Soil Sampling

and

Methods

of

Analysis

Edited by
Martin R. Carter
for
CANADIAN SOCIETY OF SOIL SCIENCE

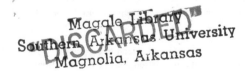

LEWIS PUBLISHERS
Boca Raton   Ann Arbor   London   Tokyo

Library of Congress Cataloging-in-Publication Data

Soil sampling and methods of analysis / edited by M. R. Carter.
    p. cm.
  Includes bibliographical references and index.
  ISBN 0-87371-861-5
  1. Soils--Analysis. 2. Soils--Sampling. I. Carter, M. R.
(Martin R.)
S593.S7425 1993
631.4'028'7--dc20   92-38583
                                                CIP

PRINTED IN THE UNITED STATES OF AMERICA
2 3 4 5 6 7 8 9 0

Printed on acid-free paper

PREFACE

Soil as a natural resource is a complex body made up of interacting mineral, organic, water, and air components with both biotic and abiotic features. From a practical viewpoint, soil is nonrenewable; thus, its characterization is of prime importance in regard to conservation strategies. Soil science, the study of soil as a natural resource, is interdisciplinary in character and has important fields of study in pedology, agriculture, plant science, forestry, engineering, geology, geography, and biology. The need to describe and understand soil material requires the continued development of suitable analytical methods to characterize soil chemical, physical, and biological properties.

Canadian soil scientists have been concerned with soil analysis methodology for several decades. The Canada Soil Survey Committee endorsed the preparation of *Manual on Soil Sampling and Methods of Analysis* which was published in 1976. A revised, second edition was promoted and published by the Canadian Society of Soil Science (CSSS) in 1978. Both editions were edited by J. A. McKeague. In 1989, the governing Council of the CSSS recommended the preparation of a new soil analysis publication. *Soil Sampling and Methods of Analysis* is the fruit of this new endeavor. The 75 chapters, prepared by 94 authors and co-authors, covers a wide range of recommended and updated methods for soil chemical, biological, and physical analysis, including methods for characterization of organic and frozen soils. Each chapter was reviewed by two members of a nine member Editorial Committee.

Overall, the book aims to establish a middle ground between the so called "cook-book" approach and the comprehensive, in-depth type of manual. In general, methods which allow some measure of standardization have been selected based on their commonness of use and ease of duplication, and their facility for accuracy and speed. Where possible, the range, limitations and potential for each method is characterized. Within any one volume all-inclusive coverage of any one method is impractical, thus sufficient references are supplied to inform the reader about the complexity and application of the methods and provide a source for further reading.

M. R. Carter
Editor

CANADIAN SOCIETY OF SOIL SCIENCE

The Canadian Society of Soil Science (CSSS) is a non-government, non-profit organization for scientists, engineers, technologists, administrators and students involved in soil science. It is affiliated with the International Society of Soil Science (ISSS) and the Agricultural Institute of Canada (AIC); its members are engaged in a wide variety of activities, including agriculture, horticulture, forestry, geography, geology, remote sensing, environmental science, and land-use planning.

The CSSS has three main objectives: to promote the wise use of soil for the benefit of all society; to facilitate information exchange among people involved in soil science; and to promote research and practical application of findings in soil science. Efforts to achieve these objectives are both national and international.

The official technical publication of the CSSS is *The Canadian Journal of Soil Science* which is published quarterly. It is recognized internationally and contains approximately 70 papers and 800 pages annually. The society also publishes a newsletter and hosts an annual scientific meeting. Recently, the CSSS has played a leading role in the AIC's Training Partnership Program. The objective of this program is to facilitate exchange of agricultural knowledge and expertise between Canada and developing countries so that agriculture is improved, professional institutions are strengthened, and sustainable links are developed with AIC member organizations. Currently, the CSSS is involved in Partnership Programs with the Soil Science Society of Nigeria, the Soil and Water Conservation Society of Thailand, and the Costa Rican Society of Soil Science.

Further information about the CSSS can be obtained at the following address:

Canadian Society of Soil Science
907-151 Slater Street
Ottawa, Ontario K1P 5H4
Canada

THE EDITOR

M. R. Carter, Ph.D. obtained a doctorate in Soil Science from the University of Saskatchewan in 1983. Since 1977, he has been involved with methodology to characterize soil properties and is currently a research scientist with the Research Branch, Agricultural Canada, Charlottetown, Prince Edward Island. A member of the Canadian Society of Soil Science, Dr. Carter holds an appointment as Adjunct Professor in the Department of Chemistry and Soil Science, Nova Scotia Agricultural College, and currently serves on the Editorial Advisory Board of *Soil and Tillage Research*.

CONTRIBUTORS

S. Abboud
Alberta Research Council
Box 8330, Station F
Edmonton, Alberta
Canada T6H 5X2

D. W. Anderson
Department of Soil Science
University of Saskatchewan
Saskatoon, Saskatchewan
Canada S7N 0W0

D. A. Angers
Agriculture Canada
Research Station
2560 Hochelaga Blvd.
Sainte-Foy, Quebec
Canada G1V 2J3

G. H. Baker
Division of Entomology
CSIRO, Glen Osmond
SA Australia 5064

B. C. Ball
Scottish Agricultural College
West Mains Road
Edinburgh
United Kingdom EH9 3JG

Thomas E. Bates
303 Edinburgh Road
Guelph, Ontario
Canada N1G 2K3

E. G. Beauchamp
Department of Land Resource Science
University of Guelph
Guelph, Ontario
Canada N1G 2W1

D. W. Bergstrom
Department of Land Resource Science
University of Guelph
Guelph, Ontario
Canada N1G 2W1

R. P. Beyaert
Agriculture Canada
Research Station
P. O. Box 186
Delhi, Ontario
Canada N4B 2W9

C. A. Campbell
Agriculture Canada
Research Station
P. O. Box 1030
Swift Current, Saskatchewan
Canada S9H 3X2

J. Caron
Soil Science Department
Laval University
Sainte-Foy, Quebec
Canada G1K 7P4

M. R. Carter
Agriculture Canada
Research Station
P. O. Box 1210
Charlottetown
Prince Edward Island
Canada C1A 7M8

Yeh-Moon Chae
Wastes and Chemicals Division
Alberta Environment
10405 Jasper Avenue
Edmonton, Alberta
Canada T5J 3N7

F. J. Cook
CSIRO, Centre for Environmental
  Mechanics
GPO Box 821
Canberra ACT
Australia
2601

Jean Crepin
Norwest Labs
9938 67th Avenue
Edmonton, Alberta
Canada T6E 0P5

J. L. B. Culley
Agriculture Canada
Centre for Land and Biological Resources
  Research
Ottawa, Ontario
Canada K1A 0C6

Y. Dalpé
Agriculture Canada
Centre for Land and Biological Resources
  Research
Ottawa, Ontario
Canada K1A 0C6

E. de Jong
Department of Soil Science
University of Saskatchewan
Saskatoon, Saskatchewan
Canada S7N 0W0

R. de Jong
Agriculture Canada
Centre for Land and Biological Resources
  Research
Ottawa, Ontario
Canada K1A 0C6

C. R. de Kimpe
Agriculture Canada
Research Coordination Directorate
Ottawa, Ontario
Canada K1A 0C5

E. Dickson
Department of Land Resource Science
University of Guelph
Guelph, Ontario
Canada N1G 2W1

M. Duquette
SNC-Lavalin
2 Place Felix-Martin
Montreal, Quebec
Canada H2Z 1Z3

B. H. Ellert
Agriculture Canada
Centre for Land and Biological Resources
  Research
Ottawa, Ontario
Canada K1A 0C6

J. A. Elliott
Department of Soil Science
University of Saskatchewan
Saskatoon, Saskatchewan
Canada S7N 0W0

R. E. Farrell
Department of Soil Science
University of Saskatchewan
Saskatoon, Saskatchewan
Canada S7N 0W0

C. T. Figueiredo
Department of Soil Science
University of Alberta
Edmonton, Alberta
Canada T6G 2E3

C. A. Fox
Agriculture Canada
Centre for Land and Biological Resources
  Research
Ottawa, Ontario
Canada K1A 0C6

Y. T. Galganov
Agriculture Canada
Centre for Land and Biological Resources
  Research
Ottawa, Ontario
Canada K1A 0C6

J. J. Germida
Department of Soil Science
University of Saskatchewan
Saskatoon, Saskatchewan
Canada S7N 0W0

Tee Boon Goh
Department of Soil Science
University of Manitoba
Winnipeg, Manitoba
Canada R3T 2N2

C. D. Grant
Department of Soil Science
University of Adelaide
Waite Campus
PMB 1 Glen Osmond
5A Australia 5069

E. G. Gregorich
Agriculture Canada
Centre for Land and Biological Resources
 Research
Ottawa, Ontario
Canada K1A 0C6

P. H. Groenevelt
Department of Land Resource Science
University of Guelph
Guelph, Ontario
Canada N1G 2W1

R. K. Guertin
Agriculture Canada
Centre for Land and Biological Resources
 Research
Ottawa, Ontario
Canada K1A 0C6

U. C. Gupta
Agriculture Canada
Research Station
P. O. Box 1210
Charlottetown, P.E.I.
Canada C1A 7M8

W. H. Hendershot
Department of Renewable Resources
Macdonald Campus, McGill University
Ste. Anne de Bellevue, Quebec
Canada H9X 1C0

Milan Ihnat
Agriculture Canada
Centre for Land and Biological Resources
 Research
Ottawa, Ontario
Canada K1A 0C6

Y. W. Jame
Agriculture Canada
Research Station
P. O. Box 1030
Swift Current, Saskatchewan
Canada S9H 3X2

H. H. Janzen
Agriculture Canada
Research Station
P. O. Box 3000, Main
Lethbridge, Alberta
Canada T1J 4B1

Richard L. Johnson
Alberta Environmental Centre
Bag 4000
Vegreville, Alberta
Canada T9C 1T4

Y. P. Kalra
Northern Forestry Centre
Forestry Canada
5320-122 Street
Edmonton, Alberta
Canada T6H 3S5

A. Karam
Soil Science Department
Laval University
Sainte-Foy, Quebec
Canada G1K 7P4

R. E. Karamanos
Department of Soil Science
University of Saskatchewan
Saskatoon, Saskatchewan
Canada S7N 0W0

B. D. Kay
Department of Land Resource Science
University of Guelph
Guelph, Ontario
Canada N1G 2W1

J. Kimpinski
Agriculture Canada
Research Station
P. O. Box 1210
Charlottetown, Prince Edward Island
Canada C1A 7M8

J.-M. Konrad
Department of Civil Engineering
Laval University
Sainte-Foy, Quebec
Canada G1K 7P4

C. G. Kowalenko
Agriculture Canada
Research Station
P. O. Box 1000
Agassiz, British Columbia
Canada V0M 1A0

H. Lalande
Department of Renewable Resources
Macdonald Campus, McGill University
Ste. Anne de Bellevue, Quebec
Canada H9X 1C0

K. E. Lee
Division of Soils
CSIRO Glen Osmond, SA
Australia 5064

J. Liang
Saskatchewan Soil Testing Laboratory
University of Saskatchewan
Saskatoon, Saskatchewan
Canada S7N 0W0

G. P. Lilley
Landcare Research New Zealand
Private Bag 31-902
Lower Hutt
New Zealand

N. J. Livingston
Department of Biology
University of Victoria
Victoria, British Columbia
Canada V8W 2Y2

L. E. Lowe
Department of Soil Science
University of British Columbia
Vancouver, British Columbia
Canada V6T 1Z4

D. G. Maynard
Northern Forestry Centre
Forestry Canada
5320-122nd Street
Edmonton, Alberta
Canada T6H 3S5

R. A. McBride
Department of Land Resource Science
University of Guelph
Guelph, Ontario
Canada N1G 2W1

W. B. McGill
Department of Soil Science
University of Alberta
Edmonton, Alberta
Canada T6G 2E3

G. R. Mehuys
Department of Renewable Resources
Macdonald Campus, McGill University
Sainte Anne de Bellevue, Quebec
Canada H9X 1C0

A. R. Mermut
Department of Soil Science
University of Saskatchewan
Saskatoon, Saskatchewan
Canada S7N 0W0

J. O. Moir
Department of Soil Science
University of Saskatchewan
Saskatoon, Saskatchewan
Canada S7N 0W0

R. A. Nunns
Landcare Research New Zealand
Private Bag 31-902
Lower Hutt
New Zealand

I. P. O'Halloran
Department of Renewable Resources
Macdonald Campus, McGill University
21,111 Lakeshore Road
Sainte Anne de Bellevue, Quebec
H9X 1C0

P. E. Olsen
Agriculture Canada
Research Station
P. O. Box 29
Beaverlodge, Alberta
Canada T0H 0C0

L. E. Parent
Soil Science Department
Laval University
Sainte-Foy, Quebec
Canada G1K 7P4

G. T. Patterson
Agriculture Canada
Canada Soil Survey
Nova Scotia Agricultural College
P. O. Box 550
Truro, Nova Scotia
Canada B2N 5E3

E. Perfect
Department of Land Resource Science
University of Guelph
Guelph, Ontario
Canada N1G 2W1

R. Protz
Department of Land Resource Science
University of Guelph
Guelph, Ontario
Canada N1G 2W1

W. D. Reynolds
Agriculture Canada
Centre for Land and Biological Resources
  Research
Ottawa, Ontario
Canada K1A 0C6

W. A. Rice
Agriculture Canada
Research Station
P. O. Box 29
Beaverlodge, Alberta
Canada T0H 0C0

Pierre J. H. Richard
Department of Geography
University of Montreal
Montreal, Quebec
Canada H3C 3J7

John E. Richards
Agriculture Canada
Research Station
P. O. Box 20280
Fredericton, New Brunswick
Canada E3B 4Z7

G. J. Ross
Agriculture Canada
Centre for Land and Biological Resources
  Research
Ottawa, Ontario
Canada K1A 0C6

Fons J. Schellekens
Geotechnical Science Laboratories
Carleton University
Ottawa, Ontario
Canada K1S 5B6

J. J. Schoenau
Department of Soil Science
University of Saskatchewan
Saskatoon, Saskatchewan
Canada S7N 0W0

B. H. Sheldrick
Agriculture Canada
Centre for Land and Biological Resources
  Research
Ottawa, Ontario
Canada K1A 0C6

Marsha I. Sheppard
Ecological Research Section
AECL Research
Whiteshell Laboratories
Pinawa, Manitoba
Canada R0E 1L0

R. R. Simard
Agriculture Canada
Research Station
2560 Hochelaga Blvd.
Sainte-Foy, Quebec
Canada G1V 2J3

Y. K. Soon
Agriculture Canada
Research Station
P. O. Box 29
Beaverlodge, Alberta
Canada T0H 0C0

R. J. St. Arnaud
Department of Soil Science
University of Saskatchewan
Saskatoon, Saskatchewan
Canada S7N 0W0

S. Sweeney
Department of Land Resource Science
University of Guelph
Guelph, Ontario
Canada N1G 2W1

Charles Tarnocai
Agriculture Canada
Centre for Land and Biological Resources
  Research
Ottawa, Ontario
Canada K1A 0C6

Denis H. Thibault
AECL
Whiteshell Laboratories
Pinawa, Manitoba
Canada R0E 1L0

H. Tiessen
Department of Soil Science
University of Saskatchewan
Saskatoon, Saskatchewan
Canada S7N 0W0

G. C. Topp
Agriculture Canada
Centre for Land and Biological Resources
  Research
Ottawa, Ontario
Canada K1A 0C6

T. Sen Tran
Service des Sols
Ministere de l'Agriculture
Service de Recherche en Sols
Pecheries et de L'Alimentation du
  Quebec
2700 Einstein
Sainte Foy, Quebec
Canada G1P 3W8

I. A. van Haneghem
Department of Agricultural
  Engineering and Physics, Agrotechnion
Bomenweg 4, 6703 HD Wageningen
The Netherlands

W. van Lierop
555 Chemin Dermott
Cookshire, Quebec
Canada J0B 1M0

W. K. P. van Loon
Department of Agricultural
  Engineering and Physics, Agrotechnion
Bomenweg 4, 6703 HD Wageningen
The Netherlands

R. P. Voroney
Department of Land Resource Science
University of Guelph
Guelph, Ontario
Canada N1G 2W1

C. Wang
Agriculture Canada
Centre for Land and Biological Resources
  Research
Ottawa, Ontario
Canada K1A 0C6

B. P. Warkentin
Water Resources Research Institute
Oregon State University
Corvallis OR
U.S. 97331-2213

C. J. Warren
Department of Land Resource Science
University of Guelph
Guelph, Ontario
Canada N1G 2W1

W. M. White
1521 Northmount Drive, N.W.
Calgary, Alberta
Canada T2L 0G8

Peter J. Williams
Geotechnical Science Laboratories
Carleton University
Ottawa, Ontario
Canada K1S 5B6

J. P. Winter
Department of Land Resource Science
University of Guelph
Guelph, Ontario
Canada N1G 2W1

TABLE OF CONTENTS

COLLECTION AND PREPARATION OF SOIL SAMPLES

SOIL TEST ANALYSES

SOIL CHEMICAL ANALYSES

SOIL BIOLOGICAL ANALYSES

SOIL BIOCHEMICAL ANALYSES

ANALYSIS OF ORGANIC SOILS

SOIL PHYSICAL ANALYSES

SOIL MINERALOGICAL ANALYSES

ANALYSIS OF FROZEN SOILS

Collection and Preparation of Soil Samples

Chapter 1
Site Description

G. T. Patterson

Agriculture Canada
Truro, Nova Scotia, Canada

1.1 INTRODUCTION

Information about the soil sampling site serves as a link between the soil sampling point and the soil horizon and landscape. In this way, a description of site attributes provides a context for the various soil properties determined on the sample. Site information may also aid in the final evaluation and judgment of soil analytical results

Site information can be classified into three categories: basic sampling information data, such as who did the sampling, where, when, and why; information about the landscape; and a summary of soil horizon data. It is not possible to compile a definitive list of site attributes that should be collected for every soil-related project. Various methods are available for measuring the value of site attributes. Generally, the purpose or scope of the project will determine which site attributes and methods should be utilized. For example, latitude-longitude measurements of location may be appropriate at the national level, but legal descriptions may be more useful at the farm level. All site data are not necessarily measured in the field. For example, elevation may be taken from topographic maps, and soil map unit names from soil survey reports.

1.2 SITE ATTRIBUTES

The following tables provide a list of site attributes applicable to soil-related studies. In Table 1.1 a list of basic information related to sample site and sampling method is provided. Landscape and soil horizon attributes, as useful accessory information, are listed in Tables 1.2 and 1.3, respectively. A bibliography of references concerned with soil sampling site description is given at the end of the chapter.

Soil Sampling and Methods of Analysis, M. R. Carter, Ed.,
Canadian Society of Soil Science. © 1993 Lewis Publishers.

TABLE 1.1 A List of Basic Sampling Data

1	Soil name
2	Soil map unit name
3	Horizon or depth sampled
4	Sampling date
5	Name of sampler
6	Location e.g.,
	Legal description
	Latitude-longitude
7	Sampling plan e.g.,
	Random
	Purposeful
8	Sampling method e.g.,
	Probe or auger
	Core
9	Vegetation or crop

TABLE 1.2 A List of Landscape Attributes

1	Landform	
2	Climate	
3	Land use	
4	Drainage	
5	Stoniness class	
6	Rockiness class	
7	Flooding events	
8	Parent material	
	1	Particle size
	2	Mode of deposition
9	Topography	
	1	Steepness of slope
	2	Length of slope
	3	Shape
	4	Site position
	5	Slope pattern
	6	Elevation
	7	Aspect

TABLE 1.3 A List of Soil Profile Attributes

1	Depth to water table	
2	Rooting zone	
	1	Thickness
	2	Particle size
3	Root-restricting layer	
	1	Thickness
	2	Kind
	3	% of area affected
4	Depth to bedrock	
5	Depth to free carbonates	
6	Depth to saline conditions	

REFERENCES

Day, J. R., Ed. 1983. The Canada Soil Information System (CanSIS). Manual for describing soils in the field. Expert Committee on Soil Survey. LRCC #82–52. Ottawa. 97 pp.

Knapik, L. J., Russell, W. B., Riddell, K. M., and Stevens, N. 1988. Forest Ecosystem Classification and Land System Mapping Pilot Project. Duck Mountain, Manitoba. Can. Forest Serv. and MB Forest. Br. 129 pp.

Soil Survey Staff. 1951. Soil Survey Manual. SCS. U.S. Department of Agriculture Handbook 18. 503 pp.

Soil Survey Staff. 1975. Soil Taxonomy. SCS. U.S. Department of Agriculture Handbook 436. 754 pp.

Taylor, N. H. and Pohlen, I. J. 1962. Soil Survey Method. A New Zealand Handbook for the Field Study of Soils. Soil Bureau Bull. 25. Taita Exp. Sta. 241 pp.

Walmsley, M., Utzig, G., Vold, T., Mood, D., and van Barnveld, J. 1980. Describing Ecosystems in the Field. RAB Tech. Paper 2. Land Management Report 7. BC Min. Env. 209 pp.

Webster, R. and Butler, B. E. 1976. Soil classification and survey studies and Ginninderra. Aust. J. Soil Res. 14: 1–24.

Chapter 2
Soil Sampling for Environmental Assessment

Jean Crépin

Norwest Labs

Edmonton, Alberta, Canada

Richard L. Johnson

Alberta Environmental Centre

Vegreville, Alberta, Canada

2.1 INTRODUCTION

The objective of soil sampling is to obtain reliable information about a particular soil. Although samples are collected to get information about the larger soil body, called the "population", the sample may or may not be representative of the population, depending on how the sample is selected and collected. In the past, sampling has been the weakest feature of resource survey and field research. Many of the data obtained only with difficulty in laboratory and field were of little use because the original sampling was unsatisfactory (Webster and Oliver 1990).

There are excellent reviews covering the general subject of sampling for chemical analysis (Kratochvil et al. 1984), sampling for soil surveys (Webster and Oliver 1990), and sampling for agriculture or forestry laboratory analysis (Petersen and Calvin 1986). In this chapter we concentrate on sampling designs and procedures with particular emphasis on evaluating environmental changes in soil.

All soils are naturally variable: their properties change horizontally across the landscape and vertically down the soil profile. The soil should be subdivided into classes which are as homogeneous as possible. Soil map units derived from changes in topography, underlying geology, and dominant vegetation type can be used for horizontal subdivisions. Soil horizons are excellent subdivisions of vertical change.

Environmental disturbances, such as mechanical mixing, gaseous deposition, liquid spills, and solid waste application, introduce additional variation to natural landscapes. Historical data on machinery, chemical and amendment use, and the evaluation of migration pathways can be used to further subdivide the soil population into smaller classes. All major sources of variation within the population should be sampled if valid inferences are to be made about the population from the sample.

Sampling and the resulting analytical work appear to be very expensive. But the reclamation of environmentally damaged soils or their disposal when damage exceeds acceptable limits can be even more costly. Sampling can save vast amounts of work if precise boundaries and limits can be established. Work in land reclamation or soil disposal or litigation translates into large financial costs.

2.2 STATISTICAL CONCEPTS IN SOIL SAMPLING

The most familiar and important statistic used in soil science is the mean, $\bar{x}$, the arithmetic average of a variable among a collection of samples.

This dispersion of individual samples around the mean is measured as the variance, s^2. A wide variance indicates wide dispersion, a small variance indicates little dispersion. The *standard deviation,* s, is the square root of the variance.

Sample estimates are subject to variation. The mean amount of polychlorinated biphenyls (pcb) in soil as estimated from a sample of three will seldom be the same as estimates that would have been obtained from other samples of three. An estimate of pcb concentration in soil that ranges from 8 to 14 ppm inspires more confidence and allows a more careful design of clean-up than an estimate ranging from 2 to 20 ppm. The standard error of the estimate (sample), more commonly just called the *standard error,* s_x, indicates the reliability of the mean, $\bar{x}$.

For large sample numbers (>50) we can be 95% confident that the actual population mean will be within two standard errors of the sample mean. Thus, for a pcb-contaminated soil with a sample mean of 11 ppm and a standard error of 1.5 ppm, the true population mean will be within 8 to 14 ppm. This estimate will be wrong, because of natural sampling variation, only 5% of the time.

The values obtained by adding or subtracting two standard errors to the sample mean are the 95% *confidence limits,* or CL. Confidence limits for any level, 67, 80, 90, 95, or 99%, can be calculated from the sample mean and standard error.

Variance, the measure of variability among units, is often related to the size of the units; large items tend to have a larger variance than small items. Plant root depth has a larger variance than average pore diameter. The *coefficient of variation,* CV, puts the expression of variability on a relative basis by dividing the standard deviation by the mean:

$$CV\% \ = \ 100 \, \frac{s}{\bar{x}} \tag{2.1}$$

Sampling and analytical work are expensive. But costs can be cut by taking only as many samples as are needed for a given level of precision. For example, let us say that unless a

5% chance occurs we want our sample mean to be within ± 1.5 ppm of the population mean. Using the following formula we can calculate n, the size of the sample needed:

$$n = \frac{t^2 s^2}{D^2} \tag{2.2}$$

where: t = a number chosen from a ''t'' table for a chosen level of precision, in this case, 95%; the degrees of freedom for ''t'' are first chosen arbitrarily, say 10, and then modified by reiteration.

s^2 = The variance which is known beforehand from other studies or estimated by $s^2 = (R/4)^2$, where R is the estimated range likely to be encountered in sampling (Freese 1962).

D = the variability in mean estimation of pcb we are willing to accept.

Therefore, for a soil with a pcb concentration ranging from 0 to 13 ppm, the estimated sample number at 95% probability and within 1.5 ppm of the true mean (CL) is

$$n = \frac{(2.23)^2(3.25)^2}{(1.5)^2} = 23 \tag{2.3}$$

Since 23 samples are many more than the 10 we used to obtain a ''t'' value, we must run a reiteration using a ''t'' value equal to our new estimate:

$$n = \frac{(2.069)^2(3.25)^2}{(1.5)^2} = 20 \tag{2.4}$$

Testing 20 samples for pcb is expensive, but the estimated variance is high (a range from 0 to 13 ppm). Any sampling scheme that reduces the variance will lower the number of samples needed (see the next section). The number can also be lowered by relaxing the probability from 95 to 90% or by allowing the confidence limits to increase to 2 ppm or higher.

The ultimate goal is to sample with the precision necessary to satisfy the sampler's objectives. One can cut down on sample numbers if the variability is low, if less precision is needed, or if a higher probability of sampling error is acceptable (say 5% as opposed to 10%). However, in cases where more than one property will be measured on each soil sample, the total number of samples should be based on the property requiring the highest sample number.

2.3 SOIL SAMPLING PLANS

Soil sampling can be judgmental or random. The former consists of sampling typical or visible differences. In environmental studies judgmental sampling often forms the basis of exploratory sampling. In soil survey judgmental sampling is the rule rather than the exception in spite of the following objections to this sampling procedure: the results rely heavily on the personal judgment of the surveyor; there is no way to verify except by checking random transects in the field the quality of the surveyor's judgment (Arnold 1979); there is no way of communicating the expert's confidence in his choice — that is, no estimates of error are possible and objective conclusions cannot be drawn about the soil population (Mapping Systems Working Group 1981); and the survey can be biased in that sampling preference is given to one part of the population (Webster 1977).

Random sampling is probability sampling and benefits from the advantages accruing to the use of probability theory (McCammon 1975). Random sampling does not mean "casual" or "haphazard". Paradoxically, random sampling is achieved in practice only by following strict rules. This means that there must be ways of identifying the individual in a population and of selecting the individuals without bias.

2.3.1 Exploratory Sampling

Exploratory samples may be used for qualitative assessment of soils where an impact or damage is visible or anticipated.

The investigator will design the sampling program according to a presumed cause of a disturbance, taking into account the nature of the disturbance, the mobility of any contaminant and the natural division of the landscape. Control samples are required; the number will depend on amount of variability in the natural landscape under investigation. The minimum number of control sites is two. Exploratory sampling makes good use of field and laboratory resources, using the preliminary information to design more efficient sampling plans. The exploratory sample can also deal with known impact sites where quantitative data on soil properties are needed ("hot spots").

Some examples of exploratory sampling are

1   *Single-industry waste sites.* The waste composition is known and the affected area is known. The sampling may be limited to a few samples to confirm the information and to determine concentrations.

2   *Small waste sites.* The area affected is known, but the types of contaminants may be unknown. The sampling is designed to support qualitative analyses. (What substances are present?) This information may lead to further sampling, but it may also limit the number of properties to be tested.

3   *Chemical spills.* The area affected is usually visible; either the soil has changed color or there is no vegetation. The sampling may concentrate on confirming maximum concentrations so that neutralizing agents can be applied in adequate amounts.

4   *Physical disturbance of previously undisturbed soils.* Surface horizon mixing, dilution of organic matter, or compaction of subsoils may require initial measurements before remedial measures are designed or further sampling is contemplated.

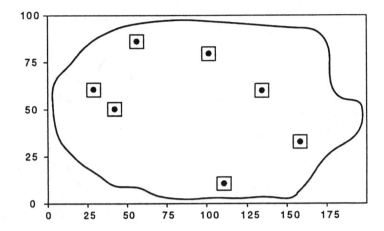

FIGURE 2.1. Map of sample area showing random sampling locations.

2.3.2 Simple Random Sampling

A simple random sample is one that allows every possible combination of sample units to be selected. The possible combinations are limited only by the sample size. Random sampling is accomplished by ensuring that at any stage of sampling the selection of a particular unit is not influenced by other units that have already been selected. The most common means of ensuring this is to assign every unit in the population a number and draw sample units from a table of random digits.

The determination of random sampling sites is best carried out prior to going to the field (Figure 2.1). All personal bias relating to landscape position can be avoided in this way. The number of samples is determined using Equation 2.2.

On small sites (<0.5 ha) that have been affected in a uniform way, as few as five to ten samples may be sufficient. Larger areas or ones that vary more may need up to 25 samples. There is little gain in precision when sample numbers exceed 25 (Webster and Oliver 1990). If a better estimate of the mean is needed the affected area should be divided into less variable units (see following Section 2.3.3).

Random sampling provides estimates of a mean and CL, but may not provide sufficient information on the pattern of disturbance. The simple random sampling plan is often used for: postreclamation assessment of mine and gravel pit sites, land-farming operations, assessment of logging blocks, and soil storage piles. All of these tend to be uniform because the soil has been treated similarly within the disturbance type. Variability should be low, but a statistical analysis will confirm or refute that hypothesis.

2.3.3 Stratified Random Sampling

In stratified random sampling the total area is broken into a number of strata or subpopulations and a random sample is taken from each stratum (Figure 2.2).

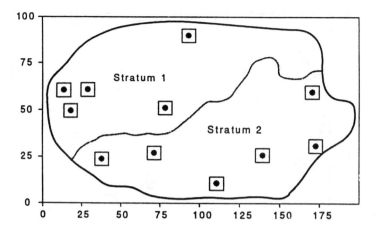

FIGURE 2.2. Stratified sampling.

This method is used: (1) to make statements about each stratum or subpopulation separately and (2) to increase the precision of estimates over the entire area (Petersen and Calvin 1965, Webster 1977). In order for stratified random sampling to be more precise than simple random sampling, the stratification must eliminate some variation attributable to sampling error. Stratification must make the sampling units (strata) more homogeneous than the population as a whole. The basis of effective stratification may be topography, type of vegetation cover, type of soil, estimated exposure to or concentration of contaminants, or the difference in type of clean-up procedure.

Soil map units are often used to define strata. The gain in efficiency of stratified over simple random sampling is due primarily to better coverage of a soil population. Stratified sampling also allows for a statistical analysis of variability within and between strata.

Stratified sampling can require a significant amount of presampling work to define the strata and their physical boundaries. However, the benefits are often worthwhile. First, the increased homogeneity of the soils in each stratum increases sampling precision. The design simplifies the interpretation of the information while increasing the possibility of damage assessment. Second, depending on the type of investigation, the cost of clean-up, reclamation, or mitigation can be significantly reduced.

This approach lends itself to a variety of statistical treatments. The simplest case may be the comparison of an affected site vs. an unaffected site, resulting in only two strata. Variability within each stratum can be quantified and differences between means can be statistically evaluated. In more complicated cases, where many strata exist, comparisons are made among all units. Within-stratum variability can also be quantified.

The design of a sampling plan for a stratified random scheme should ensure that: the strata do not overlap; the sum of the sizes of the strata equals the area of the sample area; there are no purposefully excluded populations; and the random sample locations within each stratum are selected as described in Section 2.3.2.

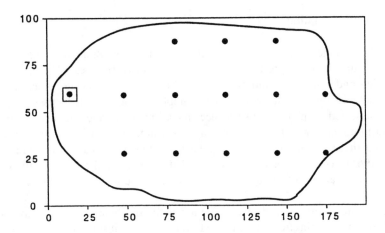

FIGURE 2.3. Map of a sample site showing systematic sampling locations.

2.3.4 Systematic or Grid Sampling

Systematic sampling, a scheme in which selected units are at regular distances from each other, attempts to guarantee complete coverage of a soil population. Sampling points are usually located at regular intervals on a grid as shown in Figure 2.3.

The sampling points follow a simple pattern and are separated by a fixed distance. Therefore, locating sampling points in the field will be easier than in random sampling. Although the sampling pattern is predetermined the first sample location should be selected at random. All the sample sites follow the pattern from that point on. In some cases, however, there may be a need for stratified systematic sampling. This is encountered when the strata are well-defined entities, such as field research plots.

Systematic sampling results in a gain in precision (Finney 1948, Williams 1956) and is preferred by some for mapping procedures. The production of high-quality maps requires complete and even coverage, unbiased sampling, and the highest possible precision within the limits determined by the number of samples.

There are two common criticisms leveled against systematic sampling: a loss in precision in comparison to simple random or stratified random methods when the population has a periodic trend in variation which corresponds to the sampling interval; and the difficulty in estimating sampling error from the sample itself. In response to the first criticism. it is recommended that a grid pattern be chosen that is unrelated in spacing and orientation and that sample numbers be increased (Webster 1977). Petersen and Calvin (1965) show several methods for estimating sampling error from systematic samples and point out what assumptions are made for each.

Systematic sampling can always replace random sampling. It can also replace stratified random sampling for spill sites where there is a likely concentration gradient. Systematic sampling also allows for the use of three-dimensional data analyses, as well as the newer tools of geostatistics, such as kriging (U.S. EPA 1989).

2.3.5 Composite Sampling

When only the average value of a soil property is needed, a substantial saving in analytical costs can be realized by compositing samples. A number of field samples representing the soil population under study are thoroughly mixed to form a composite which is then sub-sampled for submission to the laboratory. This sampling plan can only be used for properties which are unaffected by physical disturbance (most biological and chemical properties). It is important to remember that only the mean is measured; there will be no estimate of the variance of the mean nor an estimate of the precision with which it can be measured (Petersen and Calvin 1965). The number of samples needed to estimate the mean of the composite depends on the variability of the property and the desirable level of confidence.

Composite sampling can be used in conjunction with stratification, that is, the landscape can be divided into meaningful units and good averages of each soil property can be obtained by compositing samples within each unit. The average of the units can then be used to calculate the mean, standard deviation, and other required statistics of the entire soil population in the landscape.

2.3.6 Geostatistics

Most soil properties vary continuously in space. The values at sites close together are more similar than those further apart; they are not independent of one another. This property is known as spatial dependence. Geostatistics is a science that applies to a family of methods that enables the analysis, evaluation, or characterization of spatially correlated data.

A geostatistical analysis is a two-step process. First, a model is developed that describes the spatial relationship between a location where the value of a soil property will be estimated and the existing data obtained from sample points which are various distances away from the location. Data points near to the location will be closely related and have a large influence on the estimate; points far away will tend to be less related and have less influence. The model describing the influence of the spatial relationship of nearby and faraway sample points is called a semi-variogram. The second step of the geostatistical analysis is kriging. This involves estimating the value for each point in the area under study. For each point to be estimated the surrounding points provide a weighted contribution to the estimate, depending upon the semi-variogram function.

The real value of a geostatistical analysis is the estimation of soil properties for locations within the area that were not sampled. In addition, kriging allows estimation of the precision associated with the estimate.

Figure 2.4 is an example of a map of topsoil depths produced by the kriging procedure. Environmental scientists interested in sulfur impact might want the sample point to represent sulfur deposition rates or soil pH status; the principles remain the same: soil properties can be systematically sampled, a semi-variogram constructed and interpreted, and a map produced by kriging in which interpolated isopleths are used to represent environmental gradients.

For a general discussion of geostatistical theory, see Clark (1979) or Davis (1986). Journel and Huijbregts (1978) and David (1984) provide detailed, advanced explanations of the theory and practice of geostatistics. Recent publications on the application of geostatistics to soil science can be found in Oliver (1987), Warrick et al. (1986), and Webster and Oliver (1990).

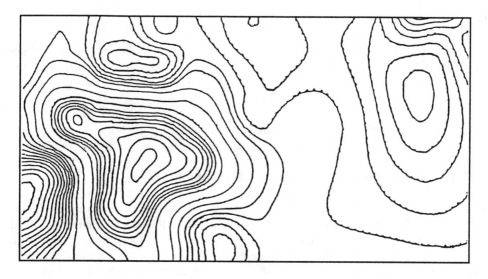

FIGURE 2.4. A computer-interpolated map of topsoil depths with isopleths at 5-cm intervals.

2.3.7 Sampling for Linear Disturbances

While there is no special sampling plan for disturbances caused by roads, power lines, or pipelines, their ubiquity and the increased demand for impact assessment make it necessary to mention linear disturbances separately. The main characteristics are

1   Linear disturbances in many landforms, soil types, land uses, and climatic zones

2   Environmental damage is often related to the loss of topsoil, soil horizon mixing, and changes in the soil physical characteristics, such as compaction, soil structure, and permeability (Culley et al. 1981)

The sampling plan requires the division of the study area into strata and the implementation of a stratified random sampling plan (Section 2.3.3). Strata are usually defined and can be restricted on the basis of soil map units or soil classes, topography, land use, moisture regime, or level of disturbance, including the way the soils and parent material were handled.

Linear disturbances can occur in repeating strata. For example, cultivated chernozemic soils may be interspersed with gleysols or luvisols. The intensity of the sampling will depend upon the requirements of the assessment as well as the variability between and within strata.

2.4 SOIL SAMPLING PROCEDURES

Environmental assessment concerns itself with undisturbed and mechanically disturbed soils. In this chapter a soil is considered to be mechanically disturbed when more than the topsoil horizon is removed.

2.4.1 Undisturbed Soils

Most soils are mechanically undisturbed. These soils have recognizable horizons that are the result of soil genesis. The Canadian System of Soil Classification (ECSS 1987) and the Canada Soil Information System (ECSS 1983) provide excellent accounts of how to recognize and describe the profiles of these undisturbed soils.

Undisturbed soils should be sampled by horizon. Samples taken by horizon have reduced variances for all soil properties. Essentially, the horizon can serve as a means of stratification.

2.4.2 Procedures Appropriate to Mechanically Disturbed Soils

Soils that are physically disturbed often show no visible stratification (no horizons). In this case the soil should be sampled by depth. The recommended depths for sampling depend on the purpose of assessment and history of the soil. Soil biological factors, such as microbial biomass, microbial activity, soil fauna, and enzyme activity, are most important in surface samples. The active root zone, frequently the 0- to 30-cm depth under Canadian conditions, but ultimately depending on the soil, the climate, and the vegetation, always deserves special attention. The subsoil, from 30 to 100 cm, controls water storage, salt movement, and compaction status. The parent material, lying below the subsoil, can be sampled by depth intervals or by type if it is stratified (in layers). The kind of sampling pattern chosen for disturbed soils will depend on the purpose of the investigation and the nature of the soil after disturbance.

2.4.2.1 *Random Depth Sampling*

This method is used where the material has no visible or known stratification and where an assessment of the variability of the whole soil mass is necessary (Figure 2.5). Each sample is taken from a random depth which is chosen prior to entering the field and determined by the following formula:

$$\text{Depth} = \text{Total soil depth} \times \text{RN} \qquad (2.5)$$

where RN = a random number between 0 and 1.

2.4.2.2 *Stratified Random Depth Sampling*

Where there are visible strata or where strata are defined arbitrarily for purposes of the study (e.g., the upper 1 m of parent material or the entire vadose zone below the topsoil), samples can be acquired at random within each stratum. Equation 2.5 is used to select the location, except that the depth of the stratum is substituted for the "total soil depth". Figure 2.6 illustrates this sampling procedure.

2.4.2.3 *Discrete Depth Sampling*

This method uses predetermined sampling depths (Figure 2.7). This is the preferred method for sampling for volatile organics. If the actual thickness of the sample is the entire depth interval which is then homogenized to yield a representative subsample, discrete depth sampling will not be appropriate for volatile organics.

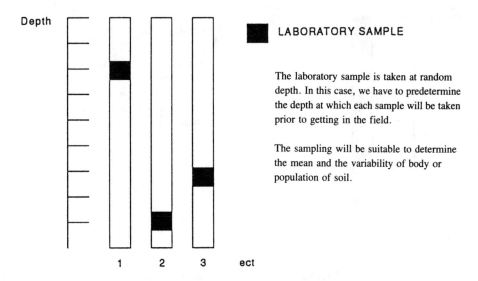

FIGURE 2.5. Random depth sampling.

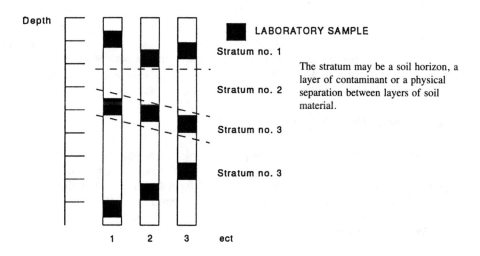

FIGURE 2.6. Stratified random depth sampling.

Normally, sampling depths are evenly spaced, but when assessing a site after clean-up or reclamation, it may be advantageous to design the sampling depths so that narrow intervals occur near the surface and wider intervals are used below. In this way more detailed information can be collected from the depths contributing most to human, animal, or plant response.

2.4.3 Procedures Appropriate to All Soils

There are some sampling methods which are used on both mechanically disturbed and undisturbed soils. The decision to employ these procedures depends on the nature of the impact, the anticipated response, and the statistical analysis to be applied to the data.

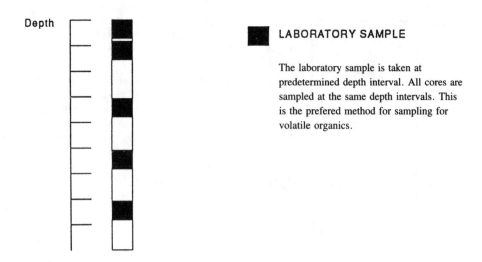

FIGURE 2.7. Discrete depth sampling.

2.4.3.1 Sampling Thin Surface Horizons

Sampling the soil surface, especially for atmospheric deposition where a contaminant impinges as a gas, a mist, a dust, or is deposited in rain, necessitates assessing a very shallow (<5 cm) depth. Care must be taken to avoid unnecessary dilution with nonimpacted, deeper soil; otherwise, both the sampling and analytical error may be increased. Often, the bulk density of the shallow surface layer must also be measured.

2.4.3.2 Composite Depth Sampling

In this method the whole soil core is homogenized and a subsample is taken for laboratory analysis. Using this method good information can be obtained very inexpensively about average values of soil properties at each site, but all information about variability with depth will be lost. Mixing the sample precludes its use for the measurement of volatile compounds.

2.4.3.3 Equivalent Depth Sampling

Often, a physically disturbed soil needs to be compared to an undisturbed "control" soil. One recommended sampling procedure is termed "equivalent depth sampling". The following is modified from a description by Powlson and Jenkinson (1981).

First, one decides which soil (disturbed or undisturbed) is most compacted. Second, the less compacted soil is sampled slightly deeper than the more compacted soil. Third, the compacted soil is sampled to a depth that yields a weight of oven-dried (O.D.) soil equal to the O.D. weight of the less compacted soil. This is done by sampling the more compacted soil both to the depth of interest and from an extra depth immediately below. A portion of this extra soil is added so that the total weights of compacted and uncompacted soils are the same. The "equivalent sampling depth" (E.D.) of compacted soil is calculated from:

$$\text{E.D.} = \text{Depth of interest} + \text{extra depth} \times \frac{\text{weight of portion of extra soil added}}{\text{total weight of extra soil}}$$

The same approach can be modified for field sampling where precision is not needed. One decides which soil is most compacted and by approximately what amount. The less compacted soil is sampled deeper than the more compacted soil by the amount of compaction that has been estimated. For example, if one thinks that soil A has an average bulk density of 1.0 Mg/m^3 and soil B has an average bulk density of 1.5 Mg/m^3, then the sample of soil A will be taken 50% deeper than the E.D. of soil B. Yet, the properties analyzed from the two samples will be considered to be from comparable depths.

2.5 QA/QC IN SOIL SAMPLING

Quality assurance (QA) and quality control (QC) are used to assess and minimize errors in all phases of environmental studies. Perhaps the most exact and concise description of the goal of a QA/QC program was made by Taylor (1987):

> "When the uncertainty of a measured value is one-third or less of the permissible tolerance for its use, it can be considered as essentially errorless for that use."

Error or variability of results can occur randomly or be due to bias. The two major sources of variability are population variability and measurement variability. The former was discussed in some detail in Section 2.2; the latter can be arbitrarily divided into sampling variability, handling, transport and preparation variability, subsampling variability, lab analysis variability, and between-batch variability. ("Batch" is defined as a group of samples, collected, shipped, and analyzed under similar conditions.) Typically, the error arising from field sampling is much larger than that associated with sample preparation, handling, or analysis, yet the majority of effort has been spent at quantifying and controlling error in the laboratory (see Chapter 26).

The U.S. Environmental Protection Agency (EPA) has proposed a systematic approach to the identification and control of error in the sampling of soil for environmental studies (van Ee et al. 1990). This approach adopts concepts traditionally used for QA/QC in analytical laboratories; for example, the use of duplicate, split, spiked, evaluation, and calibration samples. Furthermore, they have published a manual and software computer program (U.S. EPA 1991) that provides the foundation for answering the two essential questions in QA/QC programs in field sampling: (1) How many and what type of samples are required to assess the quality of data in a field sampling effort? (2) How can the information from QA samples be used to identify and control the sources of error and uncertainty in measurement programs?

REFERENCES

Arnold, R. W. 1979. Strategies for field resource inventories. Agronomy mimeo 79–20. Cornell University, Ithaca, NY. 36 pp.

Clark, I. 1979. Practical Geostatistics. Elsevier Scientific Publishing, New York. 202 pp.

Culley, J. L. B., Dow, B. K., Presant, E. W., and Maclean, A. J. 1981. Impact of Installation of Oil Pipeline on the Productivity of Ontario Cropland. Land Resource Research Institute Contribution No. 66. Research Branch, Agriculture Canada, Ottawa, ON. 88 pp.

David, M. 1984. Geostatistical Ore Reserve Estimation. Elsevier Scientific Publishing, New York. 364 pp.

Davis, J. C. 1986. Statistical Data Analysis in Geology. John Wiley & Sons, New York. 550 pp.

ECSS. 1987. The Canadian System of Soil Classification. Expert Committee on Soil Survey. Research Branch, Agriculture Canada. Publication 1646. 2nd ed. Ottawa, ON. 164 pp.

ECSS. 1983. Canada Soil Information System (CanSIS). Expert Committee on Soil Survey. Research Branch, Agriculture Canada. Contribution No. 82–52. J. H. Day, ed. Ottawa, ON. 97 pp.

Finney, D. J. 1948. Random and systematic sampling in timber surveys. J. Forestry 22: 64–99.

Freese, F. 1962. Elementary forest sampling. Agriculture Handbook No. 232. U.S. Department of Agriculture. Reprinted in 1981 by Oregon State University, Corvallis. 91 pp.

Journel, A. G. and Huijbregts, C. J. 1978. Mining Geostatistics. Academic Press, New York, 600 pp.

Kratochvil, B., Walker, D., and Taylor, J. K. 1984. Sampling for chemical analysis. Anal. Chem. 56 (5): 13R–129R.

Mapping Systems Working Group. 1981. A Soil Mapping System for Canada. Revised. Land Resource Research Institute Contribution No. 142. Research Branch, Agriculture Canada. Ottawa, ON. 94 pp.

McCammon, R. B. 1975. Statistics and probability. Pages 1–20 in Concepts in Geostatistics. R. B. McCammon, Ed. Springer-Verlag, New York.

Oliver, M. A. 1987. Geostatistics and its application to soil science. Soil Use Manage. 3: 8–20.

Petersen, R. G. and Calvin, L. D. 1986. Sampling. Pages 33–52 in A. Klute, Ed. Methods of soil analysis. Part I. Physical and mineralogical methods. Agronomy No. 9, 2nd ed. American Society of Agronomy, Madison, WI.

Petersen, R. G. and Calvin L. D. 1965. Sampling. Pages 44–72 in C. A. Black, Ed. Methods of soil analysis. Part I. Physical and mineralogical properties. Agronomy No. 9. American Society of Agronomy, Madison, WI.

Powlson, D. S. and Jenkinson, D. S. 1981. A comparison of the organic matter, biomass, adenosine triphosphate and mineralizable nitrogen contents of ploughed and direct drilled soils. J. Agric. Sci. 97: 713–721.

Taylor, J. K. 1987. Principles of calibration. American Society for Testing and Materials. Special technical publication. Sampling calibration atmospheric measurements, 957 pp., 14–18.

U.S. EPA. 1991. ASSESS Users Guide. U.S. Environmental Protection Agency, Environmental Monitoring Systems Laboratory. EPA/600/8–91001. Las Vegas, NV.

U.S. EPA. 1989. Methods of Evaluating the Attainment of Clean Up Standards. Volume I: Soils and Solid Media. U.S. Environmental Protection Agency, Washington, DC.

van Ee, J. J., Blume, L. J., and Starks, T. H. 1990. A Rationale for the Assessment of Errors in the Sampling of Soils. U.S. Environmental Protection Agency, Environmental Monitoring Systems Laboratory. EPA/600/4–90/013. Las Vegas, NV. 57 pp. EPA/600/X-89/203

Warrick, A. W., Myers, D. E., and Neilsen, D. E. 1986. Geostatistical methods applied to soil science. Pages 53–82 in A. Klute, Ed. Methods of soil analysis. Part I. Physical and mineralogical methods. Agronomy No. 9, 2nd ed. American Society of Agronomy, Madison, WI.

Webster, R. and Oliver, M. A. 1990. Statistical Methods in Soil and Land Resource Survey. Oxford University Press, Oxford. 315 pp.

Webster, R. 1977. Quantitative and Numerical Methods in Soil Classification and Survey. Clarendon Press, Oxford. 269 pp.

Williams, R. M. 1956. The variance of the mean of systematic samples. Biometrika 43: 137–148.

Chapter 3
Soil Handling and Preparation

Thomas E. Bates

University of Guelph

Guelph, Ontario, Canada

Much of this discussion deals with soil handling and preparation for determination of plant-available nutrients. Other analyses such as total nitrogen content, X-ray diffraction analysis to determine clay mineral content can be expected to require different handling and preparation.

Except for the particular aspect under study, research on soil testing for plant-available nutrients should use exactly the same procedures as are used in the routine soil test laboratory. This includes measuring instead of weighing samples in most cases. Much research has to be redone before it can be applied in a soil test laboratory because of failure to do this.

Sampling of field soils is one of the more difficult and more important aspects of prediction of nutrient requirements by soil analysis, but will be discussed elsewhere. No amount of care in preparation and analysis can overcome the problems of careless or inappropriate sampling in the field.

3.1 DRYING

Most soil samples for testing arrive at the laboratory in "field-moist" conditions. This means that the samples may range from air dry to saturated.

Usually the first operation when moist samples are received is to dry them. For nitrate and ammonium-N determination the samples are usually air dried before sending to the laboratory or transported in a frozen or refrigerated state to prevent changes in nitrate and ammonium content. Freezing and drying have both been found by some workers to affect exchangeable ammonium-N and fixed ammonium-N (Allen and Grimshaw 1962, Harding and Ross 1964). Allen and Grimshaw also found that freezing affected extractable P. Bremner and Mulvaney (1982) provide a useful discussion of the pretreatment of soils for total and mineral N content and conclude that on drying of soil samples having a low percentage of total N (1–3%) in the form of inorganic N, drying is unlikely to significantly affect the results for total N analysis.

Soil Sampling and Methods of Analysis, M. R. Carter, Ed.,

Canadian Society of Soil Science. © 1993 Lewis Publishers.

TABLE 3.1 A Survey of Soil-Handling Methods for Routine Soil Testing, 1989

	Drying	Grinding method	Sieve size	Measuring
Newfoundland	40°C	Mortar & pestle	1 mm & 2 mm[a]	Volume
P.E.I.	35°C	"Mechanical"	2 mm	Weight
Nova Scotia	35°C	Mortar & pestle	2 mm	Volume
New Brunswick	Air dry	Mortar & pestle	2 mm	Volume[b]
Quebec	<37°C	Mortar & pestle	2 mm	Volume
Ontario	35°C	Flail	2 mm	Volume
Manitoba	<35°C	"Heavy-duty grinder"	2 mm	Volume
Saskatchewan	30°C	Flail (except for micro)	2 mm	Volume
Alberta	60°C	Flail	2 mm	Volume[c]
British Columbia	55°C	Wood blocks	2 mm	Volume

[a] 1 mm for mineral soils, 2 mm for organic.
[b] Organic soils weighed.
[c] Alberta allows a choice in whether laboratories weigh or measure samples.

For routine soil testing all Canadian provinces recommend drying the samples before analysis (Table 3.1). Analysis of undried samples is not recommended because of the effect of moisture content on soil to extractant ratio and the need for determination of moisture content.

A surprising range of drying temperatures is used across Canada (Table 3.1). The degree of drying or the temperature of drying is well known to have marked and important effects on the amount of potassium extracted on soils containing micaceous clays and/or vermiculite. Shongwe (1979) obtained closer correlations of ammonium acetate-extractable K with plant K uptake on moist Ontario soils than on soils dried at 35°C, and numerous authors have shown similar results. The amount of potassium released or fixed by drying increases with the degree of drying or increase in drying temperature. An extreme example of this is shown with the Marshall subsoil from Iowa in Figure 3.1 (Bates 1961).

Soils containing micaceous minerals and/or vermiculite should be dried at as low an air temperature as is practical. Temperatures below 35°C are probably not practical in Ontario, as we have air temperatures as high as 30°C with very high relative humidity from time to time. There is little doubt that relative humidity will affect K release and fixation. However, with the differences in drying temperatures shown in Table 3.1, it should not be expected that the same amount of K will be extracted on soils analyzed in different provincial laboratories, even if identical extractions are used. On soils which do not contain micaceous minerals and/or vermiculite the temperature or degree of drying may not be critical.

The best that can be done is to use the drying method recommended for the province from which the soils are taken.

3.2 GRINDING AND SCREENING

3.2.1 Routine Testing

For routine soil tests almost all provinces grind samples to pass a 2-mm screen, and there seems little reason to change from this. Use of a finer screen will remove the very coarse

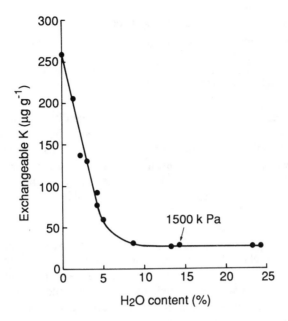

FIGURE 3.1. Exchangeable K as affected by drying on a soil known to fix large amounts of K (Bates 1961).

sand, 1.0–2.0 mm diameter, unless the sand particles are broken by grinding. Nylon or stainless screens are recommended for micronutrient samples.

Canadian laboratories use a range of grinding methods from a wood block or agate mortar and pestle to flail-type and other mechanical grinders. The soil should not be subjected to sufficient force or abrasion to break up the individual sand, silt, or clay particles. Where micronutrient and metal tests are to be run, particular care should be taken in the choice and use of grinders to prevent contamination. Some provinces avoid the use of metal grinders for micronutrients and others do not. Stainless steel is the usual choice where metal is used for screening and grinding. It seems to be generally satisfactory, but, considering the range of metals used in stainless steel, this should not be taken for granted.

3.2.2 Special Tests

For some tests where very small samples are used in the analysis, grinding finer than 2 mm is recommended. This is usually the case for total nitrogen (Bremner and Mulvaney 1982), suggesting that for macro-Kjeldahl analysis, soils containing $<0.5\%$ N should be ground to pass through a 32-mesh screen and soils containing $>1\%$ N should be ground to pass a 100-mesh screen. It is frequently recommended to grind to from 100 to 300 mesh.

3.3 VOLUME VS. WEIGHT

For routine soil testing most provinces measure a volume of dried, screened sample before placing it in a flask to which a specific volume of extractant is added (Table 3.1), but some laboratories weigh each sample instead. The sample is measured by scooping it up with a "scoop" of measured volume. It is common to tap the heaping scoopful of soil with a spatula, or on a bench top, after which a spatula, a glass rod in one laboratory, is used to scrape the excess soil off the scoop to provide a consistent volume. A spatula held vertically is recommended for scraping.

There has been a great deal of discussion and somewhat less research on weighing vs. measuring samples (Mehlich 1972 and 1973; van Lierop 1981 and 1989).

There is no doubt that soils after drying and sieving have different bulk densities than the same soils in the field. Ideally, one would like to represent the rooting volume of the crop. Over a range of soils neither a weighed nor a screened measured sample does this well. The discrepancy is very large when high organic mineral soils and organic soils with bulk densities ranging to lower than 0.200 (van Lierop 1981) are tested by weight along with low organic mineral soils. It is also large, but somewhat less, when samples are measured after drying and screening.

The question whether to weigh samples or to use measured samples should be resolved by determining which correlates best with plant uptake of nutrients. In Ontario, studies in the greenhouse, where a very wide range of bulk densities was used, including mineral and organic soils, and samples were measured, there was no tendency for the organic soils to react differently in terms of plant nutrient uptake than the soils with a high bulk density. An examination of the bulk densities makes it obvious that if weighed samples were used, the organic soils would require a completely different calibration. This was the case in separate studies with several phosphorus, potassium, zinc, and manganese soil tests. Where the soils to be tested do not have a wide range in bulk density, weighing or measuring samples are likely to provide similar results.

Weighed samples can be expected to provide better laboratory precision, but in individual research projects it is essential that the decision to weigh or measure samples agree with what is done in actual testing where the research will be applied. This applies to soil test calibration work and also to development and comparison of extractants. If this is not done the research will probably have to be repeated before it can be put to use. Far too many researchers have weighed their samples, but expected the soil test laboratories that measures samples to apply the results.

3.3.1 Measurement Units

Many of the arguments as to whether soils should be measured or weighed for soil test analysis have carried over into what units should be used to express the results. There seems little reason to express the results of soil tests in which the sample has been measured in other than volume units ($mg\ L^{-1}$). Likewise, results from weighed samples should logically be expressed in $mg\ kg^{-1}$. There is little need or logic to converting the results from measured samples from $mg\ L^{-1}$ to $mg\ kg^{-1}$ by assuming an average bulk density (Mehlich 1973).

Either of the above, $mg\ L^{-1}$ or $mg\ kg^{-1}$, can be converted to $kg\ ha^{-1}$ or $lb\ ac^{-1}$ if desired. The volume conversion from $mg\ L^{-1}$ to $kg\ ha^{-1}$ should be done on a strictly volume basis, assuming a depth of 15 or 20 cm.

3.4 STORAGE

Dried soil samples are frequently stored for up to 3 months after testing, presumably for retesting if required. Soil test laboratory check samples are frequently stored for one or more years and tested periodically. Storage of research samples for a few weeks or many years is also common.

Storage is probably most important for research samples from field calibration trials. These samples should be saved, probably for 10 to 20 years. Where they are saved, a new soil test extractant or method can be calibrated simply by testing the old samples with the new method and using the old response data. Likewise, if crop uptake is measured on a number of soils in the greenhouse to evaluate extractants, a new extractant can be evaluated very inexpensively several years later by simply extracting the stored soils used in the previous experiment with the new extractant.

If soils are stored for any of the above reasons it is important to know whether or not the parameter of interest has changed in storage.

The Saskatchewan provincial laboratory has stored dry soils in sealed jars in the laboratory for over 15 years and Alberta for 4 years without any apparent changes in the test values (personal communication). In Ontario, the author has stored samples in cardboard containers for 9 years without any change in test values. In this case storage was in a building with just sufficient winter heat to prevent freezing and with some dehumidification in the summer. Most soil test laboratories save their check samples for months without apparent problems, although the New Brunswick laboratory reports a decline in soil pH after 1 month (personal communication). It seems probable that if the storage is at a reasonably low humidity it has little effect on most soil tests. None of the above reports are believed to include forms of N. There is certainly opinion to the contrary (Bartlett and James 1980, Shuman 1980), so it should not be taken for granted. It may depend on the element and/or the extractant. Allen and Grimshaw (1962) found that samples air dried at 40°C for 24 h released ammonium-N and P in storage.

The author found little information on storage of frozen samples. Nelson and Bremner (1972) report small changes in exchangeable and fixed ammonium and nitrate-N in samples stored at $-5°C$ for 9 months. Allen and Grimshaw (1962) did not recommend frozen $(-15°C)$ storage because of changes in ammonium-N and phosphorus.

REFERENCES

Allen, S. E. and Grimshaw, H. M. 1962. Effect of low temperature storage on the extractable nutrient ions in soils. J. Sci. Food Agric. 13: 525–529.

Bartlett, R. and James, B. 1980. Studying dried, stored soil samples — some pitfalls. Soil Sci. Soc. Am. J. 44: 721–724.

Bates, T. E. 1961. Potassium Release in Soils as Affected by Drying, Ph.D. thesis, Iowa State University, Ames.

Bremner, J. M. and Mulvaney, C. S. 1982. Pages 600–601 in A. L. Page et al., Eds. Methods of

soil analysis. Part 2. 2nd ed. American Society of Agronomy, Madison, WI.

Harding, D. E. and Ross, D. J. 1964. Some factors in low temperature storage influencing the mineralisable-nitrogen of soils. J. Sci. Food Agric. 15: 829–834.

Mehlich, A. 1972. Uniformity of expressing soil test results: a case for calculating results on a volume basis. Commun. Soil Sci. Plant Anal. 3: 417–424.

Mehlich, A. 1973. Uniformity of soil test results as influenced by volume weight. Commun. Soil Sci. Plant Anal. 4: 475–486.

Nelson, D. W. and Bremner, J. M. 1972. Preservation of soil samples for inorganic nitrogen analysis. Agron. J. 64: 196–199.

Shongwe, N. G. 1979. Comparison of Methods of Measuring Plant-Available Potassium in Southern Ontario Soils. M.Sc. thesis, University of Guelph, Guelph, Ont.

Shuman, L. M. 1980. Effects of soil temperature, moisture and air-drying on extractable manganese, iron, copper and zinc. Soil Sci. 130: 336–343.

van Lierop, W. 1981. Laboratory determination of field bulk density for improving fertilizer recommendations of organic soils. Can. J. Soil Sci. 61: 475–482.

van Lierop, W. 1989. Effect of assumptions on accuracy of analytical results and liming recommendations when testing a volume or weight of soil. Commun. Soil Sci. Plant Anal. 20: 121–137.

Soil Test Analyses

Chapter 4
Nitrate and Exchangeable Ammonium Nitrogen

D. G. Maynard and Y. P. Kalra

Forestry Canada
Edmonton, Alberta, Canada

4.1 INTRODUCTION

Inorganic N in soils is predominantly NO_3 and NH_4. Nitrite is seldom present in detectable amounts, and its determination is normally unwarranted except in neutral to alkaline soils receiving NH_4 or NH_4-producing fertilizers (Keeney and Nelson 1982). In Canada, provincial soil testing laboratories usually determine NO_3 to estimate available N in agricultural soils, while laboratories analyzing tree nursery and forest soils often determine both NO_3 and NH_4.

There is considerable diversity among laboratories in the extraction and determination of NO_3 and NH_4. In addition, incubation methods (both aerobic and anaerobic) have been used to determine the potential mineralization of N. These are outlined in Chapter 33.

Nitrate is water soluble and a number of solutions including water have been used as extractants. These include saturated (0.35%) $CaSO_4 \cdot 2H_2O$ solution 0.03 M NH_4F and 0.015 M H_2SO_4, 0.01 M $CaCl_2$, 0.5 M $NaHCO_3$, pH 8.5, 0.01 M $CuSO_4$, 0.01 M $CuSO_4$ containing Ag_2SO_4, and 2.0 M KCl (Soltanpour and Workman 1981; Keeney and Nelson 1982; Ministry of Agriculture, Fisheries and Food 1986; Blakemore et al. 1987; Houba et al. 1988; Laverty and Bollo-Kamara 1988). Exchangeable NH_4 is defined as NH_4 that can be extracted at room temperature with a neutral K salt solution. Various molarities have been used, such as 0.05 M K_2SO_4 (Keeney and Nelson 1982), 0.1 M KCl (Houba et al. 1988), 1.0 M KCl, and 2.0 M KCl (Keeney and Nelson 1982). The most common extractant for NO_3 and NH_4, however, is 2.0 M KCl.

The methods of determination for NO_3 and NH_4 are even more diverse than the methods of extraction (Keeney and Nelson 1982). These range from specific ion electrode to manual colorimetric techniques, microdiffusion, steam distillation, and flow injection analysis. Steam distillation is still a preferred method when using ^{15}N; however, for routine analysis automated colorimetric techniques are preferred. They are rapid, free from most soil interferences, and very sensitive.

Soil Sampling and Methods of Analysis, M. R. Carter, Ed.,
Canadian Society of Soil Science. © 1993 Lewis Publishers.

The methods for the most commonly used extractant (2.0 M KCl) and the automated colorimetric methods for the determination of NO_3 and NH_4 are presented here. The steam distillation methods for determination of NO_3 and NH_4 have not been included, since they have not changed much over the last several years. Detailed descriptions of them are included elsewhere (Bremner 1965; Keeney and Nelson 1982). For NO_3, extraction with 0.01 M $CuSO_4$ containing Ag_2SO_4 and determination by phenoldisulfonic acid methods are also presented.

4.2 EXTRACTION OF NO_3-N AND NH_4-N WITH 2.0 M KCl

4.2.1 Principle

Ammonium is held in an exchangeable form in soils in the same manner as exchangeable metallic cations. Fixed or nonexchangeable NH_4 can make up a significant portion of soil N; however, fixed NH_4 is defined as the NH_4 in soil that cannot be replaced by a neutral K salt solution (Kenney and Nelson 1982). Exchangeable NH_4 is extracted by shaking with 2.0 M KCl. Nitrate is water soluble and hence can also be extracted by the same 2.0 M KCl extract. Nitrite is seldom present in detectable amounts in soil and therefore is usually not determined.

The extraction of NO_3 and NH_4 is carried out on air-dried soil for most soil test procedures; however, air drying may result in large changes to the extraction of NO_3 and NH_4 concentrations. Extracting soils moist immediately after sampling is the ideal situation; however, this may cause problems with respect to storage and in obtaining a representative subsample. The use of moist soils would be preferred on samples related to biological studies (Haynes and Swift 1989). For most test procedures and fertilizer recommendations, air-dried samples at low temperature (e.g., room temperature) in an NH_4-free environment are used.

4.2.2 MATERIALS AND REAGENTS

1   Reciprocating shaker.

2   Dispensing bottle.

3   Erlenmeyer flasks, 125 mL.

4   Filter funnels.

5   Whatman® No. 42 filter papers.

6   Aluminum dishes.

7   Potassium chloride (2.0 M KCl): dissolve 149 g KCl in approximately 800 mL NH_3-free double-distilled water and dilute to 1 L.

4.2.3 PROCEDURE

A. Moisture determination.

1   Weigh 5.00 g of moist soil in a preweighed aluminum dish.

2   Dry overnight in an oven at 105°C.

3   Cool in a desiccator and weigh.

B. Extraction procedure.

1   Weigh (5.0 g) field-moist soil (or moist soil incubated for mineralization experiments) into an Erlenmeyer flask.

2   Add 50 mL 2.0 M KCl solution using the dispensing bottle. (If the sample is limited, it can be reduced to a minimum of 1.0 g and 10 mL 2.0 M KCl to keep 1:10 ratio.)

3   Carry a reagent blank throughout the procedure.

4   Stopper the flasks and shake for 30 min.

5   Filter into 60-mL Nalgene bottles.

6   Analyze for NO_3-N and NH_4-N within 24 h (see comment 3).

4.2.4 COMMENTS

1   Significant changes in the amounts of NH_4 and NO_3 can take place on prolonged storage at room temperature of air-dried samples. A study conducted by the Western Enviro-Agricultural Laboratory Association showed that the NO_3 content of soils decreased significantly after a 3-year storage of air-dried samples at room temperature (unpublished results). Increases in NH_4 content have also been reported by Bremner (1965) and Selmer-Olsen (1971).

2   Filter paper can contain significant amounts of NO_3 and NH_4 that can potentially contaminate extracts (Muneta 1980; Heffernan 1985; Sparrow and Masiak 1987).

3   Ammonium and nitrate in KCl extracts should be done within 24 h of extraction (Keeney and Nelson 1982). If the extracts cannot be analyzed immediately they should be frozen. Potassium chloride extracts keep indefinitely when frozen (Heffernan 1985).

4   This method yields highly reproducible results.

4.3 DETERMINATION OF NO$_3$-N IN 2.0 *M* KCl EXTRACTS BY AUTOANALYZER (CADMIUM REDUCTION PROCEDURE)

4.3.1 Principle

Nitrate is determined by an automated spectrophotometric method. Nitrates are reduced to nitrite by a copper cadmium reductor column. The nitrite ion reacts with sulfanilamide under acidic conditions to form a diazo compound. This couples with *N*-1-naphthyl-ethylenediamine dihydrochloride to form a reddish purple azo dye (Technicon® Instrument Corporation 1971).

4.3.2 MATERIALS AND REAGENTS

1   Technicon® AutoAnalyzer® consisting of sampler, manifold, proportioning pump, cadmium reductor column, colorimeter, and recorder.

2   Reductor column preparation.

   a. Cadmium metal is ground and sized. Particles used in the column must be between 25 and 60 mesh size.

   b. New or used cadmium particles (10 g) are cleaned with 50 mL of 6 *M* HCl for 1 min. Decant HCl and wash the cadmium with another 50 mL of 6 *M* HCl for 1 min.

   c. Decant HCl and wash cadmium several times with distilled water.

   d. Decant distilled water and add 50 mL of 2% CuSO$_4$·5H$_2$O. Wash cadmium until no blue color remains in solution.

   e. Rinse cadmium several times with double-distilled water, then decant.

   f. Add 50 mL of 2% CuSO$_4$·5H$_2$O and wash until no blue color remains in the solution.

   g. Decant and wash thoroughly with distilled water.

   h. Fill reductor column with NH$_4$Cl reagent (or water) and transfer prepared cadmium particles to column using a Pasteur pipet. Be careful not to allow any air bubbles to be trapped in the column. (Note: in place of the Reductor Tube 189–0000, a 35-cm length of 2.00 mm I.D. Tygon® tubing can be used.)

   i. Prior to sample analysis, condition the column with 100 mg N (nitrate) L^{-1} for 5 min followed by 100 mg N (nitrite) L^{-1} for 10 min.

3　Standards:

   a. Stock solution (100 μg NO_3-N mL^{-1}): dissolve 0.7218 g of KNO_3 (dried overnight at 105°C) in double-distilled H_2O and dilute to 1 L. Add 1 mL of chloroform to preserve the solution.

   b. Working standards: pipet 0.5, 1.0, 1.5, and 2.0 mL of stock solution into a 100-mL volumetric flask and make to volume with 2.0 M KCl solution to obtain 0.5, 1.0, 1.5, and 2.0 μg NO_3-N mL^{-1} standard solution, respectively.

4　Ammonium chloride reagent: dissolve 10 g NH_4Cl in alkaline water and dilute to 1 L. (Alkaline water is prepared by adding just enough dilute NH_4OH to double-distilled H_2O to attain a pH of 8.5.) Add 0.5 mL of Brij-35. (Note: it takes only two drops of dilute NH_4OH. Dilute NH_4OH is prepared by adding four to five drops of concentrated NH_4OH to about 30 mL H_2O.)

5　Color reagent: to about 750 mL of double-distilled H_2O, carefully add 100 mL of concentrated H_3PO_4 and 10 g of sulfanilamide. Dissolve completely. Add 0.5 g of *N*-1-naphthyl-ethylenediamine dihydrochloride (Marshall's reagent), and dissolve. Dilute to 1 L volume. Add 0.5 mL of Brij-35. Store in an amber glass bottle. This reagent is stable for 1 month.

4.3.3 PROCEDURE

1　If refrigerated, bring the soil extracts to room temperature.

2　Shake extracts well.

3　Set up AutoAnalyzer® as shown in flow diagram (Figure 4.1, Technicon® Instrument Corporation 1971; Kalra and Maynard 1991) and allow the colorimeter to warm up for at least 30 min.

4　Place all reagent tubes in double-distilled H_2O and run for 10 min.

5　Insert tubes in correct reagents and run for 20 min to ensure thorough flushing of the system (feed 2.0 M KCl through the wash line).

6　Set the recorder baseline.

7　Place the sample tube in high standard for 5 min.

8　When the intensity appears on chart, set to 85–90% absorption.

9　Reset the baseline, if necessary.

10　Transfer standard solutions (in duplicate) to sample cups and arrange on the tray in descending order. Switch on the sampler.

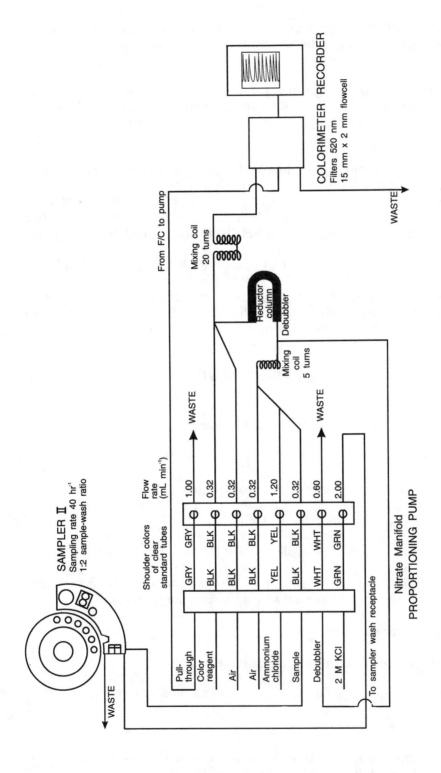

FIGURE 4.1. Flow diagram for the determination of NO_3-N in 2.0 *M* KCl extracts by autoanalyzer. (From Technicon Instrument Corporation [No. 32-69W and No. 154-71W], Bran and Luebbe, Inc., Buffalo Grove, IL. With permission.)

11 After running the standards, run the samples.

12 After run is complete, rerun the standards to ensure that there has been no drifting. Reestablish baseline.

13 Place the tubes in double-distilled H_2O, rinse and run for 20 min before turning the proportionating pump off.

4.3.4 Calculation (See Comment 5)

Prepare a standard curve from recorded readings (absorption vs. concentration) of standards and read off as μg NO_3-N mL^{-1} in KCl extract. For a 1:10 soil-to-solution ratio:

$$NO_3\text{-N in moist soil } (\mu g\ g^{-1}) = NO_3\text{-N } \mu g\ mL^{-1} \times 10 \qquad (4.1)$$

4.3.5 COMMENTS

1 Use double-distilled water throughout the procedure.

2 All reagent bottles, sample cups, and new pump tubes should be rinsed with approxiately 1 *M* HCl.

3 Extracts high in NO_3 should be diluted with 2.0 *M* KCl solution prior to analysis.

4 Range: 0.01–2.00 μg NO_3-N g^{-1} soil.

5 Results can also be calculated as follows:

$$\text{Moisture factor} = \frac{\text{moist soil (g)}}{\text{oven-dried soil (g)}} \qquad (4.2)$$

$$NO_3\text{-N in oven-dried soil } (\mu g\ g^{-1}) = NO_3\text{-N in moist soil } (\mu g\ g^{-1}) \times \text{moisture factor} \qquad (4.3)$$

6 The analysis of KCl extracts can also be performed by Bran and Luebbe TRAACS 800 continuous-flow AutoAnalyzer® system. The methods given for the AA-2 system have been converted for use on the TRAACS 800 (Tel and Heseltine 1990).

7 The method includes NO_3-N plus NO_2-N; therefore, samples containing significant amounts of NO_2-N will result in the overestimation of NO_3-N.

4.3.6 Precision and Accuracy

There are no standard reference samples for accuracy determination. Precision measurements for NO_3-N carried out for soil test quality assurance program of the Alberta Institute of

Pedology (Heaney et al. 1988) indicated that NO_3-N was one of the most variable parameters measured. Coefficient of variation ranged from 4.8 to 30.4% for samples with 67.3 ± 3.2 (S.D.) and 3.3 ± 1.0 (S.D.) μg NO_3-N g^{-1}, respectively.

4.4 DETERMINATION OF NH_4-N IN 2.0-*M* KCl EXTRACTS BY AUTOANALYZER INDOPHENOL BLUE PROCEDURE (PHENATE METHOD)

4.4.1 Principle

Ammonium is determined by an automated spectrophotometric method utilizing the Berthelot reaction (Searle 1984). Phenol and NH_4 react to form an intense blue color. The intensity of color is proportional to the NH_4 present. Sodium hypochlorite and sodium nitroprusside solutions are used as oxidant and catalyst, respectively.

4.4.2 MATERIALS AND REAGENTS

1   Technicon® AutoAnalyzer® consisting of sampler, manifold, proportioning pump, heating bath, colorimeter, and recorder.

2   Standard solutions:

   a. Stock solution #1 (1000 μg NH_4-N mL^{-1}): dissolve 4.7170 g $(NH_4)_2SO_4$ (dried at 105°C) and dilute to 1 L. Store the solution in a refrigerator.

   b. Stock solution #2 (100 μg NH_4-N mL^{-1}): dilute 10 mL of stock solution #1 to 100 mL with 2.0 *M* KCl solution. Store the solution in a refrigerator.

   c. Working standards: transfer 0, 1, 2, 5, 7, and 10 mL of stock solution #2 to 100-mL volumetric flasks. Make to volume with 2.0 *M* KCl. This will provide 0, 1, 5, 7, and 10 μg NH_4-N mL^{-1} standard solutions, respectively. Prepare daily.

3   Complexing reagent: dissolve 33 g of potassium sodium tartrate, KNa-$C_4H_4O_6 \cdot H_2O$, and 24 g of sodium citrate, $HOC(COONa)(CH_2COONa)_2 \cdot H_2O$, in 950 mL of H_2O. Adjust to pH 5.0 with concentrated H_2SO_4. Dilute to 1 L. Add 0.5 mL of Brij-35.

4   Alkaline phenol: using a 1-L Erlenmeyer flask, dissolve 83 g of phenol in 50 mL of H_2O. Cautiously add, in small increments with agitation, 180 mL of 20% (5 *M*) NaOH. Dilute to 1 L H_2O. Store alkaline phenol reagent in an amber bottle. (To make 20% NaOH, dissolve 200 g of NaOH and dilute to 1 L.)

5   Sodium hypochlorite (NaOCl): dilute 200 mL of household bleach (5.25% NaOCl) to 1 L. This reagent must be prepared daily, immediately before use to obtain optimum results because the NaOCl concentration in this reagent decreases on standing.

6   Sodium nitroprusside: dissolve 0.5 g of sodium nitroprusside (disodium nitroferricyanide) $Na_2Fe(CN)_5NO \cdot 2H_2O$, in 900 mL of H_2O and dilute to 1 L. Store in dark-colored bottle in a refrigerator.

4.4.3 Procedure

Follow the procedure (4.3.3) outlined for NO_3-N using the flow diagram given in Figure 4.2 (Technicon® Instrument Corporation 1973; Kalra and Maynard 1991).

4.4.4 Calculation

Same as given in 4.3.4.

4.4.5 COMMENTS

1   Use NH_4-free double-distilled water throughout the procedure.

2   See comments 2, 3, 5, and 6 in Section 4.3.5.

3   Range: 0.01–2.00 μ NH_4-N g^{-1} soil.

4   It is critical that the operating temperature is 50 ± 1°C.

4.4.6 Precision and Accuracy

There are no standard reference samples for accuracy determination. Long-term analyses of laboratory samples gave coefficient of variations of 21–24% for several samples over a wide range of concentrations.

4.5 EXTRACTION OF NO_3-N WITH 0.01 M $CuSO_4$ AND MANUAL COLORIMETRIC DETERMINATION

4.5.1 Principle

Nitrate is extracted with 0.01 M $CuSO_4$ containing Ag_2SO_4. A clear soil extract is obtained by using $CuSO_4$, $Ca(OH)_2$, and $MgCO_3$. Chloride interference is prevented by the use of Ag_2SO_4. Nitrate is determined colorimetrically by the nitrophenoldisulfonic yellow color method. The colorimetric method depends upon the nitration of position 6 of 2,4-phenol-disulfonic acid in fuming H_2SO_4 (Jackson 1958):

$$C_6H_3OH(HSO_3)_2 + HNO_3 \rightarrow C_6H_2OH(HSO_3)_2NO_2 + H_2O \qquad (4.4)$$

The product behaves as a nitrophenolic-type indicator; it is colorless in acid and yellow when neutralized or in alkaline solution, e.g., NH_4OH, due to the formation of triammonium salt (ammonium nitrophenoldisulfonic acid).

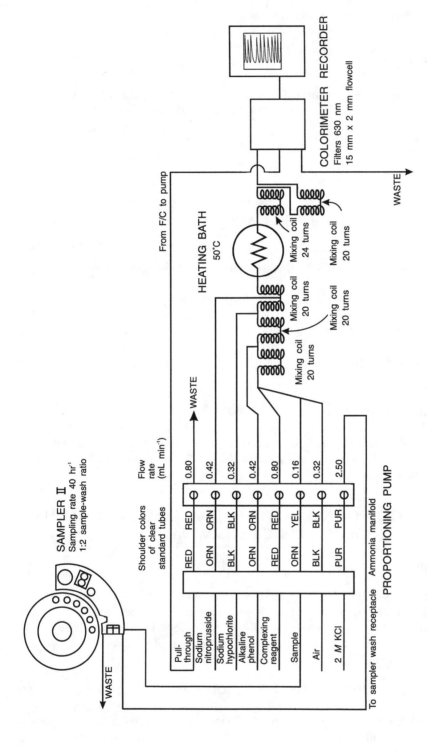

FIGURE 4.2. Flow diagram for the determination of NH$_4$-N in 2.0 M KCl extracts by autoanalyzer. (From Technicon Instrument Corporation [No. 32-69W and No. 154-71W], Bran and Luebbe, Inc., Buffalo Grove, IL. With permission.)

4.5.2 MATERIALS AND REAGENTS

1   Reciprocating shaker.

2   Heavy-duty hot plate.

3   Spectrophotometer.

4   Dispenser.

5   Phenoldisulfonic acid (phenol 2,4-disulfonic acid): transfer 70 mL pure liquid phenol (carbolic acid) to an 800-mL Kjeldahl flask. Add 450 mL concentrated H_2SO_4 while shaking. Add 225 mL fuming H_2SO_4 (13–15% SO_3). Mix well. Place Kjeldahl flask (loosely stoppered) in boiling water in a beaker and heat for 2 h. Store resulting phenoldisulfonic acid [$C_6H_3OH(HSO_3)_2$] solution in a glass-stoppered bottle.

6   Dilute ammonium hydroxide (about 7.5 M NH_4OH): mix one part NH_4OH (sp gr 0.90) with one part H_2O.

7   Copper sulfate solution (0.5 M): 125 g $CuSO_4{\cdot}5H_2O$ L^{-1}.

8   Silver sulfate solution (0.6% solution): 6.0 g Ag_2SO_4 L^{-1}. Heat or shake well until all salt is dissolved.

9   Nitrate-extracting solution: mix 200 mL of 0.5 M copper sulfate solution and 1 L 0.6% silver sulfate solution and dilute to 10 L with water. Mix well.

10 Standard nitrate solution (100 μg mL^{-1} stock solution): dissolve 0.7221 g KNO_3 (oven dried at 105°C) in water and dilute to 1 L. Mix thoroughly.

11 Standard nitrate solution (10 μg N mL^{-1}; working solution): dilute 100 mL of 100 μg N mL^{-1} stock solution to 1 L with water. Mix well.

12 Calcium hydroxide, reagent-grade powder.

13 Magnesium carbonate, reagent-grade powder.

4.5.3 PROCEDURE

1   Place moist soil equivalent to 5.0 g oven-dried soil (2.5 g of peat) in an Erlenmeyer flask.

2   Add 25 mL nitrate-extracting solution.

3   Shake contents for 10 min.

4   Add about 0.2 g $Ca(OH)_2$ and shake for 5 min.

5   Add about 0.5 g $MgCO_3$ and shake for 10–15 min.

6   Allow to settle for a few minutes.

7   Filter through a Whatman® filter paper No. 42.

8   Pipet 10 mL of clear filtrate into a 100-mL beaker. Evaporate to dryness on a hot plate at low heat in a fumehood free from HNO_3 fumes. Do not continue heating beyond dryness.

9   When completely dry, cool residue, add 2 mL phenoldisulfonic acid rapidly (from a buret having the tip cut off or a dispensette) covering the residue quickly. Rotate beaker so that reagent comes in contact with all residual salt. (Caution: *phenoldisulfonic acid is very corrosive.*) Allow to stand for 10–15 min.

10 Add 16.5 mL cold water. Rotate beaker to dissolve residue (stir with a glass rod until all residue is in solution).

11 After beakers are cool, add dilute NH_4OH slowly until solution is distinctly alkaline as indicated by the development of a stable yellow color (15 mL).

12 After beakers are cool, add 16.5 mL water (volume at the end of step 12 is 50 mL). Mix thoroughly.

13 Read concentration of NO_3-N at 415 nm.

14 Standards: evaporate 0, 2, 5, 8, and 10 mL of the 10-µg NO_3-N mL^{-1} working solution after adding 10 mL NO_3-extracting solution in 100-mL beakers and evaporate to dryness. Follow steps 9–13. After step 12, these solutions will have 0, 0.40, 1.00, 1.60, and 2.00 µg NO_3-N mL^{-1}.

4.5.4 Calculation

NO_3-N in soil (µg g^{-1}) =

$$
NO_3\text{-N in test solution} \times \frac{\text{Volume after color development (mL)}}{\text{Volume evaporated (mL)}} \times \frac{\text{Volume of extracting solution (mL)}}{\text{Weight oven-dried soil (g)}} \quad (4.5)
$$
(µg mL^{-1})

4.5.5 COMMENTS

1   Depending upon the expected NO_3 content of soil, 5 to 25 mL of soil extract should be evaporated.

2   Calcium hydroxide, $MgCO_3$ and Ag_2SO_4, should be free of nitrate.

3   A perfectly clear and colorless soil extract must be obtained in order to secure accurate results with this method. Colored soil extracts should be decolorized with activated charcoal, or blanks must be prepared from the extracts to zero spectrophotometer.

4   Phenoldisulfonic acid should be added when samples have completely dried.

5   Ammonium hydroxide loses strength after long storage. Therefore, a fresh reagent should be used.

6   This method is used when only NO_3 is needed and only a limited number of samples are to be analyzed.

4.5.6 Precision and Accuracy

Insufficient data available.

REFERENCES

Blakemore, L. C., Searle, P. L., and Daly, B. K. 1987. Methods for chemical analysis of soils. N.Z. Soil Bureau Scientific Report 80, Lower Hutt, N.Z.

Bremner, J. M. 1965. Inorganic forms of nitrogen. Pages 1179–1237 in C. A. Black et al., Eds. Methods of soil analysis. Part 2. Agronomy No. 9. American Society of Agronomy, Madison, WI.

Haynes, R. J. and Swift, R. S. 1989. Effect of rewetting air-dried soils on pH and accumulation of mineral nitrogen. J. Soil Sci. 40: 340–347.

Heaney, D. J., McGill, W. B., and Nguyen, C. 1988. Soil test quality assurance program, final report. Alberta Institute of Pedology, Edmonton, Alberta.

Heffernan, B. 1985. A handbook of methods of inorganic chemical analysis for forest soils, foliage and water. CSIRO, Canberra, Australia.

Houba, V. J. G., Lee, J. J. van der, Novozamsky, I., and Walinga, I. 1988. Soil and plant analysis. Part 5: Soil analysis procedures. Wageningen Agricultural University, The Netherlands.

Jackson, M. L. 1958. Soil chemical analysis. Prentice-Hall, Englewood Cliffs, NJ.

Kalra, Y. P. and Maynard, D. G. 1991. Methods manual for forest soil and plant analysis. Northern Forestry Centre, Northwest Region, Forestry Canada, Edmonton, Alberta. Information Report NOR-X-319.

Keeney, D. R. and Nelson, D. W. 1982. Nitrogen in organic forms. Pages 643–698 in A. L. Page et al., Eds. Methods of soil analysis. Part 2. Agronomy No. 9, American Society of Agronomy, Madison, WI.

Laverty, D. H. and Bollo-Kamara, A. 1988. Recommended methods of soil analyses for Canadian

Prairie agricultural soils. Alberta Agriculture, Edmonton, Alberta. Rep. 300.

Ministry of Agriculture, Fisheries and Food. 1986. The analysis of agricultural materials; a manual of the analytical methods used by the Agricultural Development and Advisory Service, London. Reference Book 427.

Muneta, P. 1980. Analytical errors resulting from nitrate contamination of filter paper. J. Assoc. Off. Anal. Chem. 63: 937–938.

Searle, P. L. 1984. The Berthelot or indophenol reaction and its use in the analytical chemistry of nitrogen: a review. Analyst 109: 549–568.

Selmer-Olsen, A. R. 1971. Determination of ammonium in soil extracts by an automated indophenol method. Analyst 96: 565–568.

Soltanpour, P. N. and Workman, S. M. 1981. Soil-testing methods used at Colorado State University soil testing laboratory for the evaluation of fertility, salinity, sodicity and trace element toxicity. Technical Bulletin 142. Colorado State University, Fort Collins, CO.

Sparrow, S. D. and Masiak, D. T. 1987. Errors in analysis for ammonium and nitrate caused by contamination from filter papers. Soil Sci. Soc. Am. J. 51: 107–110.

Technicon Instrument Corporation. 1971. Nitrate + nitrite in water. Industrial method No. 32–69W. Technicon Instrument Corporation, Tarrytown, NY.

Technicon Instrument Corporation. 1973. Ammonia in water and seawater. Industrial method No. 154–71W, Technicon Instrument Corporation, Tarrytown, NY.

Tel, D. A. and Heseltine, C. 1990. The analyses of KCl soil extracts for nitrate, nitrite and ammonium using a TRAACS 800 analyzer. Commun. Soil Sci. Plant Anal. 21: 1681–1688.

Chapter 5
Ammonium Acetate-Extractable Elements

R. R. Simard

Agriculture Canada
Sainte-Foy, Quebec, Canada

5.1 INTRODUCTION

The alkali and alkaline earth metals in soils can be broadly classified in four different fractions which are in dynamic equilibrium with each other: solution, rapidly exchangeable, slowly exchangeable, and structural (Simard et al. 1990). The neutral 1 M ammonium acetate (NH_4OAc) extraction method is the most widely used procedure to extract the water-soluble and rapidly exchangeable fractions (Knudsen et al. 1982, Lanyon and Heald 1982). This method is used to assess the amount of ''plant-available'' K (McLean and Watson 1985), Mg (Metson 1974), and Ca (Doll and Lucas 1973). Na, Li, Ba, and Sr, which are also extracted by 1 M NH_4OAc, are not essential to plant growth. Soils are tested for Na to diagnose sodic and sodic-saline problems. Nutritional studies might require information about the amounts of plant-available Li, Sr, and Ba.

The amount of plant-available K, Ca, and Mg has also been estimated by extractions in water, dilute salts (Tran et al. 1987), dilute acids (Richards and Bates 1989), and by multielement extraction procedures such as the Kelowna (Van Lierop and Gouch 1989), Mehlich III (Simard et al. 1990), and 0.02 M $SrCl_2^-$-citric acid 0.05 M (Simard et al. 1989b) methods. The amounts extracted by these procedures are different, but are often very highly correlated with each other (Van Lierop and Tran 1985, Van Lierop and Gouch 1989).

Extractions with boiling 1 M HNO_3 or 1 M HCl are used to determine the amount of slowly exchangeable K and Mg (Simard et al. 1989a). This fraction has been used with the NH_4OAc-extractable fractions to predict the amount of plant-available K for crops with large requirements, such as silage corn and forage legumes (Smith and Matthews 1957, Richards and Bates 1988).

5.2 AMMONIUM ACETATE EXTRACTION

This method extracts the water-soluble and rapidly exchangeable fractions of the alkali and alkaline earth cations by displacement with NH_4^+ from the exchange sites. This is done by shaking the soil with a solution of 1 *M* NH$_4$OAc adjusted to pH 7. After filtration, the amounts of K, Na, and Li are determined preferably by flame emission and Ca, Mg, Sr, and Ba by atomic absorption spectroscopy (AAS).

5.2.1 MATERIALS AND REAGENTS

1   Reciprocating shaker.

2   Erlenmeyer flasks, 125 mL.

3   Filter funnels.

4   Filter paper Whatman® No. 2 or equivalent.

5   Disposable plastic vials.

6   Flame photometer, atomic absorption spectrophotometer (AAS), or inductively coupled plasma spectrophotometer.

7   A 3-cm^3 calibrated soil scoop.

8   Ammonium acetate (NH$_4$OAc), 1.0 *M* adjusted to pH 7. For each liter of solution desired, dissolve 77.08 g of NH$_4$OAc with distilled water or add 58 mL of glacial acetic acid (99%) to about 600 mL of water, and then add 70 mL of concentrated ammonium hydroxide (NH$_4$OH). The NH$_4$OH is best added under a fume hood through a long-stemmed funnel so that it is introduced into the bottom of the acid solution; let the solution cool down before adjusting the pH to 7.00 with NH$_4$OH or acetic acid using a pH meter.

9   Lanthanum solution. For each liter of solution desired, dissolve carefully 46.88 g of La$_2$O$_3$ in about 200 mL of 12 *M* HCl and make to volume with distilled water. This solution should contain 4% La.

10 Standard solutions. Certified atomic absorption standards or dilute separately the following quantities of oven-dried, reagent-grade compounds in 1 L of distilled water: K: 1.908 g KCl; Na: 2.542 g NaCl; Li: 6.108 g LiCl; Ca: 3.6681 g CaCl$_2$·2H$_2$O; Mg: 8.3632 g MgCl$_2$·6H$_2$O; Sr: 3.043 g SrCl$_2$·6H$_2$O; Ba:1.779 g BaCl$_2$·2H$_2$O. This will give a 1000-ppm stock solution for each metal. Several metals can be prepared in combination in a single solution. Standards are then prepared in the desired range, depending on the specifications of the spectrophotometer. Prepare solutions of 100 ppm by diluting the 1000-ppm stock solutions in the ammonium acetate extracting solution. An example for a 10-ppm standard add: 10 mL of the 100-ppm solution containing each of the cations and add 25 mL of the 4% lanthanum solution and make to volume with distilled water in a 100-mL volumetric flask.

5.3 PROCEDURE

5.3.1 Extraction and Analysis

1 Weigh 3.00 g of air-dried <2 mm soil.

2 Add the soil sample to a 125-mL Erlenmeyer flask and add 30 mL of the extracting solution with a repipet dispenser.

3 Prepare the standards in the concentration range recommended for your spectrometer, in the same way as your unknowns. The matrix of your samples and standards should contain the same quantities of La and NH_4OAc.

4 Stopper the flask and shake for a period of 15 min on a reciprocating shaker set at 120 oscillations per minute.

5 Filter extracts through Whatman® No. 2 or similar grade paper. Refilter if the extract is not clear. Save the filtrate in a plastic vial for analysis.

6 Dilute an aliquot of the extract, with a repipet device, with distilled water containing 1% La (wt/vol); a dilution of 1:10 would be appropriate for K, Na, Sr, and Ba and 1:40 for Ca and Mg.

5.3.2 Calculations

Extractable amount (kg/ha) = Reading (ppm) × 22.4 × dilution factor.

5.3.3 Comments

In areas where commercial laboratories use the scoop technique, a mass for the 3.0 mL of soil should be determined if the results are to be used in soil test calibration. This method does not differentiate Ca and Mg extracted from their carbonates; accordingly, calcareous soils should not be analyzed for their NH_4OAc-extractable Ca and Mg contents. The amounts of rapidly exchangeable cations are obtained by subtracting their water-soluble contents. Sodic or gypsiferous soils are not normally analyzed by this procedure. Since NH_4^+ forms inner-sphere surface complexes with 2:1 layer-type clay minerals and since it can even dislodge cations from easily weathered primary soil minerals, the use of this method to measure soil cation exchange capacity has significant potential for inaccuracy (Sposito 1989).

In Canada, the critical levels for the ammonium acetate-extractable K are 100–150 ppm for canola; 120–160 ppm for cereals, corn, and soya; 125–200 ppm for grass; and 150–200 for alfalfa (Neufeld 1980, Bates 1989, CPVQ 1989). Magnesium soil tests are generally based on the amounts of ammonium acetate-extractable Mg, although certain regions also consider the amounts of extractable K and Ca in their recommendations. Soils are rarely deficient in Ca, per se, but might require additions of lime to correct possible elemental (Al, Mn) toxicities.

REFERENCES

Bates, T. E. 1989. Fertilizer requirement tables for field crops in Ontario. 1988 revision. University of Guelph, Guelph, Ont. 27 pp.

Conseil des productions végétales du Québec. 1989. Grilles de fertilization 1989. Gouv. Québec. 18 pp.

Doll, E. C. and Lucas, R. E. 1973. Testing soils for potassium, calcium and magnesium. Pages 133–151 in L. M. Walsh and J. D. Beaton, Eds. Soil testing and plant analysis. Soil Sci. Soc. Am., Madison, WI.

Knudsen, D., Paterson, G. A., and Pratt, P. F. 1982. Lithium sodium and potassium. Pages 225–246 in A. L. Page et al., Eds. Methods of soil analysis. Part 2. 2nd ed. Agronomy No. 9. American Society of Agronomy, Madison, WI.

Lanyon, L. E. and Heald, W. R. 1982. Magnesium, calcium, strontium and barium. Pages 247–262 in A. L. Page et al., Eds. Methods of soil analysis. Part 2. 2nd ed. Agronomy No. 9. American Society of Agronomy, Madison, WI.

McLean, E. O. and Watson, M. E. 1985. Soil measurements of plant-available potassium. Pages 277–308 in R. D. Munson, Ed. Potassium in Agriculture. American Society of Agronomy, Madison, WI.

Metson, A. J. 1974. Magnesium in New Zealand soils. I. Some factors governing the availability of soil magnesium. A review, N.Z. J. Exp. Agric. 2: 329–347.

Neufeld, J. H. 1980. Soil testing methods and interpretations. British Columbia Ministry of Agriculture Publ. 80–2. 29 pp.

Richards, J. E. and Bates, T. E. 1988. Studies on the potassium-supplying capacities of southern Ontario soils. II. Nitric acid extraction of non-exchangeable K and its availability to crops. Can. J. Soil Sci. 68: 199–208.

Richards, J. E. and Bates, T. E. 1989. Studies on the potassium-supplying capacities of southern Ontario soils. III. Measurement of available K. Can. J. Soil Sci. 69: 597–610.

Simard, R. R., De Kimpe, C. R., and Zizka, J. 1989a. The kinetics of non-exchangeable potassium and magnesium release from Québec soils. Can. J. Soil Sci. 69: 663–675.

Simard, R. R., Deschênes, M., and Zizka, J. 1989b. Sr-citrate, une nouvelle solution d'extraction des éléments nutritifs des sols. Proc. 35th Annu. Meet. Can. Soc. Soil Sci. p. 34.

Simard, R. R., Zizka, J., and De Kimpe, C. R. 1990. Le prélèvement du K par la luzerne (*Medicago sativa* L.) et sa dynamique dans 30 sols du Québec. Can. J. Soil Sci. 70: 379–393.

Smith, J. A. and Matthews, B. C. 1957. Release of potassium by 18 Ontario soils during continuous cropping in the greenhouse. Can. J. Soil Sci. 37: 1–10.

Sposito, G. 1989. The chemistry of soils. Oxford University Press, New York. 277 pp.

Tran, T. S., Tabi, M., and De Kimpe, C. R. 1987. Relation du potassium extrait par EUF et quelques méthodes chimiques avec les propriétés du sol et le rendement des plantes. Can. J. Soil Sci. 67: 17–31.

Van Lierop, W. and Tran, T. S. 1985. Comparative potassium levels removed from soils by electro-ultrafiltration and some chemical extractants. Can. J. Soil Sci. 65: 25–34.

Van Lierop, W. and Gouch, N. A. 1989. Extraction of potassium and sodium from acid and calcareous soils with the Kelowna multiple element extractant. Can. J. Soil Sci. 69: 235–242.

Chapter 6
Mehlich III-Extractable Elements

T. Sen Tran

Ministère de l'Agriculture, Sainte-Foy, Quebec, Canada

R. R. Simard

Agriculture Canada, Sainte-Foy, Quebec, Canada

6.1 INTRODUCTION

The Mehlich III method was developed by Mehlich (1984) as multielement soil extraction. This extractant is composed of 0.2 M CH$_3$COOH, 0.25 M NH$_4$NO$_3$, 0.015 M NH$_4$F, 0.013 M HNO$_3$, and 0.001 M ethylene diamine tetraacetic acid (EDTA). In the Mehlich III procedure, phosphorus is extracted by reaction with acetic acid and fluoride compounds. Exchangeable K, Ca, Mg, and Na are extracted by the action of ammonium nitrate and nitric acid. Finally, the micronutrients (Cu, Zn, Mn, and Fe) are extracted by NH$_4$ and the chelating agent EDTA (Figure 6.1).

The amount of Mehlich III-extractable P was found to be closely related to that determined by other conventional extractants such as Bray 1, Bray 2, Olsen, Mehlich II, and bicarbonate resin (Mehlich 1984, Wolf and Baker 1985, Michaelson et al. 1987, Ness et al. 1988, Tran et al. 1990). The Mehlich III extractant is neutralized less by carbonate compounds in soil than the double acid (Mehlich I) and the Bray 1, and is less aggressive towards apatite or other calcium phosphate than the double acid and Bray 2 extractants. The amount of Mehlich III-extractable P is highly correlated with plant P uptake in a wide range of soils in Quebec (Tran and Giroux 1985, 1989).

The amounts of exchangeable bases (K, Ca, Mg, Na) determined by the Mehlich III method are nearly identical to those obtained by the ammonium acetate method (Hanlon and Johnson 1984, Michaelson et al. 1987, Tran and Giroux 1989). Moreover, the Mehlich III-extractable amounts of Cu, Zn, Mn, and Fe are also closely related to those obtained by the double acid, diethylene triamine pentaacetic acid-triethanolamine (DTPA-TEA), or 0.1 M HCl methods and can be used in soil testing for micronutrients (Mehlich 1984; Makarim and Cox 1983, Mascagni and Cox 1985, Tran 1989).

Soil Sampling and Methods of Analysis, M. R. Carter, Ed.,
Canadian Society of Soil Science. © 1993 Lewis Publishers.

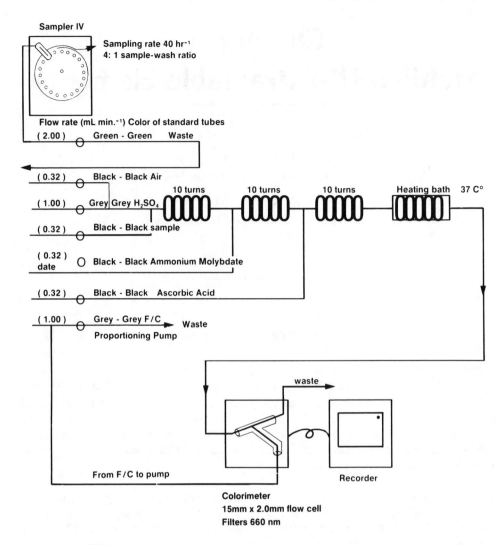

FIGURE 6.1. Flow diagram for the automated analysis of P in Mehlich III soil extract.

6.2 MATERIALS AND REAGENTS

1   Reciprocating shaker.

2   Erlenmeyer flasks, 125 mL.

3   Filter funnels.

4   Filter paper (Whatman® No. 42).

5   Disposable plastic vials.

6 Spectrophotometer for conventional colorimetry at 882 μg or automatic colorimetric (Technicon® AutoAnalyzer®), flame photometer, or atomic absorption spectrophotometer or inductively coupled plasma atomic-emission spectrophotometer (ICP).

7 Mehlich III extracting solution

a. Stock solution M-3: (1.5 M NH_4F + 0.1 M EDTA). Dissolve 55.56 g of ammonium fluoride (NH_4F) in 600 mL of distilled water. Add 29.23 g of EDTA (F.W. 292.24) to this mixture, dissolve, bring to 1 L, mix thoroughly, and store in plastic bottle.

b. In a plastic carboy containing 8 L of distilled water, add 200.1 g of ammonium nitrate (NH_4NO_3), 100 mL of stock solution M-3, 115 mL acetic acid (CH_3COOH), 82 mL of 10% v/v nitric acid (10 mL concentrate 70% HNO_3 in 100 mL of water), dissolve, bring to 10 L with distilled water, and mix thoroughly.

8 Solutions for the manual determination of phosphorus.

a. Solution A: dissolve 12 g of ammonium molybdate ($NH_4)_6$ Mo_7O_{24}·$4H_2O$ in 250 mL of distilled water. In a 100-mL flask, dissolve 0.2908 g of potassium antimony tartrate in 80 mL of water. Transfer these two solutions into a 2-L volumetric flask containing 1000 mL of 2.5 M H_2SO_4 (141 mL concentrated H_2SO_4/L), bring to 2 L with water, mix thoroughly, and store in the dark at 4°C.

b. Solution B: dissolve 1.056 g of ascorbic acid in 200 mL of solution A, prepare this solution daily.

c. Standard solution of P: use certified P standard or prepare a solution of 100 μg g^{-1} P by dissolving 0.4393 g of KH_2PO_4 in 1 L of distilled water. Prepare standard solutions of 0, 2, 4, 6, 8, and 10 μg g^{-1} P in diluted Mehlich III extractant. The P concentration in the Mehlich III extractant can be determined by other colorimetric methods as in the case of Bray 1-extractable P.

9 Solutions for automated determination of phosphorus Technicon® II AutoAnalyzer®, modified industrial method No. 94–70W: Technicon® AutoAnalyzer II 1973).

a. Molybdate-antimony solutions: dissolve 30 g of ammonium molybdate ($NH_4)_6$ Mo_7O_{24}·$4H_2O$ in 600 mL of water. Add 0.15 g of potassium antimony tartrate ($KSbO$·$C_4H_4O_6$) and bring to 1 L volume with water.

b. 1 M sulfuric acid: dilute 56 mL of concentrated sulfuric acid in 400 mL of water, cool the solution. Add 1 mL of Aerosol® 22 agent and dilute to 1 L with water. This solution has to be prepared daily.

c. Ascorbic acid solution: dissolve 12 g of ascorbic acid in 200 mL of water, add 1 mL of "Levor IV" wetting agent, and mix altogether.

d. Standard solutions of P: as shown in 8.c.

10 Solutions for K, Ca, Mg, Na determination by atomic absorption:

a. 10% w/v lanthanum chloride solution.

b. Solution 0.06% CsCl + 0.2% $LaCl_3$: dissolve 3.16 g of cesium chloride in 100 mL of a lanthanum chloride solution ($LaCl_3$ 10%). To avoid the precipitation of this solution in contact with Mehlich III soil extract, it is recommended to use diluted extract with water (1:10).

c. Combined K and Na standard solutions: use certified atomic absorption standard and prepare solutions of 0.5, 1.0, 1.5, 2.0, and 0.3, 0.6, 0.9, and 1.2 ppm of K and Na, respectively.

d. Combined Ca and Mg standard solutions. Prepare 2, 4, 6, 8, 10, and 0.2, 0.4, 0.6, 0.8, and 1.0 $\mu g\ g^{-1}$ of Ca and Mg, respectively.

11 Standard solution for Cu, Zn, Mn determination by atomic absorption.

a. Combined Cu and Zn standard solution: 0, 0.2, 0.4, 0.8, 1.2 to 2.0 $\mu g\ g^{-1}$ of Cu and of Zn in Mehlich III extractant.

b. Mn standard solutions: prepare 0, 0.4, 0.8, 1.2 to 4 $\mu g\ g^{-1}$ of Mn in diluted Mehlich III extractant.

6.3 PROCEDURE

6.3.1 Extraction

1 Weigh 3 g of soil passed through a 2-mm sieve into a 125-mL Erlenmeyer flask.

2 Add 30 mL of the Mehlich III extracting solution (soil:solution ratio 1:10).

3 Shake immediately on rotative shaker for 5 min (120 oscillations/min).

4 Filter through No. 42 Whatman filter paper and save the filtrate in plastic vials. Analyses should be made as soon as possible.

6.3.2 Determination of P by Manual Colorimetric Method

1 Pipet 2 mL of the clear filtrate into a 25-mL volumetric flask.

2   Add 15 mL of distilled water and 4 mL of solution B, dilute to 25 mL with distilled water, and mix.

3   After 10 min for color development, measure the absorbance at 882 μm.

6.3.3 Determination of P by Automated Method (Technicon® AutoAnalyzer®)

1   Turn on the different modules of AutoAnalyzer® for at least 30 min to warm up.

2   Put each tube into its reagent as indicated in the following diagram (Figure 6.1). Run pump for 20 min to equilibrate the system.

3   Adjust the base line with the Mehlich III extractant and the maximum absorbance level of the recorder with standard solution containing highest P concentration.

4   Arrange standards and samples on sampler and run the system. Use water with some "Levor IV" wetting agent added as a washing solution.

6.3.4 Determination of K, Ca, Mg, and Na by Atomic Absorption or by Flame Emission

1   Pipet 1 to 5 mL of filtrate into a 50-mL volumetric flask.

2   Add distilled water and mix.

3   Add 1 mL of CsCl + LaCl$_3$ solution, bring to volume with distilled water.

4   Determine Ca, Mg by atomic absorption and K, Na by flame emission.

6.3.5 Determination of Cu, Zn, and Mn by Atomic Absorption

Cu and Zn concentrations in the extract are determined without dilution while the Mn concentration is determined in diluted Mehlich III extract.

6.4 COMMENTS

The Mehlich III method is used in the soil testing program in Quebec and the Bray 2 soil test levels for different crops have been converted to Mehlich III levels (CPVQ, 1989). The critical level of Mehlich III-extractable P for most common crops was about 50 to 60 μg g^{-1} (Tran and Giroux 1985; Sims 1989; Simard et al. 1991). On most noncalcareous soils, the amount of Mehlich III-extractable P is approximately the same as that determined by the Bray 1 method (Tran et al. 1990).

No conversion is needed for K and Na as the amounts of K and Na extracted by Mehlich III are equal to those determined by ammonium acetate. For exchangeable Ca and Mg, the

Mehlich III extracts about 1.10 times more than ammonium acetate, and this correction is included for the conversion of Mehlich III values to ammonium acetate values.

The critical level proposed by Lindsay and Norvell (1978) for DTPA-extractable Zn in soils varied from 0.6 to 1 μg g^{-1}; these levels correspond to about 1.2 to 1.8 of Mehlich III-extractable Zn. The critical level of DTPA-extractable Cu (0.2 μg g^{-1}) corresponds to about 0.4 μg g^{-1} of Mehlich III-extractable Cu (Tran 1989; Makarim and Cox 1983). For higher Zn and Cu levels, these regression equations can be used for mineral soils (Tran 1989):

Mehlich III Zn (ppm) = 0.64 + 1.15 DTPA-Zn $R^2 = 0.722**$
Mehlich III Zn (ppm) = 0.16 + 0.50 (0.1 N HCl)-Zn $R^2 = 0.922**$
Mehlich III Cu (ppm) = 0.29 + 1.24 DTPA-Cu $R^2 = 0.864**$
Mehlich III Cu (ppm) = 0.07 + 0.66 (0.1 N HCl)-Cu $R^2 = 0.828**$

The Mehlich III critical Mn levels proposed by Mascagni and Cox (1985) for soybean were 3.9 and 8.0 mg L^{-1} at soil pH 6 and 7, respectively. Sims (1989) reported the Mehlich III critical values of 7.7 to 16.9 mg L^{-1} for Mn and 0.3 to 2.7 mg L^{-1} for Zn, at soil pH from 6.0 to 7.0, respectively; while for Cu, Mehlich III critical values ranged from 0.5 to 1.1 mg L^{-1} for soils containing 2 to 8% of organic matter.

REFERENCES

C.P.V.Q. Grille de fertilisation 1989. Conseil des productions végétales du Québec. Ministère de l'Agriculture, des Pêcheries et de l'Alimentation du Québec.

Hanlon, E. A. and Johnson, G. V. 1984. Bray/Kurtz, Mehlich-III, AB/D and acetate extractions of P, K, and Mg in four Oklahoma soils. Comm. Soil Sci. Plant Anal. 15: 277–294.

Lindsay, W. L. and Norvell, W. A. 1978. Development of a DTPA soil test for zinc, iron, manganese, and copper. Soil Sci. Soc. Am. J. 42: 421–428.

Makarim, A. K. and Cox, F. R. 1983. Evaluation of the need for copper with several soil extractants. Agron. J. 75: 493–496.

Mascagni, H. J. and Cox, F. R. 1985. Calibration of a manganese availability index for soybean soil test data. Soil Sci. Soc. Am J. 49: 382–386.

Mehlich, A. 1984. Mehlich-3 soil test extractant: a modification of Mehlich-2 extractant. Comm. Soil Sci. Plant Anal. 15: 1409–1416.

Michaelson, G. J., Ping, C. L., and Mitchell, C. A. 1987. Correlation of Mehlich-3, Bray 1 and ammonium acetate extractable P, K, Ca, and Mg for Alaska agricultural soils. Commun. Soil Sci. Plant Anal. 18: 1003–1015.

Ness, P., Grava, J., and Bloom, P. R. 1988. Correlation of several tests for phosphorus with resin extractable phosphorus for 30 alkaline soils. Commun. Soil Sci. Plant Anal. 19: 675–689.

Simard, R. R., Tran, T. Sen, and Zizka, J. 1991. Strontium chloride-citric acid extraction evaluated as a soil-testing procedure for phosphorus. Soil Sci. Soc. Am. J. 55: 414–421.

Sims, J. T. 1989. Comparison of Mehlich-1 and Mehlich-3 extractants for P, K, Ca, Mg, Mn, Cu and Zn in Atlantic Coastal plain soils. Commun. Soil Sci. Plant Anal. 20: 1707–1726.

Technicon Auto Analyzer II. 1973. Orthophosphate in water and wastewater. Industrial method No. 94–70W.

Tran, T. Sen 1989. Détermination des minéraux et oligo-éléments par la méthode Mehlich-III. Méthodes d'analyse des sols, des fumiers, et des tissus végétaux. Conseil des productions végétales du Québec. Ministère de l'Agriculture, des Pêcheries et de l'Alimentation du Québec. Agdex. 533.

Tran, T. Sen and Giroux, M. 1985. Disponibilité du phosphore dans les sols neutres et calcaires du Québec en relation avec les propriétés chimiques et physiques. Can. J. Soil Sci. 67: 1–16.

Tran, T. Sen and Giroux, M. 1989. Evaluation de la méthode Mehlich-III pour déterminer les éléments nutritifs (P, K, Ca, Mg, Na) des sols du Québec. Agrosol. 2: 27–33.

Tran, T. Sen, Giroux, M., Guilbeault, J., and Audesse, P. 1990. Evaluation of Mehlich-III extractant to estimate the available P in Quebec soils. Commun. Soil Sci. Plant Anal. 21: 1–28.

Wolf, A. M. and Baker, D. E. 1985. Comparisons of soil test phosphorus by Olsen, Bray P1, Mehlich-I and Mehlich-III methods. Commun. Soil Sci. Plant Anal. 16: 467–484.

Chapter 7
Sodium Bicarbonate-Extractable P, K, and N

J. J. Schoenau and R. E. Karamanos

University of Saskatchewan

Saskatoon, Saskatchewan, Canada

7.1 OVERVIEW

Sodium bicarbonate ($NaHCO_3$)-extractable P is commonly used as an index of soil-available P. Sodium bicarbonate acts through a pH and ion effect to remove solution phosphate plus some labile solid phase P compounds such as phosphate adsorbed to free $CaCO_3$, slightly soluble calcium phosphate precipitates, and phosphate loosely sorbed to Al and Fe oxide surfaces. Sodium bicarbonate-extractable P, commonly termed Olsen P (Olsen et al. 1954), appears to be a good predictor of relative P availability in any given soil type, be it acidic or calcareous (Kamprath and Watson 1980, Matar et al. 1988). However, the test has been criticized as being geographically limited, since bicarbonate-extractable P levels do not indicate relative P availability when comparing soils with large differences in P chemistry and forms (Menon et al. 1988).

In a sodium bicarbonate extraction, the Na ions act through mass action to displace K ions from negatively charged sites on the soil colloids. Thus, the extract includes soil solution K plus "exchangeable" K and together they constitute the readily available K pool in soils (McLean and Watson 1985). While sodium can exchange with K ions held on the planar surface of clay minerals, it is relatively inefficient in removing K ions held in the interlayer spaces of 2:1 clay minerals due to the large hydrated radius of the sodium ion compared to potassium. While interlayer or fixed K is not considered a readily available K source, it is slowly released via weathering and in some soils, depending on the mineralogy and the environmental conditions, can make a significant contribution to plant-available K over a growing season. NH_4-acetate extractions will remove a larger portion of the slowly available K pool, since the ammonium ion is approximately the same size as the K ion and can fit into the interlayer spaces.

To assess the potential soil contribution to plant-available N over a growing season, one must take into account the inorganic N (NO_3^- and NH_4^+) present at the beginning of the growing season plus whatever is mineralized from the soil organic matter over the time period in question. Nitrate exists almost exclusively in soil solution and can be removed by

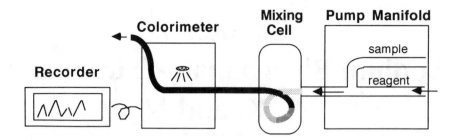

FIGURE 7.1. Schematic of typical automated analysis system.

any aqueous extractant, including $NaHCO_3$. However, water or a dilute salt solution will only remove a portion of the ammonium, since much of it exists adsorbed to negatively charged colloids. In warm, well-aerated soils of neutral pH, nitrification is rapid; therefore, nitrate is the main inorganic N form, but in acidic or uncultivated soils, the nitrification process may be inhibited so that ammonium predominates.

In high-rainfall areas, mineralization of organic N over the growing season may constitute the main source of available N, since stored inorganic N prior to seeding is often negligible. In such regions, a soil test which can provide an estimate of potentially mineralizable N is especially useful. $NaHCo_3$-extractable organic N (MacLean 1964) has been shown to be closely related to available N as assessed through plant uptake experiments (Smith 1966). In early work, the total N content of the bicarbonate extract was determined through Kjeldahl digestion; a time-consuming and difficult procedure not well suited to routine analysis. More recently, a simpler UV absorbance technique has been developed and shown to be highly correlated with N uptake by corn plants (Fox and Piekielek 1978).

Simultaneous N, P, and K determination in a single extract offers considerable time and cost savings. In this chapter, procedures are outlined for sodium bicarbonate extraction and the measurement of phosphate, potassium, and nitrate ions contained within the extract. Automated systems are frequenty employed to measure the concentration of nutrient ions in the extract and references to the materials and methodology are provided. The most commonly used automated analysis system is the Technicon® AutoAnalyzer®. In a typical automated system (Figure 7.1), a sampler automatically introduces sample into the flow stream. A proportioning pump and manifold moves the sample and reagent streams into the system. Samples and reagents are then mixed in a mixing cell, allowing a chemical reaction to proceed at constant temperature. The chemical reaction results in the formation of a colored complex which absorbs light of specific wavelengths. The colored solution is then pumped in an air-segmented stream through a colorimeter where absorbance is measured. The absorbance reading is proportional to the concentration of the ion.

Flow injection systems such as the Lachat® are a relatively new approach to automated colorimetry. Unlike the AutoAnalyzer® system in which the flow is segmented by air bubbles, in flow injection a continuous stream of reagent into which the sample is injected is used. Flow injection systems are capable of analyzing a large number of samples over a short time period.

7.2 NaHCO$_3$ EXTRACTION

The extraction step involves shaking a soil sample with 0.5 M NaHCO$_3$ and then filtering the extract to obtain a clear, particulate-free filtrate.

7.2.1 REAGENTS

1 0.5 M NaHCO$_3$ extracting solution: dissolve 42 g of NaHCO$_3$ in 1000 mL of deionized water. Then add 0.5 g of NaOH to adjust the pH of the solution to 8.5. Avoid exposure of the solution to air. Plastic is preferable to glass as a storage vessel. Olsen and Sommers (1982) recommend that if glass is used, the NaHCO$_3$ extracting solution be prepared fresh each month, since changes in the pH of the solution over time will affect the amount of P extracted.

2 Charcoal suspension: prepare by mixing 300 g of charcoal with 900 mL of deionized water.

7.2.2 PROCEDURE

1 Place A 2.5-g ($\pm$ 0.02 g) sample of air-dried soil (ground to less than 2 mm) in a 125-mL Erlenmeyer flask.

2 To each flask add 50 mL of NaHCO$_3$ solution at 25°C plus 0.4 mL of the charcoal suspension.

3 Shake for 30 min on a reciprocating shaker at 120 strokes per minute.

4 Filter the extract into sample cups using medium retention filter paper (Whatman® No. 40). If the filtrate is cloudy, refilter as necessary.

7.2.3 Comments

Shaking speed will depend on the construction of the shaker. Both shaking time and speed can affect the amount of element extracted. This is particularly true in the case of P, with a tendency towards higher amounts extracted over longer shaking times and with more vigorous shaking (Olsen and Sommers 1982). Since bicarbonate extractions are used most often as availability indices rather than absolute measurements of the available pool, keeping the shaking procedure as uniform as possible between samples is of greater importance than a specific extraction time and speed. Temperature of the NaHCO$_3$ extracting solution is another source of variability in NaHCO$_3$-P determinations. Olsen et al. (1954) found that extractable P increased 0.43 μg P g^{-1} for each degree rise in temperature between 20 and 30°C for soils testing between 5 and 40 μg P g^{-1}.

7.3 ION MEASUREMENT

Determination of phosphate, potassium, and nitrate ion concentrations in the $NaHCO_3$ extract may proceed once a clear filtrate has been obtained. Phosphate and nitrate ion concentrations are determined using colorimetry, while potassium concentration is assessed using flame emission spectrometry.

7.3.1 Phosphate

Phosphate in the extract is measured by the reaction of phosphate with ammonium molybdate in an acid medium to form molybdophosphoric acid. The molybdophosphoric acid is then reduced to a blue colored complex through reaction with ascorbic acid (Murphy and Riley 1962). Absorbance readings are taken at a 712-nm wavelength using a spectrophotometer. A standard curve constructed from absorbance readings of standards is used to convert absorbance values of samples to phosphate concentration. A manual method for color development and measurement is presented below. Automated methods (Hamm et al. 1970; Technicon® Industrial Systems 1973a) use similar chemistry.

7.3.1.1 REAGENTS

1   Ammonium molybdate solution: dissolve 40 g of ammonium molybdate $[(NH_4)_6Mo_7O_{24}\cdot4H_2O]$ in 1000 mL of deionized water.

2   Ascorbic acid solution: dissolve 26.4 g of L-ascorbic acid in 500 mL of deionized water.

3   Antimony potassium tartrate solution: dissolve 1.454 g of antimony potassium tartrate in 500 mL of deionized water.

The above solutions are stable for 2 to 3 months if well stoppered and stored under refrigeration.

4   Using the above reagents, prepare the Murphy-Riley color developing solution in a 500-mL flask as follows: add 250 mL of 2.5 M H_2SO_4, followed by 75 mL of ammonium molybdate solution, 50 mL of ascorbic acid solution, and 25 mL of antimony potassium tartrate solution. Then add 100 mL of deionized water and mix on a magnetic stirrer. Note: reagents should be added in the proper order and the contents of the flask swirled after each addition. The Murphy-Riley solution should be kept in an amber bottle in a dark location to protect from light. It is recommended that fresh Murphy-Riley solution be prepared daily.

5   Standards: prepare 100 mL of a base P standard with concentration of 5 µg P mL^{-1}.

7.3.1.2 PROCEDURE

1   Standards: 0, 0.1, 0.2, 0.3, and 0.4 μg P mL^{-1}. Add 1, 2, 3, and 4 mL of the base P standard (5 μg P mL^{-1}) to 50-mL volumetric flasks. To each flask then add 10 mL of the NaHCO$_3$ extracting solution.

2   Samples: pipette 10 mL of filtered NaHCO$_3$ extract into 50-mL volumetric flasks.

3   To both standard and sample flasks, add 10 mL of deionized water and one drop of *p*-nitrophenol indicator. Then add 0.5 *M* H$_2$SO$_4$ dropwise until the solution becomes clear. At the point where indicator color disappears, the correct pH for the color development reaction has been attained. If the end point is exceeded through addition of excessive acid, the pH may be brought back up again through the addition of NaOH.

4   To each flask add 8 mL of the Murphy-Riley solution and bring the volume up to 50 mL with deionized water. Allow 15 min for color development.

5   Read absorbance values in a 1-cm cuvette on a spectrophotometer set at 712-nm wavelength. Absorbance values for the standards (0, 0.1, 0.2, 0.3, and 0.4 μg P ml^{-1}) are used to construct a standard curve to convert absorbance values for the samples to concentration.

7.3.2 Potassium

Potassium in the NaHCO$_3$ extract is usually measured by flame emission spectrophotometry, either manually or in an automated system. An automated system for flame emission measurement of K is described by Hamm et al. (1970). In the automated procedure, a sample stream of NaHCO$_3$ is combined with acidified lithium nitrate, segmented with air, and then relayed to a constant temperature dialyzer (37°C). A debubbler is used to remove air from the stream as it enters the atomizing chamber of the flame photometer. A manual method for measurement of K concentration using an atomic absorption/emission spectrophotometer is outlined below.

7.3.2.1 REAGENTS

1   Lithium solution (167 meq Li L^{-1}): dilute 223 mL of Li stock solution (1500 meq L$^-$;1) to 2000 mL with deionized water. In flame emission spectrophotometry, lithium is added to minimize K ionization.

2   K standards: prepare 10, 20, 40, 60, 80, and 100 μg K mL^{-1} standards by adding an appropriate volume of K stock solution to a volumetric flask and making to volume with NaHCO$_3$ extracting solution.

7.3.2.2 PROCEDURE

1   Dilute standards and sample extracts 10 X using the Li solution prepared above.

2   Atomic absorption/emission spectrophotometer (e.g., Pye® Unican® SP 191) set up: Requirements in setting up the instrument for K measurement will vary, depending on instrument design, and the reader is advised to consult the operating manual for specifics. Ensure that all hollow cathode lamps are turned off and the instrument is set for the emission mode. Set the wavelength to 766.5 nm and fine tune the instrument while aspirating the high standard.

3   Aspirate samples and standards and record emission readings. Use a standard curve to convert emission readings for samples to K concentration.

7.3.3 Nitrate

The nitrate contained in bicarbonate extracts is commonly determined using an automated system in which nitrate is reduced to nitrite using alkaline copper and hydrazine, followed by reaction of the nitrite with sulfanilamide under acid conditions to form a diazonium salt. This salt then reacts with a coupling agent (*N*-1-naphthylethylenediamine dihydrochloride) to form a colored azo compound which is measured by absorbance at 520 nm (Kamphake et al. 1967, Hamm et al. 1970, Technicon® Industrial Systems 1973b). The above colorimetric procedure may be performed on an AutoAnalyzer® or flow injection system. A copper-cadmium reductor column may be used in place of the copper and hydrazine sulfate reagents to reduce the nitrate to nitrite (Technicon® Industrial Systems 1973b).

The chemistry in the automated system is similar to that in a manual colorimetric method described by Edwards et al. (1962). In the method of Edwards et al. (1962), zinc is used to reduce nitrate to nitrite. This method is outlined below.

7.3.3.1 REAGENTS

1   Zinc sulfate solution: dissolve 100 g $ZnSO_4 \cdot 7\ H_2O$ in deionized water and make to 1000 mL.

2   Sodium hydroxide solution: dissolve 240 g of NaOH in deionized water and make to 1000 mL.

3   Standards: using $NaNO_3$, prepare 0.25, 0.50, 0.75, 1.0, 1.5, 2.0, 2.5, and 3.0 μg N mL^{-1} standards in 0.5 M $NaHCO_3$.

4   Sulfanilic acid reagent: dissolve 0.60 g sulfanilic acid in deionized water and dilute to 100 mL.

5   Zinc: add 1.000 g of finely powdered zinc to 200 g NaCl in a bottle or beaker. Mix completely by shaking. Shake well before use.

6   Naphthylamine hydrochloride reagent: add 1.0 mL of concentrated HCl to about 50 mL of deionized water. Then add 0.60 g 1-naphthylamine hydrochloride and make to 100 mL volume. Store under refrigeration. After 1 to 2 weeks, a precipitate may form, which can be removed by filtration.

7   2 *M* sodium acetate solution.

7.3.3.2 PROCEDURE

1   In a clean test tube containing an appropriate volume of standard or sample solution, add 1.0 mL of 2.2 *M* HCl and 1.0 mL of sulfanilic acid. Dilute to about 45 mL and mix.

2   Add 1 mL of well-mixed Zn-NaCl mixture. Invert the sample ten times, wait 2 min, invert ten times, wait 2 min, and then invert again ten times. Filter into a 50-mL volumetric flask.

3   Add 1.0 mL naphthylamine hydrochloride, mix, and add 1.0 mL of sodium acetate solution. Allow 5 min of color development, dilute to 50 mL, and measure absorbance on a spectrophotometer set at 520 nm.

7.3.4 Comments

Edwards et al. (1962) have identified the amount of zinc, time of contact for reduction, and temperature as critical in the determination. Ferric, cupric, mercurous, silver, lead, auric, antimonous, bismuth, chlorplatinate, and metavanadate ions are stated to interfere in the reaction.

REFERENCES

Edwards, G. P., Pfafflin, J. R., Schwartz, L. H., and Lauren, P. M. 1962. Determination of nitrates in wastewater effluents and water. J. Water Pollut. Control Fed. 34: 1112–1116.

Fox, R. H. and Piekielek, W. P. 1978. A rapid method for estimating the nitrogen-supplying capability of soil. Soil Sci. Soc. Am. J. 42: 751–753.

Hamm, J. W., Radford, F. G., and Halstead, E. H. 1970. The simultaneous determination of nitrogen, phosphorus and potassium in sodium bicarbonate extracts of soils. Adv. Auto. Anal. 2: 65–69.

Kamphake, L. J., Hannah, S. A., and Cohen, J. M. 1967. Automated analysis for nitrate by hydrazine reduction. Page 205 in Water Research, Vol. 1. Pergamon Press, New York.

Kamprath, E. J. and Watson, M. E. 1980. Conventional soil and tissue tests for assessing the phosphorus status of soil. Pages 433–469 in F. E. Khasawneh, E. C. Sample, and E.J. Kamprath, Eds. The role of phosphorus in agriculture. American Society of Agronomy, Madison, WI.

MacLean, A. A. 1964. Measurement of nitrogen supplying-power of soils by extraction with sodium bicarbonate. Nature 203: 1307–1308.

Matar, A. E., Garabed, S., Riahi, S., and Mazid, A. 1988. A comparison of four soil test procedures for determination of available phosphorus in calcereous soils of the Mediterranean region. Commun. in Soil Sci. Plant Anal. 19: 127–140.

McLean, E. O. and Watson, M. E. 1985. Soil measurements of plant-available potassium. Pages 277–308 in R. D. Munson, Ed. Potassium in agriculture. American Society of Agronomy, Madison, WI.

Menon, R. G., Hammond, L. L., and Sissingh, H. A. 1988. Determination of plant-available phosphorus by the iron hydroxide-impregnated filter paper (P_i) soil test. Soil Sci. Soc. Am. J. 52: 110–115.

Murphy, J. and Riley, J. P. 1962. A modified single solution method for the determination of phosphates in natural waters. Anal. Chem. Acta 27: 31–36.

Olsen, S. R. and Sommers, L. E. 1982. Phosphorus. Pages 403–430 in A. L. Page, Ed. Methods of soil analysis. Part 2. 2nd ed. American Society of Agronomy, Madison, WI.

Olsen, S. R., Cole, C. V., Watanabe, F. S., and Dean, L. A. 1954. Estimation of available phosphorus in soils by extraction with sodium bicarbonate. U.S. Department of Agriculture Circ. 939.

Smith, J. A. 1966. An evaluation of nitrogen soil test methods for Ontario soils. Can. J. Soil Sci. 46: 185–194.

Technicon Industrial Systems. 1973a. Ortho phosphate in water and wastewater. Industrial Method No. 94–70W. Technicon Industrial Systems, Tarrytown, NY.

Technicon Industrial Systems. 1973b. Nitrate and nitrite in water and wastewater. Industrial Method No. 100–70W. Technicon Industrial Systems, Tarrytown, NY.

Chapter 8
Available Potassium

Thomas E. Bates

University of Guelph

Guelph, Ontario, Canada

John E. Richards

Agriculture Canada

Fredericton, New Brunswick, Canada

8.1 INTRODUCTION

Potassium soil tests are based on two steps: (1) extraction of a portion of the soil K and (2) correlation of the extracted K to crop response to applied K to provide fertilizer recommendations. This chapter is concerned solely with methods of extracting soil K. Reviews on the development of soil tests are given by Hanway (1973), Corey (1987), and Dahnke and Olson (1990). In addition, thorough discussions on methods for the measurement of soil K were written by Quemener (1979) and by McLean and Watson (1985).

One of the most important factors affecting soil K availability to crops is the mineralogy of the soil under consideration which can affect release of nonexchangeable K or fixation of applied K (MacLean 1961, Richards et al. 1988). Since Canadian soils vary markedly in their clay mineralogy (Kodama 1979), it should not be expected that one set of recommendations based on one extractant would be satisfactory for all of Canada. At the time of writing, there are five soil test extractants used by the ten provincial soil testing services. They are CH_3COONH_4 (Wood and DeTurk 1940), Mehlich III (Mehlich 1984), double acid or Mehlich I (Mehlich 1953), $NaHCO_3$ (Olsen et al. 1954), and the Kelowna extractant (van Lierop and Gough 1989). Details of extraction are given in Table 8.1

Other extractants have resulted in significant correlations with plant uptake. In Ontario, where the soils are dominated by micaceous materials and are capable of releasing large amounts of nonexchangeable K, recent studies (Richards and Bates 1989, Liu and Bates 1990) have shown that the following extractants are effective for predicting soil K availability in greenhouse trials: ammonium bicarbonate and diethylene triamine pentaacetic acid (AB-DTPA) (Soltanpour and Schwab 1977; Soltanpour and Workman 1979); 2 M NaCl (McKeague 1978); 0.1 M HNO_3 (Maclean 1961); and electroultrafiltration (EUF) (Nemeth 1979). In Quebec, Tran et al. (1987) reported that $CaCl_2$ and EUF provided a good estimate of plant-available K. In recent studies, EUF was superior to CH_3COONH_4 (Richards and Bates 1989, Simard et al. 1991) and to Mehlich III (Simard et al. 1991) for predicting the long-

Soil Sampling and Methods of Analysis, M. R. Carter, Ed.,

Canadian Society of Soil Science. © 1993 Lewis Publishers.

TABLE 8.1 Canadian Soil Test Extraction Methods for Potassium, 1991

Province	Extractant	Soil:solution ratio	Shaking time (min)	Analytical instrument	Additives
Newfoundland	CH_3COONH_4	1:10 v/v	15	Atomic absorption	$LaCl_3$-HNO_3
P.E.I.	Mehlich III	1:10 w/v	5	ICP-AES	
Nova Scotia	Double acid[a]	1:10 v/v	5	ICP-AES	
New Brunswick	CH_3COONH_4	1:10 v/v	30	Atomic absorption	
Quebec	Mehlich III	1:10 v/v or w/v	5	Flame emission	$CaCl + LaCl_3$
Ontario	CH_3COONH_4	1:10 v/v	15	Flame emission	
Manitoba	CH_3COONH_4	1:10 w/v	30	Flame emission	$LiNO_3$
Saskatchewan	$NaCHO_3$	1:20 v/v	30	Flame emission	
Alberta	CH_3COONH_4	1:5 v/v	5	Flame emission	$LiNo_3$
British Columbia	$CH_3COOH + NH_4$	1:10 v/v	5	ICP-AES	

[a] Also known as Mehlich I.

term supplying capacity of Quebec soils and of some Ontario soils. The EUF procedures resulted in problems on soils high in exchangeable Ca and/or carbonates. On these soils the electrical current tends to increase markedly during extraction with the result that plant-available K may be overestimated.

All the above extractants remove solution K and slightly different amounts of exchangeable K, and some may remove a portion of the soil K which is not readily exchangeable with Ca^{2+} or Mg^{2+}. Therefore, the amounts of soil K removed by the extractants differ (Richards and Bates 1988, 1989, Liu and Bates 1990, Simard et al. 1991). In the soil fertility literature, exchangeable K is often defined in terms of the amount of K extracted with $1\ M\ CH_3COONH_4$. Since NH_4^+ and K^+ have similar hydrated radii, NH_4^+ is believed to displace K held in the K-specific wedge zones of weathered micas and vermiculites, thus overestimating the amount of K which is exchangeable with Ca^{2+} and Mg^{2+} (Martin and Sparks 1985). Richards and Bates (1989) showed that this K which was not displaced by Ca^{2+} and Mg^{2+} and which was extracted by NH_4OAc was considerably less available to crops than the Ca^{2+}- and Mg^{2+}-exchangeable K. The amount of K extracted by CH_3COONH_4 minus that extracted by NaCl or $CaCl_2$ provides a measure of this less available fraction.

8.2 SOIL EXTRACTION

The degree of agitation during extraction and the extraction time can affect CH_3COONH_4-extractable K, and this effect may differ among soils (Grava 1980). It seems likely that this is true with other extractants. Grava (1980) reported that flask size and shape and the number of oscillations per minute were interrelated in this effect. Munter (1988) also found in some soils that the type of flask affected CH_3COONH_4-extractable K. Obviously, an acceptable procedure must ensure that every soil, regardless of texture, be maintained in suspension during agitation. Grava (1980) and Munter (1988) provide some guidance in that regard.

8.3 AMMONIUM ACETATE EXTRACTION (WOOD AND DETURK 1940)

Of the five Canadian provinces recommending the CH_3COONH_4 extractant, four use a 1:10 dilution and one a 1:5 ratio (Table 8.1). It seems probable that this will lead to differences in the amount of K extracted from some soils. Two laboratories use a 15-min shaking time and three use 5 min. Grava (1980) showed a difference in the amount of K extracted from some Minnesota soils when the extraction time was increased from 5 to 15 min and where 50-mL Erlenmeyer flasks were used with 10 mL of extractants. Where 30-mL Wheaton bottles were used the differences were larger.

The IM CH_3COONH_4-extraction procedure is described in Chapter 5.

8.4 MEHLICH III EXTRACTION (MEHLICH 1984)

The Mehlich III extractant is used in Quebec and Prince Edward Island for potassium and other nutrients. On a wide range of Ontario soils Mehlich III extracted slightly more K than CH_3COONH_4, but the amounts of K removed by the two were highly correlated ($R^2 = 0.977$). In a greenhouse study, correlations of the amount of K extracted with plant K uptake were similar (0.85 and 0.86) for the two extractants (Liu and Bates 1990). The Mehlich III procedure is described in Chapter 6.

8.5 KELOWNA EXTRACTION (VAN LEIROP AND GOUGH 1989)

The Kelowna extractant is used in British Columbia for several nutrients, including K.

8.5.1 Preparation of the Kelowna Extractant (0.25 *M* CH$_3$COOH + 0.015 *M* NH$_4$F)

The extracting solution is made up of 50 mL stock solution A and 50 mL of stock solution B diluted to 1 L of extractant.

> Stock solution A: 300 mL glacial (5 *M*) CH$_3$COOH L^{-1}
> Stock solution B: 11.112 g 0.3 *M* NH$_4$F L^{-1}

The authors report that addition of ethylene diamine tetraacetic acid (EDTA) or DTPA to this extracting solution to extract micronutrients does not change the amount of K extracted.

8.5.2 Extraction

Measure 2.5 mL of soil into a 50-mL Erlenmeyer flask, add 25 mL of Kelowna extracting solution (only soil:solution ratio specified by authors), and swirl for 5 min at 180 oscillations min^{-1}. Filter through a Whatman® No. 2 filter paper and determine K in the extract.

8.6 MEASURING K IN SOIL EXTRACTS

There do not appear to be problems determining K in any of the three extractants described: CH$_3$COONH$_4$, Mehlich III, and Kelowna.

8.6.1 Standards

1   Stock solution: 1000 mg K L^{-1}: dissolve 1.9073 g oven-dry reagent-grade KCl in the appropriate extracting solution, bring to a volume of 1 L with extracting solution, and mix well.

2   Prepare a 100-mg K L^{-1} standard by diluting 100 mL of the 1000 mg L^{-1} stock solution to 1 L with the appropriate extracting solution. Pipette 10, 20, 30, 40, and 50 mL of the 100 mg L^{-1} solution into 100-mL volumetric flasks and bring each to volume with extracting solution. These solutions will contain 10, 20, 30, 40 and 50 mg K L^{-1}, respectively. The extracting solution serves as the 0-mg K L^{-1} standard. Certified standard solutions may also be purchased.

8.6.2 Flame Photometric Determination of K

Potassium is commonly determined on a flame emission spectrophotometer at 766.5 nm. Some laboratories add Li or La to the extracting solution when they determine K by flame emission. More details on the use of Li and La are provided in Chapter 15, Section 15.4.3.

8.6.3 Inductively Coupled Plasma (ICP) Emission Spectrographic Determination of K

Potassium can be determined sufficiently accurately on some ICP spectrographs.

REFERENCES

Corey, R. B. 1987. Soil test procedures; correlation. Pages 15–22 in Soil testing: sampling, correlation, calibration and interpretation, J. R. Brown, Ed. Soil Science Society of America, Madison, WI.

Dahnke, W. C. and Olson, R. A. 1990. Soil test correlation calibration and recommendation. Pages 45–71 in Soil testing and plant analysis. 3rd ed. R. L. Westerman, Ed. Soil Science Society of America, Madison, WI.

Grava, J. 1980. Importance of soil extraction techniques. Pages 9–11 in Recommended soil test procedures for the north central region. Bull. 499 North Dakota Agric. Exp. Stn., North Dakota State University, Fargo.

Hanway, J. J. 1973. Experimental methods for correlating and calibrating soil tests. Pages 55–66 in L. M. Walsh and J. D. Beaton, Eds. Soil testing and plant analysis. Rev. ed. Soil Science Society of America, Madison, WI.

Kodama, H. 1979. Clay minerals in Canadian soils: their origin, distribution and alteration. Can. J. Soil Sci. 59: 37–58.

Liu, L. and Bates, T. E. 1990. Evaluation of soil extractants for the prediction of plant-available potassium in Ontario soils. Can. J. Soil Sci. 70: 607–615.

MacLean, A. J. 1961. Potassium supplying power of some Canadian soils. Can. J. Soil Sci. 41: 196–206.

Martin, H. W. and Sparks, D. L. 1985. On the behaviour of non-exchangeable potassium release from two coastal plains soils. Soil Sci. Soc. Am. J. 47: 883–887.

McKeague, J. E., ed. 1978. Manual on soil sampling and methods of analysis. Canadian Society of Soil Science, Ottawa, Ont.

McLean, E. O. and Watson, M. E. 1985. Soil measurement of plant-available potassium. Pages 277–307 in R. D. Munson, Ed. Potassium in agriculture. Soil Science Society of America, Madison, WI.

Mehlich, A. 1953. Determination of phosphorus, calcium, magnesium, potassium, sodium and ammonium by the North Carolina Soil Testing Laboratory, Mimeo Report, North Carolina State University, Raleigh.

Mehlich, A. 1984. Mehlich 3 soil test extractant. A modification of Mehlich 2 extractant. Commun. Soil Sci. Plant Anal. 15: 1409–1416.

Munter, R. C. 1988. Laboratory factors affecting the extractability of nutrients. Pages 8–12 in Recommended chemical soil test procedures for the North Central Region, North Dakota Agric. Exp. Stn. Bull. 499, North Dakota State University, Fargo.

Nemeth, K. 1979. The availability of nutrients in the soil as determined by electro-ultra filtration (EUF). Adv. Agron. 31: 155–187.

Olsen, S. R., Cole, C. W., Watanabe, F. S., and Dean, L. A. 1954. Estimation of available phosphorus in soils by extraction with sodium bicarbonate. U.S. Department of Agriculture ARC publ. 939.

Quemener, J. 1979. The measurement of soil potassium. IPI Research Topics No. 4. International Potash Institute CH-3048 Bern-Worblaufen, Switzerland. 48 pp.

Richards, J. E. and Bates, T. E. 1988. Studies on the potassium supplying capacities of Southern Ontario soils. II. Nitric acid extraction of the non-exchangeable K and its availability to crops. Can. J. Soil Sci. 68: 199–208.

Richards, J. E. and Bates, T. E. 1989. Studies on the potassium-supplying capacities of Southern

Ontario soils. III. Measurement of available K. Can. J. Soil Sci. 69: 597–610.

Richards, J. E., Bates, T. E., and Sheppard, S. C. 1988. Studies on the potassium-supplying capacities of Southern Ontario soils. I. Field and greenhouse experiments. Can. J. Soil Sci. 68: 183–197.

Soltanpour, P. N. and Schwab, A. P. 1977. A new soil test for simultaneous extraction of macro and micro-nutrients in alkaline soils. Commun. Soil Sci. Plant Anal. 8: 195–207.

Soltanpour, P. N. and Workman, S. 1979. Modification of the NH$_4$HCO$_3$-DTPA test to omit carbon black. Commun. Soil Sci. Plant Anal. 10: 1411–1420.

Simard, R. R., Tran, T. S., and Zizka, J. 1991. Evaluation of the electro-ultrafiltration technique as a measure of the K supplying power of Quebec soils. Plant Soil 132: 91–101.

Tran, T. S., Tabi, M., and DeKimpe, C. R. 1987. Relation du potassium extrait par euf et quelques méthodes chimiques avec les propriétés du sol et le rendement des plantes. Can. J. Soil Sci. 67: 17–31.

van Lierop, W. and Gough, N. A. 1989. Extraction of potassium and sodium from acid and calcareous soils with the Kelowna multiple element extraction. Can. J. Soil Sci. 69: 235–242.

Wood, L. K. and DeTurk, E. E. 1940. The adsorption of potassium in soils in replaceable form. Soil Sci. Soc. Am. Proc. 5: 152–161.

Chapter 9
Extraction of Available Sulfur

C. G. Kowalenko
Agriculture Canada Research Station
Agassiz, British Columbia, Canada

9.1 INTRODUCTION

Sulfur occurs in soils in both organic and inorganic forms, but only a fraction of that present is available for crop growth (Beaton et al. 1968, Metson 1979, Tabatabai 1982). It is normally assumed that direct uptake of sulfur by plants occurs largely as inorganic sulfate. Sulfate is the main inorganic form in most soils, although elemental and sulfide forms may be present in soils that are under predominantly anaerobic conditions. Highly reduced forms are relatively insoluble and hence not likely to be directly available to plants. It is possible that oxidized forms other than sulfate such as thiosulfate, tetrathionate, or sulfite (Nor and Tabatabai 1976, Wainwright and Johnson 1980) may be found in soils, but are probably present only as intermediates during oxidation or reduction of sulfur. Some of these forms may be toxic to plant growth. Sulfate may be present in the soil solution, adsorbed on soil surfaces, or as insoluble compounds such as gypsum (Nelson 1982) or associated with calcium carbonate (Roberts and Bettany 1985). Adsorption of sulfate occurs on positive charges that are pH dependent and these sites are negligible above pH 6.5 (Tabatabai 1982). The insoluble sulfate compounds are probably not taken up directly by the crop. On a theoretical basis, then, the solution and adsorbed forms of sulfate are the primary pools of sulfur in the soil that are immediately available for plant uptake.

Although there is a good theoretical basis for the solution and adsorbed pools being preset in the soil, there are some practical limitations to their quantification. Limitations occur in both the extraction and subsequent chemical quantification. Before choosing a method and interpreting the results, these limitations should be thoroughly understood.

The choice of the extractant will depend on analytical equipment available and type of soil to be analyzed. Numerous extractants have been used for soil sulfate (Beaton et al. 1968). Extractants may include water, acetates, carbonates, chlorides, phosphates, citrates, and oxalates (Beaton et al. 1968, Jones 1986, Kilmer and Nearpass 1960). The nature of the anion influences the ability of the extractant to displace adsorbed sulfate. The displacement order on the adsorption sites are hydroxyl > phosphate > sulfate = acetate > nitrate = chloride. A water extract would be the theoretical choice for extracting only the solution sulfate; however, a weak calcium chloride solution is more frequently used, as calcium would result in better extraction by flocculating the soil and may depress extraction of colored organic materials (Tabatabai 1982). Lithium chloride is also used frequently instead of calcium chloride, since the lithium would inhibit microbial activity during and after extraction (Tabatabai 1982). The salt concentration, soil-to-extractant ratio, and probably extraction time should be kept as low as possible so that some of the adsorbed sulfate, if present, is not extracted with the solution sulfate (Kowalenko and Lowe 1975). Adsorbed sulfate (together with solution sulfate) is usually extracted with a phosphate solution. Calcium phosphate is the most frequent choice, but sodium and potassium phosphates have also been used (Beaton et al. 1968). A concentration of 500 μg P mL^{-1} is usually adequate to displace sulfate in most soils; however, for soils that fix considerable phosphate, such as certain subsoils, 2000 μg P mL^{-1} may be required. Extraction times and ratios should be sufficiently large to remove all of the adsorbed sulfate. The pH of the extractant will also influence the removal of adsorbed sulfate, since the adsorption process is pH dependent. Although a high pH would theoretically be preferred, there is a problem of extracting additional and highly colored organic materials with high-pH extractants. Acidic extractants may extract portions of gypsum- or carbonate-associated sulfate that may be present in some soils. Buffered extractants may result in more consistent results. Preextraction sample treatment such as air drying will also influence the results (Kowalenko and Lowe 1975, Tabatabai 1982).

There are a number of methods for quantifying sulfate (Beaton et al. 1968, Patterson and Pappenhagen 1978, Tabatabai 1982), but not all can be applied to determinations of sulfate in all soil extracts. Ideally, the method should be quantitative, have adequate sensitivity, not be greatly influenced by the extractant used nor any constituent extracted from the soil, and must be specific to the sulfate forms of sulfur. Unfortunately, there does not appear to be a specific, direct colorimetric method to determine sulfate. Quantitative methods that have most frequently been applied to soil extracts include sulfate analysis using barium sulfate precipitation or sulfide analysis after hydriodic acid reduction. Numerous variations of the barium sulfate reaction have been used, including titrimetric, turbidimetric, gravimetric, and colorimetric methods (Beaton et al.1968). The methods that involve barium can be relatively easy to do, and have been automated; however, they are not very sensitive and are subject to many interferences. The analysis of sulfate as sulfide after hydriodic acid reduction is quite sensitive and relatively free from interferences, but the reduction procedure is quite time consuming, has not been automated to date, and the chemicals involved are fairly costly. Further, since the hydriodic acid reagent reduces both organic and inorganic sulfate to sulfide, the analysis is not specific to inorganic sulfate. More recently, ion chromatography, inductively coupled plasma, and X-ray fluoresence have been used for sulfur analysis of soil extracts (Beaton et al. 1968, Gibson and Giltrap 1979, Maynard et al. 1987, Tabatabai 1982). Although ion chromatography is specific for inorganic sulfate analysis, is quite sensitive, and not affected greatly by interferences, specialized instrumentation is required and special attention to the influence of the extraction salts (concentration and types) must be given. Inductively coupled plasma and X-ray fluoresence provide fast and relatively interference-free results, but the instrumentation involved is costly and the analysis is not specific. The value produced is total sulfur and will include organic and inorganic sulfate, carbon-bonded organic sulfur, etc. Various methods, such as precipitation of inorganic sulfate as barium and subsequent analysis of the sulfate in the precipitate, adsorption of organics by charcoal, or separation of organics from inorganics by ion exchange resin or molecular size (e.g., Sephedex) have been attempted to make the methods specific to inorganic sulfate, but each has distinct limitations (Kowalenko and Lowe 1975).

The choice of a method for extracting available sulfur from soils, then, should be made carefully, taking into consideration the purpose of the analysis, the soil type involved, the nature of the extractant, and the analytical method to quantify the sulfur extracted. Interpretation of the results should consider each of these factors and it should be clearly understood whether the analysis determines only inorganic sulfate or includes other forms of sulfur as well.

The methods that will be described include the extraction of "solution" and "adsorbed plus solution" sulfate. The quantification procedure described involves hydriodic acid reduction of sulfate to sulfide and subsequent determination of sulfur as bismuth sulfide in a modified reduction/distillation apparatus (Kowalenko 1985). This combination provides a much faster and more versatile method for determining sulfate sulfur than the methylene blue approach. The hydriodic acid reduction is generally specific to sulfate; therefore, the final result will include sulfate of both organic and inorganic forms.

9.2 DETERMINATION OF SOIL SOLUTION SULFATE

Although there are many variations to extraction of sulfate in the soil solution, the extractant proposed here is a weak calcium chloride solution (0.01 M L^{-1}) and a 1:2 soil weight to extractant volume ratio shaken for 30 min. The presence of calcium chloride serves to reduce extraction of organic materials and increases flocculation of the soil in the solution. The concentration of calcium chloride and the soil to extractant ratio are both kept as low as possible to minimize extraction of adsorbed sulfate if it is present in the soil.

9.2.1 MATERIALS AND REAGENTS

9.2.1.1 For Soil Extraction

1. 0.01 M CaCl$_2$ extractant: dissolve 1.47 g reagent-grade calcium chloride dihydrate per L of water. Prepare quantities sufficient for the numbers of extractions involved.

2. Extraction vessel: 250-mL Erlenmeyer with stoppers are suitable, although centrifuge tubes, etc., may also be used.

3. Shaker: reciprocating or end over end.

4. Filtration equipment: funnels, filter paper, and filtrate receiver container.

9.2.1.2 For Quantification of Sulfate as Bismuth Sulfide

1. Custom-built reduction/distillation apparatus (Kowalenko 1985): the apparatus can be fabricated from readily available glass materials (burette, ground-glass joints, Taylor tubes, and glass tubing) as illustrated in Figure 9.1. A number of the modified Taylor tubes should be made to facilitate digestion of batches of samples and preparation of standards. This apparatus must be supported in such a way that it can be moved sufficiently high to remove and replace the modified Taylor tubes above a heater.

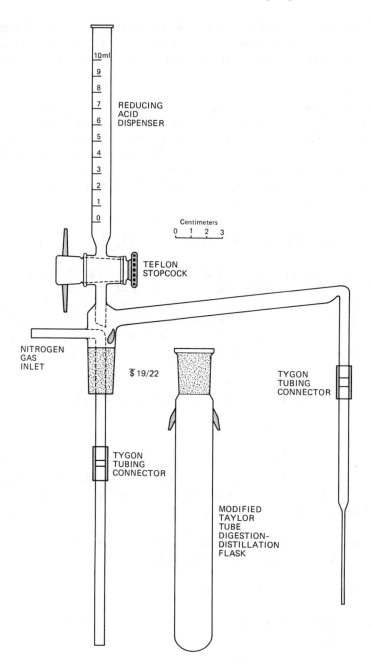

FIGURE 9.1. Simplified digestion-distillation apparatus for total or sulfate-sulfur analyses. (From Kowalenko, C. G., Commun. Soil Sci. Plant Anal., 16, 289, 1985. With permission.)

2   Heating block for hydriodic acid reduction/distillation: the reducing reagent requires heat (110–115°C, i.e., boiling point of the reagent) for the quantitative conversion of sulfate to hydrogen sulfide. A small aluminum block (approximately 7 × 7 × 4 cm with a 2-cm-deep hole into which the Taylor tube will fit) heated by a small heater has been found to be acceptable (Kowalenko 1985). Since the Taylor tube is required to reflux the reducing

reagent, the upper part of the glass apparatus should be shielded from the heater (with metal or other suitable material) as much as possible.

3   Nitrogen gas: the gas must be relatively pure and free from sulfides in particular. The gas may be purified by bubbling it through a solution containing 5 to 10 g mercuric chloride in 100 mL of 2% potassium permanganate. The flow of the nitrogen gas to the reduction/distillation apparatus should be regulated to approximately 200 mL min^{-1}. This can be done by commercially available flow meters (e.g., Rotometer [Kowalenko 1985]) or forcing the gas through an appropriate length (e.g., 30 cm) of capillary glass tubing (Kowalenko and Lowe 1972).

4   Hydriodic acid reducing reagent: mix 4 volumes (e.g., 400 mL) of hydriodic acid (e.g., 57% with 1–2.5% hypophosphorus acid preservative), 1 volume (e.g., 100 mL) of hypophosphorus acid (50%), and 2 volumes (e.g., 200 mL) formic acid (90%) and, while bubbling purified nitrogen gas through it, heat for 10 min at 115–117°C. Continue the nitrogen gas flow through the reagent while cooling. The heating should be done in a well-ventilated hood; refluxing or a special gas-trapping apparatus (Tabatabai 1982) is recommended. Since this reagent is not very stable, only sufficient reagent for several days of sample and standard analyses should be prepared. Storage in a brown bottle and refrigeration will extend its stability.

5   1 *M* sodium hydroxide: for absorbing hydrogen sulfide.

6   Bismuth reagent: dissolve 3.4 g bismuth nitrate pentahydrate in 230 mL glacial acetic acid. Also, dissolve 30 g gelatin in 500 mL water and mix thoroughly. Both solutions will require gentle heating for dissolution. Filter the bismuth solution if it is not clear. The final bismuth reagent is prepared by combining the bismuth and gelatin solutions and diluting to 1 L. This reagent is quite stable at room temperature.

7   Sulfate-sulfur standards: prepare a 1000-µg S mL^{-1} stock solution by dissolving 5.435 g dried reagent-grade potassium sulfate and diluting to 1 L. Working standards are made by appropriate dilutions.

8   Spectrophotometer: the instrument should be suitable for measurement at 400 nm and capable of accommodating small (e.g., 7.5-mL) volumes, including provision for rinsing the cuvette or analysis chamber.

9.2.2 PROCEDURE

9.2.2.1 *Soil Extraction*

Weigh soil sample (e.g., 10 g) into extraction vessel, add an appropriate volume of extractant (e.g., 20 mL) for 1:2 soil weight to extractant volume ratio, shake for 30 min, and then filter through extractant-washed filter paper. Filtration will be more efficient if the soil is allowed to settle briefly.

9.2.2.2 *Quantification of Sulfate as Bismuth Sulfide*

1   Pipette an appropriate volume (e.g., 2–5 mL) of the filtered extract into the modified Taylor tube for the hydriodic acid reduction/distillation apparatus and evaporate (up to 100°C) to dryness.

2   Assemble the custom-built reduction/distillation apparatus above the small heating block in such a way that the modified Taylor tube can be easily installed on or removed from the dispenser portion. This can be accomplished by either having the heater in a fixed position and the dispenser portion with the Taylor tube easily raised and lowered or the dispenser plus Taylor tube fixed and the heater on a jack assembly. A tube (50-mL test tube or larger, depending on the range of the standard curve) containing the sodium hydroxide solution for absorbing the hydrogen sulfide should be fixed in a position such that the nitrogen gas from the outlet of the apparatus will bubble through several centimeters of the absorbing solution. The volume of the absorbing solution is adjusted for the concentration range of sulfate to be analyzed. Adjust the nitrogen gas at the appropriate rate and fill the burette with reducing reagent. As each distillation is completed and the Taylor tube is removed, a watch glass is placed above the heater to intercept any drops of reducing reagent. The entire apparatus should be adequately ventilated.

3   Condition the reduction/distillation apparatus by attaching a modified Taylor tube containing a high (e.g., 200 μg S mL^{-1}) sulfate standard to the reduction/distillation apparatus, place the apparatus in heating position, adjust the nitrogen gas flow, and dispense 4 mL of reducing reagent into the attached Taylor tube. The apparatus requires conditioning at the beginning of each new session with high sulfate-sulfur standard to ensure quantitative initial distillation. Distill until all the sulfate has been reduced and transferred into the absorbing solution. The time required for this process will vary with the flow rate of the nitrogen and the "dead" volume within the apparatus. About 8 to 10 min should be adequate, but calibration under specific conditions is recommended (Kowalenko 1985). After distillation of the hydrogen sulfide is complete, remove the tube containing the absorption solution from the apparatus and check that the distillation process is functioning by immediately adding an appropriate volume of bismuth reagent and mix thoroughly. This volume should correspond to the volume of the absorbing solution (2:1 absorbing solution:bismuth reagent) depending on the range of the standard sulfate-sulfur required. For example, 20 mL of absorbing solution is suitable for 0–200 μg S mL^{-1} range and 5 mL for a 0–40 μg S mL^{-1} range.

4   After the initial setup and conditioning of the apparatus, digested soil samples and corresponding dried standards are distilled into subsequent volumes of absorption solution, and bismuth reagent is immediately added in preparation for quantitative measurements of the bismuth sulfide produced. Measurements are best conducted in batches and adequate standards included in each batch.

5   Measure the absorbance of the bismuth sulfide solution from the soil samples and standards at 400 nm in a spectrophotometer.

9.2.3 Calculations

Calculate the sulfate ion or sulfate-sulfur content by taking into consideration the aliquot volume of extract used and the soil to extractant ratio. Each batch of sample analyses should contain adequate standards to ensure that the standard curve is consistent from batch to batch. Conversion to account for varying soil moisture such that results are expressed on an oven-dried soil basis is recommended.

9.2.4 Comments

1   Filter paper has been found to contain variable amounts of sulfate which will be leached during the filtration. Washing the filter paper with some of the extractant prior to filtration is recommended.

2   Water has been shown to reduce the efficiency of hydriodic acid reagent to reduce sulfate to sulfide (Kowalenko and Lowe 1975); therefore the standard and sample solution aliquot volumes should either be the same throughout or all the volumes evaporated to dryness. Although evaporating the sample to dryness is time consuming, it does provide an opportunity for altering the sensitivity of the analysis (i.e., evaporate a small volume for soils with significant sulfate-sulfur content or a large volume for soil with a low sulfate-sulfur content).

3   The methylene blue color reaction is subject to interference; therefore the nitrogen gas stream must pass through a pyrogallol-sodium phosphate wash just prior to the hydrogen sulfide absorption (Johnson and Nishita 1952). The bismuth sulfide method is much less sensitive to interference; therefore the pyrogallol-sodium phosphate wash can be eliminated (Kowalenko and Lowe 1972). Use of a Taylor tube rather than a condenser to provide refluxing can shorten reduction/distillation times from 60 to 10 min (Kowalenko 1985). The apparatus is also simpler to fabricate and is versatile for different types of analyses. Although the sensitivity of the bismuth reaction is considerably lower than the methylene blue reaction, it can be adequately enhanced for most soil studies by decreasing the volume into which the hydrogen sulfide is absorbed and/or by increasing the size of the original sample being analyzed. However, as the volumes of the absorbing solution and the bismuth reagent are decreased to increase the sensitivity, increased attention must be given to the precision and reproducibility of absorbing solution and bismuth reagent volume measurements, particularly relative to the standard samples. The spectrophotometer should be capable of accommodating the small sample sizes, including appropriate rinsing between samples.

4   If more than 6 mg nitrate that is present in the sample to be analyzed is transferred into the absorption solution, a special procedure should be applied to eliminate the interference whether the bismuth or the methylene blue colorimetric method is used (Johnson and Nishita 1952, Kowalenko and

Lowe 1972). This amount of nitrate would probably occur only under unusual situations.

5   The original sulfate determination procedure (Johnson and Nishita 1952) recommended that the nitrogen gas should be purified before use. Currently available sources of nitrogen are more uniform and free from impurities; therefore, purification of the gas may be omitted. Prepurified nitrogen gas has been used successfully without further gas purification. The purity of the gas for analysis purposes can be evaluated by examining blanks. There should also be fairly good control of the flow rate of the nitrogen gas with a high enough rate to transfer the hydrogen sulfide produced into the receiver solution quickly, but slow enough that the sulfide gas can be absorbed by the sodium hydroxide.

6   Sources of contamination, such as rubber connectors or lubricants for sealing connections, should be considered, particularly in the reduction/distillation procedure. A small amount of water is adequate to seal the Taylor tube to the rest of the apparatus during the reduction/distillation.

7   Hydriodic acid is available in concentrations ranging from 48 to 66% and with or without preservative. Although these products contain varying quantities of sulfate contamination, the sulfate is removed by heating the mixed reagent. The 57% hydriodic acid with preservative has been found to be acceptable. If other products are used, the proportion of hydriodic acid to the other acids may need adjustment and the final reagent tested for effectiveness. Adequate time should be allowed for acquisition of hydriodic acid, as stocks are often limited.

8   Although the hydriodic acid reduction procedure is not influenced by a wide variety of salts, it is recommended that the standards should be made up in the extraction solutions. This precaution will also provide a check on sulfate or sulfide contamination.

9   The hydriodic acid method, although relatively specific for sulfate, includes both inorganic and organic forms. This should not be neglected when the results are being interpreted.

9.3 DETERMINATION OF SOIL SOLUTION PLUS ADSORBED SULFATE

Numerous extractants have been used to extract soil solution plus adsorbed sulfate, but phosphate-containing solutions are most commonly used. Phosphate-containing extractants made up from calcium, potassium, and sodium salts have been employed with varying phosphorus concentrations. However, 500 μg P mL^{-1} is the most commonly used concentration. A higher phosphorus content is preferred, particularly if the soils have a high capacity to fix sulfate, such as in certain subsoils. The hydriodic acid quantification method can tolerate relatively large concentrations of a variety of salts. The soil weight to solution volume should be kept as wide as possible to completely displace the adsorbed sulfate. However, as the ratio is widened, the sensitivity of the quantification will have to increase correspondingly. The extraction time should also be adequately long for complete displacement.

The method outlined here utilizes a 0.5 M sodium phosphate extractant which is buffered at pH 7. This extractant, with 15,485 μg P mL^{-1} concentration, will ensure complete sulfate displacement from soils with high fixation capacity. Extractant pH is kept above 6.5 to ensure sulfate desorption, but not above pH 7 so that extraction of organic sulfate is minimized (Bart 1969). The sulfate is determined by hydriodic acid reduction because the method has adequate sensitivity, is relatively free from interferences, and is specific for sulfate (both organic and inorganic).

9.3.1 MATERIALS AND REAGENTS

1　Phosphate buffer extractant: 0.5 M phosphate buffer at pH 7.0 is made by mixing a 3:2 proportion of 0.5 M Na$_2$HPO$_4$ (70.980 g Na$_2$HPO$_4$ L^{-1}) and 0.5 M NaH$_2$PO$_4$ (68.995 g NaH$_2$PO$_4$·H$_2$O L^{-1}), then adjusting the pH by adding one or the other of the solutions as required.

2　All other materials and reagents for extraction and quantification as outlined for soil solution sulfate above (Section 9.2.1).

9.3.2 Procedure

Extract soil as outlined for soil solution sulfate above (9.2.2), except that the phosphate buffer extractant is used instead of the calcium chloride extractant and the soil weight : extractant volume ratio is 1:5 (e.g., 5 g soil and 25 mL phosphate buffer extractant). Shake for 30 min and filter using extractant-washed filter paper. Determine the sulfate-sulfur concentration in the phosphate buffer extract as for calcium chloride extract (Section 9.2.2).

9.3.3 Calculations

Calculate the sulfate ion or preferably sulfate-sulfur content by taking into consideration the aliquot volume of extract and soil to extractant ratio. Each batch of sample analysis should contain adequate standards to ensure that the standard curve is consistent from batch to batch. Conversion to account for varying soil moisture such that results are expressed on an oven-dried soil basis is recommended.

9.3.4 COMMENTS

1　Precautions outlined above for the soil solution sulfate procedure (i.e., sulfur contamination in reagents and filter paper, effect of water on sulfate reduction, and realizing that the value includes organic and inorganic sulfate-sulfur) must be taken.

2　Adsorption of anions in soil is pH dependent and little if any adsorbed sulfate would be expected to be found in soils of pH 6.5 or greater.

REFERENCES

Bart, A. L. 1969. Some factors affecting the extraction of sulphate from selected lower Fraser Valley and Vancouver Island soils. M.Sc. thesis. University of British Columbia.

Beaton, J. D., Burns, G. R., and Platou, J. 1968. Determination of sulphur in soils and plant material. Technical Bulletin No. 14. The Sulphur Institute, Washington, D.C. 56 pp.

Gibson, A. R. and Giltrap, D. J. 1979. Measurement of extractable soil sulphur in the presence of phosphate using ion-exchange resin paper discs and XRF spectrometry. N.Z. J. Agric. Res. 22: 439–443.

Johnson, C. M. and Nishita, H. 1952. Microestimation of sulfur in plant materials, soils, and irrigation waters. Anal. Chem. 24: 736–742.

Jones, M. B. 1986. Sulfur availability indexes. Pages 549–566 in M. A. Tabatabai, Ed. Sulfur in agriculture. Agronomy No. 27. American Society of Agronomy, Madison, WI.

Kilmer, V. J. and Nearpass, D. C. 1960. The determination of available sulfur in soils. Soil Sci. Soc. Am. Proc. 24: 337–340.

Kowalenko, C. G. 1985. A modified apparatus for quick and versatile sulphate sulphur analysis using hydriodic acid reduction. Commun. Soil Sci. Plant Anal. 16: 289–300.

Kowalenko, C. G. and Lowe, L. E. 1972. Observations on the bismuth sulphide colorimetric procedure for sulfate analysis in soil. Commun. Soil Sci. Plant Anal. 3: 79–86.

Kowalenko, C. G. and Lowe, L. E. 1975. Evaluation of several extraction methods and of a closed incubation method for sulfur mineralization. Can. J. Soil Sci. 55: 1–8.

Maynard, D. G., Kalra, Y. P., and Radford, F. G. 1987. Extraction and determination of sulfur in organic horizons of forest soils. Soil Sci. Soc. Am. J. 51: 801–805.

Metson, A. J. 1979. Sulphur in New Zealand soils. I. A review of sulphur in soils with particular reference to adsorbed sulphate-sulphur. N.Z.J. Agric. Res. 22: 95–114.

Nelson, R. E. 1982. Carbonate and gypsum. Pages 181–197 in A. L. Page, R. H. Miller, and D. R. Keeney, Eds. Methods of soil analysis. Part 2. Chemical and microbiological properties. 2nd ed. Agronomy No. 9. American Society of Agronomy, Madison, WI.

Nor, Y. M. and Tabatabai, M. A. 1976. Extraction and colorimetric determination of thiosulphate and tetrathionate in soils. Soil Sci. 122: 171–178.

Patterson, G. D., Jr. and Pappenhagen, J. M. 1978. Sulfur. Pages 463–527 in D. F. Boltz and J. A. Howell, Eds. Colorimetric determination of nonmetals. John Wiley & Sons, Toronto.

Roberts, T. L. and Bettany, J. R. 1985. The influence of topography on the nature and distribution of soil sulfur across a narrow environmental gradient. Can. J. Soil Sci. 65: 419–434.

Tabatabai, M. A. 1982. Sulfur. Pages 501–538 in A. L. Page, R. H. Miller, and D. R. Keeney, Eds. Methods of soil analysis. Part 2. Chemical and microbiological properties. 2nd ed. Agronomy No. 9. American Society of Agronomy, Madison, WI.

Wainwright, M. and Johnson, J. 1980. Determination of sulphite in mineral soils. Plant Soil 54: 299–305.

Chapter 10
Characterization of Available P by Sequential Extraction

H. Tiessen and J. O. Moir

University of Saskatchewan

Saskatoon, Saskatchewan, Canada

10.1 INTRODUCTION

It is not a straightforward task to describe methods for the determination of soil-available P for two basic reasons:

1. Methods for the determination of available P in an agronomic context never measure the quantity of P available to a crop, but measure a pool of soil P that is somehow related to that portion of soil P which is plant available. The relationship is established over years of agronomic experimentation and testing of fertilizer responses through regression equations. These equations relate plant performance to measured soil P levels, or indicate levels of P deficiency expressed as potential fertilizer requirement for a crop, and typically account for 50–60% of the observed variability. Results obtained with this approach are not always transferable from one crop or soil to another, and different equations are established by soil testing services for varying crops and soils. The approach frequently breaks down entirely, when perennial plants or natural ecosystems are examined, because measurable pools are often small, and P cycling rather than pool size is a major determinant of annual productivity.

2. Phosphorus availability needs to be defined with respect to an external sink, i.e., a plant, plant community, or crop. Plants differ in their ability to extract P from soils due to differences in rooting systems, mycorrhizal associations, and growth rates. Since any "immediately available" pool of P is constantly replenished through reactions of dissolution or desorption of "less available" P, and through the mineralization of organic P, the pool size of "total available" P is strongly time dependent.

10.2 SOIL TEST METHODS FOR AVAILABLE P

Agronomic tests for available P are designed with several aims; they should:

1   Be simple enough for routine application;

Soil Sampling and Methods of Analysis, M. R. Carter, Ed.,

Canadian Society of Soil Science. © 1993 Lewis Publishers.

2   Extract sufficient P to be easily measurable;

3   Extract sufficient P to represent a significant portion of potential plant up-take, so that plant supply is represented more closely by the quantity measured rather than being dependent on P turnover and replenishment of the measured pool;

4   Not extract significant amounts of P that are not plant available.

This is achieved with solutions of moderately lowered or raised pH, which release P associated with the soil mineral phase, without significantly solubilizing minerals. Alternatively, or in combination with these pH changes, specific anions are introduced that bring P into solution by competing with P sorption sites or by lowering the solubility of cations that bind P in the soil. Based on these principles, numerous extraction methods exist, all of which have some merits and limitations and are used in various parts of the world, where their value relies on long-term correlation studies that establish the relationship between extract and crop response. An exhaustive review of extraction methods by a working group in Spain (Anon. 1982) listed 50 different methods and more than 50 publications comparing different extracts.

Worldwide, the most common methods are probably the alkaline bicarbonate extract of Olsen et al. (1954) and the acid ammonium fluoride extract (Bray and Kurtz 1945) with various modifications. In addition, an extraction using lactate (Egnér et al. 1960) is popular in Europe. A rationale for the use of bicarbonate or lactate for the extraction of available P is provided by the consideration that plant roots produce CO_2 which forms bicarbonate in the soil solution, and also produce various organic acids similar to lactate that may solubilize soil P. It is hoped therefore that these extractants somehow simulate the action of plant roots and thus give a more appropriate measure of plant-available P. Chelating extracts (Onken et al. 1980) have been proposed for similar reasons.

The bicarbonate extract (Olsen et al. 1954) has been used successfully on a wide range of acid to alkaline soils. Available P is extracted with a solution of sodium bicarbonate at a pH of 8.5 for 30 min. Interference from organic matter dissolved in the solution has frequently been eliminated by sorbing the organic matter onto activated acid-washed charcoal (carbon black) added to the extract, but it is difficult to obtain P-free charcoal. An alternative was therefore developed which eliminates organic interference with polyacrylamide (Banderis et al. 1976). When the blue phospho-molybdate complex is measured at a wavelength of 712 nm, color interference from the yellow organic matter is negligible. At high organic matter concentrations in the extract, the organic matter will precipitate upon acidification during the Murphy-Riley (1962) procedure, and interfere with P colorimetry. The extraction time of 30 min has been designed for rapid routine soil testing. A more complete extraction is obtained by extracting for 16 h (Colwell 1963). In all applications that attempt to functionally evaluate the bicarbonate-extractable P pool, and that include organic P determination, the more complete 16-h extract should be used.

The acid ammonium fluoride extract (Bray and Kurtz 1945) has been widely used on acid and neutral soils, and a large data base exists. The relatively low acid strength and importance of acidity for the extraction mechanism make the method unsuitable for calcareous or strongly alkaline soils which would partially neutralize the acidity and eliminate the standard test conditions. The exotic composition makes this a purely chemical test that cannot be interpreted in terms of plant function like the bicarbonate or some of the organic acid or chelating extracts.

10.3 APPROACHES FOR CHARACTERIZING AVAILABLE P

Since available P is a functional concept rather than a measurable quantity, no simple direct measurements are available. Plant-available P is all P that is taken up by a plant during a specific period, such as a cropping season, year, or growth cycle. Since the plant obtains P through its roots or root symbionts from the soil solution, available P is composed of solution P plus P that enters the solution during the period used to define availability. P may enter the solution by desorption or dissolution of inorganic P (P_i) associated with the solid phase of the soil or by the mineralization of organic P (P_o).

Whether desorption or dissolution are critical processes in the supply from P_i forms cannot usually be resolved. In one case the solubility product of the least soluble P compound, and in the other the saturation of sorbing surfaces would determine the P supply at equilibrium. Countless publications have dealt with the reverse of these reactions — precipitation and adsorption, and empirical data usually fit either process to some degree (Syers and Curtin 1989). There is an increasing realization, though, that solid-phase P is not static, and that sorption-desorption and precipitation-dissolution equilibria change with time due to secondary processes (Parfitt et al. 1989) such as recrystallization (Barrow 1983) or solid-state diffusion (Willet et al. 1988). A measurement of available P_i therefore needs to consider both the amounts and rates of release of P from the solid phase. Very few appropriate methods have been published. Among the approaches taken are repeated water extracts and sorption-desorption isotherms (Bache and Williams 1971, Fox and Kamprath 1970), possibly at elevated temperatures to substitute for impractically long reaction times (Barrow and Shaw 1975).

In a simpler approach, a sink for solution P_i in form of an anion-exchange resin may be used, which offsets the equilibrium between dissolved and soluble P_i. "Exchangeable" P_i as well as some of the more soluble precipitated P forms will enter the depleted solution and be absorbed by the resin. The P sorbed by the resin is subsequently measured. Several different methods for resin extractions have been developed and tested, using different anionic forms, ratios of soil:water:resin, times and methods of shaking, and enclosure in bags or mixing through the suspension (Sibbesen 1977, 1978, Barrow and Shaw 1977, Tiessen 1991). By far the simplest method uses Teflon®-based anion-exchange membranes, which can be cut into strips and used repeatedly and easily (Schoenau and Huang 1991, Saggar et al. 1990).

The pool measured by resin extraction is very similar to that assessed with isotopic dilution (Amer et al. 1955). The sorption by a resin is usually complete within 20 h, and only minor changes are observed thereafter. Similarly isotopic-exchange methods reach a relatively steady state after a few hours, followed by continuing small changes (Fardeau and Jappe 1980). These continuing changes represent the activity of less soluble or slower pools of soil P, which replenish available P at rates varying from days to years. Errors involved in the measurement of these small changes make it impossible to extrapolate the continuing release (or isotopic dilution) rates to times appropriate for cropping seasons and growth cycles. Phosphorus taking part in longer-term transformations can be examined with sequential extractions, which first remove labile P, and then the more stable forms.

A widely used sequential extraction method was developed by Chang and Jackson (1957) with many later modifications such as that of Williams et al. (1967). The sequential procedure employs NH_4Cl to extract "labile" P_i followed by NH_4F to dissolve specifically Al-associated P_i. This is followed by NaOH to extract Fe-bound P_i and by dithionite-citrate for "occluded" P_i forms. HCl dissolves Ca-bound P_i and the final residue is analyzed by Na_2CO_3 fusion for total P or it can be submitted to a more complicated analysis for P_i and P_o by ignition plus acid extraction (Williams et al. 1967). As in all other methods of P_o determination, the amount of P_o is not measured, but calculated by difference. Acid-extractable P_i is subtracted from the greater amount of P_i that is acid extractable after ignition of the soil organic matter (Saunders and Williams 1955).

The procedure presented many interpretational problems; since P_i reprecipitated during the fluoride extraction, the separation of Al- and Fe-associated P_i was not reliable, and the reductant-soluble or "occluded" P_i was an ill-defined pool (Williams and Walker 1969). However, the sequence of alkaline followed by acid extraction gives a reliable distinction between Al + Fe and Ca-associated P_i (Kurmies 1972). This distinction reflects the weathering stage of the soil and can be used to monitor the fate of rock phosphate fertilizer in weathered soils that contain little Ca-bound P. The P_o extracted by the procedure has usually been ignored, although it was shown to be important in plant nutrition (Kelly et al. 1983).

An alternative P-fractionation scheme was developed by Hedley et al. (1982a). This sequential extraction aimed at quantifying labile (plant-available) P_i, Ca-associated P_i, Fe + Al-associated P_i, as well as labile and more stable-forms of P_o using the following extracts: labile forms of P_i which are thought to consist of P_i adsorbed on surfaces of more crystalline P compounds, sesquioxides, or carbonates (Mattingly 1975) were extracted with resin and bicarbonate. Hydroxide-extractable P_i has lower plant availability (Marks 1977) and is thought to be associated with amorphous and some crystalline Al and Fe phosphates. A more precise characterization of these P_i forms is usually not possible, since mixed compounds containing Ca, Al, Fe, P, and other ions predominate in soils (Sawhney 1973). Labile bicarbonate-extractable P_o is easily mineralizable and contributes to plant-available P (Bowman and Cole 1978). More stable forms of P_o involved in the long-term transformations of P in soils are extracted with hydroxide (Batsula and Krivonosova 1973).

Each of the extracts obtained can be assigned some role in the P transformations occurring in a soil under different conditions of incubations (Hedley et al. 1982a), near plant roots (Hedley et al. 1982b), with cultivation (Tiessen et al. 1983), or soil development (Tiessen et al. 1984, Roberts et al. 1985). These empirical assignments can then be used to characterize P status of the soil relative to a conceptual model of P pools and their transformations.

This fractionation approach is currently the only one that can be used with moderate success for the evaluation of available P_o. Due to the reactivity of mineralized P with the soils mineral phase, determination of a potentially mineralizable P_o pool, analogous to the mineralizable N or S pools measured with incubation and leaching techniques (Ellert and Bettany 1988), is not feasible. The nature of different extractable P_o pools is even less well defined than that of the P_i pools (Stewart and Tiessen 1987). Their turnover and availability frequently depend on the mineralization of C during which P is released as a side product, although solubilized P_o will be rapidly mineralized by soil enzymes. This complex situation means that there is presently no satisfactory method for measuring available P_o, beyond the rather empirical sequential extraction combined with conceptual models and possibly separate organic matter studies (Tiessen et al. 1983, Stewart and Tiessen 1987).

The original fractionation (Hedley et al. 1982a) left between 20 and 60% of the P in the soil unextracted. This residue often contained significant amounts of P_o that sometimes participated in relatively short-term transformations. On relatively young, Ca-dominated soils this residual P_o can be extracted by NaOH after the acid extraction, while on more weathered soils, hot HCl (Metha et al. 1954) extracts most of the organic and inorganic residual P. The hot HCl method appears to work satisfactorily on most soils, and is presented below as part of an extensive soil P fractionation.

10.4 P FRACTIONATION PROCEDURE

10.4.1 MATERIALS

10.4.1.1 Extracting Solutions

1   0.5 M HCl: bring 88.5 mL conc. HCl to 2 L with deionized H_2O.

2   0.5 M NaHCO$_3$ (pH 8.5): dissolve 84 g NaHCO$_3$ + 1 g NaOH to 2 L with H_2O.

3   0.1 M NaOH: dissolve 4 g NaOH to 1 L with H_2O.

4   1 M HCl: bring 177 mL conc. HCl to 2 L with H_2O.

5   conc. HCl: 11.3 M.

6   H_2O_2: 30% hydrogen peroxide (P free).

7   conc. (18 M) H_2SO_4.

10.4.1.2 Reagents for P Determination

1   Make the following solutions:
      40.0 g ammonium molybdate/1000 mL
      26.4 g l-ascorbic acid/500 mL
      1.454 g antimony potassium tartrate/500 mL
      278 mL conc. H_2SO_4/2 L (2.5 M H_2SO_4)

2   Mix the developing reagent in the following sequence, swirling contents of flask after each addition: to 250 mL 2.5 M H_2SO_4, add 75 mL ammonium molybdate solution, then 50 mL ascorbate solution, and finally 25 mL of antimony potassium tartrate solution. Dilute to a total volume of 500 mL with deionized H_2O and mix.

3   For organic matter precipitation and pH adjustment make up:
     0.9 M H_2SO_4: bring 100 mL conc. H_2SO_4 to 2 L with H_2O.
     0.25 M H_2SO_4: bring 100 mL 2.5 M H_2SO_4 to 2 L with H_2O.
     4 M NaOH: dissolve 160 g NaOH to 1 L with H_2O.

10.4.1.3 *Resin Strip*

Use anion-exchange membrane (BDH No. 55164 or equivalent), cut into strips (9 × 62 mm), converted to bicarbonate form. To regenerate after the adsorbed P has been extracted with HCl, wash resin strips for 3 d with 6 batches of 0.5 M HCl, followed by washing a further 3 d with 6 batches of 0.5 M NaHCO$_3$ (pH 8.5). Then rinse well with deionized/distilled water.

10.4.2 ANALYTICAL PROCEDURES

10.4.2.1 *P Determination by the Murphy-Riley (1962) Method*

This method is used directly for the P recovered from the resin strip and for P_i determination in the two HCl extracts:

1   Pipette suitable aliquot into a 50-mL volumetric flask.
     Use paranitrophenol as an indicator. If the extract is acid, first adjust pH with 4 M NaOH to yellow and then with ≈0.25 M H_2SO_4 until the indicator just turns clear. For alkaline extracts just acidify.

2   Add 8 mL of color developing solution, make to volume, shake, and read on spectrophotometer at 712 nm after 10 min (color is stable for 24 h).

10.4.2.2 *Determination of Inorganic P on 0.5 M NaHCO$_3$ and 0.1 N NaOH*

1   Pipette 10 mL solution into a 50-mL centrifuge tube.

2   Acidify to pH 1.5 and set in fridge for 30 min;
     to acidify 0.5 M NaHCO$_3$ extract use 6 mL of 0.9 M H_2SO_4;
     to acidify 0.1 M NaOH extract use 1.6 mL of 0.9 M H_2SO_4.

3   Centrifuge at 25,000 × g for 10 min at 0°C.

4   Decant supernatant into a 50-mL volumetric flask.

5   Rinse tube carefully so as not to disturb the organic matter with a little acidified water and add to the solution in the flask (2 or 3 times).

6   Adjust pH and measure P by the Murphy-Riley method.

10.4.2.3 *Determination of Total P in 0.5 M NaHCO₃, 0.1 M NaOH, and conc. HCl Extracts (EPA 1971)*

1 Pipette 5 mL solution into a 50-mL volumetric flask.

2 To 0.5 M NaHCO$_3$ extract: add $\approx$0.5 g ammonium persulfate + 10 mL 0.9 M H$_2$SO$_4$;

to 0.1 M NaOH extract: add $\approx$0.6 g ammonium persulfate + 10 mL 0.9 M H$_2$SO$_4$;

to conc. HCl extract: add $\approx$0.4 g ammonium persulfate + 10 mL deionized water.

3 Cover with tinfoil (double layer for conc. HCl) and autoclave: NaHCO$_3$ and HCl extracts for 60 min, NaOH extract for 90 min.

4 Adjust pH and measure P by the Murphy-Riley method.

10.4.3 EXTRACTION PROCEDURE

Day 1: Weigh 0.5 g soil into a 50-mL centrifuge tube, add two resin strips + 30 mL deionized water, and shake overnight (16 h).

Day 2: Remove resin strips and wash soil back into tube. Place resin strip in a clean 50-mL tube and add 20 mL 0.5 M HCl. Set aside for 1 h to allow gas to escape, cap, and shake overnight. Determine P using Murphy-Riley method. Centrifuge soil suspension at 25,000 $\times$ g for 10 min at 0°C. Decant water through a millipore filter (pore size 0.45 μm). Discard water and wash any soil off filter back into the tube with a little 0.5 M NaHCO$_3$ (pH 8.5) solution. Add more NaHCO$_3$ solution to bring solution volume to 30 mL (done by weight) and shake suspension overnight. Make sure all soil is free from bottom of tube before putting on the shaker.

Day 3: Centrifuge soil suspension at 25,000 $\times$ g for 10 min at 0°C. Decant NaHCO$_3$ extract through a millipore filter into a clean vial. Determine inorganic and total P$_t$ on bicarbonate extract. Wash any soil off filter back into the tube using a little 0.1 M NaOH. Make volume of NaOH solution to 30 mL and shake suspension overnight.

Day 4: Centrifuge suspension at 25,000 $\times$ g for 10 min at 0°C. Decant NaOH extract through a millipore filter into a clean vial. Determine inorganic and P$_t$ on NaOH extract. Wash any soil off filter back into the tube using a little 1 M HCl. Make volume of HCl to 30 mL and shake suspension overnight.

Day 5:

1. Centrifuge soil suspension at 25,000 $\times$ g for 10 min at 0°C. Decant HCl extract through a millipore filter into a clean vial. Determine P in extract. (In this step any residue on the filter paper is not washed back into the tube; decant gently so as not to lose soil.)

2. Soil residue heated with 10 mL conc. HCl in a water bath at 80°C for 10 min. (Vortex to mix soil and HCl well and loosen caps before putting into the hot bath.

The mixture will take about 10 min to come to temperature — check with a thermometer in a tube containing HCl only — i.e., the tubes will be in the hot water for a total of 20 min.) Remove and add a further 5 mL conc. HCl, vortex, and allow to stand at room temperature for 1 h (vortex every 15 min). Tighten caps, centrifuge at 25,000 × g for 10 min at 0°C, and decant supernatant into a 50-mL volumetric flask. Wash soil with 10 mL H_2O, centrifuge, and add wash to flask twice. Make to volume, filter if necessary (i.e., soil in solution) through a No. 40 paper, and determine inorganic and P_t in HCl solution.

3.   Add 10 mL deionized water to soil residue and disperse soil. Suction suspension into 75-mL digestion tubes (use the minimum amount of water possible to transfer all the soil residue), add 5 mL conc. H_2SO_4 + one boiling chip (Hengar Granules, Hengar Co., Philadelphia, PA, Cat. No. 136C), vortex, and put on a cold digestion block. Raise the temperature very slowly to evaporate water and when 360°C is reached start treating with H_2O_2 in the following way: remove tubes from heat and let cool to hand-warm; add 0.5 mL of H_2O_2; reheat for 30 min during which H_2O_2 is used up. Repeat H_2O_2 additions until liquid is clear (usually about 10 times). Make sure there is adequate heating after the final H_2O_2 addition, since residual H_2O_2 interferes with the P determination. Cool, make to volume, shake, and transfer to vials (either filter or allow residue to settle out overnight). Determine P in solution. (This digestion is based on Thomas et al. 1967.)

10.4.4 Interpretation and Limitations

The interpretation of data obtained from this sequential fractionation is based on an understanding of the action of the individual extractants, their sequence (Figure 10.1), and their relationship to the chemical and biological properties of the soil. Some typical examples for soils from Canada, Ghana, and Brazil are given in Table 10.1.

It must be remembered that, while the fractionation is an attempt to separate P pools according to their lability, any chemical fractionation can at best only approximate biological functions. Resin P is reasonably well defined as freely exchangeable P_i, since the resin extract does not chemically modify the soil solution. Bicarbonate extracts a P_i fraction which is likely to be plant available, since the chemical changes introduced are minor and somewhat representative of root action (respiration). This fraction is not comparable to the widely used fertility test (Olsen et al. 1954) because the resin-extractable pool has already been removed at this point.

Bicarb-P_i and OH-P_i are not really completely separate pools, particularly in acid soils, but represent a continuum of Fe- and Al-associated P extractable with increasing pH (original pH of the soil to 8.5 to 13).

The P_o extracted with these two extractants is also likely to represent similar pools. Since P_o is determined by the difference between P_t and P_i in each extract, there is a source of error. The P_t determination is quite reliable, but P_i is determined in the supernatant after precipitation of organic matter with acid. Any nonprecipitated P_o (fulvic acid P) will not significantly react with the Murphy-Riley solution, so that P_i is rarely overestimated. Any P_i, though, that precipitates along with the organic matter upon matter upon acidification would be erroneously determined as P_o (P_t-P_i). This may happen with P_i associated with Fe or Al hydroxides which are soluble at high, but insoluble at low pH. It has so far been impossible to quantify the P_o overestimation. In soils with low extractable organic matter

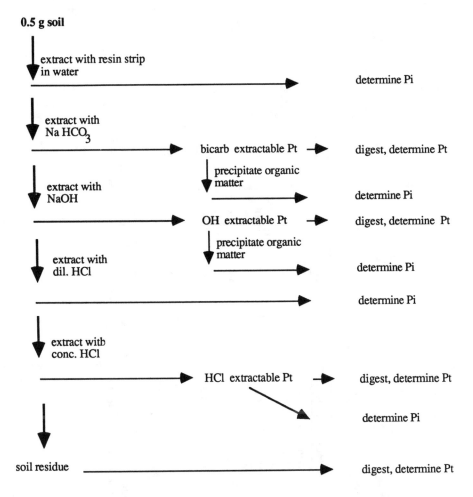

FIGURE 10.1. Flow chart of the sequential P extraction.

contents (low enough not to cause precipitation in the acid Murphy-Riley solution) it is possible to determine P_i in the extract without prior acid precipitation using a blank correction for the color of the extracts.

The dilute HCl P_i is clearly defined as Ca-associated P, since Fe- or Al-associated P, that might remain unextracted after the NaOH extraction, is insoluble in acid. There is rarely any P_o in this extract.

The hot concentrated HCl extract does not present the same problems as the other P_o extracts, since P_i is determined directly. This extract is useful for distinguishing P_i and P_o in very stable residual pools. But at the same time, P_o extracted at this step may simply come from particulate organic matter that is not alkali extractable, but may be easily bioavailable.

The residue left after the hot concentrated HCl extraction is unlikely to contain anything but highly recalcitrant P_i.

For a reliable interpretation of the P_o fractions and their role in P cycling, it is advisable to supplement the fractionation with a suitable characterization of soil organic matter, so that characteristics can be inferred from the combined results of different techniques.

TABLE 10.1 Some Examples of Sequentially Extracted Soils

Soil	P_i					P_o			Residue
	Resin	Bicarb.	OH	Dil. HCl	Conc. HCl	Bicarb.	OH	Conc. HCl	
Chernozem (Mollisol), Canada, native prairie									
	62	33	52	216	(69)[a]	32	88	(193)	0
Chernozem (Mollisol), Canada, 65 yr. cultivated									
	18	9	22	218	(64)	10	52	(140)	0
Latosol (Oxisol), Brazil, native thorn forest									
	1	6	13	2	59	5	22	2	17
Latosol (Oxisol), Brazil, 5 yr. cultivated									
	0.5	7	20	3	65	4	15	2	22
Savanna ochrosol (Alfisol), Ghana, iron-nodule free									
	3	6	11	12	—[a]	4	48	—	86
Savanna ochrosol (Alfisol), Ghana, 75% iron-nodule content									
	1	8	57	8	—	6	35	—	554

[a] These soils were analyzed without the conc. HCl step; for the Mollisol, P_i and P_o were determined in the residue, providing an approximate equivalent to the P_i/P_o distinction given by the conc. HCl extract.

REFERENCES

Amer, F., Bouldin, D. R., Black, C. A., and Duke, F. R. 1955. Characterisation of soil phosphorus by anion exchange resin adsorption and ^{32}P equilibration. Plant Soil 6: 391–408.

Anon. 1982. Groupo de trabajo de normalisacion de metodos analiticos. Revision bibliografica de metodos de extraction de phosphoro assimilable en suelos 1957–80. Pages 1085–1112 in Anales de Edafologia y Agrobiologia 41.

Bache, B. W. and Williams, E. G. 1971. A phosphate sorption index for soils. J. Soil Sci. 22: 289–301.

Banderis, A. S., Barter, D. H., and Henderson, K. 1976. The use of polyacrylamide to replace carbon in the determination of Olsen's extractable phosphate in soil. J. Soil Sci. 27: 71–74.

Barrow, N. J. 1983. On the reversibility of phosphate sorption of soils. J. Soil Sci. 34: 751–758.

Barrow, N. J. and Shaw, T. C. 1975. The slow reactions between soil and anions. 2. Effect of time and temperature on the decrease in phosphate concentration in the soil solution. Soil Sci. 119: 167–177.

Barrow, N. J. and Shaw, T. C. 1977. Factors affecting the amount of phosphate extracted from soil by anion exchange resin. Geoderma 18: 309–323.

Batsula, A. A. and Krivonosova, G. M. 1973. Phosphorus in the humic and fulvic acids of some Ukrainian soils. Soviet Soil Sci. 5: 347–350.

Bowman, R. A. and Cole, C. V. 1978. Transformations of organic phosphorus substances in soils as evaluated by NaHCO$_3$ extraction. Soil Sci. 125: 49–54.

Bray, R. H. and Kurtz, L. T. 1945. Determination of total, organic and available forms of phosphorus in soils. Soil Sci. 59: 39–45.

Chang, S. C. and Jackson, M. L. 1957. Fractionation of soil phosphorus. Soil Sci. 84: 133–144.

Colwell, J. D. 1963. The estimation of the phosphorus fertilizer requirement of wheat in southern New South Wales by soil analysis. Aust. J. Exp. Agric. Anim. Husb. 3: 190–197.

Egnér, H., Riehm, H., and Domingo, W. R. 1960. Untersuchungen über die chemische Bodenanalyse als Grundlage für die Beurteilung des Nährstoffzustandes der Böden. 2. Chemische Extraktionsmethoden zur Phosphor- und Kaliumbestimmung. Kungl. Landbrukshögskolans Annaler, Uppsala, Sweden 26: 199–215.

Ellert, B. H. and Bettany, J. R. 1988. Comparisons of kinetic models for describing net sulfur and nitrogen mineralization. Soil Sci. Soc. Am. J. 52: 1692–1702.

EPA 1971. Methods of chemical analysis for water and wastes. U.S. Environmental Protection Agency, Cincinnati, OH.

Fardeau, J. C. and Jappe, J. 1980. Choix de la fertilisation phosphorique des sols tropicaux: emploi du phosphore 32. Agron. Trop. 35:225–231.

Fox, R. L. and Kamprath, E. J. 1970. Phosphate sorption isotherms for evaluating the phosphate requirements of soils. Soil Sci. Soc. Am. Proc. 34: 902–907.

Hedley, M. J., Stewart, J. W. B., and Chauhan, B. S. 1982a. Changes in inorganic and organic soil phosphorus fractions induced by cultivation practices and by laboratory incubations. Soil Sci. Soc. Am. J. 46: 970–976.

Hedley, M. J., White, R. E., and Nye, P. H. 1982b. Plant-induced changes in the rhizosphere of rape seedlings. III. Changes in L-value, soil phosphate fractions and phosphatase activity. New Phytol. 91: 45–56.

Kelley, J., Lambert, M. J., and Turner, J. 1983. Available phosphorus forms in forest soils and their possible ecological significance. Commun. Soil. Sci. Plant Anal. 14: 1217–1234.

Kurmies, B. 1972. Zur Fraktionierung der Bodenphosphate. Die Phosphorsäure 29: 118–151.

Marks, G. 1977. Beitrag zur präzisierten Charakterisierung von pflanzen-verfügbarem Phosphat in Ackerböden. Arch. Acker Pflanzenbau Bodenkd. 21: 447–456.

Mattingly, G. E. G. 1975. Labile phosphate in soils. Soil Sci. 119: 369–375.

Metha, N. C., Legg, J. O., Goring, C. A. I., and Black, C. A. 1954. Determination of organic phosphorus in soils. I. Extraction method. Soil Sci. Soc. Am. Proc. 18: 443–449.

Murphy, J. and Riley, J. P. 1962. A modified single solution method for the determination of phosphate in natural waters. Anal. Chim. Acta 27: 31–36.

Olsen, S. R., Cole, C. V., Watanabe, F. S., and Dean, L. A. 1954. Estimation of available phosphorus in soils by extraction with sodium bicarbonate. U.S. Department of Agriculture Circular 939.

Onken, A. B., Matheson, R., and Williams, E. J. 1980. Evaluation of EDTA-extractable P as a soil test procedure. Soil Sci. Soc. Am. J. 44: 783–786.

Parfitt, R. L., Hume, L. J., and Sparling, G. P. 1989. Loss of availability of phosphate in New Zealand soils. J. Soil Sci. 40: 371–382.

Roberts, T. L., Stewart, J. W. B., and Bettany, J. R. 1985. The influence of topography on the distribution of organic and inorganic soil phosphorus across a narrow environmental gradient. Can. J. Soil Sci. 65: 651–665.

Saggar, S., Hedley, M. J., and White, R. E. 1990. A simplified resin membrane technique for extracting phosphorus from soils. Fertil. Res. 24: 173–180.

Saunders, W. M. H. and Williams, E. G. 1955. Observations on the determination of total organic phosphorus in soils. J. Soil Sci. 6: 254–267.

Sawhney, B. L. 1973. Electron microprobe analysis of phosphates in soils and sediments. Soil Sci. Soc. Am. Proc. 37: 658–660.

Schoenau, J. J. and Huang, W. Z. 1991. Anion-exchange membrane, water, and sodium bicarbonate extractions as soil tests for phosphorus. Commun. Soil Sci. Plant Anal. 22: 465–492.

Sibbesen, E. 1977. An investigation of the anion exchange resin method for soil phosphate extraction. Plant Soil 46: 665–669.

Sibbesen, E. 1978. A simple ion exchange resin procedure for extracting plant available elements from soil. Plant Soil 50: 305–321.

Stewart, J. W. B. and Tiessen, H. 1987. Dynamics of soil organic phosphorus. Biogeochemistry. 4: 41–60.

Syers, J. K. and Curtin, D. 1989. Inorganic reactions controlling phosphorus cycling. Pages 17–29 in H. Tiessen, Ed. P Cycles in Terrestrial and Aquatic Ecosystems. Regional Workshop 1. Proceedings of a workshop by the Scientific Committee on Problems of Environment (SCOPE). Published by the University of Saskatchewan, Saskatoon.

Thomas, R. L., Sheard, R. W., and Moyer, J. R. 1967. Comparison of conventional and automated procedures for nitrogen, phosphorus and potassium analysis of plant material using a single digestion. Agron. J. 59: 240–243.

Tiessen, H. 1991. Characterisation of soil phosphorus and its availability in different ecosystems, in Trends in Soil Science, Council for Scientific Research Integration, Trivandrum, India.

Tiessen, H., Stewart, J. W. B., and Moir, J. O. 1983. Changes in organic and inorganic phosphorus composition of two grassland soils and their particle size fractions during 60–70 years of cultivation. J. Soil Sci. 34: 815–823.

Tiessen, H., Stewart, J. W. B., and Cole, C. V. 1984. Pathways of phosphorus transformations in soils of differing pedogenesis. Soil Sci. Soc. Am. J. 48: 853–858.

Willett, I. R., Chartres, C. J., and Nguyen, T. T. 1988. Migration of phosphate into aggregated particles of ferrihydrite. J. Soil Sci. 39: 275–282.

Williams, J. D. H., Syers, J. K., and Walker, T. W. 1967. Fractionation of soil inorganic phosphate by a modification of Chang and Jackson's procedure. Soil Sci. Soc. Am. Proc. 31: 736–739.

Williams, J. D. H. and Walker, T. W. 1969. Fractionation of phosphate in a maturity sequence of New Zealand basaltic soil profiles. Soil Sci. 107: 22–30.

Chapter 11
DTPA-Extractable Fe, Mn, Cu, and Zn

J. Liang and R. E. Karamanos

University of Saskatchewan

Saskatoon, Saskatchewan, Canada

11.1 OVERVIEW

The methods most commonly used to extract the available micronutrient cations Fe, Mn, Cu, and Zn include the diethylene triamine pentaacetic acid (DTPA) method, ethylene diamine tetraacetic acid (EDTA) method, and 0.1 M HCl method. The DTPA soil test was originally developed by Lindsay and Norvell, as outlined in 1978, for micronutrient extraction of near-neutral and calcareous soils. The detailed theoretical basis for the DTPA soil test and reviews on its use have been published by Lindsay and Norvell (1978), Cox (1987), and O'Connor (1988). DTPA has been selected among a number of chelating agents because it has the most favorable combination of stability constants for the simultaneous complexing of Fe, Mn, Cu, and Zn. The extractant which is highly pH dependent is buffered at pH 7.30 with triethanolamine (TEA). The addition of 0.01 M $CaCl_2$ enables the extract to attain equilibrium with $CaCl_2$ so that it can minimize the dissolution of $CaCO_3$ from calcareous soils.

The DTPA method is inexpensive, reproducible, and easily adaptable to routine operations, if standardized procedures for preparation and extraction of the samples are established (Soltanpour et al. 1976). Various reports showed that DTPA extracts micronutrient metals from labile pools in soil (Wallance and Mueller 1968, Rule and Graham 1976) and that the level of DTPA-extractable metals correlate significantly with plant uptake (Haq and Miller 1972, Randall et al. 1976). However, the method is not suitable for sludge-amended soils (Barbarick and Workman 1987, Bidwell and Dowdy 1987).

The DTPA soil test was modified by Soltanpour and Schwab (1977) by shortening the shaking time and adding ammonium bicarbonate for extraction of P and other cations. These modifications have been evaluated for the simultaneous extraction of macro- and micro-nutrients in alkaline soils (Barbarick and Workman 1987).

Soil Sampling and Methods of Analysis, M. R. Carter, Ed.,

Canadian Society of Soil Science. © 1993 Lewis Publishers.

11.2 SPECIAL CONSIDERATION REGARDING SAMPLE PREPARATION

Because the levels of trace metals are very low in soil, contamination is always a problem. Special attention has to be paid to sampling and sample preparation. In the field, the tools used for sampling, such as augers, spades, or trowels, represent possible sources of contamination. Rusty metal implements should not be used, while those covered by paint are likely to give rise to considerable contamination from Zn. Soil samples should be taken with the aid of a well-polished steel shovel or an Al sample tube. Stainless steel products may be used if a qualitative test for acid-soluble Cu is negative. The composite sample should be mixed in an acid-washed plastic bucket or container. Samples should be placed in polyethylene bags, air-dried in wooden trays, ground with a wooden rolling pin, passed through a 2-mm nylon or polyethene sieve, and stored in a plastic or paper container.

11.3 DTPA EXTRACTION

The extraction step involves shaking a soil sample with 0.005 M DTPA extracting solution and then filtering the extract to obtain a clear, particulate-free filtrate.

11.3.1 Reagents

DTPA extracting solution: 0.005 M DTPA, 0.01 M calcium chloride ($CaCl_2$), and 0.1 M TEA, ($HOCH_2CH_2)_3N$ adjusted to pH 7.30 with dilute hydrochloride (HCl). To prepare 10 L of this solution dissolve 149.2 g of reagent-grade TEA, 19.67 g of DTPA, and 14.7 g of $CaCl_2·2H_2O$ in approximately 200 mL of deionized water. Allow sufficient time for the DTPA to dissolve, and dilute to 9 L. Adjust the pH to 7.30 $\pm$ 0.05 with 1:1 HCl while stirring and dilute to 10 L. This solution is stable for several months.

11.3.2 EXTRACTING PROCEDURE

1   Place 10 g of air-dry soil (2-mm) in a 125-mL Erlenmeyer flask. To each flask add 20 mL of the DTPA extracting solution.

2   Cover each flask with stretchable parafilm or polyethylene stoppers, and secure upright on a horizontal shaker with a stroke of 8.0 cm and a speed of 120 cycles min^{-1}. Shake for exactly 2 h.

3   Filter the suspension by gravity through Whatman® No.42 filter paper. If the filtrate is cloudy, refilter as necessary.

4   A blank solution containing all reagents should be used to correct for contamination.

11.3.3. Comments

The DTPA test is a nonequilibrium extraction. Therefore, factors such as shaking time, shaking speed, and shape of extraction vessel influence the quantity of metals extracted.

TABLE 11.1 Operating Parameters for Fe, Mn, Cu, and Zn

	Fe	Mn	Cu	Zn
Optimum conc. range (μg mL^{-1})	0.3–10	0.1–10	0.2–10	0.05–2
Sensitivity (μg mL^{-1})[a]	0.12	0.05	0.1	0.02
Detection limit (μg mL^{-1})[1]	0.02	0.01	0.01	0.005
Wavelength (nm)	248.3	279.5	324.7	213.9

[a] Sensitivity and detection limits vary among manufacturers and instruments.

Adapted from American Public Health Association (1985).

These factors must be standardized in each laboratory or the critical level for each of the micronutrients will be affected. Deviations from these standards will require recalibration of the test with plant growth response. A sample preservative is not necessary if analysis can be completed within a day or two. Otherwise, refrigeration may be required to retard microbial growth.

11.4 CATION MEASUREMENT

Determination of Fe, Mn, Cu, and Zn cation concentrations in the DTPA extract may proceed once a clear filtrate has been obtained. All four cation concentrations may be determined by atomic absorption spectrometry (AAS) manually or in an automated system. Inductively coupled plasma argon emission measurement of metal cations is also becoming more popular, but this equipment is more expensive and is not as widely available. A manual method for measurement of the four cations concentration using an atomic absorption spectrophotometer is outlined below.

11.4.1 REAGENTS

1   Standards containing 1000 μg mL^{-1} of the various metals are available from commercial sources.

2   Working standards: working standards should be prepared by diluting the 1000 μg mL^{-};1 stock solutions with DTPA extracting solution to include the optimum working ranges (Table 11.1). An intermediate dilution of the 1000 μg mL^{-1} stock solution to 100 μg mL^{-1} facilitates preparation of the standard solutions. The standards should be prepared with the same amounts of the reagents as contained in the samples.

11.4.2 Procedure

The concentration of the metal cations is determined using a calibration curve prepared with the standard solution or read directly from instruments equipped with microprocessor. Requirements in calibrating the instrument will vary depending on its design. Specific operating conditions and procedure for the instrument are outlined in the manufacturer's operating manual. Samples with concentration below the optimum range should be analyzed by flameless AAS while samples with concentrations above the optimum range should be diluted to bring them into the optimum range of the analytical instrument. The diluting solution should contain the same concentration of reagents as the samples and standards.

11.4.3 Calculations

The results expressed as micrograms per gram of soil are equal to twice the filtrate concentration. The filtrate concentration is determined from the sample solutions using the calibration curve or direct readout. The concentration of DTPA-extractable Fe, Mn, Cu, and Zn in the soil is calculated as follows:

$$\mu g \; g^{-1} \text{ metals in soil} = 20 \times (\mu g \; g^{-1} \text{ metals in sample solution} \atop - \; \mu g \; g^{-1} \text{ metals in blank solution})/10 \qquad (11.1)$$

11.4.4 Comments

Background correction is usually necessary to avoid interference from broad-band absorption and scattering from other chemical species in the samples. To reduce matrix interference, the standards should contain the same quantity of acid as the samples.

REFERENCES

American Public Health Association. 1985. Standard Methods for the Examination of Waste and Wastewater. 16th ed. Washington, D.C.

Barbarick, K. A. and Workman, S. M. 1987. Ammonium bicarbonate-DTPA and DTPA extractions of sludge-amended soils. J. Environ. Qual. 16: 125–130.

Bidwell, A. M. and Dowdy, R. H. 1987. Cadmium and zinc availability to corn following termination of sewage sludge applications. J. Environ. Qual. 16: 438–442.

Cox, F. R. 1987. Micronutrients soil tests: correlation and calibration. Pages 97–227 in Soil Testing: Sampling, Correlation, Calibration and Interpretation. SSSA special publication No. 21.

Haq, A. U. and Miller, M. H. 1972. Prediction of available soil Zn, Cu, and Mn using chemical extractants. Agron. J. 64: 779–782.

Lindsay, W. L. and Norvell, W. A. 1978. Development of a DTPA soil test for zinc, iron, manganese, and copper. Soil Sci. Soc. Am. J. 42: 421–428.

O'Connor, G. A. 1988. Use and misuse of the DTPA soil test. J. Environ. Qual. 17: 715–718.

Randall, G. W., Shulte, E. E., and Corey, R. B. 1976. Correlations of plant manganese with extractable soil manganese and soil factors. Soil Sci. Soc. Am. J. 40: 282–286.

Rule, J. H. and Graham, E. R. 1976. Soil labile pools of manganese, iron, and zinc as measured by plant uptake and DTPA equilibrium. Soil Sci. Soc. Am. J. 40: 853–857.

Soltanpour, P. N. and Schwab, A. P. 1977. A new soil test for simultaneous extraction of macro- and micronutrients in alkaline soils. Commun. Soil Sci. Plant Anal. 3: 195–207.

Soltanpour, P. N., Khan, A., and Lindsay, W. L. 1976. Factors affecting DTPA-extractable Zn, Fe, Mn, and Cu from soils. Commun. Soil Sci. Plant Anal. 7: 797–821.

Wallance, A. and Mueller, R. T. 1968. Effect of chelating agents on the availability of ^{54}Mn following its addition as carrier-free ^{54}Mn to three different soils. Soil Sci. Soc. Am. Proc. 32: 828–830.

Chapter 12
Boron, Molybdenum, and Selenium

U. C. Gupta

Agriculture Canada
Charlottetown, Prince Edward Island, Canada

12.1 INTRODUCTION

The availability of micronutrient anions, B, Mo, and Se to crops grown on the podzol soils of eastern Canada generally tends to be low. Because of the acidic nature of these soils the availability of Mo is greatly reduced. The fact that Mo availability to plants increases with increasing pH may possibly be explained by an anion exchange of the types given below.

$$2 \, OH^- \rightleftarrows MoO_4^- \tag{12.1}$$

Under oxidizing conditions the relatively insoluble molybdenumtrioxide is converted to highly soluble salts as shown below.

$$MoO_3 \rightleftarrows \text{soluble Mo salts} \tag{12.2}$$

The rate of these two above-mentioned reactions is largely governed by pH and is shifted to the left under acid conditions (Berger and Pratt 1963). This would account for the often observed fact that Mo availability is lowered by increasing acidity.

Boron is subject to loss by leaching, particularly on sandy soils, and thus responses to B have been observed in a variety of crops grown in eastern Canada as summarized by Gupta (1979).

Responses to Se are generally not found on crop yields. However, it is highly essential for livestock. Selenium concentration in soils in humid regions is generally inadequate to produce crops sufficient in Se to meet the needs of livestock. In acid soils, the ferric-iron selenite complex is formed, which is only slightly available to plants (National Academy of Science-National Research Council 1971). Selenium is generally present in excessive amounts only in semiarid and arid regions in soils derived from cretaceous shales, where it tends to form selenates (Rosenfeld and Beath 1964).

12.2 BORON

A number of extractants such as 0.05 M HCl (Ponnamperuma et al. 1981), 0.01 M CaCl$_2$ + 0.5 M mannitol (Cartwright et al. 1983), hot 0.02 M CaCl$_2$ (Parker and Gardner 1981),

Soil Sampling and Methods of Analysis, M. R. Carter, Ed.,
Canadian Society of Soil Science. © 1993 Lewis Publishers.

91

and 1 N NH$_4$OAC (Gupta and Stewart 1978) have been employed for determining the availability of B in soils. One advantage of using CaCl$_2$ is that it extracts little color from the soil, and predicted error due to this color is found to be low at 0.00–0.07 µg B g^{-1} soil (Parker and Gardner 1981). Such filtered extracts are also free of colloidal matter (Jeffrey and McCallum 1988).

Azotobacter chroococcum was also considered as a possible microbiological indicator for B availability in soils (Gerretsen and de Hoop 1954). However, owing to the complicated behavior of soils, the microbiological determination of B by the Azotobacter method has been considered unsatisfactory (Bradford 1966).

The most commonly used method is still the hot water extraction of soils as originally developed by Berger and Truog (1939). A number of modified versions of this procedure have since appeared. Offiah and Axley (1988) have used B-spiked hot water extraction for soils. This method is claimed to have an advantage over unspiked hot water extraction in that it removes from consideration a portion of the B fixing capacity of soils that does not relate well to plant uptake. A longer boiling time of 10 min as opposed to the normally used 5-min boiling was found to reduce error for a Typic Hapludult soil by removing enough boron to reach the plateau region of the extraction curve (Odom 1980).

Once extracted from the soil, B can be analyzed by the colorimetric methods using reagents such as carmine (Hatcher and Wilcox 1950), azomethine-H (Wolf 1971), and most recently by inductively coupled plasma (ICP) and atomic emission spectrometry (Haubold et al. 1988, Jeffrey and McCallum 1988).

The method described here is the one used in the author's laboratory in Charlottetown. The method of boiling soil with water is a simple modification (Gupta 1967) of the one developed by Berger and Truog (1939). The extracted B in the filtered extract is determined by the azomethine-H colorimetric method (Gupta 1979).

12.2.1 REAGENTS

1   Azomethine-H: dissolve 0.5 g azomethine-H and 1.0 g L-ascorbic acid in about 10 mL redistilled water with gentle heating in a water bath or under a hot water tap at about 30°C, and make the volume up to 100 mL with redistilled water. If the solution is not clear, it should be reheated again till it dissolves. Prepare fresh azomethine-H solution for everyday use.

2   Ethylene diamine tetraacetic acid (EDTA) reagent (0.025 *M*): dissolve 9.3 g EDTA in redistilled water and make the volume up to 1 L with redistilled water. Add 1 mL Brij-35 and mix.

3   Buffer solution: dissolve 250 g ammonium acetate in 500 mL redistilled water and adjust the pH to about 5.5 by slowly adding approximately 100 mL glacial acetic acid, with constant stirring. Add 0.5 mL Brij-35 and mix.

12.2.2 PROCEDURE (GUPTA 1979)

1   Weigh 25 g air-dried soil, screened through a 2-mm sieve, in a beaker and add about 0.4 g charcoal.

2   The amount of charcoal added will vary with the organic matter content of the soil and should be just sufficient to produce a colorless extract after 5 min of boiling on a hot plate.

3   The loss of weight due to boiling should be made up by adding redistilled water and the mixture should be filtered while still hot. For a blank take 50 mL redistilled water and add same quantity of charcoal as to the soil and treat in the same manner as above.

4   Take 5 mL of the clear filtrate in a test tube and add 2 mL buffer reagent, 2 mL EDTA solution, and 2 mL azomethine-H solution.

5   Mix the contents thoroughly after the addition of each chemical.

6   Let the solutions stand for 1 h and measure the absorbance at 430 nm.

7   The color thus developed has been found to be stable for up to 3–4 h.

8   The pH of the colored extract should be about 5.0.

12.2.3 COMMENTS

1   The use of azomethine-H is an improvement over that of carmine, quinalizarin, and curcumin, since the procedure involving this chemical does not require the use of a concentrated acid.

2   This method has been found to give reproducible results when compared to the carmine method (Gupta 1979, Sipola and Ervio 1977).

3   It is difficult to use an autoanalyzer because of its insensitivity at lower B concentrations generally found in the hot water extract of most soils.

4   Excess amounts of charcoal can result in loss of extractable B from soils.

12.2.4 Determination of Boron by ICP

This technique has been found to be rapid and reliable for determining B in soil extracts using the procedure described in Section 12.2.2.

12.2.5 COMMENTS

1   All glassware used in plant or soil B analyses must be washed with a 1:1 mixture of boiling HCl acid prior to use. Storage of the filtered extracts for the analysis of B by the ICP technique must be in plastic sampling cups.

2   Results for the B analysis are difficult to compare, since the National Bureau of Standards does not supply any official B values for the NBS citrus leaf sample.

12.3 MOLYBDENUM

Studies on the extraction of available Mo from soils have been limited. Further, the extremely low amounts of available Mo in soils under deficiency conditions make it difficult to determine Mo accurately. The Grigg or Tamm reagent (Grigg 1953), acid ammonium oxalate, still remains the primary extractant used, although the extracted Mo by this method on some iron-rich soils may be misleading. Some of the other extractants used for estimating the availability of Mo in soils are water (Lavy and Barber 1964, Gupta and MacKay 1965b), hot water (Lowe and Massey 1965), anion-exchange resin (Bhella and Dawson 1972), ammonium bicarbonate-diethylene triamine pentaacetic acid (AB-DTPA) (Soltanpour 1985), and 1 M neutral ammonium acetate (Advisory Soil Analysis and Interpretation 1985).

Grigg's reagent, when soil pH is considered as a factor, has yielded better correlations between soil and plant Mo as summarized by Cox (1987). The author has also used this reagent, and thus this method is described in detail here.

12.3.1 REAGENTS (BINGLEY 1963)

1   H_2SO_4, concentrated.

2   H_2O_2, 30%.

3   Tartaric acid, 50%.

4   Potassium iodide, 50%.

5   Amyl acetate (pentacetate amylacetic ester).

6   Thiourea, 10%.

7   Ascorbic acid, 5%.

8   Dithiol (Gupta and MacKay 1965a) used in our laboratory was diacetyldithiol. Currently, it is available as toluene-3,4-dithiol zinc derivative. To 0.2 g of analytical-grade melted dithiol add 100 mL 1% NaOH warmed to 38°C and stir for about 10 min on a warm plate (about 30°C). Now add 1.7 mL thioglycolic acid, stir for about 2 min and filter.

9   As described by Gupta and MacKay (1966), 0.2 *M* ammonium oxalate buffered to pH 3.3 with oxalic acid (24.9 g ammonium oxalate plus 12.605 g oxalic acid per liter).

12.3.2 METHOD

1   Shake 15 g air-dried soil with 150 mL 0.2 *M* ammonium oxalate (pH 3.3) solution for 16 h and filter the mixture.

2   Evaporate an aliquot of 120 mL to dryness and ignite for 3 to 4 h at 550°C to remove oxalate and organic matter.

3   Digest the contents using H_2SO_4 and H_2O_2 and make up to 100 mL as described by Gupta and MacKay (1965b).

4   Transfer 50 mL of this extract into a separatory funnel.

5   Because there is sufficient Fe in the extract, addition of ferric ammonium sulfate is not necessary and the Fe already present is reduced by adding 2 mL KI solution.

6   Then neutralize the liberated iodine with 2 mL ascorbic acid solution. If the color does not disappear, add a few more drops of ascorbic acid.

7   Next add 0.25 mL of 50% tartaric acid solution, shake, and add 2 mL of 10% thiourea solution, then thoroughly mix the contents.

8   Now add 4 mL of 0.2% dithiol reagent, shake for 20 s, and allow to stand for 30 min.

9   Finally, add 10 mL amylacetate, shake vigorously for 2 min, and allow to stand for 1 h for complete separation.

10   Run off and discard the aqueous phase and drain off the organic phase in a centrifuge tube and centrifuge for 15 min at 2000 rpm.

11   Measure the absorbance at 680 nm.

12.3.3 COMMENTS

1   The extracted green complex formed between Mo and dithiol is stable for at least 24 h. Most other reagents are stable for lesser periods of time.

2   Large amounts of ferric iron in soil extracts can interfere seriously by oxidative destruction of the Mo-dithiol complex (interference may result in a black precipitate). Such interference should be eliminated by completely reducing the ferric iron to the ferrous state using the KI solution.

3   It may be possible to analyze Mo in the soil extracts by ICP or using graphite furnace-atomic absorption spectroscopy with some experimentation.

12.4 SELENIUM

Total soil Se has proved to be of little value in predicting plant uptake (Lindberg and Bingefors 1970). Cary and Allaway (1969) have shown that the plant uptake of Se is closely correlated to the water-soluble Se fraction in soils rich in Se, although such a correlation has not been demonstrated for soils low in Se.

In Se-deficient soils, the extractants generally extract small quantities of Se. Unpublished results of Gupta at the Charlottetown Research Station showed that the mean water-soluble Se in acid podzol soils of Prince Edward Island was only 0.0055 μg g^{-1}. This on average constituted less than 4% of the total soil Se and was not related to plant Se.

Total Se in soils occurs in highly variable amounts, with most soils containing 0.1 to 2 μg g^{-1} (Swaine 1955). Soils in the Atlantic region of Canada, where Se deficiency in livestock is most prevalent, contain from 0.20 to 0.27 μg g^{-1} (Gupta and Winter 1975). Selenium content of lacustrine clay in Saskatchewan ranged from 0.24 to 1.92 μg g^{-1} (Doyle and Fletcher 1977).

12.4.1 EXTRACTION OF SELENIUM IN SOILS

There are five most commonly used extractants as given below:

1   AB-DPTA

2   Hot water

3   Saturated paste extractants

4   DTPA (2 h)

5   0.5 M Na$_2$CO$_3$

12.4.2 METHOD

1   The ratio of soil to the extractants listed above varies from 1:2 to 1:5 and the extraction time from 15 min to 2 h as summarized by Jump and Sabey (1989).

2   The filtered extracts can be analyzed for Se using a hydride generating system attached to an ICP emission spectrometer (Soltanpour et al. 1982a).

All the above five extractants, when tested on soils producing high Se, showed high correlation between the wheat plant Se and Se extracted from soils (Jump and Sabey 1989). However, Se extracted in saturated soil pastes and expressed as mg Se L^{-1} of extract was found to be the best predictor of Se uptake in Se-accumulating plants. The data of Jump and Sabey (1989) suggested that soil or mine spoil materials that yield more than 0.1 mg Se L^{-1} in saturated extract may produce Se-toxic plants.

The AB-DTPA extract has been found to predict availability better when Se in wheat grain was correlated with Se in 0–90 cm as opposed to 0–30 cm soil depth (Soltanpour et al. 1982b). This was found to be particularly useful to screen soils and overburden material for potential toxicity of Se.

The parent material has a significant effect upon the Se composition of plants. For example, field studies conducted on wheat in west central Saskatchewan showed higher Se values in wheat plants grown on lacustrine clay and glacial till, intermediate in plants grown on lacustrine silt, and lowest on aeolian sand (Doyle and Fletcher 1977). A similar trend characterized the C horizon soil, with highest Se values with lacustrine clay and lowest with aeolian sand. The findings of Doyle and Fletcher (1977) point to the potential usefulness of information on the Se content of soil parent materials when designing sampling programs for investigating regional variations in plant Se content.

12.4.3 COMMENTS

1   The extractants developed have been found to be suitable for predicting the availability of Se in Se-toxic areas only. Because of rather small quantities of available Se in Se-deficient areas, no reliable extractant has yet been developed for such soils. Therefore, plant Se and total soil Se will continue to serve as the best tools available for testing the Se status of Se-deficient soils.

2   The term deficiency or deficient in connection with Se has implications in livestock and human nutrition only and not in plant nutrition, since no known yield responses to Se have been found on cultivated crops.

REFERENCES

Advisory Soil Analysis and Interpretation. 1985. 2. Trace Elements, Bull. 1, Page 7–8. MacAulay Institute for Soil Research and Scottish Agricultural Colleges Liaison Group.

Berger, K. C. and Pratt, P. F. 1963. Advances in secondary and micronutrient fertilization. Pages 287–340 in M. H. McVickar et al., Eds. Fertilizer Technology and Usage. Soil Sci. Soc. Am., Madison, WI.

Berger, K. C. and Truog, K. 1939. Boron determination in soils and plants using the quinalizarin reaction. Ind. Eng. Chem. 11: 540–545.

Bhella, H. S. and Dawson, M. D. 1972. The use of anion exchange resin for determining available soil molybdenum. Soil Sci. Soc. Am. Proc. 36: 177–178.

Bingley, J. B. 1963. Determination of molybdenum in biological materials with control of copper interference dithiol. Agric. Food Chem. 11: 130–131.

Bradford, G. R. 1966. Pages 33–61 in Chapman, Ed. Diagnostic Criteria for Plants and Soils. University of California Division of Agriculture Science, Riverside.

Cartwright, B., Tiller, K. G., Zarcinas, B. A., and Spouncer, L. R. 1983. The chemical assessment of the boron status of soils. Aust. J. Soil Res. 21: 321–330.

Cary, E. E. and Allaway, W. H. 1969. The stability of different forms of selenium applied to low-selenium soils. Soil Sci. Soc. Am. Proc. 33: 571–574.

Cox, F. R. 1987. Micronutrient Soil Tests: Correlation and Calibration Soil Testing: Sampling, Correlation, Calibration, and Interpretation. SSSA Special Publ. No. 21, 97–117.

Doyle, P. J. and Fletcher, W. K. 1977. Influence of soil parent material on the selenium content of wheat from west-central Saskatchewan. Can. J. Plant Sci. 57: 859–864.

Gerretsen, F. C. and de Hoop, H. 1954. Boron, an essential micro-element for *Azotobacter chroococcum*. Plant Soil 5: 349–367.

Grigg, J. L. 1953. Determination of the available molybdenum of soils. N.Z. J. Sci. Tech. Sect. A-34: 405–414.

Gupta, S. K. and Stewart, J. W. B. 1978. An automated procedure for determination of boron in soils, plants and irrigation waters. Schweiz. Landwirtscha. Forsch. 17: 51–55.

Gupta, U. C. 1967. A simplified method for determining hot-water soluble boron in podzol soils. Soil Sci. 103: 424–428.

Gupta, U. C. 1979. Some factors affecting the determination of hot-water soluble boron from podzol soils using azomethine-H. Can. J. Soil Sci. 59: 241–247.

Gupta, U. C. and MacKay, D. C. 1965a. Determination of Mo in plant materials. Soil Sci. 99: 414–415.

Gupta, U. C. and MacKay, D. C. 1965b. Extraction of water-soluble copper and molybdenum

from podzol soils. Soil Sci. Soc. Am. Proc. 29: 323.

Gupta, U. C. and MacKay, D. C. 1966. Procedure for the determination of exchangeable copper and molybdenum in podzol soils. Soil Sci. 101: 93–97.

Gupta, U. C. and Winter, K. A. 1975. Selenium content of soils and crops and the effects of lime and sulfur on plant selenium. Can. J. Soil Sci. 55: 161–166.

Hatcher, J. T. and Wilcox, L. V. 1950. Colorimetric determination of boron using carmine. Anal. Chem. 22: 567–569.

Haubold, W., Koenig, E., and Schmid, R. 1988. Determination of the hot-water soluble boron content in soils using inductively coupled plasma atomic emission spectrometry. Fresenius Z. Anal. Chem. 33: 713–720.

Jeffrey, A. J. and McCallum, L. E. 1988. Investigation of a hot 0.01 M calcium chloride soil boron extraction procedure followed by ICP-AES analysis. Commun. Soil Sci. Plant Anal. 19: 663–673.

Jump, R. K. and Sabey, B. R. 1989. Soil test extractants for predicting selenium in plants. Pages 95–105 in L. W. Jacobs, Ed. Selenium in agriculture and the environment. Am. Soc. Agron, Inc; Soil Sci. Soc. Am., Inc, Madison, SSSA Spec. Publ. 23 Chapter 5.

Lavy, T. L. and Barber, S. A. 1964. Movement of molybdenum in the soil and its effect on availability to the plant. Proc. Soil Sci. Soc. Am. 28: 93–97.

Lindberg, P. and Bingefors, S. 1970. Selenium levels of forages and soils in different regions of Sweden. Acta Agric. Scand. 20: 133–136.

Lowe, R. H. and Massey, H. F. 1965. Hot water extraction for available soil molybdenum. Soil Sci. 100: 238–243.

National Academy of Sciences-National Research Council. 1971. Selenium in Nutrition. NAS-NRC, Washington, D.C.

Odom, J. W. 1980. Kinetics of the hot-water soluble boron soil test. Commun. Soil. Sci. Plant Anal. 11: 759–765.

Offiah, O. and Axley, J. H. 1988. Improvement of boron soil test. Commun. Soil Sci. Plant Anal. 19: 1527–1542.

Parker, D. R. and Gardner, E. H. 1981. The determination of hot-water-soluble boron in some acid Oregon soils using a modified azomethine-H procedure. Commun. Soil Sci. Plant Anal. 12: 1311–1322.

Ponnamperuma, F. N., Cayton, M. T., and Lantin, R. S. 1981. Dilute hydrochloric acid as an extractant for available zinc, copper, and boron in rice soils. Plant Soil 61: 297–310.

Rosenfeld, I. and Beath, O. A. 1964. Selenium; Geobotany, Biochemistry, Toxicity and Nutrition. Pages 1–7. Academic Press, New York.

Sippola, J. and Ervio, R. 1977. Determination of boron in soils and plants by the azomethine-H method. Finn. Chem. Lett. 138–140.

Soltanpour, P. N. 1985. Use of ammonium bicarbonate DTPA soil test to evaluate elemental availability and toxicity. Commun. Soil Sci. Plant Anal. 16: 323–338.

Soltanpour, P. N., Jones, J. B., and Workman, S. M. 1982a. Optical emission spectrometry. Pages 29–65 in A. L. Page et al., Eds. Methods of soil analysis. Part 2. 2nd ed. Agronomy No. 9. American Society of Agronomy, Madison, WI.

Soltanpour, P. N., Olsen, S. R., and Goos, R. J. 1982b. Effect of nitrogen fertilization on dryland wheat on grain selenium concentration. Soil Sci. Soc. Am. J. 46: 430–433.

Swaine, D. J. 1955. The trace-element content of soils. Common. Bur. Soil Sci. Tech. Commun. 48: 157.

Wolf, B. 1971. The determination of boron in soil extracts, plant materials, composts, manures, water, and nutrient solutions. Soil Sci. Plant Anal. 2 363–374.

Chapter 13
Cadmium, Chromium, Lead, and Nickel

Y. K. Soon

Agriculture Canada
Beaverlodge, Alberta, Canada

S. Abboud

Alberta Research Council
Edmonton, Alberta, Canada

13.1 INTRODUCTION

High concentrations of some heavy metals in soils, derived either from natural sources or through pollution, can adversely affect crops and consumers of the crops, e.g., humans and animals. This chapter will be concerned only with Cd, Cr, Pb, and Ni. Chromium, Ni, and Pb normally occur in soils in the range of 10 to 100 mg kg^{-1}, and Cd below 1 mg kg^{-1} (Mattigod and Page 1983). However, in serpentine soils derived from ultrabasic rocks, Cr and Ni can occur in concentrations of several thousand mg kg^{-1}, and are at least partly responsible for the inherent infertility of these soils (Shewry and Peterson 1976). Soils in many parts of the world have also become polluted with Cr, Cd, Ni, and Pb, among other metals, due to mining and smelting activities, disposal of sewage sludge and other wastes, and combustion emissions (Patterson 1971). Cadmium and lead pose considerable health concerns to humans and animals as environmental pollutants (Lisk 1972); chromium is an essential micronutrient and, at higher concentrations, a toxic element for mammals and other animals (Mertz 1982), while Ni is a possible carcinogen (Lisk 1972). However, Ni has also been suggested as an essential micronutrient for higher plants (e.g., Brown et al. 1987). Metals in contaminated soils enter the food chain by plant uptake, direct ingestion of soil by animals, and entering water supplies. Estuaries may be contaminated by sediments and shellfish affected. Trace element pollution is therefore a subject of increasing importance to the soil scientist. The interest in these metals in soils extends beyond that of agricultural soil testing, i.e., in relation to crop yield and nutrient deficiency. For example, lead-contaminated soil and dust appear to be responsible for blood lead levels increasing above background level in children (because of their hand-to-mouth activities and high rate of intestinal absorption) (Centers for Disease Control 1985).

With the possible exception of Cd, the soil chemistry of these metals is not adequately understood, especially that of chromium and nickel. The solubility in soils follows the order:

Cd > Ni > Pb ≥ Cr. These metals tend to be concentrated in hydrous oxides of iron and possibly aluminum (Le Riche and Weir 1963, Garcia-Miragaya et al. 1981), and organic matter (Soon and Bates 1982, Soon and Abboud 1990) components of soil. Nickel is present as a constituent of silicate clay minerals to some extent, but Cd does not appear to enter into the crystal lattice of layer silicate minerals. Among these metals, only Cd tends to occur in significant proportion in the exchangeable form in soils (Soon and Bates 1982). Chromium in soils is believed to be derived mostly from chromite, a mixed oxide of Cr(III) and Fe(II). Cr^{3+} has an ionic radius between that of Fe^{3+} and Al^{3+} and thus coprecipitates readily in iron oxide and aluminum oxide. The soil chemistry of Cr(III) is also similar in many respects to that of Fe(III) and to a lesser extent Al(III). The pH-redox relationship in most cultivated soils would favor the formation of Cr(III) over that of Cr(VI) (Cary et al. 1977), with the reduction of Cr(VI), a more toxic form, to Cr(III) being slower in alkaline than in acid soils.

Although plant availability is also a concern, in assessing metal pollution of soils, accurate quantification of accumulation of metals in soils for evaluation of the degree of contamination and potential for movement in leachates is desirable. A variety of extraction procedures are available and the choice of procedure to use will depend on the metals of interest and the objective for quantifying the metal loading in the soil.

Types of chemical extractants used for quantifying the level of extractable Cd, Cr, Ni, and Pb in soils include chelating agents (e.g., ethylene diamine tetraacetic acid [EDTA] and diethylene triamine pentaacetic acid [DTPA]), dilute acids (e.g., 1 M HNO_3 and 0.5 M acetic acid), water, and buffered and unbuffered salt solutions (e.g., 1 M ammonium acetate at pH 7 and 1 M KNO_3, respectively) (Haq et al. 1980, Garcia-Miragaya et al. 1981, Lake et al. 1984, Soon and Abboud 1990). No one extractant has been found to provide the best correlations for several heavy metals in the plant-available form under a wide range of soil properties; therefore a number of extractants may have to be employed. For example, Haq et al. (1980) found among the extractants they tested that 0.5 M acetic acid extraction gave the best correlation for Ni, while distilled water extraction was best for Cd. Correlations between plant availability and extractable amount are often improved by including other soil properties in the regression equations, especially pH, and organic matter and clay content. Thus, Haq et al. (1980) found that DTPA was most suitable for Cd, Ni, and Zn extractions if one or more of these soil properties were included in the respective regression equations relating plant metal content to extracted metal content and other properties of soil.

13.2 TOTAL SORBED CADMIUM, CHROMIUM, NICKEL, AND LEAD

A total elemental analysis that includes metals trapped within the crystal lattice of silicate minerals is usually not necessary for an assessment of soil contamination with heavy metals. However, if such an analysis is desired, a rapid and reasonably precise method that includes Cd, Cr, Ni, and Pb is an aqua regia-HF digestion for 1 h at 100°C in a Teflon®-lined bomb (Buckley and Cranston 1971). Fusion methods are generally not suitable, unless otherwise demonstrated, because of possible volatile losses of Cd and Pb. Strong acid extractants such as concentrated HNO_3 and aqua regia are often used to determine "total" metals in contaminated soils (Frank et al. 1976, Merry et al. 1983). The U.S. Environmental Protection Agency (1986) has adopted an acid digestion procedure (EPA method 3050) for total sorbed metals. These "total" analyses would exclude most trace metals trapped within silicate crystal lattices. However, the EPA method 3050 gives a reliable measure of the amounts of metals added to soils as nonsilicates from industrial sources (Risser and Baker 1990), i.e., potentially available for natural leaching and biological processes.

13.2.1 REAGENTS

1   Nitric acid HNO_3: concentrated (BDH Analar® grade) and 1:1 dilution in deionized water (>10 megaohm-cm resistivity).

2   30% hydrogen peroxide, H_2O_2.

3   Hydrochloric acid, HCl: concentrated (BDH Analar® grade).

13.2.2 PROCEDURE

1   Add 10 mL of 1:1 HNO_3 to 2 g of air-dried soil (<1 mm) in a 150-mL beaker.

2   Place the sample on a hot plate, cover with a watch glass, and heat (reflux) at 95°C for 15 min.

3   Cool the digestate and add 5 mL of conc. HNO_3. Reflux for an additional 30 min at 95°C.

4   Repeat the last step, and reduce the solution to about 5 mL without boiling (by only partially covering the beaker).

5   Cool the sample again and add 2 mL of deionized water and 3 mL of 30% H_2O_2.

6   With the beaker covered, heat the sample gently to start the peroxide re-action. If effervescence becomes excessively vigorous, remove the sample from the hot plate. Continue to add 30% H_2O_2 in 1-mL increments, followed by gentle heating until the effervescence subsides.

7   Add 5 mL of conc. HCl and 10 mL of deionized water and reflux the sample for an additional 15 min without boiling.

8   Cool and filter the sample through a Whatman® No. 42 filter paper. Dilute to 50 mL with deionized water. Analyze for Cd, Cr, Ni, and Pb by atomic absorption or inductively coupled plasma (ICP) spectroscopy (Table 13.1).

13.2.3 Comments

1. The above method was developed by the U.S. Environmental Protection Agency for heavy metals in soils, sediments, and sludges, and extracts several times more metals than the DTPA extractant (Section 13.3). It has not been validated for organic soils.

2. Some instrument handbooks recommend a luminous air-acetylene flame for maximum sensitivity for the determination of chromium by atomic absorption. It is well known that chromium analysis is subject to numerous interelement and oxidation state effects in air-

TABLE 13.1 Best Detection Limits of Cd, Cr, Ni,
and Pb by Atomic Absorption (AA)
and ICP-Emission Spectroscopy

| | | Method of analysis | |
	Flame AA	Electrothermal AA $\mu g\ L^{-1}$	ICP
Cd	0.5	0.0002	0.07
Cr	2	0.004	0.08
Ni	2	0.05	0.2
Pb	10	0.007	1.0

From Parson, M. L., Major, S., and Forster, A. R.,
Appl. Spectrosc., 37, 411, 1983. With permission.

acetylene flame. Commercially prepared standard solutions contain Cr(VI), whereas most
Cr in soil and environmental samples is probably present as Cr(III) or a mixture of oxidation
states. Some interference effects can be overcome by using a nitrous oxide-acetylene flame
or a nonluminous air-acetylene flame with considerable loss of sensitivity (Thompson 1978).
Analysis of the other elements should be more routine.

13.3 DTPA EXTRACTION
(MODIFIED FROM LINDSAY AND NORVELL 1978)

For many environmental and agricultural applications, it is not necessary to determine the
total contents of elements in soil. When it is sufficient to only quantify the labile or ''available'' fraction, extraction with DTPA is a suitable procedure. This method was originally
developed to identify inadequate supplies of the plant-available micronutrient metals, Cu,
Zn, Fe, and Mn, in near-neutral and calcareous soils (Lindsay and Norvell 1978). Therefore,
in soils heavily contaminated by metals, the extraction of metals may be limited by saturation
of the 0.005 M DTPA used at the recommended 2:1 extractant:soil ratio. A combined
maximum of 10 mmol kg^{-1} of all metals would be theoretically extractable (assuming all
Ca^{2+} would be displaced from the chelate). Norvell (1984) suggested that using the DTPA
extractant at a 5:1 extractant:soil ratio was effective in extracting metals in acid and metal-
contaminated soils. The buffered alkaline pH (7.3) was not as serious a limitation as anticipated for acid soils. Although buffering the extractant at pH 5.3 increased metal extractability
(Norvell 1984), the alkaline (pH 7.3)-buffered extractant is suitable for both acid and calcareous soil and is therefore recommended.

13.3.1 Reagents

DTPA extracting solution: the solution is 5 mM DTPA + 10 mM CaCl$_2$ + 0.1 M triethanolamine (TEA). To prepare 5 L of this solution, dissolve 74.6 g of reagent-grade TEA
in approximately 1 L of deionized water (>10 megaohm-cm resistivity). Next dissolve 9.835
g DTPA and 7.35 g CaCl$_2$2H$_2$O in the TEA solution. Allow time for DTPA to dissolve,
and dilute to approximately 4 L. Adjust the pH to 7.3 ± 0.05 with 1:1 HCl (approximately
42 mL required) and dilute to 5 L final volume. The solution is stable for several months.

13.3.2 PROCEDURES

1   Place 5 g of air-dried soil (>2 mm) in a 125-mL Erlenmeyer flask.

2   Add 25 mL of the DTPA extracting solution and shake for 2 h at a speed of 120 cycles min^{-1}.

3   After shaking, filter through a Whatman® No. 42 filter paper. Refilter if filtrate is turbid.

4   Analyze metals in filtrate by atomic spectroscopy.

13.3.3 Comments

The above procedure would also be suitable for measuring extractable Zn and Cu in contaminated soil. The procedure would be suitable particularly for Cd and Ni. Although Karamanos et al. (1976) recovered 80% of Pb added to soil with DTPA extraction, Haq et al. (1980) reported that plant tissue Pb could not be predicted from DTPA extraction of soils. Until more data are available, this extraction procedure should be used with caution for Pb. It is not effective for Cr. Very little work has been done on organic soils, therefore caution is advised.

13.4 EDTA EXTRACTION
(MODIFIED FROM MITCHELL ET AL. 1957)

Dilute EDTA has been a preferred extractant for soils long before DTPA was promoted by the Colorado SU group (Lindsay and Norvell 1978). It was first proposed by Cheng and Bray (1953) and subsequently advocated by Viro (1955a and b). Different concentrations (0.01 to 0.1 *M*) have been used in combination with buffered pH values of 4.6 to 9.0 (see Borggaard 1976), although pH 7 appears to be a favorite. Different soil:extractant ratios and extraction times have been used. Clayton and Tiller (1979) used a 7-d extraction with 0.1 *M* EDTA at pH 6.0 for classification of soils with respect to the degree of contamination with Cd, Pb, and Zn. This long extraction period is not suitable for routine analysis. The procedure described below has been in use at the Macaulay Land Use Research Institute for many of their trace element studies (e.g., Berrow and Mitchell 1980).

13.4.1 REAGENTS

1   0.05 *M* EDTA (pH 7): dissolve 93.05 g of EDTA (di-sodium salt) in approximately 4 L of distilled and deionized water (DD). Adjust to pH 7.0 with 7 *M* NH$_4$OH, and make up to 5 L with DD water.

13.4.2 PROCEDURE

1   Weigh 5 g of air-dried (<2 mm) soil into a 125-mL Erlenmeyer flask and add
     25 mL of 0.05 *M* EDTA solution.

2   Shake for 1 h at a speed of 120 cycles min^{-1}.

3   Filter through a Whatman® No. 42 filter paper after shaking, and analyze for
     metals by atomic spectroscopic methods.

13.4.3 Comments

EDTA is a generally useful extractant for many elements (Borggaard 1976, Berrow and
Mitchell 1980). The above procedure is suitable for use with acid and calcareous soils.
Merry and Tiller (1978) found that Cd, Cu, Pb, and Zn concentrations in plants growing
on contaminated soils were related to the EDTA-extracted metals from the soils. EDTA
extraction is also useful for estimating labile Mo and Se in potentially toxic soils, including
peat soils (Williams and Thornton 1973). The limited amount of data available suggest that
EDTA is also one of the slightly better extractants for Cr (Shewry and Peterson 1976, Berrow
and Mitchell 1980), even though the amount extracted is still small (<1 to 4% of total
amount). However, EDTA-extractable Cr has not been shown to be correlated with plant
availability.

13.5 1 *M* HNO$_3$ EXTRACTION

Previous studies have shown 1 *M* HNO$_3$ to be a suitable extractant for a variety of heavy
metals, including Cd, Ni, and Pb (John 1971, Soon and Bates 1982, Garcia-Miragaya et
al. 1981). Metals so extracted include those fractions complexed by organic matter and
sorbed by or coprecipitated with hydrous oxides, carbonates, and sulfides. Cary et al. (1977)
found 0.1 *M* HCl to be the most efficient among several mild extractants in recovering
applied Cr from soils. This suggests that 1 *M* HNO$_3$ may also be suitable for Cr. John (1971)
recovered 77 to 100% of added lead with 1 *M* HNO$_3$ extraction at room temperature and
considered this extraction suitable for measuring lead contamination of soils.

13.5.1 Reagents

1 *M* HNO$_3$: add 250 mL of concentrated (70–71%) HNO$_3$ to 3 L of DD water and dilute
to 4 L.

13.5.2 PROCEDURE

1   Weigh 3 g of air-dried soil (<2 mm) into a 125-mL Erlenmeyer flask and add
     30 mL of 1 *M* HNO$_3$ extractant.

2   Shake for 2 h at a speed of 120 cycles min^{-1}.

3   Filter through a Whatman® No. 42 filter paper and analyze for the metals
     by atomic spectroscopy.

13.6 ACID AMMONIUM OXALATE-EXTRACTABLE CHROMIUM

Cr(III) coprecipitates with some metal hydroxides at the pH range of most soils. Adsorption on the clay fraction is also a major retention mechanism in soils, especially for Cr(III) (Anderson 1977). Like Pb, Cr in plants tends to accumulate in root tissues and does not normally correlate well with the extractable amount in soil. This may also be partly due to the paucity of work on development of extractants for labile Cr. Recent work by Shewry and Peterson (1976) indicates that acid ammonium oxalate may be one of the more effective extractants of soil Cr and that it extracts a form of Cr different from that extracted by EDTA. Grove and Ellis (1980) also showed that ammonium oxalate was an effective extractant for Cr added to soil in sewage sludge or as $CrCl_3$ or CrO_3. This is not surprising, since acid ammonium oxalate is an effective extractant of hydrous iron and aluminum oxides.

Acid ammonium oxalate (Tamm's reagent) is quite commonly used to extract available Mo (Chapter 12, 12.4). The procedure outlined in Section 12.4 may be used to combine the extraction of Mo and Cr. An aliquot of the filtrate is used to determine Cr. Refer to Section 13.2 for possible problems with Cr analysis by atomic absorption. ICP analysis should remove most of these problems. When used to extract soil Cr, acid oxalate solution would dissolve Cr mostly coprecipitated with or occluded by amorphous iron and aluminum oxides plus some Cr associated with organic matter. It is likely that Cr so extracted would not be in readily plant-available forms.

REFERENCES

Anderson, A. 1977. Heavy metals in Swedish soils: on their retention, distribution and amounts. Swedish J. Agric. Res. 7: 7–20.

Berrow, M. L. and Mitchell, R. L. 1980. Location of trace elements in soil profiles: total and extractable contents of individual horizons. Trans. R. Soc. Edinburgh: Earth Sciences. 17: 103–121.

Borggaard, O. K. 1976. The use of EDTA in soil analysis. Acta Agric. Scand. 26: 144–156.

Brown, P. H., Welch, R. M., and Cary, E. E. 1987. Nickel: a micronutrient essential for higher plants. Plant Physiol. 85: 801–803.

Buckley, D. E. and Cranston, R. E. 1971. Atomic absorption analyses in 18 elements from a single decomposition of aluminosilicate. Chem. Geol. 7: 273–284.

Cary, E. E., Allaway, W. H., and Olsen, O. E. 1977. Control of chromium concentrations in food plants. 2. Chemistry of chromium in soils and its availability to plants. J. Agric. Food Chem. 25: 305–309.

Centers for Disease Control. 1985. Preventing lead poisoning in young children. U.S. Department of Health and Human Services, Atlanta, GA.

Cheng, K. L. and Bray, R. H. 1953. Two specific methods of determining copper in soil and plant material. Anal. Chem. 25: 655–659.

Clayton, P. M. and Tiller, K. G. 1979. A chemical method for the determination of the heavy metal content of soils in environmental studies. Div. of Soils Tech. Paper No. 41. CSIRO, Australia.

Frank, R., Ishida, K., and Suda, P. 1976. Metals in agricultural soils of Ontario. Can. J. Soil Sci. 56: 181–196.

Garcia-Miragaya, J., Castro, S., and Paolini, J. 1981. Lead and zinc levels and chemical fractionation in roadside soils of Caracas, Venezuela. Water, Air, Soil Pollut. 15: 285–297.

Grove, J. H. and Ellis, B. G. 1980. Extractable chromium as related to soil pH and applied chromium. Soil Sci. Soc. Am. J. 44: 238–242.

Haq, A. U., Bates, T. E., and Soon, Y. K. 1980. Comparison of extractants for plant-available zinc, cadmium, nickel and copper in contaminated soils. Soil Sci. Soc. Am. J. 44: 772–777.

John, M. K. 1971. Lead contamination of some agricultural soils in western Canada. Environ. Sci. Technol. 5: 1199–1203.

Karamanos, R. E., Bettany, J. R., and Rennie, D. A. 1976. Extractability of added lead in soils using lead-210. Can. J. Soil Sci. 56: 37–42.

Lake, D. L., Kirk, R. W. W., and Lester, J. N. 1984. Fractionation, characterization, and speciation of heavy metals in sewage sludge and sludge-amended soils: a review. J. Environ. Qual. 13: 175–183.

Le Riche, H. H. and Weir, A. H. 1963. A method of studying trace elements in soil fractions. J. Soil Sci. 14: 225–235.

Lindsay, W. L. and Norvell, W. A. 1978. Development of a DTPA test for zinc, iron, manganese and copper. Soil Sci. Soc. Am. J. 42: 421–428.

Lisk, D. J. 1972. Trace metals in soils, plants, and animals. Adv. Agron. 24: 267–325.

Mattigod, S. V. and Page, A. L. 1983. Assessment of metal pollution in soils. Pages 355–394 in I. Thornton, Ed. Applied environmental geochemistry. Academic Press, London.

Merry, R. H. and Tiller, K. G. 1978. The contamination of pasture by a lead smelter in a semi-arid environment. Aust. J. Exp. Agric. Anim. Husb. 18: 89–96.

Merry, R. H., Tiller, K. G., and Alston, A. M. 1983. Accumulation of copper, lead and arsenic in some Australian orchard soils. Aust. J. Soil Res. 21: 549–561.

Mertz, W. 1982. Introduction. Pages 1–4 in S. Langard, Ed. Biological and environmental aspects of chromium. Elsevier Biomedical Press, Amsterdam.

Mitchell, R. L., Reith, J. W. S., and Johnston, I. M. 1957. Soil copper status and plant uptake. Pages 249–261 in T. Wallace, Ed. Plant analysis and fertilizer problems. IRHO, Paris, France.

Norvell, W. A. 1984. Comparison of chelating agents as extractants for metals in diverse soil materials. Soil Sci. Soc. Am. J. 48: 1285–1292.

Parson, M. L., Major, S., and Forster, A. R. 1983. Trace element determination by atomic spectroscopic methods — state of the art. Appl. Spectrosc. 37: 411–418.

Patterson, J. B. E. 1971. Metal toxicities arising from industry. Pages 193–207 in Trace elements in soils and crops. MAFF Technical Bull. No. 21. Her Majesty's Stationery Office, London.

Risser, J. A. and Baker, D. E. 1990. Testing soils for toxic metals. Pages 275–298 in R. L. Westerman, Ed. Soil testing and plant analysis, 3rd ed. Soil Science Society of America, Madison, WI.

Shewry, P. R. and Peterson, P. J. 1976. Distribution of chromium and nickel in plants and soil from serpentine and other sites. J. Ecol. 64: 195–212.

Soon, Y. K. and Bates, T. E. 1982. Chemical pools of cadmium, nickel and zinc in polluted soils and some preliminary indications of their availability to plants. J. Soil Sci. 33: 477–488.

Soon, Y. K. and Abboud, S. 1990. Trace elements in agricultural soils of northwestern Alberta. Can. J. Soil Sci. 70: 277–288.

Thompson, K. C. 1978. Shape of the atomic absorption calibration graphs for chromium using an air-acetylene flame. Analyst. 103: 1258–1262.

U.S. Environmental Protection Agency. 1986. Acid digestion of sediment, sludge and soils. In Test methods for evaluating solid wastes. EPA SW-846. U.S. Government Printing Office, Washington, D.C.

Viro, P. J. 1955a. Use of ethylenediaminetetraacetic acid in soil analysis. I. Experimental. Soil Sci. 79: 459–465.

Viro, P. J. 1955b. Use of ethylenediaminetetraacetic acid in soil analysis. II. Determination of soil fertility. Soil Sci. 80: 69–74.

Williams, C. and Thornton, I. 1973. The use of soil extractants to estimate plant-available molybdenum and selenium in potentially toxic soils. Plant Soil. 39: 149–159.

Chapter 14
Lime Requirement

T. Sen Tran

Ministere de l'Agriculture
Sainte-Foy, Quebec, Canada

W. van Lierop

14.1 INTRODUCTION

Soil lime requirement (LR) is defined as the amount of limestone ($CaCO_3$) required by a plow-layer (weight or volume) to raise its pH to a desired level. The target pH is usually selected to ensure optimum crop yield. LR are occasionally selected from tables relating soil pH and texture data. Although pH indicates the intensity of soil acidity, and should be used to determine whether or not a soil needs liming, texture is only a crude indicator of buffering capacity, and thus cannot give accurate estimates of the amount of acidity to be neutralized for achieving a selected pH. LR rates so chosen are inaccurate because soil acidity, and hence LR, are related mainly to organic matter and (when pH values are <5.5) exchangeable aluminum contents (Tran and van Lierop 1981a, Curtin et al. 1984). Many rapid single-buffer pH procedures have been suggested to assess the amount of acidity that has to be neutralized to achieve a desired pH (by corollary the LR) (Woodruff 1947, Shoemaker et al. 1961, Adams and Evans 1962, Mehich 1976). More laborious double-buffer procedures have also been proposed (Yuan 1974, McLean et al. 1978), and these may be more accurate for determining the LR of soils with low buffering capacities. Other approaches have also been suggested for determining whether or not soils need lime, for example the concentration of Al and/or Mn extracted with 0.01 or 0.02 M $CaCl_2$ (Hoyt and Nyborg 1971, Hoyt and Webber 1974, Hoyt and Nyborg 1987). Although more involved approaches can probably identify soils which respond to liming by increased yields more accurately than using pH alone, a LR test is needed, nonetheless, to determine the amount of limestone required to achieve a target pH. Required liming rates for achieving several practical pH values can be determined rapidly and accurately in routine laboratory testing by using a single-buffer procedure (Tran and van Lierop 1981b, 1982, Soon and Bates 1986).

14.2 SHOEMAKER, MCLEAN, AND PRATT (SMP)
SINGLE-BUFFER METHOD

The SMP single-buffer procedure is probably the most widely used routine LR test by North American laboratories. It is utilized in Quebec, Ontario, and British Columbia, and possibly to a lesser extent in other provinces (van Lierop and Tran 1983, Soon and Bates 1986, Webber et al. 1977). The SMP single-buffer procedure is particularly well adapted for determining the LR of acid soils needing more limestone than 2 tonnes ha^{-1}. The accuracy of single-buffer procedures relies on its calibration of decreasing soil-buffer pH values with increasing LR rates. Calibration accuracy for determining low LR values of acid mineral soils was improved for the SMP method by fitting curvilinear instead of linear equations to the relationships between soil-buffer pH and incubation LR values (Tran and van Lierop 1981b, 1982, Soon and Bates 1986). LR rates can be determined accurately by using scooped soil samples (van Lierop 1989). Soil-buffer pH can be determined on the same soil sample after having measured soil pH in either H$_2$O or 0.01 M CaCl$_2$ (McLean 1982).

14.2.1 MATERIALS AND REAGENTS

1   pH meter.

2   Disposable plastic beakers.

3   Automatic pipette.

4   Glass stirring rods.

5   Rotative mechanical shaker (120 oscillations/min).

6   SMP buffer solution.

   a. A quantity of 10 L of SMP buffer suffices for determining 500 soil samples. Dissolve the following chemicals in about 6 L of distilled water:
      - 18 g p-nitrophenol (NO$_2$C$_6$H$_4$OH, FW = 139.11)
      - 30 g potassium chromate (K$_2$CrO$_4$, FW = 194.2)
      - 531 g calcium chloride dihydrate (CaCl$_2$·2H$_2$O, FW = 147.02)

   b. Dissolve 20 g of calcium acetate [(CH$_3$COO)$_2$Ca·H$_2$O, FW = 176.19] in a separate flask containing about 1 L of distilled water.

   c. Add solution *b* to solution *a* and continue stirring for about 2 h.

   d. Add 100 mL of dilute triethanolamine (TEA) solution: TEA [N (CH$_2$OH)$_3$, FW = 149.19] is very viscous and difficult to pipette accurately. It is recommended that a dilute TEA solution be prepared by mixing 250 mL (or by weighing 250 mL × 1.122 g L^{-1} = 280.15 g) of TEA to 1 L with distilled water and mixing well.

   e. Dilute to 10 L with distilled water and continue stirring or shaking periodically for 6 to 8 h.

f. Adjust pH to 7.5 $\pm$ 0.02 by titrating with either 4 M NaOH or 4 M HCl as required.

g. Filter through fiberglass membrane if necessary.

h. Verify buffering capacity of prepared SMP buffer by titrating 20 mL from pH 7.5 to 5.5 with 0.1 M HCl; it should be 0.28 $\pm$ 0.005 cmol (+) HCl/ pH unit.

14.2.2 PROCEDURE

1   Measure 10 mL or weigh 10 g of air-dried soil ($<$2 mm) soil samples in appropriate beakers.

2   Add 10 mL of distilled water and stir with glass rod and repeat stirring periodically during next 30 min.

3   Measure the soil pH (H_2O) and rinse electrodes with a minimum of distilled water.

4   If the soil pH (H_2O) is smaller than the desired pH, add 20 mL of SMP buffer to the soil-water mixture (soil-water-buffer ratio is 1:1:2 by volume) and stir with glass rod.

5   Place soil-water-buffer samples on shaker for 15 min at about 150–200 cycles min^{-1}. Remove samples from shaker and let stand for 15 min.

6   Adjust the pH meter to read 7.5 with SMP buffer.

7   Stir sample thoroughly and read the soil-water-buffer to nearest 0.05 pH unit.

8   As the SMP buffer solution can affect the accuracy of the glass electrode after approximately 200 buffer-pH determinations. It is strongly recommended to regenerate the electrodes by appropriate procedure. The combined glass electrode can be regenerated by immersing it into a plastic beaker containing a solution of 10% ammonium hydrogen fluoride ($NH_4F \cdot HF$, FW = 54.04) for 1 min. After etching, dip electrode into 1:1 H_2O-HCl solution to remove silicate. Rinse the electrode thoroughly with distilled water and immerse in hot 3 M KCl solution (50°C) for 5 h. The electrolytes in the electrode (saturated KCl or calomel) must be replaced if necessary.

14.2.3 Calculations

The rates of limestone required to attain a soil pH (H_2O) of 5.5, 6.0, or 6.5 [pH 5.4 (H_2O) for organic soils] can be calculated from soil-buffer pH values (pH_B) using the following regression equations. The equations provide reliable recommendations when soil buffer pH values are situated between 6.9 and 4.9 for mineral soils. However, soil buffer pH values located between 5.9 and 3.9 should be used with organic soils. Avoid calculating a LR for a target pH that is lower than the soil pH. The rates are expressed in tonnes limestone ha^{-1} for a plow-layer of 20 cm depth (2 million L soil).

1   Quebec soil testing (van Lierop and Tran 1983).
       a. Mineral soils:

$$LR\ (5.5) = 210.4 - 61.29\ pH_B + 4.484\ pH_B^2\ R^2 = 0.894**$$
$$LR\ (6.0) = 179.8 - 49.22\ pH_B + 3.387\ pH_B^2\ R^2 = 0.936**$$
$$LR\ (6.5) = 107.2 - 22.27\ pH_B + 0.983\ pH_B^2\ R^2 = 0.903**$$

       b. Organic soils (van Lierop 1983):

$$LR\ (5.4) = 69.3 - 11.56\ pH_B;\ R^2 = 0.915**$$

2   Ontario recommendations are based on the following regression equations (Soon and Bates 1986).

$$LR\ (5.5) = 37.7 - 5.75\ pH_B\ R^2 = 0.765**$$
$$LR\ (6.0) = 255.4 - 73.15\ pH_B + 5.26\ pH_B^2;\ R^2 = 0.827**$$
$$LR\ (6.5) = 291.6 - 80.99\ pH_B + 5.64\ pH_B^2;\ R^2 = 0.901**$$
$$LR\ (7.0) = 334.5 - 90.79\ pH_B + 6.19\ pH_B^2$$

       LR was on lime with an agricultural index of 75.

14.2.4 Comments

For a LR that is greater than about 10 tonnes limestone ha^{-1}, it is recommended that the rate be divided into two or more applications to avoid local overliming. This is important as a liming recommendation assumes that the material is homogeneously incorporated into a 20-cm plow-layer, a precept that is difficult to achieve in practice. When surface applying liming material, without significant incorporation (without tillage), the rate should be reduced to about a third. Where some tillage is practiced, but not to the 20-cm depth, the liming rate should be reduced proportionately (van Lierop 1989). There usually is no advantage to using hydrated lime, as most of it will be converted to $CaCO_3$ (limestone) by combining with CO_2 in the surrounding atmosphere if not mixed thoroughly in the soil layer soon after being applied; however, if chosen, liming rates should be reduced to $\cong 75\%$ of limestone. Generally, its cost is significantly higher.

REFERENCES

Adams, F. and Evans, C. E. 1962. A rapid method for measuring lime requirement of red-yellow podzolic soils. Soil Sci. Soc. Am. Proc. 26: 355–357.

Curtin, D., Rostad, H. P. W., and Huang, P. M. 1984. Soil acidity in relation to soil properties and lime requirement. Can. J. Soil Sci. 64: 645–654.

Hoyt, P. B. and Nyborg, M. 1971. Toxic metals in soils. 1. Estimation of plant available aluminum. Soil Sci. Soc. Am. Proc. 35: 236–240.

Hoyt, P. B. and Nyborg, M. 1987. Field calibration of liming responses of four crops using pH, Al, and Mn. Plant Soil. 102: 21–25.

Hoyt, P. B. and Webber, M. D. 1974. Rapid measurement of plant-available aluminum and manganese in acid Canadian soils. Can. J. Soil Sci. 54: 54–61.

McLean, E. O. 1982. Soil pH and lime requirement. Pages 199–224 in A. L. Page, R. H. Miller, and D. R. Keeney, Eds. Methods of soil analysis. Part 2, Agronomy No. 9.

McLean, E. O., Eckert, D. J., Reddy, G. Y., and Trierweiler, J. F. 1978. An improved SMP soil lime requirement method incorporating double buffer and quick-test features. Soil Sci. Soc. Am. J. 42: 311–316.

Mehlich, A. 1976. New buffer-pH method for rapid estimation of exchangeable acidity and lime requirements of soils. Commun. Soil Sci. Plant Anal. 7: 637–652.

Shoemaker, H. E., McLean, E. O., and Pratt, P. F. 1961. Buffer methods for determining lime requirement of soils with appreciable amounts of extractable aluminum. Soil Sci. Soc. Am. Proc. 25: 274–277.

Soon, Y. K. and Bates, T. E. 1986. Determination of the lime requirement for acid soils in Ontario using the SMP buffer methods. Can. J. Soil Sci. 66: 373–376.

Tran, T. Sen and van Lierop, W. 1981a. Evaluation des méthodes de détermination du besoin en chaux en relation avec les propriétés physiques et chimiques des sols acides. Sci. Sol 3: 253–267.

Tran, T. Sen and van Lierop, W. 1981b. Evaluation and improvement of buffer-pH lime requirement methods. Soil Sci. 131: 178–188.

Tran, T. Sen and van Lierop, W. 1982. Lime requirement determination for attaining pH 5.5 and 6.0 of coarse-textured soils using buffer-pH methods. Soil Sci. Soc. Am. J. 46: 1008–1014.

van Lierop, W. 1983. Lime requirement determination of acid organic soils using buffer-pH methods. Can. J. Soil Sci. 63: 411–423.

van Lierop, W. 1989. Effect of assumptions on accuracy of analytical results and liming recommendations when testing a volume or weight of soil. Commun. Soil Sci. Plant Anal. 30: 121–137.

van Lierop, W. and Tran, T. Sen. 1983. Détermination du besoin en chaux des sols minéraux et organiques par la méthode tampon SMP. Science et Technique Agdex 534. Min Agric. Pêch. Ali. du Québec.

Webber, M. D., Hoyt, P. B., Nyborg, M., and Corneau Dianne. 1977. A comparison of lime requirement methods for acid Canadian soils. Can. J. Soil Sci. 57: 361–370.

Woodruff, C. M. 1947. Determination of exchangeable hydrogen and lime requirement of the soil by means of the glass electrode and a buffered solution. Soil Sci. Soc. Am. Proc. 12: 141–142.

Yuan, T. L. 1974. A double buffer method for the determination of lime requirement of acid soils. Soil Sci. Soc. Am. Proc. 38: 437–441.

Chapter 15
Chemical Characterization of Plant Tissue

John E. Richards

Agriculture Canada

Fredericton, New Brunswick, Canada

15.1 INTRODUCTION

Chemical analysis of plant tissue has been used for over a century to assess the ability of soil to supply nutrients to crop plants (Martin-Prével et al. 1984). Plant analysis provides a direct means of integrating some of the complicated soil-plant mechanisms which govern nutrient uptake from soils. The basic principles underlying the use of plant analysis in soil-plant investigations have been outlined in several reviews (Bates 1971, Munson and Nelson 1973, Bouma 1983, Jones 1985, Walworth and Sumner 1986, 1987).

Perhaps one of the largest potential errors in the use of plant analysis is one of interpretation. Traditionally, one of the major uses of plant analysis has been for diagnosing nutrient deficiencies in crops during the growing season. Samples are taken from a specific plant part at a predetermined stage of growth and nutrient concentrations in these samples are compared to those obtained from plants which were not grown under nutrient stress. According to several researchers (Melstead et al. 1969, Dow and Roberts 1982), there is a "critical" concentration for each plant nutrient; when concentrations of the nutrient exceed the "critical" concentration, no response to either increasing levels of fertilizer or soil-labile forms of the nutrient should be expected. However, nutrient concentrations change markedly during the growing season and vary among plant parts (Bates 1971, Davidescu and Davidescu 1972, Munson and Nelson 1973, Jones 1985, Walworth and Sumner 1986). Therefore, for the correct interpretation to be made, it is essential that samples be taken from a specific and consistent position on the plant at an established morphological stage of growth. Beaufils (1973) advanced the diagnosis and recommendation integrated system (DRIS) as one means of circumventing the problems associated with the interpretation of plant tissue chemical analysis. The principles underlying the DRIS system were reviewed elsewhere (Beaufils 1973, Sumner 1979, 1990, Walworth and Sumner 1986, 1987).

This chapter is concerned with quantitative chemical analysis of inorganic constituents of plant tissue, and, consequently, quick test methods of analysis were not covered. The reader is referred to the methods manuals of Chapman and Pratt (1961), Davidescu and Davidescu

Soil Sampling and Methods of Analysis, M. R. Carter, Ed.,

Canadian Society of Soil Science. © 1993 Lewis Publishers.

(1972), Greweling (1976), Gaines and Mitchell (1979), and of course to the Association of Official Analytical Chemists (AOAC) methods manual (Issac 1990). In addition, methods for analyzing plant sap are not covered; see Jones (1985) for a discussion of methods of analyzing plant sap.

15.2 SAMPLING AND PROCESSING

15.2.1 Sampling

As discussed in the preceding section, it is essential that sampling is done at the prescribed morphological stage of growth and from the correct plant part (Davidescu and Davidescu 1972, Jones 1985). Jones and Steyn (1973) suggest sampling procedures for a wide range of field and vegetable crops. The plant samples must be taken from areas of the field which are representative of the soil condition being examined and a sufficient number of plants should be sampled to characterize adequately the condition (Chapman and Pratt 1961, Jones and Steyn 1973). Actual procedures used to sample tissue vary according to the chemical analyses to be conducted; procedures are discussed by Jones et al. (1971) and by Davidescu and Davidescu (1972). A thorough discussion on some of the pitfalls surrounding sampling is given by Jones (1985).

After sampling and prior to drying, plant samples may be stored in either paper or plastic bags, although some workers have reported that paper bags may cause boron contamination (Greweling 1976). If samples are stored in polyethylene bags, they should be open to the atmosphere If the plant samples cannot be dried immediately, they should be stored in polyethylene bags at 4°C to prevent losses in fresh weight (Jones and Steyn 1973).

15.2.2 Washing Plant Leaves — Decontamination

There is generally a thin layer of soil on the leaves of plant tissue taken from the field which must be removed prior to analysis for Fe (Jones and Steyn 1973, Moraghan 1991), Cr and Co (Greweling 1976), and Al (Webber 1971). Washing of the leaves is also required when the sampled crop has been sprayed by a fungicide and micronutrient analysis is to be conducted. Care must be taken to minimize exposure of plant parts to the cleansing solution, since appreciable losses of water-soluble compounds (K, Cl, NO_3, Mn) may occur. Generally, losses of water-soluble compounds are greater from dead rather than from living tissue. Greweling (1976) speculates that under some conditions, plant tissues may extract micronutrients from cleansing solutions which are grossly contaminated with soil. Prior to washing leaves, the investigator should take some sample leaves and determine if the water-soluble compounds are being leached from the plant tissue.

15.2.2.1 PROCEDURE

1   Wash plant parts in distilled water containing 0.1–0.3% detergent (phosphate free if P is to be determined). Plant tissue should be placed in container with the wash solution and should be gently moved within the container.

2   Wash in 0.01 *M* sodium ethylene diamine tetra acetic (NaEDTA).

3 Rinse in distilled water.

4 Place in paper bags and dry immediately.

15.2.2.2 *Comments*

Care should be taken not to rub the plant tissue during any of the washing steps. Only a small amount of detergent need be added in step 1; its purpose is only to break the surface tension.

15.2.3 Drying, Grinding, and Storage

Traditionally, plant samples have been dried for 24 to 36 h in a convection oven set at 70°C. Care should be taken to insure that the plant tissue is not bunched together in isolated areas of the oven, as large losses of N may occur.

Plant tissue may also be dried in a microwave oven. This procedure is rapid and can dry individual samples to a brittle state in as little as 5 min (Stephenson and Gaines 1975).

After drying, the plant tissue must be milled. The major purpose of grinding is to eliminate sampling errors when the sample is weighed and the plant tissue should be ground fine enough to achieve this objective. Generally, plant tissue is ground to pass a 1-mm sieve. Thorough discussions on grinding of plant tissue and its effects on plant analysis are given by Jones and Steyn (1973) and Greweling (1976).

15.2.4 Pre-analysis

Prior to weighing out milled plant tissue for chemical analysis, the oven-dried plant samples should be placed in a convection oven set at 70°C overnight, and then allowed to cool to room temperature (Greweling 1976). Care should be taken to prevent the samples from gathering moisture from the atmosphere. The sample should be thoroughly mixed before taking a subsample for chemical analysis, since some separation of the milled plant parts may occur during storage (Greweling 1976).

15.3 SAMPLE DIGESTION-ASHING PROCEDURES

Two major forms of sample digestion are commonly employed in the analysis of plant tissue; dry and wet ashing. In dry ashing, oxidation of the plant sample occurs at high temperatures in a muffle furnace. In wet-ashing techniques, the sample is usually digested with some combination of HNO_3, H_2SO_4, H_2O_2, and $HClO_4$. Both methods have their advantages and disadvantages, although, in many cases, the choice of method is based on the availability of equipment. More complete discussions on ashing procedures are given by Greweling (1976), Hoenig and DeBorger (1983), Watson (1984), and by Jones (1985). In addition to an excellent discussion, Greweling (1976) also outlines various ashing methods in detail.

15.3.1 Dry Ashing (Greweling 1976)

Dry ashing is, in many laboratories, the preferred method of ashing plant samples for P, K, Ca, Mg, and trace element analysis (especially B). It is relatively simple, requires very little operator attention, and avoids the use of boiling acids. Isaac and Johnson (1975) found that dry ashing in porcelain crucibles and wet ashing extracted similar quantities of Ca, Cu, Fe, Mn, Mg, K, and Zn. Complete recovery of some trace elements, however, is not always attained in dry ashing (Greweling 1976). Incomplete recovery can be due to: volatilization of elements such as Se, S, and the halogens; retention of the element on the wall of the crucible (Cu can be retained on the walls of Si crucibles); and formation of compounds which are not completely soluble in the acid used for extraction (Greweling 1976, Hoenig and DeBorger 1983, Watson 1984). The last two mechanisms usually can be overcome by using platinum crucibles, treatment of the ash with hydrofluoric acid (Greweling 1976), and carefully controlling the ashing temperature (Hoenig and DeBorger 1983). Dry ashing with hydrofluoric acid to remove silica is covered in detail by Greweling (1976).

Based on the large amount of contradictory literature regarding the success of dry ashing for organic matter destruction for the quantitative determination of some elements, Jones (1985) argues that it should be used only with the greatest of caution. According to Jones (1985), the best dry-ashing technique available is the one described by Munter and Grande (1981).

15.3.1.1 MATERIALS AND REAGENTS

1   Muffle furnace — capable of achieving 550°C ± 15°C.

2   High form fused silica or porcelain 30-mL crucibles.

3   Electric hot plate.

4   20-mL volumetric flasks.

5   Hydrochloric acid (HCl) — concentrated, reagent grade.

6   Nitric acid (HNO_3) — 5 *M*, reagent grade (dilute 320 mL concentrated HNO_3 to 1 L).

7   Nitric acid (HNO_3) — 2 *M*, reagent grade (dilute 128 mL concentrated HNO_3 to 1 L).

15.3.1.2 PROCEDURE

1   Weigh a 1-g sample of ground plant tissue into high form crucible.

2   Slowly raise the temperature of the muffle furnace to 550°C and ash at this temperature for a minimum of 4 h or until virtually no black ash remains.

3   Remove from oven.

4   If considerable carbon remains, remove by adding 2 mL of 5 *M* HNO_3 to the cool ash, evaporate the sample to dryness on a hot plate set at 100–120°C,

place in a *cool* (<200°C) muffle furnace, heat to 500°C for 1 h, and remove from oven.

5 Moisten the ash with a few drops of distilled water, add 2 mL concentrated HCl, evaporate to dryness on a hot plate, and bake for an additional 1 h.

6 Remove from hot plate and add 2.5 mL of 2 M HNO_3, ensuring salts are dissolved (if needed stir with a plastic policeman).

7 Transfer to appropriate volumetric flask (generally 20 mL), bring up to volume with distilled water, and allow sufficient time for the silica to settle out before removing aliquot for analysis.

15.3.1.3 COMMENTS

1 For most purposes, porcelain crucibles are satisfactory. If the sample is being analyzed for aluminum, however, fused silica crucibles should be used. If trace elements are to be determined, it may be necessary to make the final volume of the solution (step 7) up to 10 mL, or the amount of plant tissue used in step 1 can be increased up to 2 g. In either case, subsequent dilutions may be required. If more plant tissue is ashed, more time will be required for the digestion. If boron is to be determined, borosilicate-free glassware and laboratory ware must be used.

2 Note that in step 4 considerable care should be exercised when placing the sample containing HNO_3 in the muffle furnace, since HNO_3 can react violently with carbon. Greweling (1976) states that in most circumstances, complete destruction of carbon is unnecessary and that step 4 is optional.

3 Step 5 is intended to dehydrate the silica and prevent it from dissolving in the final acid treatment; this effectively reduces the interferences of silica in some determinations. Step 5 also solubilizes metal oxides (such as of aluminum) and, according to Greweling (1976), hydrolyzes pyrophosphates into orthophosphates. Note that for Ca, Cu, Fe, Mn, Mg, K, and Zn that step 5 is not recommended by AOAC (Issac 1990). If so desired, 6 M HNO_3 can be used to dissolve the ash in step 6 (Issac 1990).

4 Issac and Jones (1972) reported that the concentrations of Al, B, Cu, Fe, K, and Mn extracted from five different plant tissues were markedly affected by the ashing temperature. They found that ashing at 500°C for 4 or 15 h gave similar concentrations to those obtained by wet ashing.

15.3.2 Dry Ashing with Magnesium Nitrate (Greweling 1976, Issac 1990)

Sulfur volatilizes at the temperatures attained in a muffle furnace if no precautions are taken. Losses are eliminated when a basic compound, such as $Mg(NO_3)_2$, is intimately mixed with the plant sample (Greweling 1976).

15.3.2.1 MATERIALS AND REAGENTS

1   Muffle furnace — capable of achieving 550°C ± 15°C.

2   High form porcelain 50-mL crucibles.

3   Electric hot plate.

4   50-mL Volumetric flask; 1-L volumetric flask; 1-L beaker.

5   $Mg(NO_3)_2$ solution — place 955 g of $Mg(NO_3)_2$ into a large beaker and add 350 mL distilled water and 1 g of MgO. Boil solution and filter through a fast filter paper. Dilute to 1 L.

6   Hydrochloric acid (HCl) — concentrated, reagent grade.

1   Weigh a 1-g sample into a large porcelain crucible and add 7.5 mL $Mg(NO_3)_2$ solution, ensuring that all plant material is in intimate contact with the $Mg(NO_3)_2$ solution.

2   Heat on hot plate (180°C) so that solution slowly evaporates.

3   Once evaporation is complete, transfer crucible, while still hot, to cool furnace.

4   Gradually raise temperature to 500°C and ash until no carbon remains. If carbon is present, break up sample and return to oven.

5   Remove from oven when digestion is complete and let cool.

6   Carefully add 10 mL concentrated HCl, evaporate to dryness on a hot plate set at 100–120°C, and bake for 1 h to dehydrate silica.

7   Add several mL water to moisten ash, then add 1 mL concentrated HCl and warm on hot plate to dissolve salts.

8   Transfer to 50-mL volumetric flask using warm distilled water, make up to volume, and allow silica to settle out before taking aliquots for analysis. If the solution is cloudy, add HCl dropwise.

15.3.2.3 *Comments*

It is imperative that the $MgNO_3$ solution thoroughly wet the plant sample and that the temperature in the muffle furnace be increased slowly. Silica dehydration (step 6) is difficult due to a $MgCl_2$ scum (Greweling 1976) which forms on the surface of the liquid; however, excessive heat may result in expulsion of crystals, so careful attention must be paid to heating the sample in step 6. More complete comments are found in Greweling (1976).

15.3.3 Wet Digestion: Sulfuric Acid-Hydrogen Peroxide

Wet-digestion techniques are useful for the digestion of plant tissue prior to a wide variety of chemical analyses (N, P, K, Ca, Mg, S, Zn, Mn, Fe, Cu, Al, Se, and others). Wet digestions can require more operator attention than dry ashing, use more acids, and can require specialized fume hoods. However, interferences due to silica are minimized and there is no potential retention of trace elements by silica as in dry ashing.

The sulfuric acid-hydrogen peroxide method presented here was originally used for nitrogen determinations (Miller and Miller 1948). Thomas et al. (1967) showed, however, that this digestion could also be used for determination of phosphorus and potassium. The digestion is also suitable for magnesium analysis. The digest is not suitable for determination of Ca, as $CaSO_4$ may form (Greweling 1976). In this method, the plant tissue is charred with H_2SO_4 and oxidation of plant tissue is accomplished by H_2O_2 additions.

15.3.3.1 MATERIALS AND REAGENTS

1   40-position aluminum block digestor.

2   50-mL Taylor digestion tubes.

3   Sulfuric acid (H_2SO_4) — concentrated, reagent grade.

4   Hydrogen peroxide (H_2O_2) — 30%, reagent grade.

15.3.3.2 PROCEDURE

1   Weigh 0.25 g plant tissue and place in digestion tube.

2   Add 5 mL of H_2SO_4, ensuring that the acid washes any plant tissue on the side of the digestion tube down to the bottom.

3   Place in digestion block and digest at 330°C for 2 h.

4   Remove from digestion block, cool for 15 min, carefully add 4–5 drops of H_2O_2, and place tubes in digestor for a further 5 min.

5   Repeat step 4 until solutions are clear.

6   Return tubes to digestor for a further 1 h of digestion.

7   Remove tubes and upon cooling add sufficient distilled water to bring the solution up to 50 mL.

8   Allow solution to stand overnight so the silica settles out, then decant so-
     lution into storage tubes.

15.3.3.3 COMMENTS

1   The oxidation of the organic matter with H_2O_2 (step 4) is a critical part of
     the procedure. Under no circumstances should there be any splattering
     when H_2O_2 is added; the H_2O_2 should be added slowly, especially on the
     first addition, and the tubes must have cooled. In the later additions, no
     more than 5 drops of H_2O_2 should be added at one time because a large
     excess of H_2O_2 in the absence of organic matter will result in oxidation of
     some of the ammonia (Linder and Harley 1941).

2   If 75-mL digestion tubes and the phosphorus method (15.4.2) are used, the
     plant tissue may be increased to 0.30 g and the H_2SO_4 to 6 mL; this permits
     a simple dilution to be made prior to P analysis. In some cases, subsequent
     dilutions may be avoided by varying the quantity of plant tissue and acid to
     be digested. For example, in my laboratory, 0.15 g plant tissue is digested
     in 5 mL H_2SO_4 which is made up to a final volume of 75 mL (500:1 dilution).
     This permits the use of an autoanalyzer to measure simultaneously plant N
     (DeMerchant 1988) and P (DeMerchant 1989) with no final dilution required.

3   Nelson and Sommers (1980) provide a comparison between the Dumas and
     Kjeldahl procedures and the effects of varying sample to acid ratios and
     reaction vessel on the reliability of the Kjeldahl procedure.

15.3.4 Wet Digestion: Nitric Acid-Perchloric (Gaines and Mitchell 1979)

This procedure is suitable for digestion of a wide range of biological materials for analysis
of P, K, Ca, Mg, S, Al, Se, and the trace elements (with the exception of B). Some of the
reported drawbacks of dry ashing are avoided, due principally to the fact that there is no
retention of trace elements by silica.

15.3.4.1 MATERIALS AND REAGENTS

1   40-position aluminum block digestor.

2   50-mL Taylor digestion tubes.

3   Small Pyrex® funnels.

4   Perchloric acid fume hood.

5   Nitric acid (HNO_3) — reagent grade, concentrated.

6   Perchloric acid — reagent grade, concentrated.

15.3.4.2 PROCEDURE

1   Acid mixture — dilute two parts HNO_3 with one part $HClO_4$.

2   Weigh 0.5 g into 50-mL Taylor digestion tube.

3   Add 5 mL of the acid mixture.

4   Place small Pyrex® funnel into tube and put tube into block digestor.

5   Heat sample at 60°C for 15 min or until reaction is complete.

6   Increase heat to 120°C and digest for 75 min or until sample clears.

7   Remove from block digestor when clear, cool, and add sufficient distilled water to bring solution up to 50 mL.

15.3.4.3 COMMENTS

1   This digestion procedure shouldnot be used for plant tissue high in organic material or oils (Greweling 1976, Gaines and Mitchell 1979), since a vigorous reaction between perchloric acid and the oils may occur. Care must be taken in the use of perchloric acid, since a potentially explosive mixture can occur when it comes in contact with a carbon source.

2   Perchloric acid should only be used in fume hoods which are clean, free of organic materials, and can be washed down to remove perchloric acid fumes after use (Greweling 1976).

3   Zasoski and Burau (1977) describe a similar digestion for multielement analyses.

4   Blanchar et al. (1965) describe an alternate digestion in which the nitric acid is added to the plant material alone and the mixture is allowed to stand overnight prior to heating and addition of $HClO_4$. If this procedure is used, the analyst must ensure that the plant tissue receives an adequate nitric acid pretreatment prior to addition of perchloric acid.

5   Greweling (1976) describes a variation of this technique, in which sulfuric acid is added to the nitric acid to raise the boiling point of the material. This addition is primarily for increased safety, since nitric acid has a very low boiling temperature and it is quite easy to boil the nitric acid off and have foaming of the material. Considerable care must be exercised to prevent foaming of the sample during the nitric acid pretreatment.

15.3.5 Wet Digestion: Nitric Acid by Microwave (Rechcigl and Payne 1990)

This is a relatively new technique (Abu-Samra et al. 1975) and has proven useful for digestion of a wide range of biological material (White and Douthit 1985, Kingston and Jassie 1986, 1988, Kalra et al. 1989, Rechcigl and Payne 1990). Use of sealed-chamber digestion vessels in which pressure increases occur during heating have resulted in short digestion times and reduced reagent use (Kingston and Jassie 1986). Rechcigl and Payne found that, for five forage species, microwave digestion was as effective as digestion by H_2O_2-H_2SO_4 or dry ashing for the quantitative analysis of a wide range of elements. The methods and equipment employed by Rechcigl and Payne (1990) are probably the most common in use at the time of writing and will not be outlined here. There are, however, some disadvantages to the technique. The Teflon® PFA digestion vessels are expensive, as are the commercially available microwave ovens which have been modified to vent acid fumes. In addition, only a limited number of samples (usually 12) can be digested at one time, which can negate any increase in productivity due to reduced digestion times.

15.4 ANALYTICAL METHODS

Although not stated explicitly in any method, it is assumed that the analyst will use appropriate standard plant samples. Use of standard plant samples, such as those of the National Bureau of Standards (Standard Reference Materials, National Institute of Standards and Technology, Gaithersburg, MD) can help assure that the results of the analyses of scientific material are reasonable and valid. The appropriate use of reference materials is outlined in Chapter 26.

Recently, chemical analysis of a number of elements has become possible by inductively coupled plasma emission spectroscopy (ICP) (Ward et al. 1980, Blakemore and Billedeau 1981, White and Douthit 1985, Isaac and Johnson 1985) and the operating characteristics have been well defined for plant tissue. With the notable exception of N, ICP provides rapid, simultaneous analysis of most elements of interest in soil-plant investigations. Samples can be digested by the dry-ash technique, by nitric acid-perchloric acid digestion, or by other wet-ashing techniques. Since most research laboratories do not have ICP capability, more traditional methods are described below.

15.4.1 Nitrogen — by Autoanalyzer (Thomas et al. 1967)

This automated method is based on the Bertholet reaction in which a blue-colored compound is formed when ammonium is added to sodium phenoxide followed by sodium hypochlorite (Tel 1989b). The manifold described below is for a Technicon® AutoAnalyzer® II and is intended for the simultaneous determination of N and P in extract of plant tissue (Thomas et al. 1967). The manifold has been modified for the newer generation of autoanalyzers by DeMerchant (1988) and Tel (1989b).

15.4.1.1 MATERIALS AND REAGENTS

1   Digest in Section 15.3.3.

2   AutoAnalyzer® II and manifold shown in Figure 15.1.

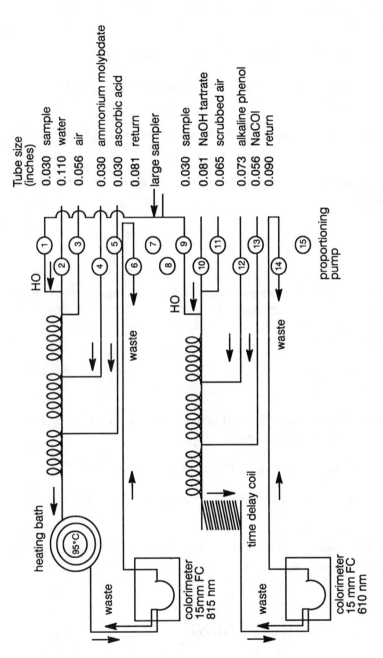

FIGURE 15.1. The autoanalyzer manifold used for simultaneous determination of N and P in extracts of plant tissue. (After Thomas, R. L. et al., Agron. J., 59, 240, 1967.)

3   Liquid phenol — reagent grade.

4   Sodium hydroxide (NaOH) — reagent grade.

5   Sodium hypochlorite (NaOCl) — commercial bleach with a minimum of 5% NaOCl may be used.

6   Sodium potassium tartrate ($NaKC_4H_4O_6\cdot4H_2O$) — reagent grade.

7   5 M NaOH — slowly add 200 g NaOH to 700 mL distilled water; when NaOH has dissolved, bring final volume up to 1 L.

8   1 M NaOH — add 40 g NaOH to 700 mL distilled water and bring final volume up to 1 L.

9   0.36 M H_2SO_4 — add 20 mL concentrated H_2SO_4 to 900 mL distilled water; bring final volume up to 1 L.

15.4.1.2 PROCEDURE

1   Sodium phenate solution — in a fume hood add 138 mL liquid phenol into ice-cold 5 M NaOH with slow stirring. Bring volume up to 1 L with cold 5 M NaOH. Store in dark bottles in a refrigerator. Warm solution to room temperature prior to using.

2   N-diluent — dissolve 50 g sodium potassium tartrate in 1 M NaOH; bring final volume up to 1 L.

3   Dilute digest 15.3.3 by 5:1 with distilled water. This can be done by hand or with a dilution loop on the autoanalyzer.

4   Make up standards ranging from 0 to 100 mg N L^{-1} in 0.36 M H_2SO_4.

5   Set up autoanalyzer as shown in Figure 15.1. Calculation of N and P in plant tissue:

$$\text{N, P (\%) in plant tissue} = \text{autoanalyzer reading (mg } L^{-1}) \times \text{dilution} \qquad (15.1)$$
$$\text{Where dilution} = \text{original dilution} \times \text{final dilution}$$
$$= 200 \times 5 = 1000$$

15.4.1.3 COMMENTS

1   Liquid phenol can be replaced with sodium salicylate ($Na_2C_7H_5O_3$) in equimolar amounts. If ammonium concentrations are low, a small amount of sodium nitroprusside ($Na_2Fe(CN)_5NO\cdot2H_2O$) may be added to the sodium salicylate solution (0.009 g per 175 g of sodium salicylate).

2 Some plant NO_3 will be reduced to ammonia during digestion; the quantity being dependent upon the amount of sugars present in the plant tissue. In mature field-grown plants, NO_3 is generally low. However, if plants are taken at an early stage of growth, or have received high rates of applied N, there may be substantial quantities of NO_3 present. In our laboratory, only 60% of the NO_3 present in young corn or added to young corn tissue was reduced to ammonia during the digestion. It should be noted that the standard Kjeldahl technique also suffers from the same problem (Preez and Bate 1989). If NO_3 is present in substantial quantities, the digestion must be modified. Nitrate can be reduced to NH_4 prior to digestion by the addition of salicylic acid (Novozamsky et al. 1983). A H_2O_2 pretreatment has been reported to eliminate NO_3 reduction during digestion, permitting quantitative analysis of NH_4 and NO_3 (Bowman et al. 1988). Tel (1992) has described a modification to the digestion, in which the plant tissue is pretreated with salicylic acid.

3 Other methods of determining ammonium in plant tissue extracts have been given by Nelson and Sommers (1973) (modification includes NO_3 reduction to NH_4), by Isaac and Johnson (1976), and by Gaines and Mitchell (1979). O'Neill and Webb (1970) described an autoanalyzer manifold for the simultaneous determination of N, P, and K. Ammonium in plant tissue extracts can, of course, be determined by the steam distillation technique described by Bremner (1965).

15.4.2 Phosphorus — by Autoanalyzer (Thomas et al. 1967)

The automated method outlined here is a companion to the method given for nitrogen. The manifold is shown in Figure 15.1. In this procedure, plant phosphorus is converted to orthophosphates during digestion with H_2SO_4-H_2O_2. The determination of orthophosphates is based on a colorimetric method where a blue color is formed by the reaction of orthophosphate and molybdate ions followed by reduction with ascorbic acid. This reaction is sensitive to minute fluctuations in pH.

15.4.2.1 MATERIALS AND REAGENTS

1 Digest in Section 15.3.3.

2 AutoAnalyzer® II and manifold shown in Figure 15.1.

3 Ammonium molybdate [$(NH_4)_6Mo_7O_{24}\cdot4H_2O$].

4 Ascorbic acid.

5 Sulfuric acid (H_2SO_4) — concentrated, reagent grade.

6 Levor V.

7 0.36 M H_2SO_4 — add 20 mL concentrated H_2SO_4 to 900 mL distilled water; bring final volume up to 1 L.

15.4.2.2 PROCEDURE

1 Ammonium molybdate solution. Dissolve 10 g ammonium molybdate in 200 mL H_2O previously heated to 60°C and cool. Add 225 mL concentrated H_2SO_4 to 400 mL distilled water and cool. Slowly add the molybdate solution to the H_2SO_4 solution with continuous stirring. Make final volume up to 2 L with distilled water.

2 Ascorbic acid solution. Dissolve 0.88 g ascorbic acid in 75 mL distilled water, bring volume up to 100 mL. Note that this solution must be prepared daily.

3 P-diluent. Add 1 mL Levor V to 1 L distilled water.

4 Dilute digest 15.3.3 by 5:1 with distilled water. This can be done by hand or with a dilution loop on the autoanalyzer.

5 Make up standards ranging from 0 to 10 mg P L^{-1} in 0.36 M H_2SO_4.

6 Set up autoanalyzer as shown in Figure 15.1.

7 Calculation of P in plant tissue — see step 5 in Section 15.4.1.

15.4.2.3 COMMENTS

1 Since this reaction is very sensitive to pH, some adjustments may be required to the ammonium molybdate solution if a dilution loop is added to the manifold. Phosphorus can also be determined on extracts 15.3.1 and 15.3.4, but some adjustments to the ammonium molybdate solution will be required.

2 Stannous chloride can be substituted for ascorbic acid with the result that the method is less sensitive to pH fluctuations. DeMerchant (1989) developed a manifold for the newer generation autoanalyzers (TRAACS 800) employing stannous chloride. Phosphorus can also be determined by the manual method of Murphy and Riley (1962).

15.4.3 Potassium, Calcium, and Magnesium — by Atomic Absorption

Potassium and magnesium can be determined on extracts 15.3.1, 15.3.3, and 15.3.4. Calcium can be determined on extracts 15.3.1 and 15.3.4. Under certain circumstances, Ca may be difficult to determine on extract 15.3.3 due to the possible formation of $CaSO_4$. Dilutions are made with a 2000-mg La L^{-1} solution, eliminating possible interferences.

15.4.3.1 MATERIALS AND REAGENTS

1   Digest 15.3.1.

2   Lanthanum nitrate [La(NO$_3$)$_3$·H$_2$O].

3   Atomic absorption spectrophotometer (AAS).

15.4.3.2 PROCEDURE

1   La-diluent: add 6.235 g La(NO$_3$)$_3$·H$_2$O in 900 mL distilled water, make up to 1 L.

2   Dilute digest 15.3.1 1:100 with the La-diluent.

3   Prepare standards either singly or in one solution: 0–3 mg K L^{-1}; 0–3 mg Ca L^{-1}; 0–0.3 mg Mg L^{-1} in a 2000-mg La L^{-1} solution.

4   Read with AAS using an acetylene-air flame and conditions given for the particular element in the operator's manual.

5   Calculation.

$$\text{K, Mg, Ca (\%) in plant tissue} = \text{AAS reading (mg L}^{-1}) \times \text{dilution} \qquad (15.2)$$
$$\text{Where dilution} = \text{original dilution} \times \text{final dilution}$$
$$= 20 \times 100 = 2000$$

15.4.3.3 COMMENTS

1   Lanthanum chloride or lanthanum oxide may be used instead of lanthanum nitrate. Although some workers do not use a suppressant, its use is essential for analytical analyses of plant tissue. Smith et al. (1983) outline some considerations for the use of interference suppressants.

2   If extract 15.3.4 (nitric-perchloric acid digest) is used, dilute it 1:20 with a 2100-mg La L^{-1} solution.

3   If extract 15.3.3 (sulfuric acid-hydrogen peroxide) is used, dilute it 1:10 with a 2200-mg La L^{-1} solution.

4   If possible, K should be read by flame emission, as results with AAS can be variable (Jones 1985).

5   Issac (1980) provides a good background regarding the use of AAS for analysis of plant tissue.

15.4.4 Sulfur — Manual Turbidimetric (Gaines and Mitchell 1979)

Plant sulfur is oxidized to sulfate during the digestion procedure. Upon introduction of $BaCl_2$, a $BaSO_4$ precipitate is formed which is determined turbidimetrically.

15.4.4.1 MATERIALS AND REAGENTS

1   Digest in Section 15.3.2.

2   Barium chloride crystals ($BaCl_2 \cdot H_2O$) — reagent grade, 60 to 80 mesh.

3   Sodium chloride (NaCl) — reagent grade.

4   Ethyl or isopropyl alcohol, reagent grade.

5   Hydrochloric acid — concentrated, reagent grade.

6   Glycerol — reagent grade.

7   Potassium sulfate crystals (K_2SO_4) — reagent grade.

8   Magnetic stirrer.

9   Colorimeter.

15.4.4.2 PROCEDURE

1   Prepare Conditioning Reagent: into 300 mL distilled water, dissolve 75 g NaCl, then add 100 mL of ethyl or isopropyl alcohol, 30 mL HCl (concentrated), and mix well. Add 50 mL glycerol, mix well, and store.

2   Prepare Standard Sulfate Solution: this should contain 250 mg S L^{-1}. Dissolve 1.3585 g potassium sulfate into distilled water and bring up to 1 L.

3   Prepare Calibration Curve: pipette aliquots of the Standard Sulfate Solution into 100-mL beakers; the aliquots should contain 100 to 1000 µg S; make up to a volume of 50 mL. Conduct steps 5 through 8.

4   Into a 100-mL beaker, pipette sufficient sample to give 100 to 1000 µg S; generally, 10-mL aliquots are sufficient. Add sufficient distilled water for a final volume of 50 mL.

5   Add 2.5 mL of the Conditioning Reagent to each beaker and mix.

6   Using a magnetic stirrer, commence mixing and add 1 g $BaCl_2 \cdot 2H_2O$. If possible, use a small spoon to measure the $BaCl_2 \cdot 2H_2O$, since it is essential that the same amount be added each time.

7 Mix for exactly 60 s, remove beaker, and let stand for exactly 4 min prior to reading.

8 Measure transmittancy vs. H_2O at 420 nm.

9 Calculations: using semilogarithmic paper or a computer, determine the standard curve of S (μg S mL^{-1}) vs. percent transmittance, and estimate the solution concentrations of S (μg S mL^{-1}).

$$\text{Sulfur in plant tissue (\%)} = \mu\text{g S per 10-mL aliquot} \times 0.0005 \qquad (15.3)$$

15.4.4.3 COMMENTS

1 If there is visible color in step 5 prior to addition of $BaCl_2 \cdot 2H_2O$, measure the "apparent" sulfate, then add $BaCl_2 \cdot 2H_2O$ and remeasure sulfur concentrations. Subtract the "apparent" concentration from the second concentration for the actual concentration of S.

2 It is essential that conditions, such as stirring speed, shape of the container, source of $BaCl_2$, and stirring and resting times (step 7), be kept as constant as possible. The size of the barium chloride crystal is critical and must be constant within 10 mesh (Greweling 1976). Sensitivity is increased by using small crystals and is decreased at the longer wavelengths; measurement at 720 nm will decrease sensitivity by a factor of 2 to 3.

3 An alternative digestion using a HNO_3-$HClO_4$ digestion is given by Blanchar et al. (1965). Other methods for determining S were described in detail by Beaton et al. (1968) and by Blanchar (1986). The BaCl turbidimetric method was automated for AutoAnalyzer® II (D. A. Tel, Department of Land Resource Science, University of Guelph, Guelph, ON).

15.4.5 Zn, Mn, Cu, and Fe — by Atomic Absorption

The dry-ash extract (15.3.1) can be used for determination of Zn, Mn, Cu, and Fe by AAS.

15.4.5.1 MATERIALS AND REAGENTS

1 Digest 15.3.1.

2 AAS.

15.4.5.2 PROCECURE

1   Preparation of a Cu-Mn-Zn multistandard. The following ranges are recommended for standards: 0–4.0 mg Cu L^{-1}; 0–20 mg Mn L^{-1}; 0–20 mg Zn L^{-1}. A stock solution of Cu-Mn-Zn containing 20 mg Cu L^{-1} and 100 mg of Mn and Zn L^{-1} may be prepared. The multistandard stock solution should possess and be diluted with exactly the same matrix as the digest 15.3.1 (0.25 M HNO_3 here).

2   Preparation of Fe standard. The following range is recommended for Fe: 0–8 mg Fe L^{-1}.

3   Measure the concentration of each element with AAS using an acetylene-air flame and conditions given for the particular element in the operator's manual.

4   Calculation.

$$\text{Zn, Cu, Mn, or Fe in plant tissue} = \text{AAS reading (mg } L^{-1}) \times \text{dilution} \quad (15.4)$$
$$\text{Where dilution} = \frac{\text{final volume (mL)}}{\text{mass plant tissue (g)}} = \frac{20}{1.0} = 20$$

15.4.5.3 COMMENTS

1   No dilution of dry ash is usually required; if the solutions must be diluted use 0.25 M HNO_3. If concentrations are too low to read accurately, make final volume of extract 15.3.1 up to 10 mL instead of 20 mL.

2   Contamination is a serious problem in the analysis of trace elements, and, consequently, all glassware and laboratory ware must be cleaned in 1 + 1 HCl, followed by a distilled water rinse. Note that it is not uncommon for distilled water to contain large concentrations of Cu, and therefore a reagent blank must be carried through the procedure. Silica may interfere with Mn determinations; this should not be a problem if the silica has been properly dehydrated.

15.4.6 Al — by Atomic Absorption

Aluminum in plant tissue can be successfully determined by AAS. Lanthanum is required to overcome interferences. The dry-ashing procedure outlined in 15.3.1 must be modified so that there is a final concentration of 2000 mg La L^{-1}.

15.4.6.1 MATERIALS AND REAGENTS

1 Lanthanum nitrate [La(NO$_3$)$_3$·H$_2$O].

2 AAS.

15.4.6.2 PROCEDURE

1 La-diluent: add 8.3133 g La(NO$_3$)$_3$·H$_2$O in 900 mL distilled water and make up to 1 L.

2 Modification to digest 15.3.1: in step 7; bring solution up to 10 mL by adding sufficient La-diluent. The final solution should have a concentration of 2000 mg La L^{-1}.

3 Prepare standards: 0–20 mg Al L^{-1} in 0.5 M HNO$_3$.

4 Read with AAS using an acetylene-air flame and conditions given for the particular element in the operator's manual.

5 Calculation.

$$\text{Al in plant tissue (mg Al L}^{-1}) = \text{AAS reading (mg L}^{-1}) \times \text{dilution} \qquad (15.5)$$
$$\text{Where dilution} = \frac{\text{final volume (mL)}}{\text{mass plant tissue (g)}} = \frac{10}{1.0} = 10$$

15.4.6.3 COMMENTS

1 Determination of Al by AAS is possible, but is difficult. Si and Fe can interfere with Al when determined with an air-acetylene flame; consequently, a nitrous oxide-acetylene flame and suppressants should be used. Webber (1974) used 2000 mg L^{-1} La to suppress enhancements from Na, K, Sr, and Fe and was able to determine Al in solution at 1 mg L^{-1} or less by AAS, although the minimum concentration most workers appear to achieve consistently is the range of 2 to 4 mg L^{-1}. Use of electrothermal excitation (graphite furnace) can greatly increase the sensitivity of analysis, but with a concomitant increase in the difficulty of analysis.

2 The temperature of dry ashing may be critical for Al determinations. Isaac and Jones (1972) found that Al concentrations of five plant species generally increased with increasing temperature of dry ashing. They speculated that this may be due to contamination from the muffle furnace itself. Webber (1971) found, however, that Al concentrations increased with increasing temperatures in plant tissue grown on soil, but not in plant tissue grown in hydroponic solution, suggesting that contamination was due to soil particles.

He found that washing the plant tissue virtually eliminated differences in Al concentrations at 500 and 1000°C.

3   Aluminum in plant tissue is also frequently determined by the aluminon method. This method is outlined by Jayman and Sivasubramaniam (1974) and is discussed in detail by Barnhisel and Bertsch (1982). This method is sensitive and does not require expensive instrumentation.

15.4.7 Boron — by Autoanalyzer (Tel, 1989a)

This method was developed by D. A. Tel (Department of Land Resource Science, University of Guelph, Guelph, ON, N1G 2W1; 1989a). Boron is determined colorimetrically after reaction with azomethine-H at a pH of 4.9.

15.4.7.1 MATERIALS AND REAGENTS

1   Azomethine-H.

2   L-Ascorbic acid.

3   Ammonium acetate.

4   Glacial acetic acid.

5   Disodium ethylene diamine tetra acetate (EDTA).

6   Brij-35.

7   Sodium hydroxide.

8   Boric acid.

9   HCl — 0.4 M HCl — (wash solution): dilute 33.3 mL concentrated HCl to 1 L with distilled water.

10   Magnetic stirrer.

11   AutoAnalyzer® II and manifold shown in Figure 15.2.

15.4.7.2 PROCEDURE

1   Buffer reagent. Dissolve 250 g ammonium acetate in 500 mL distilled water. With constant stirring, adjust pH to 5.8 by slowly adding approximately 100 mL glacial acetic acid. Add 0.5 mL Brij-35.

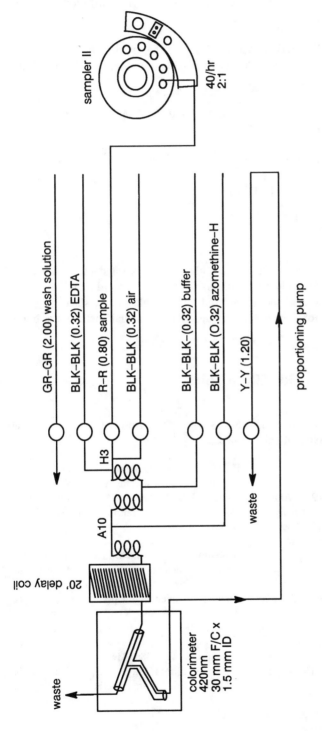

FIGURE 15.2. Manifold for boron in plant tissue. (After Tel, D. A., unpublished method, 1989a.)

2　Azomethine-H reagent. Dissolve 0.5 g azomethine-H and 1.0 g L-ascorbic acid in 10 mL distilled water with gentle heating in a water bath set at 30°C; make up to 100 mL with distilled water. If the solution is not clear, reheat. The solution is stable for 1 to 2 weeks.

3　EDTA reagent (0.025 *M* EDTA). Dissolve 9.3 g EDTA in distilled water, add 4.0 g NaOH, dissolve, make up to 1 L with distilled water, add 1 mL Brij-35, and mix.

4　Modification to digest 15.3.1: in step 6 add 2.0 mL 2 *M* HCl, dissolve the ash, and make the solution up to 10 mL.

5　Standards should contain 0 to 6.0 mg B L^{-1}.

6　Set up autoanalyzer as shown in Figure 15.2.

15.4.7.3 Comments

Borosilicate glassware should not be used; plastic "glassware" is satisfactory. Clean all containers and glassware in 6 *M* HCl and rinse with distilled water. According to Greweling (1976), neither soap nor chromic acid should be used to clean glassware, since they may result in absorption of boron. John et al. (1975) describe a manual method for boron determination in plant tissue.

REFERENCES

Abu-Samra, A., Morris, J. S., and Koirtyohann, S. R. 1975. Wet ashing of some biological samples in a microwave oven. Anal. Chem. 47: 1475–1477.

Barnhisel, R. and Bertsch, P. M. 1982. Aluminum. Pages 275–300 in Methods of soil analysis. Part 2. Chemical and microbiological properties. Agronomy No. 9. Soil Science Society of America, Madison, WI.

Bates, T. E. 1971. Factors affecting critical nutrient concentrations in plants and their evaluation: a review. Soil Sci. 112: 116–130.

Beaton, J. D., Burns, G. R., and Platou, J. 1968. Determination of sulphur in soils and plant material. Technical Bulletin Number 14. The Sulphur Institute, Washington, D.C.

Beaufils, E. R. 1973. Diagnosis and recommendation integrated system (DRIS). A general scheme for experimentation and calibration based on principles developed from research in plant nutrition.

Soil Science, Bull. No. 1, University of Natal, S.A.

Bouma, D. 1983. Diagnosis of mineral deficiencies using plant tests., in A. Läuchli and R. L. Bieleski, Eds. Encyclopedia of plant physiology, 15A: Inorganic plant nutrition. Springer-Verlag, Berlin.

Blakemore, W. E. and Billedeau, S. M. 1981. Analysis of laboratory animal feed for toxic and essential elements by atomic absorption and inductively coupled argon plasma emission spectrometry. J. Assoc. Off. Anal. Chem. 44: 1284–1290.

Blanchar, R. W. 1986. Measurement of sulfur in soils and plants. Pages 455–490 in M. A. Tabatabai, Ed. Sulfur in agriculture. Agronomy No. 27. Soil Science Society of America, Madison, WI.

Blanchar, R. W., Rehm, G., and Caldwell, A. C. 1965. Sulfur in plant materials by digestion with nitric and perchloric acid. Soil Sci. Soc. Am. Proc. 29: 71–72.

Bowman, D. C., Paul, J. L., and Carlson, R. M. 1988. A method to exclude nitrate from Kjeldahl digestion of plant tissues. Commun. Soil Sci. Plant Anal. 19: 205–213.

Bremner, J. M. 1965. Total nitrogen. In C. A. Black, Ed. Methods of soil analysis. Part II. Chemical and microbiological properties. Agronomy No. 9. American Society of Agronomy, Madison, WI.

Chapman, H. D. and Pratt, P. F. 1961. Methods of analysis for soils, plants, and waters. University of California, Riverside.

Davidescu, D. and Davidescu, V. 1972. Evaluation of fertility by plant and soil analysis. Translated by S. L. Kent, 1982. Abacus Press, Tunbridge Wells, Kent, England.

DeMerchant, G. P. 1988. Total nitrogen in plant tissue. Unpublished method. Agriculture Canada. Fredericton Research Station. Box 20280, Fredericton, NB, E3B 4Z7.

DeMerchant, G. P. 1989. Total phosphorus in plant tissue. Unpublished method. Agriculture Canada. Fredericton Research Station. Box 20280, Fredericton, NB, E3B 4Z7.

Dow, A. I. and Roberts, S. 1982. Proposal: critical nutrient ranges for crop diagnosis. Agron. J. 74: 401–403.

Gaines, T. P. and Mitchell, G. A. 1979. Chemical methods for soil and plant analysis. Agronomy handbook No. 1. University of Georgia, Coastal Plain Experimental Station.

Greweling, T. 1976. Chemical analysis of plant tissue. Agronomy No. 6. Cornell University Agricultural Experimental Station, Ithaca, NY. Search Agriculture 6(8): 1–35.

Hoenig, M. and DeBorger, R. 1983. Particular problems encountered in trace metal analysis of plant material by atomic absorption spectrometry. Spectrochem. Acta 38: 873–880.

Issac, R. A. 1980. Atomic absorption methods for analysis of soil extracts and plant tissue extracts. J. Assoc. Off. Anal. Chem. 63: 788–796.

Issac, R. A. 1990. Plants. Pages 40–60 in K. Helrich, Ed. Official methods of analysis of the Association of Official Analytical Chemists. Arlington, VA.

Isaac, R. A. and Jones, J. B. 1972. Effects of various dry ashing temperatures on the determination of 13 nutrients in five plant tissues. Commun. Soil Sci. Plant Anal. 3: 261–269.

Isaac, R. A. and Johnson, W. C. 1975. Collaborative study of wet and dry ashing techniques for the elemental analysis of plant tissue by atomic absorption spectrophotometry. J. Assoc. Off. Anal. Chem. 58: 436–440.

Isaac, R. A. and Johnson, W. C. 1976. Determination of total nitrogen in plant tissue, using a block digestor. J. Assoc. Off. Anal. Chem. 59: 98–100.

Isaac, R. A. and Johnson, W. C. 1985. Elemental analysis of plant tissue by plasma emission spectroscopy: collaborative study. J. Assoc. Off. Anal. Chem. 58: 436–440.

Jayman, T. C. Z. and Sivasubramaniam, S. 1974. The use of ascorbic acid to eliminate interference from iron in the aluminon method for determining aluminium in plant and soil extracts. Analyst 99: 296–301.

John, M. K., Chuah, H. H., and Neufeld, J. H. 1975. Application of improved azomethine-H method to the determination of boron in soils and plants. Anal. Lett. 8: 559–568.

Jones, J. B., Jr. 1985. Soil testing and plant analysis: guides to the fertilization of horticultural crops. Hortic. Rev. 7: 1–68.

Jones, J. B., Jr. and Steyn, W. J. A. 1973. Sampling, handling, and analyzing plant tissue samples. Pages 249–270 in L. M. Walsh and J. D. Beaton, Eds. Soil testing and plant analysis. Soil Science Society of America, Madison, WI.

Jones, J. B., Jr., Large, R. L., Pfleiderer, D. B., and Klosky, 1971. How to properly sample for a plant analysis. Crops Soils 23: 15–18.

Kalra, Y. P., Maynard, D. G., and Radford, F. G. 1989. Microwave digestion of tree foliage for multi-element analysis. Can. J. For. Res. 19: 981–985.

Kingston, H. M. and Jassie, L. B. 1986. Microwave energy for acid decomposition using biological and botanical samples. Anal. Chem. 58: 2534–2541.

Kingston, H. M. and Jassie, L. B., Eds. 1988. Introduction to microwave sample preparation. Am. Chem. Soc., Washington, D.C.

Linder, R. C. and Harley, C. P. 1942. A rapid method for the determination of nitrogen in plant tissue. Science 96: 565–566.

Martin-Prével, P., Gagnard, J., and Gautier, J., Eds. 1984. L'analyse végétale dans le contrôle et l'alimentation des plantes tempérées et tropicales. Lavoisier, Paris, France. (English translation) Plant analysis as a guide to the nutrient requirements of temperate and tropical crops. Lavoisier Publishing, New York.

Melstead, S. W., Motto, H. L., and Peck, T. R. 1969. Critical plant nutrient composition values useful in interpreting plant analysis data. Agron. J. 61: 17–20.

Miller, G. L. and Miller, E. E. 1948. Determination of nitrogen in biological materials. Anal. Chem. 20: 481–488.

Moraghan, J. T. 1991. Removal of endogenous iron, manganese and zinc during plant washing. Commun. Soil Sci. Plant Anal. 22: 323–330.

Munson, R. D. and Nelson, W. L. 1973. Principles and practices in plant analysis. Pages 223–270 in L. M. Walsh and J. D. Beaton, Eds. Soil testing and plant analysis. Soil Science Society of America, Madison, WI.

Munter, R. C. and Grande, R.A. 1981. Plant tissue and soil extract analysis by ICP-atomic emission spectrometry. Pages 653–672 in R. M. Barnes, Ed. Developments in atomic plasma spectrochemical analysis. Heyden and Son, London.

Murphy, J. and Riley, J. P. 1962. A modified single solution method for the determination of phosphate in natural waters. Anal. Chem. Acta 27: 31–36.

Nelson, D. W. and Sommers, L. E. 1973. Determination of total nitrogen in plant material. Agron. J. 65: 109–112.

Nelson, D. W. and Sommers, L. E. 1980. Total nitrogen analysis of soil and plant tissues. J. Assoc. Off. Anal. Chem. 63: 770–778.

Novozamsky, I., Houba, V. J. G., Eck, R. van, and Vark, W. van. 1983. A novel digestion technique for multi-element plant analysis. Commun. Soil Sci. Plant Anal. 14: 239–248.

O'Neill, J. V. and Webb, R. A. 1970. Simultaneous determination of nitrogen, phosphorus, and potassium in plant material by automatic methods. J. Sci. Food Agric. 21: 217–219.

Preez, D. R. du, and Bate, G. C. 1989. Recovery of nitrate-N in dry soil and plant samples by the standard, unmodified Kjeldahl procedure. Commun. Soil Sci. Plant Anal. 20: 1915–1931.

Rechcigl, J. E. and Payne, J. E. 1990. Comparison of a microwave digestion system to other digestion methods for plant tissue analysis. Commun. Soil Sci. Plant Anal. 21: 2209–2218.

Smith, R., Bezuidenhout, E. M., and Heerden, A. M. van. 1983. The use of interference suppressants in the direct flame atomic absorption determination of metals in water — some practical considerations. Water Res. 17: 1483–1489.

Stephenson, M. G. and Gaines, T. P. 1975. Microwave drying as a rapid means on sample preparation. Tobacco Sci. 19: 51–52.

Sumner, M. E. 1979. Interpretation of foliar analyses for diagnostic purposes. Agron. J. 71: 343–348.

Sumner, M. E. 1990. Advances in the use and application of plant analysis. Commun. Soil Sci. Plant Anal. 21: 1409–1430.

Tel, D. A. 1989a. An automated procedure for the determination of boron in 0.4 N HCl plant extract. Unpublished method. Department of Land Resource Science, University of Guelph, Guelph, ON.

Tel, D. A. 1989b. Ammonia in soil and plant extracts, water and waste water. Technicon TRAACS 800 Industrial Method No. 780–89T. Technicon Instruments Corporation, Tarrytown, NY.

Tel, D. A., Liu, L., Shelp, B. J. 1992. Determination of total nitrogen in plant digests using a TRAACS 800 autoanalyzer. Commun. Soil Sci. Plant Anal. 23: 2771–2779.

Thomas, R. L., Sheard, R. W., and Moyer, J. R. 1967. Comparison of conventional and automated procedures for nitrogen, phosphorus, and potassium analyses of plant tissue using a single digestion. Agron. J. 59: 240–243.

Walworth, J. L. and Sumner, M. E. 1986. Foliar diagnosis: a review. Pages 193–241 in B. Tinker and Läuchi, Eds. Advances in plant nutrition. Part 3. Praeger, New York.

Walworth, J. L. and Sumner, M. E. 1987. The diagnosis and recommendation integrated system. In B. A. Stewart, Ed. Advances in soil science, Part 6. Springer-Verlag, New York.

Ward, A. F., Marciello, L. F., Carrara, L., and Luciano, V. J. 1980. Simultaneous determination of major, minor, and trace elements in agricultural and biological samples by inductively coupled argon plasma spectrometry. Spectrosc. Lett. 13: 803–831.

Watson, C. 1984. Sample preparation for the analysis of heavy metals in foods. Trends Anal. Chem. 3: 25–28.

Webber, M. D. 1971. Aluminum in plant tissue grown in soil. Can. J. Soil Sci. 51: 471–476.

Webber, M. D. 1974. Atomic absorption measurements of Al in plant digests and neutral salt extracts of soil. Can. J. Soil Sci. 54: 81–87.

White, R. H. and Douthit, G. E. 1985. Use of microwave oven and nitric acid-hydrogen peroxide digestion to prepare botanical materials for elemental analysis by inductively coupled argon plasma emission spectroscopy. J. Assoc. Off. Anal. Chem. 68: 766–769.

Zasoski, R. J. and Burau, R. G. 1977. A rapid nitric-perchloric acid digestion method for multielement analysis. Commun. Soil Sci. Plant Anal. 8: 425–436.

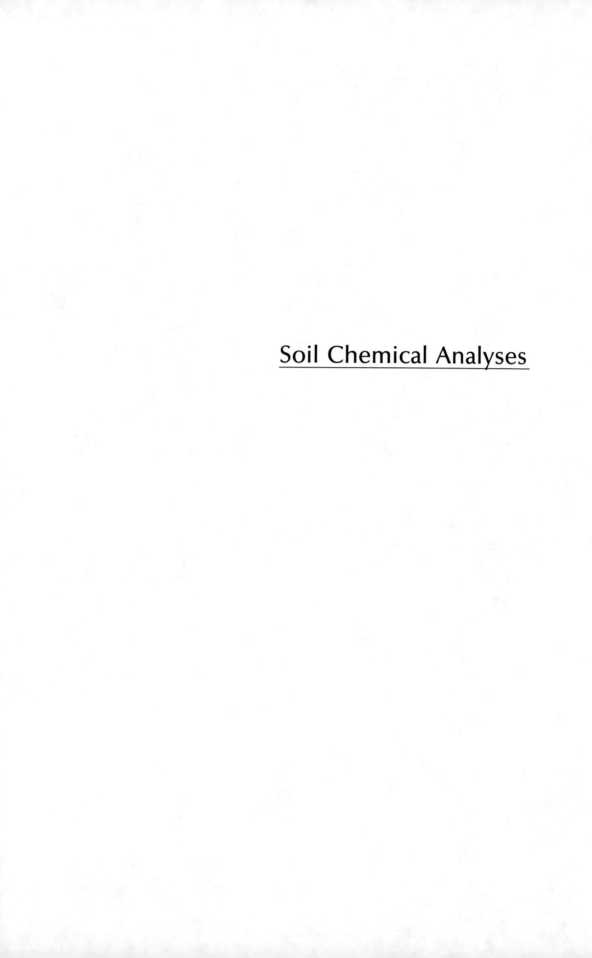

Soil Chemical Analyses

Chapter 16
Soil Reaction and Exchangeable Acidity

W. H. Hendershot and H. Lalande

McGill University

Ste. Anne de Bellevue, Quebec, Canada

M. Duquette

SNC-Lavalin

Montreal, Quebec, Canada

16.1 INTRODUCTION

Soil pH is one of the most common and important measurements in standard soil analyses. Many soil chemical and biological reactions are controlled by the pH of the soil solution in equilibrium with the soil particle surfaces.

Soil pH is measured in an aqueous matrix such as water or a dilute salt solution. Soil pH measured in water is the pH closest to the pH of soil solution in the field (this is true for soils with low electrical conductivity and for soils that are not fertilized), but is dependent on the degree of dilution (the soil-to-solution ratio). Measuring soil pH in a matrix of 0.01 M $CaCl_2$, as opposed to water, has certain advantages for agricultural soils, but the addition of the salt does lower the pH by about 0.5 pH units compared to soil pH in water (Schofield and Taylor 1955). In soil correlation work the use of pH in $CaCl_2$ is preferred because the measurement will be less dependent on the recent fertilizer history. Other methods for soil pH measurement, such as pH in 1 M KCl, are presented elsewhere (Peech 1965); these methods are not commonly used in Canada for routine analysis and are not included in this section.

16.2 SOIL pH IN WATER

When measuring soil pH in water, the main concern is that an increase in the amount of water added will cause an increase in pH; it is therefore important to keep the ratio constant and as low as possible. However, the supernatant solution must be sufficient to immerse the electrode properly without causing too much stress when inserting the tip of the electrode into the soil and to allow the porous pin on the electrode to remain in the solution above the soil.

Soil Sampling and Methods of Analysis, M. R. Carter, Ed.,

Canadian Society of Soil Science. © 1993 Lewis Publishers.

16.2.1 MATERIAL

1   pH meter: an appropriate instrument provided with two calibration points should be used.

2   Combined electrode: since the volume of soil is generally limited and the soil-to-solution ratio kept as low as possible, a combination electrode is a valuable asset.

3   30-mL long form beakers (Pyrex® or disposable plastic): beakers that have a narrow shape help to immerse the electrode in the supernatant without introducing the tip into the soil.

4   Stirrers: disposable plastic stirrers or glass rods can be used.

16.2.2 PROCEDURE

1   Weigh 10 g of air-dried mineral soil (<2 mm) into a beaker and add 20 mL of distilled/deionized (DD) water. For organic samples, use 2 g of soil in 20 mL of DD water. Record the soil-to-solution ratio used. Include duplicate quality control (QC) samples in each batch.

2   Stir the suspension intermittently for 30 min.

3   Let stand about 1 h.

4   Immerse the electrode into the clear supernatant and record the pH once the reading is constant. Note: both the glass membrane and the porous salt bridge must be immersed.

16.2.3 COMMENTS

1   Soil samples containing high amounts of organic matter tend to form a thick dry paste when the ratio is kept the same as for mineral samples; therefore a decreased ratio of sample to water must be accepted (1:5 or 1:10).

2   Two pH standards should be used to calibrate the pH meter. They must be chosen in accordance with the pH range expected for the soils analyzed (pH 4.0 and 7.0 or pH 7.0 and 10.0).

3   A large amount of a soil similar to the samples being analyzed should be kept as an indicator of the variability of pH results over time; duplicate subsamples of this QC sample should be run with each batch of samples measured. Failure of the QC to fall within acceptable limits means that the whole batch should be reanalyzed.

16.3 SOIL pH IN 0.01 *M* CaCl₂

Standard measurementof soil pH in CaCl₂ is probably the most commonly used method to characterize soil pH. As mentioned by Peech (1965), Davey and Conyers (1988), and Conyers and Davey (1988), the use of CaCl₂ has some advantages for pH measurement: (1) the pH is not affected within a range of the soil-to-solution ratios used; (2) the pH is almost independent of the soluble salt concentration for nonsaline soils; (3) this method is a fairly good approximation of the field pH for agricultural soils; (4) because the suspension remains flocculated, errors due to the liquid junction potential are minimized; (5) no significant differences in soil pH determination are observed for moist or air-dried soil; and (6) 1 year of storage of air-dried soil does not affect the pH.

16.3.1 MATERIAL AND REAGENTS

1 pH meter: an appropriate instrument provided with two calibration points should be used.

2 Combined electrode: since the volume of soil is generally limited and the soil-to-solution ratio kept to a minimum, a combination electrode is a valuable asset.

3 30-mL long form beakers (Pyrex® or disposable plastic): beakers that have a narrow shape help to immerse the electrode in the supernatant without introducing the tip of the electrode in the soil, thus avoiding breakage.

4 Stirrers: disposable plastic stirrers or glass rods can be used.

5 Calcium chloride, 0.01 *M:* dissolve 2.940 g of calcium chloride dihydrate (CaCl₂·2H₂O) with DD water in a 2000-mL volumetric flask. The electrical conductivity of the CaCl₂ solution must be between 2.24 and 2.40 mS cm⁻¹ at 25°C.

16.3.2 PROCEDURE

1 Weigh 10 g of air-dried mineral soil (<2 mm), or 2 g of organic soil into a 20-mL beaker and add 20 mL of 0.01 *M* CaCl₂. Note the soil-to-solution ratio used. Include duplicate QC samples in each batch.

2 Stir the suspension intermittently for 30 min.

3 Let stand about 1 h.

4 Immerse a combination electrode into the clear supernatant and record the pH once the reading is constant. Note: both the glass membrane and the porous salt bridge must be immersed.

16.3.3 COMMENTS

1   The pH and electrical conductivity of the $CaCl_2$ should be fairly constant, i.e., pH in the range of 5.5 to 6.5 and the electrical conductivity around 2.4 mS cm^{-1} at 25°C. If the pH is outside this range, it should be adjusted with HCl or $Ca(OH)_2$ solution. If the electrical conductivity is not within the acceptable range, a new solution must be prepared.

2   The comments given in Section 16.2.3 apply to this method as well.

16.4 EXCHANGEABLE ACIDITY (THOMAS 1982)

In addition to bases (e.g., Ca, Mg, K, Na), there is also an amount of acidity that can be displaced from the exchange complex of a soil. The amount of this acidity is largely a function of soil pH and the exchange capacity. In most soils the exchangeable acidity will be composed of (1) exchangeable H^+, (2) exchangeable Al as either Al^{3+} or partially neutralized Al-OH compounds such as $AlOH^{2+}$ or $Al(OH)_2^+$, and (3) weak organic acids.

When a soil is limed the exchangeable acidity is neutralized as the pH rises. Hence, exchangeable acidity is one measure of the amount of lime that will be needed to buffer soil pH. The measurement of exchangeable acidity also allows for the measurement of exchangeable Al by first forcing the dissociation of all Al-OH complexes by the addition of KF (thus increasing pH) followed by a back titration. Alternatively, exchangeable Al can be measured directly by atomic absorption spectrometry.

16.4.1 MATERIALS AND REAGENTS

1   50-mL beakers and plastic stirring sticks.

2   Buchner funnels (55-mm diameter) and 250-mL filtering flasks connected to low-pressure vacuum line.

3   Replacing solution, 1 *M* potassium chloride: dissolve 149.2 g of KCl in a 2000-mL volumetric flask and make to volume with DD water.

4   Aluminum complexing solution, 1 *M* potassium fluoride: dissolve 11.62 g of KF in about 180 mL of DD water in a 200-mL beaker and then titrate to a phenolphthalein endpoint with sodium hydroxide (NaOH). Transfer to a 200-mL volumetric flask and make to volume.

5   Hydrochloric acid (HCl), approximately 0.1 *M*, standardized.

6   Sodium hydroxide (NaOH), approximately 0.1 *M*, standardized.

7   Phenolphthalein solution: dissolve 1 g of phenolphthalein in 100 mL of ethanol.

16.4.2 PROCEDURES

1 Weigh a 10.00-g sample of soil into a 50-mL beaker, add 25 mL of 1 *M* KCl solution, mix, and let stand for 30 min. Transfer to a Buchner funnel fitted with Whatman® No. 42 filter paper and mounted on a 250-mL vacuum flask. Rinse with an additional volume of 125 mL of KCl in 25-mL increments to a total of 150 mL; the exact volume is not critical.

2 To obtain KCl acidity, add 4 or 5 drops of phenolphthalein to the filtrate in the flask and titrate with 0.1 *M* NaOH to the first permanent pink endpoint. (Note: a deep pink is too far.) Titrate a blank (150 mL of KCl solution) to the endpoint and record the amount. Centimoles of KCl-extracted acidity per kg of soil (cmol (+) kg^{-1}) are calculated as shown below.

3 To estimate amounts of Al and H$^+$, record the titer for NaOH, then add 10 mL of 1 *M* KF, and titrate with 0.1 *M* HCl until pink color disappears, and record the amount of HCl added. Wait for 30 min, and add additional HCl to a clear endpoint; add this amount of HCl to that recorded above. Aluminum and hydrogen are calculated as shown below.

16.4.3 Calculations

$$\text{cmol}(+) \text{ kg}^{-1} \text{ KCl acidity} = \frac{(\text{mL NaOH sample} - \text{mL NaOH blank}) \times M \times 100}{\text{g sample}} \quad (16.1)$$

$$\text{cmol}(+) \text{ kg}^{-1} \text{ KCl exchangeable Al} = \frac{\text{mL HCl} \times M \times 100}{\text{g sample}} \quad (16.2)$$

$$\text{cmol}(+) \text{ kg}^{-1} \text{ H}^+ = \text{KCl acidity} - \text{KCl exchangeable Al} \quad (16.3)$$

where M is the exact molarity of the acid or base solution.

REFERENCES

Conyers, M. K. and Davey, B. G. 1988. Observations on some routine methods for soil pH determination. Soil Sci. 145: 29–36.

Davey, B. G. and Conyers, M. K. 1988. Determining the pH of acid soils. Soil Sci. 146: 141–150.

Peech, M. 1965. Hydrogen-ion activity. Pages 914–926 in C. A. Black, Ed. Methods of soil analysis Part 2. Agronomy No. 9. American Society of Agronomy, Madison, WI.

Schofield, R. K. and Taylor, A. W. 1955. The measurement of soil pH. Soil Sci. Soc. Am. Proc. 19: 164–167.

Thomas, G. W. 1982. Exchangeable cations. Pages 159–165 in A. L. Page et al., Eds. Methods of soil analysis. Agronomy No. 9, 2nd ed. American Society of Agronomy, Madison, WI.

Chapter 17
Soil Solution

Y. K. Soon

Agriculture Canada
Beaverlodge, Alberta, Canada

C. J. Warren

University of Guelph
Guelph, Ontario, Canada

17.1 INTRODUCTION

Isolation of the soil solution and its subsequent chemical characterization have useful applications in studies of plant nutrition, soil chemistry and biochemistry, mineral stability, and assessment of environmental contaminants (such as toxic metals and pesticides) and their movement in soils.

The soil solution may be defined as the aqueous liquid phase of the soil and its solutes (SSSA 1975). The solutes consist of dissolved electrolytes and gases, conveniently considered at equilibrium or quasiequilibrium with definable solid and gas phases of the soil, and small quantities of other water-soluble compounds such as organic substances and metabolites. Two implications emerge from its designation as a phase: (1) it has uniform macroscopic properties like electrolyte concentration and temperature, and (2) it can be isolated from the soil and subsequently studied in the laboratory (Sposito 1989).

The concept of the soil solution and the development of techniques to isolate it have been reviewed (Fried and Broeshart 1967, Pearson 1971, Adams 1974). The review by Adams should be considered essential reading by anyone initiating a study of the soil solution.

The microscopic concept of the soil-water interface is that of a usually net-negatively charged particle surface surrounded by a diffuse layer of hydrated counter-ions, called the micellar or inner solution, merging into a homogeneous intermicellar or outer solution. It is generally accepted that it is the outer solution which constitutes the soil solution. Since it is not presently feasible to quantify the chemical composition of the soil solution *in situ*, the question becomes one of attempting to isolate the "true" soil solution.

Methods of studying the chemistry of soil solutions may be direct or indirect. Soil solution studies involving only a few ions may be attempted indirectly by the null point quantity/

Soil Sampling and Methods of Analysis, M. R. Carter, Ed.,
Canadian Society of Soil Science. © 1993 Lewis Publishers.

intensity (Q/I) equilibration method (Beckett 1964, White and Beckett 1964). The activity ratio $a_K/(a_{Ca+Mg})^{0.5}$ of soils obtained by the null point Q/I method has been found to be in good agreement with that determined in displaced soil solutions (Moss 1969).

Direct methods of studying the soil solution require that the solution be isolated without changing its chemical composition. (1) Extraction of soil solution by compaction is not satisfactory because of interacting double layers and salt sieving (Bolt 1961, Appelo 1977). (2) Soil solution isolated by absorption on filter paper (Schuffelen et al. 1964, Hinkley and Patterson 1973) or by the pressure memebrane apparatus (Fried and Broeshart 1967) is not reliable because of retention of ions and release of contaminants from filter materials. (3) *In situ* soil solution sampling by suction into ceramic porous cup containers is used extensively in hydrology and envionmental monitoring (Reeve and Doering 1965, Wood 1973, Michael et al. 1989); however, several studies show that the chemical composition of the extracted solution is altered by sorption or contamination (Grover and Lamborn 1970, Hansen and Harris 1975, Rauland-Rasmussen 1989). In a variation of the technique, Hossner and Phillips (1973) used a porous plastic material for saturated soil conditions, and Nielsen (1972) used a silica-gel filter sampler for a greenhouse study. However, the reliability of these latter filter materials under a wider range of conditions has yet to be demonstrated.

Currently acceptable techniques of isolating soil solutions from soil include the classical column displacement technique and two centrifugal displacement techniques.

The classical column displacement method was introduced in 1866 by Schloessing (cited by Nye and Tinker 1977). Schloessing effected the displacement of soil solution from a column of soil with carmine-colored water. The dye was used to identify the displacent, i.e., the displacing solution, in the effluent. If the soil solution has been equilibrated with the soil and the initial moisture content is not too low, there should be little or no disturbance of the diffuse double layer, and successive fractions of the soil solution should be similar in chemical composition. The reliability of the column displacement method has been demonstrated by Parker (1921) and Burd and Martin (1923), and more recently by Adams and co-workers (see Adams 1974).

The disadvantages of the column displacement technique are (1) the necessity for experience and skill on the part of the operator; (2) the long time period required, and consequently the low throughput of sample; and (3) the need for large soil samples.

The appeal of the centrifugal methods is that no skill or previous experience is required, small amounts of soil are sufficient (as little as 10 g, Reynolds 1984), and the soil solution can be extracted within a period from half an hour to an hour. Previously, the small volume of soil solution extracted, typically about 3 to 17 mL per 100 g of dry soil, was considered a serious disadvantage of the centrifugal methods. With modern analytical instruments capable of multielemental analysis with a single run, such as inductively coupled plasma emission spectroscopy (ICP-ES) and ion chromatography, this limitation is no longer a serious problem: a typical uptake rate of ICP-ES is about 2.4 mL per minute, and sample size for ion chromatography is typically 20 to 100 μL. Third-generation continuous-flow analyzers can analyze for NH_4^+, phosphate, nitrate, sulfate, and chloride using only about 1 mL of soil solution. Adams et al. (1980) reported that the column displacement technique, the centrifugal filtration method, and the immiscible displacement-centrifugal method gave similar results for soils ranging in texture from loamy sand to clay.

A problem common to all methods of soil solution isolation using air-dried soil as the starting material concerns the length of time the soil needs to be incubated with water to approach equilibrium with respect to the soil solution. Larsen and Widdowson (1968) observed increases in ion concentrations in the soil solution over a period of 53 d of incubation, particularly with respect to Ca^{2+}, Mg^{2+}, and NO_3^-. They attributed the increases to mineralization and nitrification. Since lag periods in nitrification upon incubation of soil commonly run to about 10 d, an incubation period of about 10 d may be a good compromise. This is supported by Larsen and Widdowson's data. However, investigators should determine the optimum incubation period for minimizing these effects in their soils. Different patterns of mineral N production and pH changes on rewetting soils appear to depend on initial soil pH values (Haynes and Swift 1989). These chemical changes can be minimized by using field-moist soil, whenever possible, as the starting material. Edmeades et al. (1985) recommended extraction of the soil solution within 24 h of sampling. However, field-moist soils do not always yield sufficient quantities of soil solution for analysis. To ensure an adequate amount of extracted soil solution, Linehan et al. (1989) remoistened field-moist soil samples to 90% of field capacity and reequilibrated the samples for 48 h before attempting to extract the soil solution.

17.2 COLUMN DISPLACEMENT METHOD (ADAMS ET AL. 1980)

Soil solution displacement from a soil column by another liquid may be regarded as the result of a mechanical piston effect. The soil moisture content at the bottom of the column must be in excess of the water-holding capacity before any solution starts dripping. Various liquids such as ethanol, dilute potassium thiocyanate, 50% glycerol, and saturated $CaSO_4$ have been used successfully as displacents.

17.2.1 MATERIALS AND REAGENTS

1   Packing column: a 500-mL dispensing buret with a glass wool plug at the bottom can be conveniently used as the packing column. Alternatively, the glass column described by Adams et al. (1980) may be used. To reduce contamination from dust during solution collection, a collecting cup made from a 50-mL centrifuge tube and rubber bung with two holes (a sample inlet and an air outlet) may be attached to the outlet of the packing column.

2   Displacent solution: saturated $CaSO_4$ containing 0.4% KSCN is prepared by dissolving 4.85 g $CaSO_4 \cdot 2H_2O$ in 2 L of deionized water, filtering off undissolved $CaSO_4$, and then dissolving 8 g KSCN in the filtrate.

3   5% (w/v) $FeCl_3$ in 0.1 M HCl: dissolve 5 g of $FeCl_3$ in 100 mL of 0.1 M HCl.

4   Acid-washed quartz sand, 0.5 to 1 mm size.

17.2.2 PROCEDURE

1   Pass moist soil samples through a 6-mm screen.

2   Transfer the soil in portions of about 30 g to the packing tube and tamp lightly with a rubber bung (#7) attached to a long rod. Continue until 400 to 500 g of moist soil have been added.

3   Place the column on a suitable rack and attach the solution-collecting cup.

4   Refirm the soil at the top of the column with the rubber bung-and-rod, and add about 1 cm of quartz sand to minimize disturbance of the soil during the following step.

5   Add about 100 mL of the displacent solution to the top of the column. (The exact amount will depend on the amount of soil used and the water-holding capacity of the soil.)

6   Discard the first 2–3 mL of leachate as a precaution against minor contamination. Collect the subsequent displaced fractions in 5-mL aliquots until SCN^- appears in the effluent (detected by reacting a few drops of effluent with acidified 5% $FeCl_3$, which results in the formation of a deep red-colored ferric thiocyanate complex in the presence of SCN^-).

7   Bulk the various fractions uncontaminated with SCN^-.

8   Determine pH and EC. Analyze for other constituents as required as soon as possible, following the scheme shown in Figure 17.1. Extracted samples should be stored at 4 or $-18°C$, i.e., in a refrigerator or freezer, until analysis can be done. The soil solution may be analyzed for (1) major cations, anions, and organic ligands; (2) ions that pair significantly; and (3) minor ions of special interest, e.g., trace elements.

17.2.3 COMMENTS

1   The ferric thiocyanate method is probably the most sensitive one for detecting the presence of the displacing solution in the effluent. Pierre (1925) showed that detected amounts of SCN^- appears in the displaced solution before changes in soil solution composition were measurable.

2   Obtaining a properly displaced soil solution requires skill on the part of the operator. According to Parker (1921), "After a little experience, one can readily determine the proper degree of packing for any soil at a given moisture content". Too little packing can result in channeling and early appearance of the displacent in the effluent. Too much compacting needlessly prolongs the displacement time. When properly packed, the displaced soil solution usually appears after about 90 min or more. For fine-textured soil,

SOIL SOLUTION

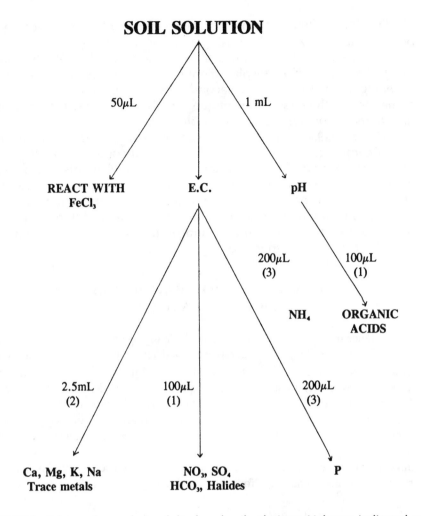

FIGURE 17.1. Scheme for analysis of displaced soil solutions. Volumes indicated are approximate and will vary with instruments and methods used (modified from Wolt and Graveel 1986). Numbers beside volume indicate analysis by (1) ion chromatography, (2) ICP-emission spectroscopy or atomic absorption, and (3) colorimetry.

mixing with dry quartz sand (0.5 to 1 mm size) in a proportion of up to 1 part sand to 3 parts soil can reduce displacement time. In samples where rates of displacement are much less than normal, i.e., less than 10 mL h^{-1}, application of air pressure of up to 35 kPa at the top of the column should be helpful.

3   The time required to complete displacement increases with column height and the yield of soil solution displaced is inversely related to the diameter of the column (Parker 1921). More time is also required for complete displacement at a lower soil moisture content. Soil solution displacement should be completed within an 8-h period to minimize changes in the soil associated with microbial activity. According to Adams (1974), 25 to 75% of the soil water can be displaced within this time period.

4   The limitations inherent to the classical column displacement technique can
     be eliminated or reduced by using a mechanical vacuum extractor. Up to
     24 samples can be displaced in 2 h (Wolt and Graveel 1986). This improve-
     ment allows the column displacement method to be used routinely by first,
     speeding up the process, and secondly, by reducing the constant attention
     demanded by the classical technique. Another advantage is that about 70 g
     only of moist soil are required per column (a 60-mL disposal syringe). Wolt
     and Graveel (1986) compared the vacuum-assisted displacement method
     with a centrifugal filtration method and found some differences between
     the two methods in the concentrations of certain elements, and a generally
     higher pH in the vacuum-extracted solution. The higher pH values may be
     attributed to the partial vacuum and degassing of the extracted solution.
     When equilibration of a $CaCO_3$-CO_2-H_2O-dominated system is critical, use
     of a vacuum extractor may not be appropriate. Changes in H^+ activity may
     also significantly alter speciation of hydrolyzable metals.

17.3 CENTRIFUGAL FILTRATION METHOD
(DAVIES AND DAVIES 1963)

Current acceptance of the centrifugal method for isolating soil solution stems from the work
of Davies and Davies (1963). The effectiveness and reliability of the method have been
demonstrated (Adams et al. 1980, Elkhatib et al. 1987). Most published techniques rely on
modifications of commercial equipment that require at least moderate workshop facilities.
Reynolds (1984) has described an apparatus that can be readily modified without workshop
facilities, but requires a high-speed centrifuge (to 10^4 rpm). The following method uses
equipment that requires minimal modification and does not need a high-speed centrifuge.

The centrifugal force developed at any point in the centrifuged soil column is conveniently
measured as the relative centrifugal force (RCF) given by the formula:

$$RCF = \frac{(2\pi n)^2 r}{g}$$

(17.1)

where n = no. of revolutions s^{-1}
      r = distance from the center of rotation in cm
      g = 981 cm s^{-2}

A relationship exists between the RCF developed and the minimum size of soil pores drained
(assumed to be equivalent to capillary pores). It can be shown that pores of approximately
1 μm diameter or greater can be drained at a RCF of $1000 \times g$; $RCF > 4500 \times g$ is
required to drain pores of less than 0.1 μm diameter (Edmunds and Bath 1976).

17.3.1 MATERIALS AND EQUIPMENT

1   The centrifugal apparatus consists of: (a) a soil-containing cylinder made by
     cutting off the top half of a 60-mL disposal syringe; and (b) a solution col-
     lecting cup, made by cutting off the top 58 mm of a 50-mL polypropylene
     or polyallomer centrifuge tube (see Figure 17.2).

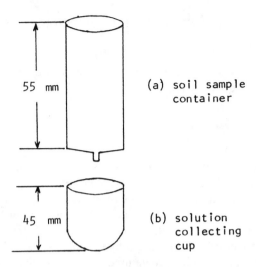

FIGURE 17.2. Centrifugal filtration apparatus.

2   Centrifuge with horizontal rotors and 50-mL centrifuge shields or adaptors, preferably with refrigeration for temperature control.

3   Glass wool.

4   Whatman® No. 42 filters cut into 27-mm-diameter disks.

5   Plastic vials, e.g., 7-dram snap cap vials.

6   Parafilm.

17.3.2 PROCEDURE

1   Cover the drainage hole of the soil container with a small plug of glass wool, and line the bottom with a disk of filter paper (Whatman® No. 42).

2   Rinse soil container, glass wool plug, and filter disk with 1 mL of 0.1 M HCl, followed by 3 mL of deionized water and dry partially by centrifuging for 5 min at 3000 rpm.

3   Place about 25 g of moist soil in the soil container and cover with a sheet or disk of parafilm to reduce evaporative loss. A separate subsample of each soil may be retained for moisture determination, if required.

4   Place a clean solution-collecting cup followed by the soil-containing cylinder into the centrifuge shield.

5   Use paired samples (corresponding to a soil or treatment) for balancing the centrifuge (to the nearest 0.01 g).

6 Centrifuge at 3000 rpm (RCF of 1500 $\times$ g at bottom of soil column) or higher for 30 min.

7 Analyze the solution for pH and other constituents as recommended in Section 17.2.2 (Figure 17.1).

17.3.3 Comments

The method described above removes about 40% of soil moisture, initially at 33 kPa tension, from a loam soil and about 21% from a silty clay loam. Each 25 g of moist soil yields from 1.2 to 2.6 mL of soil solution. This efficiency of solution extraction is greater than that of the Davies and Davies (1963) apparatus. As described, the filter disk retains about 10 μL of extracted solution. Centrifugal filtration methods using high centrifugal forces (RCF > 5000 $\times$ g) have been reported to remove up to 88% of 33-kPa soil moisture (Elkhatib et al. 1987). The wall of the disposal syringe used to construct the soil container should be similar in thickness to the wall of a 50-mL high-speed centrifuge tube. Consequently, the apparatus described above should withstand centrifugation at RCF > 1500 $\times$ g. Where field samples have been obtained using a 25-mm soil corer, this method reduces soil disturbance to a minimum, since the soil core can be cut and directly transfered to the soil container with a minimum of handling.

17.4 IMMISCIBLE DISPLACEMENT CENTRIFUGAL METHOD (WHELAN AND BARROW 1980)

This technique combines the principles involved in immiscible displacement with the convenience of centrifugation using commercially available centrifuge tubes. A dense immiscible liquid (density >1.0 kg L^{-1}) is used to displace the soil solution which, under a centrifugal force, floats on top of the denser immiscible displacent. Desirable properties of a displacent include: high density, low volatility, low toxicity, low water solubility, low price, and high chemical inertness. The method as originally proposed by Mubarak and Olsen (1976) uses CCl_4 as the displacent. However, because of the high toxicity of CCl_4, other liquids have been proposed: 1,1,1-trichloroethane and tetrachloroethylene (Whalen and Barrow 1980); 1,1,2-trichloro-trifluoroethane (Kinniburgh and Miles 1983); and ethyl benzoylacetate (Elkhatib et al. 1986). 1,1,1-Trichloroethane has not been recommended because it contains nitrate as a stabilizer (Whalen and Barrow 1980). Ethyl benzoylacetate is nontoxic, but expensive, and not so readily available. Tetrachloroethylene has a low toxicity, is economical to use, and is recommended by Whelan and Barrow (1980). It does require, however, the use of nylon, polyallomer, or Teflon® centrifuge tubes. An equally good choice of displacent is trifluoroethane (Kinniburgh and Miles 1983; Phillips and Bond 1989), which has a low toxicity and is compatible with standard polypropylene tubes, but has a low boiling point and thus requires careful handling.

17.4.1 MATERIALS AND REAGENTS

1 Centrifuge tubes made of polyallomer or Teflon®, 50-mL capacity with caps, acid washed.

2   Tetrachloroethylene, C_2Cl_4 (density = 1.63 kg L^{-1}).

3   Disposal syringes, 10 mL.

4   25-mm diameter syringe nylon filter, 0.2 or 0.45 μm.

5   High-speed refrigerated centrifuge.

6   Plastic vials.

17.4.2 PROCEDURE

1   Weigh 20 to 21 g of moist soil into paired sets of centrifuge tubes. Cap after filling to minimize evaporation. Retain a subsample for moisture determination.

2   Add 20 mL of tetrachloroethylene. Balance each pair of tubes and contents, including caps, to 0.002 g by removing or adding tetrachloroethylene with a syringe fitted with a hypodermic needle.

3   Centrifuge the capped centrifuge tubes for 1 h at about 15,000 rpm (RCF = 23,000 × g).

4   Remove the displaced soil solution with a 10-mL syringe, taking care to withdraw as little tetrachloroethylene as possible, and transfer to a plastic vial where the solution may be combined with other subsamples for bulking, if necessary.

5   If small amounts of particulate matter are present, pass the aqueous extract through a syringe filter. Prior to filtering, pass 5 to 10 mL of deionized water through the filter to minimize the risk of contamination.

6   Determine the pH and EC as soon as possible. Analyze for other constituents as soon as possible after this, according to Section 17.2.2.

17.4.3 Comments

Most of the immiscible liquids used as displacent are toxic to some degree when they are inhaled, ingested, or absorbed through the skin. Handling should be done in a fume hood as much as possible. The added expense of replenishing the displacent supply is also another consideration, although used displacents can be distilled and recycled with proper equipment. The water solubility of tetrachloroethylene is about 150 mg L^{-1} at 25°C; however, this is not a serious problem. With some soils, inorganic material may accumulate between the extracted solution and the displacent. In such cases, the displaced soil solution can be poured into a filter funnel fitted with a 15-cm phase separation filter paper (Whatman® silicone treated No. 1PS). The paper retains the aqueous phase. Allow the particulate matter to settle before pouring the solution into storage vials.

17.5 CALCULATIONS

In many applications of soil chemistry and mineralogy, it is desirable to predict the activities and chemical species of the elements of interest in the soil solution, e.g., the effect of aluminum speciation on plant growth. Several computer codes have been written for performing speciation calculations (Nordstrom et al. 1979). Computer codes differ mainly in the calculation of activity coefficients, the numerical methods used to solve the equations, the number of entries, the number of options available, and the thermodynamic data bases. The following short summary of some of the features of computer codes commonly used to perform speciation calculations on aqueous solutions is provided for information only:

1. SOLMINEQ (Kharaka and Barnes 1973) was initially written in PL/1 and designed for solution-mineral-equilibrium computations in geologic systems. The code has recently been updated and rewritten in FORTRAN for use on both mainframe and IBM®-compatible microcomputers (Kharaka et al. 1988). The code now has a data base of 260 organic and inorganic aqueous species and over 200 minerals. Acquisition inquiries to:

> Y. K. Kharaka
> U.S. Geological Survey
> Water Resources Division
> 345 Middlefield Rd, MS # 427
> Menlo Park, CA 94025, U.S.A.

or

> E. H. Perkins or W. D. Gunter
> Alberta Research Council
> Oil Sands and Hydrocarbon Recovery Dept.
> P.O. Box 8330, Postal Station F
> Edmonton, Alberta, Canada T6H 5X2

2. GEOCHEM (Sposito and Mattigod 1980) was written in FORTRAN IV and originally designed for soil systems. Its major advantage over other codes is the inclusion of equilibrium constants for ion-exchange reactions and a large number of organometallic complexes in its data base. The GEOCHEM code has been replaced with a new version for PC microcomputer called SOILCHEM (Sposito and Coves 1988). Acquisition inquiries:

> Manager, Campus Software Office
> 295 Evans Hall
> University of California
> Berkeley, CA 94720, U.S.A.

3. PHREEQE (Parkhurst et al. 1980) was written in FORTRAN IV and designed to model pH-redox equilibria in a geochemical system. The utility of the code lies in its very versatile set of options which include: a large number of redox reactions in the data base, and the ability to simulate pH titrations, evaporation, and mixing to two solutions. A version of the code (PC-PHREEQE) has been adapted to IBM®-compatible microcomputers and modified to include accurate calculation of activity coefficients to high ionic strengths (Crowe and Longstaffe 1987). Acquisition inquiries:

> D. L. Parkhurst or D. C. Thorstenson
> U.S. Geological Survey

                    Water Resources Division
                    12201 Sunrise Valley Drive
                    Reston, VA 22092, U.S.A.

        or

                    A. S. Crowe
                    National Water Research Institute
                    P.O. Box 5050
                    Burlington, Ontario
                    L7R 4A6 Canada

4. MINTEQ (Felmy et al. 1984, Peterson et al. 1987) and a PC microcomputer version, MINTEQA2 (Allison et al. 1990) are written in FORTRAN and originally developed as part of a model for assessing the fate of priority pollutant metals in aquatic systems. The main features of the code include: a very large and easily expandable data base, temperature correction using the Van't Hoff relationship, and six different adsorption algorithms. Acquisition inquiries for the PC version:

                    Center for Exposure Assessment Modelling (CEAM)
                    Environmental Research Laboratory
                    Office of Research and Development
                    U.S. Environmental Protection Agency
                    Athens, Georgia 30613-7799, U.S.A.

Some limitations of speciation calculation in equilibrium models have been considered by Baham (1984) and Sposito (1989). Perhaps the greatest problem is the quality or availability of thermodynamic constants.

Stability constants for many assumed complexes in soils may be inaccurate or unavailable. Inadequate knowledge of pertinent chemical equilibria such as oxidation-reduction or precipitation-dissolution, or of certain constituents in the soil solution (e.g., dissolved or suspended, monomeric vs. polymeric) is also a problem. A critically compiled thermodynamic data file represents, in the scientific judgment of those who have labored to assemble it, the best available estimates for the constants. No two research groups will necessarily agree on which data set represents the most reliable data and now many soluble complexes occur in solution. Other errors may be introduced by improper analytical technique or assumptions, e.g., extracting the soil solution at 5°C when the soil was initially incubated at 20°C. Properly used, however, the usefulness of equilibrium speciation calculations by a computer, especially for trace elements, has been amply demonstrated (Mattigod and Sposito 1979, Sposito and Bingham 1981).

REFERENCES

Adams, F. 1974. Soil solution. Pages 441–482 in E. W. Carson, ed. The plant root and its environment. University Press of Virginia, Charlottesville, VA.

Adams, F., Burmester, C., Hue, N. V., and Long, F. L. 1980. A comparison of column-displace-ment and centrifuge methods for obtaining soil solutions. Soil Sci. Soc. Am. J. 44: 733–735.

Allison, J. D., Brown, D. S., and Novo-Gradac, K. J. 1990. MINTEQ/PRODEFAZ, a geochemical assessment model for environmental systems. Version 3.0. User's manual. U.S. Environmental Protection Agency, Athens, GA.

Appelo, C. A. J. 1977. Chemistry of water expelled from compacting clay layers: a model based on Donnan equilibrium. Chem. Geol. 19: 91–98.

Baham, J. 1984. Prediction of ion activities in soil solution: computer equilibrium modelling. Soil Sci. Soc. Am. J. 48: 525–531.

Beckett, P. H. T. 1964. Studies on soil potassium. 1. J. Soil Sci. 15: 1–8.

Bolt, G. 1961. The pressure filtrate of colloidal suspensions. II. Experimental data on homoionic clays. Kolloid Z. 175: 144–150.

Burd, J. S. and Martin, J. C. 1923. Water displacement of soils and the soil solution. J. Agr. Sci. 13: 265–295.

Crowe, A. S. and Longstaffe, F. J. 1987. Extension of geochemical modelling techniques to brines: coupling the Pitzer equations with models, in Proc. of Solving Groundwater Problems with Models. Nat. Water Well Assoc. Conf. Feb. 10–12, 1987. Denver, CO.

Davies, B. E. and Davies, R. I. 1963. A simple centrifugation method for obtaining small samples of soil solution. Nature 198: 216–217.

Edmeades, D. C., Wheeler, D. J., and Clinton, O. E. 1985. The chemical composition and ionic strength of soil solutions from New Zealand topsoils. Aust. J. Soil Res. 23: 151–165.

Edmunds, W. M. and Bath, A. H. 1976. Centrifuge extraction and chemical analysis of interstitial waters. Environ. Sci. Technol. 10: 467–472.

Elkhatib, E. A., Bennett, O. L., Baligar, V. C., and Wright, R. J. 1986. A centrifuge method for obtaining soil solution using an immiscible liquid. Soil Sci. Soc. Am. J. 50: 297–299.

Elkhatib, E. A., Hern, J. L., and Staley, T. E. 1987. A rapid centrifugation method for obtaining soil solution. Soil Sci. Soc. Am. J. 51: 578–583.

Felmy, A. R., Girvin, D. C., and Jenne, E. A. 1984. MINTEQ — a computer program for calculating aqueous geochemical equilibria. U.S. Environmental Protection Agency, Athens, GA.

Fried, M. and Broeshart, H. 1967. The soil-plant system. Academic Press, New York.

Grover, B. L. and Lamborn, R. E. 1970. Preparation of porous ceramic cups to be used for extraction of soil water having low solute concentrations. Soil Sci. Soc. Am. Proc. 34: 706–708.

Hansen, E. A. and Harris, A. R. 1975. Validity of soil water samples collected with porous ceramic cups. Soil Sci. Soc. Am. Proc. 39: 528–536.

Haynes, R. J. and Swift, R. S. 1989. Effect of re-wetting air-dried soils on pH and accumulation of mineral nitrogen. J. Soil Sci. 40: 341–347.

Hinkley, T. and Patterson, C. 1973. Concentrations of metals in soil moisture film. Nature Phy. Sci. 246: 123–124.

Hossner, L. R. and Phillips, D. P. 1973. Extraction of soil solution from flooded soil using a porous plastic filter. Soil Sci. 115: 87–88.

Kharaka, Y. K. and Barnes, I. 1973. SOLMNEQ: solution-mineral equilibrium computations. NTIS #PB215–899. National Technical Information Service, Springfield, VA.

Kharaka, Y. K., Gunter, W. D., Agarwal, P. K., Perkins, E. H., and Debral, J. D. 1988. SOLMINEQ.88: a computer program code for geochemical modelling of water-rock interactions. U.S. Geological Survey Water Resource Investigations Rep. 88–4227.

Kinniburgh, D. G. and Miles, D. L. 1983. Extraction and chemical analysis of interstitial water from soils and rocks. Environ. Sci. Technol. 17: 362–368.

Larsen, S. and Widdowson, A. E. 1968. Chemical composition of soil solution. J. Sci. Food Agric. 19: 693–695.

Linehan, D. J., Sinclair, A. H., and Mitchell, M. C. 1989. Seasonal changes in Cu, Mn, Zn and Co concentrations in the root-zone of barley (*Hordeum vulgare* L.). J. Soil Sci. 40: 103–115.

Mattigod, S. V. and Sposito, G. 1979. Chemical modelling in trace metal equilibria in contaminated soil solutions using the computer program GEOCHEM. Pages 837–855 in Chemical modelling in aqueous systems. Am. Chem. Soc. Symp. Ser. No. 93. Washington D.C.

Michael, J. L., Neary, D. G., and Wells, M. J. M. 1989. Picloram movement in soil solution and streamflow from a coastal plain forest. J. Environ. Qual. 18: 89–95.

Moss, P. 1969. A comparison of potassium activity ratios derived from equilibrium procedures and from measurements on displaced soil solution. J. Soil Sci. 20: 297–306.

Mubarak, A. and Olson, R. A. 1976. Immiscible displacement of the soil solution by centrifugation. Soil Sci. Soc. Am. Proc. 40: 329–331.

Nielsen, N. E. 1972. A transport kinetic concept of ion uptake from soil by plants. 1. A method for isolating soil solution from soils with or without plant cover. Plant Soil 36: 505–520.

Nordstrom, D. K., Plummer, L. N., Wigley, T. J. L., et al. 1979. Comparison of computerized chemical models for equilibrium calculations in aqueous systems. Pages 875–892 in Chemical modelling in aqueous systems. Am. Chem. Soc. Symp. Ser. No. 93. Washington D.C.

Nye, P. H. and Tinker, P. B. 1977. Solute movement in the soil-root system. Blackwell Scientific, Oxford.

Parker, F. W. 1921. Methods of studying the concentration and composition of the soil solution. Soil Sci. 12: 204–232.

Parkhurst, D. L., Thorstenson, D. C., and Plummer, L. N. 1980. PHREEQE — a computer program for geochemical calculations, U.S. Geological Survey, Water Resource Investigations Rep. 80–86.

Pearson, R. W. 1971. Introduction to symposium — the soil solution. Soil Sci. Soc. Am. Proc. 35: 417–420.

Peterson, S. R., Hostetler, C. J., Deutsch, W. J., and Cowan, C. E. 1987. MINTEQ user's manual. NUREG/CR-4808 (PNL-6106). Pacific Northwest Laboratory, Richland, WA.

Phillips, I. R. and Bond, W. J. 1989. Extraction procedure for determining solution and exchangeable ions in the same sample. Soil Sci. Soc. Am. J. 53: 1294–1297.

Pierre, W. H. 1925. The H-ion concentration of soils as affected by carbonic acid and the soil-water ratio, and the nature of soil acidity as revealed by these studies. Soil Sci. 20: 285–305.

Rauland-Rasmussen, K. 1989. Aluminium contamination and other changes of acid soil solution isolated by means of porcelain suction cups. J. Soil Sci. 40: 95–101.

Reeve, R. C. and Doering, E. J. 1965. Sampling the soil solution for salinity appraisal. Soil Sci. 99: 339–344.

Reynolds, B. 1984. A simple method for the extraction of soil solution by high speed centrifugation. Plant Soil 78: 437–440.

Schuffelen, A. C., Koenigs, F. F. R., and Bolt, G. H. 1964. The isolation of the soil solution with the aid of filter paper tablets. 8th Int. Congr. Soil Sci. Trans. Vol 3: 519–528. Bucharest, Romania.

Sposito, S. 1989. The chemistry of soils. Oxford University Press, New York.

Sposito, G. and Mattigod, S. V. 1980. GEOCHEM: a computer program for the calculation of chemical equilibria in soil solutions and other natural water systems. Kearney Foundation of Soil Science, University of California, Riverside.

Sposito, G. and Bingham, F. T. 1981. Computer modelling of trace metal speciation in soil solutions: correlations with trace metal uptake by higher plants. J. Plant Nutr. 3: 35–49.

Sposito, G. and Coves, J. 1988. SOILCHEM: a program for the calculation of chemical speciation in soils. University of California, Berkeley.

SSSA. 1975. Glossary of soil science terms. Soil Science Society of America. Madison, WI.

Whelan, B. R. and Barrow, N. J. 1980. A study of a method for displacing soil solution by centrifuging with an immiscible liquid. J. Environ. Qual. 9: 315–319.

White, R. E. and Beckett, P. H. T. 1964. Studies on the phosphate potentials of soil. 1. Plant Soil 20: 1–19.

Wolt, J. and Graveel, J. G. 1986. A rapid routine method for obtaining soil solution using vacuum displacement. Soil Sci. Soc. Am. J. 50: 602–605.

Wood, W. W. 1973. A technique using porous cups for water sampling at any depth in the unsaturated zone. Water Resour. Res. 9: 486–488.

Chapter 18
Soluble Salts

H. H. Janzen

Agriculture Canada
Lethbridge, Alberta, Canada

18.1 INTRODUCTION

Soil salinity is a widespread limitation to agricultural production in semiarid and arid soils throughout the world. The accumulation of soluble salts in the soil profile curtails crop growth by increasing the osmotic potential of the soil solution and inducing specific ion toxicities or nutrient imbalances. The predominant solutes responsible for salinity include the cations sodium, calcium, and magnesium, and the anions sulfate and chloride (Richards 1954). Minor amounts of potassium, bicarbonate, carbonate, and nitrate may also be present. Aside from direct effects on crop growth, salts can also limit productivity by adverse effects on soil structure. In particular, high concentrations of sodium on cation-exchange sites will disperse soils and impede water and air movement (Bernstein 1975).

A number of approaches have been devised to characterize soil salinity. Most conventional methods employ aqueous or direct extraction of the soil solution and subsequent analysis of salt concentrations. More recently, a number of techniques, including four electrode probes, electromagnetic (EM) induction sensors, and time-domain reflectometry (TDR), have been developed for monitoring of salinity level *in situ*. Several of these field methods have been described by Dasberg and Nadler (1988) and Rhoades (1990). This chapter will focus on laboratory techniques for the quantification and characterization of soil salinity.

The choice of methods will depend to a large degree on the objective of the analysis, whether it be the classification of salt-affected soils, measurement of total salt concentration, or identification of specific ion problems. All of the laboratory methods involve extraction of salts from the soil solution, analysis of ions in the solution, and calculation of various indices for the identification of salt-related problems.

18.2 EXTRACTION

Soluble salts are removed from the soil using aqueous extraction techniques employing various extractant:soil ratios. The technique offering most accurate characterization of ionic conditions in the environment of the roots is the direct extraction of the soil solution (Chapter 17). Soil solution extraction, however, is possible only under relatively high soil moisture conditions and is time consuming in comparison to other extraction techniques. The most common method of extraction, used almost universally in the analysis of soil salinity, is the saturation paste extraction (Richards 1954). This procedure offers advantages of convenience and greater extract volume relative to direct solution extraction. As well, it can be reproducibly related to field soil water contents and compensates for variation in soil moisture retention.

Some investigators employ constant extraction ratios (e.g., 1:1, 1:5) which are more convenient to use and yield higher extract volumes without vacuum, but are not as closely related to field soil moisture contents. As well, the greater dilution of salts using these procedures may result in errors from hydrolysis, peptization, cation exchange, and dissolution of minerals such as gypsum.

The best extraction technique for a given application will depend on analytical objectives and research requirements. Precise quantification of ions in the soil solution necessitates direct extraction of the soil solution. While the saturation extraction procedure effectively measures total salt concentrations in the soil solution, it does not accurately reflect ionic composition of that solution, particularly in regard to calcium concentrations (Janzen and Chang 1988). Although the saturation extract remains the standard for measuring total salinity, fixed extraction ratios have been found to be closely correlated with the saturation extract in a wide range of soils (Hogg and Henry 1984). Extraction using a fixed ratio may be particularly useful for monitoring relative changes in solute concentrations.

18.2.1 Direct Extraction

Precise evaluation of the ionic composition and characteristics of the soil solution in saline soils requires the direct extraction of the soil solution. Several methods of extraction are described in Chapter 17 of this volume.

18.2.2 SATURATION EXTRACT (RHOADES 1982)

1 Weigh from 200 to 400 g of soil with known moisture content into a container with lid. Record the total weight of the container and the soil sample. (The weight of the soil used will depend on the volume of extract required. In general, approximately one third of the water added is recovered in the saturation extract.)

2 Add sufficient deionized water while mixing to saturate the soil sample. At saturation, the soil paste glistens, flows slightly when the container is tipped and slides cleanly from the spatula. A trench carved in the soil surface will readily close upon jarring the container.

3 Allow the sample to stand for at least 4 h and check to ensure that saturation criteria are still met. If free water has accumulated on the surface, add a weighed amount of soil and remix. If the soil has stiffened or does not glisten, add distilled water and mix thoroughly.

4 Weigh the container with contents. Record the increase in weight, which corresponds to the amount of water added. (Alternatively, the amount of

water added can be determined volumetrically by dispensing water from a burette). Calculate the saturation percentage (SP) as follows:

$$SP = \frac{(\text{weight of } H_2O \text{ added} + \text{weight of } H_2O \text{ in sample})}{\text{oven-dry weight of soil}} \times 100 \qquad (18.1)$$

5   After allowing the paste to stand for at least an additional 4 h, transfer to a Buchner funnel fitted with highly retentive filter paper. Apply vacuum and collect the extract until air passes through the filter. Turbid filtrates should be discarded or refiltered. Add 1 drop of 0.1% $(NaPO_3)_6$ solution per 25 mL of extract to prevent precipitation of $CaCO_3$.

6   Store extracts at 4°C until analyzed.

18.2.3 Comment

If possible, organic soils should be extracted without prior drying, which affects the SP. Organic soils may require an overnight imbibition period and a second addition of water to achieve a definite saturation endpoint.

18.2.4 FIXED-RATIO EXTRACT (RHOADES 1982)

1   Weigh the appropriate amount of air-dried soil into a flask, add sufficient deionized water to achieve the desired extraction ratio, and shake for 1 h.

2   Filter the suspension using highly retentive filter paper and store the filtrate at 4°C.

18.3 ANALYSES

18.3.1 Electrical Conductivity

The total solute concentration in the various extracts is normally determined by analysis of the electrical conductivity. Although the relationship between conductivity and salt concentrations varies somewhat depending on the ionic composition of the solution, electrical conductivity provides a rapid and reasonably accurate estimate of solute concentration.

1   Calibrate conductivity meter using standard KCl solution. At 25°C, the conductivities of 0.010 M and 0.100 M KCl solutions are 1.412 and 12.90 dS m^{-1}, respectively.

2   Read conductivity of the extracts. If the meter does not provide automatic temperature compensation, correct readings to 25°C based on measured temperature of the extract and known compensation factors.

Table 18.1 Crop Response to Salinity Measured as Electrical
Conductivity (EC) of the Saturation Extract

EC (dS m^{-1} at 25°C)	Crop response
0–2	Almost negligible effects
2–4	Yields of very sensitive crops restricted
4–8	Yields of most crops restricted
8–16	Only tolerant crops yield satisfactorily
>16	Only very tolerant crops yield satisfactorily

Adapted from Bernstein, L., Annu. Rev. Phytopathol., 13, 295, 1975.

18.3.2 Ion Concentrations

From a quantitative standpoint, the most important cations in aqueous extracts of soils are Ca^{2+}, Mg^{2+}, K^+, and Na^+. These cations are normally determined by atomic absorption spectrometry, atomic emission spectrometry, or compleximetric titration methods. Standard methods are outlined by Knudsen et al. (1982) and Lanyon and Heald (1982). A relatively recent innovation has been the development of ion chromatography techniques for the analysis of these cations (Basta and Tabatabai 1985).

The primary anions in soil extracts are normally sulfate, chloride, nitrate, and carbonate or bicarbonate. Standard methods for the analysis of these ions are described by Rhoades (1982). Ion chromatography has been reported to allow simultaneous analysis of several anions in aqueous extracts (Nieto and Frankenberger 1985).

18.4 CALCULATIONS AND INTERPRETATION

18.4.1 Electrical Conductivity

The relationship between crop growth and electrical conductivity of saturation extracts for a variety of crops has been extensively reviewed (Maas and Hoffman 1977, Carter 1982, Van Genuchten and Hoffman 1984). General salinity effects are presented in Table 18.1. Crop response to salinity at a given site may vary somewhat from reported values because of the differences in salt composition, crop varieties, climatic factors, and soil properties.

18.4.2 Ion Activities

Although amounts of ions present are normally expressed in concentration units, the influence of various salts may be more accurately reflected by the activity of the ions. Because of the high ionic strength of soil solutions in saline conditions, ion activities are often very different from absolute concentrations. Activities can be estimated from concentrations using models such as that published by Sposito and Coves (1988). Alternatively, activities can be measured directly using ion-selective electrodes.

18.4.3 Sodium Adsorption Ratio (SAR)

The SAR, a useful index of the sodicity or relative sodium status of soil solutions, aqueous extracts, or water in equilibrium with soil, is calculated as follows:

$$SAR = [Na^+]/[Ca^{2+} + Mg^{2+}]^{0.5} \tag{18.2}$$

where concentrations are in mmol L^{-1}.

By convention, soils with SAR values greater than 13 are usually considered to be sodic (Soil Science Society of America 1984), although other critical values have been proposed (Bennett 1988).

18.4.4 Exchangeable Sodium Percentage (ESP)

The ESP, a measure of soil sodicity, is the molar proportion of cation-exchange sites in a soil (CEC) occupied by sodium (Na_{exch}):

$$ESP = Na_{exch}/CEC \times 100 \tag{18.3}$$

This ratio can be calculated from direct measurements of exchangeable cations, though such analyses may be subject to considerable errors arising from anion exclusion or dissolution of slightly soluble minerals (Bohn et al. 1979, p. 229). The ESP can be estimated from the SAR using the following equation (Kamphorst and Bolt 1978):

$$\frac{ESP}{100 - ESP} = 0.015 \, SAR \tag{18.4}$$

An alternative, more complex Gapon-type equation was proposed by Sposito (1977).

Traditionally, sodic soils have been defined as having an ESP >15 (Richards 1954).

18.4.5 Critical Calcium Ratio

A number of studies have shown that crop yield in a salt-affected soil is strongly influenced by the ratio of calcium to that of other cations in the soil solution (Howard and Adams 1965, Carter et al. 1979, Janzen and Chang 1987). Yield reductions are typically observed when this ratio falls below approximately 0.10.

REFERENCES

Basta, N. T. and Tabatabai, M. 1985. Determination of exchangeable bases in soils by ion chromatography. Soil Sci. Soc. Am. J. 49: 84–89.

Bennett, D. R. 1988. Soil chemical criteria for irrigation suitability classification of Brown Solonetzic soils. Can. J. Soil Sci. 68: 703–714.

Bernstein, L. 1975. Effects of salinity and sodicity on plant growth. Annu. Rev. Phytopathol. 13: 295–312.

Bohn, H., McNeal, B., and O'Connor, G. 1979. Soil chemistry. John Wiley & Sons, Toronto.

Carter, D. L. 1982. Salinity and plant productivity. Pages 117–133 in M. Rechcigl, Ed. CRC handbook of agricultural productivity. Vol. 1. Plant productivity. CRC Press, Boca Raton, FL.

Carter, M. R., Webster, G. R., and Cairns, R. R. 1979. Calcium deficiency in some Solonetzic soils of Alberta. J. Soil Sci. 30: 161–174.

Dasberg, S. and Nadler, A. 1988. Soil salinity measurements. Soil Use Manag. 4: 127–133.

Hogg, T. J. and Henry, J. L. 1984. Comparison of 1:1 and 1:2 suspensions and extracts with the saturation extract in estimating salinity in Saskatchewan soils. Can. J. Soil Sci. 64: 699–704.

Howard, D. D. and Adams, F. 1965. Calcium requirement for penetration of subsoils by primary cotton roots. Soil Sci. Soc. Am. Proc. 29: 558–562.

Janzen, H. H. and Chang, C. 1987. Cation nutrition of barley as influenced by soil solution composition in a saline soil. Can. J. Soil Sci. 67: 619–629.

Janzen, H. H. and Chang, C. 1988. Cation concentrations in the saturation extract and soil solution extract of soil salinized with various sulfate salts. Comm. Soil Sci. Plant Anal. 19: 405–430.

Kamphorst, A. and Bolt, G. H. 1978. Saline and sodic soils. Pages 171–191 in G. H. Bolt and M. G. M. Bruggenwert, Eds. Soil chemistry. A. Basic elements. 2nd ed. Elsevier Scientific, Amsterdam.

Knudsen, D., Peterson, G. A., and Pratt, P. F. 1982. Lithium, sodium, and potassium. Pages 225–246 in A. L. Page, Ed. Methods of soil analysis. Part 2. Chemical and microbiological properties. Agronomy No. 9, 2nd ed. American Society of Agronomy, Madison, WI.

Lanyon, L. E. and Heald, W. R. (1982. Magnesium, calcium, strontium, and barium.

Pages 247–262 in A. L. Page, Ed. Methods of soil analysis. Part 2. Chemical and microbiological properties. Agronomy No. 9, 2nd ed. American Society of Agronomy, Madison, WI.

Maas, E. V. and Hoffman, G. J. 1977. Crop salt tolerance — current assessment. J. Irrig. Drain. Div. Proc. Am. Soc. Civil Eng. 103: 115–134.

Nieto, K. F. and Frankenberger, W. T. 1985. Single ion chromatography. I. Analysis of inorganic anions in soils. Soil Sci. Soc. Am. J. 49: 587–592.

Rhoades, J. D. 1982. Soluble salts. Pages 167–179 in A. L. Page, Ed. Methods of soil analysis. Part 2. Chemical and microbiological properties. Agronomy No. 9, 2nd ed. American Society of Agronomy, Madison, WI.

Rhoades, J. D. 1990. Determining soil salinity from measurements of electrical conductivity. Comm. Soil Sci. Plant Anal. 21: 1887–1926.

Richards, L. E., Ed. 1954. Diagnosis and improvement of saline and alkali soils. U.S. Salinity Laboratory, U.S. Department of Agriculture Handbook 60.

Soil Science Society of America. 1984. Glossary of soil science terms. Soil Sci. Soc. Am. Inc., Madison, WI.

Sposito, G. 1977. The Gapon and the Vanselow selectivity coefficients. Soil Sci. Soc. Am. J. 41: 1205–1206.

Sposito, G. and Coves, J. 1988. SOILCHEM: a computer program for the calculation of chemical speciation in soils. University of California, Riverside and Berkeley.

Van Genuchten, M. T. and Hoffman, G. J. 1984. Analysis of crop tolerance data. Pages 258–271 in I. Shainberg and J. Shalhevet, Eds. Soil salinity under irrigation. Ecological studies. Vol. 51. Springer-Verlag, Berlin.

Chapter 19
Ion Exchange and Exchangeable Cations

W. H. Hendershot and H. Lalande

McGill University
Ste. Anne de Bellevue, Quebec, Canada

M. Duquette

SNC-Lavalin
Montreal, Quebec, Canada

19.1 INTRODUCTION

Soils possess electrostatic charge as a result of atomic substitution in the lattices of soil minerals (permanent charge) and as a result of hydrolysis reactions on broken edges of the lattices and the surfaces of oxides, hydroxides, hydrous oxides, and organic matter (pH-dependent charge). These charges attract counter (exchangeable) ions and form the exchange complex. The principle of the methods used to measure exchangeable ions is to saturate the exchange complex with some ion that forces the exchangeable ions already present on the charged surfaces into solution (law of mass action). Exchange capacity can then be calculated as the sum of the individual cations displaced from the soil (summation method); or the ion used to saturate the exchange complex, termed the index ion, can be displaced with a concentrated solution of a different salt and the exchange capacity calculated as the amount of the index ion displaced (displacement method).

The cation-exchange capacity (CEC) is a measure of the amount of ions that can be adsorbed, in an exchangeable fashion, on the negative charge sites of the soil (Bache 1976). The results are commonly expressed in centimoles of positive charge per kilogram of soil (cmol $[+]$ kg^{-1}). Anion-exchange capacity (AEC) is expressed in terms of negative charge (cmol $[-]$ kg^{-1}). In most Canadian soils, CEC is much greater than AEC; as a result, in most routine soil analysis, we only measure CEC and exchangeable cations.

The measurement of CEC is complicated by: (1) errors due to the dissolution of soluble salts, $CaCO_3$ and gypsum ($CaSO_4 \cdot H_2O$); (2) specific adsorption of K and NH_4 in the interlayer position in vermiculites and micas (including illite or hydrous mica); and (3) the specific adsorption of trivalent cations such as Al^{3+} or Fe^{3+} on the surface of soil particles. In general, the errors can be reduced by using a method of CEC determination that employs

Soil Sampling and Methods of Analysis, M. R. Carter, Ed.,
Canadian Society of Soil Science. © 1993 Lewis Publishers.

reagents of similar concentration and pH to those of the soil to be analyzed. For this reason a method buffered at pH 7.0 or 8.2 using relatively high concentrations of saturating and extracting solutions will decrease errors due to dissolution of $CaCO_3$ and gypsum in soils from arid regions (Thomas 1982). In acidic soils, solutions buffered at pH 7.0 or 8.2 are less effective in replacing trivalent cations and an unbuffered method will provide a better estimate of the CEC and exchangeable cations.

Methods using a solution at a buffered pH (Section 19.4) are commonly used with agricultural soils providing a measurement that is independent of recent fertilization and liming practices. For forest soils and other low-pH soils, it is often preferable to measure CEC at the pH of the soil (Section 19.2), thus providing a more accurate measure of exchangeable cations and CEC under field conditions.

Soils containing appreciable amounts of amorphous materials (e.g., podzols and some brunisols) will show order of magnitude changes in CEC and AEC as a result of acidification or liming. The method for measuring pH-dependent CEC and AEC (Section 19.3) is provided for those who wish to study the variation in charge properties as a function of pH. The method provides more useful information than does the potentiometric titration method. Although both can be used to give an estimate of the point of zero charge, the pH-dependent CEC and AEC methods also provide a measure of the absolute amount of exchange capacity at any pH.

19.2 EXCHANGEABLE CATIONS AND EFFECTIVE CEC BY THE $BaCl_2$ METHOD (HENDERSHOT AND DUQUETTE 1986)

The $BaCl_2$ method provides a rapid means of determining the exchangeable cations and the "effective" CEC of a wide range of soil types. In this method CEC is calculated as the sum of exchangeable cations (Ca, Mg, K, Na, Al, Fe, and Mn). The method is particularly applicable in forestry or studies of environmental problems related to soils where information on the CEC at the pH of the soil in the field is of prime importance. In soils with large amounts of pH-dependent cation exchange sites the value measured at pH 7 will be considerably higher than that measured by this method. Problems may arise if this method is used with saline soils containing very high levels of SO_4, since $BaSO_4$ will precipitate.

This method has been compared to other methods of determining the CEC at the soil pH and provides comparable results (Hendershot and Duquette 1986, Ngewoh et al. 1989). Barium is a good flocculant and is able to displace trivalent cations. The relatively low ionic strength of the equilibrating solution causes a smaller change in pH than do more concentrated salt solutions. This method is simple and rapid; however, it is recommended that exchangeable iron and manganese be measured, since they may be more abundant in some acidic soils than other commonly considered cations such as potassium and sodium.

19.2.1 MATERIALS AND REAGENTS

1   Centrifuge tubes (50 mL) with screw caps and low-speed centrifuge.

2   End-over-end shaker.

3   Barium chloride, 0.1 *M*: dissolve 24.43 g of $BaCl_2 \cdot 2H_2O$ with distilled/deionized (DD) water and make to volume in a 1000-mL volumetric flask.

4   Standards of Ca, Mg, K, Na, Al, Fe, and Mn are prepared using atomic absorption reagent-grade liquid standards of 1000 mg L^{-1}. The matrix in the standards must correspond to the $BaCl_2$ concentration of the analyzed sample (diluted or nondiluted matrix).

5   Lanthanum solution, 100 g L^{-1}: dissolve 53.49 g of $LaCl_3 \cdot 7H_2O$ in a 200-mL volumetric flask and make to volume.

19.2.2 PROCEDURE

1   Weigh out about 0.5 g of air-dried (<2 mm) organic soil or fine textured soil to 3.0 g of coarse textured soil into a 50-mL centrifuge tube. Record the exact weight of soil used to the nearest 0.001 g. Include blanks, duplicates, and quality control (QC) samples.

2   Add 30.0 mL of 0.1 *M* $BaCl_2$ to each tube and shake slowly on an end-over-end shaker (15 rpm) for 2 h.

3   Centrifuge (15 min, 700 × *g*) and filter the supernatant (SN) with Whatman® No. 41 filter paper.

4   Analyze the following cations in the SN solution with an atomic absorption spectrophotometer (AAS) or other suitable instrument: Ca, Mg, K, Na, Al, Fe, and Mn. Dilution (10- or 100-fold) is usually required for Ca, K, and Mg. The addition of 0.1 mL of La solution to a 10-mL aliquot of diluted extract is required for the determination of Ca and Mg by AAS (for detailed instructions on this and other aspects of analysis refer to the manual for your AAS).

5   If desired, the pH of the equilibrating solution can be measured on a *separate* aliquot of the $BaCl_2$ solution. Leakage of K from the KCl salt bridge of the pH electrode is significant and therefore you cannot use the same aliquot for K analysis as for pH measurement.

19.2.3 CALCULATIONS

1   Exchangeable cations.

$$M^+ \text{ cmol } (+) \text{ kg}^{-1} = C \text{ cmol } (+) \text{ L}^{-1} \times (0.03 \text{ L/wt. soil g}) \times 1000 \text{ g kg}^{-1} \times DF \quad (19.1)$$

where,
M^+: concentration of an adsorbed cation, cmol $(+)$ kg^{-1}.
C: concentration of the same cation measured in the $BaCl_2$ extract [cmol $(+)$ L^{-1}].
DF: dilution factor, if applicable.

2   Effective CEC.

$$\text{Effective CEC cmol} (+) \text{ kg}^{-1} = \Sigma \text{ M}^+ \text{ cmol} (+) \text{ kg}^{-1} \tag{19.2}$$

19.2.4 Comments

A large amount of a soil similar to the samples being analyzed should be kept as an indicator of the variability of results over time; duplicate subsamples of this QC sample should be run with each batch of samples measured. Failure of the QC to fall within acceptable limits means that the whole batch should be reanalyzed. Analysis of QC samples is also useful to verify that samples analyzed by different people in the same laboratory are comparable, and that results do not change from one year to another or from one batch of chemicals to another.

For the sake of simplicity AAS standards are usually made up by diluting 1000 mg L^{-1} concentrate to lower mg L^{-1} values suitable for the range of the instrument being used. Calibrate the machine using the corresponding cmol($+$) L^{-1} value; the conversion values are as follows:

$$1 \text{ mg } L^{-1} \text{ Ca } = 4.99 \times 10^{-3} \text{ cmol}(+) \text{ L}^{-1};$$
$$1 \text{ mg } L^{-1} \text{ Mg } = 8.23 \times 10^{-3} \text{ cmol}(+) \text{ L}^{-1};$$
$$1 \text{ mg } L^{-1} \text{ K } = 2.56 \times 10^{-3} \text{ cmol}(+) \text{ L}^{-1};$$
$$1 \text{ mg } L^{-1} \text{ Na } = 4.35 \times 10^{-3} \text{ cmol}(+) \text{ L}^{-1};$$
$$1 \text{ mg } L^{-1} \text{ Al } = 11.12 \times 10^{-3} \text{ cmol}(+) \text{ L}^{-1};$$
$$1 \text{ mg } L^{-1} \text{ Fe } = 5.37 \times 10^{-3} \text{ cmol}(+) \text{ L}^{-1};$$
$$1 \text{ mg } L^{-1} \text{ Mn } = 3.64 \times 10^{-3} \text{ cmol}(+) \text{ L}^{-1}.$$

19.3 pH-DEPENDENT AEC-CEC
(FEY AND LeROUX 1976, JAMES 1984)

In the literature, the method of Fey and LeRoux (1976) is often cited in research on pH-dependent CEC and AEC. The method is time consuming because of the multiple saturation and pH adjustment steps. The method of James (1984) is preferred because there are fewer steps and therefore it is faster with less chance of errors due to contamination or loss of soil. Originally this procedure was written using NH_4 as the saturating cation, which can result in the collapse of expanded clays such as vermiculite. The step-by-step procedure of James was retained and the same saturating and replacing salts as used by Fey and LeRoux were adopted. The only disadvantage with the modified procedure is that it is more difficult to obtain an even distribution of pH values than with the method of Fey and LeRoux, but this can be corrected by rerunning the analysis and adjusting the amounts of HNO_3 or $Ca(OH)_2$ added.

19.3.1 MATERIALS AND REAGENTS

1   Centrifuge tubes (50 mL) with screw caps and low-speed centrifuge.

2   Vortex centrifuge tube mixer and end-over-end shaker.

3   Calcium nitrate, 0.05 *M*: dissolve 23.62 g of calcium nitrate tetrahydrate [$Ca(NO_3)_2 \cdot 4H_2O$] with DD water in a 2000-mL volumetric flask.

4   Nitric acid, 0.1 *M*: dilute 6.3 mL of concentrated nitric acid (HNO_3) with DD water in a 1000-mL volumetric flask.

5   Calcium hydroxide, 0.05 *M*: dissolve 3.70 g of calcium hydroxide [$Ca(OH)_2$] with DD water in a 1000-mL volumetric flask, then filter through a Whatman® 41 filter (a prefiltration step can be done using a glass microfiber filter (Whatman® GF/C)).

6   Calcium nitrate, 0.005 *M*: dilute 200 mL of 0.05 *M* $Ca(NO_3)_2$ solution with DD water in a 2000-mL volumetric flask.

7   Potassium chloride, 1.0 *M*: dissolve 149.12 g of potassium chloride (KCl) with DD water in a 2000-mL volumetric flask.

8   Lanthanum solution, 100 g L^{-1}: dissolve 53.49 g of $LaCl_3 \cdot 7H_2O$ with DD water in a 200-mL volumetric flask and make to volume.

19.3.2 PROCEDURE

1   Weigh twenty 50-mL centrifuge tubes to the nearest 0.001 g (one set of 20 tubes for each soil sample to be analyzed).

2   Add 1.0 g subsamples of air-dried <2 mm soil and record the weight to the nearest 0.001 g. The analysis is done in duplicate with one pair of QC samples per batch. If moist soil is used, start by weighing out four additional samples into small beakers and air dry to determine the weight of moist soil to be used (1 g air-dry equivalent).

3   Add 25 mL 0.05 *M* $Ca(NO_3)_2$ solution, cap the tubes, and shake 1 h using an end-over-end shaker (15 rpm).

4   Centrifuge (10 min, 700 × *g*) and discard SN by decantation. Be careful to avoid loss of soil during decantation.

5   Add a new 25-mL aliquot of 0.05 *M* $Ca(NO_3)_2$ solution to each tube. Then add 0, 0.25, 0.5, 1.0, or 2.5 mL of 0.1 *M* HNO_3 to tubes in duplicate, and finally add 0.25, 0.5, 1.0 or 2.5 mL of 0.05 *M* $Ca(OH)_2$ to the remaining tubes in duplicate. Add 1.0 mL of 0.1 *M* HNO_3 or 0.05 *M* $Ca(OH)_2$ to the QC sample. A vortex mixer is useful to resuspend the soil after addition of the solution.

6   Cap and shake overnight on an end-over-end shaker.

7   Centrifuge (10 min, 700 × *g*) and discard SN.

8   Resuspend the soil in 25 mL of 0.005 *M* $Ca(NO_3)_2$, centrifuge (10 min, 700 × *g*), and discard SN.

9   Repeat step 8, but measure pH of SN and *keep it* for the analysis of Ca and NO_3 (after 100-fold dilution with DD water). Weigh tubes plus the soil and the interstitial soil solution.

10   Add 25 mL of 1.0 *M* KCl, shake 1 h, and centrifuge (10 min, 700 × *g*).

11   *Keep this SN* for determination of displaced Ca and NO_3. Dilute this KCl extract tenfold with DD water.

12   Measure Ca by AAS in the tenfold diluted KCl extract (saved in step 11) and in the 0.005 *M* $Ca(NO_3)_2$ equilibration solution (saved in step 9). To prevent interference from PO_4 during the measurement of Ca, the solution should contain 1000 mg L^{-1} La (0.1 mL of 100 mg L^{-1} in a 10-mL aliquot: refer to your AAS operating manual).

13   Measure NO_3 in the undiluted KCl extract (saved in step 11) and in the diluted 0.005 *M* $Ca(NO_3)_2$ equilibration solution (saved in step 9).

19.3.3 CALCULATIONS

1   Residual Ca and NO_3

   a. Volume of interstitial solution: subtract the weight of the empty tube with the soil (step 2) from weight measured in step 9 to calculate weight of residual 0.005 *M* $Ca(NO_3)_2$ solution (Vol_{res}). Assume 1 g equals 1 mL.

   b. Residual amount of Ca and NO_3 (Ca_{res} and $NO_{3\ res}$);

$$Ca_{res} \text{ (mmole)} = Vol_{res} \text{ (mL)} \times Ca_{sol} \text{ (m}M) \times 0.001 \text{ (L mL}^{-1}) \times DF \qquad (19.3)$$

$$NO_{3\ res} \text{ (mmole)} = Vol_{res} \text{ (mL)} \times NO_{3sol} \text{ (m}M) \times 0.001 \text{ (L mL}^{-1}) \times DF \qquad (19.4)$$

   where Ca_{sol} and NO_{sol} are the measured concentrations of calcium and nitrate in the 0.005 *M* $Ca(NO_3)_2$ wash solution saved in step 9 (units in m*M*). DF: dilution factor if applicable.

2   Total amount of calcium and nitrate (Ca_t $NO_{3\ t}$) in the KCl extract (including the residual):

$$Ca_t \text{ (mmole)} = Ca_{KCl} \text{ (m}M) \times 25 \text{ (mL)} \times 0.001 \text{ (L mL}^{-1}) \times DF \qquad (19.5)$$

$$NO_{3\ t} \text{ (mmole)} = NO_{3KCl} \text{ (m}M) \times 25 \text{ (mL)} \times 0.001 \text{ (L mL}^{-1}) \times DF \qquad (19.6)$$

   where Ca_{KCl}, NO_{3KCl}: calcium and nitrate concentration (m*M*) in the KCl extract saved in step 11. DF: dilution factor if applicable.

3.   Calculation of the CEC and AEC:

   CEC cmol (+) kg^{-1} =   (19.7)

$(Ca_t - Ca_{res})$ (mmole) $\times$ 0.2 (cmol $(+)$ mmole^{-1}) $\times$ 1000 g kg^{-1})/wt. soil (g)

AEC cmol $(-)$ kg^{-1} = \hfill (19.8)

$(NO_{3\ t} - NO_{3\ res})$ (mmole) $\times$ 0.1 (cmol $(-)$ mmole^{-1}) $\times$ 1000 (g kg^{-1})/wt. soil (g)

19.4 EXCHANGEABLE CATIONS AND TOTAL EXCHANGE CAPACITY BY THE AMMONIUM ACETATE METHOD AT pH 7.0 (LAVKULICH 1981)

The method described here was developed by Lavkulich (1981) for standard analysis of a wide range of soil types. It involves fewer steps than some other similar methods, such as that of McKeague (1978). Problems with this approach to measuring exchangeable cations and CEC have been discussed extensively in the literature (Chapman 1965, Bache 1976, Rhoades 1982, Thomas 1982), but we agree with the conclusion of Thomas (1982) that "There is no evidence at the present time that cations other than NH_4^+ give results that are less arbitrary than those obtained using NH_4^+".

Errors due to the dissolution of $CaCO_3$ and gypsum will result in an excess of Ca^{2+} being extracted by NH_4^+ and a decrease in the amount of NH_4^+ retained due to competition between Ca^{2+} and NH_4^+ during equilibration in the saturating step. In soils containing these minerals, exchangeable Ca will be too high and total CEC too low. The former problem cannot easily be corrected (Thomas 1982); however, more accurate measurement of CEC in this type of soil can be obtained by using the method described by Rhoades (1982).

Fixation of K^+ and NH_4^+ in phyllosilicates can result in either an over- or underestimation of exchangeable K^+ when NH_4^+ is used as an extractant, depending on whether the NH_4^+ moves through the interlayer positions replacing the K^+, or whether it causes the collapse of the edges, preventing further exchange.

Compared to the other methods presented in this chapter, this method uses a larger sample size which helps to decrease the sample to sample variability. Another advantage of this procedure is that there are no decantation steps that can cause the loss of sample, particularly in the case of organic soils.

The method described below can be used to measure either exchangeable cations and CEC or just exchangeable cations. In the latter case the sum of exchangeable cations (including Al) could be used as an estimate of CEC. Due to the high pH of the extracting solution, the amount of Al measured will usually be lower than that displaced by $BaCl_2$ or KCl.

19.4.1 MATERIALS AND REAGENTS

1   Centrifuge tubes: 100-mL centrifuge tubes and stoppers.

2   Reciprocal shaker.

3   Buchner funnels (55-mm diameter) and 500-mL filtering flasks connected to low-pressure vacuum line.

4   Ammonium acetate, 1 *M*: dissolve 77.08 g of NH_4OAc with DD water and
     make to volume in a 1000-mL volumetric flask. Adjust pH to 7.0 with am-
     monium hydroxide or acetic acid.

5   Isopropanol.

6   Potassium chloride, 1 *M*: dissolve 74.6 g of KCl with DD water and make to
     volume in a 1000-mL volumetric flask.

7   Standard ammonium solution, 200 mg L^{-1} N: dissolve 0.238 g of $(NH_4)_2SO_4$
     (dried for 3–4 h at 40°C) in 100 mL of DD water and then dilute to volume
     in a 250-mL volumetric flask. Prepare diluted standards (10, 20, 40, and 80
     mg L^{-1} from the 200-mg L^{-1} stock).

8   Prepare Ca, Mg, K, and Na standards using 1 *M* NH_4OAc as the matrix.

19.4.2 PROCEDURES

19.4.2.1 *For Exchangeable Cations*

1   a. *For samples low in organic matter:* weigh out 10.000 g of soil into a 100-
        mL centrifuge tube.
     b. *For samples high in organic matter:* weigh out 5.000 or 2.000 g.
     c. Prepare a blank and include a QC sample.

2   Add 40 mL of 1 *M* NH_4OAc to the centrifuge tube. Stopper the tube and
     shake for 5 min on a reciprocal shaker (115 rpm). Remove tubes from shaker,
     agitate to rinse down soil adhering to the sides of the tube, and let stand
     overnight.

3   Shake tube again for 15 min. Prepare Buchner funnels with Whatman® No.
     42 filter paper and place them above 500-mL filtering flasks.

4   Transfer contents of the tube to the funnel with suction applied. Rinse the
     tube and the stopper with 1 *M* NH_4OAc from a wash bottle.

5   Wash the soil in the Buchner funnel with four 30-mL portions of 1 *M* NH_4OAc.
     Let each portion drain completely before adding the next, but do not allow
     the soil to become dry or cracked.

6   Transfer the leachate to a 250-mL volumetric flask; rinse the filtering flask
     with 1 *M* NH_4OAc and make up to volume with 1 *M* NH_4OAc. Mix well and
     save a portion of the extract for analysis of Ca, Mg, K, and Na. Keep samples
     refrigerated prior to analysis.

19.4.2.2 *For Total Exchange Capacity (CEC)*

7   Replace the funnels containing the ammonium-saturated soil onto the filtering flasks. To remove the residual NH_4OAc from the soil, wash the soil in the Buchner funnel with three 40-mL portions of isopropanol, again letting each portion drain completely before adding the next. (Turn off the suction after the last washing before the soil dries out.) Discard the isopropanol washings and rinse the flask *well* with tap water followed by DD water.

8   Replace the funnels onto the flasks and leach the soil with four 50-mL portions of 1 *M* KCl, again letting each portion drain completely before adding the next. Transfer the leachate to a 250-mL volumetric flask. Rinse the filtering flask into the volumetric flask with DD water and make up to volume with DD water. Mix well and save a portion of the extract for analysis of NH_4 by autoanalyzer.

19.4.3 CALCULATIONS

1   Exchangeable cations:

$$M^+ \text{ cmol }(+) \text{ kg}^{-1} = C \text{ cmol }(+) \text{ L}^{-1} \times (0.25 \text{ L/wt soil g}) \times 1000 \text{ g kg}^{-1} \qquad (19.9)$$

where,
M^+: concentration of adsorbed cation, $\text{cmol}(+) \text{ kg}^{-1}$
C: concentration of cation in the NH_4OAc extract $[\text{cmol}(+) \text{ L}^{-1}]$

Note: see Section 19.2.4 for conversion of mg L^{-1} to $\text{cmol}(+) \text{ L}^{-1}$.

2   CEC:

$$\text{CEC cmol }(+) \text{ kg}^{-1} = \qquad\qquad\qquad (19.10)$$
$$(\text{mg L}^{-1} \text{ N} \times (1 \text{ cmol }(+)/140 \text{ mg}) \times (0.25 \text{ L/wt soil g}) \times 1000 \text{ g kg}^{-1}$$

REFERENCES

Bache, B. W. 1976. The measurement of cation exchange capacity of soils. J. Sci. Food Agric. 27: 273–280.

Chapman, H. D. 1965. Cation exchange capacity. Pages 891–901 in C. A. Black et al., Eds. Methods of soil analysis. Agronomy No. 9. American Society of Agronomy, Madison, WI.

Fey, M. V. and LeRoux, J. 1976. Electric charges on sesquioxidic soil clays. Soil Sci. Soc. Am. J. 40: 359–364.

Hendershot, W. H. and Duquette, M. 1986. A simple barium chloride method for determining cation exchange capacity and exchangeable cations. Soil Sci. Soc. Am. J. 50: 605–608.

James, B. D. 1984. Personal communication.

Lavkulich, L. M. 1981. Methods manual, pedology laboratory. Department of Soil Science, University of British Columbia, Vancouver.

McKeague, J. A. 1978. Manual on soil sampling and methods of analysis. 2nd ed. Canadian Society of Soil Science.

Ngewoh, Z. S., Taylor, R. W., and Shuford, J. W. 1989. Exchangeable cations and CEC determinations of some highly weathered soils. Commun. Soil Sci. Plant Anal. 20: 1833–1855.

Rhoades, J. D. 1982. Cation exchange capacity. Pages 149–157 in A. L. Page et al., Eds. Methods of soil analysis. Agronomy No. 9. 2nd ed. American Society of Agronomy, Madison, WI.

Thomas, G. W. 1982. Exchangeable cations. Pages 159–165 in A. L. Page et al., Eds. Methods of soil analysis. Agronomy No. 9. 2nd ed. American Society of Agronomy, Madison, WI.

Chapter 20
Carbonates

Tee Boon Goh

University of Manitoba

Winnipeg, Manitoba, Canada

R. J. St. Arnaud and A. R. Mermut

University of Saskatchewan

Saskatoon, Saskatchewan, Canada

20.1 INTRODUCTION

Inorganic carbon occurs in soils commonly as the carbonate minerals calcite ($CaCO_3$), dolomite [$CaMg(CO_3)_2$] and magnesian calcites ($Ca_{1-x}Mg_xCO_3$). Other less common forms are aragonite ($CaCO_3$) and siderite ($FeCO_3$). Carbonate in soils can be of primary (inherited from parent material) or secondary (pedogenic) origin. Secondary carbonates are usually aggregates of silt- and clay-sized calcite crystals which are easily identified in grain mounts. Larger crystals of calcite or dolomite are of primary origin (Doner and Lynn 1989). Once only routinely reported by sedimentologists, the qualitative and quantitative determinations, especially of Ca and Mg carbonates, are useful in studies of soil genesis and classification, and micronutrient and phosphorus sorption. Furthermore, the soil carbonate impacts on root and water movement, soil pH (Nelson 1982), and the nature of the exchange complex (St. Arnaud and Herbillon 1973).

A variety of methods can be used for the determination of calcite, dolomite, and magnesian calcite in soils. Chemical determinations of carbonates include the use of empirical standard curves relating pH to known carbonate content as well as the measurement of CO_2 evolved when treated with acid. These permit a measurement of inorganic C from carbonates in soil. Most procedures express the carbonate content as the calcium carbonate equivalent. Further analysis of the Ca and Mg content provides a means of estimating the kind of inorganic carbonate in soil.

20.2 CARBONATE CONTENT BY USE OF EMPIRICAL STANDARD CURVE (LOEPPERT ET AL. 1984)

The analysis of suitable for rapid and routine analysis of large numbers of samples. A known quantity of acetic acid is consumed by reaction with carbonates and the final pH following complete dissolution of $CaCO_3$ is recorded for each sample. Calcium carbonate content is

Soil Sampling and Methods of Analysis, M. R. Carter, Ed.,

Canadian Society of Soil Science. © 1993 Lewis Publishers.

determined empirically from a standard curve relating pH to weight of $CaCO_3$ according to the equation

$$pH = K + n \log [CaCO_3/(T - CaCO_3)] \qquad (20.1)$$

in which K and n are constants and T is the total amount of $CaCO_3$ which could be completely neutralized by the quantity of acetic acid used.

20.2.1 REAGENTS AND EQUIPMENT

1   Calcite standard: pure calcite such as Iceland spar calcite ground to <270 mesh in size is suitable.

2   Acetic acid, 0.400 mol L^{-1}: dilute 400 mL of 1 mol L^{-1} CH_3COOH to the mark in a 1-L volumetric flask with deionized/distilled water.

3   pH meter: a digital pH meter is recommended.

4   Ultrasonic probe: a suitable model with a probe that can be inserted into a 50-mL centrifuge tube.

20.2.2 PROCEDURE

20.2.2.1 Standard Curve

1   Weigh accurately, Iceland spar calcite, ranging from 5 to 500 mg into separate 50-mL polypropylene centrifuge tubes.

2   Add 25 mL of acetic acid, 0.400 mol L^{-1}, which is sufficient to exactly neutralize all the $CaCO_3$ in the largest sample of the standard (500 mg $CaCO_3$ in this case) according to the reaction

$$CaCO_3 + 2CH_3COOH \rightarrow Ca^{2+} + 2CH_3COO^- + H_2O + CO_2 \qquad (20.2)$$

3   Shake tubes intermittently for 8 h. Allow tubes to stand overnight with caps loosened to allow escape of CO_2.

4   Degas using ultrasonic probe at low setting to prevent excessive splashing for approximately 30 s.

5   Centrifuge and record pH of the supernatant to two decimal places after 4 min.

6   Plot standard curve of pH vs. $\log [CaCO_3/(T - CaCO_3)]$. Note: T is the weight of $CaCO_3$ (mg) used to exactly neutralize the volume of acetic acid used

and will vary if either the volume or the concentration of acetic acid is changed.

20.2.2.2 Calcium Carbonate Contents of Soil Samples

1   Weigh accurately, up to 2 g soil (<100 mesh size) containing up to 400 mg $CaCO_3$. Reduce the soil sample weight if the carbonate content exceeds 20%.

2   Repeat steps (2) through (5) as for the standard curve above.

20.2.3 Calculations

From the pH value recorded, determine the value of log $[CaCO_3/(T - CaCO_3)]$ using the standard curve and calculate the weight of $CaCO_3$ (mg) in the soil sample.

$$\text{the \% CaCO3 equivalent} = \frac{\text{mg } CaCO_3}{\text{mg sample}} \times 100 \qquad (20.3)$$

20.2.4 Comments

The total carbonate content so determined is expressed as the calcium carbonate equivalent. If dolomite is present in the soil sample, increased reaction times may be required for the dissolution to go to completion. The accuracy of results is influenced by (1) proton consumption by soil constituents, (2) acid-generating hydrolysis reactions during mineral decomposition, (3) high P_{CO2}, (4) volatilization of acetic acid, and (5) errors in pH determination. These can be minimized by standard additions of Ca^{2+}, from a solution of $CaCl_2$, to all samples and standards, grinding of samples to increase reactivity of sand-sized carbonates, and reduction of reaction time between acetic acid and other minerals, covers to reduce loss of acetic acid, degassing CO_2 and reduction of suspension effects in pH reading (Loeppert et al. 1984).

20.3 APPROXIMATE GRAVIMETRIC METHOD (ALLISON AND MOODIE 1965, RAAD 1978)

A preweighed soil sample containing carbonates is reacted with acid. The resultant loss in weight from CO_2 released is used to calculate the calcium carbonate content. Calcite and dolomite cannot be accurately distinguished, but a fair estimate of the proportion of dolomite in the sample can be obtained by checking the weight loss with time.

20.3.1 REAGENTS

1   Hydrochloric acid (HCl), 4 mol L^{-1}.

2   Hydrochloric acid (HCl)-ferrous chloride ($FeCl_2 \cdot 4H_2O$) reagent: dissolve 3 g of $FeCl_2 \cdot 4H_2O$ per 100 mL of 4 mol L^{-1} HCl immediately before use.

20.3.2 PROCEDURE

1   Weigh a stoppered, 50-mL Erlenmeyer flask containing 10 mL of the HCl-$FeCl_2$ reagent.

2   Transfer a 1- to 10-g soil sample containing between 100 to 300 mg of carbonate to the flask gradually to avoid excessive frothing.

3   After effervescence has subsided, replace the stopper loosely and allow the carbonate to further decompose in the mixture for about 30 min with occasional swirling to displace any accumulated CO_2. Replace the stopper and weigh the flask plus contents.

4   Repeat step (3) until the change in weight of the flask and its contents is no more than 2 to 3 mg. The reaction is usually complete within 2 h.

20.3.3 Calculations

Weight of CO_2 lost from carbonates = difference in initial and final weights of (flask + stopper contents)

$$\% \ CaCO_3 \ \text{equivalent} = \frac{g \ CO_2 \ \text{lost}}{g \ \text{soil}} \times 227.3 \qquad (20.4)$$

20.3.4 Comments

When dolomite is present, it is considerably less reactive to cold HCl. Therefore, if the weight is observed to decrease markedly after 30 min, some dolomite is present. The use of acid containing $FeCl_2$ as an antioxidant eliminates errors caused by oxidizing interferences due to MnO_2 in soil. The accuracy of this method depends upon the accuracy of weighing and the degree to which CO_2 retained in solution is compensated for by loss of water vapor.

20.4 QUANTITATIVE GRAVIMETRIC METHOD (U.S. DEPARTMENT OF AGRICULTURE SOIL CONSERVATION SERVICE 1967)

The loss in weight of a soil sample is measured accurately after reaction between carbonates in the soil and acid. In this method, the loss of water vapor evolved with CO_2 is eliminated by a trap containing anhydrone. The addition of a CO_2 trap to the apparatus is an alternative to the method, by measuring the gain rather than loss in weight, and provides a check against any leaks in the connections to the glassware. With several units in operation the method is quite rapid and accurate.

A. Glass wool plugs
B. Anhydrone, $Mg(ClO_4)_2$
C. Vial containing 6 mol L^{-1} HCl
D. Stopcock
E. Stopcock
F. 125-mL Erlenmeyer flask
G. Stopper
H. U-tube
I. Calcium chloride tube (shortened)
J. Glass tube

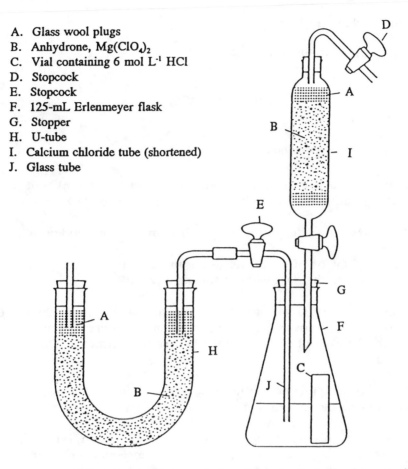

FIGURE 20.1. Apparatus for accurate quantitative determination of calcium carbonate equivalent. (After Raad, A. A., Manual on Soil Sampling and Methods of Analysis, 2nd ed. Can. Soc. Soil Sci., Ottawa, ON.)

20.4.1 Apparatus

The apparatus is assembled as depicted in Figure 20.1 for the weight loss method. A polyethylene drying tube packed with Ascarite II® to trap CO_2 can also be attached to the end of the gas train after stopcock D in the weight gain method.

20.4.2 REAGENTS

1 Hydrochloric acid (HCl), 6 mol L^{-1}.

2 Hydrochloric acid (HCl)-ferrous chloride ($FeCl_2 \cdot 4H_2O$) reagent: dissolve 3 g of $FeCl_2 \cdot 4H_2O$ per 100 mL of 6 mol L^{-1} HCl immediately before use.

3   Anhydrone [Mg(ClO$_4$)$_2$], drying agent.

4   Ascarite II®: 20–30 mesh, optional.

20.4.3 PROCEDURE

20.4.3.1 *Weight Loss Method*

1   Weigh a 1- to 10-g sample of oven-dried soil (<100 mesh) containing less than 1 g CaCO$_3$ equivalent into a 125-mL Erlenmeyer flask.

2   Wash down the sides of the flask with 10 mL of deionized/destilled water.

3   Transfer 7 mL of HCl-FeCl$_2$ reagent into vial C (Figure 20.1) and place the vial upright in the flask without spilling any acid.

4   Moisten stopper G with glycerin, sprinkle it with a small amount of 180-mesh abrasive to overcome slipperiness, and assemble the apparatus as in Figure 20.1 without connecting the U-tube to stopcock E. Close stopcocks D and E.

5   Place the apparatus beside the balance and allow the temperature in the flask to equilibrate with that of the air in the balance.

6   Using tongs, place the apparatus on the weighing pan, open stopcock D, and record weight to the nearest 0.1 mg. Close stopcock D immediately. Weigh again after 10 min to ensure that weight has stabilized.

7   Open stopcock D and shake the flask to upset vial C, thus allowing the acid to react with the soil.

8   After 10 min attach the U-tube H to the apparatus, open stopcock E, and apply gentle suction at stopcock D at a rate of 5 to 10 bubbles per second at tube J to sweep out CO$_2$ with dry air. Shake the flask at 10-min intervals.

9   Stop the suction when the reaction is complete (usually 30 min; 1 h if dolomite is present). Close stopcocks D and E. Disconnect the U-tube H. Wait for 1 h and weigh apparatus and contents with stopcock D open. Check the weight after 10 min.

20.4.3.2 *Weight Gain Method*

1   Weigh drying tube containing Ascarite II® to the nearest 0.1 mg. Attach to the apparatus depicted in Figure 20.1 at stopcock D.

2   Proceed as described in the weight loss method, but apply suction at the end of the polyethylene drying tube so that the gas train passes through the CO$_2$ trap.

3   Disconnect from the CO_2 trap and weigh drying tube and contents.

20.4.4 Calculations

20.4.4.1 *Weight Loss Method*

$$\% \ CaCO_3 \ equivalent = \frac{(Initial \ weight, \ g \ - \ Final \ weight, \ g)}{sample \ weight, \ g} \times 227.3 \quad (20.5)$$

20.4.4.2 *Weight Gain Method*

$$\% \ CaCO_3 \ equivalent = \frac{(Final \ weight, \ g \ - \ Initial \ weight, \ g)}{sample \ weight, \ g} \times 227.3 \quad (20.6)$$

20.4.5 Comments

The results obtained by the two methods should agree within the limits of weighing error. Larger discrepancies may indicate leaks in the connections of the apparatus.

20.5 QUANTIFICATION OF CALCITE AND DOLOMITE (PETERSON ET AL. 1966, ROSTAD AND ST. ARNAUD 1970)

The citrate buffer method described by Raad (1978) is presented here. Calcite and dolomite are selectively dissolved in a citrate buffer solution, Ca and Mg in solution are determined, and the calcite content of the sample calculated. It is assumed that dolomite has a Ca:Mg molar ratio of 1:1 and the only sources of the Ca and Mg in the solution are calcite and dolomite (i.e., no magnesian calcite is present). The portion of dolomite dissolved in the citrate buffer is calculated from the Mg in solution; an equivalent amount of Ca is assigned to it, and the remaining Ca determines the calcite content of the sample. The total dolomite content of the sample is obtained by the difference between the total carbonate content previously determined in another subsample and the portion of carbonate from calcite. As a check of accuracy, the dolomite content of the sample can also be calculated from the Mg in solution. The method is useful if clay-sized dolomite is present in the sample.

20.5.1 REAGENTS

1   Citrate buffer: dissolve 64 g citric acid ($C_6H_8O_7$) in 1 L of deionized water. Titrate to pH 5.85 with concentrated NH_4OH.

2   Sodium chloride-ethanol: dissolve 58.5 g NaCl in 30% (v/v) ethanol and bring to 1 L with deionized water.

3   Sodium dithionite.

20.5.2 PROCEDURE

1   Weigh 50 to 500 mg oven-dried soil ground to pass a 100-mesh sieve into a 50-mL centrifuge tube.

2   Wash twice with NaCl-ethanol solution and discard washing.

3   Add 25 mL of citrate buffer solution and heat in a water bath at 80°C. Add approximately 0.5 g of sodium dithionite and continue heating with stirring for about 15 min.

4   Centrifuge and collect supernatant in a 250-mL volumetric flask. Wash the residue once with 25 mL of citrate buffer, centrifuge, and collect the supernatant. Make to volume with deionized water.

5   Determine Ca and Mg in solution by atomic absorption spectroscopy using standards made up in the same concentrations of citrate buffer and dithionite. Standards and sample solutions should contain 1 mg La mL^{-1} to minimize interference effects.

6   The total carbonate content is determined in another subsample by other quantitative methods (Section 20.2).

20.5.3 Calculations

If citrate-soluble Ca = X mmol, citrate-soluble Mg = Y mmol, and total carbonate = Z mmol,

the mmol calcite-Ca = X-Y

since 1 mmol calcite weighs 100 mg,   (20.7

$$\% \text{ calcite in sample} = \frac{(X-Y) \text{ mmol}}{\text{mg sample}} \times \frac{100 \text{ mg}}{\text{mmol}} \times 100$$

the mmol dolomite-CO$_3$ = Z − (X-Y)

mmol dolomite = $^1/_2$ (mmol dolomite-CO$_3$)   (20.8)

since 1 mmol dolomite weighs 184 mg,

$$\% \text{ dolomite in sample} = \frac{^1/_2[Z - (X-Y)] \text{ mmol}}{\text{mg sample}} \times \frac{184 \text{ mg}}{\text{mmol}} \times 100$$

Alternatively, the dolomite content can be calculated by

the mmol dolomite-Mg = Y   (20.9)

$$\% \text{ dolomite in sample} = \frac{Y \text{ mmol}}{\text{mg sample}} \times \frac{184 \text{ mg}}{\text{mmol}} \times 100$$

The % CaCO$_3$ equivalent of the dolomite present

$$= \frac{[Z - (X\text{-}Y)] \text{ mmol}}{\text{mg sample}} \times \frac{100 \text{ mg}}{\text{mmol}} \times 100 \quad (20.10)$$

20.5.4 Comments

The dolomite content calculated from dolomite-CO$_3$ should agree with that calculated from the Mg in solution unless some source of citrate-soluble Mg other than dolomite or magnesian calcite is present. If magnesian calcite is present, the calcite content of the sample will be underestimated. Therefore, identification and quantification of magnesian calcite (Chapter 69) should be conducted for more specialized research.

REFERENCES

Allison, L. E. and Moodie, C. D. 1965. Carbonate. Pages 1379–1396 in C. A. Black et al., Eds. Methods of soil analysis. Part 2. Chemical and microbiological properties. 1st ed. American Society of Agronomy, Madison, WI.

Doner, H. E. and Lynn, W. C. 1989. Carbonate, halide, sulfate and sulfide minerals. Pages 279–330 in J. B. Dixon and S. B. Weed, Eds. Minerals in soil environments. 2nd ed. Soil Sci. Soc. Am., Madison, WI.

Loeppert, R. H., Hallmark, C. T., and Koshy, M. M. 1984. Routine procedure for rapid determination of soil carbonates. Soil Sci. Soc. Am. J. 48: 1030–1033.

Nelson, R. E. 1982. Carbonate and gypsum. Pages 181–197 in A. L. Page et al., Eds. Methods of soil analysis. Part 2. Chemical and microbiological properties. 2nd ed. American Society of Agronomy, Madison, WI.

Peterson, G. W., Chesters, G., and Lee, G. B. 1966. Quantitative determination of calcite and dolomite in soils. J. Soil Sci. 17: 328–338.

Raad, A. A. 1978. Carbonates, Pages 86–98 in J. A. McKeague, Ed. Manual on soil sampling and methods of analysis. 2nd ed. Can. Soc. Soil Sci., Ottawa, ON.

Rostad, H. P. W. and St. Arnaud, R. J. 1970. Nature of carbonate minerals in two Saskatchewan soils. Can. J. Soil Sci. 50: 65–70.

St. Arnaud, R. J. and Herbillon, A. J. 1973. Occurrence and genesis of secondary magnesium-bearing calcites in soils. Geoderma 9: 279–298.

U.S. Department of Agriculture Soil Conservation Service. 1967. Calcium carbonate. Pages 28–30 in Soil survey investigations report No. 1. Soil survey laboratory methods and procedures for collecting soil samples. U.S. Government Printing Office, Washington, D.C.

Chapter 21
Total and Organic Carbon

H. Tiessen and J. O. Moir

University of Saskatchewan

Saskatoon, Saskatchewan, Canada

21.1 INTRODUCTION

Numerous methods are available for the determination of organic carbon (OC), all of which operate on one of three basic principles:

1   Wet oxidation of OC in an acid dichromate solution followed by back titration of the remaining dichromate with ferrous ammonium sulfate and a suitable indicator, or by photometric determination of Cr^{III}.

2   Wet oxidation of OC in an acid dichromate solution and collection and determination of the evolved CO_2. If no pretreatment of calcareous soils is used, carbonates dissolved by the acid medium will be included in the CO_2 determination to give a measure of total C.

3   Dry oxidation of OC in a furnace, followed by direct determination or collection and determination of the evolved CO_2. At high furnace temperatures carbonates will also decompose, giving a measure of total C.

Limitations and applicability of methods in these three categories are discussed below:

21.1.1 Titrimetric Dichromate Redox Methods

Dichromate redox methods potentially suffer from a number of interferences and low OC recoveries. One of the first versions was introduced by Schollenberger (1927) as a rapid, approximate method, but during subsequent years of widespread use this aspect has often been ignored. In one modification of the method (Walkley and Black 1934), OC is oxidized solely by the action of acid dichromate, heated by the exothermic mixing of aqueous dichromate and concentrated sulfuric acid. Oxidation of OC by this method is notoriously incomplete due to the low temperatures ($\approx 120°C$) reached, but can easily be improved by supplementary heating on a hot plate or block digestor (Schollenberger 1945, Nelson and Sommers 1975). The length and intensity of external heating affect OC recoveries, so that conditions must be closely controlled. Correction factors for incomplete oxidation are an integral part of these methods (Walkley and Black 1934), but these factors are not valid across different soils (Gillman et al. 1986). Elemental carbon, e.g., charcoal, is to a large extent excluded (Metson et al. 1979), which may be desirable for some studies.

Upon oxidation, OC is calculated from the titer of ferrous ammonium sulfate obtained in the back titration of unused dichromate. This assumes a specific oxidation state of the C in the sample, and published methods all assume that C^0 is oxidized to C^{IV}. On average, OC contained in soils is near 0 oxidation state, but, if specific fractions of soil organic matter are to be analyzed, different states may occur. Carbohydrates have an oxidation state of C^0, while C in aromatic compounds has an oxidation state below zero.

The acid dichromate digestion solution decomposes at temperatures above 150°C, limiting the temperature that can be employed in any redox method (Charles and Simmons 1986). Clays have been reported to cause decomposition of dichromate, leading to overestimated OC values (Metson et al. 1979).

The reduction of dichromate is also affected by the presence of other redox-active components of the soil. The presence of Fe^{II}, Mn^{II}, and Cl^- causes an overestimation of soil OC, since all three are oxidized during the digestion. Higher oxides of Mn cause an underestimation of OC. With careful work many of these interferences can be eliminated. Thorough air drying of soil samples for a few days will oxidize most Fe^{II}, but oxidation of Mn^{II} may be slow at acid or neutral pH, and hydromorphic soils may present problems. Higher oxides of Mn are usually negligible, particularly since crystalline minerals are virtually inactive, and only fresh pedogenic precipitates would interfere seriously with the method. In soils or sediments where MnO_2 may be a problem, it can be determined by a separate redox titration with cold acidified $FeSO_4$ and dichromate using diphenylamine as an indicator (Jackson 1962). Interference from the oxidation of Cl^- can be reduced by adding Ag_2SO_4 to the reaction mixture, or, alternatively, Cl^- can be determined separately and the soil C content corrected using the stoichiometric equivalent of 1 mol C = 4 mol Cl^- (Jackson 1962).

An additional problem is presented by the often low visibility of the color change during back titration. Indicators used are *o*-phenanthroline (green to reddish brown) or *N*-phenylanthranilic acid (dark violet-green to light green). Some workers prefer phenylanthranilic acid, but in clayey soils with high chroma, the color change may be difficult to observe. Both indicators require some experience before overtitration can be avoided. An alternative to the back-titration with redox indicators is the direct determination of Cr^{3+} by spectrophotometry at 660 nm with calibration against carbohydrate standards (Heanes 1984).

Despite the difficulties and inaccuracies, the dichromate redox method is widely used because it requires a minimum of equipment, can be adapted to handle large numbers of samples, and it is suitable for comparative work on similar soils. For this reason one selected method is described in the following pages.

21.1.2 Dichromate Oxidation-CO_2 Determination

When wet oxidation of OC in an acid dichromate solution is followed by collection and determination of the evolved CO_2, many of the interferences of the titrimetric dichromate redox method are eliminated, and digestion temperatures can be raised above 150°C. Recoveries of OC are usually better with these oxidation-CO_2 collection methods than with titration methods, since differences in overall oxidation states presented by soil samples have no influence on the result, there are no interferences during back-titration, and higher oxidation temperatures can be employed. A 3:2 mixture of H_2SO_4 and H_3PO_4 with dichromate boils at 210°C. The addition of HIO_3 and fuming H_2SO_4 appears to offer little further advantage (Allison 1960).

Trapping of the evolved CO_2 in solid absorbers followed by weighing, or in alkali solutions followed by back titration of the alkali have been used with CO_2 absorption trains similar to those used in dry combustion procedures (Allison 1960). These methods usually only handle one sample at a time. An adaptation of the wet-digestion principle to a block digestor allowing the processing of 40 samples at a time has been described by Snyder and Trofymow (1984). The method by Snyder and Trofymow (1984) has been selected for presentation because of its ease of operation and facility to handle larger sample numbers, although the authors used a temperature of only 120°C.

21.1.3 Dry Combustion-CO$_2$ Determination

Several options are available for the dry oxidation of OC in a furnace, followed by direct determination or collection and determination of the evolved CO_2.

McKeague (1976) described a method using a tube resistance furnace for combusting soil C, followed by collection of evolved CO_2 in a NaOH absorption tower and back titration of the NaOH. The method, or similar ones, have been widely used as a standard for OC analysis. Its main limitations are low sensitivity resulting from the back titration of a relatively large volume of NaOH in which often only small quantities of CO_2 have been absorbed, and incomplete combustion of soil C. The low sensitivity (>5 mg C) can be overcome by increasing sample sizes, or, where this is not possible, by using the modified method (>0.5 mg C) of Tiessen et al. (1981). The latter method employs a direct titration of the trapped CO_2 between the 2 pKs of carbonic acid. The endpoints of the titration are stabilized against the effects of dissolved (not chemically bound) CO_2 in the solution by addition of carbonic anhydrase (Underwood 1961). The direct titration of carbonic acid also means that the absorbent NaOH need not be standardized.

Most resistance furnace methods suffer from potentially incomplete oxidation of C. Frequently, combustion temperatures below 500°C are used in order to allow the determination of OC in the presence of carbonates which decompose only at higher temperatures. Gallardo and Saavedra (1987) reported that some humic materials resist combustion at temperatures below 600°C, particularly in the presence of gibbsite. The use of a stream of O_2 in the tube furnace at a temperature of 900°C provides for near complete combustion of OC, and addition of MnO_2 and CuO as catalysts convert any evolved CO to CO_2 (Tiessen et al. 1981). Samples containing carbonates must be pretreated with acid, since combustion at 900°C will otherwise give a measure of total carbon. A problem encountered in routine work at the high temperature is the formation of hairline cracks in the ceramic tube of the furnace, causing leaks and low CO_2 recoveries. This can be minimized by reducing heat-cool cycles of the furnace and running analyses on a more continuous basis. The method by Tiessen et al. (1981) is described below, with the understanding that back titration of the NaOH is an option if sample sizes are large enough.

The most convenient combustion methods use automated instruments, which heat the sample by induction and determine the evolved CO_2 by infrared absorption. Various microprocessor-controlled instruments are available which oxidize the sample at 950°C in oxygen. At these high temperatures, total carbon is measured, and a pretreatment or separate determination of carbonate is required to arrive at organic C. Many of the commercially available units suffer from low sensitivity for C because of the design compromises required by the practice of combining C with H, N, or S analytical facilities.

21.2 WET OXIDATION-REDOX TITRATION METHOD

The procedure involves the oxidation of OC with a dichromate solution and titration with an Fe^{II} solution to determine the amount of dichromate remaining. It is based on the original description by Schollenberger (1945) and the modification by Jackson (1962) for the *o*-phenanthroline indicator. *N*-Phenylanthranilic acid can be substituted if desired (Nelson and Sommers 1975). The concentration of sulfuric acid and dichromate used are those of Tyurin (1931) and have given satisfactory results at relatively low consumption of reagents. Numerous alternative concentrations have been used (Nelson and Sommers 1982). Carbonates are not measured by this method.

21.2.1 REAGENTS

1   Digestion mixture: dissolve 39.22 g $K_2Cr_2O_7$ (oven dried at 90°C) (FW 294.19 g) in 800–900 mL deionized water. Slowly add 1000 mL conc H_2SO_4. Cool and make to 2000 mL volume with H_2O. This results in a 0.066 M (0.4 N) potassium dichromate, 9 M sulfuric acid solution. This solution is the only primary standard in the method, and its concentration is therefore critical.

2   Ferrous ammonium sulfate: dissolve 157 g $Fe (NH_4)_2(SO_4)_2 \cdot 6H_2O$ in about 1000 mL H_2O containing 100 mL conc H_2SO_4. Make to 2000 mL volume with H_2O, to give a ≈ 0.2 M (≈ 0.2 N) ferrous ammonium sulfate solution. It is not possible to make this solution to an exact concentration of Fe^{II}, since it oxidizes during preparation and storage. The solution does not store well, and must be standardized against chromic acid at each use.

3   Phosphoric acid: 85% conc. H_3PO_4.

4   Indicator solution: dissolve 3.00 g of *o*-phenanthroline monohydrate and 1.40 g of ferrous sulfate heptahydrate ($FeSO_4 \cdot 7H_2O$) in water. Dilute the solution to a volume of 200 mL. The indicator is commercially available under the name of Ferroin from the G. Frederick Smith Chemical Co. (Columbus, OH). Alternatively, dissolve 0.1 g of *N*-phenylanthranilic acid and 0.1 g of Na_2CO_3 in 100 mL of H_2O.

21.2.2 PROCEDURE

1   Weigh a sample (100 mesh, 0.15 mm) of 0.5–1.0 g that contains between 1 and 10 mg C. Place weighed sample into a 250-mL digestion tube, add 15 mL digestion mixture, and place on a 150°C preheated block digestor for 45 min.

2   Allow digested samples to cool before adding 50 mL H_2O, 5 mL 85% H_3PO_4, and four drops of indicator. The phosphoric acid eliminates interference from ferric iron. Then titrate with ferrous ammonium sulfate to a color change from dark violet-green to light green with *N*-phenylanthranilic acid or from green to reddish brown with *o*-phenanthroline. Each set should include two heated and two unheated blanks.

3   In the absence of a block digestor samples may be boiled gently on a hot plate for 45 min using a reflux condensor. Many variations of the dichromate method exist using different concentrations of acid or dichromate and different redox indicators. Since recoveries of OC are variable, the method should be rigorously standardized for comparative studies.

21.2.3 CALCULATIONS

The reactions involved are

$$4\ Cr^{VI} + 3\ C^0 \rightarrow 4\ Cr^{III} + 3\ C^{IV} \tag{21.1}$$

and for the back titration:

$$Cr^{VI} + 3\ Fe^{II} \rightarrow Cr^{III} + 3\ Fe^{III} \tag{21.2}$$

1   Calculate the molarity (= normality) of the ferrous ammonium sulfate solution based on titration against the unheated blanks. Equation 21.2 applies for this calculation, and approximately 2 mL of ferrous solution are required per 1 mL of dichromate solution.

2   Calculate total mg C in sample as:

$$\text{mg C in sample} = (B - T)(M)\ (0.003)(1000) \tag{21.3}$$

where B and T are the titers of the heated blank and the sample, respectively, and M is the molarity of the ferrous solution.

Note: if a significant difference between heated and unheated blanks is apparent, a correction may be desirable.

21.3 WET OXIDATION-CO_2 TRAP METHOD (SNYDER AND TROFYMOW 1984)

The sample is oxidized inside a closed digestion tube with a H_2SO_4-dichromate mixture, the evolved CO_2 is allowed to diffuse inside the digestion tube to a shell vial containing NaOH or ethanolamine, and the trapped CO_2 is subsequently titrated. The method as described by Snyder and Trofymow (1984) can handle solid or liquid samples, but uses a temperature of only 120°C. If only solid samples are processed, i.e., no water vapors are present, the digestion can be performed at higher temperatures, as long as it remains below the boiling point of the acid mixture used. If boiling occurs the caps may leak, or the tubes

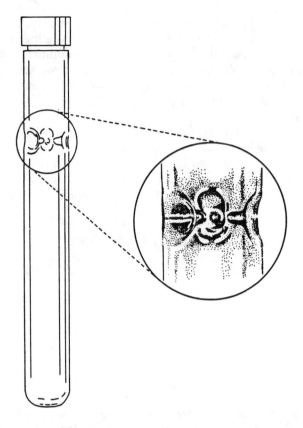

FIGURE 21.1. Culture tube, modified for the digestion of OC or carbonates, and diffusion and trapping of evolved CO_2. (Drawing courtesy of J. Deubner.)

may even break. For the determination of OC only, in soils that contain carbonates an acid pretreatment is required. The pretreatment solution is that suggested by Allison (1960), containing antioxidant to prevent OC oxidation during this step, although Snyder and Trofymow (1984) report that no significant losses of OC were observed when the antioxidant was omitted. An adaptation of the method for the separate determination of carbonates (Snyder and Trofymow 1984) is also given.

21.3.1 Preparation of Tubes

Standard culture tubes (Pyrex® No. 9825-25X, or similar) capped with screw caps containing a conical polyseal (size 24-400, e.g., VWR No. 16217-089) are modified with two indentations near the top capable of supporting an inserted glass shell vial (1 dram, 15 × 45 mm). The indentations can be easily made with a sharpened graphite rod by a glass blower (Figure 21.1).

For the modified procedure in which carbonates can also be determined, additional septum caps (e.g., Mininert valve cap, Precision Sampling No. 614163), and smaller shell vials (1/2 dram, 12 × 35 mm), allowing clearance for injection of acid into the closed tubes, are required.

21.3.2 REAGENTS

1   Acid pretreatment for elimination or determination of carbonates: dilute 57 mL conc H_2SO_4 in 600 mL H_2O and add 92 g of $FeSO_4 \cdot 7H_2O$. Dissolve and make to 1000 mL to give approximately 1 M H_2SO_4 containing 5% antioxidant.

2   Digestion mixture: These are kept separate and only combined in the reaction tube: a. $K_2Cr_2O_7$ (oven dried at 90°C) (FW 294.19 g), and b. a mixture of three parts conc H_2SO_4 and two parts 85% H_3PO_4.

3   CO_2 absorption solution: dissolve 16.0 g of NaOH and bring to 200 mL with distilled water to give a ≈ 2 M solution. This should be kept in an airtight flask, or under a CO_2 trap.

4   Indicator solution: dissolve 0.4 g thymolphthalein in 100 mL of a mixture of 1:1 ethanol:distilled water.

5   Barium chloride solution: dissolve 41.66 g $BaCl_2$ in distilled water, make to 200 mL, to give a ≈ 1 M solution

6   Titrant: use exactly 1.000 M HCl (e.g., Titrisol®).

21.3.3 PROCEDURE

21.3.3.1 Sample Preparation

Soil or plant samples should be crushed to pass through a 100-mesh sieve. Sample size is limited to 0.5–2.0 g of soil or 0.001–0.02 g of plant material. Liquid samples up to about 5 mL can be digested without preparation; larger samples should be evaporated to less than 5 mL in the digestion tubes at 100°C. When liquid samples are processed, temperature of the digestion must be limited to 120°C.

For samples containing up to 10% carbonates, 3 mL of pretreatment acid are added per gram of soil. The pretreatment is done in the digestion tubes by shaking them uncapped for 60 min on a reciprocal shaker at low speed. The water added with the dilute acid limits the digestion temperature. Alternatively, 1 M HCl, which can be gently evaporated, can be used for the pretreatment. The method can be modified to allow the quantification of inorganic carbon at this point, by injecting the acid through a septum into the closed culture tubes containing the CO_2-absorption vials. Modifications required are described in detail below.

21.3.3.2 Organic Carbon Oxidation

1   Place samples into the bottom of the digestion tubes with a long spatula and then pretreat to remove carbonates.

2   Approximately 1 g of $K_2Cr_2O_7$ is added using a long spatula or glass funnel. Add 25 mL of a 3:2 (v/v) mixture of conc H_2SO_4:85% H_3PO_4 and quickly insert the CO_2 trap (shell vial containing NaOH).

3   Tightly cap the tubes and place in a 150°C digestion block for 2 h.

4   Remove the tubes from the block, and after 12 h, remove and titrate the CO_2 trap.

The amount of NaOH in the trap limits the amount of CO_2 that can be absorbed. Using 1 mL of 2 M NaOH in a 6-mL capacity shell vial allows titration directly in the vial. Use of 1 mL of 2 M NaOH will trap a theoretical maximum of 12 mg CO_2-C, but absorption efficiency drops before this maximum is reached.

21.3.3.3 *Titration Procedure*

Carbonic acid trapped in the NaOH can be titrated by the direct, two endpoint method described in the following section, or by back titration. If back titration is used, the NaOH used for CO_2 absorption must be standardized.

1   To back titrate, add an excess of $BaCl_2$ (2 mL of 1 M $BaCl_2$) to the NaOH to precipitate $BaCO_3$.

2   Add approximately five drops of 0.4% thymolpthalein, and titrate the NaOH with 1.000 M HCl using a microburette accurate to 0.001 mL. The burette used and the concentration of the acid can be varied, depending on the level of accuracy desired. Four blanks per 40-tube digestion batch should be included.

21.3.3.4 *Modification for the Determination of Carbonates*

1   After placing the sample into the digestion tube, place a small 1/2-dram shell vial containing 1 mL of 2 M NaOH in the tube and close with a Mininert valve cap.

2   Inject 3 mL of 1 M H_2SO_4/$FeSO_4$ pretreatment solution past the CO_2 trap using a 12-cm 20-gauge Luer lock needle and syringe.

3   Remove the needle and close the Mininert valve. The tubes remain capped for at least 12 h to allow complete carbonate decomposition and CO_2 absorption.

4   Remove the trap for titration and process the samples further for OC as above.

5   Remove the trap for titration and process the samples further for OC as above.

21.3.4 Calculation of C

The reaction for CO_2 absorption is

$$H_2CO_3 + 2\,NaOH + BaCl_2 \rightarrow BaCO_3 + 2\,NaCl + 2\,H_2O \qquad (21.4)$$

Thus 2 mol of OH^- are used for each 1 mol (12 g) of C, and C is calculated as:

mg C = (mean mL HCl of blanks − mL HCl of sample) * M of HCl * (12/2)

When 1.00 M HCl is used in the titration, this formula reduces to:

mg C trapped = (mL HCl blank − mL HCl sample) * 6

21.4 DRY COMBUSTION METHOD FOR TOTAL OR ORGANIC CARBON

The method uses a tube furnace at a temperature of 900°C, trapping of evolved CO_2 in NaOH, and direct two-endpoint titration of the CO_2. It was described by Tiessen et al. (1981). At 900°C carbonates are also volatilized, so that total carbon is determined. The determination of OC in soils containing carbonates requires an acid pretreatment.

21.4.1 REAGENTS

1 Potassium iodide solution, 50 g/100 mL.

2 Sodium hydroxide, approximately 0.1 *M*.

3 Standardized 0.100 *M* HCl or H_2SO_4 (Titrisol®).

4 Approximately 0.1 *M* hydrochloric or 0.05 *M* sulfuric acid for the first end-point titration; this is optional.

5 Approximately 1 *M* hydrochloric or 0.5 *M* sulfuric acid.

6 Cupric oxide (CuO) wire.

7 Manganese dioxide powder (MnO_2).

8 Carbonic anhydrase solution, 10 mg bovine carbonic anhydrase (Sigma®) in 10 mL distilled water.

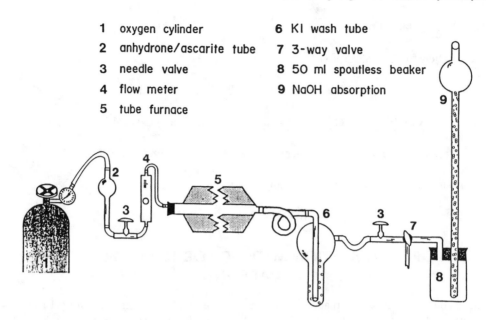

1	oxygen cylinder	6	KI wash tube
2	anhydrone/ascarite tube	7	3-way valve
3	needle valve	8	50 ml spoutless beaker
4	flow meter	9	NaOH absorption
5	tube furnace		

FIGURE 21.2. Tube furnace and gas-absorption train for the determination of total or organic carbon. (Modified from Tiessen, H. et al. Commun. Soil Sci. Plant Anal. 12(3): 211–218.)

21.4.2 Combustion Apparatus

A tube furnace capable of reaching 1000°C is used (several models and manufacturers are available), in combination with ceramic tubes ("high-temperature mulite" with an I.D. of 7/8 in from McDanel, Beaver Falls, PA). The ends of the tubes remain cool enough to make all gas lead connections with ordinary rubber stoppers and Tygon® tubing. Standard ceramic combustion boats are used to carry the samples.

To ensure uniform combustion conditions the tube furnace is supplied with oxygen from a pressure tank (technical grade, purified through drying agent and CO_2-absorbant "Ascarite®"). Alternatively, air from a compressor may be used and purified. Blanks may then be slightly higher.

A few grams of cupric oxide (CuO) in wire form are placed downstream of the samples in the hot part of the combustion tube to effect complete conversion of C to CO_2. This catalyst can be used for many weeks. Gases are led from the combustion tube through a loop of Tygon® tubing into a purification trap (Figure 21.2). The loop of transparent tubing reduces the risk of back flush of liquid into the hot combustion tube with pressure changes. The purification trap contains a 50% potassium iodide solution which has to be renewed after every full day of analyses. This removes oxides of N and S, which interfere with the later titrations.

The KI wash is followed by a CO_2-absorption tower containing 25 mL of 0.1 M NaOH. The absorption tower is filled with 3-mm glass beads to maximize gas-liquid contact. Gas flow is maintained at 40–60 mL min^{-1} by means of two flow values, one before the furnace and one before the absorption tower. Two valves are used to prevent pulse surging of the flow. Gas pressure is kept below 7 g cm^{-2} (1 psi). The entire apparatus (Figure 21.2) is designed to minimize inner dead volume to achieve fast flushing of the system with low flow rates.

All glass components are standard laboratory supply, except for the 45-cm-tall absorption tower, which was prepared by a glass blower.

21.4.3 PROCEDURE

21.4.3.1 Sample Preparation

Soil samples are prepared by crushing them to pass a 0.5-mm sieve. For whole soils or similar materials, between 0.5 and 1.0 g of sample (0.5–10 mg C) is placed into a combustion boat. For the determination of OC in soils containing carbonates, the combustion boat containing the soil sample is placed on a hot plate at 80°C and the sample wetted with about 0.5 mL of about 2 M HCl. Upon drying, the process is repeated until no more effervescence is seen.

21.4.3.2 Combustion and Titration

1   Preheat the tube furnace to about 900°C and turn on the gas flow, directing it through the absorption tower. Check with water in the tower that the flow gives an even, slow stream of bubbles. Open the three-way valve venting the gas to the air. Attach a beaker containing 25 mL of 0.1 M NaOH to the tower (the gas is still vented, flowing through the furnace to keep a steady temperature).

   At all times the system must be kept free of leaks. This can be checked by monitoring the flow with the exit of the tower plugged or some other suitable method.

2   The dry sample (pretreated or not) is lightly covered with MnO_2 (approximately 0.3 g) and the boat placed in the preheated furnace. Immediately close the three-way valve directing the gas stream through the tower. The combustion is usually complete after 4 min and a combustion time of 10 min per sample gives a wide safety margin.

3   After 10 min the system is vented and the NaOH is washed from the tower into the 50-mL beaker. If the wash volume is kept to a minimum, the titration can be performed in the same beaker without transfers. Completion of the washdown can be checked with indicator paper on the last effluent from the tower.

4   The solution is then titrated:
    1. Quickly bring to a pH of about 10 with 1 M HCl.
    2. Add five drops of carbonic anhydrase solution.
    3. Adjust the pH to exactly 8.30 with 0.1 M HCl.
    4. Then lower the pH to 3.70 and the volume of 0.100 M HCl required for this step is recorded. An autotitrator is useful for step 4, and if a two-endpoint titrator is available and the expense of standardized acid is not

critical, step 3 can also be performed automatically. The titration can be performed while the next sample is in the tube furnace.

21.4.4 Calculations

A blank combustion has to be performed and the amount of carbon in a sample is calculated as:

$$\text{mg C in sample} = (\text{mL } 0.1\ M \text{ HCl used for sample} - \text{mL } 0.1\ M \text{ HCl used for blank}) * 1.2 \quad (21.5)$$

This is a very simple relationship, since 0.1 mol of H^+ is used by the system for each 0.1 mol of CO_2 (1.2 g of C). If the sample was pretreated for the removal of carbonates, this value represents OC; without pretreatment, total C is obtained. Inorganic C could be calculated as the difference, or determined directly using a modified digestion-absorption train as described by Tiessen et al. (1983).

REFERENCES

Allison, L. E. 1960. Wet-combustion apparatus and procedure for organic and inorganic carbon in soil. Soil Sci. Soc. Am. Proc. 24: 36–40.

Charles, M. J. and Simmons, M. S. 1986. Methods for the determination of carbon in soils and sediments. A review. Analyst 111: 385–390.

Gallardo, J. F. and Saavedra, J. 1987. Soil organic matter determination. Commun. Soil Sci. Plant Anal. 18(6): 699–707.

Gillman, G. P., Sinclair, D. F., and Beech, T. A. 1986. Recovery of organic carbon by the Walkley and Black procedure in highly weathered soils. Commun. Soil Sci. Plant Anal. 17(8): 885–892.

Heanes, D. L. 1984. Determination of total and organic C in soils by an improved chromic acid digestion and spectrophotometric procedure. Commun. Soil Sci. Plant Anal. 15(10): 1191–1213.

Jackson, M. L. 1962. Soil chemical analysis, Constable & Co. Ltd., London.

McKeague, J. A. (Ed.) 1976. Manual on soil sampling and methods of analysis. Soil Research Institute of Canada.

Metson, A. J., Blakemore, L. C., and Rhoades, D. A. 1979. Methods for the determination of soil organic carbon: a review, and application to New Zealand soils. N.Z. J. Sci. 22: 205–28.

Nelson, D. W. and Sommers, L. E. 1975. A rapid and accurate procedure for estimation of organic carbon in soils. Proc. Indiana Acad. Sci. (for 1974) 84: 456–62.

Nelson, D. W. and Sommers, L. E. 1982. Total carbon, organic carbon and organic matter. Chapter 29 in Methods of soil analysis. Part 2. 2nd ed. American Society of Agronomy, Soil Science Society of America, Madison, WI.

Schollenberger, C. J. 1927. A rapid approximate method for determining soil organic matter. Soil Sci. 24: 65–68.

Schollenberger, C. J. 1945. Determination of soil organic matter, Soil Sci. 59: 53–56.

Snyder, J. D. and Trofymow, J. A. 1984. A rapid accurate wet oxidation diffusion procedure for determining organic and inorganic carbon in plant and soil samples. Commun. Soil Sci. Plant Anal. 15(5): 587–597.

Tiessen, H., Bettany, J. R., and Stewart, J. W. B. 1981. An improved method for the determination of carbon in soils and soil extracts by dry combustion. Commun. Soil Sci. Plant Anal. 12(3): 211–218.

Tiessen, H., Roberts, T. L., and Stewart, J. W. B. 1983. Carbonate analysis in soils and minerals by acid digestion and two-endpoint titration. Commun. Soil Sci. Plant Anal. 14(2): 161–166.

Tyurin, I. V. 1931. A modification of the volumetric method of determining soil organic matter by means of chromic acid. Pochvovodenye 26: 36–47.

Underwood, A. L. 1961. Carbonic anhydrase in the titration of carbon dioxide solutions. Anal. Chem. 33: 955–956.

Walkley, A. and Black, I. A. 1934. An examination of the Degtjareff method for determining soil organic matter, and a proposed modification of the chromic acid titration method. Soil Sci. 34: 29–38.

Chapter 22
Total Nitrogen

W. B. McGill and C. T. Figueiredo

University of Alberta
Edmonton, Alberta, Canada

22.1 INTRODUCTION

Total N analyses may be divided into two categories: (1) wet-oxidation (e.g., Kjeldahl method) or (2) dry-combustion (e.g., Dumas method). Wet-oxidation techniques are the most common and involve conversion of organic and inorganic N to NH_4^+ and its subsequent measurement. The Dumas method normally involves an initial oxidation step followed by passage of the gases through a reduction furnace to reduce NO_x to N_2. The quantity of N_2 is normally determined using a thermal conductivity detector.

The Dumas method is being revived due to the increasing availability of automated analytical instruments with which to determine C, H, N, and S on the same sample, and O with a simple modification. Sample variability is a concern with combustion techniques because of the small sample size required. Bellomonte et al. (1987) concluded that the automatic Dumas procedure was comparable to Kjeldahl analyses for heterogeneous substrates. They reported c.vs. of 0.79% for analyses of total N content on cereal flour and 1.08% on meats when using a commercial Dumas system. Kirsten and Jansson (1986), Marshall and Whiteway (1985), and Minagawa et al. (1984) have investigated the use of dry-combustion techniques for total N, and when connected in line with a mass spectrometer, for ^{15}N analyses of heterogeneous substrates.

Kjeldahl procedures are the most widely used for total N and continue to be modified. Although fixed NH_4^+ is normally included in the Kjeldahl digestion, it may not be quantitatively measured in some soils with a high proportion (e.g., 30–60%) of their N as fixed NH_4^+ (Bremner and Mulvaney 1982). In such cases, an HF-HCl modification, as described by Bremner and Mulvaney (1982), may be necessary to destroy clay minerals. Recent literature has emphasized methods to analyze samples containing nitrate, and revisions to digestion and ammonium measurement protocols.

Inclusion of NO_3^- is a problem because Kjeldahl digestion includes some, but not all, NO_3^-, thereby precluding the addition of NO_3^- from a separate analysis to the Kjeldahl total N values (Goh 1972, Bremner and Mulvaney 1982, Wikoff and Moraghan 1985). Hence, efforts to simplify the inclusion of NO_3^- in Kjeldahl digestions continue. Pruden et al. (1985)

Soil Sampling and Methods of Analysis, M. R. Carter, Ed.,
Canadian Society of Soil Science. © 1993 Lewis Publishers.

proposed the use of Zn to reduce Cr(III) to Cr(II), which in turn reduces NO_3^- and NO_2^-. Zn is added to the soil sample in the digestion tube, followed by $CrK(SO_4)_2$ and H_2SO_4, and allowed to stand at room temperature for 2 h. Normal Kjeldahl digestion follows. A modification by Dalal et al. (1984) uses sodium thiosulfate to reduce NO_3^- or NO_2^- and requires no pretreatment. The method is satisfactory with wet or dry samples; recovery of ^{15}N-labeled NO_3^- or NO_2^- was complete from soil or aqueous solutions DuPreez and Bate (1989) reported that phenyl acetate added to dry samples resulted in quantitative recovery of NO_3^- or NO_2^-. In the presence of NO_3^- or NO_2^-, and under acidic conditions, phenyl acetate yields nitrophenolic compounds. They found there was no need for subsequent addition of a reductant, and the method required no pretreatment, but it was suitable for dry samples only.

Rather than reduce all the nitrate, an alternative approach is to prevent nitrate reduction during Kjeldahl digestion. Bowman et al. (1988) concluded that pretreatment of plant tissues (100 mg) with 3 mL of H_2O_2 at 100°C for 15 min prior to addition of H_2SO_4, K_2SO_4, and catalyst prevented reduction and inclusion of nitrate in the analysis of plant material containing up to 0.73% nitrate-N.

Recent modifications to the Kjeldahl procedure include "peroxy" methods of digestion, and modifications of the Berthelot reaction and of diffusion techniques for ammonium measurement. The "peroxy" method replaces added salts and metal catalysts with peroxymonosulfuric acid (H_2SO_5, Caro's acid), which is a more powerful oxidant than hydrogen peroxide. This method involves carbonizing the sample in H_2SO_4 prior to adding the hydrogen peroxide-sulfuric acid reagent ("peroxy" reagent). It works as much as 25 times faster than conventional Kjeldahl procedures and provides complete recovery of N when tested on a variety of plant materials and one of the most refractory organic compounds, nicotinic acid (Hach et al. 1985). To enhance safety and improve speed, Hach et al. (1987) developed a system using a Vigreux fractionation head to simplify addition of the digestion solution, containing only H_2SO_4 and H_2O_2, and to maintain constant residual H_2O_2 in the digestion system. Their procedure required only 17.5 min to digest nicotinic acid, compared to 3 h for standard Kjeldahl techniques. Further investigation on soil materials would be warranted.

Lange et al. (1983) provided calibration details for block digestion using either micro- or macro-Kjeldahl systems in combination with the Berthelot reaction for colorimetric determination of NH_4^+. They tested their method on several pure compounds, including nicotinic acid (as an example of a refractory compound). A manual version of the Berthelot reaction has been used to quantify NH_4^+ from either Kjeldahl digestion or 2 M KCl extracts of soils (Wang and Oien 1986).

MacKown et al. (1987) reported on a diffusion technique to replace distillation for determination of NH_4^+ following digestion. It is simple and convenient; and because it uses disposable containers, it eliminates problems of cross contamination in studies using ^{15}N.

Selection of the most suitable combination of variables for the Kjeldahl method must be based on local requirements and facilities. Digestion options include: H_2SO_4 or H_2O_2; heating mantles or digestion blocks; macro- or semimicro digestion; inclusion or omission of NO_2^- plus NO_3^-. Subsequent measurement of NH_4^+ may use the Berthelot reaction, NH_4^+ electrode, diffusion in digestion tubes or in conway dishes, steam distillation directly from digestion tubes or from standard taper flasks, macro- or semimicro distillation, titration with an indicator or using automated titrators.

The choice here has been to present details of modern methods that could be widely used, to introduce special purpose methods by way of comments, and to provide citations for older or classical methods. Digestion blocks with the temperature controlled electronically and with digestion tubes are now common. Blocks are replacing heating mantles and tubes are replacing standard taper semimicro- or 0.8-L macro-Kjeldahl digestion flasks for total N analyses by the Kjeldahl procedure.

Micro-Kjeldahl digestion procedures are given with and without steps to include nitrate and nitrite. In many soils, N as NO_2^- plus NO_3^- is a negligible component of total N, and procedure 22.2 is satisfactory.

Macro-Kjeldahl procedures are not reported here. They are little used now because of the cost of equipment and chemicals and the high precision of semimicro-Kjeldahl and Dumas procedures. Refer to McKeague (1978) for macro-Kjeldahl techniques.

Soil samples should be ground to pass a 100-mesh sieve for all Kjeldahl procedures. Finer grinding is normally required for analyses using automated Dumas procedures.

22.2 MICRO-KJELDAHL DIGESTION FOLLOWED BY STEAM DISTILLATION: WITHOUT PRETREATMENT TO INCLUDE NO_2^- AND NO_3^- QUANTITATIVELY

This method is appropriate for total N determination on samples of surface soil horizons in which the nitrate and nitrite contents are negligible. If used with samples containing significant amounts of nitrate or nitrite, the results will be higher than for the fixed NH_4^+ plus organic N content alone, but lower than for total N including nitrate and nitrite. This method is not recommended for analysis of total N in soil samples from ^{15}N tracer studies because of the significant influence of highly labeled NO_2^- or NO_3^- on ^{15}N analyses. Method 22.3 is recommended for such samples.

22.2.1 MATERIALS AND REAGENTS: DIGESTION

1   Heating block with digestion tubes, timer, and temperature controller. The block must be capable of maintaining a temperature of 360°C for up to 5 h. Blocks holding 40 tubes (20 mm O.D. × 350 mm long) calibrated to hold 0.1 L are commonly used for micro-Kjeldahl digestions.

2   Air condenser designed to fit over the digestion tubes in the block (see Comments 22.2.5).

3   Concentrated H_2SO_4.

4   K_2SO_4, $CuSO_4$: Mixed in ratio of 8.8:1 (K_2SO_4:$CuSO_4 \cdot 5H_2O$).

5   Hengar granules, both selenized and nonselenized.

22.2.2 MATERIALS AND REAGENTS: DISTILLATION AND TITRATION

1 Micro-Kjeldahl steam distillation apparatus (Figure 22.1). See Comments 22.2.5.

2 Steam distillation flasks: 0.5-L round bottom, with 19/38 standard taper ground glass joint.

3 NaOH: 10 M and 0.1 M, prepared in CO_2-free deionized water.

4 Boric acid (2%) plus indicator: place 80 g of boric acid (H_3BO_3) powder into a 0.25-L beaker. Add ≈20–40 mL of H_2O and mix with a glass rod to wet all the H_3BO_3. Pour into ≈3 L of H_2O in a 4-L flask and stir with an electric stir rod. Once wet, the H_3BO_3 dissolves readily. Add 80 mL of mixed indicator prepared as follows: 0.099 g bromocresol green and 0.066 g methyl red dissolved in 100 mL ethanol. Add 0.1 M NaOH cautiously until the solution turns reddish-purple (pH 4.8–5.0). Make up to 4 L with deionized H_2O and mix thoroughly.

5 Graduated beakers: 100 mL.

6 Burette: 10 mL graduated at 0.02- or 0.01-mL intervals. A magnetic stirrer is desirable.

7 H_2SO_4: 0.01 M (standardized).

22.2.3 PROCEDURE: DIGESTION, DISTILLATION, AND TITRATION

1 Place sample, containing about 1 mg N, in a dry digestion tube. This will vary from 0.25–2.0 g for soil and from 0.02–0.05 g for plant tissue.

2 Add 2 mL deionized H_2O (3 mL if using 2 g soil) and swirl to wet all the soil.

3 To each tube add 3.5 g of K_2SO_4:$CuSO_4$ mix.

4 Add one selenized and one nonselenized Hengar granule.

5 Add 10 mL concentrated H_2SO_4.

6 Place the digestion tubes into the digestion block.

7 Program the block to raise the temperature to 220°C and keep it there for 1.5 h. Digestion will start and water will be removed during this time.

8 After 1.5 h at 220°C, put the air condensers onto the digestion tubes in the block.

9   Program the block to raise the temperature to 360°C and hold it there for 3.5 h.

10   After digestion is complete, cool the samples overnight in the block or on a fiberglass pad.

11   Remove the air condenser and rinse it with water.

12   Slowly and with swirling add 25 mL deionized water to each cooled digestion tube. Vortex the sample to dislodge all particles that may have solidified during cooling. If all the material does not enter into suspension, warm gently until it does. Transfer the sample quantitatively with three washes of deionized water to a 0.5-L round-bottom distillation flask.

13   With the condensers on, connect the distillation flask to the steam distillation apparatus; secure it with a clamp.

14   Open the steam supply to the distillation head to allow steam into the tubing. Be sure the drain line is already open so that at this point steam exits to the drain.

15   Place a 100-mL graduated beaker with 5 mL of 2% H_3BO_3 under the condenser so that the tip of the condenser is immersed in the H_3BO_3.

16   Very slowly add an excess (usually 30 mL) of 10 M NaOH through the distillation head. Do not completely empty the NaOH reservoir, otherwise NH_3 may be lost through the stopcock.

17   Close the pinch clamp, or stopcock, going to the steam drain; this directs steam into the distillation flask. The steam generation rate should be such that the distillate is collected at about 6 mL/min. Collect 40 mL of distillate.

18   Open the pinch clamp to the steam drain and remove the distillation flask then close the pinch clamp to the distillation head. This sequence is important to prevent steam burns and suck-back of fluid from the distillation flask into the steam line.

19   Wash the tip of the condenser into the beaker.

20   Titrate the distillate with 0.01 M H_2SO_4. The color change at the endpoint is from green to pink (pH ≈ 5.4).

22.2.4 Calculations

One mL of 0.01 M H_2SO_4 is equivalent to 0.28 mg of N.

$$\text{Total N \%} = \frac{\text{mL (Sample - Blank)} \times \text{M} \times 2.8}{\text{mass of sample (g)}} \tag{22.1}$$

22.2.5 COMMENTS

1   Heating blocks and tubes supplied by Tecator or Technicon® have been found satisfactory for Kjeldahl digestions in our laboratory. The air condenser has been described by Panasiuk and Redshaw (1977) and can be constructed by a competent glassblower. Equivalent devices are available from Tecator.

2   The K_2SO_4:$CuSO_4$ mix can be prepared in the laboratory, obtained as a loose mix or in appropriately-sized packages from commercial suppliers (e.g., Kjeltabs® — trademark of Tecator, Inc — from Fisher Scientific). Bulk mixes should be kept tightly sealed during storage to avoid adsorption of water and caking.

3   Distillation systems similar to those in Figure 22.1 are described by Hesse (1971), Bremner and Mulvaney (1982), and Bremner and Breitenback (1983). Commercial systems specifically designed for use with digestion tubes are also available (e.g., Tecator, available from Fisher Scientific). We use two distillation heads, each supplied with steam from a 5-L round-bottom boiling flask heated by a 600-W heating mantle. Several dozen Hengar granules are in each flask. H_3PO_4 (about 2 mL) is added to each boiling flask to absorb NH_3.

4   We have eliminated sample transfer from the digestion tube to a distillation flask by doing digestions in a 0.25-L digestion tube designed for a block that holds 20, rather than 40, tubes. The distillation head was modified by attaching a rubber stopper with a hole through which the standard taper joint of the distillation head fits. The 0.25-L digestion tube is attached to the distillation unit by fitting the end over the rubber stopper. It is held securely in place with a clamp, allowing distillation directly from the digestion tube. The distillation system described by Bremner and Breitenback (1983) uses tubes designed for the 40-tube blocks.

5   The NaOH must be added slowly and carefully to avoid violent bubbling that would force the liquid into the condenser and contaminate the distillation head. The amount of NaOH needed varies with the amount of H_2SO_4 consumed during digestion of the sample. Consumption of H_2SO_4 varies with the amount of soil organic matter and reduced minerals present; e.g., 1 g of C consumes 10 mL of H_2SO_4 (Bremner and Mulvaney 1982).

6   Distillation can be replaced by autoanalyzer protocols on the digested sample. When we use this approach, the digested sample is diluted to 0.1 L followed by an autoanalyzer method for measurement of NH_4^+ (Smith and Scott 1991).

7   Use of an automatic titrator can improve consistency and eliminates the need to mix an indicator into the boric acid solution.

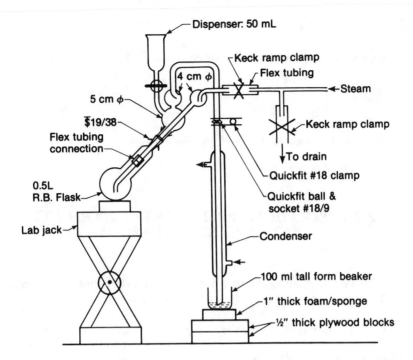

FIGURE 22.1. Steam distillation apparatus.

8   The above Kjeldahl digestion method does not recover fixed NH_4^+ in some soils. For total N analysis of soils with a high proportion of their N as fixed NH_4^+, an HF-HCl modification as described by Bremner and Mulvaney (1982) may be necessary to destroy clay minerals.

22.3 MICRO-KJELDAHL DIGESTION FOLLOWED BY STEAM DISTILLATION: NO_2^- AND NO_3^- INCLUDED QUANTITATIVELY

This is the method of choice for total N analysis of samples of surface soil horizons containing an appreciable quantity of N as NO_2^- or NO_3^-. Because of the significant influence of highly labeled NO_2^- or NO_3^- on ^{15}N analyses, this method is recommended for analysis of total N in all soil samples from ^{15}N tracer studies. This method is the same as in Section 22.2, except for the addition of a pretreatment to oxidize NO_2^- to NO_3^- and then to reduce the NO_3^- to NH_4^+.

22.3.1 MATERIALS AND REAGENTS: PRETREATMENT AND DIGESTION

1   All items from Section 22.2.1, plus the following:

2   Potassium permanganate solution: dissolve 50 g $KMnO_4$ in 1 L deionized water; store in an amber bottle.

3   Dilute H_2SO_4: (1:1 [v/v] with H_2O).

4   Fe powder; finer than 100-mesh sieve (e.g., Fisher Scientific).

5   *N*-Octyl alcohol.

22.3.2 Materials and Reagents: Distillation and Titration

All items from Section 22.2.2.

22.3.3 PROCEDURE: PRETREATMENT TO REDUCE NO_2^- AND NO_3^- TO NH_4^+

1   Place sample, containing about 1 mg N, in a dry digestion tube. Usually this is 0.25–2.0 g of soil and from 0.02–0.05 g of plant tissue.

2   Add 2 mL deionized H_2O (3 mL if using 2 g soil) and swirl to wet all the soil.

3   Add 1 mL $KMnO_4$ and swirl for 30 s.

4   Hold the digestion tube at a 45° angle and very slowly pipette 2 mL dilute H_2SO_4.

5   Allow to stand 5 min.

6   Add one drop *N*-octyl alcohol (to control frothing).

7   Add 0.5 g reduced Fe using a scoop, through a dry, long-stemmed funnel or thistle funnel tube.

8   Immediately cover the digestion tube with an inverted 25-mL beaker or 50-mL Erlenmeyer flask inverted to prevent loss of water.

9   Swirl to bring the Fe into contact with the acid.

10   Allow to stand (about 15 min) until strong effervescence has ceased.

11   Place digestion tubes into the digestion block and program it to raise the temperature to 100°C and hold it there for 1 h.

12   Cool the tubes before proceeding to digestion.

22.3.4 Procedure: Digestion, Distillation, and Titration

Follow steps 3 to 20 inclusive, of Section 22.2.3.

22.3.5 Calculations

One mL of 0.01 M H_2SO_4 is equivalent to 0.28 mg of N.

$$\text{Total N \%} = \frac{\text{mL(Sample-Blank)} \times M \times 2.8}{\text{mass of sample (g)}} \tag{22.2}$$

22.3.6 COMMENTS

1. The $KMnO_4$ oxidizes NO_2^- to NO_3^-, which is reduced to NH_4^+ by the reduced Fe.

2. *N*-Octyl alcohol is to reduce frothing.

3. Goh (1972) concluded that it was not necessary to include the permanganate pretreatment when reduced iron is used as a reductant in the procedure to include nitrate.

4. Please see Section 22.2.5 for additional important information.

22.4 DUMAS METHOD

Several automated Dumas systems are available (Fiedler et al. 1973, Kirsten and Jansson 1986; Bellomonte et al. 1987, Tabatabai and Bremner 1991). They are suitable for ^{15}N tracer studies when linked by a continuous flow interface from the nitrogen analyzer to an isotope ratio mass spectrometer (Fiedler and Proksch 1975, Minagawa et al. 1984, Marshall and Whiteway 1985).

Dumas combustion techniques can be performed manually in sealed tubes, or automatically with automatic nitrogen analysis (ANA) in a continuous process using helium carrier gas with total nitrogen content usually measured using a thermal conductivity detector. ANA, based on the Dumas technique, as now used is microscale, and very fine grinding is critical to ensure a representative sample. When an ANA is being used, samples (1–100 mg) of the dry, finely ground soil are weighed into a tin cup and loaded into an autosampler. The samples are sequentially introduced into an oxidation furnace at $>1000°C$, where flash combustion in a stream of pure oxygen generates temperatures of 1800–2000°C, to yield CO_2, H_2O, N_2, and NO_x. These gases enter a reduction furnace containing Cu at 650°C, which removes excess O_2 and converts NO_x to N_2. All CO_2 and H_2O vapors are absorbed in chemical traps and N_2 is separated from residual impurities with an integral gas chromatograph (Pella and Colombo 1973). Quantification normally requires about 3 min.

Dumas techniques have the advantage of requiring less laboratory space, do not produce noxious fumes, and include all forms of N without lengthy pretreatments (Bellomonte et al. 1987). For tracer studies, they avoid digestion, distillation, titration, evaporation, and subsequent oxidation of NH_3 to N_2. Such systems may combine both C and N analyses in one instrument.

REFERENCES

Bellomonte, G., Costantini, A., and Giammarioli, S. 1987. Comparison of modified automatic Dumas method and the traditional Kjeldahl method for nitrogen determination in infant food. J. Assoc. Offic. Anal. Chem. 70: 227–229.

Bowman, D. C., Paul, J. L., and Carlson, R. M. 1988. A method to exclude nitrate from Kjeldahl digestion of plant tissues. Commun. Soil Sci. Plant Anal. 19: 205–213.

Bremner, J. M. and Breitenback, G. A. 1983. A simple method for determination of ammonium in semimicro-Kjeldahl analysis of soils and plant materials using a block digester. Commun. Soil Sci. Plant Anal. 14: 905–913.

Bremner, J. M. and Mulvaney, C. S. 1982. Nitrogen — total. Pages 595–624 in A. L. Page, Ed. Methods of soil analysis, 2nd ed. Agronomy No. 9. American Society of Agronomy, Madison, WI.

Dalal, R. C., Sahrawat, K. L., and Myers, R. J. K. 1984. Inclusion of nitrate and nitrite in the Kjeldahl nitrogen determination of soils and plant materials using sodium thiosulphate. Commun. Soil Sci. Plant Anal. 15: 1453–1461.

DuPreez, D. R. and Bate, G. C. 1989. A simple method for the quantitative recovery of nitrate-N during Kjeldahl analysis of dry soil and plant samples. Commun. Soil Sci. Plant Anal. 20: 345–357.

Fiedler, R. and Proksch, G. 1975. The determination of nitrogen-15 by emission and mass spectrometry in biochemical analysis: a review. Anal. Chim. Acta 78: 1–62.

Fiedler, R., Proksch, G., and Koepf, A. 1973. The determination of total nitrogen in plant materials with an automatic nitrogen analyzer. Anal. Chim. Acta 63: 435–443.

Goh, K. M. 1972. Comparison and evaluation of methods for including nitrate in the total determination of soils. J. Sci. Food Agric. 23: 275–284.

Hach, C. C., Brayton, S. V., and Kopelove, A. B. 1985. A powerful Kjeldahl nitrogen method using peroxymonosulfuric acid. J. Agric. Food Chem. 33: 1117–1123.

Hach, C. C., Bowden, B. K., Kopelove, A. B., and Brayton, S. V. 1987. More powerful peroxide Kjeldahl digestion method. J. Assoc. Offic. Anal. Chem. 70: 783–787.

Hesse, P. R. 1971. A textbook of soil chemical analysis. John Murray, London.

Kirsten, W. J. and Jansson, K. H. 1986. Rapid and automatic determination of nitrogen using flash combustion of large samples. Anal. Chem. 58: 2109–2112.

Lange, R., Friebe, R., and Linow, F. 1983. Zur anwendung der methodenkombination Kjeldahl-Aufschluss/Berthelot reaktion bei der stickstoffbestimmung in biologischen materialien. 6. Die blockaufschlusstechnik im halbmikro- und makromasstab bei biologischen feststoffen. Nahrung 27: 645–658.

MacKown, C. T., Brooks, P. D., and Smith, M. S. 1987. Diffusion of nitrogen-15 Kjeldahl digests for isotope analysis. Soil Sci. Soc. Am. J. 51: 87–90.

Marshall, R. B. and Whiteway, J. N. 1985. Automation of an interface between a nitrogen analyzer and an isotope ratio mass spectrometer. Analyst 110: 867–871.

McKeague, J. A., Ed., 1978. Manual on soil sampling and methods of analysis. 2nd ed. Canadian Society of Soil Science, Ottawa, ON.

Minagawa, M., Winter, D. A., and Kaplan, I. R. 1984. Comparison of Kjeldahl and combustion methods for measurement of nitrogen isotope ratios in organic matter. Anal. Chem. 56: 1859–1861.

Panasiuk, R. and Redshaw, E. S. 1977. A simple apparatus used for effective fume control during plant tissue digestion using a heating block. Commun. Soil Sci. Plant Anal. 8: 411–416.

Pella, E. and Colombo, B. 1973. Study of carbon, hydrogen and nitrogen determination by combustion-gas chromatography. Mikrochim. Acta (Wein) 697–719.

Pruden, G., Kalembasa, S. J., and Jenkinson, D. S. 1985. Reduction of nitrate prior to Kjeldahl digestion. J. Sci. Food Agric. 36: 71–73.

Smith, K. A. and Scott, A. 1991. Continuous-flow, flow-injection, and discrete analysis. Pages 183–227 in K. A. Smith, Ed. Soil analysis: modern instrumental techniques. 2nd ed. Marcel Dekker, New York.

Tabatabai, M. A. and Bremner, J. M. 1991. Automated instruments for determination of total carbon, nitrogen, and sulfur in soils by combustion techniques. Pages 261–286 in K. A. Smith, Ed. Soil analysis: modern instrumental techniques. 2nd ed. Marcel Dekker, New York.

Wang, L. and Oien, A. 1986. Determination of Kjeldahl nitrogen and exchangeable ammonium in soil by the indophenol method. Acta Agric. Scand. 36: 60–70.

Wikoff, L. and Moraghan, J. T. 1985. Recovery of soil nitrate by Kjeldahl analysis. Commun. Soil Sci. Plant Anal. 16: 923–929.

Chapter 23
Total and Organic Phosphorus

I. P. O'Halloran

McGill University
Ste. Anne de Bellevue, Quebec, Canada

23.1 INTRODUCTION

This chapter describes several methods used for the determination of either total phosphorus (P_t) or organic phosphorus (P_o) in soils.

The two most widely recognized procedures for the determination of soil P_t are the sodium carbonate (Na_2CO_3) fusion method and the perchloric acid ($HClO_4$) digestion method (Olsen and Sommers 1982). Two additional procedures, an alkaline oxidation method (Dick and Tabatabai 1977) using sodium hypobromite (NaOBr)/sodium hydroxide (NaOH), and a digestion method (Bowman 1988) using sulfuric acid (H_2SO_4)/hydrogen peroxide (H_2O_2)/hydrofluoric acid (HF), are also presented.

Soil P_o can be determined either by extraction (Mehta et al. 1954) or ignition techniques (Saunders and Williams 1955), and numerous variations of these procedures exist. In each case P_o is not determined directly, but rather by the difference between a measure of total soil P and inorganic P (P_i). The ignition technique is less laborious than the extraction technique, but it is also subject to a greater number of errors, depending on the properties of the soil analyzed. In general, for most soils the two techniques give comparable values for soil P_o, although the ignition method tends to give higher P_o values for highly weathered soils. A third method, developed by Bowman (1989), involves the sequential extraction of soil P_o using H_2SO_4 and NaOH. This extraction procedure gives similar results as the Mehta extraction procedure, and is fastest of the three procedures presented.

Several methods exist for the determination of phosphate in solutions. Inductively coupled plasma spectroscopy is one method which can be used, although many researchers may not have ready access to these spectrophotometers. Conventionally, colorimetric techniques have been used for determinations of P concentrations in various extracts. Numerous colorimetric methods exist, and the reader is cautioned that slight deviations from the proposed procedures can lead to erroneous results. For further information concerning the colorimetric determination of P the reader is referred to Jackson (1958) and Olsen and Sommers (1982).

The ascorbic acid method of Murphy and Riley (1962) is suitable for the determination of orthophosphate in all digests and extracts collected in Sections 23.2 and 23.3. This method uses the blue color developed by the molybdophosphate complex reduced by ascorbic acid

Soil Sampling and Methods of Analysis, M. R. Carter, Ed.,
Canadian Society of Soil Science. © 1993 Lewis Publishers.

in the presence of antimony (Sb) to estimate the concentration of P in solution. The addition of Sb eliminates the necessity of heating the solution as described by Fogg and Wilkinson (1958). This procedure is fairly simple, less susceptible to interferences than procedures using $SnCl_2$ as a reductant, and is capable of being used manually or adapted for automated systems. This method is presented in Chapter 10.

Note: the aliquot size of the various digests and extracts collected in Sections 23.2 and 23.3 is extremely important. If the spectrophotometer used has a 1-cm cuvette, the aliquot should not contain more than 40 μg P as orthophosphate. If the spectrophotometer has a 4-cm cuvette, the aliquot should not contain more than 10 μg P as orthophosphate.

23.2 TOTAL PHOSPHORUS

23.2.1 Introduction

Although the Na_2CO_2 fusion method is the most reliable procedure, it is also very time consuming, and generally unsuitable for the analyses of large numbers of samples. The procedure requires the use of platinum crucibles which are expensive and require special handling to avoid damage. Digestion with $HClO_4$ is more adaptable as a routine laboratory procedure, although special perchloric acid fumehoods are required, and care must be taken to prevent the buildup of $HClO_4$, which can cause explosions. It has been reported that the $HClO_4$ digestion method gives relatively low P_t values in highly weathered materials and with samples containing apatite inclusions (Syers et al. 1967, 1968), but this method should be suitable for most Canadian soils. Caution must also be exerted when dealing with samples high in organic matter, and the routine oxidation of organic matter in samples with nitric acid (HNO_3) and heat is recommended. The NaOBr-NaOH oxidation method gives similar results as the $HClO_4$ digestion, and has the advantage that it does not require the use of a special fumehood. The $H_2SO_4/H_2O_2/HF$ digestion method is relatively fast and gives results approximately equal to the $HClO_4$ digestion and NaOBr-NaOH oxidation methods, and is ideally suited for small numbers of samples. This procedure, however, requires HF-resistant materials, since HF attacks glass.

Regardless of the procedure selected, it is recommended that finely ground soil, 0.15 to 0.18 mm (100 to 80 mesh) be used allow for efficient removal of P from the soil material, and to improve the reproducibility. The moisture content of the soil should be known so as to allow expression of P content on an oven-dry basis. Blank samples containing no soil should also be included to assess the possibility of P contamination and to serve as a suitable reagent blank for the colorimetric determination of P.

23.2.2 Sodium Carbonate Fusion (Olsen and Sommers 1982)

This method involves the preparation of a soil-Na_2CO_3 melt in a platinum crucible, and thus the user should be aware of the proper methods of handling and caring for platinum utensils. This method is not well suited for the analysis of large numbers of samples due to the amount of time required per sample. Results are very reproducible, with C.V.s normally less than 4%.

23.2.2.1 MATERIALS AND REAGENTS

1   Platinum crucible and platinum-tipped crucible tongs.

2   Sodium carbonate (Na_2CO_3), anhydrous.

3   Sulfuric acid (H_2SO_4), 4.5 *M*: add 250 mL concentrated H_2SO_4 to 600 mL distilled/deionized water in a 1-L volumetric flask. Allow to cool, and make to volume using distilled/deionized water.

4   Sulfuric acid (H_2SO_4), 1 *M*: add 56 mL concentrated H_2SO_4 to 600 mL distilled/deionized water in a 1-L. volumetric flask. Allow to cool, and make to volume using distilled/deionized water.

5   Color developing solutions: see Chapter 10.

23.2.2.2 PROCEDURE

1   Thoroughly mix 0.5 g of finely ground (0.15 to 0.18 mm) soil with 4 to 5 g of Na_2CO_3 in a platinum crucible. Cover mixture with a thin layer of Na_2CO_3 (approximately 1 g). Blank samples containing no soil should also be analyzed.

2   Heat the crucible gently (ensuring that the crucible is not in the reducing part of the flame) with a Meker burner to drive off any moisture. Place the lid on the crucible so that about one fifth of the crucible top is uncovered. Continue heating with a low flame for the first 10 min so as to gently fuse the mass.

3   Continue heating at full blast of Meker burner for 15 to 20 min. During this time there may be considerable spattering of the melt. Cautiously lift the lid periodically to provide an oxidizing environment when the spattering has ceased.

4   Remove the flame and allow the crucible to cool momentarily before handling. This will prevent the easy deformation of the crucible. Gently rotate the crucible as it cools so as to distribute the fused mass as a thin layer over the interior.

5   When the crucible is cool enough to touch, roll it gently between hands to facilitate removal of the melt. Remove the melt with 30 mL of 4.5 *M* H_2SO_4, taking care that no material is lost due to effervescence, and place it in a beaker. Place the crucible and lid in a separate small beaker containing 25

mL of 1 M H_2SO_4, and heat the contents to boiling. Add this portion to the main solution.

6 Transfer solution and suspended material to a 250-mL volumetric flask, make to volume with distilled/deionized water, and mix thoroughly. The final solution is approximately 0.6 M H_2SO_4.

7 Either filter two or three times with suction through the same Whatman® No. 42 filter paper to ensure clear filtrate, or allow the sediment to settle.

8 Determine P concentration in an aliquot (see Section 23.1) of either the clear filtrate or the supernatant and in soil sample as indicated in Chapter 10 (10.4.2).

23.2.3 Perchloric Acid Digestion (Olsen and Sommers, 1982)

The method described by Olsen and Sommers (1982) indicates that before digesting in 60% $HClO_4$, samples high in organic matter should be digested in concentrated HNO_3 (to oxidize organic matter). As a safety precaution, it is recommended that samples routinely be digested in concentrated HNO_3 before proceeding with the $HClO_4$ digestion. This method is suitable for the determination of phosphorus in a large number of samples, although the use of a *perchloric acid fumehood is essential*. The digestion can be carried out using 250-mL Erlenmeyer flasks and a hot plate, or by using an aluminum block digestor with 75-, 100-, or 250-mL digestion tubes. (See comments for the method describing the procedure for using 75- or 100-mL digestion tubes.)

23.2.3.1 MATERIALS AND REAGENTS

1 Perchloric acid fumehood.

2 Hot plate (with 250-mL Erlenmeyer flasks) or block digestor (with 250-mL digestion tubes).

3 Nitric acid (HNO_3), concentrated.

4 Perchloric acid ($HClO_4$), 60%.

5 Color developing solutions: see Chapter 10.

23.2.3.2 PROCEDURE

1 Accurately weigh approximately 2.0 g of finely ground soil (0.15 to 0.18 mm) into a 250-mL volumetric or Erlenmeyer flask. Blank samples containing no soil should also be analyzed.

2   Add 20 mL concentrated HNO_3 to flask and mix well. Heat (approximately 130°C) to oxidize the organic matter in the sample. Organic matter oxidation is complete when the dark color due to the organic matter in the sample disappears.

3   Allow the soil-HNO_3 mixture to cool slightly. In the perchloric acid fume-hood add 30 mL of $HClO_4$ and digest the sample at the boiling temperature (approximately 200°C) for 20 min. During this time dense white fumes should appear and the insoluble solid material left in the bottom of flask/digestion tube should appear like white sand. If necessary, use a little (less than 2 mL) extra $HClO_4$ to wash down any black particles that have stuck to the sides of the flask/digestion tube. Heat for another 10–15 min.

4   Allow the mixture to cool. With distilled/deionized water transfer the mixture to a 250-mL volumetric flask and make to volume with distilled/deionized water. Mix thoroughly.

5   Allow sediment to settle before taking an aliquot for analysis.

6   Determine P concentration in an aliquot (see Section 23.1) of the clear supernatant as indicated in Chapter 10 (10.4.2).

23.2.3.3 COMMENTS

1   A 40-tube block digestion system with volumetric 75- or 100-mL digestion tubes can also be used in this procedure by using half the amounts of sample, HNO_3, and $HClO_4$ described above. The digested material is diluted to a final volume of 75 or 100 mL (step 4).

23.2.4 Sodium Hypobromite/Sodium Hydroxide Alkaline Oxidation Method (Dick and Tabatabai 1977)

This method involves the boiling to dryness of a mixture of soil and NaOBr-NaOH solution using a sand bath. Formic acid is added after completion of the NaOBr-NaOH treatment to destroy residual NaOBr remaining after the oxidation of the sample. P_t is then extracted from the sample using 0.5 M H_2SO_4. The results of this method agree closely with those obtained by the $HClO_4$ method. The method permits the digestion of a large number of samples at one time, although more manipulation of the sample is required compared to the $HClO_4$ method.

23.2.4.1 MATERIALS AND REAGENTS

1   Sand bath for which the temperature of the sand can be regulated at 260–280°C. A sand bath may be prepared by placing 3–4 cm of silica sand on a hot plate.

2   Fumehood.

3   Boiling flask (50 mL) with stoppers.

4   Centrifuge with 50-mL centrifuge tubes.

5   Sodium hydroxide (NaOH), 2 *M*: dissolve 80 g NaOH in a 1-L volumetric
     flask containing 600 mL of distilled/deionized water. Allow to cool, and make
     to volume with distilled/deionized water.

6   Sodium hypobromite (NaOBr)/sodium hydroxide (NaOH) solution: prepare
     this solution in a fumehood by adding 3 mL of bromine slowly (0.5 mL per
     min) and with constant stirring to 100 mL of 2 *M* NaOH. Prepare the NaOBr-
     NaOH solution immediately before use.

7   Formic acid (HCOOH), 90%.

8   Sulfuric acid (H_2SO_4), 0.5 *M*: add 28 mL concentrated H_2SO_4 to 600 mL
     distilled/deionized water in a 1-L volumetric flask. Mix, allow to cool and
     make to volume using distilled/deionized water.

9   Color developing solutions: see Chapter 10 (10.4.2).

23.2.4.2 PROCEDURES

1   Accurately weigh between 0.10–0.20 g of finely ground (0.15 to 0.18 mm)
     soil into a dry 50-mL boiling flask. Blanks containing no soil should also be
     analyzed.

2   Add 3 mL of NaOBr-NaOH solution to the boiling flask, and swirl the flask
     for a few seconds to mix the contents. Allow the flask to stand for 5 min,
     and then swirl the flask again for a few seconds.

3   Place the flask upright in a sand bath (temperature regulated between 260
     and 280°C) situated in a fumehood. Heat the flask for 10–15 min until its
     contents are evaporated to dryness, and continue heating for an additional
     30 min.

4   Remove the flask from the sand bath, cool for about 5 min, add 4 mL of
     distilled/deionized water and 1 mL of 90% HCOOH. Mix the contents, and
     then add 25 mL of 0.5 *M* H_2SO_4. Stopper the flask and mix the contents.

5   Transfer the mixture to a 50-mL plastic centrifuge tube and centrifuge at
     15,000 × *g* for 1 min.

6   Determine P concentration in an aliquot (see Section 23.1) of the clear
     supernatant as indicated in Chapter 10 (10.4.2).

23.2.4.3 COMMENTS

1   It is very important that the NaOBr-NaOH solution be prepared immediately before use. Dick and Tabatabai (1977) reported that storing the NaOBr-NaOH at 4°C for 24 h reduced P_t values by 2–4%.

2   Sample sizes up to 0.5 g can be analyzed for most soils. However, samples containing high amounts of Fe show large decreases in the value of P_t when sample sizes are increased above 0.2 g.

23.2.5   Sulfuric Acid/Hydrogen Peroxide/Hydrofluoric Acid Digestion (Bowman 1988)

This method involves the digestion of the soil sample by the sequential additions of concentrated H_2SO_4, H_2O_2, and HF. The precision and accuracy is similar to those of the $HClO_4$ and NaOBr-NaOH methods. The method is suited for the analysis of a small number of samples at one time. The time required for manual additions of the H_2O_2 and HF is similar to the manipulations required for the NaOBr-NaOH method, which makes the procedure slightly more labor intensive than the $HClO_4$ method.

23.2.5.1 MATERIALS AND REAGENTS

1   Fluoropolymer beaker (100 mL) of known weight.

2   Fumehood.

3   Balance or 50-mL volumetric flask.

4   Sulfuric acid (H_2SO_4), concentrated.

5   Hydrogen peroxide (H_2O_2), 30%.

6   Hydrofluoric acid (HF), concentrated (see comments).

7   Color developing solutions: see Chapter 10 (10.4.2).

23.2.5.2 PROCEDURES

1   Accurately weigh 0.5 g of finely ground (0.15 to 0.18 mm) soil into a 100-mL fluoropolymer beaker of known weight. For soil high in organic matter use 0.25 g. Blank samples containing no soil should also be analyzed.

2   In the fumehood add 5 mL (9.2 g) of H_2SO_4 to the soil and gently swirl to remove solid materials adhering to the bottom of the beaker.

3   In the fumehood, slowly add 0.5 mL of H_2O_2 and mix vigorously to promote the oxidation of organic materials. Repeat this step until 3 mL of H_2O_2 has been added to the beaker. Let sample sit until the reaction with H_2O_2 has subsided.

4   Add 0.5 mL of HF to the beaker and mix. Repeat this step again so that a total of 1 mL of HF has been added.

5   Place beaker on a preheated hot plate (approximately 150°C) for 10 to 12 min to eliminate excess H_2O_2.

6   Remove beaker, and while sample is still warm, wash down the sides of the beaker with 10 to 20 mL of distilled/deionized water. Mix and cool to room temperature.

7   Weigh the beaker and its contents and add sufficient distilled/deionized water to bring the final weight to 55 g (equivalent to 50-mL volume). Alternatively, the material in the beaker can be quantitatively transferred to a 50-mL volumetric flask and made to volume using distilled/deionized water.

8   Mix and filter the extract through quantitative fine filter paper.

9   Determine P concentration in an aliquot (see Section 23.1) of the clear filtrate as indicated in Chapter 10 (10.4.2).

23.2.5.3 COMMENTS

1   HF acid attacks glass and it is very important that HF-resistant materials are used. An example of such material is Teflon®. The use of this trade name is provided for the benefit of the reader and does not imply endorsement by the CSSS.

2   Excess H_2O_2 will interfere with the colorimetric determination of P in the digested sample. The formation of a yellow color instead of the blue color normally associated with the reduced phospho-molybdate complex indicates the presence of excess H_2O_2.

23.2 ORGANIC PHOSPHORUS

23.3.1 Introduction

Soil P_o is not measured directly, but rather estimated by the difference between P_t and P_i determined either by extraction or ignition techniques. Differences between techniques or soil types may reflect a change in the efficiency of the procedure, rather than a true change in the amount of P_o in the soil. The extraction techniques (Mehta et al. 1954, Anderson 1960, Bowman 1989) involve the use of various acid and base treatments with the subsequent determination of the P_i and P_t in the extractants. The two major problems with these techniques are the incomplete extraction of soil P_o and the possible hydrolysis of P_o by the extractants.

In general, these techniques tend to give the lower range of soil P_o values. The ignition technique uses either high (550°C, Saunders and Williams 1955) or low (250°C, Legg and Black 1955) temperatures to oxidize soil P_o to P_i. Matched ignited and unignited samples are then extracted with either weak or strong acids, and the difference between P_t (ignited sample) and P_i (unignited sample) is considered P_o. This technique may result in erroneous estimates of P_o due to changes in the solubilities of P minerals by ignition at either high or low temperatures, while ignition at higher temperatures may also cause volatilization of P. Each technique has its advantages and disadvantages, depending on the situation and the purpose of the study in question. As indicated by Bowman (1989), the extraction techniques are more suited for comparisons of P_o levels across different soil types, while ignition techniques are more suitable for comparisons among treatments within a soil type. Due to the errors which may be associated with P_o determinations and since the P_o is determined by difference, little significance can be given to treatments which differ by less than 20 μg P g^{-1} soil (Olsen and Sommers 1982).

Further information regarding comparisons of various methods for the determination of P_o in soils can be obtained by referring to Condron et al. 1990, Dormaar (1964), Dormaar and Webster (1963), (1964), Hance and Anderson (1962), Sommers et al. (1970), Steward and Oades (1972), and Williams et al. (1970).

23.3.2 Hydrochloric Acid/Sodium Hydroxide Extraction Method (Mehta et al. 1954)

In this method soils are sequentially treated with concentrated HCl, 0.5 M NaOH at room temperature, and 0.5 M NaOH at 90°C. P_i in extracts is determined immediately after extraction, and P_t determined after the oxidation of the organic matter with HClO$_4$.

23.3.2.1 MATERIALS AND REAGENTS

1   Polypropylene centrifuge tubes (100 mL).

2   Steam bath.

3   Oven for NaOH extraction of samples at 90°C.

4   Perchloric acid fumehood.

5   Aluminum block digestor and digestion tubes.

6   Concentrated sulfuric acid (H_2SO_4).

7   Concentrated hydrochloric acid (HCl).

8   Sodium hydroxide (NaOH), 0.5 M: dissolve 20 g NaOH in a 1-L volumetric flask containing 600 mL of distilled/deionized water. Make to volume with distilled/deionized water.

9   Perchloric acid (HClO$_4$), 72%.

10   Color developing solutions: see Chapter 10 (10.4.2).

23.3.2.2 PROCEDURES

1   Weigh 1.0 g of finely ground (<0.15 mm) soil into a 100-mL polypropylene centrifuge tube. Blank samples containing no soil should also be analyzed.

2   Add 10 mL of concentrated HCl, mix thoroughly, and then place the tube in a steam bath for approximately 10 min (final solution temperature should be about 70°C). Remove the tube from the steam bath, add 10 mL concentrated HCl, and allow to stand at room temperature for 1 h. Add approximately 50 mL of distilled/deionized water to the tube, mix thoroughly, and centrifuge the suspension (approximately 1500 × *g*) for 15 min. Decant the supernatant into a 250-mL volumetric flask containing 50 mL of distilled/deionized water, taking care to keep the residue in the tube.

3   Add 30 mL of 0.5 *M* NaOH to the soil residue in the tube. Mix well and allow to stand at room temperature for 1 h. Centrifuge (1500 × *g*) the suspension for 15 min, and decant the supernatant into the 250-mL volumetric flask containing the concentrated HCl extract, taking care to keep the residue in the tube.

4   Add 60 mL of 0.5 *M* NaOH to the soil residue in the tube, mix well, and place the tube in a 90°C oven for 8 h. Tubes should be loosely capped (e.g., 50-mL beaker, funnel, etc.) during heating. Remove tubes from oven, allow to cool, and then centrifuge (1500 × *g*) for 15 min. Transfer supernatant to the 250-mL volumetric flask containing the concentrated HCl and cold NaOH extracts.

5   Dilute the contents of the flask to volume with distilled/deionized water and mix thoroughly to suspend flocculated organic matter. Note: determination of P_i (step 9) in this material should be made as soon as possible (see comments). Pipette a 5-mL aliquot to a digestion tube.

6   To determine P_t in the extract add one drop of concentrated H_2SO_4 and 1 mL of 72% $HClO_4$ to the digestion tube (in the fumehood), and place the tube in the digestion block. Raise the temperature of the block slowly until dense white fumes of $HClO_4$ appear. Place a 25-mm funnel atop the tube. Increase the temperature to 205°C and digest the sample for 30 min. Remove the tube from the digestion block, allow to cool, and quantitatively transfer the digest to a 50-mL volumetric flask, rinsing the tube with <35 mL of distilled/deionized water. If this solution is to be used directly for colorimetric determination, the volume of liquid in the flask must be less than 42 mL.

7   If the material digested contains more than 2000 μg P g^{-1} soil, dilute the digest to 50 mL with distilled/deionized water, and use a suitable aliquot (see Section 23.1) for the colorimetric determination of P (see Chapter 10).

8   To determine P_i in the extract, allow the flocculated organic material to settle, and then remove a suitable aliquot (see Section 23.1) for P determination (see Chapter 10).

23.3.2.3 *Calculations*

After determining the concentration of P in the digests and extracts and converting each to a soil weight basis (e.g., mg P kg^{-1} soil), P_o is calculated as:

$$Po = Pt \text{ (digest)} - Pi \text{ (extract)} \qquad (23.1)$$

23.3.2.4 COMMENTS

1   Care must be taken when handling perchloric acid, and specially designed perchloric acid fumehoods should be used when using perchloric acid to reduce the potential for explosions.

2   P_i should be determined in the extracts as soon as possible to reduce the chance of P hydrolysis resulting in an underestimation of soil P_o.

23.3.3 Sulfuric Acid/Sodium Hydroxide Extraction Method (Bowman 1989)

In this procedure, P_o is sequentially extracted from the soil using concentrated H_2SO_4 and 0.5 M NaOH. This method is simple, faster than either the HCl/NaOH extraction method of Mehta et al. (1954) or the ignition method of Saunders and Williams (1955), and is equally precise. Results are similar to those obtained by the HCl/NaOH extraction method, and less than those of the ignition method. The procedure appears to cause very little hydrolysis of P_o. Condron et al. (1989) found it to be a suitable technique for tropical soils.

23.3.3.1 MATERIALS AND REAGENTS

1   Volumetric flask (50 mL).

2   Erlenmeyer flask (250 mL).

3   Graduated pipette.

4   Sulfuric acid (H_2SO_4), concentrated.

5   Sulfuric acid (H_2SO_4), 5.5 M: add 306 mL of concentrated H_2SO_4 to a 1-L volumetric flask containing 500 mL of distilled/deionized water. Mix, allow to cool, and make to volume using distilled/deionized water.

6   Sodium hydroxide (NaOH), 0.5 M: dissolve 20 g NaOH in a 1-L volumetric flask containing 600 mL of distilled/deionized water. Make to volume with distilled/deionized water.

7　　Potassium persulfate ($K_2S_2O_8$).

8　　Color developing solutions: see Chapter 10.

23.3.3.2 PROCEDURES

1　　Weigh 2.0 g of finely ground (<0.15 mm) soil into a 50-mL volumetric flask. Blank samples containing no soil should also be analyzed.

2　　Add 3 mL of concentrated H_2SO_4 (4 mL if soil is calcareous) to the flask and swirl gently to mix the acid and soil together.

3　　With a graduated pipette, add 4 mL of distilled/deionized water 1 mL at a time. Swirl the flask vigorously for 5 to 10 s after each addition.

4　　Wash down the sides of the flask with about 5 to 10 mL of distilled/deionized water, and mix vigorously. Add distilled/deionized water to approximately 1 cm below the graduation line, let cool, and make to volume with distilled/deionized water and mix thoroughly. The final volume of this extract is slightly less than 50 mL.

5　　Filter 20 to 30 mL of suspension through Whatman® No. 1 filter paper, and recover and save the extract. This extract contains the acid-extractable P_o (Acid-P_o).

6　　Quantitatively transfer soil remaining in the volumetric flask to the filter paper. Wash soil on filter paper with approximately 5 mL of distilled/deionized water using a fine-tip water bottle and let drain completely. Discard this second filtrate, but save the filter paper with soil residue.

7　　Tear filter paper carefully so as not to lose any soil and place filter paper and soil residue in a 250-mL Erlenmeyer flask.

8　　Add 100 mL of 0.5 M NaOH to the flask and shake 2 h. Filter through Whatman® No. 1 filter paper, and recover and save the extract. This extract contains the base-extractable P_o (Base-P_o).

9　　For acid-extractable total P (Acid-P_t) and base-extractable total P (Base-P_t), pipette a suitable aliquot (e.g., 1 mL; see Section 23.1) into a 25-mL volumetric flask, add about 0.5 g of $K_2S_2O_8$ with a calibrated scoop, and 1 mL of 5.5 M H_2SO_4. Digest the sample at high temperature (>150°C) on a hot plate for 20 to 30 min. Lack of vigorous boiling is usually an indication that the digestion has terminated. Develop color as indicated in Chapter 10, noting that if a 25-mL volumetric flask is used, only 4 mL of color developing solution is used.

10　　Determine Acid-P_i and Base-P_i in a suitable aliquot (see Section 23.1) of the extracts as indicated in Chapter 10 (10.4.2).

23.3.3.3 Calculations

After determining the concentration of P in the digests and extracts and converting each to a soil weight basis (e.g., mg P kg^{-1} soil), P$_o$ is calculated as follows.

$$\text{Acid-Po} = \text{Acid-Pt (digest)} - \text{Acid-Pi (extract)} \tag{23.2}$$

$$\text{Base-Po} = \text{Base-Pt (digest)} - \text{Base-Pi (extract)} \tag{23.3}$$

$$\text{Total Po} = \text{Acid-Po} + \text{Base-Po} \tag{23.4}$$

23.3.3.4 COMMENTS

1 Bowman (1989) suggested that if specific detail of Acid-P$_o$ and Base-P$_o$ is not required, the extracts could be combined (10 mL acid extract + 20 mL base extract) and analyses for P$_t$ and P$_i$ conducted on the combined sample. If the extracts are combined in this manner, the aliquots for P determination should be taken immediately to minimize the effects of humic-matter precipitations.

2 The acid extraction conducted in the 50-mL volumetric flask results in an extractant volume slightly less than 50 mL due to the volume of the soil in the flask. For most mineral soils this error would be less than 2%. Condron et al. (1990) modified the procedure by placing the soil in a 250-mL plastic centrifuge tube, proceeding with steps 2 and 3 as above, and then adding 41 mL of distilled/deionized water. This would give a final volume of 48 mL for the extractant. One could just as easily add 43 mL of distilled/deionized water and have a final extractant volume of 50 mL.

3 Using the 250-mL centrifuge tubes as suggested by Condron et al. (1990) has two advantages: (1) the soil-acid mixture and the soil-water mixture can be centrifuged before filtering, and (2) there is no need to transfer the soil to another vessel, since the 100 mL of 0.5 M NaOH can be added directly to the soil and filter paper in the 250-mL centrifuge tube.

4 Bowman (1989) extracted the soil and filter paper with 98 mL of 0.5 M NaOH. Considering the amount of water retained by the soil and filter paper (usually less than 2 mL), the final extracting solution volume would be approximately 100 mL. By weighing the filter paper and soil, the amount of water retained can easily be measured for a few representative samples, and the actual extracting volume (volume water retained + volume 0.5 M NaOH added) calculated. The addition of 100 mL of 0.5 M NaOH solution would be more convenient than adding 98 mL, and without the calculation of water retained by the soil and filter paper, result in an error in volume of less than 2%.

5 Condron et al. (1990) modified the filtering step by first centrifuging the cool acid-soil mixture, and then filtering through a 45-μm Millipore® (Millipore® Corporation, Bedford, MA) filter under suction. The soil was washed

with distilled/deionized water, centrifuged, and the supernatant discarded. The Millipore® filter was added to the 250-mL centrifuge containing the soil and shaken with 98 mL of 0.5 M NaOH (again it would be easier to add 100 mL of 0.5 M NaOH). Although the amount of liquid retained by the Millipore® filter is less than that retained by the Whatman® No. 1 filter paper (approximately 1 mL less), the error would be less than 1%.

6   P_t in the acid and base extracts can also be determined by digesting the sample using ammonium persulfate and an autoclave as indicated in Chapter 10 for the determination of P_t in the $NaHCO_3$ and NaOH extracts. Digest the aliquot (usually 1 to 2 mL) of the acid extract following the procedure for P_t in the $NaHCO_3$ extracts, and the aliquot of base extract following the procedure for the NaOH extract.

23.3.4 Ignition Method (Saunders and Williams 1955, as modified by Walker and Adams 1958)

In this method, P_o is estimated by the difference between 0.5 M H_2SO_4-extractable P in a soil sample ignited at 550°C and an unignited sample. The method is suitable for the determination of soil P_o in a large number of samples. Dormaar and Webster (1964) have indicated that significant volatile of losses of P may occur at temperatures above 400°C, especially with peat soils.

23.3.4.1 MATERIALS AND REAGENTS

1   Muffle furnace and porcelain crucibles for igniting soils at 550°C.

2   Polypropylene centrifuge tubes (100 mL) with caps or stoppers.

3   Shaker capable of holding the above tubes.

4   Centrifuge.

5   Sulfuric acid (H_2SO_4), 0.5 M: add 28 mL concentrated H_2SO_4 to 600 mL distilled/deionized water in a 1-L volumetric flask. Allow to cool, and make to volume using distilled/deionized water.

6   Color developing solutions: see Chapter 10 (10.4.2).

23.3.4.2 PROCEDURES

1   Weigh 1.0 g of finely ground (0.15 to 0.18 mm) soil in a porcelain crucible, and place the crucible in a cool muffle furnace. Blank samples containing no soil should also be analyzed.

2   Slowly raise the temperature of the muffle furnace to 550°C over a period of approximately 2 h. Continue to heat the samples at 550°C for 1 h, then remove the samples and allow them to cool.

3   Transfer the ignited soil to a 100-mL polypropylene centrifuge tube for extraction of P_t.

4   To a separate 100-mL polypropylene centrifuge tube, weigh 1.0 g of unignited soil for the extraction of P_i.

5   Add 50 mL of 0.5 M H_2SO_4 to both samples, mix well, and allow to sit lightly stoppered for a few minutes to avoid pressure from CO_2 released from any carbonates that may be present in the soil sample. Tightly stopper the tubes and place them on a shaker for 16 h. Blank samples containing only 0.5 M H_2SO_4 should also be included.

6   Centrifuge the samples at approximately 1500 $\times$ g for 15 min. If the extract is not clear, filtration using acid-resistant filter paper may be required.

7   Determine P concentration in an aliquot (see Section 23.1) of clear supernatant or filtrate as indicated in Chapter 10 (10.4.2).

23.3.4.3 Calculations

After determining the concentration of P in the extracts of the ignited and unignited soil samples and converting to a soil weight basis (e.g., mg P kg^{-1} soil), P_o is calculated as follows:

$$Po = P \text{ ignited sample} - P \text{ unignited sample} \tag{23.5}$$

23.3.4.4 COMMENTS

1   To prevent volatilization of P from the sample, care must be taken to not allow temperatures in the muffle furnace to exceed 550°C when using mineral soils (Sommers et al. 1970, Williams et al. 1970).

REFERENCES

Anderson, G. 1960. Factors affecting the estimation of phosphate esters in soil. J. Sci. Food Agric. 11: 497–503.

Bowman, R. A. 1988. A rapid method to determine total phosphorus in soils. Soil Sci. Soc. Am. J. 52: 1301–1304.

Bowman, R. A. 1989. A sequential extraction procedure with concentrated sulfuric acid and dilute base for soil organic phosphorus. Soil Sci Soc. Am. J. 53: 362–366.

Condron, L. M., Moir, J. O., Tiessen, H., and Stewart, J. W. B. 1990. Critical evaluation of methods for determining total organic phosphorus in tropical soils. Soil Sci. Soc. Am. J. 54: 1261–1266.

Dick, W. A. and Tabatabai, M. A. 1977. An alkaline oxidation method for determination of total phosphorus in soils. Soil Sci. Soc. Am. J. 41: 511–514.

Dormaar, J. F. 1964. Evaluation of methods for determination of total organic phosphorus in chernozemic soils of southern Alberta. Can. J. Soil Sci. 44: 265–271.

Dormaar, J. F. and Webster, G. R. 1963. Determination of total organic phosphorus in soil by extraction methods. Can. J. Soil Sci. 43: 35–43.

Dormaar, J. F. and Webster, G. R. 1964. Losses inherent in ignition procedures for determining total organic phosphorus. Can. J. Soil Sci. 44: 1–6.

Fogg, D. N. and Wilkinson, N. T. 1958. The colorimetric determination of phosphorus. Analyst (London) 83: 406–414.

Hance, R. J. and Anderson, G. 1962. A comparative study of methods of estimating soil organic phosphate. J. Soil Sci. 13: 225–230.

Jackson, M. L. 1958. Soil chemical analysis. Prentice-Hall, Englewood Cliffs, NJ.

Legg, J. O. and Black, C. A. 1955. Determination of organic phosphorus in soils. II. Soil Sci. Soc. Am. Proc. 19: 139–143.

Mehta, N. C., Legg, J. O., Goring, C. A. I., and Black, C. A. 1954. Determination of organic phosphorus in soils. I. Extraction methods. Soil Sci. Soc. Am. Proc. 18: 443–449.

Murphy, J. and Riley, J. P. 1962. A modified single solution method for the determination of phosphate in natural waters. Anal. Chim. Acta 27: 31–36.

Olsen, S. R. and Sommers, L. E. 1982. Phosphorus. Pages 403–430 in A. L. Page, R. H. Miller, and D. R. Keeney, Eds. Methods of soil analysis. Part 2. 2nd ed. Agronomy No. 9. American Society of Agronomy Madison, WI.

Saunders, W. M. and Williams, E. G. 1955. Observations on the determination of organic phosphorus in soils. J. Soil Sci. 6: 254–267.

Sommers, L. E., Harris, R. F., Williams, J. D. H., Armstrong, D. E., and Syers, J. K. 1970. Determination of total organic phosphorus in lake sediments. Limnol. Oceanogr. 15: 301–304.

Steward, J. H. and Oades, J. M. 1972. The determination of organic phosphorus in soils. J. Soil Sci. 23: 38–49.

Syers, J. K., Williams, J. D. H., and Walker, T. W. 1967. The significance of apatite inclusions in soil phosphorus studies. Soil Sci. Soc. Am. Proc. 31: 752–756.

Syers, J. K., Williams, J. D. H., and Walker, T. W. 1968. The determination of total phosphorus in soils and parent materials. N.Z. J. Agric. Res. 11: 757–762.

Walker, T. W. and Adams, A. F. R. 1958. Studies on soil organic matter. I. Influence of phosphorus content of parent materials on accumulations of carbon, nitrogen, sulfur and organic phosphorus in grassland soils. Soil Sci. 85: 307–318.

Williams, J. D. H., Syers, J. K., Walker, T. W., and Rex, R. W. 1970. A comparison of methods for the determination of soil organic phosphorus. Soil Sci. 110: 13–18.

Chapter 24
Total and Fractions of Sulfur

C. G. Kowalenko

Agriculture Canada Research Station
Agassiz, British Columbia, Canada

24.1 INTRODUCTION

Sulfur is present in the soil in a variety of chemical forms, both organic and inorganic, and in a variety of valence states (Blanchar 1986). Numerous methods have been used to determine total sulfur in the soil, but none appears to have found universal acceptance (Beaton et al. 1968, Blanchar 1986, Tabatabai 1982), probably because the forms of sulfur and their distribution varies considerably with soil type.

Although the analytical methods for total soil sulfur vary considerably, almost all methods require two steps: (1) the conversion of all of the various sulfur forms in the soil into a single form and (2) the quantification of that form of sulfur. The two steps must be compatible with each other. Methods available for the conversion step include various types of ashing or dry combustion and wet digestion (Beaton et al. 1968, Blanchar 1986, Tabatabai 1982). Dry ashing would include the use of ovens, heating elements, open flames (e.g., fusion), enclosed flames (e.g., oxygen flask), or high-temperature combustion using induction or resistance furnaces. Wet digestion may be either alkaline or acidic. Numerous quantification methods are available for both gases and solutions in either oxidized or reduced forms. Gases can be quantified directly by infrared, chemiluminescent, coulometric, flame photometric, and other methods. Solutions (either absorbed/dissolved gases, liquid digests, or solubilized solids) are usually analyzed in the sulfide or sulfate forms by spectrometry (colorimetry, flame emission, atomic absorption, etc.), ion-selective electrode, titration, gravimetric, and chromatographic methods. Both or either of the conversion and the quantification steps can be automated. X-ray fluorescence (Jenkins 1984) is different in that quantification is done directly without a conversion step.

Comparisons of various methods for determining total sulfur of soils have been conducted (e.g., Gerzabek and Schaffer 1986, Tabatabai and Bremner 1970b); however, the comparisons have been quite limited with respect to the scope of soils analyzed and the type of methods used, nor were the methods optimized for specific types of samples (Matrai 1989, Searle 1968, Tabatabai and Bremner 1970b). Variable results have been shown among the methods; however, no one method could clearly be said to give a true estimate of total sulfur (Hogan and Maynard 1984). The highest value cannot necessarily be taken as the true value.

Each method for determining total sulfur has its own particular advantages and disadvantages. High-temperature combustion and X-ray methods are relatively fast and easy to do, but both require specialized and often costly instrumentation. High-temperature combustion has not been found to yield particularly satisfactory results for some soil samples (Gerzabek and Schaffer 1986, Lowe 1969, Tabatabai and Bremner 1970b, Tiedeman and Anderson 1971). Analysis of a wide range of soil samples showed that determinations by X-ray fluorescence requires an adjustment for the organic matter content in the sample (Brown and Kanaris-Sotiriou 1969). Low-temperature ashing can be done with more readily available laboratory equipment. Closed combustion is possible, but further development and evaluation are required. A method proposed by Killham and Wainwright (1981) did not result in complete recovery of elemental sulfur nor methionine. Tabatabai and Bremner (1970a) showed that an alkaline low-temperature ashing and an acid digestion produced results that were similar and precise for seven soils, and they concluded that the results were satisfactory. Acid digestion should be used with caution, as gaseous losses of sulfur are possible (Randall and Spencer 1980). Recently, Tabatabai et al. (1988) have shown good results using dry ashing with sodium bicarbonate and silver oxide followed by ion chromatographic analysis of the sulfate produced. More study of this oxidation may be required, since an earlier study (Tabatabai and Bremner 1970b) found that the sodium bicarbonate/silver oxide ashing coupled with sulfur determination by hydriodic acid reduction was not comparable to nor as precise as sodium hypobromite or acid digestion. The sodium bicarbonate/silver oxide ashing appeared to be suitable for plant samples (Tabatabai et al. 1988) and therefore may be suitable for high organic matter-containing soils. However, this has not been tested. A fusion technique proposed for geological samples using sodium peroxide and ion chromatography was not completely satisfactory (Stallings et al. 1988). However, Hordijk et al. (1989) found good agreement for total sulfur measurement in freshwater sediments by ion chromatography after Na_2CO_3/KNO_3 fusion with inductively coupled plasma (ICP)- and fluorescent-based methods. The sodium bicarbonate/silver oxide ash should be compatible with inductively coupled plasma sulfur quantification (Hogan and Maynard 1984, Pritchard and Lee 1984), but tests should be run.

Most of the sulfur present in surface horizons of soils is present in the organic form (Biederbeck 1978, Blanchar 1986, Tabatabai 1982). Inorganic sulfur is present largely as sulfate in aerobic soils, and in reduced forms (sulfide, sulfite, elemental sulfur, etc.) in anaerobic soils (Beaton et al. 1968, Blanchar 1986, Tabatabai 1982). Methods for the determination of inorganic sulfate in soils using varying extractants have been studied extensively and discussed in another section of this manual (see Chapter 9, Extraction of Available Sulfur). Specialized methods to determine total inorganic sulfate are required for certain soils and some subsurface horizons, such as gypsiferous (Khan and Webster 1968, Nelson 1982) or acid sulfate (Begheijn et al. 1978) soils. Methods for determining reduced forms of inorganic sulfur have been proposed (Beaton et al. 1968, Blanchar 1986, Tabatabai 1982); however, the methods have not been thoroughly evaluated (Barrow 1970, Watkinson et al. (1987) because these fractions have limited agricultural significance. The reduced forms are of greater significance in marine and freshwater sediments (e.g., see Davison and Lishman 1983). Determination of reduced inorganic forms of sulfur by zinc-hydrochloric acid distillation has been proposed for lake sediments (Aspiras et al. 1972). The method appears to be relatively specific to reduced inorganic sulfur and apparently does not include sulfate, persulfate, and organic sulfur. However, it does not completely recover all reduced forms of inorganic sulfur. This method has been applied to soils (David et al. 1982, 1983, Roberts and Bettany 1985), but more work is required to understand the limitations and significance of this analysis.

Melville et al. (1971) found that although digestion of soil with tin and hydrochloric acid resulted in measurable sulfide release, the method was not specific to inorganic sulfur. Pirela and Tabatabai (1988) found that a tin and phosphoric acid digestion reduced more soil sulfur than tin and hydrochloric acid digestion; however, the phosphoric acid method was not specific to inorganic nor organic forms, since various "model" sulfur compounds were not always quantitatively reduced. The method will have limited applicability to soil studies because of its lack of specificity, but the procedure should be studied further. In a study on the application of tin and phosphoric acid digestion to geologic materials, Mizoguchi et al. (1980) showed that copper caused some interference and that addition of iodide eliminated the interference.

The most common fractionation of soil sulfur has been the reduction of sulfate (including both organic and inorganic) with a reagent containing hydriodic acid (Beaton et al. 1968, Biederbeck 1978, Tabatabai 1982) based on original work by Johnson and Nishita (1952). The method appears to be relatively free from interferences and is quite specific to the sulfate form (Tabatabai 1982). This method, then, is usually assumed to give a good estimate of total sulfate-sulfur in most agricultural soils. This assumption should be treated with caution on certain soils, as considerable elemental sulfur may be reduced by this reagent (Pirela and Tabatabai 1988). Lowe and DeLong (1963) proposed the determination of carbon-bonded sulfur using a digestion with Raney nickel in sodium hydroxide. Although this method was shown to be quite specific for carbon-bonded sulfur, the method is not quantitative due to interference problems in soils and soil extracts (Freney et al. 1970, Scott et al. 1981). The difference between total sulfur and hydriodic acid-reducible sulfur appears to provide a better estimate of carbon-bonded sulfur than direct determination by Raney-nickel digestion.

Many other methods have been attempted to determine organic sulfur compounds or fractions in soils either directly or on extracts (Kowalenko 1978), but none have been universally accepted. Determination of sulfur-containing amino acids such as methionine, cystine, etc., has been done, but special precautions are required. These do not appear to account for a very large portion of the total sulfur present. There has been some success in determining the sulfur content of lipid extracts (Kowalenko 1978, Chae and Lowe 1981, Chae and Tabatabai 1981), but this fraction also accounts for only a small portion of the total sulfur in the soil. Microbial sulfur comprises only a small portion of the total sulfur (Chapman 1987, Saggar et al. 1981, Strick and Nakas 1984), but may have considerable biological significance. Since estimation of microbial sulfur is dependent on the quantitative determination of inorganic sulfate, the limitations of extracting and specifically determining inorganic sulfate (see Chapter 9, Extraction of Available Sulfur) must be taken into account. Measurements of microbial sulfur will be more difficult in soils than can adsorb sulfate-sulfur because the high background sulfate will make it difficult to determine the small changes in sulfate content.

The hydriodic acid reduction procedure has been widely accepted as a method to fractionate total soil sulfate into at least a chemically accepted form. The method is quite specific for sulfate. Any of the known exceptions (portions of elemental sulfur, etc.) to this specificity are usually not present in normal (i.e., aerated) agricultural soils in very large quantities. The chemical reduction method does not distinguish between inorganic and organic forms of sulfate-sulfur. An estimate of the organic sulfate-sulfur content is frequently determined by difference (i.e., subtracting the quantity of solution and adsorbed sulfate-sulfur estimated by an appropriate extraction procedure from the total sulfate-sulfur determined directly on the soil). Since the hydriodic acid reduction is considered to be a relatively good determination, the method is also assumed to give the best estimate for carbon-bonded sulfur by difference between total sulfur and total sulfate-sulfur. In poorly aerated soils, such as wetlands, correction for reduced inorganic forms of sulfur which may be present should be made.

24.2 ALKALINE DIGESTION WITH HYDRIODIC ACID REAGENT AND BISMUTH SULFIDE DETERMINATION (TABATABAI AND BREMNER 1970a, KOWALENKO 1985)

The procedure recommended is an alkaline digestion of sulfur to sulfate followed by reduction to sulfide. The sulfide is then determined colorimetrically. The sulfate is converted to hydrogen sulfide with a hydriodic acid reagent (Tabatabai and Bremner 1970a) using a modified reduction/distillation apparatus (Kowalenko 1985). The hydrogen sulfide is determined as bismuth sulfide rather than by the methylene blue reaction (Kowalenko and Lowe 1972). The bismuth sulfide method using the modified apparatus has the advantages of short analysis time, versatility for types of sulfur analyses, and the equipment requirements are small. The procedure has been used satisfactorily on a range of soils and on both organic and inorganic compounds. Application of the same method for tree foliage analysis (Guthrie and Lowe 1984) was not as satisfactory; therefore, the method should be used with caution on soils with high organic matter content. The digestion and quantification must be done in the same vessel (Tabatabai and Bremner 1970a); therefore, an alternate quantification method (e.g., ICP) cannot easily be substituted for the quantification of hydrogen sulfide released by the hydriodic acid digestion/distillation.

24.2.1 MATERIALS AND REAGENTS

1 Sodium hypobromite digestion reagent: in a fumehood, add slowly with constant stirring 3 mL of bromine to 100 mL 2 M sodium hydroxide. This reagent is not very stable, and should be prepared immediately before use or at least daily.

2 Formic acid (90%).

3 Temperature-controlled digestion block: the block heater (commercially available or can be custom built) must accommodate the Taylor tubes and be capable of maintaining a temperature of 250–260°C.

4 Materials and reagents as for quantification of sulfate as bismuth sulfide (see Chapter 9, Section 9.2.1).

24.2.2 PROCEDURE

1 Weigh a finely ground (<100 mesh) soil sample or measure an aliquot of sulfate-sulfur directly into a dry modified Taylor tube. The standard solution should be dried at approximately 100°C in a forced-air oven or, alternatively,

under an infrared lamp (Kowalenko and Lowe 1972). The weight of the sample should be adjusted so that the sulfur content is within the range of the standards that are chosen. The soil weight should be large enough to adequately represent the sample, yet not too large for the digestion solution. For example, digest 0.1–0.4 g of Ah horizon of mineral soil or 0.1 g or less of organic horizon with a 30-mL bismuth sulfide final volume that provides a 0–200 µg S L^{-1} range of analysis.

2   Add 3 mL of the sodium hypobromite digestion reagent and thoroughly wet the soil sample with the reagent by swirling the tube. After letting the sample stand for 5 min, evaporate the mixture to dryness in the digestion block at 250–260°C, then continue heating for an additional 30 min. Remove from heat and allow the tube to cool for about 5 min.

3   Resuspend the digested sample in 1 mL of water by swirling and heating briefly. After cooling, add 1 mL of formic acid to eliminate any excess bromine that may be present.

4   Perform quantification for the sulfur content as for sulfate in a dried soil extract (see Chapter 9, Section 9.2.2).

24.2.3 Calculations

Calculate the sulfur content of the sample by taking into consideration the weight of the sample and using the standard curve produced.

24.2.4 Comments

The sulfur content of soil samples containing a large amount of organic material may be underestimated by this method as described, and sequential digestions and/or longer heating times may be required (Guthrie and Lowe 1984).

24.3 TOTAL SULFATE BY HYDRIODIC ACID REAGENT REDUCTION

The method for determining total sulfate described here utilizes bismuth sulfide colorimetry and a modified apparatus (Kowalenko 1985). This modification to the original method proposed (Johnson and Nishita 1952) contributes to a much reduced reduction/distillation time for each analysis (10 vs. 60 min). The apparatus is easier to build and adaptable for easy preanalysis sample digestion for total sulfur determinations.

24.3.1 Materials and Reagents

The custom-built apparatus, materials, and reagents are the same as outlined previously (see Chapter 9, Section 9.2.1, for quantification of sulfate as bismuth sulfide).

24.3.2 Procedure

The hydriodic acid procedure is applied directly on a weighed soil sample as outlined for determination of a dried aliquot of soil solution (see Chapter 9, Section 9.2.2.2, quantification of sulfate as bismuth sulfide).

24.3.3 Calculations

Calculations are made in a manner similar to that used for total sulfur by taking into account the weight of the soil used in relation to the curve determined with dried sulfate standards. The quantity can be calculated as sulfate ion or preferably sulfate-sulfur.

24.3.4 Comments

When this method is used to determine total organic sulfate (or total carbon-bonded sulfur by difference) the capability of accurately determining total inorganic sulfate or total sulfur must be considered, particularly when unusual samples (e.g., subsurface, anaerobic, organic, etc. samples) are being examined. The difference value will involve the error or variability associated with two analyses rather than one.

REFERENCES

Aspiras, R. B., Keeney, D. R., and Chesters, G. 1972. Determination of reduced inorganic sulfur forms as sulfide by zinc-hydrochloric acid distillation. Anal. Lett. 5: 425–432.

Barrow, N. J. 1970. Note on incomplete extraction of elemental sulphur from wet soil by chloroform. J. Sci. Food Agric. 21: 439–440.

Beaton, J. D., Burns, G. R., and Platou, J. 1968. Determination of sulphur in soils and plant material. Technical Bulletin No. 14. The Sulphur Institute, Washington, D.C. 56 pp.

Begheijn, L. Th., van Breeman, N., and Velthorst, E. J. 1978. Analysis of sulfur compounds in acid sulphate soils and other recent marine soils. Comm. Soil Sci. Plant Anal. 9: 873–882.

Biederbeck, V. O. 1978. Soil organic sulfur and fertility. Pages 273–310 in M. Schnitzer and S. U. Khan, Eds. Soil organic matter. Developments in Soil Science 8, Elsevier, New York.

Blanchar, R. W. 1986. Measurement of sulfur in soils and plants. Pages 465–490 in M. A. Tabatabai, Ed. Sulfur in agriculture. Agronomy No. 27. American Society of Agronomy, Madison, WI.

Brown, G. and Kanaris-Sotiriou, R. 1969. Determination of sulphur in soils by X-ray fluorescence analysis. Analyst 94: 782–786.

Chae, Y. M. and Lowe, L. E. 1981. Fractionation by column chromatography of lipids and lipid

sulphur extracted from soils. Soil Biol. Biochem. 13: 257–260.

Chae, Y. M. and Tabatabai, M. A. 1981. Sulfolipid and phospholipid in soils and sewage sludges in Iowa. Soil Sci. Soc. Am. J. 45: 20–25.

Chapman, S. J. 1987. Microbial sulfur in some Scottish soils. Soil Biol. Biochem. 19: 301–305.

David, M. B., Mitchell, M. J., and Nakas, J. P. 1982. Organic and inorganic sulfur constituents of a forest soil and their relationship to microbial activity. Soil Sci. Soc. Am. J. 46: 847–852.

David, M. B., Schindler, S. C., Mitchell, M. J., and Strick, J. E. 1983. Importance of organic and inorganic sulfur to mineralization processes in a forest soil. Soil Biol. Biochem. 15: 671–677.

Davison, W. and Lishman, J. P. 1983. Rapid colorimetric procedure for the determination of acid volatile sulphide in sediments. Analyst 108: 1235–1239.

Freney, J. R., Melville, G. E., and Williams, C. H. 1970. The determination of carbon bonded sulfur in soil. Soil Sci. 109: 310–318.

Gerzabek, Von M. H. and Schaffer, K. 1986. [Determination of total sulfur in soil — a comparison of methods]. Die Bodenkultur 37: 1–6.

Guthrie, T. F. and Lowe, L. E. 1984. A comparison of methods for total sulphur analysis of tree foliage. Can. J. For. Res. 14: 470–473.

Hogan, G. D. and Maynard, D. G. 1984. Sulphur analysis of environmental materials by vacuum inductively coupled plasma emission spectrometry (ICP-AES). Pages 676–683 in Proceedings of sulphur-84. Sulphur Development Institute of Canada (SUDIC), Calgary.

Hordijk, C. A., van Engelen, J. J. M., Jonker, F. A., and Cappenberg, T. E. 1989. Determination of total sulfur in freshwater sediments by ion chromatography. Water Res. 23: 853–859.

Jenkins, R. 1984. X-ray fluorescence analysis. Anal. Chem. 56: 1099A–1106A.

Johnson, C. M. and Nishita, H. 1952. Microestimation of sulfur in plant materials, soils, and irrigation waters. Anal. Chem. 24: 736–742.

Khan, S. U. and Webster, G. R. 1968. Determination of gypsum in solonetzic soils by an X-ray technique. Analyst 93: 400–402.

Killham, K. and Wainwright, M. 1981. Closed combustion method for the rapid determination of total S in atmospheric polluted soils and vegetation. Environ. Pollut. (Ser. B) 2: 81–85.

Kowalenko, C. G. 1978. Organic nitrogen, phosphorus and sulfur in soils. Pages 95–136 in M. Schnitzer and S. U. Khan, Eds. Soil organic matter. Developments in Soil Science 8, Elsevier, New York.

Kowalenko, C. G. 1985. A modified apparatus for quick and versatile sulphate sulphur analysis using hydriodic acid reduction. Commun. Soil Sci. Plant Anal. 16: 289–300.

Kowalenko, C. G. and Lowe, L. E. 1972. Observations on the bismuth sulfide colorimetric procedure for sulfate analysis in soil. Commun. Soil Sci. Plant Anal. 3: 79–86.

Lowe, L. E. 1969. Sulfur fractions of selected Alberta profiles of the Gleysolic order. Can. J. Soil Sci. 49: 375–381.

Lowe, L. E. and DeLong, W. A. 1963. Carbon bonded sulfur in selected Quebec soils. Can. J. Soil Sci. 43: 151–155.

Matrai, P. A. 1989. Determination of sulfur in ocean particulates by combustion-fluorescence. Marine Chem. 26: 227–238.

Melville, G. E., Freney, J. R., and Williams, C. H. 1971. Reaction of organic sulfur compounds in soil with tin and hydrochloric acid. Soil Sci. 112: 245–248.

Mizoguchi, T., Iwahori, H., and Ishii, H. 1980. Analytical applications of condensed phosphoric acid. III. Iodometric determination of sulphur after reduction of sulphate with sodium hypophosphite and either tin metal or potassium iodide in condensed phosphoric acid. Talanta 27: 519–524.

Nelson, R. E. 1982. Carbonate and gypsum. Pages 181–197 in A. L. Page, R. H. Miller, and D. R. Keeney, Eds. Methods of soil analysis. Part 2. Chemical and microbiological properties. 2nd ed. Agronomy No. 9. American Society of Agronomy, Madison, WI.

Pirela, H. J. and Tabatabai, M. A. 1988. Reduction of organic sulfur in soils with tin and phosphoric acid. Soil Sci. Soc. Am. J. 52: 959–964.

Pritchard, M. W. and Lee, J. 1984. Simultaneous determination of boron, phosphorus and sulphur in some biological and soil materials by inductively-coupled plasma emission spectrometry. Anal. Chim. Acta 157: 313–316.

Randall, P. J. and Spencer, K. 1980. Sulfur content of plant material: a comparison of methods of oxidation prior to determination. Commun. Soil Sci. Plant Anal. 11: 257–266.

Roberts, T. L. and Bettany, J. R. 1985. The influence of topography on the nature and distribution of soil sulfur across a narrow environmental gradient. Can. J. Soil Sci. 65: 419–434.

Saggar, S., Bettany, J. R., and Stewart, J. W. B. 1981. Measurement of microbial sulfur in soil. Soil Biol. Biochem. 13: 493–498.

Scott, N. M., Bick, W., and Anderson, H. A. 1981. The measurement of sulfur-containing amino acids in some Scottish soils. J. Sci. Food Agric. 32: 21–24.

Searle, P. L. 1968. Determination of total sulfur in soil by using high-frequency induction furnace equipment. Analyst 93: 540–545.

Stallings, E. A., Candelaria, L. M., and Gladney, E. S. 1988. Investigation of a fusion technique for the determination of total sulfur in geological samples by ion chromatography. Anal. Chem. 60: 1246–1248.

Strick, J. E. and Nakas, J. P. 1984. Calibration of a microbial sulfur technique for use in forest soils. Soil Biol. Biochem. 16: 289–291.

Tabatabai, M. A. 1982. Sulfur. Pages 501–538 in A. L. Page, R. H. Miller, and D. R. Keeney, Eds. Methods of soil analysis. Part 2. Chemical and microbiological properties. 2nd ed. Agronomy No. 9. American Society of Agronomy, Madison, WI.

Tabatabai, M. A., Basta, N. T., and Pirela, H. J. 1988. Determination of total sulfur in soils and plant materials by ion chromatography. Commun. Soil Sci. Plant Anal. 19: 1701–1714.

Tabatabai, M. A. and Bremner, J. M. 1970a. An alkaline oxidation method for determination of total sulfur in soils. Soil Sci. Soc. Am. Proc. 34: 62–65.

Tabatabai, M. A. and Bremner, J. M. 1970b. Comparison of some methods for determination of total sulfur in soils. Soil Sci. Soc. Am. Proc. 34: 417–420.

Tiedeman, A. R. and Anderson, T. D. 1971. Rapid analysis of total sulphur in soils and plant material. Plant Soil 35: 197–200.

Watkinson, J. L., Lee, A., and Lauren, D. R. 1987. Measurement of elemental sulfur in soil and sediments: field sampling, sample storage, pretreatment, extraction and analysis by high-performance liquid chromatography. Aust. J. Soil Res. 25: 167–178.

Chapter 25
Extractable Al, Fe, Mn, and Si

G. J. Ross and C. Wang

Agriculture Canada

Ottawa, Ontario, Canada

25.1 INTRODUCTION

The amounts of Al, Fe, Mn, and Si extracted from soils by different dissolution methods indicate the forms of these elements present. The results are useful in studies of soil classification, soil genesis, and soil behavior. For example, the nature and amounts of extractable inorganic and organic Fe and Al may show the pathway of soil genesis. Also, extractable soil constituents are generally fine grained with large specific surface area and therefore have a marked effect on physical and chemical soil properties and behavior. For these reasons, dissolution data are commonly used in chemical criteria for soil classification. Five extraction methods may be used to provide a basis for approximating the amounts and forms of these elements in soils.

Dithionite citrate removes finely divided hematite, goethite, lepidocrocite, ferrihydrite, and noncrystalline iron oxides as well as organic-complexed Fe and Al (Mehra and Jackson 1960). It is less effective in removing noncrystalline inorganic forms of Al. The method extracts virtually no Fe (or Al) from most crystalline silicate minerals, and thus provides an estimate of "free" (nonsilicate) Fe in soils. The procedure may have to be repeated to dissolve silt- and sand-size goethite and hematite completely (Kodama and Ross 1991). Magnetite is not dissolved. In the Agriculture Canada, Centre of Land and Biology Resources Research (CLBRR) analytical service laboratory in Ottawa, the coefficients of variation at Fe levels of 1.4% and Al levels of 0.45% are 6.3% and 7.8%, respectively.

Acid ammonium oxalate removes noncrystalline inorganic forms of Fe and Al and organic-complexed Fe and Al from soils (McKeague 1967). It only slightly attacks most silicate minerals, goethite, hematite, and lepidocrocite, but it dissolves considerable amounts of magnetite (Baril and Bitton 1967) and finely divided, easily weathered silicates, such as olivine. In the CLBRR analytical service laboratory, the coefficients of variation at Fe levels of 0.67% and Al levels of 0.67% are 7.2% and 4.1%, respectively.

Hydroxylamine is closely similar to oxalate in its extraction capacity (Chao and Zhou 1983). Unlike ammonium oxalate, however, hydroxylamine does not dissolve magnetite and can therefore be used as an alternative to ammonium oxalate for soils containing magnetite (Ross

Soil Sampling and Methods of Analysis, M. R. Carter, Ed.,

Canadian Society of Soil Science. © 1993 Lewis Publishers.

et al. 1985). In the CLBRR mineralogical laboratory, the coefficients of variation at Fe levels of 0.63% and Al levels of 0.62% are 4.5% and 3.0%, respectively.

Tiron, 4,5-dihydroxy-1,3-benzene-disulfonic acid (disodium salt), does not dissolve magnetite either and can also be used instead of oxalate (Kodama and Ross 1991). Furthermore, Tiron dissolves opaline silica, whereas neither oxalate nor hydroxylamine dissolve this soil component effectively. Tiron is currently mainly used to extract clays and the method is therefore described in Chapter 70. However, it should also be suitable for soils ground to pass a 100-mesh sieve.

Pyrophosphate extracts organic-complexed Fe and Al from soils. It only slightly dissolves noncrystalline inorganic forms, and it does not significantly attack silicate minerals and crystalline Fe and Al oxides and hydroxides (McKeague et al. 1971). In the CLBRR analytical service laboratory, the coefficients of variation at Fe levels of 0.64% and Al levels of 0.69% are 5.9% and 6.0%, respectively.

From the results of these methods the following estimates can be made:

a. Finely divided goethite, hematite, and lepidocrocite: (dithionite Fe − oxalate Fe) or (dithionite Fe − hydroxylamine Fe) or (dithionite Fe − Tiron Fe).
b. Noncrystalline inorganic forms of Fe (including ferrihydrite): (oxalate Fe − pyrophosphate Fe) or (hydroxylamine Fe − pyrophosphate Fe) or (Tiron Fe − pyrophosphate Fe).
c. Organic complexed Fe: (pyrophosphate Fe).

Relationships b and c hold approximately also for Al. In the case of Mn, both dithionite and oxalate attack some crystalline oxide forms. The noncrystalline forms of Si, such as opaline silica, are completely extracted only by Tiron (Kodama and Ross 1991). They are not extracted by oxalate and only partly by dithionite and hydroxylamine. Poorly crystalline and noncrystalline aluminosilicates, including allophane and imogolite, are extracted by oxalate, hydroxylamine, and Tiron. Dithionite is less effective in extracting these compounds.

25.2 DITHIONITE-CITRATE METHOD (SOIL CONSERVATION SERVICE, U.S. DEPARTMENT OF AGRICULTURE 1972)

This method consists of shaking the samples overnight at ambient conditions in a reducing and complexing solution. This treatment is particularly useful for dissolving the "free" (nonsilicate) Fe in soils. The overnight shaking procedure is simpler than the dithionite-citrate-bicarbonate method of Mehra and Jackson (1960), and it gives closely similar results (Sheldrick and McKeague 1975).

25.2.1 REAGENTS

1   Sodium hydrosulfite (dithionite), $Na_2S_2O_4$.

2   Certified atomic absorption standards, ±1%.

3   Sodium citrate ($Na_3C_6H_5O_7 \cdot 2H_2O$), 0.68 M (200 g/L).

25.2.2 PROCEDURE

1   Weigh 0.500 g of soil, ground to pass a 35-mesh sieve, into a 50-mL plastic centrifuge tube (use 0.2 g for clays and 1 g for coarse soils).

2   Add 25 mL of the sodium citrate solution.

3   Add 0.4 g of dithionite (a calibrated scoop may be used).

4   Stopper tightly and shake in an end-over-end shaker overnight (use a horizontal shaker if end-over-end shaker is not available).

5   Remove stoppers and centrifuge for 20 min at $510 \times g$ (centrifuge at higher speed for clays). Filter extracts containing suspended material.

6   For determining Fe, Al, Mn, and Si by atomic absorption, prepare standard solutions of these elements in a matrix of the extracting solution.

7   An air-acetylene flame is suitable for the determination of Fe and Mn, and a nitrous oxide-acetylene flame is used for Al and Si.

8   If it is necessary to dilute the extracts, either dilute them with the extracting solution or prepare standards containing the same concentration of extracting solution as the diluted extracts.

9   Colorimetric methods may be used to determine the extracted elements if desired (McKeague 1978).

25.2.3 CALCULATIONS

1
$$\% \text{ Fe, Al, Mn, Si} = \frac{\mu\text{g/mL in final sol'n} \times \text{extractant (mL)} \times \text{dil.} \times 100}{\text{sample wt (mg)} \times 1000}$$

(25.1)

2   For example, for 0.500 g of sample, 25 mL of extractant, $5 \times$ dilution and 48 μg/mL of Fe determined:

$$\% \text{ Fe in sample} = \frac{48 \times 25 \times 5 \times 100}{500 \times 1000} = 1.2$$

25.3 ACID AMMONIUM OXALATE METHOD (McKEAGUE AND DAY 1966)

The oxalate procedure was used 70 years ago by Tamm to remove the sesquioxide weathering products from soils. It was revised by Schwertmann (1959), who showed that it could be used to estimate noncrystalline and poorly crystalline Fe and Al forms in soils. It extracts the Fe and Al accumulated in podzol B horizons (McKeague and Day 1966) and is useful to identify podzolic B horizons. Amounts of oxalate-extractable Fe and Al were recently proposed as criteria for the spodic horizon (ICOMOD 1989).

Oxalate selectively dissolves allophane and imogolite (Wada 1988). Amounts of these minerals can therefore be estimated from the amounts of oxalate-extractable Al and Si, taking into account that oxalate also extracts organic-complexed Al (Parfitt and Henmi 1982).

25.3.1 REAGENTS

1   Oxalate solution $(NH_4)_2 C_2O_4 \cdot H_2O$, 0.2 M (28.3 g/L).

2   Oxalic acid solution $(H_2C_2O_4 \cdot 2H_2O)$, 0.2 M (25.2 g/L).

3   Mix 700 mL of A and 535 mL of B, check pH, and adjust to 3.0 by adding either A or B.

25.3.2 PROCEDURE

1   Weigh 0.250 g of soil, ground to pass a 100-mesh sieve, into a 15-mL test tube (weigh 0.125 g for samples with more than 2% extractable Fe or Al).

2   Add 10 mL of the acid oxalate solution and stopper the tube tightly.

3   Place the tubes in an end-over-end shaker (or, if not available, in a horizontal shaker) and shake for 4 h (the extraction has to be done in the dark).

4   Centrifuge the tubes for 20 min at 510 $\times$ g (centrifuge at higher speed for clays), decant the clear supernatant into a suitable container, and analyze within a few days.

5   For determining Fe, Al, Mn, and Si by atomic absorption, follow standard atomic absorption procedures.

25.3.3 CALCULATIONS

1   $$\% \text{ Fe, Al, Mn, Si} = \frac{\mu g/mL \text{ in final sol'n} \times \text{extractant (mL)} \times \text{dil.} \times 100}{\text{sample wt (mg)} \times 1000}$$

(25.2)

2   For example, for 0.250 g of sample, 10 mL of extractant, 5 × dilution, and 12 µg/mL of Fe determined:

$$\% \text{ Fe in sample} = \frac{12 \times 10 \times 5 \times 100}{250 \times 1000} = 0.24$$

25.4 ACID HYDROXYLAMINE METHOD (ROSS ET AL. 1985, WANG ET AL. 1987)

Acid hydroxylamine extraction is used in geochemical studies for removing noncrystalline material from crystalline iron oxides, including magnetite (Chao and Zhou 1983). Ross et al. (1985) and Wang et al. (1987) modified this procedure and tested it on soil samples. For Fe and Al, the results were similar to those obtained by oxalate extraction with the advantage that hydroxylamine did not dissolve magnetite. There was less agreement between the Si results obtained by the two methods. The suitability of hydroxylamine as an extractant for Mn in soils has not been tested yet. Hydroxylamine solutions also were more easily analyzed than oxalate solutions by atomic absorption analysis.

25.4.1 REAGENTS

1   Prepare a hydroxylamine hydrochloride-hydrochloric acid solution (0.25 M $NH_2OH \cdot HCl$, 0.25 M HCl) by adding 0.25 M HCl to 17.37 g $NH_2OH \cdot HCl$ to a volume of 1 L.

25.4.2 PROCEDURE

1   Weigh 0.100 g of soil, ground to pass a 100-mesh sieve, into a 50-mL plastic centrifuge tube.

2   Add 25 mL of the hydroxylamine solution and stopper the tube tightly.

3   Place the tubes in an end-over-end shaker (or, if not available, in a horizontal shaker) and shake overnight (16 h).

4   Centrifuge the tubes for 20 min at 510 × g (centrifuge at higher speed for clays), decant the clear supernatant into a suitable container, and analyze within a few days.

5   For determining Fe, Al, Mn, and Si by atomic absorption, follow standard atomic absorption procedures. Note the points mentioned in Section 25.2.2.

25.4.3 CALCULATIONS

1

$$\% \text{ Fe, Al, Mn, Si} = \frac{\mu g/mL \text{ in final sol'n} \times \text{extractant (mL)} \times \text{dil.} \times 100}{\text{sample wt (mg)} \times 1000}$$

(25.3)

2. For example, for 0.100 g of sample, 25 mL of extractant, 5 × dilution, and 6 µg/mL of Fe determined:

$$\% \text{ Fe in sample} = \frac{6 \times 25 \times 5 \times 100}{100 \times 1000} = 0.75$$

25.5 SODIUM PYROPHOSPHATE METHOD (McKEAGUE 1967)

This method is used in Canada and elsewhere as a chemical basis for identifying podzolic B horizons (McKeague 1967). In some cases finely divided silicates and iron oxides remain suspended after low-speed centrifugation, and high-speed centrifugation or ultrafiltration is necessary to clear the extracts (McKeague and Schuppli 1982, Schuppli et al. 1983).

25.5.1 REAGENTS

1   Sodium pyrophosphate solution ($Na_4P_2O_7, \cdot 10\ H_2O$), 0.1 *M* (44.6 g/L).

2   Superfloc® (N-100), available from Cyanamid® of Canada Ltd., P.O. Box 1038, Montreal, Que. H3C 2X4.

25.5.2 PROCEDURE

1   Weigh 0.300 g of soil, ground to pass a 100-mesh sieve, into a 50-mL plastic centrifuge tube (use 1 g for samples low in extractable Fe and Al).

2   Add 30 mL of 0.1 *M* sodium pyrophosphate solution, stopper, and shake overnight in an end-over-end shaker (or, if not available, in a horizontal shaker).

3   Centrifuge at 20,000 × *g* for 10 min or, alternatively, add 0.5 mL of 0.1% superfloc solution and centrifuge at 510 × *g* for 10 min. Note the following points:

   a. Concentrations of Fe and Al in sodium pyrophosphate extracts of some soils may decrease progressively by centrifugation for longer times or at higher speeds.

   b. Ultrafiltration through a 0.025-μm Millipore® filter is recommended for tropical soils and for soils giving doubtful results by the centrifugation method.

4   Decant a portion of the clear supernatant into a suitable container and analyze within a few days.

5   For determining Fe, Al, and Mn by atomic absorption, follow standard atomic absorption procedures. Note the points mentioned in Section 25.2.2. Extracts containing suspended material should be filtered.

25.5.3 CALCULATIONS

1

$$\% \text{ Fe, Al, Mn} = \frac{\mu\text{g/mL in final sol'n} \times \text{extractant (mL)} \times 100}{\text{sample wt (mg)} \times 1000} \quad (25.4)$$

2.   For example, for 0.300 g of sample, 30 mL of extractant, 75 μg/mL of Fe determined:

$$\% \text{ Fe in sample} = \frac{75 \times 30 \times 100}{300 \times 1000} = 0.75$$

REFERENCES

Baril, R. and Bitton, G. 1967. Anomalous values of free iron in some Quebec soils containing magnetite. Can. J. Soil Sci. 47: 261.

Chao, T. T. and Zhou, L. 1983. Extraction techniques for selective dissolution of amorphous iron oxides from soils and sediments. Soil Sci. Soc. Am. J. 47: 225–232.

ICOMOD, International Committee on the Classification of Spodosols. 1989. Pages 1–7 in Circular letter No. 8, Soil Conservation Service, U.S. Department of Agriculture, Washington, D.C.

Kodama, H. and Ross, G. J. 1991. Tiron dissolution method used to remove and characterize inorganic compounds in soils. Soil Sci. Soc. Am. J. 55: 1180–1187.

McKeague, J. A. 1967. An evaluation of 0.1 *M* pyrophosphate and pyrophosphate-dithionite in comparison with oxalate as extractants of the accumulation products in Podzols and some other soils. Can. J. Soil Sci. 47: 95–99.

McKeague, J. A., Ed. 1978. Manual on soil sampling and methods of analysis. 2nd ed. Can. Soc. Soil Sci., Suite 907, 151 Slater St., Ottawa, Ont.

McKeague, J. A. and Day, J. H. 1966. Dithionite and oxalate-extractable Fe and Al as aids in differentiating various classes of soils. Can. J. Soil Sci. 46: 13–22.

McKeague, J. A. and Schuppli, P. A. 1982. Changes in concentration of Fe and Al in pyrophosphate extracts of soil, and composition of sediment resulting from ultra centrifugation in relation to spodic horizons. Soil Sci. 134: 265–270.

McKeague, J. A., Brydon, J. E., and Miles, N. M. 1971. Differentiation of forms of extractable iron and aluminum in soils. Soil Sci. Soc. Am. Proc. 35: 33–38.

Mehra, O. P. and Jackson, M. L. 1960. Iron oxide removal from soils and clays by a dithionite-citrate system buffered with sodium bicarbonate. Pages 317–327 in 7th Natl. Conf. Clays and Clay Minerals.

Parfitt, R. L. and Henmi, T. 1982. Comparison of an oxalate-extraction method and an infrared spectroscopic method for determining allophane in soil clays. Soil Sci. Plant Nutr. 28: 183–190.

Ross, G. J., Wang, C., and Schuppli, P. A. 1985. Hydroxylamine and ammonium oxalate solutions as extractants for Fe and Al from soil. Soil Sci. Soc. Am. J. 49: 783–785.

Schuppli, P. A., Ross, G. J., and McKeague, J. A. 1983. The effective removal of suspended materials from pyrophosphate extracts of soils from tropical and temperate regions. Soil Sci. Soc. Am. J. 47: 1026–1032.

Schwertmann, U. 1959. Die fractionierte Extraction der freien Eisenoxide in Böden, ihre mineralogischen Formen und ihre Entstehung-sweisen. Z. Pflanzenernaehr. Dueng. Bodenkd. 84: 194–204.

Sheldrick, B. H. and McKeague, J. A. 1975. A comparison of extractable Fe and Al data using methods followed in the U.S.A. and Canada. Can. J. Soil Sci. 55: 77–78.

Soil Conservation Service, U.S. Department of Agriculture. 1972. Soil survey laboratory methods and procedures for collecting soil samples. Soil Survey Investigations Report No. 1 (revised), U.S. Government Printing Office, Washington, D.C.

Wada, K. 1988. Allophane and Imogolite. Pages 1051–1081 in J. B. Dixon and S. B. Weed, Eds. Minerals in soil environments. Soil Science Society of America, Madison, WI.

Wang, C., Schuppli, P. A., and Ross, G. J. 1987. A comparison of hydroxylamine and ammonium oxalate solutions as extractants for Al, Fe and Si from Spodosols and Spodosol-like soils in Canada. Geoderma 40: 345–355.

Chapter 26
Reference Materials for
Data Quality

Milan Ihnat

Agriculture Canada
Ottawa, Ontario, Canada

26.1 INTRODUCTION AND SCOPE

The vital requirement for reliable analytical information and its bearing on the outcome of research and monitoring activities is obvious to practitioners of analytical science. Incorporation of reference materials (RM) into the overall measurement scheme is one cost-effective facet of a data quality program to ensure accuracy of results. In this chapter, the concept and role of RM are summarized, and procedures for their selection and utilization are presented. The thrust of this chapter is soil chemical analyses. While extractable or bioavailable elemental concentrations are frequently of interest, only the control of *total elemental concentrations* will be addressed, as RM are available only for total concentrations. Plant and related RM are listed and discussed by Muramatsu and Parr (1985), Ihnat (1988), and Cortes Toro et al. (1990).

26.2 CONCEPT AND ROLE OF REFERENCE MATERIALS

The mechanisms of arriving at and maintaining reliable analytical data falls into the category of a quality assurance and quality control system. Works on theoretical treatments and practical applications of quality assurance and control by Taylor (1981), Currie (1982), Kateman and Pijpers (1981), Dux (1986), and Keith et al. (1980), should be consulted for details.

The three usual mechanisms for achieving compatibility and transferring accuracy among laboratories are standard reference data, RM, and reference methods (Uriano and Gravatt 1977, Coleman 1980), with the latter two the more widely used. The concept and role of RM as a cornerstone in a data quality program is the thrust of this presentation. Incorporation of appropriate RM into the scheme of analysis utilizing good methods and other aspects of a quality control program is the most convenient, cost-effective mechanism by which to assess and maintain analytical data quality.

Soil Sampling and Methods of Analysis, M. R. Carter, Ed.,
Canadian Society of Soil Science. © 1993 Lewis Publishers.

A RM is considered to be any material, device, or physical system for which definitive numerical values can be associated with specific properties and that is used to calibrate a measurement process. Uriano and Gravatt (1977) use the term reference materials to describe a generic class of well-characterized, stable, homogeneous materials, produced in quantity and having one or more physical or chemical properties experimentally determined within stated measurement uncertainties. Primary RM are defined as RM having properties certified by a recognized standard laboratory or standards agency, such as the National Bureau of Standards (NBS), now known as National Institute of Standards and Technology (NIST).

Primary RM produced by NIST are denoted standard reference materials (SRM) by that organization, a term used synonymously with the term certified reference materials (CRM) recommended by the International Union of Pure and Applied Chemistry and the International Organization for Standardization (ISO). The authors further mention that a primary RM is normally produced by a national standards laboratory or other organization having legal authority to issue such materials.

NIST SRM are well-characterized materials produced in quantity to improve measurement science. SRM are certified for specific chemical or physical properties, and are issued by NIST with certificates that report the results of the characterization and indicate the intended use of the material. They are prepared and used for three main purposes: (1) to help develop accurate methods of analysis (reference methods), (2) to calibrate measurement systems, and (3) to assure the long-term adequacy and integrity of measurement quality assurance programs.

According to the ISO, a RM is defined as a material or substance one or more properties of which are sufficiently well established to be used for the calibration of an apparatus, the assessment of a measurement method, or for assigning values to materials. A CRM is defined as a RM one or more of whose property values are certified by a technically valid procedure, accompanied by or traceable to a certificate or other documentation which is issued by a certifying body.

26.3 PREREQUISITES FOR USE OF REFERENCE MATERIALS

The generation of valid analytical data and the proper use of RM dictates compliance with several prerequisites.

1   The correct analytical method must be applied by appropriately qualified and trained personnel in a suitable physical and administrative environment.

2   Suitable quality control procedures are routinely in use and the analytical system must be in a state of statistical control.

3   When dealing with the determination of *total* concentrations of elements, that is, the sum of all the element concentrations in all material (sample) phases and molecular species, it must be ascertained that the method is in fact measuring all of the element. The sample decomposition procedure must bring into selection all of the material: no grains or insoluble fraction must be left behind (e.g., Ihnat 1982). In addition, the element must be in the correct oxidation state required by the procedure.

26.4 PROCEDURES FOR REFERENCE MATERIAL SELECTION AND USE

In order to facilitate application of selection and use procedures, RM information is presented in tables at the end of the chapter. Table 26.1 lists suppliers of RM treated and mentioned in this chapter, giving codes used here and addresses for contact and ordering. Soil RM are presented in Table 26.2. Other RM relevant and complementary to soil and environmental sciences including sediments, rocks, minerals, sludges, ashes, aerosols, and water are given in the literature (e.g., Cortes Toro et al. 1990). Central to the selection of soil RM is Table 26.3, listing soil elemental concentration values.

In addition, a series of eight Canadian soil control samples (ECSS-1 to ECSS-8), together with a report of analysis (Sheldrick and Wang 1987) are available from C. Wang, Centre for Land and Biological Resources Research, Ottawa, ON K1A 0C6. These have been analyzed for a number of parameters, including several extractable elements. Values are available for pH, total and organic C, N, sand and clay content, extractable Fe and Al (citrate-dithionate, ammonium oxalate, sodium pyrophosphate), exchangeable cations (Ca, Mg, K, Na), cation exchange capacity, and Bray-extractable P. These materials replace the CSSC-1 to CSSC-28 soil series now depleted, but still in use in some laboratories and described by McKeague et al. (1978).

26.4.1 Procedures for Reference Material Selection

The consideration of material and data reliability is one of utmost importance and usually difficult to assess. In principle, the user should be able to acquire a *certified reference material* with a corresponding certificate of analysis and use it with confidence. A certified or recommended value from one producer may elicit a different confidence than that from another producer. Professional judgment of an analytical chemist must be used in the selection of appropriate materials.

Proper RM selection must be appropriate to the task. The product must resemble as closely as possible in all respects, the actual, real-life materials being processed; also, the form of the analyte, e.g., the native form and its concentration must be similar in both commodities. Furthermore, the RM must be sufficiently homogeneous so that test portions of size commensurate with the analytical method can be used. Ideally (but often impossible), two or three RM should be chosen to bracket the analyte composition of the sample. The following steps could be followed to select appropriate RM.

1   In selecting a soil RM, consult Table 26.3 for materials which have concentration values listed for the element of interest. In conjunction with Table 26.2, giving a descriptive name of the product, select the material approximating the laboratory sample to be controlled with respect to general soil type (i.e., matrix) as well as the analyte level expected.

2   Follow the same approach to choose a second or third RM to match (or bracket) the sample with respect to concentration of the given element.

3   For multielement analyses, that is, the determination of more than one element on the same laboratory subsample, go through the identical material selection steps for the second, third, etc. element to choose appropriate materials for each of these respective analytes. Maximize the number of elements to monitor by a given RM, i.e., minimize the number of RM required, by reducing the strictness of matrix and analyte matching criteria.

TABLE 26.1 Suppliers of Reference and Control Materials for Chemical Composition Quality Control in Soil and Related Environmental Sciences

Code	Name and address	Code	Name and address
AMM	Academy of Mining and Metallurgy Institute of Physics and Nuclear Techniques Al. Mickiewicza 30 PL-30-059 Krakow, Poland	IAEA	International Atomic Energy Agency Analytical Quality Control Services Laboratory Seibersdorf P.O. Box 100 A-1400 Vienna, Austria
ARC	Dr. Jorma Kumpulainen Agricultural Research Centre Central Laboratory SF-31600 Jokioinen, Finland	ICHT	Commission of Trace Analysis of the Committee for Analytical Chemistry of the Polish Academy of Sciences ul. Dorodna 16 030195 Warszawa, Poland
BCR	Community Bureau of Reference Commission of the European Communities 200 Rue de la Loi B-1049 Brussels, Belgium	IGGE	Institute of Geo-physical and Geo-chemical Exploration Langfang, Hebei 102801 People's Republic of China
BOWEN	Dr. H. J. M. Bowen West Down West Street Winterborne Kingston Dorset DT11 9AT, U.K.	IRANT	Institute of Radio-ecology and Applied Nuclear Techniques Komenského 9 P.O. Box A-41, 040 61, Kosice, Czechoslovakia
CANMET	Canada Centre for Mineral and Energy Technology Coordinator, CCRMP CANMET, EMR 555 Booth Street Ottawa, Ontario Canada K1A 0G1	NIES	National Institute for Environmental Studies Japan Environment Agency Yatabe-machi, Tsukuba Ibaraki 305, Japan
EPA	U.S. Environmental Protection Agency Environmental Monitoring and Support Laboratory Cincinnati, OH 45268. U.S.	NRCC	National Research Council Canada Institute for Environmental Chemistry Ottawa, Ontario Canada K1A 0R6
NIST	Standard Reference Materials Room 205, Building 202 National Institute of Standards and Technology Gaithersburg, MD 20899, U.S.		Dr. C. Wang Centre for Land and Biological Resources Research Agriculture Canada Ottawa, Ontario Canada K1A 0C6
FISHER	Fisher Scientific International Division Headquarters 50 Fadem Road Springfield, NJ 07081, U.S.		

TABLE 26.2 Soil Reference Materials for Chemical Composition

Material[a]	Code[b]	Selected quoted elements[c]	Ref.
Soil	AMM-SO-1	Ca Cd Cr Fe K Mn Na Pb Zn	Cortes Toro et al. (1990) Holynska et al. (1988)
Soil (calcareous loam)	BCR-CRM-141	Cd Cu Hg Pb Zn	BCR (1983, 1985)
Soil (light sandy)	BCR-CRM-142	Cd Cu Hg Pb Zn	BCR (1983, 1985)
Soil (amended with sewage sludge)	BCR-CRM-143	Cd Cu Hg Pb Zn	BCR (1983, 1985)
Soil (ferro-humic podzol)	CANMET-SO-2	Al Ba Ca Cr Cu Fe Hg K Mg Mn Na P Pb Si Sr Ti V Zn	Bowman (1990) Bowman et al. (1979)
Soil (gray-brown luvisol)	CANMET-SO-3	Al Ba Ca Cr Cu Fe Hg K Mg Mn Na P Pb Si Sr Ti V Zn	Bowman (1990) Bowman et al. (1979)
Soil (black chernozemic)	CANMET-SO-4	Al Ca Cr Cu Fe Hg K Mg Mn Na P Pb Si Sr Ti V Zn	Bowman (1990) Bowman et al. (1979)
Soil	IAEA-SOIL-7	As Cr Cu Mn Pb Sr V Zn	IAEA (1984, 1990)
Soil (dark brown podzolitic)	IGGE-GSS-1	Al B Ba Ca Cd Cr Cu Fe Hg K Mg Mn Mo Na P Pb Si Sr Ti V Zn	Xie et al. (1985, 1989)
Soil (chestnut)	IGGE-GSS-2	Al B Ba Ca Cd Cr Cu Fe Hg K Mg Mn Mo Na P Pb Si Sr Ti V Zn	Xie et al. (1985, 1989)
Soil (yellow-brown)	IGGE-GSS-3	Al B Ba Ca Cd Cr Cu Fe Hg K Mg Mn Mo Na P Pb Si Sr Ti V Zn	Xie et al. (1985, 1989)
Soil (limey-yellow)	IGGE-GSS-4	Al B Ba Ca Cd Cr Cu Fe Hg K Mg Mn Mo Na P Pb Si Sr Ti V Zn	Xie et al. (1985, 1989)
Soil (yellow-red)	IGGE-GSS-5	Al B Ba Ca Cd Cr Cu Fe Hg K Mg Mn Mo Na P Pb Si Sr Ti V Zn	Xie et al. (1985, 1989)
Soil (yellow-red)	IGGE-GSS-6	Al B Ba Ca Cd Cr Cu Fe Hg K Mg Mn Mo Na P Pb Si Sr Ti V Zn	Xie et al. (1985, 1989)
Soil (laterite)	IGGE-GSS-7	Al Ba Ca Cd Cr Cu Fe Hg K Mg Mn Mo Na P Pb Si Sr Ti V Zn	Xie et al. (1985, 1989)
Soil (loess)	IGGE-GSS-8	Al B Ba Ca Cd Cr Cu Fe Hg K Mg Mn Mo Na P Pb Si Sr Ti V Zn	Xie et al. (1985, 1989)

TABLE 26.2 (continued) Soil Reference Materials for Chemical Composition

Material[a]	Code[b]	Selected quoted elements[c]	Ref.
Soil	NIST-RM-8406	Hg	McKenzie (1990)
Soil	NIST-RM-8407	Hg	McKenzie (1990)
Soil	NIST-RM-8408	Hg	McKenzie (1990)

Note: Adapted from Muramatsu and Parr (1985), Ihnat (1988), and the specific references listed above with the exclusion of elements with noncertified (informational) values. Refer to the reference citations and appropriate certificates or reports of analysis for detailed coverage.

[a] Material descriptions are as provided by issuers or reported in other publications (listed above). Soils IGGE-GSS-3 to IGGE-GSS-7 are of tropical or subtropical origin and not directly related to Canadian soils. They are included because of their extensive elemental characterization to complement the extremely short collection of soil reference materials.

[b] Code is a combination of issuer code for the material, together with a code for the issuer adopted for this article (Cortes Toro et al. 1990).

[c] Some common elements selected from the list Al B Ba Ca Cd Cr Cu Fe Hg K Mg Mn Mo Na P Pb Si Sr Ti V Zn, for which the issuing organization or the producer of the material indicates to have certified or recommended concentration values. The reader is referred to the original certificates/reports for details regarding other elemental concentrations and their status. Only those elements with total concentration values are listed (refer to text for definition and discussion). The reader should refer to the original certificates or reports of analysis for additional details and listing of other elements.

4   Follow the same approach to choose a second or third RM with elemental concentration matching or bracketing the samples undergoing analysis.

26.4.1.1 COMMENTS

1   The foregoing is not to imply that the number of samples undergoing analysis is to be matched or exceeded by the number of RM. In the usual case, encountered in a large routine analysis operation, many similar samples are analyzed concurrently in a batch or run. One aliquot of a suitable RM will thus suffice to monitor the performance of the method for quite a number of samples.

2   A preliminary analysis of the sample would be advantageous to facilitate selection of a closely matching control material, but this is not usually feasible unless one has access to high-throughput, multielement analytical techniques or is carrying out high-reliability determinations. The certificate or other publications may reveal more details regarding the reference soils, including sampling locations to aid in reference soil selection.

3   Given the scarcity of reference soils it is generally unlikely to secure two or three RM, and the analyst may in fact be hard pressed to find even one.

TABLE 26.3 Certified and Recommended Elemental Concentration Values for Some Common Elements Reported for Soil Reference Materials Listed in Order of Element and in Increasing Order of Concentration

Material Code[a]	Concentration (mg/kg[b])	Material code	Concentration (mg/kg)
Al — aluminum		Ca (continued)	
CANMET-SO-3	30,500	CANMET-SO-4	11,100
IGGE-GSS-2	54,570	IGGE-GSS-1	12,290
CANMET-SO-4	54,600	IGGE-GSS-2	16,870
IGGE-GSS-8	63,100	IGGE-GSS-8	59,100
IGGE-GSS-3	64,790	CANMET-SO-3	146,300
IGGE-GSS-1	75,060		
CANMET-SO-2	80,700		
IGGE-GSS-6	112,380	Cd — cadmium	
IGGE-GSS-5	114,230		
IGGE-GSS-4	124,130	IGGE-GSS-3	0.059
IGGE-GSS-7	154,880	IGGE-GSS-2	0.071
		IGGE-GGS-7	0.08
		IGGE-GSS-8	0.13
B — boron		BCR-CRM-142	0.25
		AMM-SO-1	0.3
IGGE-GSS-3	23	IGGE-GGS-4	0.35
IGGE-GSS-2	36	BCR-CRM-141	0.36
IGGE-GSS-1	50	IGGE-GSS-5	0.45
IGGE-GSS-5	53	IGGE-GGS-1	4.3
IGGE-GSS-8	54	BCR-CRM-143	31.1
IGGE-GSS-6	57		
IGGE-GSS-4	97		
		Cr — chromium	
Ba — barium		CANMET-SO-2	16
		CANMET-SO-3	26
IGGE-GSS-6	118	IGGE-GSS-3	32
IGGE-GSS-7	180	AMM-SO-1	38
IGGE-GSS-4	213	IGGE-GSS-2	47
CANMET-SO-3	296	IAEA-SOIL-7	60
IGGE-GSS-5	296	CANMET-SO-4	61
IGGE-GSS-8	480	IGGE-GSS-1	62
IGGE-GSS-1	590	IGGE-GSS-8	68
IGGE-GSS-2	930	IGGE-GSS-6	75
CANMET-SO-2	966	IGGE-GSS-5	118
IGGE-GSS-3	1,210	IGGE-GSS-4	370
		IGGE-GSS-7	410
Ca — calcium			
IGGE-GSS-7	1,140	Cu — copper	
IGGE-GSS-6	1,570		
IGGE-GSS-4	1,860	CANMET-SO-2	7
AMM-SO-1	2,600	IAEA-SOIL-7	11
IGGE-GSS-3	9,080	IGGE-GSS-3	11.4
CANMET-SO-2	9,600	IGGE-GSS-2	16.3

TABLE 26.3 (continued) Certified and Recommended Elemental Concentration Values for Some Common Elements Reported for Soil Reference Materials Listed in Order of Element and in Increasing Order of Concentration

Material Code[a]	Concentration (mg/kg[b])	Material code	Concentration (mg/kg)
Cu (continued)	17	K — potassium	
CANMET-SO-3	21		
IGGE-GSS-1	22	IGGE-GSS-7	1,660
CANMET-SO-4	24.3	IGGE-GSS-4	8,550
IGGE-GSS-8	27.5	AMM-SO-1	12,050
BCR-CRM-142	32.6	IGGE-GSS-5	12,450
BCR-CRM-141	40.5	IGGE-GSS-6	14,110
IGGE-GSS-4	97	CANMET-SO-3	16,100
IGGE-GSS-7	144	CANMET-SO-4	17,300
IGGE-GSS-5	236.5	IGGE-GSS-8	20,090
BCR-CRM-143	390	IGGE-GSS-2	21,090
IGGE-GSS-6		IGGE-GSS-1	21,500
		CANMET-SO-2	24,500
Fe — iron		IGGE-GSS-3	25,240
AMM-SO-1	9,880		
IGGE-GSS-3	13,990		
CANMET-SO-3	15,100	Mg — magnesium	
CANMET-SO-4	23,700		
IGGE-GSS-2	24,620	IGGE-GSS-7	1,570
IGGE-GSS-8	31,340	IGGE-GSS-6	2,050
IGGE-GSS-1	36,300	IGGE-GSS-4	2,960
CANMET-SO-2	55,600	IGGE-GSS-3	3,500
IGGE-GSS-6	56,590	IGGE-GSS-5	3,680
IGGE-GSS-5	88,270	CANMET-SO-2	5,400
IGGE-GSS-7	131,200	CANMET-SO-4	5,600
		IGGE-GSS-2	6,270
		IGGE-GSS-1	10,920
		IGGE-GSS-8	14,360
Hg — mercury		CANMET-SO-3	49,800
IGGE-GSS-2	0.015		
IGGE-GSS-8	0.0166		
CANMET-SO-3	0.017	Mn — manganese	
CANMET-SO-4	0.030		
IGGE-GSS-1	0.032	AMM-SO-1	266
IGGE-GSS-3	0.060	IGGE-GSS-3	304
IGGE-GSS-7	0.061	IGGE-GSS-2	510
NIST-SRM-8408	0.07	CANMET-SO-3	520
IGGE-GSS-6	0.072	CANMET-SO-4	600
CANMET-SO-2	0.082	IAEA-SOIL-7	631
BCR-CRM-142	0.104	IGGE-GSS-8	650
IGGE-GSS-5	0.294	CANMET-SO-2	720
BCR-CRM-141	0.568	IGGE-GSS-5	1,360
IGGE-GSS-4	0.590	IGGE-GSS-4	1,420
BCR-CRM-143	3.92	IGGE-GSS-6	1,450
NIST-SRM-8407	50	IGGE-GSS-1	1,760
NIST-SRM-8406	110	IGGE-GSS-7	1,780

TABLE 26.3 (continued) Certified and Recommended Elemental Concentration Values for Some Common Elements Reported for Soil Reference Materials Listed in Order of Element and in Increasing Order of Concentration

Material Code[a]	Concentration (mg/kg[b])	Material code	Concentration (mg/kg)
Mo — molybdenum		**Pb (continued)**	
IGGE-GSS-3	0.30	AMM-SO-1	15
IGGE-GSS-2	0.98	CANMET-SO-4	16
IGGE-GSS-8	1.16	IGGE-GSS-2	20.2
IGGE-GSS-1	1.4	IGGE-GSS-8	21
IGGE-GSS-4	2.6	CANMET-SO-2	21
IGGE-GSS-7	2.9	IGGE-GSS-3	26.0
IGGE-GSS-5	4.6	BCR-CRM-141	29.4
IGGE-GSS-6	18	BCR-CRM-142	37.8
		IGGE-GSS-4	58.5
Na — sodium		AEA-SOIL-7	60
		IGGE-GSS-1	98
IGGE-GSS-7	550	IGGE-GSS-6	314
IGGE-GSS-4	830	IGGE-GSS-5	552
IGGE-GSS-5	905	BCR-CRM-143	1,333
IGGE-GGS-6	1,410		
AMM-SO-1	4,440	**Si — silicon**	
CANMET-SO-3	7,400		
CANMET-SO-4	10,000	IGGE-GSS-7	152,900
IGGE-GSS-2	12,020	CANMET-SO-3	158,600
IGGE-GSS-1	12,310	IGGE-GSS-4	238,300
IGGE-GSS-8	12,760	IGGE-GSS-5	245,900
CANMET-SO-2	19,000	CANMET-SO-2	249,900
IGGE-GSS-3	20,100	IGGE-GSS-6	266,300
		IGGE-GSS-8	274,100
		IGGE-GSS-1	292,800
		CANMET-SO-4	319,700
P — phosphorus		IGGE-GSS-2	343,100
		IGGE-GSS-3	349,500
IGGE-GSS-6	303		
IGGE-GSS-3	320		
IGGE-GSS-5	390		
IGGE-GSS-2	446		
CANMET-SO-3	480	**Sr — strontium**	
IGGE-GSS-4	695		
IGGE-GSS-1	735	IGGE-GSS-7	26
IGGE-GSS-8	775	IGGE-GSS-6	39
CANMET-SO-4	900	IGGE-GSS-5	41.5
IGGE-GSS-7	1,150	IGGE-GSS-4	77
CANMET-SO-2	3,000	IAEA-SOIL-7	108
		IGGE-GSS-1	155
		CANMET-SO-4	170
		IGGE-GSS-2	187
		CANMET-SO-3	217
Pb — lead		IGGE-GSS-8	236
		CANMET-SO-2	340
IGGE-GSS-7	13.6	IGGE-GSS-3	380
CANMET-SO-3	14		

TABLE 26.3 (continued) Certified and Recommended Elemental Concentration Values for Some Common Elements Reported for Soil Reference Materials Listed in Order of Element and in Increasing Order of Concentration

Material Code[a]	Concentration (mg/kg[b])	Material code	Concentration (mg/kg)
Ti — titanium			
		V (continued)	
CANMET-SO-3	2,000	IGGE-GSS-5	166
IGGE-GSS-3	2,240	IGGE-GSS-7	245
IGGE-GSS-2	2,710	IGGE-GSS-4	247
CANMET-SO-4	3,400		
IGGE-GSS-8	3,800	Zn — zinc	
IGGE-GSS-6	4,390		
IGGE-GSS-1	4,830	IGGE-GSS-3	31.4
IGGE-GSS-5	6,290	AMM-SO-1	35
CANMET-SO-2	8,600	IGGE-GSS-2	42.3
IGGE-GSS-4	10,800	CANMET-SO-3	52
IGGE-GSS-7	20,100	IGGE-GSS-8	68
		BCR-CRM-141	81.3
V — vanadium		BCR-CRM-142	92.4
		CANMET-SO-4	94
IGGE-GSS-3	36.5	IGGE-GSS-6	96.6
CANMET-SO-3	38	IAEA-SOIL-7	104
IGGE-GSS-2	62	CANMET-SO-2	124
CANMET-SO-2	64	IGGE-GSS-7	142
IAEA-SOIL-7	66	IGGE-GSS-4	210
IGGE-GSS-8	81.4	IGGE-GSS-5	494
IGGE-GSS-1	86	IGGE-GSS-1	680
CANMET-SO-4	90	BCR-CRM-143	1,272

Note: Information is for materials listed in Table 26.2, adapted from Muramatsu and Parr (1985), Ihnat (1988), and the references listed in Table 26.2 with retention only of analytical data denoted as certified or recommended.

[a] Code is a combination of issuer code, together with a code for the issuing organization adopted from Cortes Toro et al. (1990).

[b] Concentrations are in mg/kg on a dry basis or "as is" basis as recorded in the relevant reports. Concentrations for major elements reported as oxides for the IGGE materials have been converted to the elemental basis. This table contains only some of the more common elements and is for information only for use as a guide in selecting appropriate reference materials. The reader is advised to consult the relevant certificates, reports, or information sheets supplied with the materials for latest official certified or recommended concentration values and uncertainties for these and other elements.

4   For multielement analyses, it would certainly be efficient and cost effective to be able to use the same aliquot of the selected RM for quality control. Feasibility of this approach depends on whether or not that RM has certified analytical values for the elements of interest (reference to Table 26.2) and whether or not the concentration levels in the reference and submitted materials are in reasonable concordance. Suitable matching is left to the analyst's discretion and will require consideration of the level of control desired, the number and availability of RM, and the amount of RM work to be included in the analytical scheme.

26.4.2 Procedures for Reference Material Utilization

RM with highly reliable certified or recommended values can be called upon to fulfill a variety of demands within the measurement process. The major uses of such materials can be grouped into the four broad categories: (1) analytical calibration, (2) quality control, (3) analytical method development and evaluation, and (4) production and evaluation of other RM. Good coverage of the roles and uses of RM is provided by Cali et al. (1975), Uriano and Gravatt (1977), and Taylor (1985).

The use of RM for establishing calibration functions (analytical response as a function of analyte concentration) has been proposed. However, this mode of usage of natural matrix RM is not generally recommended due to uncertainties in certified elemental concentrations. These uncertainties, resulting from material inhomogeneity and measurement imprecisions and biases, are many times greater than uncertainties of pure element or pure compound assays. Since the concentrations of standard solutions used for calibration should be known with greater certainty than is required in the analysis of an actual sample, the use of high-purity, highly reliable pure elements and compounds rather than natural matrix RM is preferred for calibration.

Analytical data quality control to establish method performance (bias) is by far the most common use of RM and the one recommended by this author; a good treatment of this is given by Taylor (1985). The main use of these materials is to monitor and maintain data quality. Errors in measurement can arise in the three component steps of an analytical method: sampling (including presampling considerations), sample manipulation, and measurement. Thus, when using the RM for data quality control, the aggregate of all steps subsequent to the point at which the RM is introduced into the scheme of analysis will be monitored for performance.

The following steps are suggested for RM utilization.

1   Ensure that the analytical procedure is in a state of statistical control prior to testing, in order for credible data to be generated and errors identified and corrected.

2   Follow certificate instructions for material handling. Incorporate the RM(s) into the laboratory scheme of analysis, introducing it at earliest stage, i.e., prior to the beginning of sample decomposition. Take it through the entire analytical procedure at the same time and conditions as the actual analytical samples in order to monitor all steps within the sample manipulation and measurement stages.

3   In the case of multiple element determinations, should different sample preparation and measurement procedures (i.e., different analytical methods) be indicated for the different elements, take separate aliquots of the RM through the entire appropriate analytical scheme for proper quality control.

26.4.2.1 COMMENTS

1   The information in the tables in this chapter should be used only as guides for RM selection and use. The latest appropriate certificates or reports of analysis or other relevant publications issued with the RM when acquired must be consulted and used. The most important source of information for the RM is the certificate or report of analysis issued with the material. This document is an integral part of the RM, as it provides the analytical (certified) information, estimates of uncertainties, instructions for the correct use of the material, and other relevant information.

2   RM can monitor the performance of laboratory procedures subsequent to the point of introduction of the RM. Activities occurring prior to this, such as sampling, preservation, storage, and presampling considerations, are generally impossible to monitor by use of RM.

3   It is important that both the reference and actual samples undergo identical, simultaneous handling; if feasible, the RM could be submitted as a blind material to the analyst. It is also important that the reference and actual sample analyte concentrations be reasonably close, since method performance can vary dramatically with concentration, and conclusions at one level may not be applicable to other levels.

4   RM are best used on a regular basis. The sporadic use of RM when trouble is suspected is a legitimate use, but systematic measurement in a control chart mode of operation will generally be more informative and is highly recommended. RM may be used as the sole quality control material or they may be used in conjunction with in-house or locally produced control materials in a systematic manner in order to conserve the former.

26.4.3 Interpretation, Calculation, and Corrective Action

When possible, the analysis of several reference samples, spanning the concentration range of interest, is the most useful way to investigate measurement bias. The three-sample approach — analysis of a low, middle, and upper range sample — is practical, provided that the reference samples are sufficiently homogeneous and that the range of analytical interest is covered. However, acquiring the necessary RM from the world repertoire of materials may not always be possible. When supported by other data, the measurement of even a single reference sample can be meaningful.

Typically, the method under test should give a precision (standard deviation) with the RM and other homogeneous materials equal to or better than the overall uncertainty reported for the RM by the issuer. Results from the analysis of the RM are then compared with the certified value; rarely will the two agree exactly, due to measurement errors in each. Whether the two differ significantly is ascertained by comparing the two values and their uncertainties using simple statistical tests. If the confidence intervals intersect, the measured concentration value agrees with the certified value, and the analyst can apply his method with some confidence to the analysis of materials of similar composition. Otherwise there is disagreement and the method as applied exhibits a bias. One of the following approaches can be followed to estimate agreement.

1   **Case with all parameters available:** compare the 95% confidence levels calculated from the standard deviation, number of analyses, and the student *t* statistic with the confidence or tolerance interval of the RM using the following equations:

$$\bar{X}_1 - \bar{X}_2 = ts\left(\frac{1}{n_1} + \frac{1}{n_2}\right)^{1/2} \tag{26.1}$$

$$s^2 = [(n_1 - 1)\, s_1^2 + (n_2 - 1)\, s_2^2]/(n_1 + n_2 - 2) \tag{26.2}$$

where:
$\bar{X}_1$ is the mean concentration found by the user for the RM,
$\bar{X}_2$ is the certified, recommended, or reference value for the RM,
S_1 is the standard deviation estimated from n_1 determinations by the user,
S_2 is the standard deviation reported for the RM in the certificate of report of analysis based on n_2 determinations, *t* is the student *t* statistic

The difference $\bar{X}_1 - \bar{X}_2$ is compared to the right-hand side of equation 26.1 using the *t* value for 95% confidence ($p = 0.05$). Should it be greater (positive or negative), a discrepancy exists between the measured and certified concentration value, which indicates the analytical procedure not to be operating well and that the source of error should be sought and rectified. Should it be ascertained that an unacceptable bias exists, a correction for it should *not* be applied. Instead, diagnostic steps should be taken to identify sources of unacceptable bias or imprecision and corrective action should be taken.

2   **Case with missing n_2 and negligible uncertainty in the RM certified value:** compare the absolute value of the estimated bias $\bar{X}_1 - \bar{X}_2$ with a critical value based on

$$\bar{X}_1 - \bar{X}_2 = t\, s_1/(n_1)^{1/2} \tag{26.3}$$

using uncertainty parameters only for the measurements carried out by the analyst. Proceed further as in case 1.

3   **Case with missing n_2 and specified uncertainty in the RM:** compare the absolute value of the estimated bias $\bar{X}_1 - \bar{X}_2$ with a critical value based on

$$\bar{X}_1 - \bar{X}_2 = t\, s_1/(n_1)^{1/2} + u \tag{26.4}$$

where *u* is the uncertainty of the RM certified concentration reported in the certificate of analysis. Proceed further as in case 1.

Reference to collaborative and research data published in the literature and consultations with experienced users of the methodology may be helpful for suggested approaches. Reference to the "Handbook for SRM users" by Taylor (1985) is recommended for detailed discussion of RM use. He lists the following factors to be considered in reducing operational (nonstate-of-the-art) bias: quality of calibrants, calibration technique, contamination, mechanical losses, solution extraction efficiencies, and interferences. He gives the following factors to be considered in reducing operational random errors: technical skill, manipulative skill, environmental control, tolerances in operational parameters, instrumentation, and blank variability. It may be of interest to the user of the RM to consult the report of Gladney et al. (1987), giving a compilation of elemental concentration data reported in the literature for NIST SRM.

26.4.3.1 COMMENTS

1 It cannot be too strongly emphasized that reference must be made to the appropriate certificate or report of analysis, and not to other sources of information, for correct usage of a RM and interpretation of results.

2 It must be emphasized that the measurement system must be under statistical control before the analytical data can be used and errors identified and corrected. Identification of causes of unacceptable bias, or precision, a necessary first step prior to corrective action is not easy. Whenever excessive bias or imprecision is found to be present, corrective action must be taken, otherwise the measurement process will have limited usefulness.

3 There are not too many instances where the uncertainty for the RM is characterized by a standard deviation s_2 and corresponding number of determinations, n_2. Thus cases 2 and, particularly, 3 will most often be the ones of necessity. The uncertainty, u, in case 3 is not necessarily a standard deviation or standard error, but can reflect symmetric or asymmetric estimates of imprecision and possible systematic errors among methods used in certification.

26.5 CONCLUDING COMMENTS

In this presentation, no treatment or listing has been included of in-house, local, regional, or other "uncertified" reference products. Apart from such information being generally unavailable, the overriding reason being the requirement in reference work of materials with unassailable, certified, correct property values. Reliability and absolute confidence in the stated characteristics of RM is a basic critical criterion for their use for quality control. Homogeneous in-house materials standardized with respect to approximate analyte concentration can be incorporated into the quality control scheme for day-to-day or even more frequent monitoring of *precision*. Certified, bona fide RM are reserved for both *precision and analytical bias, monitoring, and control*. No reference soil materials are available certified for physical properties.

No RM certified for extractable elemental concentrations are available to monitor the usual procedures in soil science based on extraction. The philosophy of certification rests on the concept of independent methodology, that is, the application of theoretically and experimentally different measurement techniques and procedures to generate concordant results leading to one reliable value for the property. Such values are thus method independent. Extractable concentrations are generated by specific procedures and are thus method dependent, an idea which has to be rationalized with the fundamental method-independent concept in RM certification work.

The complexity of modern chemical analysis provides many sources of error and opportunities for introduction of bias and imprecision. Such systems must therefore be operated under a rigid quality assurance system if results are to be meaningful. The performance of the entire system needs to be monitored on a regular basis. RM are finding increasing use as test materials to monitor system performance.

It is hoped that this chapter will increase the awareness of the soil scientist and analyst of the concept, role, and utility of RM and that it will increase RM use as a cornerstone of the quality control program to establish, monitor, and maintain analytical data quality.

Contribution No. 92-44 from Centre for Land and Biological Resources Research.

REFERENCES

Bowman, W. S. 1990. Certified reference materials. CCRMP 90-1E. Canada Centre for Mineral and Energy Technology, Ottawa.

Bowman, W. S., Faye, G. H., Sutarno, R., McKeague, J. A., and Kodama, H. 1979. Soil samples SO-1, SO-2, SO-3 and SO-4 — certified references materials. CANMET Report 79-3. Canada Centre for Mineral and Energy Technology, Ottawa.

Cali, J. P., Mears, T. W., Michaelis, R. E., Reed, W. P., Seward, R. W., Stanley, C. L., Yolken, H. T., and Ku, H. H. 1975. The role of standard reference materials in measurement systems, NBS Monograph 148, National Bureau of Standards, Washington, D.C.

Coleman, R. F. 1980. Improved accuracy in automated chemistry through the use of reference materials. J. Automatic Chem. 2: 183–186.

Community Bureau of Reference (BCR). 1983. Catalogue of BCR reference materials. BCR, Brussels, Belgium.

Community Bureau of Reference (BCR). 1985. BCR reference materials addendum to catalogue 1983. BCR, Brussels, Belgium.

Cortes Toro, E., Parr, R. M., and Clements, S. A. 1990. Biological and environmental reference materials for trace elements, nuclides and organic microcontaminants, IAEA/RL/128 (Rev. 1), International Atomic Energy Agency, Vienna, Austria.

Currie, L. A. 1982. Quality of analytical results, with special reference to trace analysis and sociochemical problems. Pure Appl. Chem. 54: 715–754.

Dux, J. P. 1986. Handbook of quality assurance for the analytical chemistry laboratory. Van Nostrand Reinhold, New York.

Gladney, E. S., O'Malley, B. T., Roelandts, I., and Gills, T. E. 1987. Standard reference materials: compilation of elemental concentration data for NBS clinical, biological, geological and environmental standard reference materials. NBS Spec. Publ. 260-111. National Bureau of Standards, Gaithersburg, MD.

Holynska, B., Jasion, J., Lankosz, M., Markowicz, A., and Baran, W. 1988. Soil SO-1 reference material for trace analysis. Fresenius Z. Anal. Chem. 332: 250–254.

Ihnat, M. 1982. Importance of acid-insoluble residue in plant analysis for total macro and micro elements. Commun. Soil Sci. Plant Anal. 13: 969–979.

Ihnat, M. 1988. Biological and related reference materials for determination of elements (Appendix 1). Pages 739–760 in H. A. McKenzie and L. E. Smythe, Eds. Quantitative trace analysis of biological materials. Elsevier, Amsterdam.

International Atomic Energy Agency. 1984. Information sheet for reference material SOIL-7, Soil. IAEA, Vienna, Austria.

International Atomic Energy Agency. 1990. Intercomparison runs, reference materials. International Atomic Energy Agency, Vienna, Austria.

Kateman, G. and Pijpers, F. W. 1981. Quality control in analytical chemistry. Wiley, New York.

Keith, L. H., Crummett, W., Deegan, J. Jr., Libby, R. A., Taylor, J. K., and Wentler, G. 1980. Principles of environmental analysis. Anal. Chem. 55: 2210–2218. Also appeared as Principles of Environmental Analysis (1983), American Chemical Society, Washington, D.C.

McKeague, J. A., Sheldrick, B. H., and Desjardins, J. G. 1978. Compilation of data for CSSC reference soil samples. Soil Research Institute, Ottawa (unpublished report).

McKenzie, R. L. 1990. NIST standard reference materials catalog 1990–1991. NIST Spec. Publ. 260. National Institute of Standards and Technology, Gaithersburg, MD.

Muramatsu, Y. and Parr, R. M. 1985. Survey of currently available reference materials for use in connection with the determination of trace elements in biological and environmental materials. IAEA/RL/128. International Atomic Energy Agency, Vienna, Austria.

Sheldrick, B. H. and Wang, C. 1987. Compilation of data for ECSS reference soil samples. Land Resource Research Centre, Ottawa (unpublished report).

Taylor, J. K. 1981. Quality assurance of chemical measurements. Anal. Chem. 53: 1588A–1596A.

Taylor, J. K. 1985. Standard reference materials: handbook for SRM users. UBS Spec. Publ. 260-100. National Bureau of Standards, Gaithersburg, MD.

Uriano, G. A. and Gravatt, C. C. 1977. The role of reference materials and reference methods in chemical analysis. CRC Crit. Rev. Anal. Chem. 6: 361–411.

Xie, X., Yan, M., Li, L., and Shen, H. 1985. Usable values for Chinese standard reference samples of stream sediments, soils, and rocks: GSD 9-12, GSS 1-8 and GSR 1-6. Geostand. Newslett. 9: 277–280.

Xie, X., Yan, M., Wang, C., Li, L., and Shen, H. 1989. Geochemical standard reference samples GSD 9-12, GSS 1-8 and GSR 1-6. Geostand. Newslett. 13: 83–179.

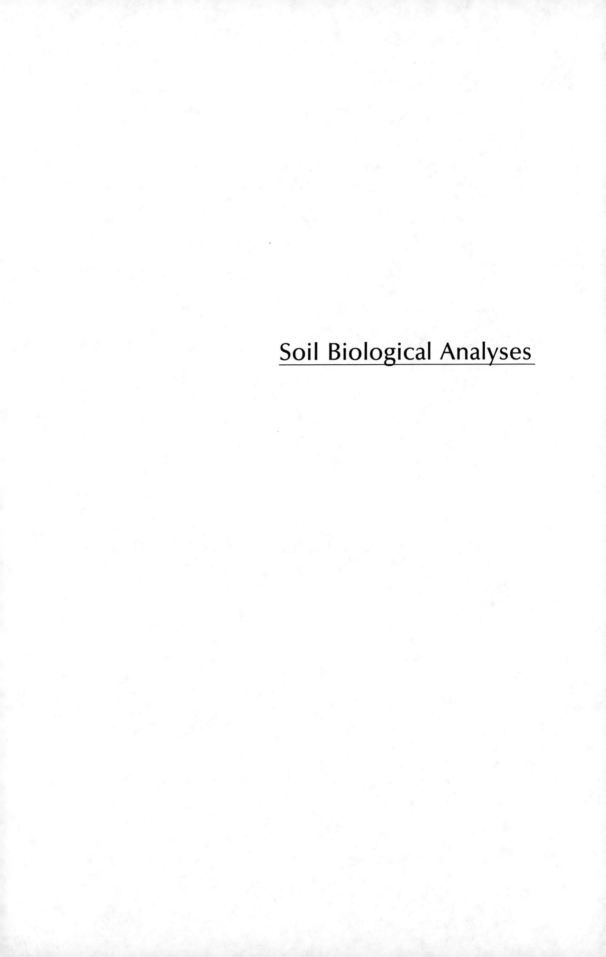

Soil Biological Analyses

Chapter 27
Cultural Methods for Soil Microorganisms

J. J. Germida

University of Saskatchewan

Saskatoon, Saskatchewan, Canada

27.1 INTRODUCTION

Soil is an ecosystem that contains a variety of microbial populations whose members represent many physiological types. For example, some microorganisms, such as fungi, are aerobic chemoorganotrophs (heterotrophs) and use organic compounds as a source of carbon and energy. Others, such as nitrifying bacteria, are aerobic chemolithotrophs (autotrophs), using CO_2 as a carbon source and oxidizing reduced inorganic N compounds to obtain energy. Some microorganisms require special growth factors, a specific environmental pH, low O_2 levels or the absence of O_2 (i.e., anaerobes) for optimum growth. The chemical, physical, and biological characteristics of a particular soil, as well as the presence of growing plants, will influence the numbers and activities of its various microbial components. Furthermore, because of the heterogeneous nature of soil, many different physiological types of organisms will be found in close proximity to one another. The microbial community in soil is important because of its relationship to soil fertility and the biogeochemical cycling of elements, and the potential use of specific members for industrial applications. Thus the need to enumerate and isolate major and minor members of the microbial community in soils.

The nonselective enumeration and isolation of soil microorganisms is relatively straightforward, but the final result is not necessarily meaningful (Casida 1968, Jensen 1968, Parkinson et al. 1971, Wollum 1982). On the other hand, selective enumeration and isolation of specific physiological types of microorganisms can provide useful and meaningful data (see, e.g., Lochhead and Chase 1943, Brown et al. 1962, Casida 1965, Simon et al. 1973, Rennie 1981, Lawrence and Germida 1988, Buyer et al. 1989). Methods to enumerate and isolate soil microorganisms are constantly changing as our knowledge of the types of microorganisms present in soil expands (Clark 1965a, Vincent 1970, Parkinson et al. 1971, Alexander 1982, Wollum 1982). This chapter provides basic principles and references on enumeration procedures and culture media for representative types of soil microorganisms.

27.2 PRINCIPLES

Enumeration of viable soil microorganisms may be accomplished by the plate count technique or most-probable number (MPN) technique. The underlying principles are (1) dispersing of a sample in a suitable diluent, (2) distributing an aliquot to an appropriate growth medium, (3) incubating inoculated plates under suitable conditions, and (4) counting the developed colonies or MPN tubes. These procedures are fairly standard and may be used to enumerate populations in bulk (Waksman 1922, Smith and Worden 1925, James and Sutherland 1939, Lochhead and Chase 1943, James 1959, Casida 1965) and rhizosphere (Katznelson and Bose 1959, Goodfellow et al. 1968, Kucey 1983) soils.

The composition of the growth medium used to enumerate microbial populations is important, as it will affect the final result. Growth media may be selective or nonselective, although no medium is truly nonselective (James 1958, Jensen 1968). Selective media contain components which allow or favor the growth of a desired group of organisms. Nonselective media should encourage growth of as many diverse groups of organisms as possible. To enumerate a specific physiological type of microorganism it is usually possible to design a medium which, when incubated under appropriate conditions of atmosphere and temperature, reduces interference from nondesired populations. For example, chitin degrading actinomycetes may be enumerated on medium containing chitin as the sole source of carbon and nitrogen (Lingappa and Lockwood 1962, Hsu and Lockwood 1975), and inclusion of specific antibiotics prevents growth of undesired organisms (Williams and Davies 1965). Similar selective media are available for many different soil microorganisms, e.g., phosphate-solubilizing bacteria (Kucey 1983), siderophore-producing microorganisms (Buyer et al. 1989), free-living, nitrogen-fixing bacteria (Brown et al. 1962, Clark 1965b, Rennie 1981), nitrifying bacteria (Schmidt and Belser 1982), or sulfur-oxidizing organisms (Postgate 1966, Germida 1985, Lawrence and Germida 1988).

Although the plate count and MPN techniques are simple to perform, their usefulness will be limited by a number of key factors (Casida 1968, Jensen 1968, Alexander 1982, Wollum 1982). Many times choice of media, problems with dispersion, and even adsorption of microbes to pipette walls can interfere with standardization of these procedures. It should be pointed out that consistency and adequate use of replicate samples will help to minimize some of these problems. Investigators must realize that microorganisms are not uniformly dispersed within the soil environment, and that numbers of any particular microorganism are not synonymous with its importance.

27.3 SPREAD PLATE COUNTING METHOD

Enumeration of microbial populations by the spread plate method is a simple and rapid method to count viable microbial cells in soil. However, counts obtained are generally 10- to 100-fold less than those determined by microscopic counts of soil smears (Skinner et al. 1952). Reasons for this discrepancy include measurement of viable and nonviable counts in soil smears, and the inability to provide adequate or appropriate nutrients in the growth media for spread plate counts (James and Sutherland 1939, Lochhead and Chase 1943, Jensen 1968). Basically, this method consists of preparing a serial dilution (e.g., 1:10 dilutions) of a soil sample in an appropriate diluent, spreading an aliquot of a dilution on the surface of an agar medium, and incubating the agar plate under appropriate environmental conditions. These spread plates may be used not only for counting microbial populations, but also as a starting source for isolation of organisms. In this latter case, an isolated colony is picked and repeatedly streaked on a suitable growth medium to check for purity. After several such transfers it may be cultured and preserved for future study and identification. Selective or nonselective media may be used, depending on the nature of the desired microorganisms. Soil extract agar (James 1958) is commonly used as a nonselective medium for enumerating soil bacteria. Recently proposed alternatives to soil extract agar include a defined "soil solution equivalent medium" (Angle et al. 1991) and "trypticase soy agar" — a commercially available complex medium (Martin 1975).

27.3.1 MATERIALS

1 Twelve petri plates containing about 20 mL of an appropriate agar medium, e.g., soil extract agar.

2 Dilution bottles (e.g., 50 × 160 mm; 200-mL capacity) and (or) test tubes (e.g., 18 × 150 mm) containing appropriate diluent, such as sterile tap water.

3 Sterile 1- and 10-mL pipettes.

4 Glass spreader (i.e., glass rod shaped like a hockey stick).

5 Glass beaker containing 95% ethanol (ETOH).

27.3.2 PREPARATION OF AGAR PETRI PLATES

1 Agar media may be prepared from commercially available dehydrated components or from recipes found in the literature. The American Type Culture Collection Catalog is an excellent source of media recipes and relevant references (ATCC 1989). Prepare media according to directions, sterilize in the autoclave at 1.05 kg/cm^2 and 121°C for 20 min, and cool to a pouring temperature of about 48°C. Some components of a medium may be heat labile and must be filter-sterilized and then added to the autoclaved agar medium (cooled to about 48–49°C) just prior to pouring plates.

2 Distribute about 20 mL of media into sterile glass or disposable, presterilized plastic petri plates and allow the agar to solidify. The plates should be allowed to sit at room temperature for 24–48 h, allowing excess surface moisture to be absorbed into the agar; this helps prevent microbial colonies from spreading over the agar surface. Poured plates not used immediately may be stored under refrigeration (2–5°C) for up to 2 weeks. These stored plates should be removed and allowed to warm to room temperature prior to use.

27.3.3 PREPARATION OF SOIL DILUTIONS

1   Samples should be collected, handled, and stored with due consideration to their ultimate use, and the effects these steps will have on microbial populations (see Chapters 1–3).

2   Pass representative soil samples through a 2-mm mesh sieve and mix thoroughly.

3   Weigh out a 10-g portion of the soil into a dilution bottle containing 95 mL of a diluent. Glass beads (about 25 2-mm beads) may be added to this dilution blank to facilitate mixing. Cap the bottle, place on a mechanical shaker, and shake for 10 min. Alternatively, shake by hand, moving the bottle through a specified arc a number of times (e.g., a 45° arc at least 50–100 times). This first dilution represents a 1:10 or a 10^{-1} dilution.

4   After removing the bottle from the shaker, shake vigorously prior to removing aliquots. To prepare a serial 1:10 dilution series of the soil sample, transfer a 10-mL sample to a 90-mL dilution blank cap and shake the dilution bottle (alternatively, it is possible to transfer 1-mL samples to 9-mL dilution blanks [prepared in 18 × 150-mm test tubes]). Continue this sequence until a dilution of 10^{-7} is reached. Subsequent spread plating of a 0.1-mL aliquot of this dilution will allow enumeration of up to 3×10^{10} colony-forming units (cfu) per g soil. Experience will indicate the appropriate range of dilutions for samples being analyzed.

27.3.4 PREPARATION OF AGAR SPREAD PLATES

1   Select a range of four dilutions that will adequately characterize the microorganisms in the sample. Transfer 0.1-mL aliquots to a separate plate from the highest dilution. Note that a 0.1-mL aliquot from a 10^{-7} dilution corresponds to an actual dilution of 10^{-8} on the plate. Repeat the process, transferring 0.1-mL aliquots from each of the next three successive and lower dilutions onto each of triplicate plates for each dilution.

2   Spread the suspension on the agar surface using a sterile glass spreader for each plate. The glass spreader is kept submerged in a beaker of ETOH, and excess ETOH is burned off prior to use. In the spreading step start with the highest dilution and progress to the next lower dilution, continuing the sequence until all the plates have been spread. Alcohol flame the glass spreader between each plate. Invert the plates and place in an incubator at an appropriate temperature.

3   Incubation conditions will depend on the facilities and the purpose of the study. When possible, one should try to mimic environmental conditions. Generally, spread plates samples are incubated in the dark, in an aerobic environment at a temperature between 24 and 28°C. Incubation periods and

conditions will vary, depending on the nature of the organisms being enumerated.

4   After a suitable incubation interval, plates are removed from the incubator and those containing 30 to 300 colonies counted. Plates with spreading or swarming organisms should be excluded from the final count. The colonies can be counted manually or by an automated laser colony counter.

27.3.5 Calculations

Average the number of colonies per plate for the dilution giving between 30 and 300 colonies. Determine the number of cfu per gram of soil as follows:

$$\text{Number of cfu g}^{-1} \text{ soil}_{DW} = \frac{(\text{Mean plate count}) (\text{Dilution factor})}{\text{Dry weight soil, initial dilution}} \quad (27.1)$$

where:

$$\text{Dry weight soil} = (\text{Weight moist soil, initial dilution} \times (1 - \frac{\% \text{ Moisture, soil sample}}{100}) \quad (27.2)$$

27.3.6 Comments

Because bacteria may exist in soil as groups or clumps of cells, it is often desirable to disperse these cells so that colonies on spread plates arise from one cell. This may be accomplished by shaking on a mechanical shaker or by hand, through application of a high shearing force as with a Waring® blender (Casida 1965), by sonic vibration (Stevenson 1959), mechanical vibration (Thornton 1922), and through use of deflocculating agents (Jensen 1968).

27.3.7 Types of Dilutions

A number of diluents may be used. In most cases tap or distilled water is adequate. Other diluents routinely used include:

1   Physiological saline: NaCl, 8.5 g; 1 L of distilled water.

2   Phosphate-buffered saline: NaCl, 8.0 g; KH_2PO_4, 0.34 g; K_2HPO_4, 1.21 g; 1 L of distilled water. Adjust pH to 7.3 with 0.1 M NaOH or HCl.

3   Peptone water: peptone, 1.0 g; 1 L of distilled water.

27.3.8 Types of Media

The choice of medium depends on the type of organism desired. Media may be made selective by omitting or altering a component, or by incubation conditions. For enumeration of total heterotrophic populations in soil, a general nonselective medium is usually employed. The following are examples of media commonly used to enumerate total soil bacteria, actinomycetes, and fungi. Additional examples of specific media are described in Section 27.5.

27.3.8.1 MEDIA FOR TOTAL HETEROTROPHIC BACTERIA

1   Soil extract agar (James 1958): 1 kg of soil is autoclaved with 1 L water for 20 min at 1.05 kg/cm^2. The liquid is strained and restored to 1 L in volume. If it is cloudy, a little $CaSO_4$ is added, and, after being allowed to stand, it is filtered through Whatman® No. 5 paper. The extract may be sterilized and solidified with agar (1.5%) as it is, or after the addition of other nutrients, e.g., 0.025% K_2HPO_4 or 0.1% glucose, 0.5% yeast extract and 0.02% K_2HPO_4.

2   Tryptic soy agar (Martin 1975): add 3 g of tryptic soy broth and 15 g of agar to 1 L distilled water. Sterilize the medium by autoclaving.

27.3.8.2 Media for Actinomycetes

Starch-casein agar (Küster and Williams 1966): starch, 10.0 g; casein (vitamin free), 0.3 g; KNO_3, 2.0 g; NaCl, 2.0 g; K_2HPO_4, 2.0 g; $MgSO_4 \cdot 7H_2O$, 0.05 g; $CaCO_3$, 0.02 g; $FeSO_4 \cdot 7H_2O$, 0.01 g; agar, 15 g; distilled water, 1 L; pH 7.2. Sterilize in autoclave as described above. Can be improved by the addition of fungistatic agents (Williams and Davies 1965).

27.3.8.3 MEDIA FOR FUNGI

1   Czapek-Dox agar: sucrose, 30.0 g; $NaNO_3$, 2.0 g; K_2HPO_4, 0.1 g; KCl, 0.5 g; $MgSO_4 \cdot 7H_2O$, 0.5 g; $FeSO_4$, trace; agar, 15 g; distilled water, 1 L; pH, 5.5. Dissolve separately and add $FeSO_4$ last. If desired, 6.5 g of yeast extract may be added (Gray and Parkinson 1968). Sterilize by autoclaving. Cool to pouring temperature, and acidify to pH 3.5 with a presterilized solution of 10% lactic acid. Use immediately.

2   Streptomycin-rose bengal agar (Martin 1950): glucose, 10 g; peptone, 5 g; KH_2PO_4, 1.0 g; $MgSO_4 \cdot 7H_2O$, 0.5 g; rose bengal, 0.03 g; agar, 20 g; tap water, 1 L. Autoclave, cool medium to about 48°C, and add 1 mL of a solution of streptomycin (0.3 g/10 mL sterile water). The final concentration of streptomycin medium should be about 30 µg/mL.

27.4 MOST PROBABLE NUMBER METHOD (MPN)

This method employs the concept of dilution to extinction in order to estimate the number of microorganisms in a given sample (Taylor 1962, Casida 1968, Alexander 1982). It is based on the presence or absence of microorganisms in replicate samples in each of several consecutive dilutions of soil. For example, if a series of test tubes containing broth medium are inoculated with aliquots representing a dilution series from 10^{-4} through 10^{-7}, and the highest dilution exhibiting growth is 10^{-5}, then the number of cells present may be estimated to be between 10^4 to 10^5. The key is that the desired organism must possess a unique characteristic or metabolic trait which can be detected. Thus, this technique can be used to count microorganisms based on growth (i.e., turbidity), metabolic activity such as substrate disappearance, and product formation. Other uses for the MPN technique include enumeration of nodule-forming rhizobia in soil (see Chapter 30) or infective vesicular-arbuscular mycorrhizal propagules in soil (see Chapter 29).

To estimate total heterotrophic counts of a soil sample the MPN procedure is similar to the spread plate count method, except that aliquots of dilutions are inoculated into test tubes of liquid medium. Alternatively, multiwell microtiter plates (e.g., 24 wells per plate) may be used in place of test tubes; this allows a savings of materials, reagents, and incubator space and allows for increased replication.

27.4.1 MATERIALS

1   25 test tubes containing appropriate culture medium or 1 disposable sterile microtiter plates — 24 or 96 wells per plate.

2   Water dilution blanks (27.3.1).

3   Sterile 1- and 10-mL pipettes (27.3.1).

27.4.2 PROCEDURES

1   Prepare medium appropriate for the desired organism.

2   Dispense aliquots of medium in test tubes and sterilize or dispense aliquots of sterile medium into presterile microtiter plates.

3   Prepare a serial 1:10 dilution sequence of the soil sample (27.3.3).

4   Select a range of dilutions that will adequately characterize the organisms in the sample. Transfer 0.1-mL aliquots to each separate well in five replicate microtiter plate wells, starting with the highest dilution. Repeat the procedure, transferring 0.1-mL aliquots from each of the next successive and lower dilutions into each of the five replicate wells for each dilution.

5   Incubate MPN assay tubes and/or plates under appropriate conditions.

6   After suitable incubation, score wells positive for growth or physiological reaction.

27.4.3 Calculations

The MPN of organisms in the original sample is calculated by reference to a MPN Table (e.g., Cochran 1950). Designate as p_1 the number of positive tubes in the least concentrated dilution in which all tubes are positive or in which the greatest number of tubes are positive. Let p_2 and p_3 represent the numbers of positive tubes in the next two higher dilutions. Refer to Table 27.1 and find the row of numbers in which p_1 and p_2 correspond to the values observed experimentally. Follow that row of numbers across the table to the column headed by the observed value of p_3. The figure at the point of intersection is the MPN or organisms in the quantity of the original sample represented in the inoculum added in the second dilution. This figure is multiplied by the appropriate dilution factor to obtain the MPN value for the original sample.

As an example, consider the instance in which a tenfold dilution with five tubes at each dilution yielded the following numbers of positive tubes after incubation: five at 10^{-4}, five at 10^{-5}, four at 10^{-6}, two at 10^{-7}, and zero at 10^{-8}. In this series, $p_1 = 5$, $p_2 = 4$, and $p_3 = 2$. For this combination of p_1, p_2, and p_3, Table 27.1 gives a 2.2 as the MPN of organisms in the quantity of inoculum applied in the 10^{-6} dilution. Multiplying this number by the dilution factor 10^6 gives 2.2 million as the MPN value for the original sample.

As a second example, consider the same situation as above except that the most concentrated dilution is 10^{-6}. Under these circumstances, $p_1 = 4$, $p_2 = 2$, and $p_3 = 0$. For this combination of p_1, p_2, and p_3, Table 27.1 gives 0.22 as the MPN of organisms in the quantity of inoculum applied in the 10^{-7} dilution. Multiplying 0.22 by 10^7 yields 2.2 million organisms as the MPN value for the original sample, as before.

The 95% confidence limits for MPN values can be calculated from prepared tables. A compilation of factors keyed to rate of dilution and to number of tubes per dilution is shown in Table 27.2. To find the upper confidence limit at the 95% level, multiply the MPN value by the appropriate factor from the table. To find the lower limit, divide the MPN value by the factor. In the first example above, the factor is 3.30, and the confidence limits are

$$(2.2)(3.30) = 7.26$$

and

$$(2.2)/3.30 = 0.66$$

27.4.4 Comments

The MPN method is usually employed to enumerate and isolate organisms that will not readily grow on solid agar medium, or those that cannot be readily identified from the background community. For example, autotrophic nitrifying bacteria (Schmidt and Belser 1982) and sulfur-oxidizing bacteria (Postgate 1966, Germida 1985, Lawrence and Germida 1988) are routinely enumerated using the MPN assay. The medium employed may be used to measure total growth, and hence optical density is satisfactory measurement. Alternatively, a physiological reaction may be monitored. For example, oxidation of sulfur to an acidic end product will alter pH and the difference may be recorded by using an appropriate pH indicator. The procedure may be used to provide a relative estimate of the numbers of many diverse physiological groups of organisms in soils. Choice of media and incubation conditions are limited only by our knowledge of specific physiological groups.

TABLE 27.1 Table of Most Probable Numbers for Use with Tenfold Dilutions and Five Tubes per Dilution (Cochran 1950)

		Most probable number for indicated values of p_3					
p_1	p_2	0	1	2	3	4	5
0	0	—	0.018	0.036	0.054	0.072	0.090
	1	0.018	0.036	0.055	0.073	0.091	0.11
	2	0.037	0.055	0.074	0.092	0.11	0.13
	3	0.056	0.074	0.093	0.11	0.13	0.15
	4	0.075	0.094	0.11	0.13	0.15	0.17
	5	0.094	0.11	0.13	0.15	0.17	0.19
1	0	0.020	0.040	0.060	0.080	0.10	0.12
	1	0.040	0.061	0.081	0.10	0.12	0.14
	2	0.061	0.082	0.10	0.12	0.15	0.17
	3	0.083	0.10	0.13	0.15	0.17	0.19
	4	0.11	0.13	0.15	0.17	0.19	0.22
	5	0.13	0.15	0.17	0.19	0.22	0.24
2	0	0.045	0.068	0.091	0.12	0.14	0.16
	1	0.068	0.092	0.12	0.14	0.17	0.19
	2	0.093	0.12	0.14	0.17	0.19	0.22
	3	0.12	0.14	0.17	0.20	0.22	0.25
	4	0.15	0.17	0.20	0.23	0.25	0.28
	5	0.17	0.20	0.23	0.26	0.29	0.32
3	0	0.078	0.11	0.13	0.16	0.20	0.23
	1	0.11	0.14	0.17	0.20	0.23	0.27
	2	0.14	0.17	0.20	0.24	0.27	0.31
	3	0.17	0.21	0.24	0.28	0.31	0.35
	4	0.21	0.24	0.28	0.32	0.36	0.40
	5	0.25	0.29	0.32	0.37	0.41	0.45
4	0	0.13	0.17	0.21	0.25	0.30	0.36
	1	0.17	0.21	0.26	0.31	0.36	0.42
	2	0.22	0.26	0.32	0.38	0.44	0.50
	3	0.27	0.33	0.39	0.45	0.52	0.59
	4	0.34	0.40	0.47	0.54	0.62	0.69
	5	0.41	0.48	0.56	0.64	0.72	0.81
5	0	0.23	0.31	0.43	0.58	0.76	0.95
	1	0.33	0.46	0.64	0.84	1.1	1.3
	2	0.49	0.70	0.95	1.2	1.5	1.8
	3	0.79	1.1	1.4	1.8	2.1	2.5
	4	1.3	1.7	2.2	2.8	3.5	4.3
	5	2.4	3.5	5.4	9.2	16	—

27.5 MEDIA FOR THE ENUMERATION AND ISOLATION OF SOIL MICROORGANISMS (GRAY AND PARKINSON 1968)

27.5.1 MEDIA FOR ISOLATION OF HETEROTROPHIC BACTERIA

1   Peptone yeast extract agar (Goodfellow et al. 1968): peptone, 5.0 g; yeast extract, 1.0 g; $FePO_4$, 0.01 g; agar, 15 g; distilled water, 1L; pH 7.2.

TABLE 27.2 Factors for Calculating the
Confidence Limits for the Most
Probable Number Count
(Cochran 1950)

Number of tubes per dilution (n)	Factor for 95% confidence limits with indicated dilution ratios			
	2	4	5	10
1	4.00	7.14	8.32	14.45
2	2.67	4.00	4.47	6.61
3	2.23	3.10	3.39	4.68
4	2.00	2.68	2.88	3.80
5	1.86	2.41	2.58	3.30
6	1.76	2.23	2.38	2.98
7	1.69	2.10	2.23	2.74
8	1.64	2.00	2.12	2.57
9	1.58	1.92	2.02	2.43
10	1.55	1.86	1.95	2.32

2   Nutrient agar: yeast extract, 1.0 g; beef extract, 3.0 g; peptone, 5.0 g; sodium chloride, 5.0 g; agar, 15 g; distilled water, 1 L; pH, 7.3.

3   Fluorescent pseudomonads (Sands and Rovira 1970, Simon et al. 1973): proteose peptone, 20.0 g; agar, 12.0 g; glycerol, 18 mL; K_2SO_4, 1.5 g; $MgSO_4 \cdot 7H_2O$, 1.5 g; distilled water, 940 mL. Adjust pH to 7.2 with 0.1 M NaOH before autoclaving. Sterilize by autoclaving. Add 150,000 units of penicillin G, 45 mg of novobiocin, 75 mg of cycloheximide, and 5 mg of chloramphenicol to 3 mL of 95% ethanol. Dilute to 60 mL with sterile distilled water, and add (filter-sterilized) to the cooled (48°C) medium prior to pouring. Prepared plates should be dried overnight before using and may be stored in the refrigerator for several weeks before use.

27.5.2 MEDIA FOR ISOLATION OF SPECIFIC PHYSIOLOGICAL GROUPS OF ORGANISMS

27.5.2.1 MICROORGANISMS INVOLVED IN CARBON TRANSFORMATIONS

1   Cellulose agar (Eggins and Pugh 1961): $NaNO_3$, 0.5 g; K_2HPO_4, 1.0 g; $MgSO_4 \cdot 7H_2O$, 0.5 g; $FeSO_4 \cdot 7H_2O$, 0.01 g; cellulose (ball-milled), 12.0 g; agar, 15 g; distilled water, 1 L.

2   Chitin agar: ball-milled, purified chitin, 10.0 g; $MgSO_4 \cdot 7H_2O$, 1.0 g; K_2HPO_4, 1.0 g; agar, 15 g; distilled water, 1 L.

3   Starch agar: 0.2% soluble starch may be added to any suitable growth medium as an alternative or additional carbohydrate. Starch hydrolysis is shown by flooding incubated plates with an iodine solution and noting clear zones.

27.5.2.2 MICROORGANISMS INVOLVED IN NITROGEN TRANSFORMATIONS

1   Combined carbon medium (Rennie 1981): solution I: K_2HPO_4, 0.8 g; KH_2PO_4, 0.2 g; NaCl, 0.1 g; Na_2Fe ethylene diamine tetraacetic acid, 28.0 mg; $Na_2MoO_4 \cdot 2H_2O$, 25.0 mg; yeast extract, 0.1 g; mannitol, 5 g; sucrose, 5 g; Na-lactate (60% v/v), 0.5 mL; distilled water, 900 mL; agar, 15 g. Solution II: $MgSO_4 \cdot 7H_2O$, 0.2 g; $CaCl_2$, 0.06 g; distilled water, 100 mL. Solution III: biotin, 5 µg/mL; PABA, 10.0 µg/mL. Autoclave solutions I and II, cool to 48°C, and mix thoroughly; then add (filter-sterilized) 1 mL/L of solution III.

2   Yeast extract mannitol medium (Allen 1957): mannitol, 10.0 g; K_2HPO_4, 0.5 g; $MgSO_4 \cdot 7H_2O$, 0.2 g; NaCl, 0.1 g; $CaCO_3$, 3.0 g; yeast extract (10%), 100 mL; agar, 15 g; distilled water, 1 L.

3   Nitrifying bacteria (Lewis and Pramer 1958): Na_2HPO_4, 13.5 g; KH_2PO_4, 0.7 g; $MgSO_4 \cdot 7H_2O$, 0.1 g; $NaHCO_3$, 0.5 g; $(NH_4)_2SO_4$, 2.5 g; $FeCl_3 \cdot 6H_2O$, 14.4 mg; $CaCl_2 \cdot 7H_2O$, 18.4 mg; distilled water, 1 L; pH, 8.0.

27.5.2.3 MICROORGANISMS INVOLVED IN SULFUR TRANSFORMATIONS

1   *Thiobacillus thiooxidans* or *T. thioparus* (Postgate 1966): $(NH_4)_2SO_4$, 0.4 g; KH_2PO_4, 4.0 g; $MgSO_4 \cdot 7H_2O$, 0.5 g; $CaCl_2$, 0.25 g; $FeSO_4$, 0.01 g; powdered sulfur, 10.0 g or $Na_2S_2O_7$, 5.0 g; distilled water, 1 L; pH, 7.0. This medium can be made selective for *T. thiooxidans*-like bacteria by using S^0 as the sulfur source and adjusting the initial pH to <3.5.

2   *T. denitrificans* (Postgate 1966): KNO_3, 1.0 g; Na_2HPO_4, 0.1 g; $Na_2S_2O_7$, 2.0 g; $NaHCO_3$, 0.1 g; $MgCl_2$, 0.1 g; distilled water, 1 L; pH, 7.0. This medium may be used for agar plates or dispensed into test tubes containing small Durham fermentation tubes to capture gas. Incubate under anaerobic conditions.

REFERENCES

Allen, O. N. 1957. Experiments in soil bacteriology. 3rd ed. Burgess, Minneapolis, MN.

Alexander, M. 1982. Most probable number method for microbial populations. Pages 815–820 in A. L. Page et al., Eds. Methods of soil analysis. Part 2. Agronomy No. 9, 2nd ed. American Society of Agronomy, Madison, WI.

American Type Culture Collection. 1989. Catalogue of bacteria and bacteriophages, 17th ed. ATCC, Rockville, MD.

Angle, J. S., McGrath, S. P., and Chaney, R. L. 1991. New culture medium containing ionic concentrations of nutrients similar to concentrations found in soil solutions. Appl. Environ. Microbiol. 57: 3674–3676.

Brown, M. E., Burlingham, S. K., and Jackson, R. M. 1962. Studies on Azotobacteria species in soil. I. Comparison of media and techniques for counting *Azotobacter* in soil. Plant Soil 17: 309–319.

Buyer, J. S., Sikora, L. J., and Chaney, R. L. 1989. A new growth medium for the study of siderophore-mediated interactions. Biol. Fertil. Soils 8: 97–101.

Casida, L. E., Jr. 1965. An abundant microorganism in soil. Appl. Microbiol. 13: 327–334.

Casida, L. E., Jr. 1968. Methods for the isolation and estimation of activity of soil bacteria. Pages 97–122 in T. R. G. Gray and D. Parkinson, Eds. The ecology of soil bacteria. University of Toronto Press, Toronto, ON, Canada.

Clark, F. E. 1965a. Agar-plate method for total microbial count. Pages 1460–1466 in C. A. Black et al., Eds. Methods of soil analysis. Part 2. Agronomy No. 9. American Society of Agronomy, Madison, WI.

Clark, F. E. 1965b. Azotobacter. Pages 1493–1497 in C. A. Black et al., Eds. Methods of soil analysis. Part 2. Agronomy No. 9. American Society of Agronomy, Madison, WI.

Cochran, W. G. 1950. Estimation of bacterial densities by means of the "most probable number". Biometrics 5: 105–116.

Eggins, H. O. W. and Pugh, G. J. F. 1961. Isolation of cellulose-decomposing fungi from soil. Nature (London) 193: 94–95.

Germida, J. J. 1985. Modified sulfur containing media for studying sulfur oxidizing microorganisms. Pages 333–344 in D. E. Caldwell, J. A. Brierley, and C. L. Brierly, Eds. Planetary ecology. Van Nostrand Reinhold, New York.

Goodfellow, M., Hill, I. R., and Gray, T. R. G. 1968. Bacteria in a pine forest soil. Pages 500–515 in T. R. G. Gray and D. Parkinson, Eds. The ecology of soil bacteria. University of Toronto Press, Toronto, ON, Canada.

Gray, T. R. G. and Parkinson, D., Eds. 1968. The ecology of soil bacteria, an international symposium. University of Toronto Press, Toronto, ON, Canada.

Hsu, S. C. and Lockwood, J. L. 1975. Powdered chitin as a selective medium for enumeration of actinomycetes in water and soil. Appl. Microbiol. 29: 422–426.

James, N. 1958. Soil extract in soil microbiology. Can. J. Microbiol. 4: 363–370.

James, N. 1959. Plate counts of bacteria and fungi in a saline soil. Can. J. Microbiol. 5: 431–439.

James, N. and Sutherland, M. L. 1939. The accuracy of the plating method for estimating the numbers of soil bacteria, actinomycetes and fungi in the dilution plated. Can. J. Res. 17: 72–86.

Jensen, V. 1968. The plate count technique. Pages 158–170 in T. R. G. Gray and D. Parkinson, Eds. The ecology of soil bacteria. University of Toronto Press, Toronto, ON, Canada.

Katznelson, H. and Bose, B. 1959. Metabolic activity and phosphate-dissolving capability of bacterial isolates from wheat roots, rhizosphere and non-rhizosphere soil. Can. J. Microbiol. 5: 79–85.

Kucey, R. M. N. 1983. Phosphate-solubilizing bacteria and fungi in various cultivated and virgin Alberta soils. Can. J. Soil. Sci. 63: 671–678.

Küster, E. and Williams, S. T. 1966. Selection of media for isolation of streptomycetes. Nature (London) 202: 928–929.

Lawrence, J. R. and Germida, J. J. 1988. Most-probable number procedure to enumerate S$^\circ$-oxidizing, thiosulfate producing heterotrophics in soil. Soil Biol. Biochem. 20: 577–578.

Lewis, R. F. and Pramer, D. 1958. Isolation of Nitrosomonas in pure culture. J. Bacteriol. 76: 524–528.

Lingappa, Y. and Lockwood, J. L. 1962. Chitin media for selective isolation and culture of actinomycetes. Phytopathology 52: 317–323.

Lochhead, A. G. and Chase, F. E. 1943. Qualitative studies of soil microorganisms. V. Nutritional requirements of the predominant bacterial flora. Soil Sci. 55: 185–195.

Martin, J. K. 1975. Comparison of agar media for counts of viable bacteria. Soil Biol. Biochem. 7: 401–402.

Martin, J. P. 1950. Use of acid, rose bengal and streptomycin in the plate method for estimating soil fungi. Soil Sci. 69: 215–232.

Parkinson, D., Gray, T. R. G., and Williams, S. T. 1971. Methods for studying the ecology of soil microorganisms. IBP Handbook No. 19. Blackwell Scientific, Oxford.

Postgate, J. R. 1966. Media for sulphur bacteria. Lab. Pract. 15: 1239–1244.

Rennie, R. J. 1981. A single medium for the isolation of acetylene-reducing (dinitrogen-fixing) bacteria from soils. Can. J. Microbiol. 27: 8–14.

Sands, D. C. and Rovira, A. D. 1970. Isolation of fluorescent pseudomonads with a selective medium. Appl. Microbiol. 20: 513–514.

Schmidt, E. L. and Belser, L. W. 1982. Nitrifying bacteria. Pages 1027–1042 in A. L. Page et al., Eds. Methods of soil analysis. Part 2. Agronomy No. 9, 2nd ed. American Society of Agronomy, Madison, WI.

Simon, A., Rovira, A. D., and Sands, D. C. 1973. An improved selective medium for the isolation of fluorescent pseudomonads. J. Appl. Bacteriol. 36: 141–145.

Skinner, F. A., Jones, P. C. T., and Mollison, J. E. 1952. A comparison of a direct- and a plating-counting technique for the quantitative estimation of soil microorganisms. J. Gen. Microbiol. 6: 261–271.

Smith, N. R. and Worden, S. 1925. Plate counts of soil micro-organisms. J. Agric. Res. 31: 501–517.

Stevenson, I. L. 1959. The effect of sonic vibration on the bacterial plate count of soil. Plant Soil 10: 1–8.

Taylor, J. 1962. The estimation of numbers of bacteria by ten-fold dilution series. J. Appl. Bacteriol. 25: 54–61.

Thornton, H. G. 1922. On the development of a standardized agar medium for counting soil bacteria with special regard to the repression of spreading colonies. Ann. Appl. Biol. 9: 241–274.

Vincent, J. M. 1970. A manual for the practical study of the root-nodule bacteria. IBP Handbook No. 15. Blackwell Scientific, Oxford.

Waksman, S. A. 1922. Microbiological analysis of soil as an index of soil fertility. II. Methods of the study of numbers of micro-organisms in the soil. Soil Sci. 14: 283–298.

Williams, S. T. and Davies, F. L. 1965. Use of antibiotics for selective isolation and enumeration of actinomycetes in soil. J. Gen. Microbiol. 38: 251–261.

Wollum, A. G., II. 1982. Culture methods for soil microorganisms. Pages 781–802 in A. L. Page et al., Eds. Methods of soil analysis. Part 2. Agronomy No. 9, 2nd ed. American Society of Agronomy, Madison, WI.

Chapter 28
Soil Microbial Biomass C and N

R. P. Voroney and J. P. Winter

University of Guelph

Guelph, Ontario, Canada

R. P. Beyaert

Agriculture Canada

Delhi, Ontario, Canada

28.1 INTRODUCTION

Measurements of soil microbial biomass have been used in studies of the flow of carbon, cycling of nutrients, and plant productivity in a variety of terrestrial ecosystems. They provide a measure of the quantity of living microbial biomass present in the soil at a particular point in time, i.e., the "standing crop". The data can be used for assessing changes in soil organic matter caused by soil management (Powlson et al. 1987, McGill et al. 1986) and tillage practices (Carter 1986), for assessing the impact of management on soil strength and porosity (Carter and White 1986), soil structure and aggregate stability (Perfect et al. 1990, Gupta and Germida 1988), for assessing soil nitrogen fertility status (Bonde et al. 1988, Doran 1987), for estimating seasonal fluctuations in microbial biomass (Ross 1990), and for serving as an indicator of the presence of toxins, e.g., metal toxicity (Chander and Brookes 1991, Brookes et al. 1986). In addition, measurements of the carbon and nutrients contained in the microbial biomass provide a basis for studies of the formation and turnover of soil organic matter, as the microbial biomass is one of the key definable fractions (Ocio et al. 1991, Brookes et al. 1990, Chaussod and Nicolardot 1982). The significance of microbial biomass measurements has recently been reviewed by Smith and Paul (1990) and Jenkinson (1988). Crossley et al. (1991) have reviewed methods for measurement of the microbial biomass as well as other components of the soil biota.

For general measurements of the soil microbial biomass, indices of microbial biomass such as assay of soil adenosine 5'-triphosphate (ATP) (Webster et al. 1984, Tate and Jenkinson 1982), anthrone-reactive carbon (Badalucco et al. 1990), ninhydrin-reactive N (Carter 1991, Amato and Ladd 1988), or of the substrate-induced respiratory response (Anderson and Domsch 1978) can be useful. However, for studies of the role of the soil microbial biomass in organic matter dynamics and nutrient cycling, a direct measurement of the C and nutrients contained in the microbial biomass is essential.

Soil Sampling and Methods of Analysis, M. R. Carter, Ed.,

Canadian Society of Soil Science. © 1993 Lewis Publishers.

The latter approach was used first by Jenkinson (1966) to study the decomposition of ryegrass residues in the field, and microbial biomass was measured using the classical $CHCl_3$ fumigation-incubation method. More recently, others have used this technique, also coupled with tracer techniques, to follow the formation of microbial biomass accompanying decomposition (Voroney and Paul 1984, Ladd et al. 1981, Amato and Ladd 1980). The fumigation-incubation method has proven to be useful on a wide variety of soils. Exceptions are for soils with a pH <4.2 (Vance et al. 1987) or for those containing significant levels of labile substrates (Ocio and Brookes 1990a, Jenkinson and Powlson 1976).

Fumigation of soil samples with $CHCl_3$ vapor causes a flush of decomposition during a subsequent 10-d incubation compared with an unfumigated soil (Jenkinson 1966). The flush of CO_2 evolution and ammonium accumulation, due to mineralization of microbial cells killed by the fumigation treatment, is directly proportional to the sizes of the microbial biomass C and N pools.

Alternatively, the microbial biomass constituents released by $CHCl_3$ fumigation treatment can be extracted directly. Microbial biomass C (Tate et al. 1988, Sparling and West 1988a, Merckx et al. 1988, Vance et al. 1987), N (Azam et al. 1989, Sparling and West 1988b, Brookes et al. 1985), S (Saggar et al. 1981), and P (Brookes et al. 1982, Hedley and Stewart 1982) have been measured by fumigation-extraction methods. Since time is usually a critical constraint for an analytical technique, methods based on direct extraction have become popular in routine soil analysis (Wu et al. 1990, Sparling and West 1988b). The fumigation-extraction method will be described in detail because it is relatively easy and much more rapid than the other methods.

28.2 CHLOROFORM FUMIGATION-EXTRACTION METHOD

Estimation of soil microbial C and N by fumigation-extraction methods has several advantages over the fumigation-incubation method. Besides being rapid, the fumigation-extraction method is applicable to soils of low pH (Couteaux and Henkinet 1990), and avoids the requirement for microbial mineralization of the C and N during an incubation (Gallardo and Schlesinger 1990, Amato and Ladd 1988, Ladd and Amato 1988, Tate et al. 1988, Merckx and Martin 1987, Vance et al. 1987, Brookes et al. 1985). A further advantage is that, provided the soils are wetted to between -5 and -10 kPa moisture potential during the fumigation stage, the fumigation-extraction method can be applied to soils of low initial water content (Sparling et al. 1990, Ross 1989, Sparling and West 1989) and to waterlogged soils (Inubushi et al. 1991). Diaz-Ravina et al. (1992) have used the method in soils heated to 600°C.

28.2.1 MATERIALS AND REAGENTS

1   $CHCl_3$: reagent-grade $CHCl_3$ contains a stabilizing agent, ethanol, which must be removed just prior to its use. The purification procedure consists of double distillation of $CHCl_3$ at 55°C, followed by washing three times, first with 18 M H_2SO_4 and then with deionized/distilled water (Jenkinson and Powlson 1976). Purified $CHCl_3$ is kept in a brown bottle containing anhydrous K_2CO_3 and used within 6 h of preparation.

Commercially available $CHCl_3$, purified, dried, distilled in glass, and stabilized with heptachlor epoxide @10 pg mL^{-1} (Caledon Laboratories, Georgetown, ON) can be used as an alternative to freshly purified $CHCl_3$. The use of $CHCl_3$ sources such as this simplifies the procedure and greatly reduces handling of $CHCl_3$ and exposure to hazardous fumes.

2   Large desiccator: the desiccator should be inert to $CHCl_3$ vapor, be of a dry-seal type, and be able to withstand a high vacuum without implosion; a thick-walled glass vacuum desiccator is suitable. A large desiccator may hold about 25 samples simultaneously.

3   Miscellaneous materials and reagents:

   a. Fumehood.

   b. Boiling chips.

   c. Two 100-mL glass containers with lids per assay (mark the sample jars with a marker insoluble to $CHCl_3$, i.e., pencil).

   d. One weighing container per assay for water content determination.

   e. 0.5 M K_2SO_4 extraction solution.

   f. Two Whatman® No. 5 filter papers per assay for filtration.

   g. Two 50-mL vials per assay for filtrate.

28.2.2 DESCRIPTION OF PROCEDURES

28.2.2.1 *Preparation of the Soil*

1   Pass the soil sample through a 6-mm mesh sieve and mix thoroughly (Ocio and Brookes 1990b).

2   Weigh out three portions of the soil, 15–50 g each, one portion into a weighing container for water content determination, and two portions into 100-mL glass bottles, one sample to be fumigated for 24 h and then extracted, and one control sample to be extracted *immediately*.

3   Dry soil in the weighing container (with the lid off) in an oven at 105°C for at least 24 h or until a constant oven-dry weight is achieved. Cool in a desiccator, reweigh, and determine the water content of the soil sample.

28.2.2.2 *Fumigation Treatment*

Procedures releasing $CHCl_3$ fumes should be conducted in a fumehood.

1   The desiccator should be lined with freshly moistened paper towels just prior to fumigation treatment. For the fumigation treatment, place the glass

sample bottles containing the soil into a desiccator together with a 100-mL beaker containing 50 mL $CHCl_3$ and a few boiling chips. Seal and evacuate the desiccator, taking care to vent the fumes released by the vacuum pump, until the $CHCl_3$ boils vigorously, and continue evacuating for ~1 min. Seal the desiccator under vacuum and place it in the dark at 25°C for 24 h.

2   After the fumigation treatment, release the vacuum, open the desiccator, and remove the beaker of $CHCl_3$ and moistened paper towels. (The waste $CHCl_3$ should be kept in a sealed bottle and disposed as a hazardous waste; the paper towels should be placed in sealed bags and disposed in the regular waste stream).

3   Remove residual $CHCl_3$ vapor from the soil samples by repeated evacuation, usually three to six times, using first a water aspirator pump, followed by a two-stage rotary oil pump capable of drawing a vacuum of 10^{-5} kPa, 15–20 min evacuation.

28.2.2.3 *Extraction of Microbial Biomass C and N*

1   Add 0.5 M K_2SO_4 to the bottles containing the unfumigated control and fumigated subsamples, using the equivalent oven-dry soil weight (g):extractant volume (mL) ratio of 1:2 to 1:5.

Soils which do not readily disperse in the extracting solution should be homogenized before shaking (Winter et al. 1993). Soil samples can be homogenized using a Brinkmann® PT 10-35 tissue homogenizer equipped with a sawtooth PT 20ST probe generator, machine power level set to 4, and homogenized for about 5–10 s.

2   Cap the jars and place on a shaker (oscillating or rotary) for 1 h. After shaking, pass the soil suspension through Whatman® No. 5 filter paper; avoid excessive evaporation during the filtration procedure. Measurements of organic C and total N can require from 5 to 30 mL of extract, depending on the methods of analysis used. Cap and store the filtrate at 4°C for not more than 2–3 d, otherwise, freeze until ready for analysis.

28.2.2.4 *Determination of Organic C, Organic N + Exchangeable NH_4^+ in the Extract*

1   Organic C, organic N + exchangeable NH_4^+ dissolved in the K_2SO_4 extracts can be determined colorimetrically using automated equipment or by standard wet chemistry techniques, e.g., organic C by dichromate digestion (Jenkinson and Powlson 1976) and organic N by semimicro-Kjeldahl digestion (Bremner and Mulvaney 1982).

2   For digestion by dichromate, an 8-mL aliquot of the K_2SO_4 extract is added to a mixture of 0.2 M $K_2Cr_2O_7$ (2 mL), 18 M H_2SO_4 (10 mL), 14.7 M H_3PO_4 (5 mL), and HgO (70 mg) and boiled under refluxing conditions for 30 min.

The excess dichromate remaining is determined by titration with 0.017 M ferrous ammonium sulfate using ferroin as an indicator.

3   Organic C can be measured with an AutoAnalyzer® II system fitted with manifold No. 116-D660-01 and utilizing Method No. 455-76W/A, Technicon® Industrial Systems. The determination involves a pretreatment to remove inorganic carbon by entraining the acidified sample with a high-velocity stream of nitrogen or carbon-free air to purge carbonate-derived CO_2. The sample is then mixed with 0.5 M H_2SO_4 and 4% (w/v) potassium persulfate and subjected to UV radiation. The resultant CO_2 generated from the organic C present in the sample is then dialyzed through a silicone rubber membrane and reacted with a weakly buffered phenolphthalein indicator. The decrease in the intensity of the color of the indicator measured at 550 nm is proportional to the organic C content. Standard solutions ranging in concentration from 0 to 150 μg C mL^{-1} are prepared by dilution of a stock potassium biphthalate solution containing 1000 mg C mL^{-1}.

4   Organic N in soil extracts (5–25 mL) can be converted to NH_4^+ by Kjeldahl digestion for 4 h at 360°C. Each sample is digested in a mixture containing 10 mL conc H_2SO_4, 3 g powdered K_2SO_4:$CuSO_4 \cdot 5H_2O$ (30:1 by weight), and a selenium catalyst in the form of one selenized Hengar® granule. After dilution, the ammonium present in the digest can either be measured directly using autoanalyzer techniques, e.g., (Method No. 325-7W, Technicon® Industrial Systems), or liberated from digests by the addition of 40% (w/v) NaOH and steam distilled into 4% (w/v) boric acid. Distillates are titrated with 0.01–0.05 M H_2SO_4 (standard) to pH 4.8. This method also includes the free and exchangeable NH_4^+-N. Measurements of microbial biomass N by the fumigation direct-extraction method do not require analysis of NO_3^- and NO_2^--N.

5   Total N (organic + inorganic) in the extract can be determined colorimetrically with an AutoAnalyzer® II system fitted with manifold No. 116-D887-01 utilizing Method No. 759-841, Technicon® Industrial Systems. This procedure oxidizes nitrogen-containing compounds to nitrate by digestion in acidic and basic conditions with UV radiation. The nitrate generated is then reduced to nitrite by reaction with a copper-hydrazine solution. Nitrite is reacted with sulfanilamide to form a diazo compound which is then coupled with N-(1-naphthyl)-ethylene diamine to form a reddish-purple color. The intensity of the color measured at 520 nm is proportional to the total nitrogen content in the extract. Standards solutions ranging in concentration from 0 to 30 μg N mL^{-1} are prepared by dilution of a stock $(NH_4)_2SO_4$ solution containing 200 mg N mL^{-1}.

28.2.3 CALCULATION OF MICROBIAL BIOMASS C AND N

1   Soil water content (WS):

$$WS\ (\%) = \frac{\text{soil wet weight (g)} - \text{soil oven-dry weight (g)}}{\text{soil oven-dry weight (g)}} \times 100 \qquad (28.1)$$

2   Weight of soil sample (oven-dry weight equivalent) taken for microbial bio-mass measurements (MS):

$$MS \text{ (g)} = \frac{\text{soil wet weight (g)} \times 100}{(100 + WS \text{ (\%)})} \tag{28.2}$$

3   Total volume of solution in the extracted soil (VS):

$$VS \text{ (mL)} = \text{soil wet weight (g)} - \text{soil oven-dry weight (g)} + \text{extractant volume (mL)} \tag{28.3}$$

4   Total weight of extractable C and N in the fumigated (O_F) and unfumigated (O_{UF}) soil samples:

$$OC_F, OC_{UF} \text{ (\mu g/g soil)} = \text{extractable C (\mu g/mL)} \times \frac{VS \text{ (mL)}}{MS \text{ (g)}} \tag{28.4}$$

$$ON_F, ON_{UF} \text{ (\mu g/g soil)} = \text{extractable N (\mu g/mL)} \times \frac{VS \text{ (mL)}}{MS \text{ (g)}} \tag{28.5}$$

5   Microbial biomass C and N in the soil (MB-C, MB-N):

a. $$MB\text{-}C \text{ (\mu g/g soil)} = (OC_F - OC_{UF})/k_{EC} \tag{28.6}$$

where: $k_{EC} = 0.25 \pm 0.05$ and represents the efficiency of extraction of microbial biomass C.

b. $$MB\text{-}N \text{ (\mu g/g soil)} = (ON_F - ON_{UF})/k_{EN} \tag{28.7}$$

where: $k_{EN} = 0.18 \pm 0.04$ and represents the efficiency of extraction of microbial biomass N.

28.2.4 COMMENTS

1   Measurements of microbial biomass should be done on fresh soil samples as quickly as possible (within hours) after sampling, without subjecting the samples to changes in temperature or moisture content. If it is not possible to complete the sample analysis immediately, soil samples should be quickly frozen and kept frozen ($-18°C$), and then thawed just prior to analysis. However, freezing soil samples may cause changes to the microbial biomass and to the extractability of nonbiomass organic matter (Winter et al. 1993). Ross (1991) has stored soil samples at 4°C with minimal changes in microbial biomass.

2   The fumigation-extraction method is relatively precise; four replicate determinations on the same soil sample should be able to detect differences in microbial biomass of 5–10% at a 0.05% level of probability. Across different soil types, the method may be less accurate due to variations in k_{EC} and k_{EN}.

3   Here, we have proposed our estimates of k_{EC} and k_{EN} which are based on *in situ* calibration techniques (Bremer and van Kessel 1990). While we are confident in these estimates of microbial biomass extraction efficiency, these values are substantially lower than those found in the literature, which have largely relied on the use of laboratory-grown microbial cultures (mostly bacteria) and have assumed a microbial C:N ratio of 8, or have been calibrated against alternative methods to estimate microbial biomass. For example, values of k_{EC} reported in the literature include: 0.38, for ten soils and calibrated using the fumigation-incubation method and ATP content (Vance et al. 1987); 0.33, for mineral soils from New Zealand and using substrate-induced respiration, *in situ* labeling techniques, and fumigation-incubation methods for calibration (Sparling and West 1988a, Tate et al. 1988); and recently revised to 0.35 (Sparling et al. 1990). Until more research is done to determine extraction efficiencies for microbial C and N, we recommend that the k_{EC} and k_{EN} values used for estimates of microbial biomass C and N be included.

4   Microbial biomass assays can be made even more rapid by substitution of the 24-h $CHCl_3$ fumigation procedure with direct addition of liquid $CHCl_3$. To suspensions of soil in the K_2SO_4 extracting solution, add 1 mL of $CHCl_3$ prior to shaking. After filtration, $CHCl_3$ is expelled by bubbling CO_2-free air through the filtrate for 30–45 s. While this method substantially reduces assay time and greatly simplifies the method, it is as accurate as the fumigation direct-extraction method, but less microbial biomass is recovered in the extract ($k_{EC} = 0.18$) (Gregorich et al. 1990). This may be an important factor if microbial biomass levels are low.

REFERENCES

Amato, M. and Ladd, J. N. 1980. Studies of nitrogen immobilization and mineralization in calcareous soils. V. Formation and distribution of isotope-labelled biomass during decomposition of ^{14}C- and ^{15}N-labelled plant material. Soil Biol. Biochem. 12: 405–411.

Amato, M. and Ladd, J. N. 1988. Assay for microbial biomass based on ninhydrin reactive nitrogen in extracts of fumigated soils. Soil Biol. Biochem. 20: 107–114.

Anderson, J. P. E. and Domsch, K. H. 1978. A physiological method for the quantitative measurement of microbial biomass in soils. Soil Biol. Biochem. 10: 215–221.

Azam, F., Mulvaney, R. L., and Stevenson, F. 1989. Synthesis of ^{15}N-labelled microbial biomass in soil *in situ* and extraction of biomass N. Biol. Fertil. Soils 7: 180–185.

Badalucco, L., Nannipieri, P., and Grego, S. 1990. Microbial biomass and anthrone-reactive carbon in soils with different organic matter contents. Soil Biol. Biochem. 22: 899–904.

Bonde, T. A., Schnurer, J., and Rosswall, T. 1988. Microbial biomass as a fraction of potentially mineralizable nitrogen in soils from long-term field experiments. Soil Biol. Biochem. 20: 447–452.

Bremer, E. and van Kessel, C. 1990. Extractability of microbial ^{14}C and ^{15}N following addition of variable rates of labelled glucose and $(NH_4)_2SO_4$ to soil. Soil Biol. Biochem. 22: 707–713.

Bremner, J. M. and Mulvaney, C. S. 1982. Nitrogen-total. Pages 595–624 in A. L. Page, Ed. Methods of soil analysis. Part 2. Chemical and microbiological properties. American Society of Agronomy, Madison, WI.

Brookes, P. C., Powlson, D. S., and Jenkinson, D. S. 1982. Measurement of microbial biomass phosphorus in soil. Soil Biol. Biochem. 14: 319–329.

Brookes, P. C., Heihen, C. E., McGrath, and Vance, E. D. 1986. Soil microbial biomass estimates in soils contaminated with metals. Soil Biol. Biochem. 18: 383–388.

Brookes, P. C., Landman, A., Pruden, G., and Jenkinson, D. S. 1985. Chloroform fumigation and the release of soil nitrogen, a rapid direct extraction method to measure microbial biomass nitrogen in soil. Soil Biol. Biochem. 17: 837–842.

Brookes, P. C., Ocio, J. A., and Wu, J. 1990. The soil microbial biomass: its measurement, properties and role in soil nitrogen and carbon dynamics following substrate incorporation. Soil Microorganisms 35: 39–51.

Carter, M. R. 1986. Microbial biomass as an index for tillage-induced changes in soil biological properties. Soil Tillage Res. 7: 29–40.

Carter, M. R. 1991. Ninhydrin-reactive N released by the fumigation-extraction method as a measure of microbial biomass under field conditions. Soil Biol. Biochem. 23: 139–143.

Carter, M. R. and White, R. P. 1986. Determination of variability in soil physical properties and microbial biomass under continuous direct-planted corn. Can. J. Soil Sci. 66: 747–750.

Chander, K. and Brookes, P. C. 1991. Plant inputs of carbon to metal-contaminated soil and effects on the microbial biomass. Soil Biol. Biochem. 23: 1169–1177.

Chaussod, R. and Nicolardot, B. 1982. Mesure de la biomasse microbienne dans les sols cultives. I. Approche cinetique et estimation simplifiee du carbone facilement mineralisable. Rev. Ecol. Biol. Sol 19: 501–512.

Couteaux, M. M. and Henkinet, R. 1990. Anomalies in microbial biomass measurements in acid organic soils using extractable carbon following chloroform fumigation. Soil Biol. Biochem. 22: 955–957.

Crossley, D. A., Jr., Coleman, D. C., Hendrix, P. F., Cheng, W., Wright, D. H., Beare, M. H., and Edwards, C. A. 1991. Modern techniques in soil ecology. Elsevier Science, New York.

Diaz-Ravina, M., Prueto, Acea, M. J., and Carballas, T. 1992. Fumigation-extraction method to estimate microbial biomass in heated soils. Soil. Biol. Biochem. 24: 259–264.

Doran, J. W. 1987. Microbial biomass and mineralizable nitrogen distributions in no-tillage and plowed soils. Biol. Fertil. Soils 5: 68–75.

Gallardo, A. and Schlesinger, W. H. 1990. Estimating microbial biomass nitrogen using the fumigation-incubation and fumigation-extraction methods in a warm-temperate forest soil. Soil Biol. Biochem. 22: 927–932.

Gregorich, E. G., Wen, G., Voroney, R. P., and Kachanoski, R. G. 1990. Calibration of a rapid chloroform extraction method for measuring soil microbial biomass C. Soil Biol. Biochem. 22: 1009–1011.

Gupta, V. V. S. R. and Germida, J. J. 1988. Distribution of microbial biomass and its activity in different soil aggregate size fractions as affected by cultivation. Soil Biol. Biochem. 21: 777–787.

Hedley, M. J. and Stewart, J. W. B. 1982. Method to measure microbial phosphate in soils. Soil Biol. Biochem. 14: 377–385.

Inubushi, K., Brookes, P. C., and Jenkinson, D. S. 1991. Soil microbial biomass C, N and ninhydrin-N in aerobic and anaerobic soils measured by the fumigation-extraction method. Soil Biol. Biochem. 23: 737–741.

Jenkinson, D. S. 1966. Studies on the decomposition of plant material in soil. II. Partial sterilization of soil and the soil biomass. J. Soil Sci. 17: 280–302.

Jenkinson, D. S. 1988. Determination of microbial biomass carbon and nitrogen in soil. Pages 368–386 in J. R. Wilson, Ed. Advances in nitrogen cycling in agricultural ecosystems. CAB International, Wallingford, U.K.

Jenkinson, D. S. and Powlson, D. S. 1976. The effects of biocidal treatment on metabolism in soil. V. A method for measuring soil biomass. Soil Biol. Biochem. 8: 167–177.

Ladd, J. N. and Amato, M. 1988. Relationships between biomass ^{14}C and soluble organic ^{14}C of a range of fumigated soils. Soil Biol. Biochem. 20: 115–116.

Ladd, J. N., Oades, J. M., and Amato, M. 1981. Microbial biomass formed from ^{14}C-^{15}N-labelled plant material decomposing in soils in the field. Soil Biol. Biochem. 13: 119–126.

McGill, W. B., Cannon, K. R., Robertson, J. A., and Cook, F. D. 1986. Dynamics of soil microbial biomass and water-soluble organic C in Breton L after 50 years of cropping to two rotations. Can. J. Soil Sci. 66: 1–19.

Merckx, R. and Martin, J. K. 1987. Extraction of microbial biomass components from rhizosphere soils. Soil Biol. Biochem. 19: 371–376.

Merckx, R., Van der Linden, A. M. A., and Leuren, K. V. 1988. The extraction of microbial biomass components from soils. Pages 327–339 in D. S. Jenkinson and K. A. Smith, Eds. Nitrogen efficiency in agricultural soils. Elsevier Applied Science, London.

Ocio, J. A. and Brookes, P. C. 1990a. An evaluation of methods for measuring the microbial biomass in soils following recent additions of wheat straw and the characterization of the biomass that develops. Soil Biol. Biochem. 22: 685–694.

Ocio, J. A. and Brookes, P. C. 1990b. Soil microbial biomass measurements in sieved and unsieved soil. Soil Biol. Biochem. 22: 999–1000.

Ocio, J. A., Martinez, J., and Brookes, P. C. 1991. Contribution of straw-derived N to total microbial biomass N following incorporation of cereal straw to soil. Soil Biol. Biochem. 23: 655–659.

Perfect, E., Kay, B. D., van Loon, W. K. P., Sheard, R. W., and Pojasok, T. 1990. Factors influencing soil structural stability within a growing season. Soil Sci. Soc. Am. J. 54: 173–179.

Powlson, D. S., Brookes, P. C., and Christensen, B. T. 1987. Measurement of soil microbial biomass provides an early indication of changes in total soil organic matter due to straw incorporation. Soil Biol. Biochem. 19: 159–164.

Ross, D. J. 1989. Estimation of soil microbial C by a fumigation-extraction procedure: influence of soil moisture content. Soil Biol. Biochem. 21: 767–772.

Ross, D. J. 1990. Estimation of soil microbial C by a fumigation-extraction method: influence of seasons, soils and calibration with the fumigation-incubation procedure. Soil Biol. Biochem. 22: 295–300.

Ross, D. J. 1991. Microbial biomass in a stored soil: a comparison of different estimation procedures. Soil Biol. Biochem. 23: 1005–1007.

Saggar, S., Bettany, J. R., and Stewart, J. W. B. 1981. Measurement of microbial sulphur in soil. Soil Biol. Biochem. 13: 493–498.

Smith, J. L. and Paul, E. A. 1990. The significance of soil microbial biomass estimations. Pages 357–396 in J.-M. Bollag and G. Stotzky, Eds. Soil biochemistry, Volume 6. Marcel Dekker, New York.

Sparling, G. P. and West, A. W. 1988a. A direct extraction method to estimate soil microbial C: calibration *in situ* using microbial respiration and ^{14}C-labelled cells. Soil Biol. Biochem. 20: 337–343.

Sparling, G. P. and West, A. W. 1988b. Modifications to the fumigation-extraction technique to permit simultaneous extraction and estimation of soil microbial biomass C and N. Commun. Soil Sci. Plant Anal. 19: 327–344.

Sparling, G. P. and West, A. W. 1989. Importance of soil water content when estimating soil microbial C, N and P by the fumigation-extraction methods. Soil Biol. Biochem. 21: 245–253.

Sparling, G. P., Felthan, C. W., Reynolds, J., West, A. W., and Singleton, P. 1990. Estimation of soil microbial C by a fumigation-extraction method: use on soils of high organic matter content, and reassessment of the k_{EC}-factor. Soil Biol. Biochem. 22: 301–307.

Tate, K. R., Ross, D. J., and Feltham, C. W. 1988. A direct extraction method to estimate soil microbial C: effects of experimental variables and some different calibration procedures. Soil Biol. Biochem. 20: 329–335.

Tate, K. R. and Jenkinson, D. S. 1982. Adenosine triphosphate measurement in soil: an improved method. Soil Biol. Biochem. 14: 331–335.

Vance, E. D., Brookes, P. C., and Jenkinson, D. S. 1987. An extraction method for measuring soil microbial biomass C. Soil Biol. Biochem. 19: 703–707.

Voroney, R. P. and Paul, E. A. 1984. Determination of k_C and k_N *in situ* for calibration of the chloroform fumigation incubation method. Soil Biol. Biochem. 16: 9–14.

Webster, J. J., Hampton, G. J., and Leach, F. R. 1984. ATP in soil: a new extractant and extraction procedure. Soil Biol. Biochem. 16: 335–345.

Winter, J. P., Zhang, Z., Tenuta, M., and Voroney, R. P. 1993. Measurement of microbial biomass by fumigation-extraction in soil stored frozen. Soil Sci. Soc. Am. J. (in press).

Wu, J., Joergensen, R. G., Pommerening, B., Chaussod, R., and Brookes, P. C. 1990. Measurement of soil microbial biomass C by fumigation-extraction — an automated procedure. Soil Biol. Biochem. 22: 1167–1169.

Chapter 29
Vesicular-Arbuscular Mycorrhiza

Y. Dalpé

Agriculture Canada
Ottawa, Ontario, Canada

29.1 INTRODUCTION

Vesicular-arbuscular mycorrhizal associations (VAM) are one of several types of symbioses formed between plant roots and soil microorganisms. This worldwide phenomenon occurs in a great majority of herbaceous and lignicolous Angiospermeae representatives (Trappe 1987); it favors growth and development of infected plants and is increasingly considered as a significant factor for improving productivity in modern agriculture. Other fungal symbioses, such as ecto- and ectendomycorrhizae, ericoid, and orchid mycorrhizae, are morphologically different and associated with trees and shrubs, ericaceous plants, and orchids, respectively.

The fungal partners of VAM symbiosis are Zygomycetes, which belong to Acaulosporaceae family (*Acaulospora, Enthrophospora*), Gigasporaceae (*Gigaspora, Scutellospora*), and Glomaceae (*Glomus, Sclerocystis*). Their morphology consists of a network of coenocytic hyphae, dispersed in both soil and roots, bearing thick-walled spores, auxiliary cells, intraradical vesicles, and intracellular arbuscules (Bonfante-Fasolo 1984). Symbiotic mycorrhizal fungi cohabitate in equilibrium with the complex of rhizosphere organisms. Specific study of this association gave rise to the concept of "mycorrhizosphere", a term suggested by Rambelli (1973) for ectomycorrhizae, and later exploited for VAM fungi. Any study of mycorrhizae must consider the mycorrhizosphere, that zone of soil where physical, chemical, and microbiological processes are influenced by plant roots and their associated mycorrhizal fungi. Because of the extensive extra- and intramatrical distribution of VAM fungi, the techniques for mycorrhizosphere analysis must include those suitable for investigations on both plant roots and soils. Root colonization, propagule numbers, and VAM capacity of a soil to colonize plants are approaches used to estimate mycorrhization. Some good general reviews dealing with soil analysis for VAM may be found in Hayman (1982), Daniels and Skipper (1982), and Kormanik and McGraw (1982). Important steps and issues for a complete analysis of VAM in soil would consist of: (1) harvesting and storage of soil and root samples, (2) extraction and estimation of fungal propagules, (3) evaluation of mycorrhizal infection potential, and (4) estimation of root colonization.

Soil Sampling and Methods of Analysis, M. R. Carter, Ed.,
Canadian Society of Soil Science. © 1993 Lewis Publishers.

29.2 HARVESTING AND STORAGE OF SOIL AND ROOT SAMPLES

As VAM structures inhabit both cortical root cells and bulk soil, their harvest consists of gently excavating and preserving roots of the chosen plant and the recovering of the first few centimeters of the mycorrhizosphere. The VAM fungal population of a site may vary considerably over a small area (up to a few m²). Therefore, precise soil sampling strategies have to be adapted to the requirements of every experiment (Hayman 1982, Reich and Barnard 1984). The procedure outlined below is commonly used and is suitable for both field surveys and sample collection from small experimental plots or containers.

29.2.1 MATERIAL

1   Plastic bags, vials.

2   Fixative solution: formaldehyde 37%, glacial acetic acid, ethanol 50 or 70%; ratio 5:5:90.

3   Waterproof marking pens, trowel, field book.

4   Refrigerator.

29.2.2 PROCEDURE

1   Choose collecting site from a special habitat or plant species. Avoid non-mycorrhizal Angiospermeae families such as Chenopodiaceae, Amaranthaceae, Cistaceae, and most Ranunculaceae and Crucifereae.

2   Subsample (three to five times) to increase precision of analysis. Keep in mind that one VAM soil analysis may require several hours to be completed.

3   Remove the first cm of surface soil to eliminate plant debris and sample soil from the top 25–30 cm or where plant roots are most common using a trowel, a borer, or an auger.

4   Dig up part of the root system as gently as possible to avoid breaking off young feeder roots. Use a knife or a trowel to break up heavy soils and to facilitate root extraction.

5   Place roots in a plastic bag for short-term preservation (a few weeks) or in a vial filled with a fixative solution for long-term preservation.

6   Put soil in a plastic bag together with the root sample.

7   Number the samples and transfer the complete information (collecting number, harvesting date, habitat, plant species, type of soil, and any other useful data) to a reference field book.

8   Store samples under low temperature (2–5°C) until analysis.

29.2.3 COMMENTS

1   Spore number and root colonization decline with depth; few propagules are usually found lower than 40–50 cm (Jakobsen and Nielsen 1983) except in some xeric habitats. In the temperate climatic zone, higher root colonization levels and spore populations are detected on mature plants, i.e., from the middle to the end of the growing season (Hayman 1970, 1982, Jakobsen and Nielsen 1983). Spore numbers were found to increase as roots senesced (Mason 1964). Fine-textured soils sometimes contain fewer VAM propagules due to the low level of soil aeration resulting in the restriction of fungal structures to the inside of root tissues (Dakessian et al. 1986).

2   Optimum storage conditions vary from 2–10°C and 1–2% relative humidity; too high temperatures or relative humidity levels increase hyperparasite activity and stimulate the germination process, thereby reducing long-term VAM propagule viability (Nemec 1987).

29.3 EXTRACTION AND ESTIMATION OF VAM PROPAGULES

Quantitative and efficient evaluation of VAM fungi is very difficult because of the diversity of soil propagules. Spore and sporocarp counts have been widely used, but this approach does not consider the VAM inoculum potential of other types of propagules, such as fragments of colonized roots, hyphae, or networks of mycelium. The estimation of the spore population, even though it may not represent the real VAM potential of a soil, has the advantages of facilitating the separation of VAM species for identification and the study of species diversity, and the isolation of spore inocula for VAM propagation or pure culture synthesis. Unfortunately, it considers neither spore viability, dormancy, competitiveness, nor the inoculum potential of other VAM propagules.

There is no unique technique available to adequately quantify VAM soil propagules. Spore numbers can be evaluated by direct counts, by using a counting slide, or by the plate method (Smith and Skipper 1979). Identification keys exclusively use morphological characters of spores. Quantification of extramatrical hyphae can be measured by estimating soil attachment of mycorrhizal fungi to roots, by soil chitin determination (Pacovsky and Bethlenfalvay 1982), or by microscopic measurements (Ames et al. 1983). Estimation of VAM fungi in fragments of colonized roots can be obtained by extracting, clearing, staining, and mounting root fragments for microscopic observation (see Section 29.5).

Extraction methods have to be adapted to the specific needs of an experiment according to the amount of soil to be analyzed, soil texture, organic matter content, and taxonomical approach. They usually consist of sieving (Gerdemann and Nicolson 1963), followed by flotation-bubbling in glycerol (Furlan and Fortin 1975), differential sedimentation in gelatin (Mosse and Jones 1968), separation by density-gradient centrifugation in sucrose (Ohms 1957), radiopaque media (Furlan et al. 1980), or centrifugation in solutions, such as sucrose, having a higher specific gravity than the spores (Allen et al. 1979) or in colloidal silica solutions (Verkade 1988). A mix of these techniques and a number of replications are often required to obtain representative results. Other techniques have been developed such as spore adhesion on a glass surface (Sutton and Barron 1972), the plate method (Smith and Skipper 1979), and dry separation of spores (Tommerup and Carter 1982). Wet sieving followed by flotation-bubbling and/or flotation-centrifugation techniques have been selected here for their ease of use and accuracy in a large number of soil types.

29.3.1 WET-SIEVING TECHNIQUE (GERDEMANN AND NICOLSON 1963)

29.3.1.1 MATERIAL

1   Soil samples.

2   Balance.

3   Flasks or beakers (500 mL, 1 L).

4   Sieves of 1000, 500, 250, 125, 50 μm, or any series between 2 mm and 35 μm.

5   Oven (90°C).

29.3.1.2 PROCEDURE

1   Estimate soil moisture by weighing 100 g of fresh soil after drying.

2   Suspend 100 g of fresh soil in water.

3   Agitate the soil vigorously to detach soil debris and aggregates, and to provide a uniform mix.

4   Pour the prepared soil through 1-mm sieve to screen out larger debris and wash abundantly.

5   Recover all liquid and soil suspension that passed through the first sieving.

6   Agitate the soil suspension and pass it gradually through the series of superposed sieves (e.g., 500, 250, 125 and 50 μm). The number of sieves can be reduced or increased according to the soil texture or the purpose of extraction.

7   Sieve under a gentle stream of water when shaking.

8   Avoid overflooding of finest sieves, as many small-diameter spores could be lost.

9   Recover each soil fraction in a flask with a minimum of water; absorb excess of water in soil by pressing a sponge under the sieve, then transfer soil to a flask with a spatula.

29.3.2 FLOTATION-BUBBLING TECHNIQUE (FURLAN AND FORTIN 1975)

29.3.2.1 MATERIAL

1   Fractions of sieved soil.

2   Filtering columns (6–10 cm wide; 50–75 cm high) with fritted disk (about 5-μm pore size) mounted on a solid base or hooked up in series with tubing and screw clamps.

3   Glycerol solution 50% (v/v) in water (specific gravity at 25°C = 1.13).

4   Compressed air outlet with manometer or inverted vacuum pump.

5   Vacuum flask and pump.

6   Flasks (1 L), rubber stoppers, glass and rubber tubes, glass rod.

7   Filter papers (Whatman® No. 2).

8   Ringer physiological solution: NaCl, 6 g; KCl, 0.1 g; CaCl$_2$, 0.1 g; distilled H$_2$O, 1 L; adjust pH to 7.2 with NaOH, filtered and autoclaved.

29.3.2.2 PROCEDURE

1   Connect the filtering column to the compressed air system with tubing. Include the manometer and a screw clamp at the base of each filtering column. Close screw clamp.

2   Turn on compressed air, unscrew clamp, and regulate air entry into each filtering column at a pressure of about 12 psi (83 kPa).

3   Pour glycerol solution (1 L or less) in the column (pressure rises slightly).

4   Add soil fraction, wash down soil with the glycerol solution.

5   Bubble the mixture for 1 to 2 min.

6   Close screw clamp, then turn off the compressed air system.

7   Wash down the soil particles adhering to the column with glycerol solution and let the soil suspension settle down for about 30 min.

8   Recover spores in a vacuum trap by aspirating the glycerol solution using a glass tube attached to a vacuum system.

9   Maintain the glass tube as close as possible to the surface of the glycerol solution to pick up floating spores.

10   Regularly wash down spores adhering to the cylinder and the glass tube with the glycerol solution and get as close as possible to the settled soil.

11   Add glycerol solution, manually agitate the soil suspension with a glass rod to resuspend soil, and repeat steps 2 to 10 two more times. The majority of spores will be recovered during the first two extractions.

12   Filter spore suspension under vacuum on several filter papers to avoid thick layers of filtered material and to facilitate the subsequent microscopic observations.

13   Store filter papers at (3–5°C) or resuspend spores in Ringer solution before using.

29.3.3 FLOTATION-CENTRIFUGATION TECHNIQUE (OHMS 1957, ALLEN ET AL. 1979)

29.3.3.1 MATERIAL

1   Fractions of sieved soils.

2   Centrifuge apparatus and 50-mL centrifuge tubes.

3   Extraction solution: 2 M sucrose with 2% Calgon® or 50% sucrose or colloidal silica solution.

4   Separatory funnel.

5   Vacuum flask and pump.

6   Filter papers (Whatman® No. 2).

29.3.3.2 PROCEDURE

1   Make a soil suspension (water:soil, 2:1, v/v) with a soil fraction obtained from the wet-sieving technique.

2   Centrifuge at 2000 rpm (6 × g) for 10 min to remove small organic debris.

3   Pour off supernatant.

4   Resuspend soil in an extraction solution (see Section 29.3.1) and agitate.

5   Centrifuge at 2000 rpm (6 × *g*) for 10 min.

6   Decant supernatant in a flask, repeat steps 5–7 two or three times.

7   Vacuum filter the spore suspension on filter paper.

8   Rinse out spores and any debris adhering to the filter paper with distilled water or Ringer solution. Rehydrate spores to avoid plasmolysis.

9   Store spore suspension in refrigerator for preservation or vacuum filter on filter papers to isolate them.

29.3.4 DENSITY GRADIENT CENTRIFUGATION (WITH RENOGRAFIN-60)

29.3.4.1 MATERIAL

1   Fractions of sieved soils.

2   Centrifuge apparatus and 50-mL centrifuge tubes.

3   Graded concentration of Renografin-60 (10, 20, 40, and 60%).

4   Pipets and syringes.

5   Cold chamber or ice buckets.

6   Fine sieve (45 μm).

29.3.4.2 PROCEDURE

1   Work under cold temperature (cold chamber or over ice).

2   Make a 10-mL soil suspension (water:soil, 2:1, v/v) obtained from the wet-sieving technique in a 50-mL centrifuge tube.

3   Prepare Renografin-60 solutions of 10, 20, 40, and 60%.

4   Delicately inject the 10% solution in the tube, maintaining the syringe needle at the bottom without disturbing the previous phase.

5   Continue in the same manner with the 20, 40, and 60% solutions.

6   Equilibrate tubes by pair and centrifuge 10 min at 1800 rpm (5 × *g*).

7   Remove spore suspension at the 20% Renografin interface with a Pasteur pipet.

8   Wash over a 45-μm sieve with water, recover in a tube, and maintain at 5°C until use.

29.3.5 ISOLATION OF SPORES FROM SUSPENSIONS

29.3.5.1 MATERIAL

1   Filter papers (Whatman® No. 2).

2   Vacuum pump.

3   Dissecting microscope.

4   Microtip® forceps No. 5.

5   Cavity slides, microscopy slides, and coverslips.

29.3.5.2 PROCEDURE

1   Vacuum filter spores from the Ringer solution or directly use the filter papers (see Section 29.1).

2   Work under a dissecting microscope (10–30×) for counting spores.

3   Use microtip forceps, micropipettes, or a vacuum-connected pipette to hand-pick spores or sporocarps if necessary.

4   Separate spores in categories by depositing them on cavity or microscopy slides for their identification or inoculum production.

29.3.6 COMMENTS

1   Spore-extraction procedures may be varied according to soil types. In soils with high organic matter content, agitation in a blender may be required to separate spores from debris and roots before extraction. On the other hand, pure sand samples or coarse-textured soils only require vigorous agitation followed by decantation before vacuum filtration. The efficiency of a chosen technique depends on both the quantity and the quality of extracted spores.

2   Extraction solutions, especially sucrose, may influence spore morphology and viability; in such cases a rapid rehydration process is necessary to recover spore membrane integrity. The use of Renografin-60 for spore extraction does not deteriorate or alter spore quality.

3   Two published methods do not involve the sieving procedure: the plate method (Smith and Skipper 1979) and the fixation-adhesion method (Sutton and Barron 1972). Both are quite fast compared with previously described methods, but are only useful with very small quantities of soil. The plate method is only efficient with spore populations greater than 20 spores per g of soil (Daniels and Skipper 1982). With the fixation-adhesion method a high number of small-size spores can be lost with the discarded portion of the sample and a large amount of unwanted organic debris is usually recovered during the process.

29.4 ESTIMATION OF VA MYCORRHIZAL INFECTIVITY OF SOILS

The VAM infection potential of a soil can be obtained by indirect estimations of propagule number. Estimates of VAM propagule populations in soils can be carried out by the "most probable number" (MPN) method (Porter 1979) or by the "mycorrhizal soil infectivity" (MSI) method (Plenchette et al. 1989). These methods consider all sources of viable VAM propagules in the soil, but do not indicate species diversity, microbial interactions, fungal identity, or type of propagule. They evaluate the number of viable propagules based on the proportion of plants found to be mycorrhizal after growth in a series of soil dilutions. These methods eliminate single-spore counts, which are very time consuming. However, they require several weeks for plant growth before evaluation of plant root colonization. For the purpose of this chapter, only the MPN method is presented, as the recent MSI method has not yet been extensively evaluated.

29.4.1 MATERIAL

1   Plastic bags.

2   Sterilized soil or sand (steamed, autoclaved, fumigated, or irradiated).

3   Seeds or pregerminated seedlings; *Allium porrum* L. plants are receptive to VAM fungi; however, many plant species have been used.

4   Garden pots (10–15 cm) or plastic tubes (nursery cone-tainers).

5   Greenhouse space or growth chamber.

6   Fertilizer solution.

7   Materials and reagents for the estimation of root colonization (see Section 29.5).

29.4.2 PROCEDURE

1   Prepare a tenfold dilution series of soil from 10^{-1} to 10^{-5} using sterilized soil as diluent.

2   Mix soils of each dilution in a plastic bag or dilution bottle.

3   Mix control soil by adding sterile soil only to diluted sample.

4   Fill pots or tubes with a constant amount of soil. Prepare five replicates for each dilution.

5   Grow plants under controlled growth conditions for 6 weeks with regular watering and fertilization (e.g., Long Ashton).

6   Harvest complete root systems, wash gently to remove soil particles, clear, bleach, and stain for the estimation of root or root length colonization (see Section 29.5).

7   Observe under dissecting or light microscope for any sign of the presence of VAM endophytes in each root system.

8   Determine the MPN for the original soil sample using tables from Fisher and Yates (1963).

29.4.3 COMMENTS

1   No method to determine VAM infectivity of soils has yet achieved universal acceptance. Direct measurements of spore population, as an indirect estimation of mycorrhizal infection potential, are widely used as valuable criteria for the evaluation of VAM populations in soil.

2   The MPN method has a low level of precision and, in order to reduce variability, the dilution power has to be reduced and the number of replicates increased. It has also been recognized that MPN estimates are dependent upon experimental conditions such as temperature and time of harvesting (Wilson and Trinick 1982), and in some cases may be underestimates due to unstained mycorrhizal structures (Morton 1985).

3   The recent MSI method (Plenchette et al. 1989) is based on a biological assay that measures the capacity of a natural soil to induce mycorrhizal infection. The MSI unit measures the minimum dry mass (g) of soil required to infect 50% of the population of a plant under the bioassay conditions (MSI_{50}). This

new approach considers both inoculum potential and soil factors and may be less time consuming and statistically more precise. This approach has to be considered when choosing an accurate method for the mycorrhizal evaluation of a soil.

29.5 ESTIMATION OF THE ROOT COLONIZATION LEVEL

The level of root colonization is estimated by the frequency of root colonization and by evaluating the intensity of each colonization. Colonization by VAM fungi is restricted to the root cortex and is usually most prevalent in young feeder roots. The intramatrical morphology of VAM fungi consists of a network of intercellular coenocytic hyphae, running through the layer of cortical cells, usually bearing intra- or extracellular vesicles, sometimes spores, and differentiating intracellular arbuscules (Bonfante-Fasolo 1984). There is no visible gross morphological change to roots due to infection, except with certain plants (e.g., *A. porrum*) which develop a yellow pigmentation once highly colonized (Becker and Gerdemann 1977). As a result, the most common method used to detect VAM root colonization is microscope examination of stained fungal structures. Several staining techniques have been tested; acid fuchsin (Kormanik and McGraw 1982) and trypan blue (Phillips and Hayman 1970, Koske and Gemma 1989) are the most utilized ones. Other quantitative and qualitative bioassays such as chitin estimation (Hepper 1977) and electrophoresis (Hepper et al. 1986) use unstained roots, but have not yet been extensively tested on a variety of plant and fungal species. Quantification of the root colonization can be done by the slide method, which indicates the length of colonized roots; the grid line intersect method, which estimates the percent of root colonized; and nonsystematic root scanning under the dissecting microscope, which rapidly gives broad categories of root colonization (0–5%, 5–25%, etc.) and the intensity of their infection (e.g., 1 = few entry points, 2 = uniform, 3 = coalesce distribution, etc.). A review of those methods has been done by Giovanetti and Mosse (1980). Use of trypan blue staining followed by the grid line intersect method to evaluate percent root colonization and the slide method for an estimation of the length of colonized root are described below.

29.5.1 MATERIAL

1   Aqueous solution of 2.5 to 10% KOH (w/v) for root clearing.

2   Aqueous solution of 3% H_2O_2 solution or aqueous solution of 20% NH_4OH for root bleaching.

3   Trypan blue solution (Phillips and Hayman 1970, Koske and Gemma 1989): trypan blue (0.5 g) in glycerol (500 mL), H_2O (450 mL), and HCl 1% (50 mL) for root coloration.

4   Destaining solutions prepared as in previous step, but without the stain.

5   PVLG mounting media (Omar et al. 1979): polyvinylalcohol, high viscosity
    (24–32 cP), 1.66 g; H_2O, 10 mL; lactic acid, 10 mL; glycerol, 1.0 mL. Dissolve
    PVA in water; stir vigorously while adding lactic acid and glycerol.

29.5.2 PROCEDURE

29.5.2.1 ROOT PREPARATION

1   Use fresh or fixed roots (see Section 29.2.2); gently wash them with water
    to remove attached soil particles and soil debris. Cut them in short pieces
    to facilitate their staining and mounting on microscope slides.

2   Clear roots in KOH solution (according to the root sensitivity) by autoclaving
    for 3 min at 121°C or heat at 90°C for 10 to 45 min (use fumehood to avoid
    breathing harmful vapor of KOH).

3   Bleach roots (used only for highly pigmented material) in H_2O_2 or NH_4OH
    for 10 to 45 min at room temperature.

4   Rinse roots in several changes of water.

5   Acidify roots in 1% HCl prior to the staining (trypan blue staining is effective
    under low pH).

6   Stain roots in trypan blue reagent by either autoclaving roots for 3 min at
    121°C or in a 90°C water bath for 15 to 60 min. Root material stains completely
    blue.

7   Destain roots in destaining solution at room temperature or at 50 to 65°C.
    Destained roots can be stored for a long term (up to several months) at
    room temperature.

8   Mount root pieces in PVLG on microscopic slides.

29.5.2.2 SCORING ROOT COLONIZATION

1   With the grid line intersect method (Giovanetti and Mosse 1980), a grid
    made of 1.27-cm (0.5-in) squares is etched on a plastic petri dish lid. Stained
    roots are deposited in the lid, flattened with the petri dish bottom, and
    examined under the microscope at 100×. Colonized root sections that in-
    tersect the lines of the grid are counted and compared with the total number
    of roots intersecting the grid lines. Results are expressed in percent of
    observed that are colonized.

2   With the slide method (Giovanetti and Mosse 1980): large numbers (50–100)
    of 1-cm root sections are mounted on slides. The length of colonized root

tissue is measured and compared to the total length of root observed. Results are expressed as a percent.

29.5.3 COMMENTS

1   Intramatrical vesicles and hyphae are indicators of the presence of VAM fungi in the root system, but only the presence of intracellular arbuscules which are the site of metabolite exchanges between plant and fungi confirm the symbiotic behavior of the fungal partner. Depending upon the physiological state of the observed mycorrhizae, arbuscules may appear as highly ramified structures originating from a single hyphae or as an irregular mass of stained material without recognizable ramification.

2   As different plant roots react differently to clearing, bleaching, and staining procedures, preliminary tests followed by adaptations of techniques are often necessary to optimize staining results. Before staining, extramatrical hyphae can be detected by examining the surface of gently washed roots under a dissecting microscope; their presence will not necessarily indicate root infection. The removal of phenol from the lactophenol/trypan blue solution (Phillips and Hayman 1970, Kormanik and McGraw 1982), as used by Koske and Gemma (1989), can provide adequate contrast with several hosts and VAM partners.

REFERENCES

Allen, M. F., Moore, T. S., and Christensen, M. 1979. Growth of vesicular-arbuscular-mycorrhizal and nonmycorrhizal *Bouteloua gracilis* in defined medium. Mycologia 71: 666–669.

Ames, R. N., Reid, C. P. P., Porter, L. K., and Campardella, C. 1983. Hyphal uptake and transport of nitrogen from two ^{15}N-labelled sources by *Glomus mosseae*, a vesicular-arbuscular mycorrhizal fungus. New Phytol. 95: 381–396.

Becker, W. N. and Gerdemann, J. W. 1977. Colorimetric quantification of vesicular-arbuscular mycorrhizal infection in onion. New Phytol. 78: 289–295.

Bonfante-Fasolo, P. 1984. Anatomy and morphology of VA mycorrhizae. Pages 5–33 in C. L. I. Powell and D. J. Bagyaraj, Eds. VA mycorrhiza. CRC Press, Boca Raton, FL.

Dakessian, S., Brown, M. S., and Bethenfalvay, G. J. 1986. Relationship of mycorrhizal growth enhancement and plant growth with soil water and texture. Plant Soil 94: 439–444.

Daniels, B. A. and Skipper, H. D. 1982. Methods of the recovery and quantitative estimation of propagules from soil. Pages 29–35 in N. C. Schenck, Ed. Methods and principles of mycorrhizal research. American Phytopathological Society.

Fisher, R. A. and Yates, F. 1963. Statistical tables for biological, agricultural and medical research. Oliver and Boyd Edinburgh Eds.

Furlan, V. and Fortin, J. A. 1975. A flotation-bubbling system for collecting Endogonaceae spores from sieved soil. Nat. Can. 102: 663–667.

Furlan, V., Bartschi, H., and Fortin, J. A. 1980. Media for density gradient extraction of endomycorrhizal spores. Trans. Br. Mycol. Soc. 75: 336–338.

Gerdemann, J. W. and Nicolson, T. H. 1963. Spores of mycorrhizal *Endogone* species extracted from soil by wet sieving and decanting. Trans. Br. Mycol. Soc. 46: 235–244.

Giovanetti, M. and Mosse, B. 1980. An evaluation of techniques for measuring vesicular-arbuscular mycorrhizal infection in roots. New Phytol. 84: 489–500.

Hayman, D. S. 1970. *Endogone* spore numbers in soil and vesicular-arbuscular mycorrhiza in wheat as influenced by season and soil treatment. Trans. Br. Mycol. Soc. 54: 53–63.

Hayman, D. S. 1982. Practical aspects of vesicular-arbuscular mycorrhiza. Pages 325–373 in N. S. S. Rao, Ed. Advances in agricultural microbiology. Butterworths.

Hepper, C. M. 1977. A colorimetric method for estimating vesicular-arbuscular mycorrhizal infection in roots. Soil Biol. Biochem. 9: 15–18.

Hepper, C. M., Sen, R., and Maskall, C. S. 1986. Identification of vesicular-arbuscular mycorrhizal fungi in roots of leek (*Allium porrum* L.) and maize (*Zea mays* L.) on the basis of enzyme mobility during polyacrylamide gel electrophoresis. New Phytol. 102: 529–539.

Jakobsen, I. and Nielsen, N. E. 1983. Vesicular-arbuscular mycorrhiza in field grown crops. 1. Mycorrhizal infection in cereals and peas at various times and soil depths. New Phytol. 93: 401–414.

Kormanik, P. P. and McGraw, A. C. 1982. Quantification of vesicular-arbuscular mycorrhizae in plant roots. Pages 37–45 in N. D. Schenck, Ed. Methods and principles of mycorrhizal research. American Phytopathological Society.

Koske, R. E. and Gemma, J. N. 1989. A modified procedure for staining roots to detect VA mycorrhizas. Mycol. Res. 92: 486–505.

Mason, D. T. 1964. A survey of numbers of *Endogone* spores in soil cropped with barley, raspberry and strawberry. Hortic. Res. 4: 98–103.

Morton, J. B. 1985. Underestimation of most probable numbers of vesicular-arbuscular endophytes because of non-staining mycorrhizae. Soil Biol. Biochem. 17: 383–384.

Mosse, B. and Jones, G. W. 1968. Separation of *Endogone* spores from organic soils debris by differential sedimentation of gelatin columns Trans. Br. Mycol. Soc. 51: 604–608.

Nemec, S. 1987. Effect of storage temperature and moisture on *Glomus* species and their subsequent effect on *Citrus* rootstock seedling growth and mycorrhiza development. Trans. Br. Mycol. Soc. 89: 205–212.

Ohms, R. E. 1957. A flotation method for collecting spores of a Phycomycetous mycorrhizal parasite from soil. Phytopathology 47: 751–752.

Omar, M. B., Bolland, L., and Heather, W. A. 1979. A permanent mounting medium for fungi. Bull. Br. Mycol. Soc. 13: 31–32.

Pacovsky, R. S. and Bethlenfalvay, G. J. 1982. Measurement of the extraradical mycelium of a vesicular-arbuscular mycorrhizal fungus in soil by chitin determination. Plant Soil 68: 143–147.

Phillips, J. M. and Hayman, D. S. 1970. Improved procedures for clearing and staining parasitic and vesicular-arbuscular mycorrhizal fungi for rapid assessment of infection. Trans. Br. Mycol. Soc. 55: 158–161.

Plenchette, C., Perrin, R., and Duvert, P. 1989. The concept of soil infectivity and a method for its determination as applied to endomycorrhizas. Can. J. Bot. 67: 112–115.

Porter, W. M. 1979. The "Most Probable Number" method for enumerating infective propagules of vesicular-arbuscular mycorrhizal fungi in soil. Aust. J. Soil Res. 17: 515–519.

Rambelli, A. 1973. The rhizosphere of mycorrhizae. Pages 299–343 in G. L. Marks and T. T. Koslowski, Eds. Ectomycorrhizae. Academic Press, New York.

Reich, L. and Barnard, J. 1984. Sampling strategies for mycorrhizal research. New Phytol. 98: 475–479.

Smith, G. W. and Skipper, H. D. 1979. Comparison of methods to extract spores of vesicular-arbuscular mycorrhizal fungi. Soil Sci. Soc. Am. J. 43: 722–725.

Sutton, J. C. and Barron, G. L. 1972. Population dynamics of *Endogone* spores in soil. Can. J. Bot. 50: 1909–1914.

Tommerup, I. C. and Carter, D. J. 1982. Dry separation of microorganisms from soil. Soil Biol. Biochem. 14: 69–71.

Trappe, J. M. 1987. Phylogenetic and ecologic aspects of mycotrophy in the angiosperms from an evolutionary standpoint. Pages 5–25 in G. R. Safir, Ed. Ecophysiology of VA mycorrhizal plants. CRC Press, Boca Raton, FL.

Verkade, S. D. 1988. Use of colloidal silica solution in the isolation of spores of *Glomus mosseae*. Mycologia 80: 109–110.

Wilson, J. M. and Trinick, M. J. 1982. Factors affecting the estimation of numbers of infective propagules of vesicular-arbuscular mycorrhizal fungi by the most probable number method. Aust. J. Soil Res. 21: 73–81.

Chapter 30
Root Nodule Bacteria and Nitrogen Fixation

W. A. Rice and P. E. Olsen

Agriculture Canada
Beaverlodge, Alberta, Canada

30.1 INTRODUCTION

The phenomenon of symbiotic nitrogen fixation is of interest to microbiologists, soil scientists, plant physiologists, plant breeders, and agronomists. Research on the subject involves a number of other cross-disciplinary interactions, but specific details of methodology of related cross-disciplinary areas will not be discussed.

The macrosymbiont of greatest agricultural significance involved in the nitrogen fixation association is the legume plant, which is nodulated by species of the genera *Rhizobium* and *Bradyrhizobium*. The separation of the legume-nodulating microorganisms into these two genera occurred in 1984 (Jordan 1984), and, therefore, literature prior to that date does not make the differentiation. In this chapter, the term rhizobia will be used to refer to species of both genera.

30.2 ISOLATION OF RHIZOBIA

30.2.1 Nodules

The practical problems involved and the guidelines for collecting nodules, preserving them, and isolating rhizobia from them have been described in detail by Date and Halliday (1987). There are many variations in the approach, depending on the extent and purpose of collecting nodules and isolating rhizobia. The procedure given here is a general description of the basic steps to be followed in obtaining a culture of the rhizobia inhabiting the nodules of a selected plant.

30.2.1.1 MATERIALS

1   Tools for excavating plants and removing roots — spade, garden trowel, knife, etc.

Soil Sampling and Methods of Analysis, M. R. Carter, Ed.,
Canadian Society of Soil Science. © 1993 Lewis Publishers.

2   Collection vessel: glass container (10- to 20-mL capacity) containing a desiccant (anhydrous calcium chloride or silica gel) occupying one fourth the volume of the container, held in place by a cotton wool plug. The container must have an airtight cap.

3   95% (v/v) ethanol.

4   Sterilant solution: 8% (w/v) sodium or calcium hypochlorite solution; or 3% (v/v) hydrogen peroxide solution.

5   Sterile water.

6   Petri dishes containing 15 mL yeast-extract mannitol agar (YEMA): 10 g mannitol, 0.5 g K_2HPO_4, 0.2 g $MgSO_4 \cdot 7H_2O$, 0.2 g NaCl, 0.5 g yeast extract, 15 g agar, 1 L water.

30.2.1.2 PROCEDURE

1   Using digging tools, remove a soil block approximately 25 cm in diameter containing roots to a depth of 10 to 15 cm.

2   Expose nodules by carefully removing soil from the roots. Exposure of the root system is greatly facilitated by immersion of the soil block in water and allowing the soil to fall away.

3   Collect exposed nodules with forceps and a sharp knife or fine scissors, severing the root about 0.5 cm on either side of the site of attachment.

4   Place nodules in collection vessel. All nodules from a single plant represent one unit of collected material and may be stored in the same container. Nodules should be processed within 3 weeks.

5   Place dry nodules in water for 30 to 60 min to allow them to fully imbibe.

6   Wash the nodules in running tap water, if necessary, to remove gross soil contamination and drain or transfer to tissue paper to surface dry.

7   Immerse the nodules for 5 to 10 s in 95% ethanol, and then in sterilant and leave for 3 to 4 min.

8   Remove sterilant and wash nodules at least five times in sterile water.

9   Crush each nodule in 1 to 2 mL of sterile water to give a turbid suspension and transfer a drop to a YEMA plate with a flamed loop or glass rod. Small nodules can be crushed with tweezers and spotted directly onto the agar plate.

10   Streak the drop of suspension onto the agar surface so that the suspension is progressively diluted.

11 Incubate the plates in an inverted position at 30°C and check daily for growth typical of *Rhizobium* or *Bradyrhizobium* along the streak lines. Those unfamiliar with the cultural characteristics of rhizobia should consult published descriptions (Jordan 1984, Date and Halliday 1987, Vincent 1970), and obtain authentic cultures of the species of interest in order to study its cultural characteristics in the investigator's own laboratory.

12 Pick off and restreak well-isolated single colonies onto fresh plates to obtain pure cultures. If more than one typical colony type appears on a plate, each of these types should be taken to pure culture.

13 Transfer pure culture isolates to screw-cap agar-slant tubes for short-term storage (up to 1 year). Alternate methods should be used for longer storage periods (Date and Halliday 1987). A very successful method is low-temperature (-135°C or lower) storage of broth cultures suspended in 10% glycerol.

14 Authenticate each pure culture isolate by confirming nodule-forming ability on test host plants grown under axenic conditions in growth pouches (see Section 30.3.1) or in Leonard jar assemblies (Vincent 1970, Gibson 1980).

30.2.1.3 COMMENTS

Several problems may be encountered in obtaining discrete single colonies, and subsequently in obtaining pure culture isolates:

1 Contamination of the agar plate is a frequent occurrence, and is particularly a problem when the contamination is due to fungi. This problem can be reduced by increasing the sterilization time or by using a different sterilant solution. The growth of fungi on the YEMA can be controlled by addition of cyclohexamide at 20 or 40 μg mL^{-1}. Dissolve 0.5 g cyclohexamide in 25 mL 95% (v/v) ethanol and add 1.2 mL L^{-1} media when cooled to 50 to 55°C (Vincent 1970).

2 Bacterial contaminants can be a particular problem because of the inability to distinguish contaminants from rhizobia. Congo red (10 mL of solution containing 1.0 to 2.5 mg mL^{-1}, sterilized separately and added to each liter of melted medium cooled to 50 to 55°C) can sometimes assist the recognition of rhizobia among other bacteria. Many contaminating bacteria absorb the dye strongly, whereas most rhizobia take it up weakly.

3 Frequently, more than one colony type, with all the morphological characteristics of rhizobia, will appear on one plate from one nodule. In some cases these colony types will form nodules on the test host, and in other cases one or more of the types may be incapable of forming nodules. These types of contaminants are often "latent" in that they appear to be carried along in subsequent transfers without detection and will suddenly appear, particularly if the culture is grown under nutritional or physical stress. It is therefore extremely important that new isolates and cultures undergoing

frequent transfer be thoroughly checked by inoculation onto test plants and reisolation from fresh nodules.

30.2.2 Recovery by Plant Infection

If satisfactory isolation is not achieved from surface-sterilized nodules, it may be possible to recover the rhizobia by inoculating the appropriate host plant with a homogenate prepared from the nodule sample. The host plant can be grown aseptically in growth pouches (see Section 30.3.1) or in Leonard jar assemblies (Vincent 1970, Gibson 1980). Isolation can then be made from fresh nodules using the technique described in Section 30.2.1. The plant-infection technique can also be used to isolate rhizobia from soil samples. In this case a 1:10 (soil:diluent) (see Section 30.3.1 for buffer diluent solution) suspension is used to inoculate the appropriate host plant species.

30.3 ENUMERATION OF RHIZOBIA

The determination of the number of living rhizobia in soil is usually done by the most probable number (MPN) plant grow-out technique, based on the recovery of rhizobia within a series of dilutions as evidenced by formation of nodules on the appropriate host plant.

30.3.1 MPN Plant Grow Out

The analysis involves determination of the number of live rhizobia of appropriate species per unit of soil. The analysis is performed by means of the MPN plant grow out technique in which serially diluted soil is applied to aseptic host plants and the viable rhizobia number is calculated from the point of nodule extinction after 4 weeks of growth. This method is based on four major assumptions: if a viable *Rhizobium* cell is inoculated on its specific host (see Table 30.1) in a nitrogen-free medium, nodules will develop in the roots of the inoculated legume plant; nodulation in the roots of the inoculated plant is proof of the presence of infective rhizobia; test cleanliness is demonstrated by the absence of nodules in plants grown from uninoculated seeds; and absence of nodules is proof of the absence of infective rhizobia. The following procedure is adapted from the official procedure as approved under authority of the Fertilizers Act (Anon. 1975).

30.3.1.1 MATERIALS

1   Disposable growth pouches and racks — the growth pouch (Canlab No. B1220) is made of strong transparent polyester film capable of withstanding steam sterilization at 100 kPa up to 20 min. In practice, sterilization is not required for most purposes, since the pouches do not contain rhizobia and any microorganisms present do not interfere with normal counts. Inserted in the pouch is a paper germination towel that is folded along the top edge into a trough and perforated to permit roots to escape from the seeding area. Suitable racks are required to provide support during seeding, inoculation, and growth.

2   Seed — undamaged certified seed of the appropriate species and cultivar.

TABLE 30.1 Scientific Name, Common Name, Number of Seeds per Kilogram, and Nodule Bacteria for Inoculation of Important Legumes Cultivated in Canada

Scientific and common name of legumes	Number of seeds per kg	Nodule bacteria
Glycine max, Merr. (Glycine soja) soybean	6,667	Bradyrhizobium japonicum
Lens culinaris lentil	6,622	R. leguminosarum
Lotus corniculatus, L. (tenuis) birdsfoot trefoil	826,875	R. loti
Medicago sativa, L. alfalfa	441,000	R. meliloti
Melilotus alba. Desr. white sweetclover	573,340	R. meliloti
Melilotus officinalis, Lam. yellow sweetclover	573,300	R. meliloti
Onobrychis viciaefolia, Scop. sainfoin	66,150	Rhizobium sp.[a]
Phaseolus vulgaris common beans	2,481	R. leguminosarum biovar phaseoli
Pisum arvense, L. field or garden peas	4,310	R. leguminosarum biovar viceae
Trifolium hybridum, L. alsike clover	1,543,500	R. leguminosarum biovar trifolii
Trifolium incarnatum, L. crimson clover	308,700	R. leguminosarum biovar trifolii
Trifolium pratense, L. red clover	606,375	R. leguminosarum biovar trifolii
Trifolium repens, L. white clover, ladino clover	1,764,000	R. leguminosarum biovar trifolii
Vicia sativa, L. common vetch	15,435	R. leguminosarum biovar viceae
Vicia faba var. equina, L. Faba bean	2,326	R. leguminosarum biovar viceae
Vicia faba var. major, L. broadbean	1,103	R. leguminosarum biovar viceae

[a] Rhizobium species designation are unknown for these legumes and specific strains are required.

3   95% (v/v) ethanol.

4   Sterilant solution — see Section 30.2.1.

5   Plant nutrient solution: 0.004 mg $CoCl_2 \cdot 6H_2O$, 2.86 mg H_3BO_3, 1.81 mg $MnCl_2 \cdot 4H_2O$, 0.22 mg $ZnSO_4 \cdot 7H_2O$, 0.08 mg $CuSO_4 \cdot 5H_2O$, 0.09 mg $H_2MoO_4 \cdot H_2O$, 493 mg $MgSO_4 \cdot 7H_2O$, 174 mg K_2HPO_4, 136 mg KH_2PO_4, 111 mg $CaCl_2$, 5 mg $FeC_6H_5O_7 \cdot xH_2O$, 1 L distilled water, 1.0 N solution of HCl or NaOH to adjust pH to 6.8. Sterilize by autoclaving at 100 kPa for 20 min.

6   Buffer diluent solution: 1.0 g peptone, 0.34 g KH_2PO_4, 1.21 g K_2HPO_4, 1 L distilled water, pH 7.0. Sterilize by autoclaving at 100 kPa for 20 min.

7   Growth chamber or room providing 550 $\mu mol\ s^{-1}m^{-2}$, 20°C during 12-h light period for clovers, or 16-h light period for other legumes and 15°C during the dark period, and relative humidity at 65 to 70%.

8   Waring® blender.

30.3.1.2 PROCEDURE

1   Add 30 mL of sterile nutrient solution to each pouch and place in rack.

2   Surface sterilize seeds by immersion in 95% ethanol for 30 s, followed by 10 min in hydrogen peroxide solution or 10 min in hypochlorite solution. Wash seeds thoroughly in at least five changes of sterile distilled water. Alternatively, small-seeded legumes may be surface sterilized by immersion in conc H_2SO_4 for 10 min, followed by at least five washes in sterile distilled water.

3   Place 20 small- ($>2 \times 10^5$ seed kg^{-1}) or 15 intermediate- (3×10^4 to 2×10^5 seeds kg^{-1}) size surface-sterilized seeds directly into the trough of each growth pouch, allow them to germinate in the dark at 20°C, then move to the growth chamber or room.

4   For large-size seeds ($<3 \times 10^4$ seeds kg^{-1}), place seeds on three layers of germination paper, and place the rolls in a humid chamber with the seed radicles pointing downward. Germinate at 20°C until radicals have elongated to 0.5 to 1.0 cm. Enlarge five holes in the seed trough of the growth pouch and place the radical of a germinated seed in each hole and cover with a sterilized, moist cotton pad for 3 to 4 d to maintain moistness until shoots have emerged. Transfer growth units to growth chamber or room.

5   Place 10 g of soil into 90 mL of cold sterilized buffer diluent in a Waring® blender and disperse for 2 min at low speed.

6   Transfer a 10-mL aliquot of the buffer-soil suspension to a bottle containing 90 mL of cold diluent and shake for 5 min. Make tenfold serial dilutions as required, depending on the expected number of rhizobia in the soil sample.

7   Prepare a subsequent fivefold dilution series (1 mL suspension into 4 mL of diluent) for six consecutive dilution steps.

8   Inoculate four growth pouches with 1.0 mL of the suspension from each fivefold dilution level, except for the last dilution level, where five pouches are inoculated. Leave an uninoculated control pouch between each set of four or five inoculated pouches. (Steps 7 and 8 can be done conveniently with 10-mL syringes by drawing up 1 mL of suspension followed by 4 mL of diluent, and then 5 mL of air. Shake the syringe to thoroughly mix the suspension and diluent, then express the air and inoculate four pouches with 1 mL of suspension each. The remaining 1 mL of suspension is diluted with an additional 4 mL of diluent and the procedure repeated for the required number of fivefold dilutions.)

9   Place racks of growth pouches in the growth chamber or room, and examine after 2 weeks for the presence of nodules. Record the results as "positive" for nodulated plants in a growth pouch and "negative" for unnodulated plants in a growth pouch. The test should be continued for at least 3 weeks to confirm "negative" readings. Controls must be free of nodules for the test to be meaningful.

30.3.1.3 CALCULATIONS

1   Obtain the MPN of rhizobia in the starting fivefold dilution series from the MPN table (Table 30.2) by locating the corresponding six-digit code representing the numbers of positive growth units at each of the fivefold dilution levels.

2   To determine the number of rhizobia per gram of soil, multiply the MPN in the starting fivefold dilution series by the tenfold dilution from which the first fivefold dilution was made.

30.3.1.4 COMMENTS

1   Soils generally contain in the order of 10^4 or less rhizobia per gram, and therefore require the tenfold dilution series to be carried to the 10^{-1} or 10^{-2} level before starting the fivefold dilution series.

2   A computer program (MPNES) (Woomer et al. 1990) is available which calculates the MPN from a variety of MPN codes and eliminates the need for Table 30.2. The program avoids interpolation of codes that fall between the codes listed in Table 30.2 and is available from NifTAL Project, 1000 Holomua Ave., Paia, HA 96779.

TABLE 30.2 Most Probable Number (MPN) of Nodule Bacteria Calculated from the Distribution of Positive (Nodulated) Growth Unit in a Plant-Infection Test Based on a Fivefold Dilution Series

No. of positive (nodulated) growth units out of four resulting from inoculation with 1-mL aliquots

Fivefold dilution series from the starting tenfold dilution or from the 1:1 dilution of preinoculated seeds						MPN of nodule bacteria in starting fivefold dilution series	
1:5	1:25	1:125	1:625	1:3,125	1:15,625[a]	Estimate	Confidence limits (95%)
0	1	0	0	0	0	1.0	0.1–7.7
0	2	0	0	0	0	2.1	0.5–9.2
0	3	0	0	0	0	3.0	0.9–10.6
1	0	0	0	0	0	1.1	0.2–7.9
1	1	0	0	0	0	2.3	0.6–9.6
1	2	0	0	0	0	3.5	1.1–11.9
1	3	0	0	0	0	4.9	1.6–14.6
2	0	0	0	0	0	2.6	0.6–10.1
2	1	0	0	0	0	4.0	1.2–12.8
2	2	0	0	0	0	5.5	1.9–16.0
2	3	0	0	0	0	7.2	2.7–19.6
3	0	0	0	0	0	4.6	1.5–14.1
3	1	0	0	0	0	6.5	2.3–18.0
3	2	0	0	0	0	8.7	3.3–23.0
3	3	0	0	0	0	11.3	4.4–29.2
4	0	0	0	0	0	8.0	3.0–21.5
4	1	0	0	0	0	11.4	4.4–29.5
4	2	0	0	0	0	16.2	6.2–42.4
4	3	0	0	0	0	24.2	9.0–64.9
4	4	0	0	0	0	40.4	15.3–106.6
4	0	1	0	0	0	10.8	4.2–28.1
4	1	1	0	0	0	15.1	5.8–39.2
4	2	1	0	0	0	21.5	8.1–57.4
4	3	1	0	0	0	32.8	12.2–87.9
4	0	2	0	0	0	14.1	5.4–36.6
4	1	2	0	0	0	19.6	7.4–51.9
4	2	2	0	0	0	28.3	10.5–76.1
4	3	2	0	0	0	43.6	16.6–114.2
4	0	3	0	0	0	18.1	6.9–47.7
4	1	3	0	0	0	25.2	9.4–67.6
4	2	3	0	0	0	36.4	13.7–96.8
4	3	3	0	0	0	56.5	21.9–146.0
4	4	1	0	0	0	5.7 × 10	2.2–14.7 × 10
4	4	2	0	0	0	8.1 × 10	3.1–21.2 × 10
4	4	3	0	0	0	12.1 × 10	4.5–32.4 × 10
4	4	4	0	0	0	20.2 × 10	7.6–53.3 × 10
4	4	0	1	0	0	5.4 × 10	2.1–14.0 × 10
4	4	1	1	0	0	7.5 × 10	2.9–19.6 × 10
4	4	2	1	0	0	10.8 × 10	4.0–28.7 × 10
4	4	3	1	0	0	16.4 × 10	6.1–43.9 × 10
4	4	0	2	0	0	7.1 × 10	2.7–18.3 × 10
4	4	1	2	0	0	9.8 × 10	3.7–26.0 × 10
4	4	2	2	0	0	14.1 × 10	5.3–38.1 × 10
4	4	3	2	0	0	21.8 × 10	8.3–57.1 × 10
4	4	0	3	0	0	9.1 × 10	3.4–23.8 × 10

TABLE 30.2 (continued) Most Probable Number (MPN) of Nodule Bacteria Calculated from the Distribution of Positive (Nodulated) Growth Unit in a Plant-Infection Test Based on a Fivefold Dilution Series

No. of positive (nodulated) growth units out of four resulting from inoculation with 1-mL aliquots

Fivefold dilution series from the starting tenfold dilution or from the 1:1 dilution of preinoculated seeds						MPN of nodule bacteria in starting fivefold dilution series	
							Confidence limits (95%)
1:5	1:25	1:125	1:625	1:3,125	1:15,625[a]	**Estimate**	
4	4	1	3	0	0	12.6×10	$4.7–33.8 \times 10$
4	4	2	3	0	0	18.2×10	$6.9–48.4 \times 10$
4	4	3	3	0	0	28.2×10	$10.9–73.0 \times 10$
4	4	4	1	0	0	2.9×10^2	$1.1–7.3 \times 10^2$
4	4	4	2	0	0	4.1×10^2	$1.6–10.6 \times 10^2$
4	4	4	3	0	0	6.0×10^2	$2.3–16.2 \times 10^2$
4	4	4	4	0	0	10.1×10^2	$3.8–26.6 \times 10^2$
4	4	4	0	1	0	2.7×10^2	$1.0–7.0 \times 10^2$
4	4	4	1	1	0	3.8×10^2	$1.5–9.8 \times 10^2$
4	4	4	2	1	0	5.4×10^2	$2.0–14.4 \times 10^2$
4	4	4	3	1	0	8.2×10^2	$3.1–22.0 \times 10^2$
4	4	4	0	2	0	3.5×10^2	$1.4–9.2 \times 10^2$
4	4	4	1	2	0	4.9×10^2	$1.8–13.0 \times 10^2$
4	4	4	2	2	0	7.1×10^2	$2.6–19.0 \times 10^2$
4	4	4	3	2	0	10.9×10^2	$4.2–28.6 \times 10^2$
4	4	4	0	3	0	4.5×10^2	$7.7–11.9 \times 10^2$
4	4	4	1	3	0	6.3×10^2	$2.3–16.9 \times 10^2$
4	4	4	2	3	0	9.1×10^2	$3.4–24.2 \times 10^2$
4	4	4	3	3	0	14.1×10^2	$5.4–36.7 \times 10^2$
4	4	4	4	1	0	14.3×10^2	$5.5–36.9 \times 10^2$
4	4	4	4	2	0	20.3×10^2	$7.8–53.0 \times 10^2$
4	4	4	4	3	0	30.2×10^2	$11.2–81.3 \times 10^2$
4	4	4	4	4	0	50.5×10^2	$19.0–133.8 \times 10^2$
4	4	4	4	0	1	13.5×10^2	$5.2–35.3 \times 10^2$
4	4	4	4	1	1	18.8×10^2	$7.2–49.0 \times 10^2$
4	4	4	4	2	1	26.9×10^2	$10.1–71.8 \times 10^2$
4	4	4	4	3	1	41.0×10^2	$15.3–110.2 \times 10^2$
4	4	4	4	0	2	17.7×10^2	$6.8–45.9 \times 10^2$
4	4	4	4	1	2	24.5×10^2	$9.2–65.0 \times 10^2$
4	4	4	4	2	2	35.3×10^2	$13.1–95.4 \times 10^2$
4	4	4	4	3	2	54.4×10^2	$20.6–143.8 \times 10^2$
4	4	4	4	0	3	22.6×10^2	$8.6–59.7 \times 10^2$
4	4	4	4	1	3	31.4×10^2	$11.7–84.7 \times 10^2$
4	4	4	4	2	3	45.5×10^2	$17.0–121.4 \times 10^2$
4	4	4	4	3	3	70.6×10^2	$27.1–184.2 \times 10^2$
4	4	4	4	4	1	7.1×10^3	$2.7–18.6 \times 10^3$
4	4	4	4	4	2	10.1×10^3	$3.8–27.0 \times 10^3$
4	4	4	4	4	3	15.1×10^3	$5.4–42.6 \times 10^3$
4	4	4	4	4	4	25.2×10^3	$8.6–74.0 \times 10^3$
4	4	4	4	4	5	$>35.5 \times 10^3$	

[a] Five growth units inoculated with 1-mL aliquots from this dilution level.

30.3.2 Plate Count

Plate counting is most suitable for determining rhizobia numbers in inoculants with a sterile base, and in inoculants with a nonsterile base, provided the rhizobia level is sufficiently high to provide readily distinguishable rhizobial colonies. This requires considerable experience to ensure reliable results. This method is less suitable for enumeration of rhizobia in soil, since other bacteria outnumber the rhizobia by a large enough margin so as to crowd them out or inhibit their development on the plate.

The spread-plate count method (Somasegaran and Hoben 1985, Vincent 1970) is often used when it is desirable to have all the colonies on the surface for easy recognition of rhizobia in the presence of other bacteria. This procedure requires the preparation of a set of serial dilutions (usually tenfold) so as to provide 30–300 colonies at some step in the series. A small volume (usually 0.1 mL) at the appropriate dilution is spread over the surface of the agar, which has been freed from excess water by holding already poured plates overnight at 37°C in an incubator.

A similar technique called the pour-plate method is also commonly used (Somasegaran and Hoben 1985, Vincent 1970). The same serial dilutions are used as for the spread-plate method. In this method, 1 mL of the selected dilutions is placed in petri dishes and within a few minutes the sample is covered with 15 mL liquefied and cooled (50°C) YEMA and mixed by rotation several times clockwise, counter clockwise, etc., to ensure uniform dispersion of the cells in the medium.

Both the spread-plate and the pour-plate methods are lengthy and require a large number of petri dishes. A variation known as the drop-plate method (Vincent 1970) is more rapid and uses less materials. However, it is limited to population estimates in pure cultures and sterile-based inoculants. This method requires the use of pipettes which have been calibrated to deliver uniform drop volumes. Uniform drops from several tenfold serial dilutions are applied to one plate and counts are made in the drop areas which show the largest number of distinct colonies.

30.3.3 Plate-Count Colony Immunoblot

A major problem with any plate-count method is the inability to positively identify rhizobia colonies when the count is being conducted on inoculants made with nonsterile carriers or on soil. A problem with both plate-count and MPN procedures is an inability to identify rhizobia at the subspecies or strain level. Identification of rhizobia at the strain level has been accomplished by the enzyme-linked immunosorbent assay (ELISA) (Olsen et al. 1983). This procedure involves the individual picking of colonies from plate counts, followed by suspension in buffer, adjustment of cell densities, individual transfer of cell suspensions to microdilution plate wells, and manipulation of microdilution plates with specialized immunoassay equipment. The colony immunoblot method is a simplification of the ELISA procedure, whereby samples are processed by the spread-plate count method and a template of the resulting colonies is obtained by laying on and then peeling off a circular nitrocellulose membrane (Olsen and Rice 1989). The membrane is then processed with a specific primary antibody, an enzyme-conjugated second antibody, and a precipitating substrate. The result is colored spots on the membrane corresponding to specific colony-antibody interactions. The immunoblot procedure simultaneously tests all colonies on a petri dish and eliminates the requirement for specialized ELISA equipment. It should be noted that the ELISA and immunoblot ELISA are strain specific, and will not provide a total count where multiple strains are present unless antibodies specific to all strains are included in the tests.

Competitive ability is an important characteristic which should be evaluated when selecting improved strains of rhizobia for specific soil and climatic conditions in Canada. Competitive ability can be assessed in the field by inoculating with test strains and then determining the proportion of nodules containing the inoculating strain. Identification of strains within nodules can be done by several methods, including the agglutination reaction (Wollum 1987), the ELISA (Kishinevsky and Jones 1987), the immunofluorescent (Bohlool 1987), and the intrinsic antibiotic resistance (Eaglesham 1987) techniques. Researchers interested in using these techniques should consult the appropriate references.

30.4 NITROGEN FIXATION

Field testing of specific plant-rhizobia associations can be conducted to determine nitrogen-fixing capacities, to assess rhizobia strain competitiveness, to determine the need for inoculation, and to select improved strains of rhizobia or host plants. It is not within the scope of this chapter to deal with the many complexities of experimental design, site selection, site preparation, and specific methods to avoid cross contamination. Information on these topics has been summarized in recent reviews (Vincent 1970, Brockwell 1980, Wynne et al. 1987).

Measurement of the fixation of atmospheric nitrogen is a key factor in determining the actual nitrogen contribution of the legume-rhizobia association to the N cycle under various soil, climatic, cultural, and management conditions. Several methods are available for estimating nitrogen fixation by legume crops in the field, but only the total-N difference, the acetylene reduction, the ^{15}N isotope dilution, and the natural ^{15}N abundance methods will be mentioned here.

30.4.1 Total-N Difference

This method is a relatively simple procedure, commonly used when only total-N analysis is available. The amount of N fixed by a legume is estimated from the difference in N yield between the legume and a non-N-fixing (reference) plant grown in the same soil under the same conditions as the legume. The most suitable reference plant is an unnodulated plant of the legume being tested (Bremer et al. 1990). This can be achieved only when the soil in which the experiment is being conducted contains no rhizobia which form effective nodules on the legume. The use of a nonnodulating isoline of the legume provides a suitable reference plant that gives reliable results (Smith and Hume 1987), but nonlegume plants have also been used successfully (Bell and Nutman 1971, Rennie 1984). The major assumption with this method is that the legume and the reference plant assimilate the same amount of soil N. However, differences in soil N uptake because of differences in root morphologies may result in erroneous estimates of N_2 fixation.

The procedure is simple in that all that is required is the inclusion of the reference crop treatment in the design of the field experiment. Care must be used in sampling to ensure that as much of the above-ground portion of the plants as possible are harvested without contaminating the plant material with soil. The plant samples are then processed and analyzed for total N by the Kjeldahl or other suitable method. Calculate the total N yield from the percent total N in the plant material and the dry matter yield of the crop, and obtain the difference in total N yield between the legume and the reference crop, which will give the estimate of N_2 fixation.

30.4.2 Acetylene Reduction Assay

In biological systems the reduction of N_2 to NH_3 is catalyzed by the enzyme nitrogenase. The total-N difference method estimates N_2 fixation by the direct measurement of the product, NH_3, and the products of its further assimilation. Nitrogenase not only reduces N_2 to NH_3, but also catalyzes the reduction of acetylene to ethylene. The acetylene reduction assay (ARA) involves the enclosure of excised nodules, detached root systems, or whole plants in a closed container containing 10% acetylene. Following specific incubation periods during which ethylene accumulates, the ethylene content of the gas phase is determined by gas chromatography (Upchurch 1987). The ARA provides a determination of the instantaneous N_2 fixation rate under the prevailing conditions. Long-term estimates, e.g., growing season estimates, require the integration of a series of measurements to cover diurnal, daily, and seasonal variations in N_2 fixation. This method, because of its simplicity, high sensitivity, rapidity, and low cost, is used in many diverse biological systems ranging from the detection of nitrogenase activity in purified enzyme components to the estimation of N_2 fixation potential of legume-rhizobia associations under controlled environment and field conditions.

There are many variations in the ARA, depending on the N_2 fixing system being measured. The procedure described here is for assessing nitrogenase activity of field-grown legumes (Rice 1980). Recent reviews should be consulted for other variations and more specific applications of the assay (Upchurch 1987, Ledgard and Peoples 1988, Turner and Gibson 1980).

30.4.2.1 MATERIALS

1   Compressed, purified-grade acetylene (C_2H_2) in a gas cylinder fitted with regulator and flow control to allow handling of gas with gas-tight syringes.

2   1-L polypropylene mason jars, with lid fitted with serum stopper.

3   1-, 30-, and 50-mL gas-tight syringes (disposable plastic syringes are suitable for this purpose).

4   60-mL serum bottles.

5   Ethylene (C_2H_4) standards: may be either premixes of known concentrations of C_2H_4 in argon or nitrogen, or samples made from injections of pure C_2H_4 into vessels of known volumes. The standards should be in the same range as the expected concentrations in the assay samples.

6   Gas chromatograph equipped with hydrogen flame ionization detector, helium or nitrogen carrier gas flow rate at 30 mL/min (276 kPa), and stainless steel column (1.8 m × 4 mm I.D.) packed with Poropak R (100–200 mesh) maintained at 50°C oven temperature.

30.4.2.2 PROCEDURE

1   Excavate either a given number of plants, or the plants from a given area or length of row. The type and size of plant and root system should be considered in selecting the approach to obtaining a representative sample of N_2-fixing nodules.

2   Cut the tops from the roots, carefully shake the soil from the roots, and place roots in 1-L mason jars. Close lid tightly to ensure gas-tight fit.

3   With 50-mL syringe, remove 100 mL air and replace with 100 mL of C_2H_2.

4   Place mason jar in the hole from which plants were excavated, cover with soil, and incubate at soil temperature for 30 min.

5   At the end of the incubation period, remove jar from soil, insert needle of 30-mL syringe through serum stopper in lid, pump syringe several times to thoroughly mix gas phase, withdraw a 30-mL gas sample, and transfer to evacuated 60-mL serum bottle. Samples should be analyzed for C_2H_4 within 72 h.

6   Just before analysis insert a syringe needle through the stopper of the 60-mL serum bottle to equilibrate contents to atmospheric pressure. Using a 1-mL syringe withdraw 0.5 mL of gas from the serum bottle and inject into gas chromatograph.

7   Check the linearity of response to a range of C_2H_4 concentrations before using the gas chromatograph for assays, and subsequently run a C_2H_4 standard each day to confirm stability.

30.4.2.3 Calculations

Measurement of peak height provides a satisfactory estimate of the content of C_2H_4, but the use of an integrator reduces the time and effort involved in obtaining the results. Programmable integrators that will give the final results in nmol C_2H_4 per plant per second, using the results of the analysis of standards and corrected for controls, are particularly time saving. Calculations will vary, depending on the approach taken to the analysis. Examples have been given by Turner and Gibson (1980).

30.4.2.4 COMMENTS

1   The ARA has been used to estimate rates of N_2 fixation, but caution must be exercised in converting between C_2H_2 reduction and N_2 fixation. In theory, the C_2H_2 to N_2 ratio is 3:1, but in practice, the ratio has been found to vary from 1.5:1 to 25:1 (Hardy et al. 1973). If this calculation is made, the conversion ratio used should be given and the limitations considered in the interpretation of the data.

2   In field studies, the ARA technique is best used for determining the comparative nitrogenase activity of various treatments without attempting to use the data to establish absolute N_2 fixation rates.

30.4.3 ^{15}N Isotope Dilution

The nitrogen isotope ^{15}N occurs in atmospheric N_2 at a constant abundance of 0.3663 atom percent. The proportions of legume plant N derived from N_2 and from the soil N can be determined, if the isotopic abundances of the atmospheric N_2 and the soil N are sufficiently different. This difference can be achieved by incorporation into the soil of small amounts of fertilizer-N enriched in ^{15}N. The procedure for determining N_2 fixation by annual legume crops has been published (Rennie 1982, Bremer et al. 1988, Kucey 1989).

There are many possible variations in field techniques and methods for determining total N and ^{15}N atom percent. More detailed information is available in recent reviews (Ledgard and Peoples 1988, Focht and Poth 1987).

30.4.4 Natural ^{15}N Abundance

Most soils are naturally enriched in ^{15}N relative to atmospheric N ($\delta^{15}N$) due to discrimination between ^{14}N and ^{15}N during biological, chemical, and physical processes (Focht and Poth 1987, Ledgard and Peoples 1988). The measurement of $\delta^{15}N$ requires the use of highly stable and specialized dual collector mass spectrometers, which require great skill and care to operate. The use of the $\delta^{15}N$ method has been questioned because of the variability of soil $\delta^{15}N$ and small differences between $\delta^{15}N$ of soil and atmospheric N. Bremer and Van Kessel (1990) compared the $\delta^{15}N$ and ^{15}N isotope dilution methods for estimating N_2 fixation in field-grown lentil (*Lens culinaris*). They concluded that both methods provided similar estimates of N_2 fixation. This reference also provides a useful discussion on the limitations and assumptions of the $\delta^{15}N$ method.

REFERENCES

Anon. 1975. Fertilizers Act and Fertilizers Regulations, Office Consolidation. C.R.C., c. 666 amended by P.C. 1979–1271. Government of Canada, Ottawa.

Bell, F. and Nutman, P. S. 1971. Experiments on nitrogen fixation by nodulated lucerne. Plant Soil Special Vol.: 231–264.

Bohlool, B. B. 1987. Fluorescence methods for study of *Rhizobium* in culture and in situ. Pages 127–147 in G. H. Elkan, Ed. Symbiotic nitrogen fixation technology. Marcel Dekker, New York.

Bremer, E., Rennie, R. J., and Rennie, D. A. 1988. Dinitrogen fixation of lentil, field pea and fababean under dryland conditions. Can. J. Soil Sci. 68: 553–562.

Bremer, E. and Van Kessel, C. 1990. Appraisal of the nitrogen-15 natural-abundance method for quantifying dinitrogen fixation. Soil Sci. Soc. Am. J. 54: 404–411.

Bremer, E., Van Kessel, C., Nelson, L., Rennie, R. J., and Rennie, D. A. 1990. Selection of *Rhizobium leguminosarum* strains for lentil (*Lens culinaris*) under growth room and field conditions. Plant Soil 121: 47–56.

Brockwell, J. 1980. Experiments with crop and pasture legumes — principles and practice. Pages 417–488 in F. J. Bergersen, Ed. Methods for evaluating biological nitrogen fixation. John Wiley & Sons, New York.

Date, R. A. and Halliday, J. 1987. Collection, isolation, cultivation, and maintenance of rhizobia. Pages 1–27 in G. H. Elkan, Ed., Symbiotic nitrogen fixation technology. Marcel Dekker, New York.

Eaglesham, A. R. J. 1987. The use of antibiotic resistance for *Rhizobium* study. Pages 185–204 in G. H. Elkan, Ed. Symbiotic nitrogen fixation technology. Marcel Dekker, New York.

Focht, D. D. and Poth, M. 1987. Measurement of biological nitrogen fixation by ^{15}N techniques. Pages 257–288 in G. H. Elkan, Ed. Symbiotic nitrogen fixation technology. Marcel Dekker, New York.

Gibson, A. H. 1980. Methods for legumes in glasshouses and controlled environment cabinets. Pages 139–184 in F. J. Bergersen, Ed. Methods for evaluating biological nitrogen fixation. John Wiley & Sons, New York.

Hardy, R. W. F., Burns, R. C., and Holsten, R. D. 1973. Applications of the acetylene-ethylene assay for the measurement of nitrogen fixation. Soil Biol. Biochem. 5: 47–81.

Jordan, D. C. 1984. III. Rhizobiaceae Conn 1938.321. Page 234 in N. R. Krieg and J. G. Holt, Eds. Bergey's manual of systematic bacteriology. 9th ed. Williams & Wilkins, Baltimore.

Kishinevsky, B. D. and Jones, D. G. 1987. Enzyme-linked immunosorbent assay (ELISA) for the detection and identification of *Rhizobium* strains. Pages 157–184 in G. H. Elkan, Ed. Symbiotic nitrogen fixation technology. Marcel Dekker, New York.

Kucey, R. M. N. 1989. Contribution of N_2 fixation to field bean and pea N uptake over the growing season under field conditions in southern Alberta. Can. J. Soil Sci. 69: 695–699.

Ledgard, S. F. and Peoples, M. B. 1988. Measurement of nitrogen fixation in the field. Pages 351–367 in J. R. Wilson, Ed. Advances in nitrogen cycling in agricultural ecosystems. C. A. B. International, Wallingford.

Olsen, P. E., Rice, W. A., Stemke, G. W., and Page, W. J. 1983. Strain-specific serological techniques for identification of *Rhizobium meliloti* in commercial alfalfa inoculants. Can. J. Microbiol. 29: 225–230.

Olsen, P. E. and Rice, W. A. 1989. *Rhizobium* strain identification and quantification in com-

mercial inoculants by immunoblot analysis. Appl. Environ. Microbiol. 55: 520–522.

Rennie, R. J. 1982. Quantifying dinitrogen (N_2) fixation in soybeans by ^{15}N isotopic dilution: the question of the nonfixing control plant. Can. J. Bot. 60: 856–861.

Rennie, R. J. 1984. Comparison of N balance and ^{15}N isotope dilution to quantify N_2 fixation in field grown legumes. Agron. J. 76: 785–790.

Rice, W. A. 1980. Seasonal patterns of nitrogen fixation and dry matter production by clovers grown in the Peace River region. Can. J. Plant Sci. 60: 847–858.

Smith, D. L. and Hume, D. J. 1987. Comparison of assay methods for N_2 fixation utilizing white bean and soybean. Can. J. Plant Sci. 67: 11–19.

Somasegaran, P. and Hoben, H. J. 1985. Methods in legume-*Rhizobium* technology. NifTAL, University of Hawaii, Honolulu.

Turner, G. L. and Gibson, A. H. 1980. Measurement of nitrogen fixation by indirect means. Pages 111–138 in F. J. Bergersen, Ed. Methods for evaluating biological nitrogen fixation. John Wiley & Sons, New York.

Upchurch, R. G. 1987. Estimation of bacterial nitrogen fixation using acetylene reduction. Pages 289–320 in G. H. Elkan, Ed. Symbiotic nitrogen fixation technology. Marcel Dekker, New York.

Vincent, J. M. 1970. A manual for the practical study of root-nodule bacteria. Blackwell Scientific, Oxford.

Wollum, A. G., II. 1987. Serological techniques for *Bradyrhizobium* and *Rhizobium* identification. Pages 149–155 in G. H. Elkan, Ed. Symbiotic nitrogen fixation technology. Marcel Dekker, New York.

Woomer, P., Bennett, J., and Yost, R. 1990. Overcoming the inflexibility of most probable number procedures. Agron. J. 82: 349–353.

Wynne, J. C., Bliss, F. A., and Rosas, J. C. 1987. Principles and practice of field designs to evaluate symbiotic nitrogen fixation. Pages 371–389 in G. H. Elkan, Ed. Symbiotic nitrogen fixation technology. Marcel Dekker, New York.

Chapter 31
Microarthropods in Soil and Litter

J. P. Winter and R. P. Voroney

University of Guelph

Guelph, Ontario, Canada

31.1 INTRODUCTION

In most soils, about 90% of the microarthropod population is composed of Collembola (springtails) and Acarina (mites), while the remainder includes Protura, Diplura, Pauropoda, and Symphyla (Wallwork 1976). While an understanding of the importance of these organisms to soil ecology is still in its infancy, microarthropods may play a significant role in accelerating plant residue decomposition through their interactions with the microflora (Seastedt 1984, Norton 1985a, Moore and Walter 1988, Verhoef and Brussaard 1990). The flow of energy and nutrients through the soil may be accelerated by microarthropods grazing on microflora, causing increased rates of microbial biomass turnover. However, their effect on decomposition through fragmentation and comminution may not be so important, since they ingest only approximately 2% of the annual plant residue production (Norton 1985a). Collembola and the acarina suborders Cryptostigmata and Astigmata are predominantly saprophytic and mycophytic. The acarina suborders Prostigmata and Mesostigmata are mainly predatory, but some Prostigmata and Mesostigmata (Uropodina) are mycophytic (Wallwork 1976, Norton 1985a).

Within a climatic region, the main factors determining the abundance of soil microarthropods include: (1) the type and quantity of decomposing organic residues and their effects on the microfloral population, (2) the structural stability of the soil and resulting porosity, and (3) the soil water regime (Wallwork 1976). They are especially abundant in the litter of boreal forest floors (e.g., 300,000 individuals/m^2), and are much less numerous in cultivated soils (50,000 individuals/m^2).

The major development of methods to extract microarthropods from plant litter and soil occurred in the first half of the 20th century, and have been reviewed by Evans et al. (1961), Macfadyen (1962), Murphy, (1962a, b), Edwards and Fletcher (1971), and Edwards (1991). Essentially the extraction methods for microarthropods fall into two categories: (1) dynamic (or active), requires the participation of the animals to move through the sample medium, away from repellant stimuli toward attractant stimuli. The typical procedure is to place a sample of soil or litter upon a sieve, and warm and dry the top of the sample, while maintaining the bottom cool and moist. As the sample dries from top down, fauna move downward to escape desiccation, and finally drop from the lower surface through the screen, to be collected in a container below. Fauna will fall from the bottom of the sample throughout the duration of extraction, but the greatest exodus occurs as the lower surface of the sample dries to -1500 kPa and the temperature rises to above 30°C (Takeda 1979, Petersen 1978). Fauna fall into a collecting jar containing a preservative solution producing neither noxious nor repellant vapors. (2) Mechanical separation uses the physical and chemical properties of the animals, such as body size, density, and hydrophobicity to mechanically extract them. In the common mechanical procedures soil is suspended by stirring in a saline solution or an oil-water mixture. When stirring is stopped, microarthropods float to the liquid surface. In the case of oil-water suspension, when stirring stops, oil floats to the surface and carries within it lipophilic microarthropods, but less plant residue having hydrophilic surfaces.

Two methods for extracting soil microarthropods will be described in this chapter: dynamic extraction, using a high- (temperature and moisture) gradient extractor, and mechanical separation in a water-heptane mixture. The high-gradient extractor has the following advantages and disadvantages when compared to heptane extraction.

1   Advantages:

   a. Less costly to operate and less labor intensive.

   b. Can extract a greater mass of substrate.

   c. Animals are extracted in better condition for identification. Mechanical methods damage many specimens.

   d. More effective in separating microarthropods from plant litter.

2   Disadvantages:

   a. Different species may not respond equally to the applied stimuli. For example, light may be an attractant for some, a repellant for others. In response to desiccation, some species may become quiescent, or attach themselves to large fauna, expecting to be carried to a more favorable environment.

   b. The extractability of different taxa may vary with the time of year (Leinaas 1978, Takeda 1979), weather events, and climate. High-gradient extractors may be less effective for hot, arid climates (Walter et al. 1987).

   c. The extractability of different taxa may be affected by changes in population age structure (Tamura 1976, Takeda 1979). Small-sized juveniles are usually less easily extracted than more mobile adults.

   d. The extractability of different taxa may vary with soil depth and between different soils. Factors such as soil texture and low organic matter, which can decrease soil porosity or continuity of pores, can affect extraction efficiency.

   e. During extraction, eggs can hatch and mortality can occur.

   f. Efficiency is affected by apparatus construction and operation.

It may be prudent to use both dynamic and mechanical methods to account for differences in extraction efficiency between species, species in different soils, and in a changing environment. A useful way of verifying the performance of a high-gradient extractor is to subject samples also to the heptane technique (Walter et al. 1987). Other methods for calibration of high-gradient extractors have been critically reviewed by Petersen (1978).

This chapter provides suggestions for soil sampling, and gives a recommended example of both the active and the mechanical extraction methods. Techniques for handling and storing microarthropods are briefly described.

31.2 SAMPLING

When using dynamic methods for extraction, it is essential that the fauna are viable and that every effort is made to facilitate their escape from the sample. Soft-bodied animals, such as Collembola (Lussenhop 1971), are very easily damaged by rough handling, and it is essential that there be no compaction of the sample if the animals are to escape successfully through the pore space and be counted.

Soil samples for a dynamic extractor may be taken by pressing a metal cylinder or "corer" into the soil. If a series of depths down a column of soil is to be sampled, a sampling tool such as illustrated in Figure 31.1 may be used. The soil is retained within metal or heat-resistant plastic rings which are 1-mm wider in internal diameter than the cutting edge of the sampling tool. Plant litter may be sampled by cutting around the exterior of a corer with a sharp knife while exerting downward pressure on the corer.

31.2.1 COMMENTS

1   The leading edge of the corer should be kept sharpened with a bevel on the lower outside edge to minimize compaction of the sample while the corer is being pressed into the soil.

2   Use a core ≥5 cm diameter to avoid sample compaction.

3   The height of the soil core has an important effect on the ability of fauna to escape. A height of 2.5–5 cm is satisfactory for most well-structured soils. Extraction of fauna from soils with low macroporosity may be best from cores 2.5 cm in height.

4   Compaction of the soil during sampling may occur even if a sampling tool is used (Figure 31.1). Subsurface soil may have to be sampled separately.

5   If using a sampling tool such as in Figure 31.1, separate the series of cylinders with fine needles. Slicing apart the cores with a knife can smear the soil and block the pores.

6   Place soil cores into plastic bags and transport in a chest preferably cooled to 5–10°C. Protect cores from vibrations during travel. Escape of fauna from the cores may be reduced if caps for the cores can be obtained.

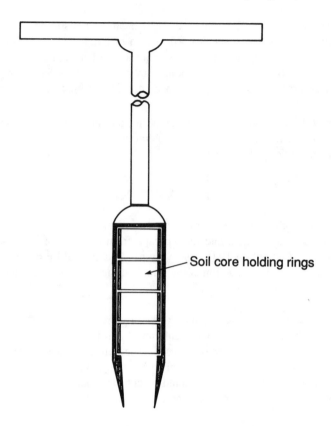

FIGURE 31.1. A coring tool for sampling depths of soil while minimizing compaction.

7   It is best to extract samples as soon after collection from the field as possible, otherwise store at 5°C. Storage may cause changes in the microarthropod population structure due to predation, breeding, molting, or mortality (Murphy 1962a).

31.3 EXTRACTION METHODS

31.3.1 High-Gradient Dynamic Extraction

31.3.1.1 *Materials and Methods*

The basic method using heat to separate microarthropods from soil was pioneered by Berlese in 1905 and Tullgren in 1918 (Murphy 1962a). Macfadyen (1955) greatly improved the method with the development of the first high-gradient extractor. The extractor described here (Figure 31.2) is for intact soil cores or loose litter samples. It is similar to that described by Norton (1985b) and was chosen for its simple construction and because it incorporates most of the refinements made to the original design by Macfadyen (1955). This extractor can operate in a refrigerator or climatically controlled cool room. It could be built to extract fauna from up to 20 samples simultaneously, but its actual dimensions will depend on the circumstances of the user.

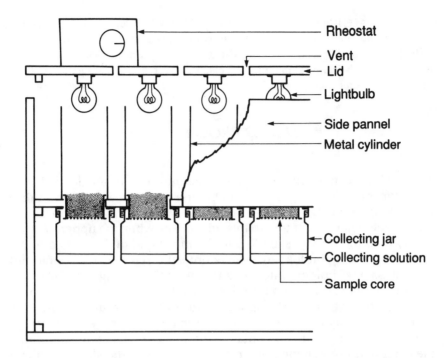

FIGURE 31.2. The front view of a high-gradient extractor for microarthropods.

The frame of the extractor consists of a varnished plywood (1.25-cm thickness) box (Figure 31.2) that is divided into upper and lower zones by a horizontal shelf. The upper zone is where heat is applied and is enclosed on all sides. Below the shelf, the lower zone is left open, front and back, so that refrigerated air can be circulated with a fan. The back panel, the side panels, and the shelf are screwed together. The front panel and the lid containing the bank of lights are removable so that samples can be loaded.

The horizontal shelf is used to hold sample cores during extraction. An array of holes are bored into the shelf through which cylindrical sample holders can fit. The holes should be 2 mm greater than the diameter of the sample holders, and 3–4 cm apart. The sample holders are prevented from dropping through these holes by rubber bands stretched over their upper ends (Figure 31.2).

Incandescent light bulbs are fitted onto the lid of the extractor to provide heat over the sample holders. These light bulbs (15 W) may be most effective if painted white. The energy output from the light bulbs is regulated by a domestic light-dimmer switch. When the extractor is in operation, heat from the light bulbs is concentrated by metal cylinders (e.g., 355-mL beverage cans, open both ends) placed on the shelf and above the sample holders. These metal cylinders may also serve to discourage cross contamination of fauna between samples in the extractor. When in operation, the bulbs are suspended at the top of the metal cylinders, but with a small space at the top of the cylinders to allow for ventilation. Holes (1-cm diam) bored in the lid between light fixtures ventilate the extractor.

During extraction, a sample is kept within its holder by fitting a cap with a 2- to 4-mm mesh screen onto the downward facing end of the holder. The mesh holes should not be too small (<2 mm) to prevent the exit of fauna, but not so large (>4 mm) as to allow an excess of soil or litter to drop into the collecting solution. Fauna escape the sample and fall into a collecting jar ($\approx$100-mL capacity) positioned below the screen. The jar should have a screw cap lid into which a hole has been cut with a diameter as large as the holes in the sample shelf. The lid of the jar is cemented onto the bottom of the shelf (Figure 31.2) so that the jar can easily be screwed into place to catch fauna. Glass jars are preferable, because they conduct heat better than plastic.

31.3.1.2 PROCEDURE

1   Soil samples contained within their sample holders are loaded into the extractor shelf. Soil samples should be oriented so that the end of the core that was uppermost in the field faces downward in the extractor (see Comments, below). An exception to this rule may be when the uppermost surface is dry; then the screen is applied to the moist end of the soil core. A plant litter sample may be loaded loosely onto the screen in a sample holder. Load all sample holders into the shelf before attaching the collecting jars so as to minimize the amount of soil or litter that could drop into the jar and make finding microarthropods in the collecting solution more difficult. A very light spray of tap water on the sample surface before loading may reduce the amount of soil contamination of the collecting solution. Keep the amount of the sample holder protruding above the shelf to a minimum so as to prevent too rapid drying of the sample and ensure the formation of a good moisture-temperature gradient.

2   When all samples are in place, position the metal cylinders over the sample holders. Close unused holes in the sample shelf with foam rubber plugs. Position the extractor lid onto the extractor.

3   Pour a collecting solution (polyethylene glycol, technical grade) into the collecting jars to a depth of 2 mm ($\approx$5 mL). Attach the canister to the sample holder, being careful not to vibrate the apparatus and cause soil or litter to fall.

4   Initiate the extraction procedure by turning on the light bulbs and setting the rheostat to give the initial temperature. The objective is to keep the temperature below the sample at 15°C, initially, while putting the top surface of the sample through a temperature regime of 20, 25, 30, 35, and 40°C for 0–2, 3–4, 5–6, 7–8, and 9–10 d, respectively. These temperatures are achieved by trial and error adjustments of the cool room or refrigerator, and the energy output of the light bulbs. Less heat energy may be required by litter samples so as not to dry them too rapidly. Soil samples with low macro-porosity, such as poorly structured heavy clay soils, may have to be dried more gradually owing to slower downward migration of fauna. Operate the extractor in a vibration-free environment.

5   At the end of the extraction run, turn off the lights, and with minimum vibration remove all the collecting jars. Pour the contents of the jars into storage vials and carefully rinse down the jars with 70% ethanol to remove any attached fauna. Remove and clean sample holders and screens.

31.3.1.3 COMMENTS

1   The extractor described above has advantages in that cross contamination of samples by fauna and drying of the sample from below are minimized. The samples experience relatively similar environmental conditions during the extraction regime.

2   In general, electric light bulbs serve as a source of drying heat, not light. Light may repel microarthropods, but it has been regarded as somewhat of a neutral stimulus (Murphy 1962a) and may in some cases be an attractant. Merchant and Crossley (1970) observed that light emitted from bulbs painted white were more efficient for extracting microarthropods than clear bulbs or those of various colors. Chemical repellents have been used, but they are not as effective as light bulbs (Murphy 1962a).

3   Inverting soil cores when placing them in a high-gradient extractor is recommended, since escape through soil pores may be facilitated. For cores taken at the soil surface, fauna are often most abundant at the top of the core, and soil voids tend to increase in size toward the top. Inverting surface cores reduces the volume of soil through which most fauna must pass.

4   Higher numbers of fauna are usually recovered from intact soil cores rather than cores that have been removed from corers and crumbled over the extractor screen (Macfadyen 1961). No conclusive explanation for this has been demonstrated, but disrupting the soil may injure the fauna or reduce the extraction efficiency by altering the rate of soil drying.

5   In earlier extractors, a collecting solution of 70% ethanol and 5% glycerol was used. While this fluid was successful for collecting fauna through a funnel into a vial, the efficiency for collecting into a jar or canister could be seriously reduced, especially for collembola (Seastedt and Crossley 1978). In the case of canisters, the alcohol was believed to act as either a repellant or an intoxicant, and many microarthropods died in the sample. For this reason, the nonvolatile polyethylene glycol collecting solution is recommended. Some workers have successfully used a solution of 50% picric acid instead of alcohol, but this may stain specimens yellow. At low concentrations, ethanol may act as an attractant, but there are no published data to suggest at what concentration it should be used. An extractor design, similar to that in Figure 31.2, was recommended by Crossley and Blair (1991), but theirs was open below the sample, minimizing effects of collecting solution fumes.

6   Little information exists on the effects of sample water content on the rate and efficiency of fauna extraction. Greater uniformity in moisture content across soil cores may produce more uniform extraction efficiency. However, finding a method to rapidly and uniformly wet all cores without affecting fauna is difficult. Most researchers do not adjust cores for moisture content. Field moisture content at the time of sampling should be measured to explain vertical migrations of microarthrods in response to rainfall.

7   Contamination of fauna in the collection fluid with soil particles can greatly increase the difficulty of counting specimens. Physical barriers have been described that prevent plant or soil from falling (Murphy 1962a, Woolley 1982), but unfortunately, they usually decrease the extraction efficiency. If necessary, microarthropods in a dirty collecting solution may be floated off using a concentrated salt solution (see Section 31.3.2).

8   The efficiency of a high-gradient extractor is affected by its construction, operation, and the nature of the samples (see Section 31.1). Many observations of their efficiency report values greater than 75% for acarina and collembola (Block 1966, Lussenhop 1971, Marshall 1972, Petersen 1978). However, when extracting microarthropods from semiarid soils, Walter et al. (1987) found efficiencies ranging from 26 to 66%. Takeda (1979) found the extraction efficiency for collembola to vary from 28% in the summer to 88% in the winter. Therefore, precise ecological research requires frequent calibration of the high-gradient extractor. For this purpose, one may use the heptane extraction method described below.

31.3.2 Mechanical Extraction: Heptane Flotation

31.3.2.1 *Introduction*

The simplest flotation method uses a saturated solution of sodium chloride or magnesium sulfate. The soil sample is disrupted and stirred in the solution (specific gravities of ≈ 1.2 Mg m^{-3}), and the less dense fauna (1.0–1.1 Mg m^{-3}) float to the surface. After leaving the solution to stand for 15 min, the fauna are decanted off the surface of the solution and stored in 70% ethanol solution. This technique has proven successful for Acari and Collembola, but not for large Cryptostigmatic mites, especially those with adhering soil, and some larger insects (Murphy 1962b). This technique is unsuitable for samples containing large amounts of organic matter and it is not as efficient as using a high-gradient extractor (Edwards and Fletcher 1971).

Better extraction of fauna may be achieved by stirring the sample in a mixture of oil and water. When left undisturbed, oil droplets rise to the surface. Microarthropods enter the oil phase, since their cuticles are lipophilic. The surface of plant residues tend to be hydrophilic and are not coated in oil. Both plant residues and microarthropods float to the surface, leaving the mineral components to settle. Plant residues reside at the top of the aqueous phase, while microarthropods reside in the oil above, at the oil-water interface. The oil phase containing microarthropods and a little of the aqueous phase are decanted off. Plant fragments can be found in the oil if small droplets of oil adhere to their surfaces or if they contain substantial amounts of entrapped air. Adding the sample to water and subjecting to a partial vacuum before stirring with oil can help to force out entrapped air.

A method of oil-water extraction of soil microarthropods is described below. It is an adaptation by Walter et al. (1987) of the standard flotation procedure of the Association of Official Analytical Chemists (Williams 1984). Recently, simplifications to this procedure have been proposed (Geurs et al. 1991), and it has been adapted for processing large (18-L) volumes of soil (Kethley 1991).

31.3.2.2 MATERIALS AND REAGENTS

1   A magnetic stirrer and a 50-mm Teflon®-coated stirring bar, or a variable-speed rotary shaker.

2   A stainless steel sieve, finer than 50 μm.

3   A Wildman trap flask (Figure 31.3), consisting of a 1- to 2-L-wide mouthed Erlenmeyer flask into which is inserted a close-fitting neoprene stopper supported on a metal rod 5 mm in diameter and about 10 cm longer than the height of the flask. The rod is threaded at the lower end and furnished with a nut and washers to hold the stopper in place. A commercially available wafer on the end of a metal rod is available from Entomological Supply Co., Inc., 2411 S. Harbor City Blvd., Melbourne, FL 32901.

4   95–100% ethanol, not denatured. Dilutions are made with distilled water on a percent volume basis.

5   *n*-Heptane containing <8% toluene.

31.3.2.3 PROCEDURE (WALTER ET AL. 1987)

1   Stage I is the immersion of the soil sample in ethanol to kill the microarthropods and fix the populations at the time of sampling. The entire soil core (about 100 g dry mass) is placed in a 2-L beaker and 500 mL of 95% ethanol is added. Parafilm is placed over the beaker to prevent evaporation. Samples may be stored in this condition for at least a few weeks with no obvious decrease in extraction efficiency. Samples are allowed to sit for at least 24 h before flotation to help free microarthropods from soil aggregates.

2   Stage II involves the clearing of organic material from the sample, the establishment of an aqueous phase, and the division of the soil sample into volumes of material that may be efficiently extracted in 2-L Erlenmeyer flasks. The mixture is placed in a vacuum chamber to eliminate air from the plant debris (time required, 2–10 min, varies with the amount of plant material). The bulk of the alcohol is decanted through a fine-mesh, stainless steel sieve (finer than 50 μm). Care must be taken to rinse oils and floating surface debris from the beaker. The material in the sieve is briefly rinsed sequentially with 95% (or 100%) ethanol, 70% ethanol, and finally with distilled water, before backrinsing the contents of the sieve into the beaker with distilled water. Distilled (or degassed) water is added to the beaker to bring the total volume to approximately 800 mL. The sample is allowed to stand for 3–5 min, then the surface debris and most of the water are decanted through the sieve, before rinsing the sieve with distilled water and finally back rinsing the sieve with distilled water into the beaker. The soil sample, now in an aqueous phase, is divided between two wide-mouth 2-L Erlenmeyer flasks (if smaller soil cores are used, a single flask is adequate), and each is placed

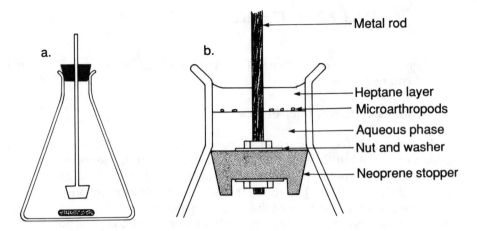

FIGURE 31.3. Equipment for heptane flotation of microarthropods.

on a magnetic stirring plate and provided with a magnetic stirring bar and a metal rod with a neoprene stopper or wafer at one end (Figure 31.3).

3   Stage III is the movement of the microarthropods from the soil sample (aqueous phase) into the heptane layer, and the isolation of the heptane layer from the aqueous phase. The neoprene stopper is placed below the water surface and approximately 25 mL of heptane is slowly poured down the rod. A clamp is used to hold the neoprene stopper above the level of the magnet, but below the water-heptane interface. A notched square of parafilm placed over the flask mouth reduces evaporation of the heptane. Alternatively, the rod can be slid through a tight-fitting hole in a neoprene stopper which fits the mouth of the flask. In this way, the flask can be sealed and the neoprene stopper held off the flask bottom (Figure 31.3). The mixture is stirred at the lowest speed that effectively suspends the sample heptane for 10–15 min (to ensure that the microarthropods in the sample have time to be trapped by the heptane layer). The mixture can also be stirred on a variable-speed rotary shaker at the lowest speed necessary. The sample is then allowed to stand for about 5 min to allow soil particles to settle. Distilled water is slowly poured along the rod to raise the heptane layer into the neck of the flask, and then allowed to stand for 2–3 min. The neoprene stopper is gently rotated to dislodge soil debris, and then slowly raised into the neck of the flask, isolating the heptane layer and a small volume of water (Figure 31.3) from the bulk of the mixture volume. The trapped volume is then rinsed into the sieve with 100% ethanol (Figure 31.3) to cut the heptane. Material in the sieve is further rinsed with 95% ethanol and finally back rinsed with 95% ethanol before rinsing into a suitable container with 70% ethanol.

31.3.2.4 COMMENTS

1   It is often useful to repeat the flotation process when microarthropods are abundant. The neoprene stopper is gently lowered to the bottom of the flask. The bulk of the aqueous phase is decanted through the sieve. The

walls of the flask are rinsed with 100% ethanol to dissolve the film of heptane, and approximately 100 mL of distilled water is added to the flask. This new volume of water is decanted through the sieve. Material in the sieve is rinsed with 100% and then 95% ethanol before a final back rinsing with distilled water into the flask. Distilled water is added to the flask to bring the volume to approximately 400 mL and the flotation procedure is repeated as above.

2   Not all mechanical methods are more efficient than using a high-gradient extractor (Edwards and Fletcher 1971). However, Walter et al. (1987) found the heptane flotation method extracted four times as many individuals and twice as many species as a high-gradient extractor. Microarthropods from laboratory cultures were added to dry sieved soil, then using a single flotation cycle the extraction efficiency of heptane flotation was estimated: for the Acarina suborders: Prostigmata (39%), Mesostigmata (69%), Cryptostigmata (84%), and Astigmata (95%); and for Collembola (89%). A second flotation cycle increased overall extraction efficiency from 78 to 88%.

3   A difficulty that can arise with mechanical methods is that both living and dead specimens are extracted. In the case of some hard-bodied mites, the presence of anal or genital shields and the absence of fungal mycelia within a body shell may be helpful in identifying mites that were viable at the time of sample collection. With other categories of fauna this may not be possible, and an overestimation of the active animal population can result.

31.4 HANDLING AND IDENTIFICATION

31.4.1 Introduction

It is not extremely difficult to identify soil microarthropods at a low level of taxonomy, such as the Acarina suborders or Collembolan families. A good general key and a few days' consultation with a specialist in soil fauna may be all that is required to acquaint a novice with the basic level of classification. On the other hand, detailed classification to the genus and species level requires a high level of skill. Taking a course in classification (short, intensive courses of Acarology: International Course in Acarology, University of Reading, Reading, England; Ohio State University Summer Institute of Acarology, Columbus, OH) is recommended. A basic key to soil fauna has recently been published by Dindal (1990). Dindal (1990) and Beham-Pelletier et al. (1985) supply major bibliographies on the biology and taxonomy of soil invertebrates.

For basic information on handling and storing specimens and for making slide mounts for examination under a light microscope, the reader is referred to Evans et al. (1961, 1985), Krantz (1978), and Woolley (1982). Preparation of specimens by serial sectioning and for electron microscopy is discussed by Krantz (1978). Methodology for laboratory culturing of microarthropods is given by Evans et al. (1961) and Krantz (1978).

31.4.2 Storage Solutions

Microarthropods collected in polyethylene glycol can be stored in that solution for a few months; however, it has not been proven for prolonged storage. A 70% ethanol solution is commonly used to preserve a collection of fauna. To guard personal health, the ethanol should be free of denaturing additives. Adding 5% glycerol to the alcohol is recommended to prevent desiccation of the specimens should the alcohol evaporate. Concentrations of ethanol greater than 80% are not recommended, since shrinkage of specimens can occur, and they may become brittle. Some fauna, such as Collembola, may have a waxy cuticle and be difficult to wet with storage solution. Collembola can be dewaxed by heating in 70% alcohol at 60°C for 1 h. This should cause them to be thoroughly wetted by the preservative solution and sink.

Specimens can be stored in the ethanol solution in small (2-mL) vials. Enclosing small vials in larger vials of ethanol solution is recommended when archiving a collection. If specimens are to be sent to a taxonomist for identification, they should be kept in ethanol solution rather than mounted on slides.

31.4.3 Preliminary Sorting and Clearing

Handling of microarthropods is carried out in liquid, usually the collecting or storage solutions contained in a Syracuse watch glass or perspex dish, or in lactic acid on a cavity slide. The animals can be manipulated by an assortment of wire loops, minute spatulas, or Pasteur pipettes. Sorting at low levels of taxonomic resolution can be done at $>60\times$ magnification under a dissecting microscope, with occasional reference to higher magnification under the stereomicroscope using bright field illumination. For higher magnification, specimens can be temporarily mounted in lactic acid on cavity microscope slides.

31.5 BRIEF STATISTICAL CONSIDERATIONS

Microarthropods frequently exhibit a "nugget" distribution with high numbers associated with fragments of plant residue, and low numbers a few centimeters away. When using 5-cm-diameter soil cores, coefficients of variation approaching 100% are not unusual and the frequency distribution of numbers per core is usually skewed downward. Therefore, many samples are necessary for an accurate estimate of the mean. Many samples may be necessary when comparing agricultural management practices, since these soils are often not as radically different in overall numbers of fauna as are soils from more diverse ecosystems. Variability may be reduced by using heptane extraction on a subsample from several well-mixed soil cores.

REFERENCES

Behan-Pelletier, V. M., Hill, S. B., Fjelberg, A., Norton, R. A., and Tomlin, A. 1985. Soil invertebrates: major reference texts. Quaest. Entomol. 21: 675–687.

Block, W. 1966. Some characteristics of the Macfadyen high gradient extractor for soil microarthropods. Oikos 17: 1–9.

Crossley, D. A., Jr. and Blair, J. M. 1991. A high-efficiency, "low-technology" Tullgren-type extractor for soil microarthropods. Agric. Ecosystems Environ. 34: 187–192.

Dindal, D. L. 1990. Soil biology guide. John Wiley & Sons, New York. 1349. pp.

Edwards, C. A. 1991. The assessment of populations of soil-inhabiting invertebrates. Agric. Ecosystems Environ. 34: 145–176.

Edwards, C. A. and Fletcher, K. E. 1971. A comparison of extraction methods for terrestrial arthropods. Pages 150–185 in J. Phillipson, Ed. Methods of study in quantitative soil ecology: population, production, and energy flow. IPB Handbook No. 18. Blackwood Scientific, London.

Evans, G. O., Sheals, J. G., and Macfarlane, D. 1961. The Terrestrial Acari of the British Isles. Vol. 1. Introduction and biology. British Museum (Natural History), London. 219 pp.

Evans, G. O., Griffiths, D. A., Macfarlane, D., Murphy, P. W., and Till, W. M. 1985. The Acari: a practical manual. Vol. 1. Morphology and classification. University of Notingham School of Agriculture, Sutton Bonington, Loughborough, Leicestershire, England.

Geurs, M., Bongers, J., and Brussaard, L. 1991. Improvements to the heptane flotation method for collecting microarthropods from silt loam soil. Agric. Ecosystems Environ. 34: 213–221.

Kethley, J. 1991. A procedure for extraction of microarthropods from bulk soil samples with emphasis on inactive stages. Agric. Ecosystems Environ. 34: 193–200.

Krantz, G. W. 1978. A manual of acarology. Oregon State University Press, Corvallis, 509 pp.

Leinaas, H. P. 1978. Seasonal variation in sampling efficiency of Collembola and Protura. Oikos 31: 307–312.

Lussenhop, J. 1971. A simplified canister-type soil arthropod extractor. Pedobiologia 11: 40–45.

Macfadyen, A. 1955. A Comparison of methods for extracting soil arthropods. Pages 315–332 in Kevan, D. K. McE. Ed. Soil zoology. Butterworths, London.

Macfadyen, A. 1961. Improved funnel-type extractors for soil arthropods. J. Anim. Ecol. 30: 171–184.

Macfadyen, A. 1962. Soil arthropod sampling. Adv. Ecol. Res. Pages 1–34. Academic Press, New York.

Marshall, V. G. 1972. Comparison of two methods of estimating efficiency of funnel extractors for soil microarthropods. Soil. Biol. Biochem. 4: 417–426.

Merchant, V. A. and Crossley, D. A., Jr. 1970. An inexpensive high-efficiency Tullgren extractor for soil microarthropods. J. Georgia Entomol. Soc. 5: 83–87.

Moore, J. C. and Walter, D. E. 1988. Arthropod regulation of micro- and mesobiota in below-ground detrital food webs. Annu. Rev. Entomol. 33: 419–439.

Murphy, P. W. 1962a. Extraction methods for soil animals. I. Dynamic methods with particular reference to funnel processes. Pages 75–114 in P. W. Murphy, Ed. Progress in soil zoology. Butterworths, London.

Murphy, P. W. 1962b. Extraction methods for soil animals. II. Mechanical methods. Pages 115–155 in P. W. Murphy, Ed. Progress in soil zoology. Butterworths, London.

Norton, R. A. 1985a. Aspects of the biology of soil arachnids, particularly saprophagous and mycophagous mites. Quaest. Entomol. 21: 523–541.

Norton, R. A. 1985b. A variation of the Merchant-Crossley soil microarthropod extractor. Quaest. Entomol. 21: 669–671.

Petersen, H. 1978. Some properties of two high gradient extractors for social microarthropods and an attempt to evaluate their extraction efficiency. Nat. Jutl. 20: 95–122.

Seastedt, T. R. 1984. The role of microarthropods in decomposition and mineralization processes. Annu. Rev. Entomol. 29: 25–46.

Seastedt, T. R. and Crossley, D. A., Jr. 1978. Further investigations of microarthropod populations using the Merchant-Crossley high-gradient extractor. J. Georgia Entomol. Soc. 13: 333–338.

Takeda, H. 1979. On the extraction process and efficiency of MacFadyen's high-gradient extractor. Pedobiologia 19: 106–112.

Tamura, H. 1976. Biases in extracting Collembola through Tullgren funnel. Rev. Ecol. Biol. Sol. 13: 21–34.

Verhoef, H. A. and Brussaard, L. 1990. Decomposition and nitrogen mineralization in natural and agro-ecosystems: the contribution of soil animals. Biogeochem. 11: 175–211.

Walter, D. E., Ketheley, J., and Moore, J. C. 1987. A heptane flotation method for recovering microarthropods from semiarid soils, with comparison to the Merchant-Crossley high-gradient extraction method and estimates of microarthropod biomass. Pedobiologia 30: 221–232.

Wallwork, J. A. 1976. The distribution and diversity of soil fauna. Academic Press, London, 335 pp.

Williams, S., Ed. 1984, Pages 887–890 in Official methods of analysis of the Association of Official Analytical Chemists, XIV ed.

Woolley, T. A. 1982. Mites and other soil microarthropods. Pages 1131–1142 in Methods of soil analysis. Part 2. Chemical and microbiological properties. American Society of Agronomy, Madison, WI.

Chapter 32
Nematodes

J. Kimpinski

Agriculture Canada
Charlottetown, Prince Edward Island, Canada

32.1 INTRODUCTION

Every laboratory develops its own modifications of existing techniques for extracting nematodes from soil. The following are brief descriptions of three common methods. The extraction of nematodes from plant tissue is not included, but this topic, as well as methods for fixing, staining, preserving, and culturing nematodes can be found in several excellent laboratory and training manuals (Ayoub 1980, Zuckerman et al. 1985, Southey 1986). Theoretical and practical aspects of sampling for nematodes can be found in Barker and Campbell (1981).

32.2 BAERMANN FUNNEL TECHNIQUE (BAERMANN 1917)

The Baermann funnel was a technique used by parasitologists to recover nematodes from feces of animals, and was adapted and modified by plant nematologists to recover nematodes from soil or plant tissue (Hooper 1986). The basic method is to wrap the sample in cloth or paper tissue and partly submerge it in water in a funnel. Nematodes move out of the saturated material into the water and settle to the bottom in the neck of the funnel. A short rubber tube fitted with a clamp is attached to the funnel neck. After a while, the clamp is loosened to release a few milliliters of water containing the nematodes into a counting dish for examination under a microscope. The equipment and procedures outlined below are similar to Townshend's (1963) method.

32.2.1 MATERIALS AND REAGENTS

1. Teflon®-coated aluminum cake pans, approximately 20 cm in diameter and 5 cm deep.

2. Saran screens (18 cm in diameter) glued with neoprene cement to acrylic rings (15 mm high and 4 mm thick). Three acrylic legs (30 mm long and 18 mm high) are cemented to the outside of each ring, and a small piece of acrylic is glued to the center of each screen to keep it about 4 mm above

the bottom of the cake pan. Screens constructed of metal gauges should not be used, since minute amounts of metallic ions can be toxic to nematodes (Pitcher and Flegg 1968).

3   Sieve (20 cm diameter) with 2-mm openings (10 mesh).

4   3-ply "Man-size Kleenex®" paper tissues, preferably cut in circles 20 cm in diameter.

5   Large test tubes (100- to 200-mL size).

32.2.2 PROCEDURE

1   Mix soil thoroughly, but gently so as not to injure nematodes.

2   Sift 50 to 100 g of soil through 10-mesh sieve to remove coarse material.

3   Place 20-cm diameter 3-ply paper tissue on each Saran screen.

4   Spread 50 g of soil in a thin layer over the paper tissue. If tissue has not been cut into a circle, fold corners over soil sample.

5   Place one screen with soil in each cake pan. Each cake pan should contain enough water to make contact with and saturate soil sample.

6   Stack pans and cover with a plastic hood or place in a large plastic bag to reduce evaporation. Add water to pans every few days if needed.

7   Incubate samples at room temperature for several days. Barker (1985) recommends 14 d for maximum recovery of nematodes.

8   At the end of the incubation period, lift the screen and allow water to drain from the soil into the pan for 10 to 15 s. Rinse the bottom of the screen into the pan with a wash bottle.

9   Swirl water in the pan and pour the contents into a large test tube. Rinse pan with a bit more water and add to test tube.

10 Allow 1 h for nematodes to settle to the bottom of test tubes.

11 Siphon down to 10 to 20 mL and pour contents carefully into a counting dish, allow a few minutes for nematodes to settle, and examine with a stereomicroscope at 10–70×.

32.2.3 Comments

The advantages of the Baermann funnel technique in its original or modified form are that the equipment is simple and easy to set up, and the method can be used to process many samples. Most of the nematode species that are suitable for this method are recovered after 24–48 h, but larger or less active types may be missed. One disadvantage of the original Baermann funnel technique is the lack of oxygen, especially in the base of the funnel where the nematodes collect. To overcome this, Stoller (1957) attached a thin polyethylene tube and bubbled oxygen or fresh air into the water. The use of a pan or tray in place of the funnel allows oxygen to diffuse rapidly into the shallow water and thin soil layer (Whitehead and Hemming 1965).

32.3 CENTRIFUGAL-FLOTATION METHOD (CAVENESS AND JENSEN 1955, JENKINS 1964)

This is a quick and easy method for extracting nematodes from soil. Specimens can be obtained in a few minutes, and recovery of inactive species and nematode eggs is greater than from most other methods of extraction. The details outlined below are primarily from a modified version by Barker (1985).

32.3.1 MATERIAL AND REAGENTS

1 Centrifuge with swinging bucket head that will hold 50-mL or larger tubes and can operate to 420 × g.

2 Several 50-mL centrifuge tubes.

3 Mechanical stirrer or vibrator mixer.

4 Sieves (20 cm diameter) with 500-, 250-, 180-, 38-, and 26-μm openings (respectively, 35, 60, 80, 400, and 500 mesh).

5 Beakers of 150- and 1000-mL size.

6 Several large test tubes (100- to 200-mL size).

7 Sucrose solution with a specific gravity of 1.18 (454 g of sugar in water to make 1 L of solution), though a specific gravity range of 1.10 to 1.18 is satisfactory for most soil-inhabiting nematodes (Thistlethwagte and Riedel 1969).

32.3.2 PROCEDURE

1 Mix soil thoroughly, but gently so as not to injure nematodes, and remove very coarse material.

2 Place 100 cm³ of soil in a 1000-mL beaker and add water to bring the volume to 600 mL.

3 Stir for 20 to 30 s and allow soil to settle for 60 s. A settling time of 20 to 30 s may be better for inactive nematode species.

4   Decant onto a 35-mesh sieve situated over a 400-mesh sieve. Hold sieves at an angle of about 45° to reduce the chance of small nematodes passing through the 400-mesh sieve.

5   Rinse the 35-mesh sieve situated over the 400-mesh sieve with water from a wash bottle. Rinse lightly, as excessive water will carry small nematodes through the 400-mesh sieve. If recovery of cysts is an objective, add a 60- or 80-mesh sieve between the 35- and 400-mesh sieves (a cyst is the body of a dead female of the family Heteroderidae, and is a spherical, moisture-resistant structure that may contain several hundred eggs).

6   Wash the debris, soil, and nematodes from the 400-mesh sieve into a 150-mL beaker.

7   Pour material from 150-mL beaker into 50-mL centrifuge tubes.

8   Place tubes in centrifuge. Be sure to balance the tubes.

9   Centrifuge at 420 × g for 5 min.

10   Discard water from tubes (nematodes are in soil at the bottom of tubes).

11   Fill centrifuge tubes with sugar solution and mix using a vibrator mixer.

12   Centrifuge for 60 s at 420 × g. Nematodes should remain suspended in the sugar solution. Do not use the brake to stop the centrifuge, as this may dislodge the soil from the bottom of the tubes.

13   Pour sucrose solution from each tube into a 600-mL beaker containing about 500 mL of water to reduce the high osmotic concentration.

14   Pour contents of 600-mL beakers slowly onto a 500-mesh sieve held at a 45° angle.

15   Gently rinse the nematodes from the 500-mesh sieve into a 150-mL beaker.

16   Pour nematode suspension into a large (100- to 200-mL) test tube.

17   Allow 1 to 2 h for nematodes to settle to bottom of test tube.

18   Siphon down to 10 to 20 mL, and pour carefully into a counting dish, allow a few minutes for nematodes to settle, and examine with a stereomicroscope at 10–70×.

32.3.3 Comments

The centrifugal flotation method usually recovers a higher percentage of inactive specimens than the Baermann funnel (Kimpinski and Welch 1971, Viglierchio and Schmitt 1983). However, this technique may not be convenient for large numbers of samples, since each soil aliquot requires a fair degree of handling and manipulation. The high osmotic concentration of the sugar solution sometimes damages or distorts specimens (Viglierchio and Yamashita 1983), and the extraction efficiency of the method may be less than 50% of the nematodes in a sample (Viglierchio and Schmitt 1983).

32.4 FENWICK CAN (FENWICK 1940)

This relatively inexpensive apparatus is widely used for extracting cysts from soil samples weighing several hundred grams. The basic principle is that cysts contain air and will float to the surface. Details outlined below are taken from Ayoub (1980) and Shepherd (1986).

32.4.1 MATERIALS AND REAGENTS

1   A 30-cm-high can, usually made of brass, tapered towards the top and having a sloped base. A drain hole (2.5 cm diam) closed with a rubber stopper is located in the side of the can at the low end of the slope. Just below the top of the can is a sloping collar with a rim 6 cm high that tapers towards a 4-cm-wide outlet.

2   Large brass funnel 20.5 cm in diameter, with a 20.5-cm-long stem.

3   Sieves (20-cm diam) with 1000-, 850-, and 250-µm openings (respectively, 18, 20, and 60 mesh), and one 60-mesh sieve about 8 cm in diameter.

4   Beaker, 600 mL in size.

32.4.2 PROCEDURE

1   Mix soil thoroughly, remove large stones and very coarse materials, and air dry for several hours.

2   Fit large brass funnel containing 18-mesh sieve into top of can.

3   Fit 20-mesh sieve over 60-mesh sieve and place under collar outlet of can to collect overflow of water.

4   Fill can with water and wet 20- and 60-mesh sieves.

5   Place soil sample on 18-mesh sieve and wash through into can with a jet of water. This will cause organic material, some soil, and many of the cysts in the soil to overflow into the collar and pass down onto 20- and 60-mesh sieves.

6   Rinse debris on the 20-mesh sieve to facilitate movement of cysts through openings down to the 60-mesh sieve.

7   Discard coarse material on 20-mesh sieve.

8   Wash material on 60-mesh sieve into a 600-mL beaker.

9   Pour floating material from 600-mL beaker through the 8-cm-diameter, 60-mesh sieve. Rotate beaker to remove material adhering to sides and discard sediment on bottom.

10   Using a small lab scoop or spoon, transfer material from sieve onto one or more counting dishes.

11   Add enough water to counting dishes to float material and examine with a stereomicroscope.

12   Transfer cysts or suspected cysts using a dissecting needle to a sample vial containing a few milliliters of water.

32.4.3 Comments

This technique is quite efficient, and Shepherd (1986) claims that 70% of cysts in a soil sample are floated up and captured on the sieves. The equipment is inexpensive and easy to set up, and the extraction takes little time. However, special equipment may be necessary when large numbers of samples are being processed (Ayoub 1980). An alternate procedure (after step 9) is to flush the material from the 60-mesh sieve with a wash bottle onto rapid-flow filter paper in a funnel, let drain, and examine the filter paper surface for cysts (John Potter, personal communication).

REFERENCES

Ayoub, S. M. 1980. Plant nematology, an agricultural training aid. NemaAid Publications, Sacramento, CA. 195 pp.

Barker, K. R. 1985. Soil sampling methods and procedures for field diagnosis. Pages 11–20 in B. M. Zuckerman, W. F. Mai, and M. B. Harrison, Eds. Plant nematology laboratory manual. University of Massachusetts Agricultural Experiment Station, Amherst.

Barker, K. R. and Campbell, C. L. 1981. Sampling nematode populations. Pages 451–474 in B. M. Zuckerman and R. A. Rohde, Eds. Plant parasitic nematodes. Vol. III. Academic Press, New York.

Baermann, G. 1917. Eine einfache Methode zur Auffindung von *Anchylostomum* (Nematoden) Larven in Erdproben. Geneesk Tijdschr. Ned. Ind. 57: 131–137.

Caveness, F. E. and Jensen, H. J. 1955. Modification of the centrifugal-flotation technique for the isolation and concentration of nematodes and their eggs from soil and plant tissue. Proc. Helminth. Soc. Wash. 22: 87–89.

Fenwick, D. W. 1940. Methods for the recovery and counting of cysts of *Heterodera schachtii* from soil. J. Helminthol. 18: 155–172.

Hooper, D. J. 1986. Extraction of free-living stages from soil. Pages 5–30 in J. F. Southey, Ed. Laboratory methods for work with plant and soil nematodes. Ministry of Agriculture, Fisheries and Food, Reference Book 402, Her Majesty's Stationery Office, London.

Jenkins, W. R. 1964. A rapid centrifugal-flotation technique for separating nematodes from soil. Plant Dis. Rep. 48: 692.

Kimpinski, J. and Welch, H. E. 1971. Comparison of Baermann funnel and sugar flotation extraction from compacted and non-compacted soils. Nematologica 17: 319–320.

Pitcher, R. S. and Flegg, J. J. M. 1968. An improved final separation sieve for the extraction of plant-parasitic nematodes from soil debris. Nematologica 14: 123–127.

Shepherd, A. M. 1986. Extraction and estimation of cyst nematodes. Pages 31–49 in J. F. Southey, Ed. Laboratory methods for work with plant and soil nematodes. Ministry of Agriculture, Fisheries and Food, Reference Book 402, Her Majesty's Stationery Office, London.

Southey, J. F., Ed. 1986. Laboratory methods for work with plant and soil nematodes. Ministry of Agriculture, Fisheries and Food, Reference Book 402, 202 pp. Her Majesty's Stationery Office, London.

Stoller, B. B. 1957. An improved test for nematodes in soil. Plant Dis. Rep. 41: 531–532.

Thistlethwagte B. and Riedel, R. M. 1969. Expressing sucrose concentration in solutions used for extracting nematodes. J. Nematol. 1: 387–388.

Townshend, J. L. 1963. A modification and evaluation of the apparatus for the Oostenbrink direct cottonwool filter extraction method. Nematologica 9: 106–110.

Viglierchio, D. R. and Schmitt, R. V. 1983. On the methodology of nematode extraction from field samples: comparison of methods for soil extraction. J. Nematol. 15: 450–454.

Viglierchio, D. R. and Yamashita, T. T. 1983. On the methodology of nematode extraction from field samples: density flotation techniques. J. Nematol. 15: 444–449.

Whitehead, A. G. and Hemming, J. R. 1965. A comparison of some quantitative methods of extracting small vermiform nematodes from soil. Ann. Appl. Biol. 55: 25–38.

Zuckerman, B. M., Mai, W. F., and Harrison, M. B., Eds. 1985. Plant nematology laboratory manual. University of Massachusetts Agricultural Experiment Station, Amherst. 212 pp.

Chapter 33
Nitrogen Mineralization Potential in Soils

C. A. Campbell
Agriculture Canada
Swift Current, Saskatchewan, Canada

B. H. Ellert
Agriculture Canada
Ottawa, Ontario, Canada

Y. W. Jame
Agriculture Canada
Swift Current, Saskatchewan, Canada

33.1 INTRODUCTION

Nitrogen is one of the major nutrients influencing crop growth, but its misuse can lead to pollution of the environment via leaching, volatilization, and erosion. Furthermore, as a costly crop production input it has great influence on net returns to the producer. The availability of soil N to the crop is therefore of great importance in agricultural systems. Although there are several thousand kilograms N per hectare in most agricultural soils of temperate climates, only 1 to 2% of this is available to plants each year. The N released does not come uniformly from the gross organic N, which is composed of a heterogeneous pool of components of varying stability (Campbell 1978). The organic components in soil include fresh crop and animal residues, microbial biomass, microbial metabolites and cell wall constituents absorbed to colloids, and the very stable humus. Although fresh residues and microbial biomass are important as mineralizable substrates, specific compounds in the mineralizable pool have not been directly identified. Therefore, incubation-leaching techniques are used to quantify the mineralizable pool of soil organic N.

Interest in laboratory methods of testing soils for their ability to supply N to crops has existed for a long time. Field and greenhouse tests provide the most accurate means, but are more expensive and time consuming relative to laboratory methods (Allison 1973). Quick chemical tests would be ideal, but most, such as the nitrate test used on the Canadian prairies, consider neither the potential mineralization over the entire growing season, nor variation due to weather conditions which are known to influence the amount of N mineralized by soils.

Soil Sampling and Methods of Analysis, M. R. Carter, Ed.,
Canadian Society of Soil Science. © 1993 Lewis Publishers.

The concept of an organic N fraction that is readily mineralized has been used to assess soil N availability in cropland, forests, and waste-disposal sites (Campbell et al. 1984, Fyles and McGill 1987, Chae and Tabatabai 1986). In the 1970s Stanford and co-workers (Stanford and Smith 1972, Stanford et al. 1973, Stanford and Epstein 1974) advanced the concept of potentially mineralizable N, denoted as N_o, and a related mineralization rate constant (k) for use in characterizing soil-available N. Since then this concept has been used, modified, and discussed in great detail by numerous scientists (Campbell 1978, Campbell et al. 1981, 1984, 1988, Juma et al. 1984, Bonde and Rosswall 1987, Ellert and Bettany 1988, Myers et al. 1982, Olness 1984, Deans et al. 1986, Fyles and McGill 1987, Chae and Tabatabai 1986, and Paustian and Bonde 1987).

33.2 THEORY

Stanford and Smith (1972) suggested that the potentially mineralizable N (N_o) of a soil and its rate constant (k) could be estimated by incubating the soil at optimum conditions and measuring the nitrogen mineralized (N_m) and time of incubation (t). They assumed that organic nitrogen mineralization at optimum temperature and moisture followed first-order kinetics. Thus:

$$\longleftarrow \text{ Total Soil Organic N } \longrightarrow \qquad\qquad \text{Inorganic N}$$

$$\boxed{\text{Resistant Soil Organic N}} \quad \boxed{N_o} \overset{k}{\longrightarrow} \boxed{N_{min}} \qquad (33.1)$$

first-order rate equation:

$$- dN/dt = kN \qquad (33.2)$$

integrate from time 0 to t:

$$\ln N_t - \ln N_o = - kt \qquad (33.3)$$

take antilogarithms:

$$N_t = N_o e^{-kt} \qquad (33.4)$$

where N_t = mineralizable organic substrate remaining at time t. At time 0:

$$\text{potentially mineralizable N} = N_o e^{-k0} = N_o \qquad (33.5)$$

The amount of inorganic N mineralized (N_{min}) is defined as the difference between the amount of mineralizable organic N at time 0 (i.e., the "potentially mineralizable N") and at time t (i.e., N_t):

$$N_{min} = N_o - N_t \qquad (33.6)$$

substitute Equation 33.4 into Equation 33.6:

$$N_{min} = N_o - N_o e^{-kt} = N_o(1 - e^{-kt}) \qquad (33.7)$$

The amounts of N_{min} accumulated after various times (t) are measured directly in laboratory incubations, whereas statistical techniques are used to estimate N_o and k from the analytical data. Stanford and Smith (1972) originally obtained successive approximations of N_o and k by regressing $\ln(N_o - N_{min})$ on t (rearrange Equation 33.6 and substitute into Equation 33.3). This method of calculation, however, is cumbersome and inaccurate (Smith et al. 1980, Talpaz et al. 1981). Estimation of N_o and k is more accurately performed by using nonlinear regression programs, such as those available in SAS (SAS, Inc., 1985) and Biomedical Department Programs (BMDP) (Ralston 1988) software, to fit Equation 33.7.

The above calculation only holds when the mineralization process is conducted under optimum conditions for mineralization. The rate constant must be adjusted for suboptimal environmental conditions to predict the actual amount of N mineralized. For North American soils optimum temperature for mineralization is about 35°C and optimum moisture content is field capacity (-0.03 to -0.01 MPa). The k is temperature dependent (Stanford et al. 1973, Campbell et al. 1981, 1984). Stanford et al. (1973) suggested $Q_{10} = 2$, but Campbell et al. (1984) showed that this value varied with climate. Methods of adjusting N_{min} for moisture content have been suggested by Stanford and Epstein (1974) and Myers et al. (1982). Campbell et al. (1984, 1988) tested their model vs. laboratory and field data by adjusting k for both temperature and moisture content and obtained a reasonable fit to their data. However, this model proved inadequate when used to estimate N mineralization under conditions of frequent wetting and drying, such as is common to semiarid conditions.

Ellert and Bettany (1988) found that the model proposed by Stanford and Smith (1972) did not adequately describe mineralization in native and cultivated forest soils which had fluctuating mineralization rates. They found that fitting nonlinear first-order kinetic models to the N mineralized during the various time intervals (incremental data) was superior to the cumulative approach (Equation 33.7) commonly used. Incremental models reduce the interdependence among observation errors and emphasize mineralization dynamics which tend to be obscured when the data are accumulated (Ellert and Bettany 1988, Paustian and Bonde 1987).

33.3 DETERMINATION OF N_o AND k

Soil (air dried or field moist) is mixed with washed quartz sand and placed in leaching apparatus. It is leached free of mineral N and then incubated at a temperature that is optimum for N mineralization. Periodically, thereafter, the NO_3^- plus NH_4-N formed during the interval is leached, collected, and determined. The mineralized N is accumulated and, together with the time of incubation, is used with Equation 33.7 to calculate N_o and k.

33.3.1 MATERIALS AND REAGENTS

1 Incubation chamber capable of providing temperatures up to 40°C and humidities near 100% so that soils will not dry out during incubation.

2 Vacuum pump for evacuating leachate at about 60 cm of mercury.

3 Soils may be incubated in leaching tubes (assembled in the laboratory, as in Figure 33.1), commercially available filter units (e.g., 150-mL membrane

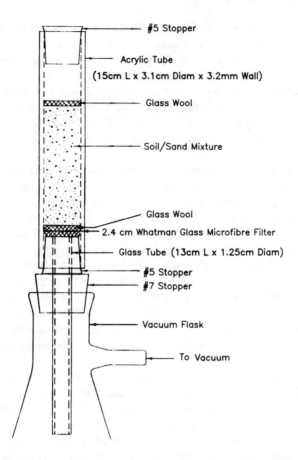

FIGURE 33.1. Diagram of home-made leaching apparatus used by the authors.

filter units, as used by MacKay and Carefoot 1981), or Buchner funnels (e.g., polypropylene funnels with detachable tops, as used by Ellert and Bettany 1988).

4   Glass wool for making pad about 6 mm thick at the bottom and 3 mm thick at the top of soil-sand mixture (Figure 33.1).

5   Acid-washed 20-mesh Ottawa sand.

6   Stock solution of $CaCl_2$ used to make up the 0.01 M solution used for each leaching operation.

7   An N-minus nutrient solution containing 0.002 M $CaSO_4$; 0.002 M $MgSO_4$; 0.005 M $Ca(H_2PO_4)_2$; and 0.0025 M K_2SO_4, for use in replacing nutrients probably lost from soil during leaching.

33.3.2 PROCEDURE

1   Soils usually are air dried and sieved prior to incubation, but the amounts of N mineralized are very sensitive to soil handling prior to incubation (see comments).

2   Mix 15 to 50 g soil with Ottawa sand at soil:sand ratios of 1:1 for medium-textured soils and 1:2 for fine-textured soils.

3   The air-dried soil and sand can be mixed (Campbell et al. 1981) or it can be moistened with a thin spray of water to give a homogenous mixture and prevent particle size segregation (Stanford and Smith 1972) during transfer to leaching tubes.

4   The soil-sand mixture is supported in the tube on a glass wool pad (Stanford and Smith 1972) or by a sandwich of glass wool/Whatman® glass microfiber filter/glass wool (Figure 33.1). The effective retention of the microfiber filter is 1.6 μm.

5   A thin glass wool pad is placed on top of the soil to avoid soil dispersion when the leaching solution is poured into the tube.

6   The mineral N initially present is leached from the system using 100 mL of 0.01 M CaCl$_2$ in small increments (10 mL at a time) followed by 25 mL of the N-minus nutrient solution. The excess solution is removed by evacuation (60 cm Hg) and discarded.

7   The tubes are then stoppered at both ends and placed in an incubator set at 35°C. The bottom stopper, a Suba Seal sleeve stopper, is pierced with a 38-mm, 16- to 18-gauge hypodermic needle to facilitate aeration. Each day the top stopper is removed for 5 min to allow aeration.

8   The leaching process is repeated every 2 weeks for 8 or 10 weeks and every 4 weeks thereafter. The leachate is collected, filtered through Whatman® No. 42 filter paper, and a blank prepared. The leachate is analyzed for NO$_3^-$ and NH$_4$ − N. (Note: some filter papers have high and variable NH$_3$ concentrations [Oh Almhain and O'Danachair 1974]).

9   When leaching, the soils are initially allowed to drain freely; vacuum is applied only to remove excess solution.

10   Normally, the incubation can be terminated when cumulative N mineralization approaches a plateau which is usually after about 20 weeks.

33.3.3 Calculations

Nonlinear regression programs in libraries of advanced statistical software (e.g., SAS or BMDP) can be used to obtain least squares estimates of N$_o$ and k. Nonlinear regression algorithms require that you specify the partial derivatives of the first-order model (Equation 33.7) with respect to N$_o$ and k:

$$\frac{\partial N_m}{\partial N_o} = 1 - e^{-kt} \tag{33.8}$$

$$\frac{\partial N_m}{\partial k} = N_o\, t e^{-kt} \tag{33.9}$$

Rough estimates of N_o and k are needed to initiate the iterative program. Estimates of initial values can be obtained using the graphical method described by Stanford and Smith (1972) or by assuming k = about 0.10 wk^{-1} (usually it is between 0.05 and 0.20) and by using a value 50% greater than accumulated mineralized N at the end of the incubation experiment. There are also derivative-free methods of analysis now accessible in BMDP and SAS software (Ralston and Jenrich 1978, Ralston 1988).

33.3.4 COMMENTS

1   Several methods of soil pretreatment and storage have been used, including field-moist soil, coarsely sieved soil (Ellert and Bettany 1988, Janzen 1987), frozen soil (Beauchamp et al. 1986), and undisturbed cores (Fyles and McGill 1987), but the most appropriate method has not been established. Soils should be refrigerated (1 to 4°C) during the interval between sampling and incubation, even if the soils are air dried immediately after sampling (Chaudhry and Cornfield 1971). To standardize the method, we recommend that soils be air dried immediately after sampling, and then refrigerated prior to incubation (storage time should be minimized). We recommend air drying, because many soils become air dry in the field (at least at the surface), and air drying lyses the soil microbial biomass, such that biomass N is included within the potentially mineralizable pool.

2   This procedure allows a capacity factor (N_o) and rate constant (k) to be estimated on the assumption that there is only one "active" fraction of soil organic matter, or that the one estimated here is the major "active" N pool. We know that there is more than one pool, and several workers have attempted to refine this calculation to define and quantify two or three pools (e.g., Paul and Juma 1981, Richter et al. 1982, Deans et al. 1986). Although the double exponential model usually fits the laboratory N mineralization data more precisely than the single exponential model, when the double exponential model was used to analyze field data its advantage over the single exponential model was insufficient to warrant its use (Campbell et al. 1988).

3   The incubation temperature has a critical influence on the estimated N_o and k. If too low a temperature is used, a lag phase in net N mineralization may be obtained during the first 2 weeks; this does not occur at higher temperatures. For example, Campbell (unpublished data) has observed this lag phase to occur at 28°C, but not at 35°C. The lag phase curve is S-shaped and does not comply with the first-order kinetic model.

4   So far, no one has established a desired minimum time of incubation; however, recently it has been shown that k values are generally lower and less variable as incubation time increases (Paustian and Bonde 1987). At the same time, it seems impractical and unnecessary to carry out such experiments for too long a period. Figure 33.2 shows a typical curve for net N mineralization. Unfortunately, in some instances, especially for forest soils, the curves tend not to follow the desired shape, and other kinetic models must be used (Ellert and Bettany 1988).

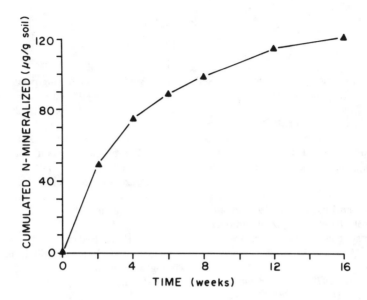

FIGURE 33.2. Typical N mineralization vs. time relationship.

5   Leaching tubes or funnels with porous disks may become clogged during incubation of fine-textured soils. Between experiments porous disks can be cleaned by soaking in dilute HCl.

6   N_o and k are inversely correlated (i.e., interdependent) when the first-order model is fitted to cumulative mineralization curves that lack distinct plateaus or horizontal asymptotes (Paustian and Bonde 1987). Campbell and co-workers (1991) reported that the product of N_o and k (N_ok), or the "initial potential rate of mineralization" is a more valid index of the availability of soil organic N than either N_o or k alone. Similarly, Dendooven and co-workers (1990) suggested that a better index of N availability would be obtained by calculating (from N_o and k) the quantity of N mineralized during 1 year at 10°C. Regardless of the index used, we encourage authors also to report the cumulative amounts of N mineralized during the incubations, because the cumulative amounts of N are definitive analytical data, whereas N_o and k are parameter estimates for a hypothetical model (first order).

7   This technique, although a most useful one, is by no means foolproof. Its most important feature is that it is the product of microbial activity. As such, then, there appears to be a need for scientists using the method to agree on methods of standardization: (a) pretreatment, (b) incubation temperature and moisture potential, (c) incubation intervals and duration, (d) soil and sand weights, (e) composition of leaching solution, (f) leaching vessels, and (g) statistical techniques. In addition to this information, published reports should include the cumulative amounts of N mineralized at the end of the incubations.

REFERENCES

Allison, F. E. 1973. Soil organic matter and its role in crop production. Developments in soil science. Vol. 3. Elsevier, Amsterdam. 637 pp.

Beauchamp, E. G., Reynolds, W. D., Brasche-Vielleneuve, D., and Kirby, K. 1986. Nitrogen mineralization kinetics with different soil pretreatments and cropping histories. Soil Sci. Soc. Am. J. 50: 1478–1483.

Bonde, T. A. and Rosswall, T. 1987. Seasonal variation of potentially mineralizable nitrogen in four cropping systems. Soil Sci. Soc. Am. J. 51: 1508–1514.

Campbell, C. A. 1978. Soil organic carbon, nitrogen and fertility. Pages 173–272 in M. Schnitzer and S. U. Khan, Eds. Soil organic matter. Developments in soil science. Vol. 8. Elsevier, Amsterdam.

Campbell, C. A., Myers, R. J. K., and Weier, K. L. 1981. Potentially mineralizable nitrogen, decomposition rates and their relationship to temperature for five Queensland soils. Aust. J. Soil Res. 19: 323–332.

Campbell, C. A., Jame, Y. W., and Winkleman, G. E. 1984. Mineralization rate constants and their use for estimating nitrogen mineralization in some Canadian prairie soils. Can. J. Soil Sci. 64: 333–343.

Campbell, C. A., Jame, Y. W., and DeJong, R. 1988. Predicting net nitrogen mineralization over a growing season: model verification. Can. J. Soil. Sci. 68: 537–552.

Campbell, C. A., Lafond, G. P., Leyshon, A. J., Zentner, R. P., and Janzen, H. H. 1991. Effect of cropping practices on the initial potential rate of N mineralization in a thin Black Chernozem. Can. J. Soil Sci. 71: 43–53.

Chae, Y. M. and Tabatabai, M. A. 1986. Mineralization of nitrogen in soils amended with organic wastes. J. Environ. Qual. 15: 193–198.

Chaudhry, I. A. and Cornfield, A. H. 1971. Low-temperature storage for preventing changes in mineralizable nitrogen and sulphur during storage of air-dried soils. Geoderma 5: 165–168.

Deans, J. R., Molina, A. E., and Clapp, C. E. 1986. Models for predicting potentially mineralizable nitrogen and decomposition rate constants. Soil Sci. Soc. Am. J. 50: 323–326.

Dendooven, L., Verstraeten, L., and Vlassak, K. 1990. The N-mineralization potential: an undefinable parameter. Pages 170–181 in R. Merckx, H. Vereecken, and K. Vlassak, Eds. Fertilizer and the environment. Leuven University Press, Leuven, Belgium.

Ellert, B. H. and Bettany, J. R. 1988. Comparison of kinetic models for describing net sulfur and nitrogen mineralization. Soil Sci. Soc. Am. J. 52: 1692–1702.

Fyles, J. W. and McGill, W. B. 1987. Nitrogen mineralization in forest soil profiles from central Alberta. Can. J. For. Res. 17: 242–249.

Janzen, H. H. 1987. Soil organic matter characteristics after long-term cropping to various spring wheat rotations. Can. J. Soil. Sci. 67: 845–856.

Juma, N. G., Paul, E. A., and Mary, B. 1984. Kinetic analysis of net nitrogen mineralization in soil. Soil Sci. Soc. Am. J. 48: 753–757.

MacKay, D. C. and Carefoot, J. M. 1981. Control of water content in laboratory determination of mineralizable nitrogen in soils. Soil Sci. Soc. Am. J. 45: 444–446.

Myers, R. J. K., Campbell, C. A., and Weier, K. L. 1982. Quantitative relationship between net nitrogen mineralization and moisture content of soils. Can. J. Soil Sci. 62: 111–124.

Oh Almhain, L. and O'Danachair, D. 1974. Filter paper as a source of error in NH_3 determinations. Analyst 99: 211–213.

Olness, A. 1984. Re: nitrogen mineralization potentials, N_o and correlations with maize response. Agron. J. 76: 171–172.

Paul, E. A. and Juma, N. G. 1981. Mineralization and immobilization of soil nitrogen by microorganisms. Pages 179–194 in F. E. Clark and T. Rosswall, Eds. Terrestrial nitrogen cycles. Processes, ecosystem strategies and management impacts. Ecol. Bull. (Stockholm) 33: 179–194.

Paustian, K. and Bonde, T. A. 1987. Interpreting incubation data on nitrogen mineralization from soil organic matter. Pages 101–112 in J. H. Cooley, Ed. Soil organic matter dynamics and soil productivity. Proceedings from an INTECOL Workshop, INTECOL Bull. 15. International Association for Ecology, Athens, GA.

Ralston, M. 1988. Derivative-free nonlinear regression. Pages 389–417 in W. J. Dixon, Ed. BioMedical Department Programs (BMDP) Statistical Software Manual, Volume 1. University of California Press, Berkeley.

Ralston, M. L. and Jenrich, R. I. 1978. DUD, a derivative-free algorithm for nonlinear least squares. Technometrics 20: 7–14.

Richter, J., Nusk, A., Hagenicht, W., and Bauer, J. 1982. Optimized N-mineralization parameters of loess soils from incubation experiments. Plant Soil 68: 379–388.

Smith, J. L., Schnable, R. R., McNeal, B. L., and Campbell, G. S. 1980. Potential errors in the first order model for estimating soil nitrogen mineralization potentials. Soil Sci. Soc. Am. J. 44: 996–1000.

Stanford, G. and Smith, S. J. 1972. Nitrogen mineralization potentials of soils. Soil Sci. Soc. Am. Proc. 36: 465–472.

Stanford, G. and Epstein, E. 1974. Nitrogen mineralization-water relations in soils. Soil Sci. Soc. Am. Proc. 38: 103–107.

Stanford, G., Frere, M. H., and Schwaninger, D. H. 1973. Temperature coefficient of soil nitrogen mineralization. Soil Sci. 115: 321–323.

Statistical Analysis System, Inc. (SAS) 1985. SAS User's Guide: Statistics, Version 5 ed. SAS Inst. Inc., Cary, NC.

Talpaz, H., Fine, P., and Bar-Yosef, B. 1981. On the estimation of N-mineralization parameters from incubation experiments. Soil Sci. Soc. Am. J. 45: 993–996.

Chapter 34
Denitrification

E. G. Beauchamp and D. W. Bergstrom

University of Guelph
Guelph, Ontario, Canada

34.1 INTRODUCTION

Anyone contemplating measurement of denitrification, especially in the field, should consult key papers (Hauck and Weaver 1986, Tiedje 1982, Tiedje et al. 1989). Although various methods have been used to measure gaseous N losses from soil during denitrification, we suggest one approach which currently appears to be most acceptable.

Measurement of biological denitrification presently involves measurement of gas concentrations. Many early attempts were made to estimate N losses from nitrogen balance determinations. This approach is much too inaccurate for present purposes. Prior to 1976, loss of N via denitrification was often determined by NO_3^- disappearance. This approach works moderately well in laboratory setups and results have been shown to compare favorably with the acetylene block method. In instances of strongly anaerobic conditions and high C substrate availability, dissimilatory nitrate reduction to ammonium may have confounded measurement. Immobilization of NO_3^- by assimilatory reduction may also have occurred. Considerable errors, however, are encountered in field measurements where nitrification and leaching occur over time.

The disappearance of ^{15}N-labeled NO_3^- has been used in laboratory and field studies to measure denitrification. The major problem in field studies is that fertilizer NO_3^- losses can be monitored, but native soil NO_3^- losses cannot be estimated. Direct measurement by mass spectrometry of $^{15}N_2$ and $^{15}N_2O$ gases produced has been used, but it has the same shortcomings as measurement of $^{15}NO_3^-$ disappearance. On the other hand, if the objective is to study the fate of applied fertilizer N, procedures involving ^{15}N labeling may be quite appropriate.

The acetylene block method is based on acetylene inhibition of reduction of nitrous oxide (N_2O) to gaseous nitrogen (N_2), the last step in the denitrification process. Accumulation of N_2O can be easily measured by gas chromatography. The acetylene block method has been widely applied in laboratory setups and also in field studies of denitrification.

A simple field approach has been to place an enclosure on the soil surface, add acetylene to the headspace or into the soil below the enclosure, and measure the accumulation of N_2O

in the headspace. A major difficulty is the diffusion of acetylene into the soil and N_2O out of the soil, especially in soils with a low air-filled porosity. Another general approach has been to place soil cores from the field into a sealable container, add acetylene, and measure N_2O accumulation. This general procedure is similar to that of laboratory setups. Based on experience and information in the literature, the acetylene block method involving "static" soil cores appears to be the most acceptable and applicable approach. As reported in the literature, measurements by this procedure have compared favorably with those by other methods. Another version of the core method is to circulate gases continuously through the core, periodically measuring N_2O accumulation. This requires an elaborate monitoring system.

The general procedure for measuring denitrification involving the acetylene block method will be described, first for a laboratory setup, and then for soil cores from the field. Subsequently, a procedure for the denitrifying enzyme activity (DEA) assay will be described.

34.2 LABORATORY PROCEDURE

34.2.1 MATERIALS AND REAGENTS

1   Erlenmeyer flasks (125 mL) with rubber serum stoppers (Suba Seal, Barnesley, England)

2   1- and 10-mL gas-tight syringes (Dynatech Precision Sampling Corp., Baton Rouge, LA)

3   Needles.

4   Inert gas source (argon or helium) connected to gas flow meter and manifold.

5   Acetylene source with sulfuric acid wash

6   Gas chromatograph.

34.2.2 PROCEDURE

1   Place 25 g of field-moist soil into a flask. Soil should be mixed sufficiently to break up larger aggregates and provide a representative subsample for analysis. Soil samples should be analyzed as soon as possible after collection.

2   Add 25 mL of water. Seal the flask with a serum stopper. Displace atmosphere from the flask by inserting two needles into the stopper, and connecting one needle to the inert gas manifold. Displacement should be complete after 15 to 20 min.

3   Remove 10 mL of headspace gas and replace with 10 mL of acetylene.

4   Within 24 h sample (0.4 mL) the headspace gas and measure the N_2O concentration using a gas chromatograph.

5   Dissolved N_2O needs to be taken into account. Dissolved N_2O (mg) in a closed system can be calculated by the formula:

$$y = \alpha \cdot x \cdot \text{(solution volume/headspace volume)} \qquad (34.1)$$

where α is the solubility of N_2O expressed as cm^3 N_2O dissolved per cm^3 of water, and x is mg of N_2O in the flask headspace. Moraghan and Buresh (1977) have provided values for α (Table 34.1). Denitrification rate is expressed as μg N_2O-N h^{-1} g^{-1} soil (dry weight basis).

34.2.3 COMMENTS

1   Water is added to saturate the soil and enhance diffusion of substrate carbon and nitrate. Gravimetric moisture content of the soil needs to be determined. Gas in the headspace may also be removed by a series of evacuations with a water tap aspirator or vacuum pump, and refilled with inert gas. This may be accomplished by two needles inserted through the serum stopper, one attached to the evacuation unit and the other to the inert gas supply. Small amounts of O_2 remaining in the flask will be consumed by microbial respiration. Approximately 10% of the headspace gas volume should be replaced with acetylene. The acetylene should be passed through two sulfuric acid washes to remove organic impurities that might serve as carbon substrate for denitrifiers.

2   A gas chromatograph equipped with a thermal conductivity detector is suitable, providing CO_2, O_2, and N_2 are separated from N_2O, and the concentrations of N_2O are relatively high. These gases can be separated by Porapak Q and Molecular Sieve 5A columns in series. Alternatively, an electron capture detector, which is especially sensitive to N_2O, could be used. Buildup of water in the column with successive sample injections is a problem. This may be overcome by equipping the chromatograph with a ten-port valve. The injected sample enters a Porapak Q precolumn and, once the N_2O has passed through, the valve is switched so that N_2O continues to the detector through another Porapak Q column; meanwhile, the water of the injected sample is back flushed and vented from the precolumn. Peak areas are measured with an integrator and N_2O concentrations are determined from a standard curve. Periodic check of chromatograph response to a standard N_2O gas sample is advised.

3   Because the structure of the soil sample as it exists in the field is not preserved and the flask headspace is replaced with inert gas, rate measurements do not represent *in situ* rates.

Table 34.1 Solubility of N_2O
in Water

Temperature (°C)	N_2O solubility (cm^3 N_2O cm^{-3} H_2O)
5	1.067
10	0.910
15	0.778
20	0.676
25	0.594
30	0.530

From Moraghan, J. T. and Buresh, R. 1977. Soil Sci. Soc. Am. J. 41: 1201–1202. With permission.

34.3 FIELD SAMPLE (STATIC CORE) TECHNIQUE

34.3.1 MATERIALS AND REAGENTS

1 Sealer jars (250 mL) with lids fitted with serum stoppers.

2 Aluminum rings 5 cm in diameter and 5 cm deep.

3 1- and 10-mL gas-tight syringes.

4 Acetylene source with sulfuric acid wash.

5 Gas chromatograph.

34.3.2 PROCEDURE

1 Collect soil cores using aluminum rings. Care should be taken to preserve the structure obtained in the field.

2 As soon as possible after sampling, weigh the soil cores and place them into glass jars. Seal the jars with lids fitted with serum stoppers. Remove 10 mL of headspace gas and replace with 10 mL of acetylene. Hold the jars at a temperature approximating that of the field site.

3 Sample (0.4 mL) headspace gas for analysis of N_2O concentration by gas chromatograph between 2 and 6 h, allowing time for acetylene to diffuse into the cores, but before NO_3^- becomes limiting. Collect a minimum of three samples and determine the rate of N_2O production by linear regression of N_2O concentration on sampling time. Determine the quantity of N_2O produced, taking air-filled porosity into account. Air-filled porosity can be estimated from total core volume, moisture content, bulk density, and an assumed particle density (see Chapter 54). Dissolved N_2O can be estimated as previously described. Denitrification rate is expressed as μg N_2O-N h^{-1} g^{-1} soil (dry weight basis).

34.3.3 COMMENTS

1   Additional detail on soil core methods is provided by Ryden et al. (1987) and Tiedje et al. (1989). Several smaller soil cores may be used. If analysis is delayed, soil cores may be stored at low temperature (e.g., 4°C) for a short period before measurement. Diffusion of acetylene into cores and N_2O out may be improved by removing intact cores from rings or using perforated rings. When acetylene is introduced into the jar, pumping with a large syringe can enhance entry of acetylene into soil pores. Acetylene is added to a concentration of approximately 10% of the headspace volume. The volume added depends upon the headspace volume.

2   Setup of gas chromatographs was described previously. Because low amounts of N_2O may be produced by soil cores, an electron capture detector is normally required. Automated gas chromatographic systems have been developed, but are not necessary unless very large numbers of samples are collected.

3   High spatial and temporal variability of denitrification rates in field soils require careful consideration of sampling strategy and replication. Parkin et al. (1987) have examined the influence of sample size on denitrification measurements. Integration of rate measurements over area is approximate. Denitrification rates frequently show a log normal distribution and require suitable statistical methods.

34.4 DENITRIFYING ENZYME ACTIVITY (MARTIN ET AL. 1988)

34.4.1 MATERIALS AND REAGENTS

1   Erlenmeyer flasks (125 mL) with rubber serum stoppers.

2   1- and 10-mL gas-tight syringes.

3   Needles.

4   Inert gas source (argon or helium) connected to gas flow meter and manifold.

5   Acetylene source with sulfuric acid wash.

6   2-mL Vacutainers® (Becton Dickinson Vacutainer® Systems, Rutherford, NJ).

7   Solution of glucose (10 mM), KNO_3 (10 mM), K_2HPO_4 (50 mM), and chloramphenicol (100 µg mL^{-1}) buffered at pH 7.0.

8   Gas chromatograph.

34.4.2 PROCEDURE

1   Place 25 g of soil into a flask. Soil should be mixed sufficiently to break up larger aggregates and provide a representative subsample for analysis. Soil samples should be analyzed as soon as possible after collection.

2   Add 25 mL of buffer solution. Seal with a rubber serum stopper and replace atmosphere in headspace with inert gas by the procedure described previously. Replace 10 mL of headspace gas with 10 mL of acetylene. Place the flask immediately onto a shaker to mix contents.

3   At 15, 30, 45, and 60 min after beginning the assay remove 0.5 mL of headspace gas with a syringe. Place the sample into a Vacutainer® filled with atmosphere at ambient pressure. Remove 0.5 mL of gas from the Vacutainers® prior to sample addition. Replace the sample volume (0.5 mL) in the flask with inert gas (from a large flask) using the syringe.

4   Determine N_2O concentration of the samples by gas chromatography, and determine the rate of N_2O production (DEA) by linear regression of N_2O concentration on sampling time. Correction for dissolved N_2O is made as described previously. DEA is expressed as $\mu g\ N_2O\text{-}N\ h^{-1}\ g^{-1}$ soil (dry weight basis).

34.4.3 Comments

DEA is an estimate of the concentration of functional denitrifying enzymes in a soil sample at the time of sampling. The measurement is based on the principle that when conditions for an enzyme-catalyzed reaction are optimized, reaction rate is proportional to enzyme concentration. In the assay, denitrifying conditions are optimized by saturation of the system with substrate carbon and NO_3^- and removal of O_2. Chloramphenicol is added to prevent protein synthesis during the assay. Acetylene is added to a concentration of approximately 10% of headspace gas volume. Adequate replication (three or four) is required. The volume of sample removed from the flask can be adjusted for amounts of N_2O produced. A 0.4-mL volume of gas is injected from the Vacutainers® into the gas chromatograph and measurement of N_2O concentration of gas in Vacutainers® is made as soon as possible (within days of collection). Standards are prepared using Vacutainers®, as in the assay procedure. Setup of gas chromatographs was previously described.

REFERENCES

Hauck, R. D. and Weaver, R. W. Eds. 1986. Field measurement of dinitrogen fixation and denitrification. SSSA Spec. Pub. No. 18, Soil Science Society of America, Madison, WI (see chapters by R. D. Hauck, J. M. Duxbury, D. E. Rolston, and D. R. Keeney).

Martin, K., Parsons, L. L., Murray, R. E., and Smith, M. S. 1988. Dynamics of soil denitrifier populations: relationships between enzyme activity, most-probable-number counts, and actual N gas loss. Appl. Environ. Microbiol. 54: 2711–2716.

Moraghan, J. T. and Buresh, R. 1977. Correction for dissolved nitrous oxide in nitrogen studies. Soil Sci. Soc. Am. J. 41: 1201–1202.

Parkin, T. B., Starr, J. L, and Meisinger, J. J. 1987. Influence of sample size on measurement of soil denitrification. Soil Sci. Soc. Am. J. 51: 1492–1501.

Ryden, J. C., Skinner, J. H., and Nixon, D. J. 1987. Soil core incubation system for the field measurement of denitrification using acetylene-inhibition. Soil Biol. Biochem. 19: 753–757.

Tiedje, J. M. 1982. Denitrification. Pages 1011–1026 in A. L. Page, R. H. Miller, and D. R. Keeney, Eds. Methods of soil analysis. Part 2. Chemical and Microbiological Properties. Agronomy No. 9, 2nd ed. American Society of Agronomy, Madison, WI.

Tiedje, J. M., Simkins, S., and Groffman, P. M. 1989. Perspectives on measurement of denitrification in the field including recommended protocols for acetylene based methods. Plant Soil 115: 261–284.

Chapter 35
Earthworms

G. H. Baker and K. E. Lee

CSIRO

Glen Osmond, S.A., Australia

35.1 INTRODUCTION

Earthworms can significantly influence soil structure, fertility, and plant productivity (Lee 1985). Several methods have therefore been developed to estimate their numbers in soils, but obtaining accurate estimates is often difficult and no one method is appropriate for all situations.

The abundance of earthworms is usually patchy within fields. Some species, and even forms within species, can differ markedly in their spatial patterns (Baker 1983). Vertical distributions in the soil can vary between species and life cycle stages, and they may also vary seasonally and even diurnally (Gerard 1963, Lee 1985, Baker et al. 1992). Some species quickly retreat deep into the soil through their burrow systems in response to digging at the surface. Many species spend part of the year inactive deep in the soil. Cocoons (eggs) are immobile. Small and cryptically colored earthworms are easily overlooked. Prior knowledge of the life history, ecology, and behavior of the species to be sampled can greatly facilitate the choice of an appropriate sampling method.

Methods used to sample earthworms can be classified as (1) physical (where the earthworms are directly removed from the soil by the operator), (2) behavioral (where they are stimulated to move out of the soil and then are collected), and (3) indirect (where numbers are estimated by trap catches or physical evidence of earthworm presence). Physical and behavioral methods have occasionally been combined to increase the accuracy of population estimates. The methods described here are the most commonly used. They include hand sorting, washing and sieving (physical methods), chemical and electrical expulsion, and trapping (behavioral methods). These methods have been reviewed more extensively in Bouche (1972a), Edwards and Lofty (1977), Bouche and Gardner (1984), and Lee (1985).

For all methods, it is important that samples should be taken at stratified random coordinates within the sampling plot and be sufficient in number and size to yield accurate estimates (Southwood 1978). The required numbers and sizes of samples will vary from one study to another (e.g., Figure 35.1). Different numbers of samples may be required to adequately measure the abundances of different species. Sample numbers may need to vary seasonally.

Soil Sampling and Methods of Analysis, M. R. Carter, Ed.,
Canadian Society of Soil Science. © 1993 Lewis Publishers.

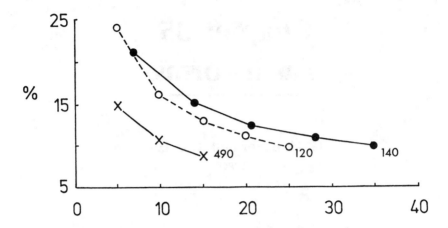

FIGURE 35.1. Relationship between the standard error, as a percentage of the mean, and the number of soil samples (each 0.1 m² in area, 30 cm deep) taken to estimate lumbricid earthworm abundance at Bracknagh, Ireland (●) and Tungkillo (○) and Springmount (X), South Australia. Numbers of earthworms m⁻² for each site are included in the figure. Data were collected in autumn (Ireland) or winter (Australia) (Baker 1983, Baker et al. 1992).

Brief comment is included at the end of this chapter on the measurement of earthworm biomass and the storage and identification of specimens once collected from the field.

35.2 HAND SORTING

Soil samples of prescribed dimensions are collected either by spade or plastic or metal corer, and the earthworms and cocoons are sorted from them by hand.

35.2.1 Materials and Reagents

Few materials are required. Spades are needed for digging the soil, or cores can be driven into the soil and the contents lifted out in one block. Trays can be used for sorting the soil. Quadrats may be used to check the dimensions of the holes dug, both top and bottom. Depth measurements can be usefully marked on the blade and handle of the spade. Quadrats may be constructed with sharp edges that can be forced into the soil to demarcate the edges of the samples to be dug out. Gloves (cloth or plastic) may prove useful for sorting earthworms from some soils (e.g., plastic gloves are useful with wet clay soils).

35.2.2 PROCEDURE

1   Samples of 0.05 to 0.25 m² in area and 0.1 to 0.3 m deep are commonly taken. Cuts made with spades are restricted as much as possible to the edges of the samples to minimize damage to the earthworms.

2   Samples are commonly stratified into layers (e.g., of 10 cm) to demonstrate vertical distribution of the worms. Total sampling depth may be varied seasonally to allow for vertical migrations.

3   Samples should be sorted against a pale-colored (e.g., white or blue) background to facilitate the detection of the earthworms.

4   Many earthworms are cut into pieces. Attempts may be made to match these pieces up when counting population numbers. Where this is not possible, it is conventional to count the head ends as individuals and to ignore other pieces.

35.2.3 Comments

Hand sorting is not practical when a substantial proportion of the population lives in deep burrows, individuals are very small and cryptic in color, or the soil is particularly hard or sticky and difficult to break into small amounts to release the worms. Hand sorting can be supplemented with washing and/or sieving to recover small earthworms and cocoons. Efficiency of hand sorting decreases with increasing sample size, probably because of the tedium of sorting through larger amounts of soil, and also with increased root content in the soil. The main disadvantage with hand sorting is that it is labor intensive. However, in contrast to several of the following methods, hand sorting can be conducted totally in the field without need of transporting large amounts of soil to the laboratory. Coring is sometimes difficult in practice because of rocks, tree roots, or other obstructions.

35.3 WASHING AND SIEVING

Soil washing and sieving may be conducted with or without previous hand sorting of the samples. The simplest methods involve washing soil through sieves using jets of water from hoses. Selection of suitable sieve sizes ensures collection of small earthworms and cocoons. Edwards et al. (1970) described a mechanized method of sieving and washing, in which soil samples were placed in rotating vertical drums, with sieves fitted to their bottoms, and subjected to high-pressure jets of water. Further developments of the washing/sieving technique have been described by Bouche (1969, 1972a, b), Bouche and Beugnot (1972), and Bouche and Gardner (1984) and form the basis for the methods described below. Recently, Judas (1988) described a washing and sieving method for the extraction of earthworms from leaf litter.

35.3.1 MATERIALS AND REAGENTS

1   Spades or corers are required for field collection of soil samples. Plastic boxes are necessary for transport of the samples to the laboratory.

2   Solutions of sodium hexametaphosphate (2%) (w/v) and formaldehyde (4%) (v/v) are used in a chemical pretreatment of the samples prior to washing them.

3   The washing machines used to extract the earthworms are best described (with figures) in Bouche (1972a, b) and Bouche and Beugnot (1972).

35.3.2 PROCEDURE

1   With Bouche's method, soil cores or blocks of soil (0.6 m² in area, 20 cm deep) are taken in an undisturbed state from the field and soaked for 2 d in a solution containing both sodium hexametaphosphate, to disperse clays, and formaldehyde to kill and fix the earthworms.

2   The soil samples are then placed in a water-filled trough, where they are broken up by paddles that are rotated mechanically on a horizontal shaft, and simultaneously flushed with water and sieved. The sieved material includes plant roots, leaves, macroarthropods, and the coarser soil particles, as well as earthworms and their cocoons.

3   Earthworms are picked by hand from the sieved material.

35.3.3 Comments

Washing/sieving methods are probably the most effective for extracting earthworms from soil. They make it possible to extract small specimens, cocoons, and inactive stages in the life cycle. Raw (1960) concluded that simply washing soil through sieves to extract earthworms was more laborious than hand sorting and probably not worth the additional effort, but with the technical advances made by Bouche and co-workers the labor involved in washing/sieving is much reduced. Provision can be made for soils that have very high clay content and are difficult to sort by the addition of clay dispersants.

35.4 CHEMICAL REPELLANTS

These methods involve the application of solutions of irritant chemicals to the soil to induce earthworms to leave. Usually, the chemicals are applied to the soil surface *in situ*, but occasionally the irritants have been applied to excavated soil samples so that the earthworms may leave the soil mass horizontally as well as vertically (Springett 1981). The chemicals that have been used include mercuric chloride, potassium permanganate, and formaldehyde (Edwards and Lofty 1977, Bouche and Gardner 1984, Lee 1985). Formaldehyde is the most commonly used chemical and is generally regarded as most efficient.

35.4.1 MATERIALS AND REAGENTS

1   A 0.22% (v/v) solution of formaldehyde (=0.55% [v/v] formalin) is most commonly used, but stronger or weaker solutions may be used as appropriate (Raw 1959, Satchell 1969, Baker 1985).

2   Equipment includes watering cans to apply solution, metal frames to contain solution within prescribed areas of soil surface, shears to clip grass in the target area, and forceps or gloves for collection of specimens from the soil surface.

35.4.2 PROCEDURE

1   Formaldehyde solution is normally applied to the soil surface using a watering can to obtain even spread.

2   The amount of solution and the period over which the surfacing earthworms are collected may vary. Raw (1959) and Baker (1985) applied 13 and 16 L of solution per square meter of soil surface, respectively, to extract lumbricid earthworms. They collected the earthworms from the soil surface for the next 20–25 min, then applied solution again and collected for a similar period. Satchell (1969) applied solution in three lots of 18 L each per square meter at 10-min intervals. He found further applications unprofitable.

3   The formaldehyde solution is best applied to an area in which the surface vegetation has been clipped short (to increase the likelihood of seeing all surfacing earthworms).

4   The area sampled (sample sizes of 0.25 m² are commonly used) should be bordered by a quadrat consisting of metal sheets (approximately 10 cm wide) which are pushed on edge into the soil surface. This helps confine the extractant to the sampling area.

5   Forceps or gloves should be used to collect the surfacing earthworms to prevent damage to the hands by the formaldehyde.

35.4.3 COMMENTS

The formaldehyde method is widely used, but it has serious limitations (Raw 1959, Bouche 1969, 1975, Satchell 1969, Lakhani and Satchell 1970, Nordstrom and Rundgren 1972, Barnes and Ellis 1979, Terhivuo 1982, Baker 1983, 1985, Bouche and Gardner 1984, Lee 1985).

1   The method is most useful for extracting some species with vertical burrows that open to the surface (e.g., *Lumbricus terrestris*). The solution is able to reach the earthworms, which then have a direct route out. Species which live very close to the surface (e.g., *Dendrobaena* and *Satchellius* spp.) may also be collected efficiently, but those that live deeper and construct mostly horizontal burrows which are rarely connected to the soil surface (e.g., *Aporrectodea* spp.) are not. Extraction is more efficient in direct-drilled cereal fields, where earthworm burrows are little disturbed, than in ploughed fields, where burrows are broken and blocked. Efficiency may vary between adult and juvenile earthworms. Cocoon numbers, of course, cannot be measured.

2   The efficiency of extraction varies with soil temperature and moisture (and hence earthworm activity), soil type (e.g., porosity and the presence or absence of compacted horizons), and concentration of formaldehyde used.

Table 35.1 Numbers of Earthworms Collected by Hand Sorting and Formaldehyde Expulsion (0.22%) from a Peat Soil at Bracknagh, Ireland. Total Area Sampled by Both Methods = 24.5 m^2.

Species	Hand sorting		Formaldehyde expulsion	
	Adults & subadults	Juveniles	Adults & subadults	Juveniles
Allolobophora chlorotica	336	302	170	107
Aporrectodea longa	21 ⎫		2 ⎫	
A. tuberculata	411 ⎬	777	95 ⎬	259
A. turgida	14 ⎭		4 ⎭	
A. rosea	239	206	61	43
Dendrobaena octaedra	15	11	10	25
Dendrodrilus rubidus	1	0	2	0
Eiseniella tetraedra	3	2	1	2
Lumbricus castaneus	8 ⎫		1 ⎫	
L. festivus	9 ⎪	482	3 ⎪	285
L. rubellus	74 ⎬		34 ⎬	
L. terrestris	52 ⎭		42 ⎭	
Octolasion cynaeum	1 ⎫	151	0 ⎫	51
O. tyrtaeum	225 ⎭		73 ⎭	
Satchellius mammalis	60	28	50	42

From Baker, G. H. Holarct. Ecol., 6, 74, 1983. With permission.

3   Comparisons of formaldehyde extraction with other methods (e.g., hand sorting, washing, and sieving) have generally shown it to be less efficient (e.g., Table 35.1). Formaldehyde is not a pleasant chemical to work with because of irritation to the nose and mouth, and health risks have been attributed to it (Feinman 1988). However, the method is relatively quick and appropriate for population estimates of some species if soil conditions and earthworm activity are optimal. If soil types vary little between experimental treatments, formaldehyde extraction is useful for comparisons of relative (cf. absolute) abundance.

4   Applications of formaldehyde solution to the soil surface have been used to encourage deep-dwelling species to move closer to the surface where they can be sampled more efficiently by hand sorting (Bouche 1969, St Remy and Daynard 1982). Formaldehyde has also been applied to the bottoms of pits to extract very deep-dwelling species (Martin 1976, Barnes and Ellis 1979).

35.5 ELECTRICAL METHODS

Some earthworms will surface if an electrode is pushed into the soil and an alternating current is discharged. This phenomenon was originally utilized for earthworm sampling by Satchell (1955). Lee (1985) considered Satchell's technique dangerous and inefficient. He could extract few earthworms relative to hand sorting. The volume of the soil affected by the current flow was dependent on soil moisture, electrolyte content of the soil water, and temperature (and hence conductivity). The volume of soil sampled was therefore variable and thus quantitative estimates of abundance were unreliable. More recently, Thielemann (1986) has developed an electrical method which appears to be more useful.

35.5.1 Materials and Reagents

The octet method of Thielemann (1986) utilizes eight electrodes (65 cm in length, 0.6 cm in diameter) mounted on a circular metal frame (52 cm in diameter). Each electrode has a separate power supply. Current spreads successively from each electrode to influence the area circumscribed by the metal frame. A switchboard includes eight switches for the electrodes, an ammeter, a voltmeter, and a rheostat for steady regulation of the current. Low currents of 30 or 60 V are sufficient. Power is supplied by a portable generator.

35.5.2 PROCEDURE

1   The metal frame is placed on the soil surface and the electrodes are individually pushed into the soil.

2   Current is varied between the electrodes such that the electrical fields within the soil change and are likely to equally affect earthworms lying in different orientations.

3   The earthworms that surface are collected by hand. Approximately 15 min are needed for adequate sampling.

35.5.3 Comments

Thielemann (1986) argued that the method was 87% efficient compared with hand sorting. A variety of species could be extracted (e.g., *Lumbricus, Aporrectodea,* and *Octolasion* spp.). The problems of an undefined sampling area have been avoided, but the method would still seem subject to variability due to soil properties influencing conductivity and earthworm activity (e.g., soil moisture). The hardness of the soil may influence the ability to insert the long, thin electrodes fully.

35.6 TRAPPING METHODS

Pitfall traps can be used to sample earthworms that are active on the surface of the soil (Boyd 1957, Bouche 1972a, 1976). The numbers of worms collected by this method reflect both abundance and activity and care must thus be taken in interpreting the data. Especially if used in conjunction with absolute measures of abundance, such as hand sorting, pitfall trapping can provide useful information on diurnal, seasonal, age, and species differences in surface activity.

Other traps, such as the baited traps described by Satchell (1980) (earthenware trays containing humus and cow dung), have also been used successfully to survey fauna in a qualitative manner.

35.6.1 MATERIALS AND REAGENTS

1   The traps usually consist of glass or plastic jars set in the soil with their open lip flush with the soil surface. Diameters of the jars may be varied (e.g., 5

to 100 + mm in diameter) according to the size of the species present and the numbers of earthworms needed (the catch will increase with the circumference of the jar aperture).

2 Preservative (e.g., formaldehyde or a saturated solution of picric acid) may be added to the traps to kill the earthworms that fall in, and also their predators, but is not essential.

3 Inner containers (e.g., plastic drink cups) can be added to the traps so that the catch can be removed easily in them without disturbing the outer limits of the trap and the soil around it.

4 Lids (e.g., ceramic tiles, sheets of metal) can be supported 2 or 3 cm above the trap apertures (e.g., on nails which are pushed into the surrounding soil) to limit contamination of the trap contents by leaves, twigs, and rainwater and to prevent access by large predators.

5 Flooding by excess water can be prevented by the inclusion of small drainage holes in the trap bottoms and walls.

35.6.2 PROCEDURE

1 The traps are set in the field and checked daily or weekly, depending on the use of preservative or not, usually at times of day which correspond to the taking of meteorological data. This timing allows easy relating of catch numbers to climatic variables.

2 Traps should be set at least several meters apart, to reduce trapping out an area too quickly and to obtain adequately accurate censuses of animals which are notoriously patchy in their distribution.

3 Traps may be run continuously or periodically, being closed with a tight-fitting lid when not in use.

35.6.3 Comments

Often it is wise to ignore the catches obtained during the first few days after the traps are set. It is common to have a "digging-in" effect, whereby the initial disturbance of setting the traps influences catch numbers. The method assumes that all earthworms are equally likely to fall into the traps, but this may not be the case. The presence of preservatives and trap lids (the latter possibly acting as shelter traps) may increase or decrease the likelihood of earthworms being caught.

35.7 OTHER METHODS

Other methods that have been used on occasion include flotation, heat extraction, mark-recapture, and counting casts made by the earthworms on the soil surface.

35.7.1 FLOTATION

1   Flotation methods involve the initial hand sorting or washing and sieving of soil samples and then stirring them in solutions (e.g., of $MgSO_4$) with specific gravities such that plant material and earthworms float off onto sieves. Earthworms can be hand picked from these sieves or may be separated from plant material by further differential flotation. Martin (1976) described a flotation technique based on methods and apparatus developed earlier by Salt and Hollick (1944) for extracting soil arthropods. Raw (1960) and Gerard (1967) describe simpler methods.

2   The flotation method may be of use where plant roots are very abundant, but the method is generally regarded as very time consuming. Bouche (1975) argued that $MgSO_4$ solutions were inefficient in floating off earthworms from soil solutions and that his mechanized washing/sieving method was more satisfactory.

35.7.2 HEAT EXTRACTION

1   Satchell (1969) and Abrahamsen (1972) described methods for extracting earthworms from soil samples based on the Baermann funnel technique, which relies on repulsion of animals from a heat source. Soil samples are partly immersed in water beneath a bank of light bulbs. The earthworms leave the heated soil and are collected by hand from the water.

2   Satchell (1969) argued that this method was more efficient than hand sorting or formalin expulsion for extracting small earthworms from the particularly dense root mats of *Deschampsia flexuosa*. Abrahamsen (1972) estimated that the method was 95% efficient. Major limitations seem to be the unsuitability of soil samples that will not hold together within water and space to house the replicate water baths needed for large numbers of samples.

35.7.3 MARKING TECHNIQUES

1   Most of the conventional marking techniques used for invertebrates are inappropriate for earthworms, principally because of their lack of a hard exoskeleton. However, Meinhardt (1976) and Mazaud and Bouche (1980) stained earthworms with nontoxic dyes, and Joyner and Harmon (1961) and Gerard (1963) labeled them with radioactive isotopes and recovered significant numbers later following release.

2   Lee (1985) suggested that the technique used by Richter (1973) for freeze-branding slugs might be quite appropriate for earthworms, but we are not aware of this having been tried.

35.7.4 Numbers of Casts

The numbers of casts produced within a prescribed area may give an indirect measure of the presence of certain species (not all species cast at the surface), but the data have to be treated with caution. The numbers of casts reflect not only abundance, but also activity. Casts will accumulate through time. They are also easily destroyed by heavy rainfall.

35.8 EARTHWORM BIOMASS

1   Earthworm biomass is often measured as well as, or instead of, earthworm numbers.

2   Some sampling methods may be quite accurate in estimating earthworm biomass because they extract the largest individuals in the population efficiently, yet inaccurate in estimating population numbers because many small individuals are missed. For example, Raw (1960) obtained 52% of the earthworm numbers and 84% of the earthworm biomass by hand sorting a wet grassland soil in which the animals were mostly confined to a thick turf mat. Small species and immature specimens of larger species were most commonly missed by the hand sorting. In contrast, Raw (1960) obtained 89% of the numbers and 95% of the biomass by hand sorting a pasture soil with no root mat, and 59% of the numbers and 90% of the biomass from a poorly structured clay soil in an old, arable field.

3   There are several sources for error in the measurement of biomass once the earthworms have been collected (e.g., state of hydration of the specimens, presence of soil in the gut in relation to recent feeding, loss of weight following preservation). For example, Piearce (1972) found that earthworms preserved in 5% formalin lost 3–5% of their body weight after 1 week. It is usual to express earthworm biomass in terms of dry weight or weight at a standard level of hydration/period in preservative. Lee (1985) describes methods for voiding gut contents from earthworms to eliminate this source of error.

35.9 STORAGE AND IDENTIFICATION

35.9.1 STORAGE

1   Earthworms collected for population or biomass estimates may be stored in either 4% formaldehyde or 70–80% ethanol (v/v). However, prolonged storage of earthworms in formaldehyde causes specimens to become brittle and unsuitable for dissection and hence for detailed taxonomy. Prolonged storage in ethanol without prior fixation causes the specimens to soften too much and decompose, which also makes them unsuitable for dissection.

2   If earthworms are required for future taxonomic purposes, the recommended procedure is to first anesthetize them in dilute ethanol (5–10%)

until they no longer respond to gentle prodding (approximately 15 min), and then to fix them for 24 h in 4% formaldehyde. Following this, the worms are stored in 70–80% ethanol.

3   Labels should be placed both inside and outside the storage containers. Paper labels should be written on in lead pencil or ink that will not dissolve.

35.9.2 IDENTIFICATION

1   Many earthworms can be identified from external characters with no need for dissection. A good hand-lens or low-power binocular microscope is sufficient to see the necessary taxonomic features. Novices should only attempt to identify adult earthworms (those possessing a clitellum, or "collar"). Juvenile specimens are more difficult.

2   Sims and Gerard (1985) give an excellent account of the general anatomy, biology, and taxonomy (including keys) of the earthworms found in Britain. These include many of the cosmopolitan species (e.g., Lumbricidae) that dominate communities in agricultural fields throughout temperate regions of the world. Ljungstrom (1970) also gives an excellent overview of earthworm taxonomy generally.

REFERENCES

Abrahamsen, G. 1972. Ecological study of Lumbricidae (Oligochaeta) in Norwegian forest soils. Pedobiologia 13: 28–39.

Baker, G. H. 1983. Distribution, abundance and species associations of earthworms (Lumbricidae) in a reclaimed peat soil in Ireland. Holarct. Ecol. 6: 74–80.

Baker, G. H. 1985. Formalin-expulsion of earthworms (Lumbricidae) from Irish peat soils. Soil Biol. Biochem. 17: 113–114.

Baker, G. H., Barrett, V. J., Grey-Gardner, R., and Buckerfield, J. C. 1992. The life history and abundance of the introduced earthworms *Aporrectodea trapezoides* and *A. caliginosa* (Annelida:Lumbricidae) in pasture soils in the Mount Lofty Ranges, South Australia. Aust. J. Ecol. 17: 177–188.

Barnes, B. T. and Ellis, F. B. 1979. Effects of different methods of cultivation and direct drilling and disposal of straw residues on populations of earthworms. J. Soil Sci. 30: 669–679.

Bouche, M. B. 1969. Comparaison critique de methodes d'evaluation des populations de Lombricides. Pedobiologia 9: 26–34.

Bouche, M. B. 1972a. Lombriciens de France. Ecologie et Systematique. INRA Publ. 72-2. Institut National des Recherches Agriculturelles, Paris.

Bouche, M. B. 1972b. Contribution a l'approche methodologique des biocenoses. I. Vers l'analyse quantitative globale des prairies. Ann. Zool.-Ecol. Anim. 44: 529–536.

Bouche, M. B. 1975. Fonctions des lombriciens. IV. Corrections et utilisations des distorsions causees par les methodes de capture. Pages 571–582 in J. Vanek, Ed. Progress in soil zoology. Proc. 5th Intl. Colloq. Soil Zool., Prague, 1973. Junk, The Hague/Academia, Prague.

Bouche, M. B. 1976. Etude de l'activite des invertebres epiges prairiaux. I. Resultats generaux et geodrilogiques (Lumbricidae:Oligochaeta). Rev. Ecol. Biol. Sol 13: 261–281.

Bouche, M. B. and Beugnot, M. 1972. Contribution a l'approche methodologique de l'etude des biocenoses. II. L'extraction des macroelements du sol par lavage-tamisage. Ann. Zool.-Ecol. Anim. 4: 537–544.

Bouche, M. B. and Gardner, R. H. 1984. Earthworm functions. VIII. Population estimation techniques. Rev. Ecol. Biol. Sol. 21: 37–63.

Boyd, J. M. 1957. Comparative aspects of the ecology of Lumbricidae on grazed and ungrazed natural maritime grassland. Oikos 8: 107–121.

Edwards, C. A. and Lofty, J. R. 1977. Biology of earthworms. 2nd ed. Chapman and Hall, London.

Edwards, C. A., Whiting, A. E., and Heath, G. W. 1970. A mechanised washing method for separation of invertebrates from soil. Pedobiologia 10: 141–148.

Feinman, S. E. 1988. Formaldehyde sensitivity and toxicity. CRC Press, Boca Raton, FL.

Gerard, B. M. 1963. The activities of some species of Lumbricidae in pasture land. Pages 49–54 in J. Doeksen and J. van der Drift, Eds. Soil organisms. North-Holland, Amsterdam.

Gerard, B. M. 1967. Factors affecting earthworms in pastures. J. Anim. Ecol. 36: 235–252.

Joyner, J. W. and Harmon, N. P. 1961. Burrows and oscillative behaviour therein of *Lumbricus terrestris*. Proc. Indiana Acad. Sci. 71: 378–384.

Judas, M. 1988. Washing-sieving extraction of earthworms from broad-leaved litter. Pedobiologia 31: 421–424.

Lakhani, K. H. and Satchell, J. E. 1970. Production of *Lumbricus terrestris* L. J. Anim. Ecol. 39: 473–492.

Lee, K. E. 1985. Earthworms. Their ecology and relationships with soils and land use. Academic Press, Sydney.

Ljungstrom, P. O. 1970. Introduction to the study of earthworm taxonomy. Pedobiologia 10: 265–285.

Martin, N. A. 1976. Effect of four insecticides on the pasture ecosystem. V. Earthworms (Oligochaeta:Lumbricidae) and Arthropoda extracted by wet sieving and salt flotation. N.Z. J. Agric. Res. 19: 111–115.

Mazaud, D. and Bouche, M. B. 1980. Introductions en surpopulation et migrations de lombriciens marques. Pages 687–701 in D. L. Dindal, Ed. Soil biology as related to land use practices. Proc. 7th Intl. Soil Zool. Colloq., Syracuse, 1979. U.S. Environmental Protection Agency, Washington, D.C.

Meinhardt, 1976. Dauerhafte Markierung von Regenwurmen durch ihre Lebendfarbung. Nachrichtenbl. Dtsch. Pflanzenschutzdienstes (Braunschweig) 28: 84–86.

Nordstrom, S. and Rundgren, S. 1972. Methods of sampling lumbricids. Oikos 23: 344–352.

Piearce, T. G. 1972. Acid intolerant and ubiquitous Lumbricidae in selected habitats in north Wales. J. Anim. Ecol. 41: 397–410.

Raw, F. 1959. Estimating earthworm populations by using formalin. Nature (London) 184: 1661–1662.

Raw, F. 1960. Earthworm population studies: a comparison of sampling methods. Nature (London) 187: 257.

Richter, K. O. 1973. Freeze-branding for individually marking the banana slug. Northwest Sci. 47: 109–113.

Salt, G. and Hollick, F. S. J. 1944. Studies of wireworm populations. I. A census of wireworms in pasture. Ann. Appl. Biol. 31: 52–64.

Satchell, J. E. 1955. An electrical method of sampling earthworm populations. Pages 356–364 in D. K. McE. Kevan, Ed. Soil zoology. Butterworths, London.

Satchell, J. E. 1969. Methods of sampling earthworm populations. Pedobiologia 9: 20–25.

Satchell, J. E. 1980. Earthworm populations of experimental birch plots on a *Calluna* podzol. Soil Biol. Biochem. 12: 311–316.

Sims, R. W. and Gerard, B. M. 1985. Earthworms. Syn. Br. Fauna No. 31. Linn. Soc. Lond, London.

Southwood, T. R. E. 1978. Ecological methods. Chapman and Hall, London.

Springett, J. A. 1981. A new method for extracting earthworms from soil cores, with a comparison of four commonly used methods for estimating earthworm populations. Pedobiologia 21: 217–222.

St Remy, E. A. de and Daynard, T. B. 1982. Effects of tillage methods on earthworm populations in monoculture corn. Can. J. Soil Sci. 62: 699–703.

Terhivuo, J. 1982. Relative efficiency of hand-sorting, formalin application and combination of both methods in extracting Lumbricidae from Finnish soils. Pedobiologia 23: 175–188.

Thielemann, U. 1986. Elektrischer Regenwurmfang mit der Oktett-Methode. Pedobiologia 29: 296–302.

Soil Biochemical Analyses

Chapter 36
Total and Labile Polysaccharide Analysis of Soils

L. E. Lowe

University of British Columbia
Vancouver, British Columbia, Canada

36.1 INTRODUCTION

There are a number of possible approaches to the estimation of carbohydrate constituents in soil. None of them is entirely satisfactory, largely because of the heterogeneity of soil polysaccharides and difficulties in achieving complete recoveries during hydrolysis of the polymers. Most procedures involve hydrolysis to release monomeric units, which are then determined individually or as a group. Various procedures for estimating total bound sugars or hexoses, pentoses, and uronic acids separately have been carefully reviewed by Cheshire (1979), including consideration of hydrolysis conditions and "clean-up" of hydrolysates.

Following release of saccharide monomers by hydrolysis, it is also possible to determine the concentration of individual monosaccharides by gas-liquid chromatography (Oades 1967) or high-performance liquid chromatography (Hounsell 1986), and estimating total polysaccharides by summation. This approach yields more detailed information, but the additional costs and equipment needed may often not be justified. Recent developments in ^{13}C-nuclear magnetic resonance methods have made it possible to obtain quantitative estimates of polysaccharide contents of solid samples without sample destruction. Such equipment is, however, rarely available to a soils laboratory, and some uncertainty exists in relation to the accuracy of such estimates.

The procedures described below are based on release of saccharide monomers by hydrolysis with sulfuric acid, followed by colorimetric estimation of total sugar content in the hydrolysates, using the "phenol-sulfuric acid" reagent. They can be used for mineral soils with relatively low polysaccharide content or for organic horizon samples with polysaccharide contents up to 40%. No very costly equipment is required. Two procedures are described, one for the estimation of total soil polysaccharides, and the other for the estimation of labile polysaccharides, which in effect means recovery of materials other than cellulose.

The procedures cannot be expected to give an accurate absolute measure of polysaccharides, because of variations in color yield by different sugar monomers, and because sugar yield

Soil Sampling and Methods of Analysis, M. R. Carter, Ed.,
Canadian Society of Soil Science. © 1993 Lewis Publishers.

varies with variation in hydrolysis procedure and with soil type. The procedures should, however, prove useful in assessing changes in polysaccharides over time, or in comparative studies. The choice of whether to determine total or labile polysaccharides, or both, will depend on whether or not the inclusion of cellulose is considered desirable in relation to the particular soil investigation involved.

36.2 TOTAL POLYSACCHARIDES

The recovery of sugars present in cellulose is achieved by pretreatment of the sample with 12 M H_2SO_4 at room temperature, prior to hydrolysis with 0.5 M H_2SO_4.

36.2.1 EQUIPMENT

1   Spectrophotometer, single beam, with optically matched cuvettes (e.g., ten 13-mm test tubes).

2   Water bath: to operate in 25–30°C range.

3   Autoclave (steam or electrical).

36.2.2 REAGENTS

1   Phenol solution, 5% w/v: dissolve 5 g phenol crystals in 100 mL distilled water. (Avoid contact with skin or breathing of vapor.)

2   Concentrated H_2SO_4, 96% w/w, reagent grade.

3   12 M H_2SO_4: prepare by dilution from concentrated H_2SO_4 (adding acid to water).

4   Stock glucose solution, 1000 $\mu g \cdot mL^{-1}$: dissolve 0.500 g glucose in 500 mL distilled water (store in refrigerator).

5   Working glucose standards: prepare 100 mL each of 20-, 30-, 40-, 50-, and 60-$\mu g \cdot mL^{-1}$ standards by dilution from stock glucose solution (prepare fresh every 3–4 d).

36.2.3 PROCEDURE

1   Weigh 0.500 to 1.000 g soil (or soil fraction) into a 250-mL Erlenmeyer flask. Add 4.0 mL 12 M H_2SO_4, ensuring that all the sample is moistened by the acid. Cover the flask with a large marble or small watch glass. Let stand for 2 h.

2   Dilute acid to 0.5 *M* by adding 92 mL distilled water. Autoclave for 1 h at 103 kPa (yielding a temperature of approximately 121°C).

3   Cool, filter into 250-mL volumetric flask, and wash residues thoroughly. Make filtrate to volume (store in refrigerator if analysis is not to be completed on the same day).

4   Prepare standard curve as follows: pipette 1 mL of each standard into separate cuvettes. To each cuvette add 1 mL phenol solution, followed by 5 mL conc H_2SO_4. Addition of the acid is made with an automatic pipette that delivers the acid rapidly to ensure good mixing (if mixing is incomplete, cuvettes can be capped with acid-resistant stoppers and mixed by inversion). After standing for 10 min, place cuvettes in a water bath at 25–30°C for 25 min. Read absorbance at 490 nm in spectrophotometer. Set zero absorbance with reagent blank prepared using 1 mL distilled water in place of a standard. Prepare calibration curve and/or calculate regression line.

5   Analysis of sample hydrolysates: follow procedure outlined above for standards, replacing 1-mL aliquot of standard with 1-mL aliquot of sample. Determine polysaccharide concentration in aliquot by reference to standard curve, and calculate percent polysaccharide by making corrections for dilutions involved. Record results as percent total polysaccharide (glucose equivalent).

36.3 LABILE POLYSACCHARIDES

This analysis is considered to recover most polysaccharides other than cellulose and should include those polymers most active in aggregation of soil particles. The pretreatment with 12 *M* H_2SO_4 is omitted.

36.3.1 Equipment and Reagents are the same as for Total Polysaccharide Analysis

36.3.2 PROCEDURE

1   Follow the same procedure as outlined for total polysaccharides, except that steps 1 and 2 are replaced by the following: weigh 0.500 g to 1.000 g soil (or soil fraction) into a 250-mL Erlenmeyer flask, and add 100 mL 0.5 *M* H_2SO_4. Autoclave for 1 h at 103 kPa (15 psi).

36.4 COMMENTS

1   The colorimetric procedure described is essentially that of Dubois et al. (1956). The hydrolysis procedures are based on those used by Ivarson and Sowden (1962) and by Cheshire (1979).

2   The procedures can be used for mineral soils, organic horizons, peats, and particle size fractions in solid state. Labile polysaccharides can also be estimated on extracted organic fractions in solution.

3   For some mineral soils, results may be improved by removing possible in-
     terferants from hydrolysates with ion-exchange resins, as reported by Doutre
     et al. (1978).

4   If an autoclave is not available, hydrolysis can be carried out by refluxing.
     Periods from 6 to 24 h have been used. Autoclaving is more convenient in
     several ways, needing less bench space and glassware, shorter hydrolysis
     times, and making it easier to handle larger numbers of samples.

5   The selection of 490 nm as a suitable wavelength for measuring absorbance
     is an arbitrary one because of the variation in absorption peaks for different
     sugar monomers. Glucose, and other hexoses, which tend to dominate the
     composition of soil polysaccharides, have absorption maxima close to 490
     nm. Colored solutions are stable for several hours.

6   The use of larger tubes for color development has the advantage of greater
     ease of mixing and the option of working with larger volumes. The use of
     small cuvettes, however, as described above, eliminates the need for trans-
     ferring strong acid solutions to cuvettes, with the attendant risks for the
     operator and for the spectrophotometer.

REFERENCES

Cheshire, M. V. 1979. Methods of analysis, pages 23–67 in Nature and origin of carbohydrates in soils. Academic Press, London.

Doutre, D. A., Hay, G. W., Hood, A., and VanLoon, G. W. 1978. Spectrophotometric meth-ods to determine carbohydrates in soil. Soil Biol. Biochem. 10: 457–462.

Dubois, M., Gilles, K. A., Hamilton, J. K., Rebers, P. A., and Smith, F. 1956. Colorimetric method for determination of sugars and related substances. Anal. Chem. 28: 350–356.

Hounsell, E. F. 1986. Carbohydrates. Chap. 4 in C. K. Lim, Ed. HPLC of small molecules.

Ivarson, K. C. and Sowden, F. J. 1962. Methods for analysis of carbohydrate material in soil. I. Colorimetric determination of uronic acids, hex-oses and pentoses. Soil Sci. 94: 245–250.

Oades, J. M. 1967. Gas-liquid chromatography of alditol acetates and application to the analysis of sugars in complex hydrolysates. J. Chromatogr. 28: 246–252.

Chapter 37
Organic Forms of Nitrogen

Yeh-Moon Chae

Alberta Environment
Edmonton, Alberta, Canada

37.1 INTRODUCTION

The organic forms of nitrogen in soils are mainly determined after treating the soils with hot mineral acids (or bases) to liberate their nitrogenous constituents. Such treatment provides reliable estimates of the amounts of total N, ammonia-N (NH_3-N), α-amino acid-N, and amino sugar-N (Bremner 1965). Current practice in the hydrolysis of the soil is to boil the samples continuously with 6 *M* HCl under reflux for 12 h (Stevenson 1982). This procedure, however, fails to extract a large proportion (about 25 to 35%) of N, which remains in the soil residue (Stevenson 1982).

The N remaining in the soil residue is usually referred to as acid-insoluble N, and the N recovered from the residue by distillation with MgO is referred as NH_3-N. The N brought down in the MgO precipitate, after distillation of NH_3-N from the acid-insoluble residue, is called "humin-N". This fraction is seldom determined and is often grouped as the unidentified or unknown N, along with other unaccounted N. This unidentified fraction was thought to be an artifact formed during the lengthy hydrolysis of N-containing and other organic components (Asami and Hara 1970). Some of this N, however, is believed to occur as a structural component of humic substances.

The main identifiable organic N compounds in soil hydrolysates are the amino acids and amino sugars. Soils contain trace quantities of nucleic acids (usually <2% of total N) and other nitrogenous biochemicals such as purine and pyrimidine bases (<1% of total N) (Anderson 1961), but specialized techniques are required for their separation and identification. Only one third to one half of the organic N in soils can be accounted for in identifiable compounds.

A significant proportion of the soil N, about 20 to 25% for surface soils, is recovered as NH_3. Some of the NH_3 is derived from indigenous fixed NH_4^+ and part comes from partial destruction of amino sugars. It is also known that NH_3 can arise from the breakdown of certain amino acids during hydrolysis (Stevenson 1982).

Methods for different forms of N in the neutralized hydrolysate are described. In the methods, different forms of N in the neutralized hydrolysate are converted to, and estimated as, NH_3.

Soil Sampling and Methods of Analysis, M. R. Carter, Ed.,
Canadian Society of Soil Science. © 1993 Lewis Publishers.

Additional colorimetric methods for amino acid-N (37.4) and amino sugar-N (37.5) are also discussed.

37.2 PREPARATION OF SOIL HYDROLYSATES FOR ORGANIC NITROGEN ANALYSES (BREMNER 1965, STEVENSON 1982)

The procedure for hydrolyzing the soil has not been standardized, and many variations in hydrolytic conditions have been employed. In general, hydrolysis is done under reflux with HCl for 6 to 24 h. In this procedure hydrolysates are prepared by hydrolysis under reflux for about 12 h using 3 mL of 6 M HCl/g of soil, and the soil hydrolysate is neutralized without prior removal of excess acid.

37.2.1 MATERIALS AND REAGENTS

1   Micro-Kjeldahl digestion unit.

2   Round-bottom flasks fitted with a standard-taper (24/40) ground-glass joint as for the quick-fit Liebig condenser.

3   Liebig condensers with 24/40 ground-glass joint.

4   Electric heating mantle.

5   Hydrochloric acid (HCl), approximately 6 M: add 513 mL of conc HCl (sp gr 1.19) to about 500 mL of water, cool, and dilute to 1 L in a volumetric flask.

6   *N*-Octyl alcohol.

7   Sodium hydroxide (NaOH), approximately 10 M: place 3.2 kg of reagent-grade NaOH in a heavy-walled 10-L Pyrex® bottle marked to indicate a volume of 8 L. Add 4 L of CO_2-free water and swirl the bottle until the alkali is dissolved. Cool the solution while the neck of bottle is closed with a rubber stopper, and then dilute it to 8 L by the addition of CO_2-free water. Swirl the bottle vigorously to mix the contents and fit the neck with some arrangement that permits the alkali to be stored and dispensed with protection from atmospheric CO_2.

8   Sodium hydroxide (NaOH), approximately 5 M: dilute 500 mL of 10 M NaOH to 1 L and store in a stoppered bottle.

9   Sodium hydroxide (NaOH), approximately 0.5 M: dilute 50 mL of 10 M NaOH to 1 L and store in a stoppered bottle.

37.2.2 PROCEDURES

1   Place a sample of finely ground (<100 mesh) soil containing about 10 mg of N in a round-bottom flask fitted with a standard-taper (24/40) ground-glass joint.

2   Add two drops of octyl alcohol and 20 mL of 6 M HCl, then swirl the flask to thoroughly mix the acid with the soil.

3   Place the flask in an electric heating mantle and connect it to a Liebig condenser fitted with 24/40 ground-glass joint.

4   Heat the soil-acid mixture so that it gently boils under reflux for 12 h.

5   Wash the condenser with a small quantity of distilled water and allow the flask to cool, then remove the flask from the condenser.

6   Filter the hydrolysis mixture through a Buchner funnel fitted with Whatman® No. 50 filter paper, using a suction filtration apparatus that allows collection of the filtrate in a 200-mL beaker.

7   Wash the residue with distilled water until the filtrate reaches the 60-mL mark on the beaker.

8   Place the bottom half of the beaker in crushed ice.

9   Neutralize to pH 6.5 ± 0.1 by slow addition of NaOH with constant stirring to ensure that the hydrolysate does not become alkaline at any stage of the neutralization process. Use 5 M NaOH to bring the pH to about 5, and complete the neutralization using 0.5 M NaOH. The hydrolysate can also be cooled in a freezer and cold NaOH used for neutralization.

10 Transfer the neutralized hydrolysate into a 100-mL volumetric flask.

11 Adjust the volume to the mark with the washings obtained by rinsing the beaker, electrodes, and stirrer several times with small quantities of distilled water.

12 Stopper the flask and invert several times to mix the contents.

13 Using an aliquot of the hydrolysate, determine different forms of N under analysis from the NH_3 liberated by steam distillation as described in Section 37.3.

37.2.3 COMMENTS

1   Poor recoveries of some forms of N may result if the quantity of acid used for hydrolysis, or the amount of N in the soil sample, is either considerably

larger or smaller than the amount specified. Accuracy may also be affected when recommended sample size (maximum sample size is 10 g of soils) for analysis is not observed (Bremner 1965).

2   If the hydrolysate is neutralized, bumping will be less likely to occur during Kjeldahl digestion. A digestion period of 30 to 60 min is usually adequate.

3   Since S-containing amino acids and amides are unstable during acid hydrolysis, the use of 6 *M* HCl has several limitations for accurate determination of soil organic N. More than 24 h were required to complete the hydrolysis of protein and chitin with boiling 6 *M* HCl (Yonebayashi and Hattori 1980). Under these conditions, decomposition of glucosamine occurred and increased with heating time, resulting in an increase of the amide-N fraction. Little glucosamine was decomposed to NH_3-N by hydrolysis with 1 *M* HCl at 100°C. However, chitin was incompletely hydrolyzed to glucosamine under these conditions. In contrast, amide-N was almost completely recovered.

4   Troublesome frothing that is sometimes encountered with calcareous soils can be eliminated by the addition of octyl alcohol to the hydrolysis mixture. Addition of octyl alcohol also increases the precision of results for soils that do not wet readily with acid. Calcareous soils should be acidified before the addition of the 20 mL of 6 *M* HCl.

5   Since hydrolysates can be stored without neutralization for long periods, the hydrolysate can be made to volume before neutralization when the hydrolysate is not to be analyzed within a day or so.

6   Hydrolysis of soil in an autoclave has been suggested as an alternative to the reflux method (Lowe 1973). The microwave digestion method (Vittori Antisari and Sequi 1988) and microwave Kjeldahl method (He et al. 1990) suggested for total N also seem to be applicable to preparation of the soil hydrolysates for organic N analyses.

37.3 DETERMINATION OF VARIOUS FORMS OF NITROGEN AS AMMONIA IN A SOIL HYDROLYSATE BY STEAM DISTILLATION

In the determination of different forms of N, all forms are converted to NH_4^+ by distillation of the neutralized soil hydrolysate (Section 37.2). The liberated NH_4^+ is collected in a H_3BO_3 indicator solution, and determined by titration with 0.0025 *M* H_2SO_4 standard. An automatic titrator with a pH-sensing electrode is much more rapid and convenient for determination of NH_4^+ in distillate collected in H_3BO_3 indicator solution.

Procedures for the determination of seven different forms of N are covered under different headings.

37.3.1 TOTAL HYDROLYZABLE NITROGEN (BREMNER AND MULVANEY 1982)

Total N in the hydrolysate is determined by the semimicro-Kjeldahl procedure described by Bremner and Mulvaney (1982) and Stevenson (1982).

37.3.1.1 MATERIALS AND REAGENTS

1   Steam distillation apparatus: this apparatus is designed so that Pyrex® Kjeldahl flasks fitted with standard-taper (19/38) ground-glass joints can be used as distillation chambers (Keeney and Nelson 1982). The steam required for distillation is generated by heating distilled water in a 5-L flask that contains glass beads to prevent vigorous boiling, and a small amount of H_2SO_4 to trap any NH_4^+ in the distilled water. Before use, the distillation apparatus should be steamed out for about 10 min to remove traces of NH_3. The rate of steam generation (7 to 8 mL of distillate per minute) should be adjusted by controlling the power supply to the mantle with a variable transformer. The temperature of the distillate obtained (less than 22°C) should be controlled by the flow of cold water through the condenser.

2   Distillation flasks: 50- and 100-mL Pyrex® Kjeldahl flasks with standard-taper (19/38) ground-glass joints and fitted with glass hooks so that they can be connected to the steam distillation apparatus by spiral steel springs (Keeney and Nelson 1982). Dimensions of the flask should be such that when the flasks are connected to the distillation apparatus, the distance between the tip of the steam inlet tube and the bottom of the flask is approximately 4 mm.

3   Microburette: 5 mL, graduated at 0.01-mL intervals.

4   Potassium sulfate-catalyst mixture: prepare the catalyst mixture by mixing 200 g of potassium sulfate (K_2SO_4), 20 g of cupric sulfate pentahydrate ($CuSO_4 \cdot 5H_2O$), and 2 g of Se. Powder the reagents separately before mixing and grind the mixture in a mortar to prevent from forming a cake during mixing.

5   Sulfuric acid (H_2SO_4), concentrated.

6   Sodium hydroxide (NaOH), approximately 10 M: prepare as described in Section 37.2.1.

7   Boric acid indicator-solution: place 80 g of boric acid (H_3BO_3) in a 5-L flask marked to indicate a volume of 4 L, add about 3800 mL of water, and heat and swirl the flask until the H_3BO_3 is dissolved. Cool the solution and add 80 mL of mixed indicator solution prepared by dissolving 0.099 g of bromocresol green and 0.066 g of methyl red in 100 mL of ethanol. Then add 0.1 M NaOH cautiously until the solution assumes a reddish purple tint (pH about 5.0), and dilute the solution to 4 L by addition of water. Mix the solution thoroughly before use.

8   Sulfuric acid (H_2SO_4), 0.0025 M standard.

9   Standard (NH_4^+ + amino sugar + amino acid)-N solution: dissolve 0.189 g of $(NH_4)_2SO_4$, 0.308 g of glucosamine HCl, and 0.254 g of alanine in water. Dilute the solution to a volume of 2 L in a volumetric flask and mix the solution thoroughly. This solution contains 20 μg of NH_4^+-N, 10 μg of amino sugar-N, and 20 μg of α-amino acid-N per mL. Store the solution in a re-

frigerator at 4°C. This reagent is stable for several months if stored in a refrigerator.

37.3.1.2 PROCEDURE

1   Place 5 mL of the neutralized hydrolysate prepared from Section 37.2.2 in a 50-mL distillation flask, and add 0.5 g of K_2SO_4-catalyst mixture and 2 mL of conc H_2SO_4.

2   Heat the flask cautiously on a micro-Kjeldahl digestion unit until the water is removed and frothing ceases. Then increase the heat until the mixture clears and complete the digestion by boiling gently for 1 h.

3   Allow the flask to cool and add about 10 mL of water slowly with shaking.

4   Cool the distillation flask under a cold-water tap and place it in a beaker containing crushed ice.

5   Transfer 5 mL of H_3BO_3-indicator solution into a 50-mL Erlenmyer flask that is marked to indicate a volume of 35 mL, and place the flask under the condenser of the steam distillation apparatus on an adjustable support.

6   Adjust the tip of the condenser to about 4 cm above the level of the H_3BO_3-indicator solution.

7   The cooled distillation flask is connected to the distillation apparatus and 10 mL of 10 *M* NaOH placed in the entry funnel. Then run the alkali slowly into the distillation flask by raising the funnel plug.

8   Rapidly rinse the funnel with 5 mL of distilled water when about 0.5 mL of alkali remains in the funnel, and allow an additional 2 mL of distilled water to run into the distillation flask before sealing the funnel.

9   Start steam distillation by closing the stopcock on the steam bypass tube of the distillation apparatus and continue the distillation for approximately 4 min.

10   Stop the distillation by opening the stopcock when the distillate reaches the 35-mL mark on the receiver flask and rinse the end of the condenser.

11   Determine NH_4^+-N in the distillate by titration with 0.0025 *M* H_2SO_4 from a microburette to the faint permanent pink endpoint color (1 mL of 0.0025 *M* H_2SO_4 equals 70 µg NH_4^+-N).

37.3.1.3 COMMENTS

1   The method can be checked by analyzing 5-mL aliquots of reagent 9.

2 In each series of analyses controls should be included to allow for any N derived from the reagents. The method can be checked by neutralization and analysis of an aliquot before and after addition of known amounts of NH_4^+-N (e.g., use $(NH_4)_2SO_4$ or NH_4Cl). The method described is not subject to interference by various inorganic and organic substances likely to be present in neutralized soil hydrolysates (Bremner 1965).

37.3.2 ACID-INSOLUBLE NITROGEN

37.3.2.1 *Procedure*

This form of N is the difference between total soil N and total hydrolyzable N.

37.3.2.2 COMMENTS

1 The percentage of the soil N recovered in acid-insoluble forms could be reduced by pretreating the soil with HF before acid hydrolysis (Cheng et al. 1975).

2 By dissolving part of the insoluble N in a dilute base followed by acid hydrolysis, the acid-insoluble fraction can be reduced to about 10% of the total N (Freney 1968, Griffith et al. 1976).

3 The quantity of acid-insoluble N can be reduced roughly by half when samples are subjected to successive hydrolyses with increasing concentrations of acid, i.e., 1 *M* 3 *M*, 6 *M*, and conc HCl containing acetic acid at 150°C (Gonzalez-Prieto and Carballas 1988).

37.3.3 AMINO ACID-N (STEVENSON 1982)

37.3.3.1 MATERIALS AND REAGENTS

1 Citric acid: grind 100 g of reagent-grade citric acid ($C_6H_8O_7 \cdot H_2O$) in a mortar and store in a small, wide-mouth bottle.

2 Ninhydrin (triketohydrindene hydrate; indane-trione hydrate): grind 10 g of reagent-grade ninhydrin in a mortar. Store in a small, wide-mouth bottle.

3 Phosphate-borate buffer, pH 11.2: place 100 g of sodium phosphate ($Na_3PO_4 \cdot 12H_2O$), 25 g of borax ($Na_2B_4O_7 \cdot 10H_2O$), and about 900 mL of water in a 1-L volumetric flask. Shake the flask until the phosphate and borate are dissolved. Dilute the solution to 1 L and store in a tightly stoppered bottle.

4 All other materials and reagents needed are described in Section 37.3.1.

37.3.3.2 PROCEDURE

1   Place 5 mL of the hydrolysate in a 50-mL distillation flask and add 1 mL of 0.5 *M* NaOH. Then heat the flask for approximatley 20 min in boiling water until the volume of the sample is reduced to 2–3 mL.

2   After cooling add 500 mg of citric acid and 100 mg of ninhydrin to the flask.

3   Place the flask in a vigorously boiling water bath (the flask bulb should be immersed completely in boiling water). After 1 min, swirl the flask for a few seconds without removing it from the bath and leave it in the bath for 9 min.

4   Cool the flask and add 10 mL of phosphate-borate buffer and 1 mL of 5 *M* NaOH.

5   Transfer 5 mL of H_3BO_3 1/N indicator solution into a 50-mL Erlenmyer flask that is marked to indicate a volume of 35 mL, and place the flask under the condenser of the steam distillation apparatus on an adjustable support.

6   Adjust the tip of the condenser to about 4 cm above the level of the H_3BO_3 1/N indicator solution.

7   Connect the flask to the steam distillation apparatus.

8   Determine the amount of NH_4^+-N liberated by steam distillation as described in Section 37.3.1.

37.3.3.3 COMMENTS

1   The quantity of α-amino acid-N recovered was halved, when samples were subjected to successive hydrolyses with increasing concentrations of acid of 1 *M*, 3 *M*, 6 *M*, and conc HCl with acetic acid at 150°C (Gonzalez-Prieto and Carballas 1988).

2   The method can be checked by analyzing 5 mL of reagent 9 (Section 37.3.1).

3   Controls should be included in each series of analyses to allow for any N derived from the reagents. The method can be checked by neutralization and analysis of aliquots before and after addition of a known amount of NH_4^+-N (as amino acid-N [as alanine]). The method described is not subject to interference by various inorganic and organic substances likely to be present in neutralized soil hydrolysates (Bremner 1965).

4   Extra ninhydrin is required sometimes for complete recovery of amino acid-N as NH_3-N (Stevenson 1982). In the determination of amino acid-N, interference from NH_4^+ and alkali-labile organic N compounds (e.g., amino sugars) is eliminated by treatment with NaOH (100°C) before the ninhydrin

reaction is carried out. An estimate of (NH_3 + amino sugar + amino acid)-N can be obtained by conducting the ninhydrin reaction without prior alkali treatment.

5   The method is best suited for the analysis of soils and sediments containing relatively large amounts of amino acids.

37.3.4 Ammonia-N (Stevenson 1982)

37.3.4.1 MATERIALS AND REAGENTS

1   Magnesium oxide (MgO): heat heavy MgO in an electric muffle furnace at 600 to 700°C for 2 h. Cool the MgO in a desiccator containing potassium hydroxide (KOH) pellets and store it in a tightly stoppered bottle.

2   All other materials and reagents needed are described in Section 37.3.1.

37.3.4.2 PROCEDURE

1   Pipette a 10-mL aliquot of the hydrolysate into a 50-mL distillation flask and add 0.07 ± 0.01 g of MgO.

2   Place 5 mL of H_3BO_3-indicator solution in a 50-mL Erlenmyer flask that is marked to indicate a volume of 20 mL and place the flask under the condenser of the steam distillation apparatus so that the end of the condenser is about 4 cm above the level of the H_3BO_3-indicator solution.

3   Connect the distillation flask to the steam distillation apparatus and immediately proceed with the distillation as described in steps 8 and 9 in Section 37.3.1.

4   Stop the distillation by opening the stopcock when the distillate reaches the 20-mL mark on the receiver flask and rinse the end of the condenser.

5   Determine the amount of NH_3 liberated in the distillate as described in Section 37.3.1.

37.3.4.3 COMMENTS

1   The method can be checked by analyzing 5 mL of reagent 9 (Section 37.3.1).

2   Controls should be included in each series of analyses to allow for any N derived from the reagents.

37.3.5 (AMMONIA + AMINO SUGAR)-N
(STEVENSON 1982)

37.3.5.1 MATERIALS AND REAGENTS

1   Phosphate-borate buffer, pH 11.2: prepared as described in Section 37.3.3.

2   All other materials and reagents are described in Section 37.3.1.

37.3.5.2 PROCEDURE

1   Place 10 mL of the hydrolysate in a 100-mL distillation flask and add 10 mL of phosphate-borate buffer. Connect the flask to the distillation apparatus.

2   Distill for approximately 4 min.

3   Determine the amount of NH_3-N liberated as described in Section 37.3.1.

37.3.5.2 COMMENTS

1   The hydrolysis method used in Section 37.2.2 causes greater decomposition of amino sugars than conventional hydrolysis procedures for amino sugar-N (Section 37.5). The correction factor to allow for hydrolysis losses is about 1.4 (Bremner 1965).

2   The recovery of hexosamine-N increased with an increase in the pH of the phosphate-borate buffer. Maximum recovery of 93.8% was attained at pH 11.5. To determine the hexosamine-N, the difference between the amounts of N recovered in (ammonia + amino sugar)-N and ammonia-N must be multiplied by a correction factor of 1.06 (Yonebayashi and Hattori 1980).

3   The method can be checked by analyzing 10 mL of reagent 9 (Section 37.3.1).

4   Controls should be included in each series of analyses to allow for any N derived from the reagents.

37.3.6 AMINO SUGAR-N

37.3.6.1 *Procedure*

This form of N is derived from the difference between the amounts of N recovered in (NH_3) + amino sugar)-N and NH_3-N.

37.3.6.2 *Comments*

The method is not suitable for analysis of samples containing low concentrations of amino sugars in the presence of relatively high concentrations of total hydrolyzable NH_3.

37.3.7 (SERINE + THREONINE)-N
(STEVENSON 1982)

37.3.7.1 MATERIALS AND REAGENTS

1   Periodic acid ($HIO_4 \cdot 2H_2O$) solution, approximately 0.2 *M*: dissolve 4.6 g of $HIO_4 \cdot 2H_2O$ in 100 mL of water and store the solution in a glass-stoppered bottle.

2   Sodium metaarsenite ($NaAsO_2$) solution, approximately 1.0 *M*: dissolve 13 g of powdered, reagent-grade $NaAsO_2$ in 100 mL of water and store the solution in a tightly stoppered bottle.

3   Standard (serine + threonine)-N solution: dissolve 0.150 g of serine and 0.170 g of threonine in water. Dilute the solution to a volume of 2 L in a volumetric flask and mix the solution thoroughly. This solution contains 10 µg of serine-N and 10 µg of threonine-N per mL. Store the solution in a refrigerator at 4°C. This reagent is stable for several months if stored in a refrigerator.

4   All other materials and reagents needed are described in Section 37.3.1.

37.3.7.2 PROCEDURE

1   After determination of (NH_3 + amino sugar)-N by steam distillation with phosphate-borate buffer (Section 37.3.5), remove the distillation flask from the distillation apparatus and rinse the steam inlet tube with 3 to 5 mL of water. The rinse water is collected in the distillation flask and the flask cooled thoroughly under a cold-water tap.

2   Add 2 mL of periodic acid solution and swirl the flask for 30 s. Add 2 mL of sodium arsenite solution.

3   Connect the flask to the distillation apparatus.

4   Distill and determine the amount of NH_3-N liberated as described in steps 8 through 11 for total hydrolyzable N (Section 37.3.1).

37.3.7.3 COMMENTS

1   The method can be checked by analyzing 10 mL of reagent 3. In the determination of (serine + threonine)-N, interference from NH_4^+ and alkali-labile N compounds in the hydrolysate is eliminated by prior steam distillation with phosphate-borate buffer at pH 11.2.

2   Controls should be included in each series of analyses to allow for any N derived from the reagents.

37.4 COLORIMETRIC METHOD FOR AMINO ACIDS

Reactions of α-amino acids with ninhydrin at pH 2.5 (Section 37.3.3) produce a stable reaction product, NH_3. However, when the reaction is carried out at pH 5, the NH_3 released reacts further with the reduced and oxidized forms of ninhydrin to form a blue-colored product. The method, based on the determination of optical density of the blue color, is described in detail by Stevenson (1982).

In the method, interference from metal cations (Fe^{3+}, Al^{3+}, Ca^{2+}, etc.) in the acid hydrolysate is eliminated by use of a chelating agent. Before the reaction of amino acids with ninhydrin, ammonia and alkali-labile amino compounds (e.g., amino sugars) are removed from the hydrolysate by an alkali treatment.

37.5 COLORIMETRIC METHOD FOR AMINO SUGARS

This method is based on the classical colorimetric procedure of Elson and Morgan (1933). In the method, a chromogen is formed by the reaction of acetylacetone with the acidic solution of *p*-dimethylaminobenzaldehyde dissolved in alcohol (Ehrlich's reagent), following which the addition of the aldehyde produces a red solution. The interference with color formation by this method can be removed by treating the hydrolysate with an anion-exchange resin, followed by a cation-exchange resin. Stevenson (1982) provides full details for this method.

REFERENCES

Anderson, G. 1961. Estimation of purines and pyrimidines in soil humic acid. Soil Sci. 91: 156–161.

Asami, T. and Hara, M. 1970. Fractionation of soil organic nitrogen after hydrolysis with hydrochloric acid. Soil Sci., Plant Nutr. 17: 222.

Bremner, J. M. 1965. Organic forms of soil nitrogen. Pages 1238–1255 in C. A. Black et al., Eds. Methods of soil analysis. Part 2. Agronomy No. 9. American Society of Agronomy, Madison, WI.

Bremner, J. M. and Mulvaney, C. S. 1982. Nitrogen-total. Pages 595–624 in A. L. Page, Ed. Methods of soil analysis. Part 2. Agronomy No. 9, 2nd ed. American Society of Agronomy, Madison, WI.

Cheng, C. N., Shufeldt, R. C., and Stevenson, F. J. 1975. Amino acid analysis of soils and sediments: extraction and desalting. Soil Biol. Biochem. 7: 143–151.

Elson, L. A. and Morgan, W. T. J. 1933. A colorimetric method for the determination of glucosamine and chondrosamine. Biochem. J. 27: 1824–1828.

Freney, J. R. 1968. The extraction and partial characterization of nonhydrolysable nitrogen in soil. Int. Congr. Soil Sci., Trans. 9th. (Aldelaide) 3: 531–539.

Gonzalez-Prieto, S. J. and Carballas, T. 1988. Modified method for the fractionation of soil organic nitrogen by successive hydrolyses. Soil Biol. Biochem. 2: 1–6.

Griffith, S. M., Sowden, F. J., and Schnitzer, M. 1976. The alkaline hydrolysis of acid-resistant soil and humid acid residues. Soil Biol. Biochem. 8: 529–531.

He, X. T., Mulvaney, R. L., and Banwart, W. L. 1990. A rapid method for total nitrogen analysis using microwave digestion. Soil Sci. Soc. Am. J. 54: 1625–1629.

Keeney, D. R. and Nelson, D. W. 1982. Nitrogen — inorganic forms. Pages 643–698 in A. L. Page, Ed. Methods of soil analysis. Part 2. Agronomy No. 9, 2nd ed. American Society of Agronomy, Madison WI.

Lowe, L. E. 1973. Amino acid distribution in forest humus layers in British Columbia. Soil Sci. Soc. Am. Proc. 37: 569–572.

Stevenson, F. J. 1982. Nitrogen-organic forms. Pages 625–641 in A. L. Page, Ed. Methods of soil analysis. Part 2. Agronomy No. 9, 2nd ed. American Society of Agronomy, Madison, WI.

Vittori Antisari, L. and Sequi, P. 1988. Comparison of total nitrogen by four procedures and sequential determination for exchangeable ammonium, organic nitrogen, and fixed ammonium in soil. Soil Sci. Soc. Am. J. 52: 1020–1023.

Yonebayashi, K. and Hattori, T. 1980. Improvements in the method for fractional determination of soil organic nitrogen. Soil Sci. Plant Nutr. 26: 469–481.

Chapter 38
Soil Humus Fractions

D. W. Anderson and J. J. Schoenau
University of Saskatchewan
Saskatoon, Saskatchewan, Canada

38.1 INTRODUCTION

The extraction of soil organic matter with alkali and separation into humic acid (HA) and fulvic acid (FA) fractions is a common technique to separate and examine soil organic matter (Kononova 1966, Anderson et al. 1974, Schnitzer 1978, Schoenau and Bettany 1987, Roberts et al. 1989, Schnitzer and Schuppli 1989). About 30 to 60% of the soil humus is removed with alkali extractants. The chemical properties of HA and FA have been extensively studied. HA is believed to be the most biologically resistant fraction of organic matter, with a core of strongly condensed aromatic structures surrounded by aliphatic side-chain components. HA generally have a higher molecular weight than FA. FA appear to be composed mainly of microbial metabolites and younger material not highly associated with the mineral fraction, but all material, old and young, can be part of the FA (McGill et al. 1975).

Characteristics of the humus fractions have proven useful in the interpretation of organic matter dynamics. For example, carbon:nitrogen:phosphorus ratios in HA and FA fractions have been used to assess the origin and turnover of nutrients in soil organic matter from different depths and zones (Schoenau and Bettany 1987, Bettany et al. 1979). HA:FA ratios are frequently used as indices of the degree of humification in a soil. Higher HA:FA ratios found in surface soils of more moist environments are believed to reflect more intense humification as a result of greater biological activity (Anderson and Coleman 1985).

Extraction of soil organic matter with NaOH is commonly used because of efficiency and simplicity, but has been criticized (Schnitzer and Schuppli 1989). The major concerns are the creation of artifacts in the extracted organic matter due to hydrolysis and oxidation (Stevenson 1982), and the arbitrary nature of the separation into HA and FA fractions. Problems with hydrolysis and oxidation appear to be at least partly overcome by extracting under an atmosphere of N_2. Low recoveries of organic matter by the NaOH extraction have also been observed. To overcome the low extraction efficiencies in soils of high base status where stable humus complexes are formed with clay and polyvalent cations, such as calcium and magnesium, anions capable of complexing cations may be included in the extractant (Anderson et al. 1974). The addition of sodium pyrophosphate has been used to increase the extraction efficiency in calcareous soils (Choudri and Stevenson 1957, Anderson et al. 1974). However, addition of $Na_4P_2O_7$ precludes the study of P in the humic materials. In

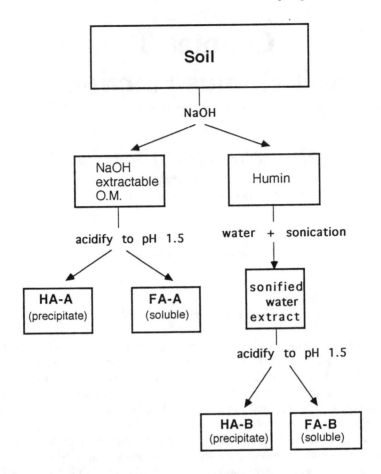

FIGURE 38.1. Flow diagram for alkali extraction and fractionation of soil humus.

this chapter a method is outlined for the extraction of soil organic matter with dilute alkali, and the subsequent separation of the extracted humus into HA and FA fractions on the basis of acid solubility (Figure 38.1).

38.2 ALKALI EXTRACTION

An alkali extraction is used to remove a portion of the soil organic matter. Alkali extractants which are commonly used include 0.1 M NaOH, 0.5 M NaOH, and 0.1 M NaOH-0.1 M Na$_4$P$_2$O$_7$. The method below describes the use of 0.5 M NaOH.

38.2.1 REAGENTS

1   0.5 M HCl: this reagent is used as a pretreatment to remove floating plant debris and inorganic forms of C, N, P, and S prior to extraction with NaOH.

2   0.5 M NaOH: dissolve 20.0 g of NaOH in 1000 mL of deionized water. The pH of the solution should be around 13.5. The NaOH solution must be prepared fresh daily and kept tightly covered, as it will absorb carbon dioxide from the atmosphere.

38.2.2 PROCEDURE

1   Place 15 g of air-dried soil (ground to less than 2 mm) into a 250-mL plastic centrifuge bottle that withstands high-speed refrigerated centrifugation.

2   Add 150 mL of 0.5 M HCl. Set aside for 1 h, stirring occasionally. Centrifuge for 15 min at 9000 $\times$ g and pour off the supernatant.

3   To wash the soil free of any remaining HCl, add 150 mL of deionized water to the centrifuge bottle, mix, then centrifuge at 9000 $\times$ g for 15 min and discard the supernatant.

4   Add 150 mL of fresh 0.5 M NaOH to the centrifuge bottle. Flush the head space of the bottle with oxygen-free N_2 gas, then quickly tighten the cap.

5   Place the bottle on an end-over-end shaker (60 turns per minute) for 18 h.

6   Following the shaking, centrifuge at 9000 $\times$ g for 15 min to separate the NaOH extract from the soil residue. Carefully decant the supernatant into a clean centrifuge bottle and retain for separation into HA and FA fractions. The residue of the extraction (humin) may be discarded or retained for analysis (see Section 38.3.2).

38.2.3 Comments

While the pretreatment step may remove a small amount of organic matter and create another artifact in the extraction, the pretreatment should be performed if organic nutrients are to be analyzed with minimal interference. Retention of the N_2 head space during extraction is important in reducing oxidation and CO_2 absorption. For this reason, centrifuge bottles should be selected with a good seal between cap and bottle. A wrist-action shaker may be substituted if an end-over-end shaker is not available.

38.3 SEPARATION INTO HUMIC FRACTIONS

The NaOH extract is fractionated into conventional HA (HA-A) and FA (FA-A) fractions by acidification. The residue of the alkali extraction, sometimes referred to as humin, is largely composed of organic material tightly bound to the mineral fraction. The humin may be discarded, retained for analysis, or sonified and fractionated into clay-associated HA (HA-B) and FA (FA-B) fractions (Anderson et al. 1974, Bettany et al. 1979).

38.3.1 REAGENTS

1   6 M HCl.

2   0.1 M NaOH: dissolve 4.0 g of NaOH in 1000 mL of deionized water. This solution is used to redissolve the HA after separation by acificiation.

38.3.2 PROCEDURE

1   Using a burette, add 6 *M* HCl to the NaOH extract in the centrifuge bottle until a pH of 1.5 is attained, checking with a pH meter. The acidification causes precipitation of a portion of the organic matter. The precipitated portion is dark brown to black in color and is termed conventional HA (HA-A), while that which remains dissolved in solution after acidification (yellowish in color) is termed conventional FA (FA-A).

2   Centrifuge at 9000 × *g* for 15 min to separate the HA and FA.

3   After centrifugation, the supernatant (FA-A) is removed and retained in vials for analysis. Then 50 mL of 0.1 *M* NaOH is added to the centrifuge bottle to redissolve the precipitate (HA-A). The HA-A may then be transferred to a vial for storage and analysis.

4   If desired, a portion of the residue of the NaOH extraction (humin) may be fractionated into HA (HA-B) and FA (FA-B) fractions. Mix the residue with 150 mL of deionized water, sonicate for 10 min with an ultrasonifier at 125 W, and then allow the suspension to stand for 48 h.

5   Centrifuge at 9000 × *g* for 15 min and remove the supernatant.

6   Acidify the supernatant to pH 1.5 as described above.

7   Remove the precipitate (HA-B) by centrifugation and retain the supernatant (FA-B).

8   Redissolve the HA-B in 100 mL of 0.1 *M* NaOH. Retain the solution for analysis.

38.3.3 Comments

Precipitated HA may be de-ashed by treatment with a hydrofluoric acid-HCl solution. However, such treatments may cause losses of nitrogenous and carbohydrate components (Schnitzer and Schuppli 1989). Samples may also be freeze-dried to provide dry humic materials for analysis. A variety of analyses may be performed on the dissolved humic constituents, including elemental analysis by digestion or automated combustion. Carbon content of dissolved humic fractions is commonly measured using a carbon analyzer such as the Beckman® model 914. The ratio of E4:E6, which gives an indication of the molecular weight of the humic substances, may be determined by dissolving an aliquot of HA or FA in 0.05 *M* $NaHCO_3$ and measuring the ratio of the absorbances at 465 and 665 nm (Kononova, 1966). The IR and ^{13}C NMR spectra of the humic substances can provide important information on the chemical structure (Schnitzer and Schuppli 1989).

Hydrolysis reactions during alkali extractions may be of concern in the study of organic S in humus. The presence of inorganic sulfate in FA extracts has led to the suggestion that organic sulfate groups are hydrolyzed to inorganic sulfate in the NaOH (Schoenau and Bettany 1987). The occurrence of such artifacts emphasizes the need for caution when interpreting the results of organic matter fractionation.

REFERENCES

Anderson, D. W. and Coleman, D. C. 1985. The dynamics of organic matter in grassland soils. J. Soil Water Conserv. 40: 211–216.

Anderson, D. W., Paul, E. A., and St. Arnaud, R. J. 1974. Extraction and characterization of humus with reference to clay-associated humus. Can. J. Soil Sci. 54: 317–323.

Bettany, J. R., Stewart, J. W. B., and Saggar, S. 1979. The nature and forms of sulfur in organic matter fractions of soils selected along an environmental gradient. Soil Sci. Soc. Am. J. 43: 481–485.

Choudri, M. B. and Stevenson, F. J. 1957. Chemical and physicochemical properties of soil humic colloids. III. Extraction of organic matter from soils. Soil Sci. Soc. Am. Proc. 21: 508–513.

Kononova, M. M. 1966. Soil organic matter. 2nd ed. Pergamon Press, London. 544 pp.

McGill, W. B., Shields, J. A., and Paul, E. A. 1975. Relation between carbon and nitrogen turnover in soil organic fractions of microbial origin. Soil Biol. Biochem. 7: 57–63.

Roberts, T. L., Bettany, J. R., and Stewart, J. W. B. 1989. A hierarchical approach to the study of organic C, N, P, and S in western Canadian soils. Can. J. Soil Sci. 69: 739–750.

Schnitzer, M. 1978. Humic substances: chemistry and reactions. Pages 1–64 in M. Schnitzer and S. U. Khan, Eds. Soil organic matter. Elsevier, Amsterdam.

Schnitzer, M. and Schuppli, P. 1989. The extraction of organic matter from selected soils and particle size fractions with 0.5 M NaOH and 0.1 M $Na_4P_2O_7$ solutions. Can. J. Soil Sci. 69: 253–262.

Schoenau, J. J. and Bettany, J. R. 1987. Organic matter leaching as a component of carbon, nitrogen, phosphorus, and sulfur cycles in a forest, grassland and gleyed soil. Soil Sci. Soc. Am. J. 51: 646–651.

Stevenson, J. F. 1982. Humus chemistry: genesis, composition, reactions. John Wiley & Sons, New York.

Chapter 39
Light Fraction and Macroorganic Matter in Mineral Soils

E. G. Gregorich and B. H. Ellert

Agriculture Canada
Ottawa, Ontario, Canada

39.1 INTRODUCTION

Soil can be separated into various physical, chemical, or biological fractions in order to identify different constituents and to elucidate the role of these constituents in soil processes. Soil organic matter can be separated by density or size; such physical fractionations are particularly useful in studying the form and function of organic constituents, which range from relatively large fragments of plant material to humified material.

Macroorganic matter can be defined as the sand-sized fraction (SSF [i.e., 53 to 2000 μm]) of organic matter. Most macroorganic matter is contained in the light fraction (LF) of soil, that is, the fraction with a density less than 2.0 g cm^{-3}. The organic matter in these fractions primarily comprises plant debris, recognized by its cellular structure, but may also contain fungal hyphae, spores, seeds, charcoal, and animal remains.

Soil scientists and ecologists separate LF and macroorganic matter from soil for a variety of reasons: to remove the nonhumified organic matter in order to characterize chemically the humified materials (e.g., Oades 1972), to isolate and measure an important reservoir of rapidly cycling organic. matter (e.g., Sollins et al. 1984), and to study the impact that transformations of plant tissue into humus have on fertility, residue persistence, and organisms (Finnell 1933, McCalla et al. 1943, Ford and Greenland 1968, Malone and Swartout 1969, Kanazawa 1979, Janzen et al. 1992).

LF organic matter generally accounts for 0.1 to 3% of the total weight of cultivated soils, but grassland and forest soils and podzols with relatively slow decomposition rates may contain considerably higher percentages (3 to 10%). LF organic matter is enriched in C and N, relative to fractions dominated by mineral constituents. Consequently, the proportion of total soil organic carbon and nitrogen that is present as the LF may reach 40 and 30%, respectively. LF organic matter comprising of plant residues contains high concentrations of oligosaccharides, polysaccharides, and hemicelluloses and thus serves as a readily decomposable substrate for microorganisms in soil. More than half the microbial populations and enzyme activities in soil may be associated with the LF (Kanazawa and Filip 1986), and the organic matter in this fraction decomposes quickly, despite a wide C:N ratio (Sollins et al. 1984, Greenland and Ford 1964). Studies with isotopic tracers indicate that LF organic matter has a faster turnover rate than other physical fractions but may consist of two or more carbon pools with different turnover rates (Bonde et al. 1992). LF organic matter has been shown to be a useful indicator of soil quality and, in particular, may be sensitive to changes in labile organic matter.

A standard technique for separation of LF and macroorganic matter from soil has not been developed but most methods generally involve flotation (density fractionation) or sieving. The use of physical fractions in organic matter studies has been reviewed (Shaymukhametov et al. 1984, Elliott and Cambardella 1991, Oades 1989, Christensen 1992), and similar methods have been used to isolate seeds or roots from soils (Ball and Miller 1989, Anderson and Ingram 1989, Böhm 1979, Barley 1955). Many early workers used various combinations of sieving in water, flotation on water, winnowing in air, dry panning, and collection of particulate organic matter that floated to the surface of a water column and adhered to the walls of a column. Since the density of soil minerals usually exceeds 2 g cm^{-3}, compared to 1.5 g cm^{-3} for organic materials, heavy liquids (density $\approx$2 g cm^{-3}) can be used to separate macroorganic matter from the mineral components. Two methods of separating organic matter are presented in this chapter: wet sieving, used to isolate the SSF, and flotation on sodium iodide at a density of 1.7 g cm^{-3}, used to isolate the LF.

39.2 MACROORGANIC MATTER

Macroorganic matter can be isolated with the sand fraction by the sieving techniques used for particle size analysis (see Chapter 47) except that pretreatments used to eliminate particulate organic matter, carbonates, and iron oxides are omitted.

39.2.1 MATERIALS AND REAGENTS

1   Reciprocating or rotary (i.e., end-over-end) shaker and several 200- to 250-mL bottles or flasks with leakproof closures. The shaker action and speed (revolutions or cycles per minute) and the flask orientation and geometry should be recorded because these variables influence the degree of soil dispersion (Stephanov 1981).

2   Sodium hexametaphosphate solution, 5 g L^{-1} (NaPO$_3$)$_6$.

3   Sieve with 53-μm openings (270 mesh, Canada Soil Survey Committee 1978), 20 cm diameter, placed in a large ($\approx$25 cm diameter) polypropylene funnel supported by a ring clamp on a laboratory stand, and several 1-L tall-form beakers to collect the silt + clay suspension that will be washed through the sieve.

4   Large wash bottle of distilled water and a spatula or rubber policeman to wash the silt + clay fraction through the sieve and to wash the sand + macroorganic matter fraction into preweighed drying tins.

39.2.2 PROCEDURE

1 Pass air-dried soil through a sieve with 2-mm openings and discard any residues retained on the sieve. Determine soil moisture content by drying at 105°C.

2 Weigh 25 g of soil into each bottle, dispense 100 mL of the sodium hexametaphosphate solution into each bottle, cap the bottles, and shake for 60 min.

3 Pour the suspension onto the 53-μm sieve, using small aliquots of water to rinse the particles from the bottle.

4 Wash the silt + clay-sized fraction, which includes mineral and fine organic matter, through the sieve using a fine jet of water from the wash bottle and gently brushing any aggregates with a rubber policeman. The sand + macroorganic matter fraction is retained on the sieve.

5 Wash the sand + macroorganic matter from the sieve to preweighed drying tins. Let excess water evaporate overnight, then place tins in the oven at 60°C to obtain the dry weight of the SSF.

6 Use a mortar and pestle to grind the SSF to pass a sieve with 250-μm openings (60 mesh). Do the same with a subsample of the whole soil. Determine the concentrations of carbon and other elements of interest in the whole soil and SSF.

39.2.3 COMMENTS

1 Soils are usually air-dried prior to dispersion to remove the effects of variations in moisture, but the amounts of macroorganic matter recovered from dried soils may differ from those recovered from field-moist soils.

2 It is important that minimal force be used during the wet-sieving step. Excessive abrasion of the soil against the sieve during this step will result in comminution and lower recovery of the macroorganic matter.

39.3 LIGHT FRACTION

LF organic matter can be isolated from the mineral part of soils by suspending the soil in a dense liquid and leaving the heavy fraction to settle to the bottom while the LF floats to the surface.

39.3.1 MATERIALS AND REAGENTS

1   Reciprocating or end-over-end shaker and several centrifuge tubes (≈100-mL capacity, polyallomer) with rubber stoppers. The shaker action and speed (revolutions or cycles per minute) and the flask orientation and geometry should be recorded because these variables influence the degree of soil dispersion.

2   Sodium iodide (USP grade is adequate), to prepare a solution with a density of 1.7 g cm^{-3}. Dispense 650 mL of water into a 1-L beaker (tall-form beakers accept the hydrometer required to measure solution density), and slowly add 795 g of NaI while the solution is stirred with a magnetic mixer. After the NaI is dissolved, cool the solution to room temperature and adjust the solution density to 1.7 g cm^{-3} according to hydrometer measurements. About 90 mL of solution, containing about 84 g of NaI, is required to separate the LF from each sample of soil.

3   Several 125-mL vacuum flasks and a vacuum hose with a disposable pipette tip cut at a 45° angle, to aspirate the light fraction.

4   A glass filtration unit consisting of a fritted glass filter support, a detachable funnel, a clamp to hold the funnel to the support, and two large (1-L) receptacles, one to collect the dense solution for reuse and the other to collect water washings for discard. Membrane filters made of nylon or "quantitative" filters, designed for easy recovery of the particles on the filter, may be used without contributing particulate carbon to the LF.

5   Three wash bottles, one containing the NaI solution (1.7 g cm^{-3}), one containing 0.01 M CaCl$_2$, and one containing distilled water.

39.3.2 PROCEDURE

1   Pass air-dried soil through a sieve with 2-mm openings, discarding the material retained on the sieve. Determine soil moisture contents by drying at 105°C.

2   Weigh 25 g of soil into each centrifuge tube, dispense 50 mL of the NaI solution into each tube, stopper the tubes, and shake for 60 min.

3   Remove the stopper from each centrifuge tube, and use the wash bottle containing NaI to wash the soil from the stopper and tube walls into the suspension (3 to 5 mL are sufficient). The tubes should be balanced for centrifugation at this time.

4   Centrifuge for 15 min at a relative centrifugal force of 1000 × g.

5   Aspirate the LF from the surface of each tube into a labeled vacuum flask. Remove enough of the dense liquid to wash the LF from the vacuum hose.

6   Pour the suspension of LF and dense liquid from the flask into the filter unit and collect the filtrate (density ≈ 1.7 g cm^{-3}) for reuse.

7   Transfer the filter unit, without disturbing the clamp or filter paper, from the receptacle containing dense solution for reuse to the receptacle used to collect the washings. Use the wash bottle containing the $CaCl_2$ to wash any LF from the walls of the vacuum flask and funnel to the filter paper. Use about 75 mL of $CaCl_2$ solution followed by 75 mL distilled water (at least 150 mL in all) to wash the NaI from the LF because halogens may interfere with some methods of carbon analysis and $CaCl_2$ will help prevent the clogging of the filter. Discard the wash water.

8   Remove the filter from the filter unit and wash the LF from the filter to preweighed drying tins. Let excess water evaporate overnight, then place tins in the oven at 60°C to obtain the dry weight of the LF.

9   To the tube containing the heavy fraction, add enough NaI solution to bring the weight of the suspension in the centrifuge tubes to 135 g (easily done by placing an empty tube on an electric balance and pressing "tare"). About 30 mL of dense solution must be added to replace the solution removed when the first aliquot of LF is aspirated.

10   Stopper the tubes, resuspend the heavy fraction, and shake for 60 min.

11   Repeat steps 3 to 8 at least once to recover another aliquot of the LF.

12   Combine the two aliquots of the LF and use a mortar and pestle to grind this fraction to pass a sieve with 250-μm openings (60 mesh). Do the same with a subsample of the whole soil. Determine the concentrations of carbon and other elements of interest in the whole soil and LF.

39.3.3 COMMENTS

1   The use of NaI to obtain the dense solutions is somewhat arbitrary, but this compound is less expensive and less toxic than most other media, is widely available, and can be used to make solutions with densities up to 1.9 g cm^{-3} at 25°C. Organic solvents have been used to fractionate soils according to density (Table 39.1), but problems with toxicity, carbon contamination, and coagulation of suspended particles were alleviated by using inorganic media. More recently, sodium metatungstate ($Na_6[H_2W_{12}O_{40}]$, Aldrich Chemical Co., Milwaukee, WI) has been used to prepare solutions with densities up to 3.1 g cm^{-3} at 25°C (Plewinsky and Kamps 1984). In addition to density, several solution properties (e.g., viscosity, surface tension, dielectric constant) may influence the results obtained by density fractionations of soils. For example, the apparent density of particulate organic matter will depend on the extent to which the dense solution occupies the cavities in the particles, and this will depend on the surface tension of the solution. More work is required to define the effects of solution properties on the density separations before results obtained from different solutions can be compared.

Table 39.1 Media Used to Prepare Heavy Solutions for Isolating the Light Fraction of Soil Organic Matter, and Density Used to Separate the Light Fraction

Media	Solution density, g cm^{-3}	Ref.
Toulet's solution (K$_2$HgI$_4$)	1.80	Khan (1959)
Bromoform/ethanol	2.00	Monnier et al. (1962)
Bromoform/petroleum spirit	2.00	Greenland and Ford (1964)
(NaPO$_3$)$_6$ + NaHCO$_3$ + MgSO$_4$	1.2?	Malone and Swartout (1969)
Tetrabromoethane/benzene	2.00	McKeague (1971)
Carbon tetrachloride	1.59	Scheffer (1977)
Calcium chloride	1.50	Al-Khafaf et al. (1977)
1,2-dibromo 3-chloropropane + surfactant	2.06	Turchenek and Oades (1979)
Sodium iodide	1.65	Spycher et al. (1983)
Sodium iodide	1.60	Sollins et al. (1984)
Zinc bromide	1.60	Ahmed and Oades (1984)
Bromoform/ethanol + surfactant	2.00	Dalal and Mayer (1986)
Sodium iodide	1.70	Strickland and Sollins (1987)
Sodium iodide	1.59	Janzen (1987)
Sodium metatungstate	1.80	Elliott et al. (1991)

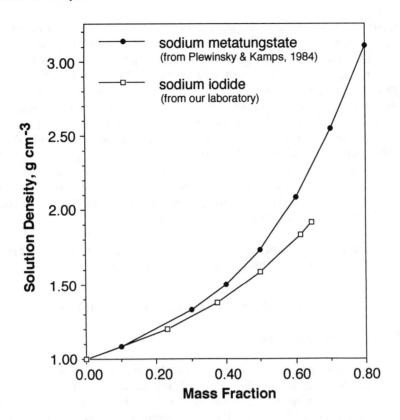

FIGURE 39.1. Densities of aqueous solutions at 20°C as a function of solute concentration.

2   The density of the solution used to isolate the LF has varied with the objectives of the researchers and the media used to prepare the dense solutions (Table 39.1). Our recommended density of 1.7 g cm^{-3} is within the range used by previous workers and is easily attained using various organic and inorganic media (e.g., the density of a saturated solution of KI is 1.72 g cm^{-3} at 20°C). If solutions of different densities are required, the mass of solute required to attain a predetermined density can be computed from plots of density against solute concentration (Figure 39.1). Concentration is expressed as mass fraction because the mass, rather than the volume, of solution components is additive:

$$F = S/(S + L) \tag{39.1}$$

$$S = FL/(1 - F) \tag{39.2}$$

$$S = F\, D_{sol'n} V_{sol'n} \tag{39.3}$$

where F = mass fraction, S = solute mass, L = solvent (water) mass, $D_{sol'n}$ = density of solution, and $V_{sol'n}$ = volume of solution. For example, a NaI solution with a density of 1.8 g cm^{-3} has a mass fraction of 0.60 (Figure 39.1) and, according to Equation 39.2, 650 g of water requires 975 g of NaI to attain a density of ≈1.8 g cm^{-3} (final solution volume ≈(975 + 650)/1.8 or 903 cm^3). Alternatively, Equation 39.3 indicates that 972 g of NaI is required to prepare 900 mL of solution at 1.8 g cm^{-3}.

3   Centrifugal force (1000 $\times$ g) is used to separate the light and heavy fractions in the above procedure, but the fractions may be separated by leaving the centrifuge tubes standing on the laboratory bench (1 $\times$ g) for 48 h. Separation without centrifugation may be more efficient when large numbers of samples are processed. However, the use of suction to remove the LF permits a better separation than that achieved by simple decantation. Centrifuge tubes allow the use of narrower solution/soil ratios (50 mL/25 g in the above procedure) compared to methods that use beakers (200 mL/30 g used by Strickland and Sollins 1987). If the solution/soil ratio is decreased further, some of the LF could become entrapped within the heavy fraction during the fractionation procedure. Therefore, it is important that the heavy fraction be resuspended and the separation procedures (steps 3 to 8 in Section 39.3.2) performed at least two or three times.

4   The use of the 125-mL vacuum flasks may be omitted by aspirating the LF directly from the centrifuge tube to the filter unit.

39.4 IMPORTANT CONSIDERATIONS

39.4.1 Methods of Disruption

Roots and decomposing plant materials are loci for the formation of aggregates in soil. Disruption of aggregates is essential to isolate these materials. Sonification, shaking, and chemical dispersion are the most common methods used to disrupt aggregates and disperse soil (Elliott and Cambardella 1991). The vibrations produced during sonification of a soil solution disrupt and disperse aggregates. Shaking soil suspensions is relatively more gentle, although shaking with glass beads or agate marbles can result in greater disruption of aggregates. Mineral salt solutions which contain cations such as sodium will facilitate the dispersion of soil.

39.4.2 Calculation of Results

Calculate the proportion of whole soil mass in the SSF or LF:

$$\text{fraction/whole soil} = (\text{SSF or LF})/\{\text{soil} /[1 + (\%H_2O/100)]\} \qquad (39.4)$$

where fraction/whole soil = proportion of whole soil mass in the SSF or LF, SSF or LF = dry weight of SSF or LF obtained, soil = original weight of soil fractionated, $\%H_2O$ = moisture content of soil fractionated.
Calculate the proportion of whole soil carbon in the LF or SSF:

$$\text{fraction C/whole soil C} = [(\text{fraction/whole soil}) \times (\text{SSF C or LF C})]/\text{soil C} \qquad (39.5)$$

where fraction C/whole soil C = proportion of whole soil carbon in the SSF or LF, fraction/whole soil = proportion of whole soil mass in the SSF or LF, SSF C or LF C = carbon concentration in the SSF or LF, soil C = carbon concentration in the whole soil.

The ash content of the SSF or LF can be determined by weighing aliquots before and after ignition in a muffle furnace for 4 h at 550°C. Since the SSF is dominated by mineral components, the ash-free portion of this fraction represents the macroorganic matter. Corrections for ash in LF may help to account for the presence of light minerals, aggregates, or phytoliths.

39.4.3 Losses During Fractionation

The mass and carbon of the whole soil can be compared with the sums of mass and carbon in the various fractions to ensure that losses during the fractionation do not introduce appreciable bias. To calculate a mass balance it is necessary to recover the silt + clay fraction in the sieving method or the heavy fraction in the flotation method. Calcium chloride (20 mL of 3 M $CaCl_2$ or some other flocculating agent) should be added so the silt + clay settles out of the suspension that passed the 53-μm sieve. After the supernatant is siphoned off, the slurry left in the bottom of the beaker can be transferred to containers that are suitable for freeze drying. In the flotation method, the heavy fraction can be recovered by siphoning off the dense solution and repeatedly resuspending the heavy fraction in wash water, followed by centrifugation and aspiration of the supernatant. When the heavy fraction fails to form a stable pellet (usually after two to three washings), it can be frozen in the centrifuge tubes and freeze dried. Samples of supernatant, wash water, and dense solutions may have to be analyzed to estimate losses of soil carbon to the liquid phase.

39.4.4 Bioassays

The medium used in separating the LF from whole soil may have deleterious effects on the viability of certain microbial populations and their activities. Studies in which the LF is characterized by bioassay may require that the LF be thoroughly washed and recolonized by inoculation with a dilute suspension (e.g., 1 g L^{-1}) of the corresponding soil.

Studies on soil biological processes affecting nutrient cycling may require fresh moist samples, because sample drying affects total microbial biomass, the type of active microbes, and enzyme activity.

REFERENCES

Ahmed, M. and Oades, J. M. 1984. Distribution of organic matter and adenosine triphosphate after fractionation of soils by physical procedures. Soil Biol. Biochem. 5: 465–470.

Al-Khafaf, S., Wierenga, P. J., and Williams, B. C. 1977. A flotation method for determining root mass in soil, Agron. J. 69: 1025–1026.

Anderson, J. M. and Ingram, S. I. 1989. Tropical soil biology and fertility: a handbook of methods. CAB International, Wallingford, England.

Ball, D. A. and Miller, S. D. 1989. A comparison of techniques for estimation of arable soil seedbanks and their relationship to weed flora. Weed Res. 29: 365–373.

Barley, K. P. 1955. The determination of macroorganic matter in soils. Agron. J. 47: 145–147.

Böhm, W. 1979. Methods of studying root systems. Ecological Studies No. 33. Springer-Verlag, New York.

Bonde, T. A., Christensen, B. T., and Cerri, C. C. 1992. Dynamics of soil organic matter as reflected by natural ^{13}C abundance in particle size fractions of forested and cultivated oxisols. Soil Biol. Biochem. 24: 275–277.

Canada, Soil Survey Committee. 1978. The Canadian system of soil classification. Publication 1646. Research Branch, Canada Department of Agriculture.

Christensen, B. T. 1992. Physical fractionation of soil and organic matter in primary particle size and density separates. Adv. Soil Sci. 20: 1–90. Springer-Verlag, New York

Dalal, R. C. and Mayer, R. J. 1986. Long-term trends in fertility of soils under continuous cultivation and cereal cropping in Southern Queensland. IV. Loss of organic carbon from different density fractions. Aust. J. Soil Res. 24: 301–309.

Elliott, E. T. and Cambardella, C. A. 1991. Physical separation of soil organic matter. Agric. Ecosys. Environ. 34: 407–419.

Elliott, E. T., Palm, C. A., Reuss, D. E., and Monz, C. A. 1991. Organic matter contained in soil aggregates from a tropical chronosequence: correction for sand and light fraction. Agric. Ecosys. Environ. 34: 443–451.

Finnell, H. H. 1933. Raw organic matter accumulations under various systems of culture. Oklahoma Exp. Station Bull. No. 216. 12 pp.

Ford, G. W. and Greenland, D. J. 1968. The dynamics of partly humified organic matter in some arable soils. Trans. 9th Int. Cong. Soil Sci. II: 403–410. Adelaide.

Greenland, D. J. and Ford, G. W. 1964. Separation of partially humified organic materials from soils by ultrasonic dispersion. Trans. 8th Int. Cong. Soil Sci. II: 137–147. Bucharest.

Janzen, H. H. 1987. Soil organic matter characteristics after long-term cropping to various spring wheat rotations. Can. J. Soil Sci. 67: 845–856.

Janzen, H. H., Campbell, C. A., Brandt, S. A., Lafond, G. P., and Townley-Smith, L. 1992. Light fraction organic matter in soils from long-term crop rotations. Soil Sci. Soc. Am. J. 56: 1799–1806.

Kanazawa, S. 1979. Studies on the plant debris in rice paddy soils. I. Morphological observation and numbers of microbes in fractionated plough layer of paddy soils. Soil Sci. Plant Nutr. 25: 59–69.

Kanazawa, S. and Filip, Z. 1986. Distribution of microorganisms, total biomass, and enzyme activities in different particles of brown soil. Microbiol. Ecol. 12: 205–215.

Khan, D. V. 1959. The composition of organic substances and their relationship to the mineral portion of the soil. Soviet Soil Sci. 1: 7–14.

Malone, C. R. and Swartout, M. B. 1969. Size, mass, and caloric content of particulate matter in old-field and forest soils. Ecology 50: 395–399.

McCalla, T. M., Duley, F. L., and Goodding, T. H. 1943. A method for measuring the plant residue fragments of the soil. Soil Sci. 55: 159–166.

McKeague, J. A. 1971. Organic matter in particulate-size and specific gravity fractions of some Ah horizons. Can. J. Soil Sci. 51: 499–505.

Monnier, G., Turc, L., and Jeanson-Luusinang, C. 1962. Une méthode de fractionnement densimétrique par centrifugation des matières organiques du sol. Ann. Agron. 13: 55–63.

Oades, J. M. 1972. Studies on soil polysaccharides. III. Composition of polysaccharides in some Australian soils. Aust. J. Soil Res. 10: 113–126.

Oades, J. M. 1989. An introduction to organic matter in mineral soils. Pages 89–159 in J. B. Dixon and S. B. Weed, Eds. Minerals in soil environments. Soil Science Society of America, Madison, WI.

Plewinsky, B. and Kamps, R. 1984. Sodium metatungstate: a new medium for binary and ternary density gradient centrifugation. Die Makromol. Chem. 185: 1429–1439.

Scheffer, B. 1977. Stabilization of organic matter in sand mixed cultures. Pages 359–363 in Soil organic matter studies. Vol. II. IAEA and FAO.

Shaymukhametov, M. S., Titova, N. A., Travnikova, L. S., and Labenets, Y. M. 1984. Use of physical fractionation methods to characterize soil organic matter. Soviet Soil Sci. 8: 131–141.

Sollins, P., Spycher, G., and Glassman, C. A. 1984. Net mineralization from light- and heavy-fraction forest soil organic matter. Soil Biol. Biochem. 16: 31–37.

Spycher, G., Sollins, P., and Rose, S. 1983. Carbon and nitrogen in the light fraction of a forest soil: vertical distribution and seasonal patterns. Soil Sci. 135: 79–87.

Stephanov, I. S. 1981. Physical methods for extracting fractions of organo-mineral substances from soils. Soviet Soil Sci. 4: 110–121.

Strickland, T. C. and Sollins, P. 1987. Improved method for separating light- and heavy-fraction organic material from soil. Soil Soc. Am. J. 51: 1390–1393.

Turchenek, L. W. and Oades, J. M. 1979. Fractionation of organo-mineral complexes by sedimentation and density techniques. Geoderma 21: 311–343.

Chapter 40
Water-Soluble Phenolic Materials

L. E. Lowe
University of British Columbia
Vancouver, British Columbia, Canada

40.1 INTRODUCTION

There are a variety of possible reasons for interest in phenolic materials in soil and a variety of forms in which such materials occur in soil. Simple phenols and phenolic acids can be present in soil solution or in soil leachates. Such compounds can also be present in adsorbed or esterified forms, as well as occurring as structural units in complex humus polymers or plant residues, which together make up a large proportion of soil organic matter. While most of the polymeric phenolic material (polyphenols) is present in the solid phase, small amounts can also be present in soil solution.

Interest in phenolic material in soils relates to known or postulated involvement in allelopathy, metal translocation, moderation of metal toxicity, role as humus precursor, and as a marker of vegetative origin of soil organic matter. Some of these concerns require information on specific phenolic compounds, while others can take advantage of information on "total phenolics" in soil solution or soil extracts. Apart from some general observations, methods for the former (analysis of specific compounds) will not be discussed here.

Of particular interest are the cinnamic acid derivatives arising from lignin breakdown and the phenolic acids arising from their further degradation. Monomeric phenolic acids have been implicated in allelopathic effects (Jalal and Read 1983), in inhibition of phosphate uptake by plants (Glass 1973), metal translocation in podzols (Vance et al. 1986), and in an important role as intermediates in humus formation (Flaig et al. 1975). The more soluble, and probably smaller, polyphenols may well also be involved in some of the same roles.

Individual phenolic compounds from soil extracts can be separated and analyzed by high-performance liquid chromatography (e.g., Vance et al. 1985, Hartley and Buchan 1979), but estimates of total water-soluble phenolic materials can be obtained by simple colorimetric procedures based on the Folin-Ciocalteau reagent. Such procedures have been used in soil studies by Morita (1980) and on peats by Kuiters and Denneman (1987).

The Folin-Ciocalteau reaction with phenols has been carefully examined by Box (1983) in relation to use on natural waters, and the colorimetric procedure described below is based on his work. It is recommended for use on water extracts of organic horizon samples or Ah

Soil Sampling and Methods of Analysis, M. R. Carter, Ed.,
Canadian Society of Soil Science. © 1993 Lewis Publishers.

horizons, which yield estimates of "free phenolics". A similar procedure was applied by Morita (1980) to alkaline pyrophosphate extracts of peats, in which polyphenolic humic substances predominate. The reported procedure has yet to be adequately evaluated for mineral soil horizons.

The method has some clear limitations, as discussed by Box (1983). It recovers a variety of individual phenolic or polyphenolic substances, which vary in their color yield in reaction with the Folin-Ciocalteau reagent. They also vary somewhat in absorption maxima. Thus the procedure should be regarded as a useful comparative measure for samples of a similar character, rather than as an absolute measure of total phenolic levels. Interferences are possible from a variety of organic and inorganic reducing agents, but are considered unlikely to have a major effect on results for most soil situations.

40.2 SOLUBLE PHENOLICS (BOX 1983)

40.2.1. EQUIPMENT

1   Spectrophotometer

40.2.2. REAGENTS

1   Na_2CO_3 solution, 20% (w/v).

2   Folin-Ciocalteau phenol reagent (available in Canada from BDH Inc.). This reagent should be stored in the dark.

3   Stock standard: vanillic acid, 100 $\mu g \cdot mL^{-1}$ (dissolution is very slow unless warmed and stirred).

4   Working standards: prepare solutions of 2, 4, 6, 8, and 10 $\mu g \cdot mL^{-1}$ vanillic acid by dilution from stock solution.

40.2.3. PROCEDURE

1   Extract 5 g soil (<35 mesh) with 50 mL water for 4 h with shaking. Centrifuge and/or filter to obtain clear extract (narrower soil:water ratios may be advisable for mineral horizon samples).

2   Place 10-mL aliquot of sample extract or standard in 200- × 25-mm test tube. Add 10 mL distilled water (if very low concentration of phenolics are encountered, a 20-mL aliquot can be used and the water omitted).

3   Add 3 mL Na_2CO_3 solution, followed by 1 mL Folin-Ciocalteau reagent. Mix well and allow to stand for 1 h at room temperature (20–25°C).

4   Read absorbance at 750 nm against blank (replacing sample with 10 mL distilled water).

5   Prepare calibration curve and calculate "water-soluble phenolics" in sample in $\mu g \cdot g^{-1}$ vanillic acid equivalent.

40.3 COMMENTS

1   For reproducible results, accurate dispensing of reagent and sample volumes must be maintained.

2   Other phenolic compounds can be used as standards where deemed more appropriate to a given application. Phenol and tannin, among others, have been used.

3   The possibility of increased positive errors in the estimate of phenolics can be anticipated for poorly drained mineral soils, where interference from Fe^{2+}, Mn^{2+}, or S^{2-} is more likely.

4   When a wavelength of 750 nm is unattainable on the available spectrophotometer, absorbance can be read at lower wavelengths down to 660 nm, with some loss of sensitivity.

5   When a "soluble phenolics" determination is made in relation to possible allelopathic or other biochemical roles, it is important to use fresh field-moist samples.

6   Kuiters and Denneman (1987) have reported a procedure in which monomeric and polymeric phenols can be determined separately. After determination of total soluble phenols, polyphenols are precipitated by addition of casein and a second analysis with Folin-Ciocalteau reagent conducted on the supernatant extract.

REFERENCES

Box, J. D. 1983. Investigation of the Folin-Ciocalteau phenol reagent for the determination of polyphenolic substances in natural waters. Water Res. 17: 511–525.

Flaig, W., Beutelspacher, H. and Rietz, E. 1975. Chemical composition and physical properties of humic substances. p. 211 in J. E. Gieseking, Ed. Soil components, Vol. I, Organic components. Springer-Verlag, New York.

Glass, A. D. M. 1973. Influence of phenolic acids in ion uptake I. Inhibition of phosphate uptake. Plant Physiol. 51: 1037–1041.

Hartley, R. D. and Buchan, H. 1979. High-performance liquid chromatography of phenolic acids and aldehydes derived from plants or from the decomposition of organic matter in soil. J. Chromatog. 180: 139–143.

Jalal, M. A. F. and Read, D. J. 1983. The organic acid composition of Calluna heathland soil with special reference to phyto- and fungitoxicity II. Monthly quantitative determination of the organic acid content of Calluna and spruce dominated soils. Plant and Soil 70: 273–286.

Kuiters, A. T. and Denneman, C. A. J. 1987. Water soluble phenolic substances in soils under several coniferous and deciduous tree species. Soil Biol. Bioch. 19: 765–769.

Morita, M. 1980. Total phenolic content in the pyrophosphate extracts of two peat soil profiles. Can. J. Soil Sci. 60: 291–297.

Vance, G. F., Mokma, D. L. and Boyd, S. A. 1986. Phenolic compounds in soils of hydrosequences and developmental sequences of spodosols. Soil Sci. Soc. Am. J. 50: 992–996.

Chapter 41
Soil Lipids

Yeh-Moon Chae

Alberta Environment

Edmonton, Alberta, Canada

41.1 INTRODUCTION

Lipids are defined as biologically derived organic compounds soluble in organic solvents. The lipid component is an important part of soil organic matter (Braids and Miller 1975). Compared to other classes of soil organics, such as carbohydrates, the lipid fraction of soils exhibits the greatest diversity of compound types, comprising a wide variety of substances ranging from rather stable paraffinic hydrocarbons to chlorophyll-degradation products, and differing from each other in chemical structure and physicochemical properties. Lipid fractions in soils, however, have received virtually no systematic attention from soil chemists, no doubt because of the difficulty of separation and identification of components.

Generally, soil lipids are believed to consist of the same compounds as plant and microbial lipids. Andreyev et al. (1981) have shown evidence that soil lipids are mostly of microbial origin, whereas Moucawi et al. (1981) state that a major proportion of soil lipids originates from plant materials. In normal aerobic soils, the lipids probably exist largely as remnants of plant and microbial tissues or other organisms.

Extensive reviews on the subject of soil lipids are given by Braids and Miller (1975), Fridland (1976), Morrison (1969), and Stevenson (1966).

41.2 EXTRACTION OF SOIL LIPIDS

Much of the lipids in soils may be linked in some form of combination with protein or carbohydrate, and these complexes are generally insoluble in organic solvents (Wagner and Muzorewa 1977). The solubility of soil lipids is likely to be further affected by the presence of large amounts of inorganic material, such as clay minerals (Wang et al. 1969) and cations of aluminum or iron (Fridland 1976). Those lipids combined with organic or inorganic soil constituents are called "bound lipids".

An acid pretreatment is required to extract the inorganic-bound lipids, such as aluminum or iron salts of fatty acids (Fridland 1976). To include those bound lipids in the extracts, soils are extracted after pretreatment with acid (Perniola et al. 1981), or the residual soils after extraction of free lipids are extracted after retreatment with acid (Ziegler 1989).

Soil Sampling and Methods of Analysis, M. R. Carter, Ed.,

Canadian Society of Soil Science. © 1993 Lewis Publishers.

It is often possible to liberate protein-bound lipid by use of ethanol. However, for most lipids, ethanol is a poor solvent. The addition of ether or benzene improves the effectiveness of ethanol, and such solvent mixtures are frequently used for extraction of soil lipids (Morrison and Bick 1967, Wagner and Muzorewa 1977).

Both the yield and the chemical nature of the material extracted by organic solvents from soils are influenced by the nature of the solvent (Sciacovelli et al. 1977) and the conditions of extraction (Morrison and Bick 1967). Schnitzer and Schuppli (1989), however, reported that both *n*-hexane and chloroform extracts contain similar aliphatic materials.

The extractability of soil lipids may be affected by the drying of the soil, since it is generally considered that drying reduces lipid solubility of producing changes in the fatty acid constituents (Hance and Anderson 1963a). The nonlipid contaminants in lipid extracts may also be a factor complicating the extraction of lipids from soils, because some lipid extractants, particularly acetone and ethanol, can extract inorganic or nonlipid organic substances. The alcohol-benzene fraction may contain some specific humic substances, such as hymatomelanic acid, and some resinoid substances and free amino acids.

The water-washing procedure also is not an entirely satisfactory way to remove nonlipid contaminants because this procedure may result in some loss of lipids and retention of nonlipid substances in the lipid extract (Nazir and Rouser 1966). The role that components of complex solvent systems play in lipid extractions has been discussed in the light of solubility parameter concepts (Schmid 1973).

If the formation of lipid artifacts caused by enzyme activity is to be avoided, some form of enzyme denaturation must precede a lipid extraction method. Methods commonly used for enzymic deactivation of tissues include the use of boiling "alcohols" such as isopropanol (Kates and Eberhardt 1957), methanol, water-saturated *n*-butanol (Colborne and Laidman 1975), or simple heat treatment at 100°C. However, water-saturated *n*-butanol is suspected of giving rise to lysophospholipid artifacts by nonenzymic hydrolysis, whereas methanol may form fatty acid methyl esters through acyl transferase activity. Extraction of soil lipids may also require attention to the potential hazards of artifact formation.

Despite much effort, the perfect extraction method has not been discovered. There are clearly many possible combinations of solvents, and their selection will depend on the type of material and the nature of the predominant lipids. The use of a variety of solvents and combinations seems essential.

A traditional method for extraction of soil lipids is described by Kononova (1966). However, this method uses benzene, which is not acceptable for use in most laboratories. The following section describes the Waring® blender extraction (Bligh and Dyer 1959), which uses a chloroform-methanol solvent system.

41.2.1 Waring® Blender Method (Bligh and Dyer 1959)

This method is based upon the ability of a chloroform-methanol mixture to form monophasic systems with tissue water, disintegrate membrane structures, and extract lipid from animal tissues. Because of its rapidity and ease of adaptability, the method has recently been employed for soil lipid extraction (Chae and Lowe 1980, Chae and Tabatabai 1981).

Fishwick and Wright (1977) have demonstrated that the chloroform-methanol solvent system based upon the principles of Bligh and Dyer (1959) is the most efficient means for use with the majority of plant tissues. For this reason, Ziegler (1989) also employed the chloroform-methanol solvent system of Bligh and Dyer (1959) for extracting unbound lipid from forest humus layers.

41.2.1.1 MATERIALS AND REAGENTS

1 Waring® blender made from stainless steel and fitted with a Teflon® gasket.

2 Infrared-red lamp.

3 Compressed nitrogen gas cylinder equipped with pressure regulator.

4 Vacuum desiccator.

5 Flash evaporator.

6 Chloroform: reagent grade.

7 Methanol: reagent grade.

41.2.1.2 PROCEDURE

1 Suspend the air-dried soil sample (100 g, <2 mm) in sufficient distilled water (around 80 mL) to make the water content of the suspension 80 ± 1%.

2 Homogenize the suspension with 300 mL of methanol-chloroform (2:1, v/v) mixture for 2 min in a Waring® blender fitted with a Teflon® gasket.

3 Add 100 mL of chloroform during blending and continue blending for a further 30 s.

4 Add 100 mL of distilled water and continue blending for another 30 s.

5 After blending, centrifuge the soil mixture in a polyethylene container at 15,000 × g to partly clarify.

6 Filter the mixture through Whatman® No. 1 filter paper on a Buchner funnel with slight suction.

7 Transfer the filtrate to a graduate cylinder (500 mL) and allow a few minutes for the chloroform layer to separate and clarify.

8 Record the volume of the chloroform layer for later use in the determination of the lipid content.

9   Remove the aqueous methanol layer by aspiration. Remove a very small volume of the chloroform layer (containing immiscible suspension) to ensure complete removal of the methanol layer. The chloroform layer contains the purified lipids.

10 Concentrate the lipid extract (the chloroform layer) to a known volume (50 mL or 100 mL) by evaporating on a flash evaporator at 35–40°C.

11 Store the concentrated lipid contained in a volumetric flask in a freezer and use a portion of this solution for fractionation and analysis.

12 Use an aliquot of this concentrated lipid for the determination of various constituents as described in the following Sections 41.3–41.6.

41.2.1.3 COMMENTS

1   There is no significant difference in lipid concentrations when soil is extracted right after soaking or after being kept in suspension overnight.

2   The lipids extracted from the soils and sewage sludges, using this method, are free of inorganic sulfate (Chae and Tabatabai 1981).

41.3 SILICIC ACID COLUMN CHROMATOGRAPHIC FRACTIONATION

The elution of the lipid extract with a silicic acid column using the sequence of chloroform, acetone, and methanol provides an essentially quantitative separation into three groups; less polar (neutral) lipids, glycolipids, and phosphatides (Rouser et al. 1967a). Mono- and diglycosyl diglycerides can be separated from each other and other lipid classes by elution with a chloroform-acetone (1:1, v/v) mixture followed by acetone (Vorbeck and Marinetti 1965). The procedure is particularly useful for brain and spinach leaf lipid extracts, in which glycolipid is high and diphosphatidyl glycerol is a very minor component (Rouser et al. 1967b). Since fairly large amounts of lipids appeared in the fraction eluted with acetone (Chae and Lowe 1981, Ziegler 1989), the procedure was considered to be efficient in the fractionation of soil lipids.

41.3.1 MATERIALS AND REAGENTS

1   Chromatographic column: 2.0-cm I.D. and 40-cm-long column equipped with a 500-mL solvent reservoir and a Teflon® stopcock and fritted disk with glass wool plug on it. Prepare a bed of 8.0-cm-high silicic acid by pouring a slurry of about 15 g of the silicic acid in chloroform into the column. Prewash the bed with three column volumes of chloroform. Allow the solvent level to ascend to the top of the bed.

2   Flash evaporator.

3 Silicic acid: 0.07- to 0.17-mm mesh. Heat activated at 120°C before use.

4 Chloroform: reagent grade.

5 Acetone: reagent grade.

6 Methanol: reagent grade.

41.3.2 PROCEDURE

1 Apply 4- to 10-mL aliquots of the concentrated lipid extract prepared in Section 41.2.1.

2 Transfer any suspended solid along with the soluble material and ensure quantitative transfer by thorough rinsing of all glasswares with chloroform.

3 Elute neutral lipids (Fraction 1) with 175 mL of chloroform at a flow rate of 3 mL per minute, and collect bulk eluent in an appropriate size of the round-bottom flask. Keep the flow rate (3 mL per minute) for the rest of four elutions.

4 Elute glycolipids containing monoglycosyl diglycerides (Fraction 2) with 90 mL of chloroform-acetone (1:1, v/v) mixture and collect the eluent.

5 Elute glycolipids containing diglyceride diglycosides and others (Fraction 3) with 700 mL of acetone and collect the eluent.

6 Elute polar lipids (Fraction 4) with 175 mL of methanol and collect the eluent.

7 Evaporate the solvent of the eluent on a flash evaporator to reduce to a small volume (usually 5–10 mL). Repeat this step for all four fractions.

8 Dilute to small known volumes in volumetric flasks (10–20 mL) with glass stoppers and store in a refrigerator for later use.

41.3.3 COMMENTS

1 The lipid composition of the four chromatographic fractions has been interpreted as follows (Rouser et al. 1967b). Fraction 1 — the chloroform eluate contains the less polar lipids including sterols, sterol esters, mono-, di-, and triglycerides, hydrocarbons, and free fatty acids. Fraction 2 — the chloroform-acetone (1:1, v/v) solvent mixture separates monoglycosyl diglycerides from diglycosyl diglycerides which is eluted with acetone. Cerebrosides are also almost completely separated from sulfatides with this solvent mixture. Other somewhat less polar lipids are eluted along with monoglycosyl diglycerides with spinach leaf lipid extracts. Fraction 3 — the acetone elutes diglycosyl diglycerides and sulfolipids with plant lipid extracts. With bacterial

lipids both neutral and acidic glycosyl glycerides are eluted, although these are not present in extracts from all microorganisms. With lipid extracts of animal organs (particularly brain), sulfatides and ceramide polyhexosides are eluted with acetone. With fecal lipid extracts a large number of uncharacterized substances devoid of phosphorus are eluted with acetone. No more than traces of phosphorus are found in acetone eluates, except with samples that contain a large amount of diphosphatidyl glycerol that is eluted in part with acetone. Fraction 4 — the methanol eluate from animal organ extracts, spinach leaves, and some bacteria contains phosphatides with, at most, traces of glycolipids. Fecal lipid extracts contain a variety of substances devoid of phosphorus that are eluted with methanol along with phosphatides.

2   After exhaustive elution with methanol, a colored residue remains on the column (Chae and Lowe 1981, Ziegler 1989), as some highly polar lipid materials are retained by the silicic acid. These highly polar lipids, however, can be eluted with 10% formic acid in methanol (Ziegler 1989).

3   Total recoveries of lipids from the columns are always over 100% due to fine particles of silicic acid eluted along with the elution solvent through the fritted glass disks (Chae and Lowe 1981). This problem can be overcome by using finer-fritted glass disks with a reduced flow rate.

4   Regeneration of the chromatographic column between runs is possible by washing the column with 70 mL methanol containing 10% formic acid and methanol, 30 mL of chloroform-methanol (1:1, v/v), and 90 mL of chloroform (Ziegler 1989).

41.4 TOTAL LIPIDS AND LIPIDS IN FOUR CHROMATOGRAPHIC FRACTIONS

Although much information in the literature on the concentrations of total lipids cannot be coordinated because of the use of different solvents and conditions for extraction, the values are comparable.

Soil lipids are only a small and variable part of organic matter. Lipid contents for 37 Canadian soils ranged from 1.0 to 50 g kg^{-1} of soil (Chae and Lowe 1980). The contents of lipids in the A1 horizon of most Russian soils ranged from between 0.6 and 14 g kg^{-1} of soil, with 2–5 g kg^{-1} being most common (Fridland 1976). This range for the lipid content is probably characteristic of most agriculturally important soils of the world.

41.4.1 MATERIALS AND REAGENTS

1   Compressed, high-purity nitrogen gas.

2   Infrared lamp.

3   Vacuum desiccator.

41.4.2 PROCEDURE

1   Transfer a portion of the concentrated lipid extracts or eluents prepared in Section 41.2 or 41.3 (containing 100 to 200 mg lipid) into a tared flask.

2   Evaporate the extract to dryness under an infrared lamp at 40–50°C under a stream of nitrogen.

3   After further drying over phosphoric anhydride in a vacuum desiccator, determine lipid content by weighing the dried residue.

41.4.3 Comments

The concentration of total lipids and the distribution of lipids in the four column chromatographic fractions are shown in Table 41.1.

41.5 PHOSPHOLIPIDS

Phospholipids are a group of biologically important organic compounds that are water insoluble, but are soluble in fat solvent. Although phospholipids occur in all living cells, the total quantity found in soils represents only <5% of organic phosphorus (Chae and Lowe 1980, Chae and Tabatabai 1981, Dormaar 1970, Hance and Anderson 1963b, Kowalenko and McKercher 1971a). Examination of lipids by thin-layer chromatographic techniques indicated that phosphatidyl choline (lecithin) and phosphatidyl ethanolamine (cephalin) are the dominant phospholipids in soils (Anderson 1980, Kowalenko and McKercher 1971b).

Current column and thin-layer chromatographic techniques for analysis of phospholipids are elaborate, time consuming, and at best semiquantitative. Recent advances in gas-liquid and high-performance liquid chromatography have provided a potential for rapid and accurate means of separating and identifying lipids in soils (Stott and Tabatabai 1985).

Lipid phosphorus (lipid-P) content in soil is determined by the method of Dick and Tabatabai (1977) for total phosphorus after an aliquot of the lipid extract is taken to dryness. Lipid-P is defined as those phosphorus-containing materials which are a component of lipids and consequently represent phospholipid. The method for total phosphorus is described in Chapter 23 (Section 23.2.4).

41.6 SULFOLIPIDS

Sulfolipids denote any sulfur-containing lipid and include sulfatides and sulfonolipids and thiolipids. Sulfolipids of diverse structure differ significantly in their chemistry and biochemistry. Due to these differences, analytical methods should accordingly follow up with a variety of techniques and methodologies, depending upon the sulfolipid to be studied, the biological source, the purpose of the analysis, and the facilities of the laboratory. The chemistry and analytical methods for sulfolipids have been extensively reviewed by Haines (1971).

Table 41.1 Concentration of Total Lipids and Lipids in Each Fraction (Fr) of Silicic Acid Column Chromatography

Soil sample		Horizon	Total lipid (%)	Less polar lipid (%) Fr #1	Glycolipid (%)		Phospholipid (%) Fr #4
Dominant vegetation					MGDG[a] Fr #2	DGDG[b] Fr #3	
Ziegler (1989)							
(1)	Norway spruce	L	7.7	41.1	38.0	6.9	11.2
(2)	Norway spruce	Of	5.8	32.6	39.1	7.0	15.1
(3)	Norway spruce	Oh	5.3	42.4	40.4	7.2	8.9
(4)	Norway spruce	Aeh	0.7	27.8	51.9	8.3	11.7
Chae and Lowe (1981)							
(5)	Western hemlock	F	4.0	33.4	57.6	8.1	3.5
(6)	Western hemlock	H	4.2	22.5	64.6	12.1	5.3
(7)	Subalpine grass	Om	1.3	34.8	54.6	9.2	4.6
(8)	Western red cedar	Ah	0.3	42.3	48.9	7.6	4.1
(9)	Grass	Ah	0.2	37.3	53.7	7.6	5.0
(10)	Grass	Ah	0.2	44.3	47.3	6.3	4.1
(11)	Grass-sedge	Oh	1.7	38.8	53.6	8.1	4.0
(12)	Grass-sedge	Oh	2.0	36.3	51.2	10.5	3.2

Note: Total lipid concentration is expressed as the percent of dry soil or leaf litter; lipid concentration in each fraction is as a percent of total lipid applied.

[a] MGDG denotes monoglycosyl diglycerides.
[b] DGDG denotes diglycosyl diglycerides.

Whereas sulfolipids are originally considered less common than phospholipids, it now appears they are ubiquitous. The retarded development of sulfolipid research has been due to the poorer analytical methods for sulfate, whereas phosphate is conveniently assayed by the molybdenum blue method.

Kean (1968) successfully applied azure A pigment for the quantitative estimation of sulfated amphiphiles extracted from animal tissues. Tadano and Ishizuka (1983) showed that the modified procedure of Kean (1968) is applicable to the accurate determination of purified or partially purified samples of the various sulfoglycolipids. However, application of the method to the soil extracts and fractions eluted from silicic acid column seems to be difficult (Chae, unpublished result).

Sulfur content in lipid extracts is determined by a method for total sulfur (Tabatabai and Bremner 1970) as described in Chapter 24 (Section 24.2).

REFERENCES

Anderson, G. 1980. Assessing organic phosphorus in soils. Pages 411–431 in F. E. Khasawneh et al., Eds. The role of phosphorus in agriculture. Soil Science Society of America, Madison, WI.

Andreyev, L. V., Nemirovskaya, I. B., Nikitin, D. I., Tomaschuk, A. Y., and Khmel'nitskiy, R. A. 1981. Lipid composition of humus. Soviet Soil Sci. 12: 406–412.

Bligh, E. G. and Dyer, W. J. 1959. A rapid method of total lipid extraction and purification. Can. J. Biochem. Physiol. 37: 911–917.

Braids, O. C. and Miller, R. H. 1975. Fats, waxes, and resins in soil. Pages 343–368 in J. E. Gieseking, Ed. Soil components. Vol. 1, Organic components. Springer-Verlag, Berlin.

Chae, Y. M. and Lowe, L. E. 1980. Distribution of lipid sulphur and total lipids in soils of British Columbia. Can. J. Soil Sci. 60: 633–640.

Chae, Y. M. and Lowe, L. E. 1981. Fractionation by column chromatography of lipids and lipid sulfur extracted from soils. Soil Biol. Biochem. 13: 257–260.

Chae, Y. M. and Tabatabai, M. A. 1981. Sulfolipid and phospholipid in soils and sewage sludges in Iowa. Soil Sci. Soc. Am. J. 45: 20–25.

Colborne, A. J. and Laidman, D. L. 1975. The extraction and analysis of wheat phospholipids. Phytochemistry 14: 2639–2645.

Dick, W. A. and Tabatabai, M. A. 1977. An alkali oxidation method for determination of total phosphorus in soils. Soil Sci. Soc. Am. J. 41: 511–514.

Dormaar, J. F. 1970. Phospholipids in chernozemic soils of southern Alberta. Soil Sci. 110: 136–139.

Fishwick, M. J. and Wright, A. J. 1977. Comparison of methods for the extraction of plant lipids. Phytochemistry 16: 1507–1510.

Fridland, Ye. V. 1976. Lipid (alcohol-benzene) fraction of organic matter in different soil groups. Soviet Soil Sci. 8: 548–557.

Haines, T. H. 1971. The chemistry of the sulfolipids. Pages 299–345 in R. T. Holman, Ed. Progress in the chemistry of fats and other lipids. Vol. II. Pergamon Press, Oxford.

Hance, R. J. and Anderson, G. 1963a. Extraction and estimation of soil phospholipids. Soil Sci. 96: 94–98.

Hance, R. J. and Anderson, G. 1963b. Identification of hydrolysis products of soil phospholipids. Soil Sci. 96: 157–161.

Kates, M. and Eberhardt, F. M. 1957. Isolation and fractionation of leaf phosphatides. Can. J. Bot. 35: 895–905.

Kean, E. L. 1968. Rapid, sensitive spectrophotometric method for quantitative determination of sulfatides. J. Lipid Res. 9: 319–327.

Kononova, M. M. 1966. Soil organic matter. Pages 382–384. Pergamon Press, New York.

Kowalenko, C. G. and McKercher, R. B. 1971a. Phospholipid P contents of Saskatchewan soils. Soil Biol. Biochem. 3: 243–247.

Kowalenko, C. G. and McKercher, R. B. 1971b. Phospholipid components extracted from Saskatchewan soils. Can. J. Soil Sci. 51: 19–22.

Morrison, R. I. 1969. Soil lipids. Pages 558–575 in G. Eglinton and M. T. J. Murphy, Eds. Organic geochemistry. Methods and results. Springer-Verlag, Berlin.

Morrison, R. I. and Bick, W. 1967. The wax fraction of soils: separation and determination of some components. J. Sci. Food Agric. 18: 351–355.

Moucawi, J., Fustec, E., and Jambu, P. 1981. Decomposition of lipids in soils: free and esterified fatty acids, alcohols and ketones. Soil Biol. Biochem. 13: 461–468.

Nazir, D. and Rouser, G. 1966. Removal of water-soluble contaminants from lipid extracts of heart. Lipids 1: 159.

Perniola, M., Ferri, G., Convertini, G., and Vannella, S. 1981. Lipid content and related fatty acid composition in fallow and cultivated soil in southern Italy. Plant Soil 59: 249–260.

Rouser, G., Kritchevsky, G., and Yamamoto, A. 1967a. Column chromatographic and associated procedure for separation and determination of phosphatides and glycolipids. Pages 99–162 in G. V. Marinetti, Ed. Lipid chromatographic analysis. Vol. I. Marcel Dekker, New York.

Rouser, G., Kritchevsky, G., Simon, G., and Nelson, G. J. 1967b. Quantitative analysis of brain and spinach leaf lipids employing silicic acid column chromatography and acetone for elution of glycolipids. Lipids 2: 37–40.

Schmid, P. 1973. Extraction and purification of lipids. II. Why is chloroform-methanol such a good lipid solvent. Physiol. Chem. Phys. 5: 141–150.

Schnitzer, M. and Schuppli, P. 1989. Method for the sequential extraction of organic matter from soils and soil fractions. Soil Sci. Soc. Am. J. 53: 1418–1424.

Sciacovelli, O., Senes, N., Solinas, V., and Testini, C. 1977. Spectroscopic studies of soil organic fractions. I. IR and NMR spectra. Soil Biol. Biochem. 9: 287–293.

Stevenson, F. J. 1966. Lipids in soil. J. Am. Oil Chem. Soc. 43: 203–210.

Stott, D. E. and Tabatabai, M. A. 1985. Identification of phospholipids in soils and sewage sludges by high-performance liquid chromatography. J. Environ. Qual. 14: 107–110.

Tabatabai, M. A. and Bremner, J. M. 1970. An alkaline oxidation method for determination of total sulfur in soils. Soil Sci. Soc. Am. Proc. 34: 62–65.

Tadano, K. and Ishizuka, I. 1983. Determination of peracetylated sulfoglycolipids using the azure A method. J. Lipid Res. 24: 1368–1375.

Vorbeck, M. L. and Marinetti, G. V. 1965. Separation of glycosyl diglycerides from phosphatides using silicic acid column chromatography. J. Lipid Res. 6: 3–6.

Wagner, G. H. and Muzorewa, E. I. 1977. Lipids of microbial origin in soil organic matter. Pages 99–104 in Soil organic matter studies. Vol. II. Proceedings of a symposium. Braunschweig, 6–10 Sept. 1976. Vienna, 1977.

Wang, T. S. C., Liang, Y. C., and Shen, W. C. 1969. Method of extraction and analysis of higher fatty acids and triglycerides in soils. Soil Sci. 107: 181–187.

Ziegler, F. 1989. Changes of lipid content and lipid composition in forest humus layers derived from Norway spruce. Soil Biol. Biochem. 21: 237–243.

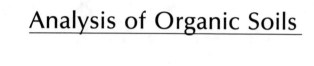

Analysis of Organic Soils

Chapter 42
Sampling Organic Soils

Marsha I. Sheppard

Atomic Energy of Canada Ltd.

Pinawa, Manitoba, Canada

Charles Tarnocai

Agriculture Canada

Ottawa, Ontario, Canada

Denis H. Thibault

Atomic Energy of Canada Ltd.

Pinawa, Manitoba, Canada

42.1 INTRODUCTION

Organic soils are sampled for very different purposes, and depending on the purpose, different methods are appropriate. Soil inventories are the most common reason for sampling organic soil. This sampling has been carried out uniformly across the country for purposes of soil characterization and classification and for resource inventory. Calculation of the dry peat yield depends on accurate knowledge of the depth, the areal extent of the peat deposit, and the moisture content (total porosity). Another need for peat sampling is to investigate the dynamics and stratigraphy and the ecology or paleoecology of peat deposits. In these cases, not only is a minimal disturbance of the structure of the peat important, but compression and contamination by material from other depths must be avoided when sampling. Similarly, in chemical and geochemical investigations of peat bogs, samples for analyses must be uncontaminated by materials from other regions within the deposit profile, and must also be free of metal contaminants. To this end, several samplers offer the use of plastic or polyethylene liners to avoid contamination from the steel core barrel. Another need is to obtain cores for controlled experiments on the physical or chemical properties of peat. In this case, uncompressed, undisturbed cores are required, preferably with their natural moisture conditions preserved.

Several compendiums have been published summarizing most of the known samplers (Radforth 1969; Day et al. 1979; Tarnocai 1980; Landva et al. 1983). The objective of this chapter is to describe one or two recommended samplers for survey sampling, "undisturbed" core sampling, and for large-core sampling. Details of the recommended corers are shown diagrammatically, and where possible, names and addresses of the manufacturers or dis-

Soil Sampling and Methods of Analysis, M. R. Carter, Ed.,

Canadian Society of Soil Science. © 1993 Lewis Publishers.

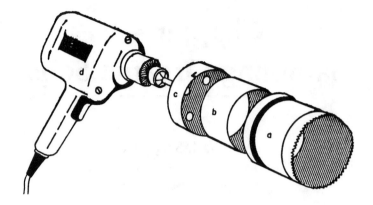

FIGURE 42.1. Peat core cutter assembly (After Stanek, W. and Silc, T., Can. J. Soil Sci., 56, 59, 1976). (a) Cylinder cutter, (b) core holder, consisting of a cylinder fitting snugly into the cylinder cutter, (c) base, (d) power drill.

tributors of the devices have been given. Some accessory tools helpful in lifting the corers and trimming the cores are also described.

42.2 SURVEY SAMPLING

Several samplers have been successfully used for obtaining a large number of samples. The Hiller borer has been used in peat studies for many decades (Radforth 1969). The borer is still used successfully for reconnaissance sampling. The Davis sampler consists of a retractable piston, or loose-fitting rod inside a 38-cm tube, 1.3 cm in diameter (Radforth 1969). The device is portable and simple, but adequate only for survey work. A peat core cutter, using the principle of the hole saw, was developed by Stanek and Silc (1976). It consists of a cylindrical cutter, a core holder, and a base attachable to a drill chuck (Figure 42.1). The coring unit could be quite portable. This device appears quite useful for survey work; however, no comparisons of the performance of this core cutter with other devices have been made. The core cutter can be operated only to a limited depth and is probably most useful for collecting bulk density samples from the wall of an open pit or eroding peat bank. Samples taken with this cutter would also be useful for subsampling for chemical and geochemical analyses and studies of depositional history.

Cuttle and Malcolm (1979) have reported an interesting and inexpensively constructed profile cutter (Figure 42.2). They indicate it is useful for coring peat soils up to 1 m deep with a 5-cm side dimension. They state it is useful for obtaining undisturbed cores. Since they include no proof or comparison of the performance of the corer with respect to compaction of the peat, we have not included it in the "undisturbed" core sampling section. A very popular cutter that is lightweight, inexpensive, and easy to operate is the manually powered cutter reported by Holowaychuk et al. (1965). The serrated cutting barrel (Figure 42.3) is pushed and rotated into the peat and then requires a spade to pry the sampler back up. A plunger assists the extrusion of the core.

Another sampler used for survey coring is the Couteaux chamber sampler (Couteaux 1962). It is 2 m long with a 10-cm inner diameter. A 2-m-long hemicylindrical probe is pushed into the peat, serving as a guide for an outer cylinder that is pushed down over it. The inner hemicylinder is turned 180° to close the chamber, then the whole apparatus is rotated and

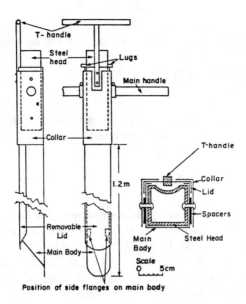

FIGURE 42.2. Profile cutter. (After Cuttle, S. P. and Malcolm, D. C., Plant Soil, 51, 297, 1979.)

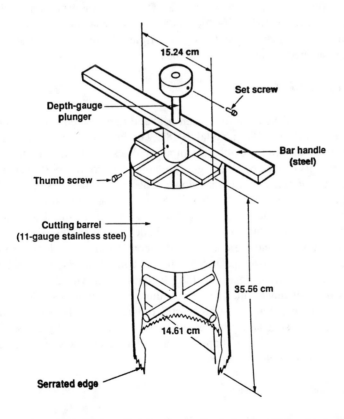

FIGURE 42.3. Core sampler for collecting at shallow depths. (After Holowaychuk, N. et al., Can. J. Soil Sci., 45, 108, 1965.)

removed from the deposit. It samples a suitable "undisturbed" core for macrofossil and chemical analyses at close intervals, and the performance and drawings for manufacture are shown in Couteaux (1962). The corer is available from P. Richard, Laboratoire Jacques-Rousseau, Department of Geography, University of Montreal, C.P. 6128 Succursale "A", Montreal, Quebec, H3C 3J7.

The Hiller borer, the modified Hoffer probe, and the Wardenaar profile cutter are probably the best samplers for survey work, and more details of their design and performance are given below.

42.2.1 Hiller Borer

This sampler is best used to get a bulk disturbed sample at a particular depth in the peat profile. The borer consists of a chamber 50 cm long and 2.8 cm in diameter, with a longitudinal slot opening (Figure 42.4). The slot is closed by a rotating sliding cover as the borer is forced to the desired depth. The insertion of the borer is aided by a rod and rod extensions attached to the upper end of the chamber. The borer is then rotated counterclockwise, thus opening the slot in the chamber, and a flange on the sliding cover scrapes the surrounding peat into the sample chamber. Water is lost from this sampler in wet peats, and in drier peats the action of the flange remolds the peat horizontally. The vertical positions of the peat components are reasonably well preserved. Woody materials and sand in the profile make sampling difficult, if not impossible. This sampler is useful for quick, qualitative evaluation of peat structure and humification; however, other samplers are better for obtaining samples for physical or chemical analysis. The original Hiller borers came from Sweden, but can be manufactured in most machine shops.

42.2.2 Modified Hoffer Sampler

A portable sampler was developed for frozen peat and fine-grained soil sampling using the Hoffer probe design (Brown 1968; Zoltai 1978). The important feature of this sampler is that it was designed to eliminate the structural weaknesses of the original Hoffer probe design and increase its efficiency. It is light, portable, easy to transport over rough terrain, and has no moving parts that could break down. The sampler consists of extension rods, a handle (30 cm long), and a durable bit that is reported to last several field seasons (Figure 42.5). The extension rods and handle are made of seamless tubing. The handle is 30 cm long with a threaded female stud welded to its middle at right angles. The bit diameter can range from 2.6 to 4.3 cm, and the core taken is about 10 to 15 cm long. An experienced person can sample and log frozen peat to a depth of 5 m. However, it takes two people to obtain cores of frozen peat at greater depths. Samples taken with this device are suitable for chemical and geochemical analysis; the authors mention its usefulness in obtaining samples for radiocarbon dating. This probe still provides the only feasible way of hand-sampling frozen organic soils at depths beyond about 0.5 m. Most of these samplers can be manufactured in ordinary welding shops. Zoltai (1978) provides sufficient detail for the manufacture of a modified Hoffer probe.

42.2.3 Wardenaar Profile Cutter

Wardenaar (1987) has described a peat profile cutter that consists of a rectangular stainless-steel box casing, divided lengthwise into two halves (Figure 42.6). The dimensions of the casing are 5 × 10 cm, with very sharp, curved cutting edges at the base. The length of the

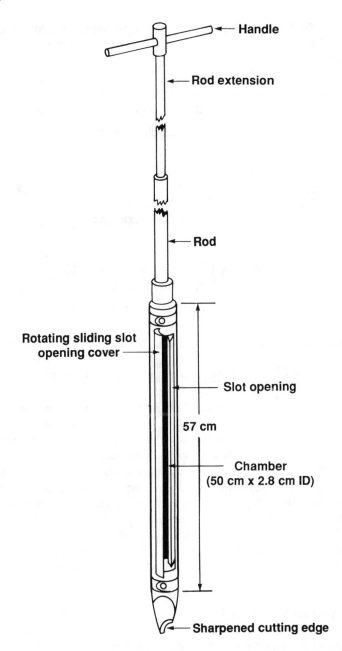

FIGURE 42.4. Hiller sampler. (After Radforth, J. R., Muskeg Engineering Handbook, I. C. MacFarlane, Ed., University of Toronto Press, 1969, chap. 5.)

corer can vary; the maximum practical length was found to be 1.7 m. One person can operate the short (0.5- to 1-m) corer: however, two people are required for the extraction of the long (1.7-m) corer. Although the author states the peat cores are essentially undisturbed, no comparisons have been made with other coring devices on critical physical parameters such as bulk density. Cores suitable for mounting, root investigations, or subsampling for chemical or geochemical investigations can be obtained quickly and easily using this profile cutter. This corer is manufactured and marketed by Eijelkamp Equipment for Soil Research B.V., Nijverheidsstraat 14, 6987 EM Giesbeek, The Netherlands.

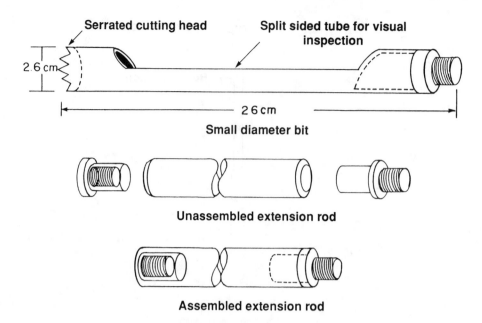

FIGURE 42.5. Modified Hoffer sampler. (After Zoltai, S. C., Can. J. Soil Sci., 58, 521, 1978.)

42.3 "UNDISTURBED-CORE" SAMPLING

Several articles state that it is difficult, if not impossible, to get undisturbed samples of peat (Hillis and Brawner 1961; Hardy 1965; Hollingshead and Raymond 1971; Helenelund et al. 1972). They indicate that the samples only appear "undisturbed". By the time they get to the test facility or laboratory, they do not have the same physical properties as the *in situ* peat (Landva et al. 1983). This may not be too worrisome for some engineering tests (Landva and Pheeney 1980); however, it may lead to gross error in investigations to study the transport of water or solutes through peat. The samplers that have had the most success in obtaining relatively undisturbed peat samples are the piston peat sampler, the Macauley/Russian, and the Mini-Mized Macauley peat samplers. The steel foil sampler is very different and is briefly described. These are all smaller samplers handled without heavy equipment. Much larger "undisturbed" blocks or cores can also be obtained with heavy equipment (see Section 42.4).

42.3.1 Piston Peat Sampler

The principle of the piston sampler came from the oceanography literature some 65 years ago (Olsson 1925), and the principle is still used for lake sediments, soft terrestrial soils, tidal marshes, and peat (Livingstone 1955; Redfield 1975; Korpijaakko 1981; Radforth 1969). During coring, the piston cone component is kept stationary with the end of the sharpened stainless-steel cylinder as it is pushed down into the peat. The piston is released and the core pushed further to the desired depth. The vacuum produced from insertion holds the peat core in place while the cylinder cuts it away from the surrounding material. Because of the strong vacuum formed, cores can also be obtained from wet deposits with most, if not all, of their moisture retained. The cylinder is generally divided into three parts (Figure 42.7). The two handles are required as operation of the sampler requires two sets of "Hiller" rods. One extension moves the sampler tube while the other is attached to the piston shaft.

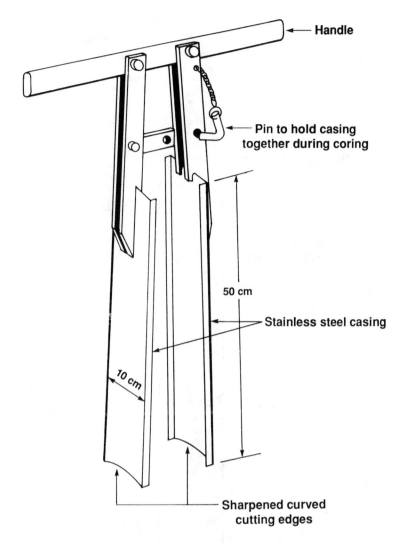

Handle

Pin to hold casing together during coring

50 cm

Stainless steel casing

10 cm

Sharpened curved cutting edges

FIGURE 42.6. Wardenaar profile cutter. (After Wardenaar, E. C. P., Can. J. Bot., 65, 1772, 1987.)

The central part is the sample cylinder. The lower portion also contains peat; however, its function is simply to hold the water in the peat in the sample cylinder when the sampler is raised. The top part of the cylinder holds the piston cone component and receives bits of peat that fall off the wall upon subsequent entries into the hole. These three components are much more discrete in the version and photos described by Korpijaakko (1981) and Pheeney (1970).

Several versions of piston peat samplers have been discussed in the literature (Landva et al. 1983; Tolonen 1967; Day et al. 1979; Korpijaakko 1981; Korpijaakko and Radforth 1972; Korpijaakko and Woolnough 1977). The samplers described vary in diameter from 4 to 10 cm and extract peat samples between 20 and 80 cm in length (Keys and Henderson 1983; Klemetti and Keys 1983; Landva et al. 1983). Wright et al. (1984) have made improvements to the traditional piston corer by adding a serrated cutting head to improve cutting of undecayed roots and woody materials. Interchangeable threaded cutting heads of hardened

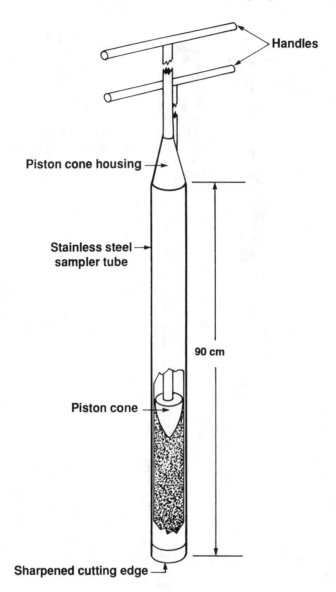

FIGURE 42.7. Piston-type sampler. (After Pheeney, P. E., Muskeg Research Institute, University of New Brunswick, N.B..)

steel sharpened to a fine edge have been used to advantage by Canada-Manitoba Soil Survey (Mills 1990, personal communication). This modification bypasses the need and expense of constructing the entire corer of special hardened steel. A comparison made between a piston sampler and the Macauley sampler indicated no statistically significant difference in mean peat densities measured with the two units (Tolonen and Ijas 1982). In this test, the piston sampler preserved field moisture better. Drawings sufficient to manufacture a piston sampler are given in Redfield (1975).

Two core samplers developed in New Zealand used the traditional piston sampler design. However, they satisfactorily combined rotary drilling action with the stationary piston to provide a more rugged survey corer (Barker and Thomas 1976). They used these corers to provide estimates of physical properties; however, no comparisons were made with other "undisturbed" corers to validate their performance. They noted that compression was a problem in dry peats. As with most samplers, difficulties were encountered with undecayed wood larger than about 50 mm. More details regarding the design and operation can be found in the original report.

42.3.2 Macauley/Russian

The most successful sampler for peat, providing samples relatively free from physical disturbance, was a Soviet device (Belokopytov and Beresnevich 1955). This sampler, with modifications, was the prototype of the Macauley sampler, which has been used successfully since its development (Jowsey 1966; Tolonen 1967). A mini-sized, one-person Macauley sampler is suitable for collecting relatively undisturbed samples (Day et al. 1979; Brown et al. 1984).

The Macauley sampler has four main components: an anchor/fin assembly, a shuttle, a head, and a nose (Figure 42.8). The shuttle is fixed rigidly to the head and nose with set bolts, welding, and brazing, and the anchor/fin assembly is pivoted loosely between the same two components so that it may turn freely. The inside diameter is about 2 cm and the core chamber is 50 cm long on this particular device, although some Macauley samplers take cores as large as 10 cm in diameter. Stainless steel is used for the shuttle, anchor, and fin, and the remainder is fabricated from mild steel. The cutting edge of the shuttle, the bottom edge of the fin, and the edge of the lower half of the anchor are sharpened to slice fibrous and woody peats. Figure 42.8 also shows cross sections of the sampler in operation to illustrate the operating principles. The chamber is pushed into the peat with the shuttle in position on the plain side of the anchor, the opposite side to the fin. Using the T-handle, the shuttle is turned 180° against the resistance of the anchor and fin, enclosing the sample completely with this one movement. On withdrawing the chamber, the sample is exposed by a reverse turn of 180°, and appears as two quarter-cylinders lying one on each side of the fin. The Macauley sampler is easy to clean and allows the "undisturbed" sample to be removed easily. Complete details of the design for the Mini-Macauley are shown in Day et al. (1979), and it can be manufactured easily.

42.3.3 Modified Mini-Mized Macauley Sampler

This sampler is a modified version of the auger described by Day et al. (1979). The modifications included redesign of the core barrel to make it much stronger, and of the fin assembly attachment to allow removal of the fin assembly from the core barrel for cleaning. Modifications were also made to strengthen the extension rod joints. In addition, two sizes of augers were made. Both augers have essentially the same design; actual drawings for manufacture can be obtained from C. Tarnocai, CLBRR, Agriculture Canada, Ottawa, ON K1A 0C6. The small-diameter version produces a 3-cm-diameter core, and, because of its light weight (it can be operated by one person), it is used mainly for reconnaissance and survey work. The large version was designed for the collection of peat samples to be used in various laboratory analyses. This large sampler produces a 5-cm-diameter peat core. Two people are required to operate the large sampler. Comparison of mean moisture contents of samples, taken over a full range of humification, using the Macauley and piston peat samplers, showed no significant difference between the two sets of data (Klemetti and Keys 1983).

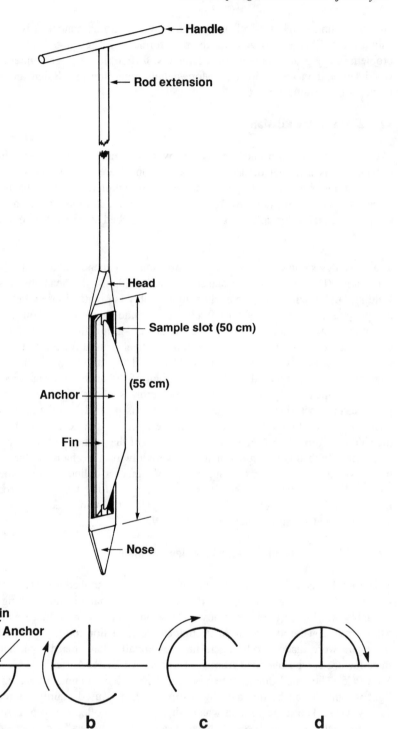

FIGURE 42.8. Mini-Mized Macauley sampler (After Day et al. 1979). Method of operation of the sampler (After Belokopytov and Bersnevich 1955). (a) Position of shuttle on entry into the peat; (b), (c), and (d) successive positions of the shuttle during sampling. The sampler is withdrawn during sampling with the shuttle in position (d).

42.3.4 Steel Foil Sampler

The steel foil sampler, similar to the piston sampler, has been used in geotechnical work (Lea and Brawner 1959) and is described in detail in Kjellman et al. (1950). Thin metal foils are coiled and fitted inside the cylinder attached to a loosely fitting piston. The piston remains stationary and the cylinder is forced into the ground. As the cylinder moves forward into the peat the foil unwinds and surrounds the peat sample as it enters the cylinder. Thus, sliding friction between the sample and the cylinder walls is minimized, preventing distortion of the sample. The manufacture of this corer is expensive because it requires high precision. The sampler also needs heavy equipment to handle it in the field.

42.4 LARGE CORE SAMPLING

Large cores or blocks of peat may be required for investigations in the field or laboratory where, to facilitate the measurements required, or to prevent contamination of the environment, the peat has to be manipulated under controlled conditions. Large block and core samplers are available for this, and three of these are described below. Freezing of the peat to maintain its integrity while it is manually lifted or cut out is also useful, and two methods based on freezing the peat will be discussed.

42.4.1 Block Samples

A 250-mm square block sampler has been used to extract peat cores up to 7 m long successfully (Landva et al. 1983). This is essentially a very large piston corer (Figure 42.9). Heavy equipment is required to position, insert, and extract the sampler, and special equipment is also required to preserve and transport the undisturbed peat block. This is definitely a preferred sampler for experimental studies on peats if work in the field cannot provide sufficiently controlled experimental conditions. Apart from disturbances at the block boundaries, this method provides a reasonably large "undisturbed" core. Drawings suitable for manufacturing this sampler are given in Landva et al. (1983).

More recently, a large core sampler has been reported that is very suitable for obtaining peat cores for many experimental purposes (Millette and Demers 1984). The advantage of the large sample size is to minimize edge effects and better represent the natural variability in physical properties occurring over short vertical and horizontal distances as a result of extreme variability in origin and composition of the peat. The outside sampling core, made of 5-mm-thick cold-rolled steel, shaped into a pipe with an inside diameter of 63 cm, is designed to hold an inner polyethylene sleeve 60 cm in diameter and 122 cm long (Figure 42.10, see also Millette and Demers [1984] for details sufficient to allow manufacture). A narrow ridge at the lower end of the steel tube holds the polyethylene insert in place. The sampler is hammered into place and is maneuvered in the field with a forklift. Blades at the bottom of the steel tube provide the cutting action. Levers attached to these blades are pulled once the depth is achieved, to close the end of the sample for extraction and transportation. Compression of the core, measured by comparing the peat levels inside and outside the core, ranged from 5 to 10%. This appears to be a very suitable technique for obtaining large soil cores for water and contaminant transport studies under controlled conditions.

A double-walled cylindrical device (Figure 42.11) that consists of an outside cylinder protector, a main inside cylinder with a lower cutting edge, and an inside cylinder assembly of one large (327 cm^3) aluminum cylinder with two end retaining rings, one above and one

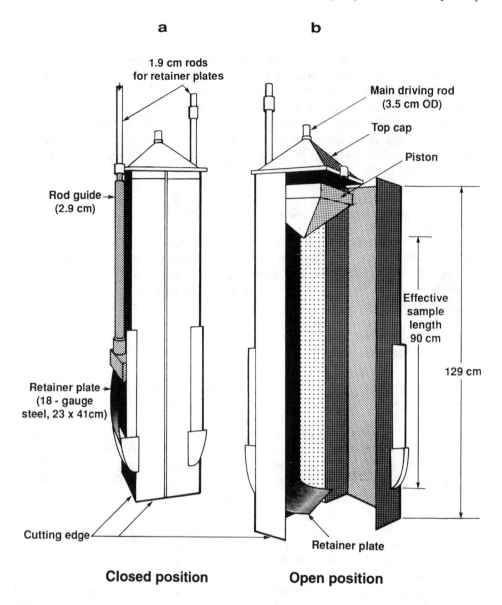

a   **b**

1.9 cm rods
for retainer plates

Main driving rod
(3.5 cm OD)

Top cap

Piston

Rod guide
(2.9 cm)

Effective
sample
length
90 cm

129 cm

Retainer plate
(18 - gauge
steel, 23 x 41cm)

Cutting edge

Retainer plate

Closed position   **Open position**

FIGURE 42.9. 25-cm-square block sampler. Sampler shown in sampling position (a) and
open (b) with one retainer plate in closed position. (After Landva, A. O. et
al., Testing of Peats and Organic Soils, ASTM STP 820, Jarrett, P. M., Ed.,
American Society for Testing Materials, 1983, 141.)

below, has also been developed for peat coring (Brown et al. 1984). The apparatus described
by Brown et al. was made of cold-rolled stainless steel with the exception of the inside
cylinder to hold the final peat sample. The sampler is pushed into the peat with a handle.
Once extracted, the retaining rings are removed and the sample is extruded. More details
can be found concerning its design, fabrication, and operation in Brown et al. (1984). Also,
details of its performance compared with the Macauley sampler and the miter box are given.
The mean density values measured with the double-walled cylinder are very similar to those
measured with the miter box (see Section 42.5), and both of these devices gave slightly
higher mean densities than the Macauley sampler, especially in hemic peats (Brown et al.

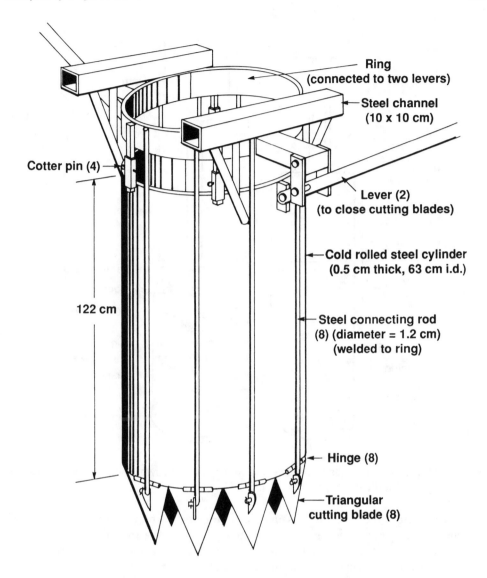

FIGURE 42.10. Large round core sampler shown upright. (After Millette, J. A. and Demers, G., Can. Agric. Eng., 26, 81, 1984.)

1984). The samples taken with the Macauley sampler were adjacent to those taken with the double-walled cylinder, however, and spatial variability in the peat deposit could explain the measured differences.

42.4.2 Frozen Core Sampling

A very effective way of obtaining undisturbed peat samples with fully retained moisture is to freeze the peat with liquid air or dry ice (Radforth 1969). Two pipes are used; the inner pipe, about 10 cm in diameter with a beveled cutting edge, is pushed into the peat; then an outer pipe is pushed in surrounding the first pipe. The liquid air is poured into the clearance space between the two pipes and left for about 30 min. This is sufficient time for the peat in the inner pipe to freeze solid. The core remains in the pipe when the pipes are extracted

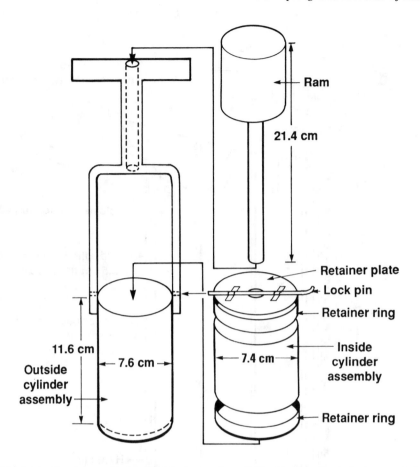

FIGURE 42.11. Double-walled cylindrical sampler. (After Brown, J. et al., Proc. of the 7th
            Int. Peat Congr., Vol. 1, 1984, 54.)

from the peat deposit. The frozen sample can be extracted by pouring hot water on the
outside of the pipe and the sample can be pushed out with a rod or plunger. Although
equipment is needed for forcing the pipes into the peat and extracting them, this method
provides an excellent sample for any detailed analysis of peat structure, as well as good
cores for contaminant transport studies.

An ordinary chain saw cuts blocks of peat extremely well when the peat is frozen *in situ*.
This requires preplanning to obtain cores for studies in the summer months; however, it
does allow the sampler to avoid insects and other vagaries of wetland sampling. The chain
saw is the best tool for sampling Folisols and is the only tool that can easily sample Folisols
with a high root content. Large-diameter cores (20 cm), 40 cm in length, have been successfully sampled for a contaminant transport study using the chain-saw method (Sheppard
and Thibault 1988). An excavation made in the peat deposit allows the chain-saw operator
to cut blocks out of the walls, similar to the approach described in the block sample method.
The blocks can be rounded or trimmed to fit the experimental containment facilities right
in the field, or can be modified in a walk-in freezer as needed.

Fibrous peats with medium to low water contents can be sampled from the walls of exca-
vations in peat deposits. Peat-cutting tools used in Europe for digging peat for fuel can be
used. These tools include a flat spade, a large knife, and a fine-toothed saw.

42.5 SAMPLE LIFTING AND TRIMMING ACCESSORIES

A simple device that enables one person to provide the lifting force necessary to extract samples from deep peat and lacustrine deposits has been presented by Blyth (1984). The lifting device has been used with a Macauley peat sampler. The sampler rods are gripped by a rope sling using a Prusik knot and are levered up using a steel beam and trestle assembly. Body weight provides the motive force and very little muscular activity is required.

A tool used to apply force to the extension tube of a piston corer upon insertion into the peat, and also to provide handles to facilitate pulling the corer up after sampling, is discussed by Redfield (1975). A small-diameter (8 mm) stainless-steel tube can be extremely useful with any coring device. Once the core is ready for extraction from the host deposit, especially in wet peats, this stainless-steel tube breaks the suction while pulling the sample out of the peat (Wardenaar 1987). The tube, with its lower end beveled to avoid getting it plugged with peat, can either be fitted permanently inside the corer or pushed down outside of the corer after the cover has been inserted.

The miter-box assembly is similar to a conventional miter box; however, this is specially designed to trim large peat blocks to fit the prescribed test facility or experimental unit (Brown et al. 1984). This device is easily manufactured; the inside base is made of wood to prevent damage to cutting tools, and the assembly consists of adjustable stainless-steel plates. A block of peat, larger than the final cut sample, is placed in the miter box, the block is secured, and the four desired cuts are made. This device was primarily designed for low-density moss peats that are easily compressed by conventional samplers. Comparisons of peat densities measured for fibric and mesic peats on miter box-trimmed samples and cylindrical-walled samples show no significant difference.

REFERENCES

Barker, P. R. and Thomas, R. F. 1976. Core samplers for fibrous and woody peats. New Zealand Soil Bureau Report No. 26, 14 pp.

Belokopytov, I. E. and Beresnevich, V. V. 1955. Giktorf's peat borers. Torf. Prom. 8: 9–10.

Blyth, A. W. 1984. A mechanical aid for use with peat samplers. Proc. of the 7th International Peat Congress. Vol. 1, 39–44.

Brown, R. J. E. 1968. Permafrost investigations in northern Ontario and northeastern Manitoba. National Research Council of Canada, Div. Bldg. Res. Tech. Paper No. 291. 72 pp.

Brown, J., Malterer, T., and Farnham, R. 1984. Surface and subsurface sampling of organic soils. Proc. of the 7th Int. Peat Congr., Vol. 1, 54–67.

Couteaux, M. 1962. Notes sur le prevelement et la preparation de certains sediments. Pollen Spores 4: 317–322.

Cuttle, S. P. and Malcolm, D. C. 1979. A corer for taking undisturbed peat samples. Plant Soil 51: 297–300.

Day, J. H., Rennie, P. J., Stanek, W., and Raymond, G. P. 1979. Peat testing manual, National Research Council of Canada, Associate Committee on Geotechnical Research, Technical Memorandum No. 125.

Hardy, R. M. 1965. Research on the shearing strength of muskeg and its application. Pages 25–32 in Proc. 10th Muskeg Research Conf. NRC ACSSM, NRC, Ottawa, Canada, Tech. Memo 85.

Helenelund, K. V., Lindqvist, L. O., and Sundman, G. 1972. Influence of sampling disturbance on the engineering properties of peat samples. Pages 229–240 in Proc. 4th Int. Peat Congress, Vol. 2.

Hillis, S. P. and Brawner, C. O. 1961. The compression of peat with reference to the construction of major highways in British Columbia. Pages 204–227 in Proc. 7th Muskeg Res. Conf., NRC, ACSSM, Tech. Memo 71.

Hollingshead, G. W. and Raymond, G. P. 1971. Prediction of undrained movements caused by embankments on muskeg. Can. Geotech. J. 8: 23–35.

Holowaychuk, N., Wildung, L. P., and Gersper, P. L. 1965. A core sampler for organic deposits. Can. J. Soil Sci. 45: 108–110.

Jowsey, P. C. 1966. An improved peat sampler. New Phytol. 65: 245–248.

Keys, D. and Henderson, R. E. 1983. Field and data compilation methods used in the inventory of the peatlands of New Brunswick, Canada. Pages 55–71 in P. M. Jarrett, Ed. Testing of Peats and Organic Soils, ASTM STP 820, American Society for Testing and Materials.

Kjellman, W., Kallstenius, T., and Wagner, O. 1950. Soil sampler with metal foils device for taking undisturbed samples of very great length. Royal Swedish Geotechnical Institute, Proc. 1.

Klemetti, V. and Keys, D. 1983. Relationships between dry density, moisture content, and decomposition of some New Brunswick peats. Pages 72–82 in P. M. Jarrett, Ed. Testing of Peats and Organic Soils, ASTP STP 820, American Society for Testing and Materials.

Korpijaakko, M. 1981. A piston sampler for undisturbed peat samples. Suo 32: 7–8 (in Finnish with an English summary).

Korpijaakko, M. and Radforth, N. W. 1972. Studies on the hydraulic conductivity of peat samples. Pages 323–334 in Proc. 4th Intern. Peat Cong., Finland, Vol. III.

Korpijaakko, M. and Woolnough, D. F. 1977. Peatland survey and inventory. Pages 63–81 in N. W. Radforth and C. O. Brawner, Eds. Muskeg

and the Northern Environment in Canada, N. W. University of Toronto Press, Toronto.

Landva, A. O. and Pheeney, P. E. 1980. Peat fabric and structure. Can. Geotech. J. 17: 416–435.

Landva, A. O., Pheeney, P. E., and Mersereau, D. E. 1983. Undisturbed sampling of peat. Pages 141–156 in P. M. Jarrett, Ed. Testing of Peats and Organic Soils, ASTM STP 820, American Society for Testing Materials.

Lea, N. D. and Brawner, C. O. 1959. Foundation and pavement design for highways on peat. Pages 406–424 in Proc. Fortieth Convention Can. Good Roads Assoc., Ottawa.

Livingstone, D. A. 1955. A light-weight piston sampler for lake deposits. Ecology 36: 137–139.

Millette, J. A. and Demers, G. 1984. The development of a large core sampler for organic soils. Can. Agric. Eng. 26: 81–83.

Olsson, J. 1925. Kolvborr, ny borrtyp for upptagning av lerprov. Tek. Tidskr. Upplaga V.V. 55: 13–16.

Pheeney, P. E. 1970. Construction diagrams and operations manual for piston-type sampler. Muskeg Research Institute, University of New Brunswick.

Radforth, J. R. 1969. Preliminary engineering investigations. Pages 141–149 in I. C. MacFarlane, Ed. Muskeg Engineering Handbook, chap. 5. University of Toronto Press.

Redfield, A. C. 1975. A piston corer for peat. Limnol. Oceanogr. 20: 1042–1045.

Sheppard, M. I. and Thibault, D. H. 1988. Migration of technetium, iodine, neptunium, and uranium in the peat of two minerotrophic mires. J. Environ. Qual. 17: 644–653.

Stanek, W. and Silc, T. 1976. An instrument and technique for collecting peat core samples in peatland surveys. Can. J. Soil Sci. 56: 59–61.

Tarnocai, C. 1980. Sampling methods. Pages 47–48 in Proc. of a Workshop on an Organic Soil Mapping and Interpretation in Newfoundland, St.

John's, Newfoundland, Agriculture Canada, Land Resource Research Institute, Ottawa.

Tolonen, K. 1967. On methods used in studies of the peatland development. II. On the peat samplers. Suo 18: 86–92 (with English summary).

Tolonen, K. and Ijas, L. 1982. Comparison of two peat samplers used in estimating the dry peat yields in field inventories. Suo 33: 33–42.

Wardenaar, E. C. P. 1987. A new hand tool for cutting peat profiles. Can. J. Bot. 65: 1772–1773.

Wright, H. E., Jr., Mann, D. H., and Glaser, P. H. 1984. Piston corers for peat and lake sediments. Ecology 65: 657–659.

Zoltai, S. C. 1978. A portable sampler for perennially frozen stone-free soils. Can. J. Soil Sci. 58: 521–523.

Chapter 43
Physical Properties of Organic Soils

L. E. Parent and J. Caron

Laval University

Sainte-Foy, Quebec, Canada

43.1 INTRODUCTION

Physical properties of peat are measured in a similar way to mineral soils, but methods developed for mineral soils are modified to facilitate standardization of peat materials. Physical properties of peat are of primary importance for the use of peat in horticulture and for studies in peat hydrology. Measurements of degree of decomposition are generally well correlated with a number of physical and chemical properties of peat materials. The degree of decomposition of peat materials is thus an important property in relation to classification and evaluation of the material for various uses (Parent 1980). Degree of decomposition or degree of humification of peat materials is assessed by measuring their fiber or humus contents. Fiber and humus contents are conjugate compositional pairs separated by passing peat materials through sieves of 60- or 100-mesh size (0.25- or 0.15-mm openings, respectively). In practice, peat decomposition is determined by a field method, the von Post pressing method (von Post and Granlund 1926, Grosse-Brauckmann 1976), and by laboratory methods, the most commonly used being the fiber volume method (Farnham and Finney 1965, Sneddon et al. 1971, Lynn et al. 1974), the mechanical dispersion and sieving method (Dinel and Levesque 1976), the centrifugation method for peat standards (Lishtvan and Kroll 1975, Malterer 1988, Malterer et al. 1992), and the Kaila colorimetric method (Kaila 1956, Schnitzer and Desjardins 1966, Lynn et al. 1974).

43.2 DEGREE OF DECOMPOSITION

43.2.1 The von Post Pressing Method

This method, introduced by the Swedish scientist Lennart von Post in 1922 (von Post and Granlund 1926), is the most reliable field method for soil and geological surveys.

43.2.1.1 *Procedure*

A fresh peat sample is first pressed in the palm of the hand. The color or turbidity of the extruded liquor or mud, as well as the proportion of extruded matter, are the only criteria for classifying peat materials. Field observations are matched to one of ten humification degrees (H1 to H10) on the von Post scale (Table 43.1).

Soil Sampling and Methods of Analysis, M. R. Carter, Ed.,

Canadian Society of Soil Science. © 1993 Lewis Publishers.

Table 43.1 The von Post Scale (H) for Assessing Degree of Peat Decomposition

H	Plant residues	Extruded matter	Residues after pressing
1	Unaltered	Clear water	Nonpasty
2	Distinct	Brown-yellow, clear water	Nonpasty
3	Distinct	Brown turbid water	Nonpasty
4	Distinct	Brown, very turbid water	Nonpasty
5	Distinct	Brown, very turbid water with plant residues	Somewhat pasty
6	Somewhat indistinct[a]	One third of the peat material extruded	Very pasty
7	Indistinct but recognizable	One half of the peat material extruded	Very pasty
8	Very indistinct	Two thirds of the peat material extruded	Few fibers
9	Almost nonrecognizable	Almost all peat material extruded	Few fibers
10	Nonrecognizable	All peat material extruded	No residue

[a] Plant residues rather indistinct, but more identifiable in the pressed residue than in the original peat material.

After Grosse-Brauckmann, G., in *Moor- und Torfkunde,* E. Scheizerbart'sche Verlag (Nägele und Obermiller), Stuttgart, 1976.

43.2.1.2 *Comments*

The von Post pressing method is not recommended for relatively dry peat materials and for the upper peat layers of drained organic soils, since the humification degree of these less compressible materials is underestimated (Grosse-Brauckmann 1976). The H values can be classified into fibric (H1 to H3), mesic (H5 to H6), and humic (H7 to H10) materials, H4 being fibric in the Canadian scheme and mesic in the German scheme. The von Post method is highly dependent on the investigator's skill. The von Post scale being an ordinal scale, the use of parametric statistical tests has been questioned (Parent et al. 1982), particularly when using the mean instead of the median value for a limited number of replicates.

43.2.2 The Fiber Volume Method (Lynn et al. 1974)

This method is an objective alternative to the von Post pressing method to classify peat materials. Fibers are separated from nonfibers by sieving rubbed or unrubbed peat materials through a 0.15-mm screen.

43.2.2.1 MATERIALS

1   A graduated 5-mL medical half-syringe adjusted for a volume of 2.5 mL (a 5-mL plastic syringe is shaved longitudinally to make the half-syringe).

2   Running tap water.

3   A 100-mesh sieve, 8 cm in diameter.

4   Absorbing tissue.

43.2.2.2 PROCEDURE

1   Place approximately 25 mL of wet sample in a piece of absorbing tissue and roll with light pressure to extract excess water. Unroll tissue and cut the sample into 6-mm pieces. Mix subsamples randomly.

2   Pack the half-syringe with randomly selected subsamples and compress just enough to saturate the material and force out any entrapped air. Do not force out any water. It is to this water content that the residue must be returned later, when the residue volume is determined.

3   For fiber determination, transfer the 2.5-mL sample to a 100-mesh sieve and wash under running tap water until the effluent appears clear. Remove excess water through the underside of the sieve by blotting with an absorbing tissue. Repack the residue into the half-syringe, and blot further with an absorbing tissue until the water content reaches the state described above. Read the residue volume on the half-syringe and record it as percent, unrubbed fiber. Transfer the residue to the 100-mesh sieve and rub between the thumb and forefinger under a stream of running tap water until the effluent is clear. Blot and repack the residue into the half-syringe as for the unrubbed fiber. Read the volume and record it as percent rubbed fiber.

43.2.2.3 *Comments*

Levesque and Mathur (1979) found the fiber content well correlated with the relative bio-degradability of peat. The fiber method appeared less repeatable than the centrifugation method (Malterer 1988, Malterer et al. 1992).

43.2.3 The Centrifugation Method (Ministry of Fuel Industry RSFSR 1976)

This method, introduced in 1965 as the Soviet standard, consists essentially in separating fibers from the so-called coagulated humus by sieving peat material through a 0.25-mm screen in a centrifuge. The degree of peat decomposition is determined graphically on a reference nomogram relating the collected sediment volumes before and after sieving. The method is applicable to all natural peat types, but peat materials containing less than 65% H_2O (w/w) require a pretreatment. This method is not calibrated for processed (milled) peat, since peat fragmentation during processing causes a two- to threefold increase in decomposition.

43.2.3.1 MATERIALS AND REAGENTS

1   Electrical centrifuge with a time relay for automatic cutoff 2 min after the switch is closed, sample tube holders and cups to fit either large or small

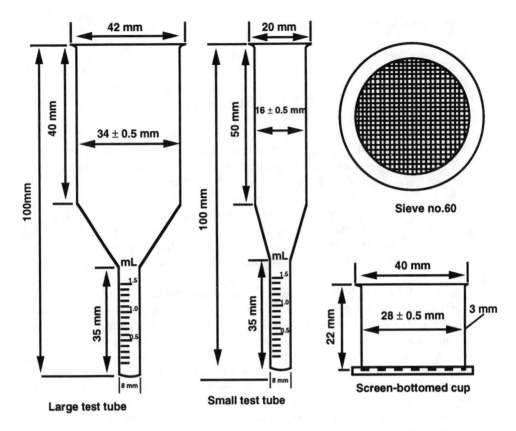

FIGURE 43.1. Large and small tubes and screen-bottomed cup used for fiber determination by the centrifugation method.

centrifuge test tubes (Figure 43.1). The tubes must be graduated by 0.1 mL up to 1.5 mL.

2   Screen-bottomed cylindrical cups made with a 0.25-mm screen fixed to a PVC ring 22 mm high and 28 mm I.D. (Figure 43.1). The 60-mesh screen is cemented to the lower wall of the PVC ring with a water-resistant adhesive. The cups must be easily inserted into the large test tubes.

3   Square plate of 625 cm², sample selector, and plunger. The sample selector is a tube, 5 mm in diameter, with an end sharpened to cut through a 3- to 4-mm peat layer spread over the plate. A plunger is used to push peat out of the sample selector. All pieces are washed after each sampling.

4   Chemical solutions: 1 M NaOH or 1 M KOH, 1 M HCl, FeCl$_3$ 10%.

43.2.3.2 PROCEDURE FOR PEAT MATERIALS CONTAINING MORE THAN 65% H$_2$O (W/W)

1   A peat sample weighing 100 to 200 g is spread out gently on the plate in order to obtain a uniform layer 3–4 mm thick. This layer is punched with

the sample selector at 10–12 points all over the surface, including areas of coarse residues. The sample, approximately 10 mm long, is pushed out with a plunger into the small test tube and covered with water up to 1 cm from the brim. One to two drops of the ferric chloride solution is added, to coagulate humus, and the sample well mixed with a rod to obtain a homogeneous suspension. At least four parallel determinations are conducted to be statistically reliable.

2 The small test tube is spun in a centrifuge for 2 min at 1000 rpm. Relative centrifugal force depends on the rotating radius from the pivot (nearly 200 $\times$ g per r = 17.5 cm). The volume of sediment, average level estimated by eye, is measured to the nearest 0.01 mL. The supernatant liquid must be clear. The volume should be 0.7 to 1.5 mL (the sample weighs approximately 0.3 to 0.5 g), but 0.7 to 1 mL for humic peats.

3 The peat material is suspended again and transferred into a screen-bottomed cup placed on top of the large tube. The small tube is rinsed with 3 to 4 mL of water. The radius of the screen-bottomed cup is 10 cm. The peat material is centrifuged for 2 min at 1000 rpm. The volume of sediment is measured to the nearest 0.01 mL. The presence of peat clumps on the screen after centrifugation indicates poor peat dispersion: in such a case the analysis should be repeated.

43.2.3.3 PROCEDURE FOR PEAT MATERIALS CONTAINING LESS THAN 65% H_2O (W/W)

1 Portions of such peat materials are placed into a porcelain dish so that two thirds to three fourths of the cup volume is occupied after peat swelling. The cup is flooded with alkali (1 *M* NaOH or 1 *M* KOH) and equilibrated for 24–30 h. Then, after stirring, peat clumps are broken up. More alkali is added as needed to obtain a uniform mixture. A sample is taken from the dish with the sample selector and put into the small test tube. The small test tube is half filled with a 1-*M* HCl solution, shaken up, and equilibrated for 2–5 min until the neutralization reaction is completed. Then the small test tube is filled with water up to 1 cm from the brim. Five to eight drops of the ferric chloride solution are added and the mixture is shaken up.

2 The small test tube is centrifuged for 2 min at 1000 rpm. After measurement of the volume of sediment to the nearest 0.01 mL, the supernatant liquid is decanted carefully without disturbing the sediment. The test tube is refilled with water and one to two drops of the ferric chloride solution are added and shaken up. The sample is further handled at step 3 as described previously.

43.2.3.4 DATA ANALYSIS

1   The degree of peat decomposition (R in percent) is obtained by reporting the volumes of sediment in the small test tube and in the large test tube on a nomogram (Figure 43.2). The degree of decomposition R is read on the right-hand side for each peat type. The degree of decomposition is higher, the higher the R value.

2   The degree of decomposition is corrected for values of ash content exceeding 15% (Table 43.2).

3   Analytical precision depends on degree of decomposition and peat water content (Table 43.3).

43.2.3.5 Comments

The centrifugation method was the official method for establishing the Soviet peat standards for peat and peatland complex utilization. The rotating speed of the centrifuge should be corrected for rotating radii differing from the 17.5 cm. The relative centrifugal force should be 190 to 200 $\times$ *g*. The centrifugation method was found more sensitive and repeatable than the fiber volume method (Malterer 1988).

43.2.4 The Colorimetric Method

This method, introduced by the Finnish peat scientist Kaila (1956), is based on the capacity of an alkaline solution of sodium pyrophosphate to extract and solubilize humic substances. The colorimetric determination of the filtrate at 550 nm or the color of the filtrate matched in a Munsell Color Chart (on 10 YR page) are measurements of the degree of decomposition.

43.2.4.1 MATERIALS AND REAGENTS

1   A 2.5-mL half syringe, a 30-mL plastic container, chromatographic paper, Erlenmeyer flasks, reciprocating shaker, funnels, and spectrophotometer.

2   Sodium pyrophosphate ($Na_4P_2O_7 \cdot 10H_2O$), as crystals or 0.025 *M* solution (dissolve 11.152 g of $Na_4P_2O_7 \cdot 10H_2O$ per L of distilled water).

43.2.4.2 PROCEDURE USING THE MUNSELL COLOR CHART (LYNN ET AL. 1974)

1   Mix a 2.5-mL half-syringe sample with 1/8 teaspoon (approximately 1 g) of sodium pyrophosphate crystals and 4 mL of water in a 30-mL plastic container, and allow the mixture to stand overnight.

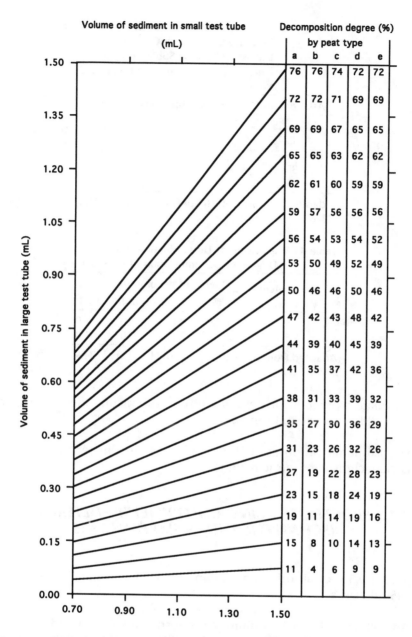

FIGURE 43.2. Nomogram for the determination of the degree of peat decomposition: a = pine-cottongrass (*Eriophorum*), cottongrass, cottongrass-sphagnum highmoor peat types; b = all other highmoor peat types; c = transitional peat types; d = woody lowmoor peat types; e = all other lowmoor peat types. (Adapted from Ministry of Fuel Industry RSFSR 1976.)

2   Mix again and insert a strip of chromatographic paper (0.5 by 3 cm) to absorb the colored solution. Allow the strip to moisten completely. Tear off the soil end, blot the strip gently on another sheet of chromatographic paper, and compare the colored strip with a Munsell Color Chart (on 10 YR Munsell page).

Table 43.2 The Degree of Decomposition May
Be Corrected for Ash Content
Exceeding 15% by Subtracting the
Following Values from the Calculated
Values:

Ash content (%)	Subtracting value (%)
≤15	0
15–25	2
25–35	3
35–45	4
45–55	5

Adapted from the Ministry of Fuel Industry RSFSR,
1976.

Table 43.3 Precision of Determination of the
Degree of Decomposition of Peat (%)

Water content	Degree of decomposition (%)			
(% w/w)	1–15	15–30	30–50	>50
>65	1.5	2.0	3.0	5.0
<65	2.0	3.0	5.0	5.0

Adapted from the Ministry of Fuel Industry RSFSR,
1976.

3 Calculate a pyrophosphate index (PI) by subtracting the chroma from the value (PI = value − chroma).

4 For taxonomic purposes, a PI of 5 indicates fibric material; a PI of 3 or less is characteristic of sapric materials.

43.2.4.3 PROCEDURE USING COLORIMETRIC DETERMINATIONS (SCHNITZER AND DESJARDINS 1965)

1 Weigh 0.5 g of air-dried and 2-mm sieved peat material and transfer into 125-mL flasks.

2 Add 50 mL of 0.025 *M* sodium pyrophosphate solution and shake for 18 h at room temperature. Filter and dilute filtrate to 250 mL with distilled water.

3 Read absorbance at 550 nm on a spectrophotometer, and multiply absorbance by 100 to give cardinal numbers for percent absorbance (PA).

4 Data on a limited number of peat materials indicate that PA <40 classifies materials as "peat", and that PA >60 classifies materials as "muck".

43.2.4.4 *Comments*

In many cases, pyrophosphate values disagree with other determinations of degree of decomposition (Kaila 1956). PI values are subject to bias due to light intensity and to color perception by different operators (Malterer et al. 1992). The colorimetric methods are semi-quantitative and best used to compare peat materials of similar origin and botanical composition, since the humification process and the original polyphenol content depend on peat genesis and peat-forming plant communities (Grosse-Brauckmann 1976).

43.3 BULK DENSITY AND TOTAL PORE SPACE

Bulk density (BD) is an easily measured property correlated to many other peat properties. Physical parameters, such as total pore space (TPS), can be calculated from the knowledge of BD and ash content.

43.3.1 MATERIALS

1   Core sampler or McCauley sampler, and sharpened knife.

2   Forced-air oven.

43.3.2 PROCEDURE

1   Trim a core of undisturbed peat with a sharpened knife to fit roughly into the core sampler. A core sampler with sharpened edge can also be used. Alternatively, peat samples can be extracted with a McCauley sampler.

2   Dry samples at 105°C over 1 or 2 d, then compute BD as:

$$BD = \text{(sample dry weight in g)}/\text{(volume of core in mL)} \qquad (43.1)$$

43.3.3 Calculations

TPS can be calculated from BD and particle density (PD). PD is estimated from ash content (Paquet et al. 1993), assuming a PD of 1.55 for the organic fraction (OM) and 2.65 for the mineral fraction (Verdonck et al. 1978):

$$\% \, OM = 100\% - (\% \text{ ash}) \qquad (43.2)$$

$$PD = (1 + F)/[(F/1.55) + (1/2.65)] \qquad (43.3)$$

where $F = (\% \, OM)/(\% \text{ ash})$

$$TPS = 100 \, [PD - BD]/PD \qquad (43.4)$$

43.3.4 Comments

Care should be taken in order to minimize peat compression by the cylinder when sampling. BD of peat materials is affected by water content. A water content of 50–60% is thus recommended for peat substrates (Verdonck et al. 1978). Macfarlane (1969) reported that estimates of PD calculated from ash content can deviate up to 18% from the actual value. The correlation between BD and ash content is high (Grigal et al. 1989). BD is also highly correlated to the degree of decomposition (Silc and Stanek 1977).

43.4 WATER-HOLDING CAPACITY (WHC)

The WHC of peat materials can be determined by a "soak and drain" method where water in saturated peat is extracted by gravity. Water can also be extracted from saturated peat over a water potential range of 0 to -10 kPa using a pressure plate (Tempe cells) or a tension table apparatus. For lower potentials (more negative values), use pressure chambers as for mineral soils.

43.4.1 "Soak and Drain" Method (Puustjarvi 1973)

43.4.1.1 MATERIALS

1   A 1-L Büchner funnel (or any cylindrical container with perforated bottom).

2   Mixing basin, at least five times as large as the container.

3   Drying oven.

43.4.1.2 PROCEDURE

1   The sample is added to the brim of the funnel and moistened, then transferred to a mixing basin.

2   Hot water (70°C) is added to suspend the peat material into a very thin suspension. After mixing with a glass rod for 1–2 min the suspension is allowed to stand for 30 min and mixed again for 1–2 min to debubble.

3   The peat suspension is transferred back into the funnel. A filter paper (high-flow) is placed on the bottom if peat is fine textured. Excess water is drained to nearly constant weight after 30 min for coarse-textured peat and 2 h for fine-textured peat.

43.4.1.3 CALCULATIONS

Peat volume (V) is calculated from peat height in the funnel. Wet peat is weighed. Gravimetric water content (Ww) is determined by weighing at least 50 g of wet peat (Wr) dried at 105°C for at least 16 h and weighing to the nearest 0.01 g (Wd). Gravimetric and volumetric WHC are computed as follows:

1   Gravimetric WHC relative to wet peat sample:

$$WHC = 100 \, [Ww - Wr]/Wr \tag{43.5}$$

2   Gravimetric WHC relative to oven-dried peat sample:

$$WHC = 100 \ [Ww - Wd]/Wd \qquad (43.6)$$

3.   Volumetric WHC:

$$WHC = 100 \ [Ww - Wd]/V \qquad (43.7)$$

43.4.1.4 Comments

The water potential at which WHC is measured is equal to half the height of the peat sample. This method allows the calculation of air capacity by difference between TPS and volumetric WHC. The above method, as well as other methods using the oven-dry weight at 105°C, might slightly overestimate the water content of peat, since peat drying at a temperature exceeding 85°C would result in some loss of organic matter (Macfarlane 1969).

43.4.2 The Pressure Plate Apparatus (Joyal et al. 1989)

In this method, saturated peat samples are equilibrated in Tempe cells at increasing potentials using a water column to control pressure.

43.4.2.1 MATERIALS

1   Tempe cells (Soil Moisture Corp.) and clamps.

2   Brass cylinders, 30 or 60 mm in height, 88 mm O.D. and 85 mm I.D.

3   Porous ceramic plates, 92 mm in diameter, 50 or 100 kPa of bubbling pressure.

4   Beakers: 250 mL.

5   Water column and pressurized air conducting system feeding the Tempe cells through a manifold.

43.4.2.2 PROCEDURE

1   Assemble the Tempe cell, brass cylinder, and ceramic plate.

2   Introduce peat material into the brass cylinder to the brim.

3   Saturate with water from the bottom up. Allow to stand for 24 h.

4   Plug the cell to the air circuit, and regulate the water pressure with reference to the water column.

5   Immerse the cell in tap water and close the bottom opening with a finger to check for air leaks on the cell side. Add vacuum grease if necessary.

6   Equilibrate at the following potentials (in cm of water column): 10, 50, and 100 (possibly other intermediates).

7   Weigh Tempe cell assembly after each equilibration period (less than 24 h for the 50-kPa ceramic plate and up to 48 h for the 100-kPa ceramic plate).

8   Weigh the Tempe cell assembly after removing the peat sample.

9   Dry the peat material at 105°C to constant weight for calculation of water retention volumetric values.

43.4.2.3 Comments

This method is derived from the original method of Reginato and van Bavel (1962) for potentials higher than -100 cm (-10 kPa). A standard pressure plate can be used to determine water retention at lower potentials (-10 to -1500 kPa).

43.4.3 The Tension Table Apparatus

In this method, a contact is established between saturated peat samples and free water in a tension table apparatus. Water potential is lowered in the range of 0 to -10 kPa.

43.4.3.1 MATERIALS

1   A tension table device made of a box filled with glass beads (30 μm) and covered with nylon (Topp and Zebchuck 1979).

2   Metal cylinders, 30 or 60 mm in height, 88 mm O.D., and 85 mm I.D.

3   Nylon mesh (53 μm) and rubber bands.

43.4.3.2 PROCEDURE

1   Fit the bottom of metal cylinders with nylon mesh and secure nylon with rubber bands.

2   Fill cylinders with peat. Slight compression may be applied (Van Dijk 1980).

3   Saturate peat samples in hot tap water (70°C) up to 1 cm from the top of the cylinder. Allow to soak for 24 h.

4 Saturate the glass beads until free water appears at the surface of the tension table. Mix to a slurry (see Topp and Zebchuck 1979). An intimate hydraulic contact should be provided between peat samples and the beads.

5 Apply water tension not exceeding -10 kPa.

6 Equilibration time depends on sample size and is in the range of 24 h for 180-mL samples to 72 h for larger samples. After equilibration is reached at a given water tension value, detach the core and weigh (total wet weight = TWW).

7 Repeat the procedure from step 4 to determine water retention at various water suction values.

8 At the end of the experiment, determine oven-dry weight (105°C) of cylinders, sample, nylon cloth, and rubber bands (total dry weight = TDW). Empty cylinder and determine dry weight of cylinder, nylon mesh, and rubber bands (C).

43.4.3.3 CALCULATIONS

1 The gravimetric WHC relative to sample oven-dry weight is calculated as:

$$\text{WHC} = 100 \, [\text{TWW} - \text{TDW}]/[\text{TDW} - \text{C}] \qquad (43.8)$$

2. The volumetric WHC is computed as:

$$\text{WHC} = 100 \, [\text{TWW} - \text{TDW}]/[\text{Volume of cylinder}] \qquad (43.9)$$

Peat shrinkage due to water extraction is usually not taken into account, but if sample volume is determined, the exact volume of peat samples could replace cylinder volume in the above equation.

43.4.3.4 Comments

The tension table method can handle more samples at a lesser cost than the pressure plate apparatus (Section 43.4.2). However, air trapping can be a problem with the tension plate method, and care should be taken to avoid breakage of the water column. Physical properties in samples of processed peat materials are very sensitive to the way samples are handled (De Kreij and De Bes 1989). In pot experiments, this problem could be circumvented by direct determination of the WHC in the pots. *In situ* water desorption characteristics of peat and peat mixes used in horticulture can be determined directly on tension tables for pots of variable geometry (Paquet et al. 1993).

43.5 PARTICLE-SIZE DISTRIBUTION OF PEAT MATERIALS

Particle size distribution extends the notion of degree of decomposition to more than two particle fractions. The procedure for obtaining particle size distribution was described by Dinel and Levesque (1976) and Levesque and Dinel (1977).

43.5.1 MATERIALS AND REAGENTS

1   Reciprocating shaker.

2   500-mL Erlenmeyer flask.

3   Filter paper (coarse).

4   A 200-mesh sieve.

5   Particle size apparatus. The apparatus consists of a cylinder with an interior diameter of 14.5 cm and a height of 46 cm with a hole near the base for entry of air and an outlet for water. A piece of rubber tubing about 1.5 cm in diameter and 45 cm long is formed into a ring and placed at the bottom of the cylinder. A nest of sieves 13.5 cm in diameter and in the following order, 10, 20, 40, 100, and 200 mesh, is assembled with a gasket made from thin (2 mm) Tygon® tubing positioned between each sieve to provide a seal. The sieves are held together with an elastic band and inserted into the cylinder. The top of the cylinder is closed with a funnel. See Figure 43.3.

6   Drying cans.

7   Oven, preferably ventilated to a fume hood.

8   Preserving solution of the following composition: 10% formaldehyde, 10% acetic acid, 45% ethanol, and 35% distilled water.

9   Glass vials.

43.5.2 PROCEDURE

1   Place 25 g of moist peat sample broken into small pieces into the 500-mL Erlenmeyer flask. Add 300 mL of water and shake the suspension for 16 h using a reciprocating shaker.

2   Determine the water content of duplicate 25-g samples by weighing when wet, drying at 70°C, and then reweighing when dry. (Note: for both Step 1 and 2, ensure that the 25-g samples are representative materials. The amount of sample that is used can be adjusted to the requirements of specific studies.)

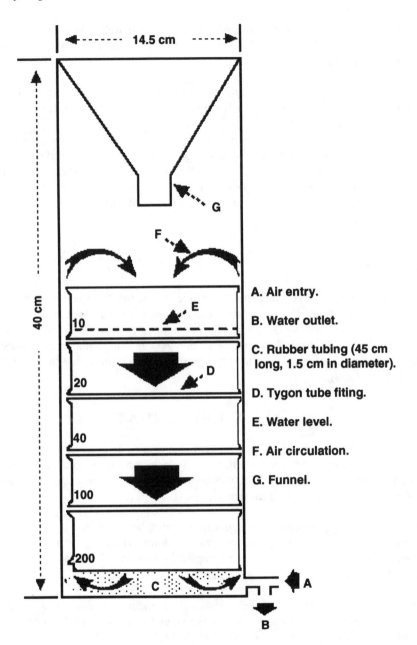

FIGURE 43.3. Sieve assembly for the determination of grain-size distribution of peat particles (sieve number indicated). (From Dinel, H. and Levesque, M. *Can. J. Soil Sci.* 56: 119–120, 1976. With permission.)

3 Pour the suspension (step 1) onto a 200-mesh sieve and wash with water to remove the material finer than 200 mesh.

4 Place the material retained on the 200-mesh sieve into the top sieve (10 mesh) of the particle size apparatus.

5 Fill the cylinder of the particle size apparatus with water to just above the level of the screen in the top sieve.

6   Introduce air into the particle size apparatus through the air inlet at the base at a rate sufficient to shake the nest of sieves and continue bubbling air vigorously for 1 h.

7   Remove the nest of sieves and recover the fibers remaining on each sieve by washing the fibers onto a funnel fitted with a coarse filter paper.

8   (a) If the fibers are to be examined for botanical composition, remove the fibers from the filter paper and place in vials. Fill the vials with the preserving solution. (b) If the fiber content is to be determined, dry the various fractions at 70°C and weigh. Calculate the proportion of the sample in each particle size range, as a percentage of the dry weight of the 25-g sample determined in step 2.

43.5.3 Comments

There seems to be no real advantage in using an electrolyte solution to disperse peat particles, and the long shaking time is not a real inconvenience, since it can be done overnight (Levesque and Dinel 1977). In Ontario Geological Survey experiments on fiber determination (100-mesh particle size), mechanical stirring for 10 min gave results similar to the unrubbed fiber method, while mechanical stirring for 16 h produced data comparable to the rubbed-fiber method (Riley 1989).

43.6 WOOD CONTENT

Wood content and stumpiness are important characteristics of peat deposits. Stumpiness is the volumetric content of stumps and other woody inclusions to the total volume of the peat deposit (Antonov and Kopenkin 1983). Stumps interfere with drainage and maintenance operations in organic soils and peat fields.

In geological surveys, a woody layer occurrence can be quantified by the frequency of intercepting such a layer over a given surface. In pedological surveys, stumpiness is defined volumetrically by estimating the surface occupied by stumps, 10 cm and more in diameter, within a 130-cm-deep control section (Mills et al. 1977).

Stumpiness is categorized as null (0%), low (0.5%), medium (0.5–2%), and high (>2%) for peat extraction (Antonov and Kopenkin 1983), and as none (<1%), moderate (1–5%), and high (>5%) for peat cultivation (Mills et al. 1977).

43.7 SATURATED HYDRAULIC CONDUCTIVITY

Hydraulic conductivity of organic soils is determined in the same manner as minerals soils. One should be aware that organic soils are composed of rather loose materials and that, consequently, the walls of augerholes may be unstable while measuring hydraulic conductivity or infiltration rate of water in the field. Measurements of hydraulic conductivity by the constant-head or the falling-head methods may be invalidated by the presence of large particles in the peat sample. Also, mathematical models developed especially to estimate saturated hydraulic conductivity from field measurements might not apply to organic soils (Hemond and Goldman 1985).

REFERENCES

Antonov, V. Ia. and Kopenkin, V. D. 1983. Technology and complex mechanization of peat production. Ed. Nedra, 2nd ed. Moscow, 288 pp. (in Russian).

De Kreij, C. and De Bes, S. S. 1989. Comparison of physical analysis of peat substrates. Acta Hortic. 238: 23–36.

Dinel, H. and Levesque, M. 1976. Une technique simple pour l'analyse granulométrique de la tourbe en milieu aqueux. Can. J. Soil Sci. 56: 119–120.

Farnham, R. S. and Finney, H. R. 1965. Classification and properties of organic soils. Adv. Agron. 17: 115–162.

Grigal, D. F., Brovold, S. L., Nord, W. S., and Ohmann, L. F. 1989. Bulk density of surface soils and peat in the North Central United States. Can. J. Soil Sci. 69: 895–900.

Grosse-Brauckmann, G. 1976. Peat stratification. Pages 91–133 in K. Göttlich, Ed. Moor- und Torfkunde. E. Scheizerbart'sche Verlag (Nägele und Obermiller), Stuttgart (in German).

Hemond, H. F. and Goldman, J. C. 1985. On non-darcian flow in peat, J. Ecol. 73: 579–584.

Joyal, P., Blain, J., and Parent, L. E. 1989. Utilization of Tempe cells in determination of physical properties of peat based substrates. Acta Hortic. 238: 63–66.

Kaila, A. 1956. Determination of the degree of humification in peat samples. Maatal. Tiet. Aikak. 28: 18–35.

Levesque, M. and Dinel, H. 1977. Fiber content, particle-size distribution and some related properties of four peat materials in eastern Canada. Can. J. Soil Sci. 57: 187–195.

Levesque, M. and Mathur, S. P. 1979. A comparison of various means of measuring the degree of decomposition of virgin peat materials in the context of their relative biodegradability. Can. J. Soil Sci. 59: 397–400.

Lishtvan, I. I. and Kroll, N. T. 1975. Basic properties of peat and methods for their deter-mination. Ed. Nauka and Tekhnika, Minsk, 320 pp. (in Russian).

Lynn, W. C., McKinzie, W. E., and Grossman, R. B. 1974. Field laboratory tests for the characterization of histosols. Pages 11–20 in A. R. Aandahl et al., Eds. Histosols: their characteri-zations, use, and classification. Soil Sci. Soc. Am. Spec. Publ. 6, Madison, WI. 136 pp.

Macfarlane, I. C. 1969. Engineering character-istics of peat. Pages 78–126 in I. C. Macfarlane, Ed. Muskeg Engineering Handbook. University of Toronto Press, Toronto. 297 pp.

Malterer, T. J. 1988. A comparative analysis of the USDA fiber volume, USSR centrifugation and von Post humification methods for determining fiber content and degree of decomposition (humification). Proc. 8th Int. Peat Congr. (Leningrad) IV: 129–139.

Malterer, T. J., Verry, E. S., and Erjavec, J. 1992. Fiber content and degree of decomposition: a review of national methods. Soil Sci. Soc. Am. J. 56: 1200–1211.

Mills, G. F., Hopkins, L. A., and Smith, R. E. 1977. Organic soils of the Roseau River Watershed in Manitoba. Agriculture Canada Monograph No. 17, 69 pp.

Ministry of Fuel Industry RSFSR. 1976. The state standard of the USSR peat. Method for the de-termination of the degree of decomposition: GOST 10650-72. Trans. Comm. I Int. Peat Soc.: Work-ing group for the classification of peat, Helsinki: 57–66.

Paquet, J., Caron, J., and Banton, O. 1992. *In situ* determination of the water desorption char-acteristics of peat substrates. Can. J. Soil Sci. 73 (in press).

Parent, L. E. 1980. Guidelines for peat utilization in Quebec and the Maritimes. Tech. Contr. J. 826, Tech. Bull. 15, Agriculture Canada Research Station, Saint-Jean-sur-Richelieu, Quebec, 41 pp. (in French).

Parent, L. E., Fanous, M. A., and Millette, J. A. 1982. Comments on parametric test limitations for analyzing data on peat decomposition. Can. J. Soil Sci. 62: 545–547.

Puustjarvi, V. 1973. A rapid method for measuring the structure of a substrate. Meded. Fac. Landbouwwet. Rijksuniv. Gent 38: 2051–2053.

Reginato, R. J. and van Bavel, C. H. M. 1962. Pressure cells for soil cores. Soil Sci. Soc. Am. Proc. 26: 1–3.

Riley, J. L. 1989. Laboratory methods for testing peat — Ontario peatland inventory project. Ontario Geological Survey Misc. Pap. 145, Ont. Min. North. Dev. Mines.

Schnitzer, M. and Desjardins, J. G. 1965. Carboxyl and phenolic hydroxyl groups in some organic soils and their relation to the degree of humification. Can. J. Soil Sci. 45: 257–264.

Schnitzer, M. and Desjardins, J. G. 1966. Oxygen-containing functional groups in organic soils and their relationship to the degree of humification as determined by solubility in sodium pyrophosphate solution. Can. J. Soil Sci. 46: 237–243.

Silc, T. and Stanek, W. 1977. Bulk density estimation of several peats in northern Ontario using the von Post humification scale. Can. J. Soil Sci. 57: 75.

Sneddon, J. I., Farstad, L., and Lavkulich, L. M. 1971. Fiber content determination and the expression of results in organic soils. Can. J. Soil Sci. 51: 138–141.

Topp, G. C. and Zebchuk, W. 1979. The determination of soil-water desorption curves for soil cores. Can. J. Soil Sci. 59: 19–26.

Van Dijk, H. 1980. Standardized methods for the physical analysis of plant substrates. Acta Hortic. 99: 221–230.

Verdonck, O. F., Cappaert, T. M., and De Boodt, M. F. 1978. Physical characterization of horticultural substrates. Acta Hortic. 82: 191–200.

von Post, L. and Granlund, E. 1926. Soedra Sveriges torvtillgangar. 1. Sver. Geol. Unders. Arsb. 19, Ser. C, No. 335, 127 s., Stockholm (in Swedish).

Chapter 44
Chemical Properties of Organic Soils

A. Karam

Laval University
Sainte-Foy, Quebec, Canada

44.1 INTRODUCTION

There is a considerable interest in the agricultural development of organic soils in many parts of the world. A wide variety of crops can be grown on peat or organic soils (van Lierop et al. 1980, Parent et al. 1985). Thus, information on the chemical properties of organic soils is of practical significance in crop production (MacLean et al. 1964). On the other hand, soil chemical properties such as mineral content, C:N ratio, soil pH, and titratable acidity may influence the rate of mineralization of organic soils (Ivarson 1977, Rayment and Mathur 1977).

Organic soils are rich in fresh or partly decomposed organic matter (White 1987). They include muck and peat soils or histosols (Canada and U.S.), the tundras, the Irish peat bogs, the moor peats (Australia), and les sols hydromorphes organiques (France). Common organic soil parent materials may include mosses (such as sphagnum), gyttja, dy, marl, volcanic ash, cattails, reeds, sedges, pondweed, grasses, and various "water-loving" deciduous and coniferous shrubs and trees (Dawson 1956, Donahue et al. 1983). Organic soils can contain layer silicate minerals from trace to appreciable amounts (Fox and Kamprath 1971). Characteristic features of many peat deposits are low pH and dark, tea-colored water (Orem 1989).

Although soil test procedures were developed originally for mineral soils, they have been widely adopted for use on organic soils (Anderson and Beverly 1985). Thus, there is no chemical extractant that is unique to organic soils. Most laboratory methods apply equally to mineral and organic soils. Authors have used procedures with slight modification for organic soil. The most frequently modified procedure is the soil:solution ratio. However, many of these methods have not been calibrated to obtain a measure of their usefulness for organic soils.

Handling and preparation of organic soils before soil analysis are critical. Chemical analyses may be conducted on fresh samples (field moisture content) or on air-dried samples. It should

Soil Sampling and Methods of Analysis, M. R. Carter, Ed.,
Canadian Society of Soil Science. © 1993 Lewis Publishers.

be noted that drying organic soils increases dry bulk density and may lead to higher soil test levels of certain nutrients, such as available P and K (Anderson and Beverly 1985). These authors postulate that screened organic soils are more easily compacted upon drying, and conversely expand upon rehydration. Anderson and Beverly (1985) recommend that organic soils be sampled on a volume weight basis in order to assure uniformity of results. According to Mehlich (1973) uniform soil test results can be obtained when expressed on a volume basis by either using a scooped sample of known volume, or a weighed sample followed by multiplication with the bulk density to express the data on a soil volume basis.

The chemical methods given for mineral soils are also applicable to organic soils, therefore, and analytical procedures are not repeated here. The analyst may choose an appropriate method from previous chapters.

44.2 SAMPLE PREPARATION (ASTM D2974 1988)

44.2.1 MATERIALS

1   Analytical balance.

2   Blender, high speed.

3   Large flat pan or equivalent.

4   Spoon or spatula.

44.2.2 PROCEDURE

1   Mix organic soil sample thoroughly and weigh a 100- to 300-g representative sample. Determine the mass of the sample and spread evenly on a large flat pan, square rubber sheet, or paper. Crush soft lumps with a spoon or spatula and let the sample come to moisture equilibrium with room air, not less than 24 h.

2   Stir occasionally to maintain maximum air exposure of the entire sample.

3   When the mass of the sample reaches a constant value, calculate the moisture removed during air drying as a percentage of the as-received mass.

4   Grind a representative portion of the air-dried sample 1 to 2 min in a high-speed blender. Determine the amount, in grams, of air-dried sample equivalent to 50 g of as-received sample as follows:

$$\text{Equivalent sample mass, g} = 50.0 - [(50 \times M)/100] \tag{44.1}$$

where M is the percent of moisture removed in air drying.

5   Place the sample in a moisture-proof container.

44.3 ASH CONTENT AND ORGANIC MATTER CONTENT (ASTM D2974 1988)

44.3.1 Introduction

The ash content of organic soils is an important component of the soil matrix which reflects the degree of mineral nutrient enrichment (Farnham and Finney 1965). Usually, as the ash content increases, the content of available nutrients also increases. The increase of ash content in peat bog soils may be due to several factors, such as mineralization of organic matter, the application of mineral fertilizers, introduction of inorganic elements by ground waters, influx of mineral elements during floods, and their deposition from atmospheric dust (Skoropanov 1968).

There are basically two procedures involved in the determination of the ash content (or inorganic fraction) of a peat or inorganic sample: dry-ashing methods and wet-ashing methods. Dry-ashing methods have been employed widely for: (1) determining ash content of organic soils (van Lierop 1981b), (2) estimating organic matter content in both noncalcareous and forest soils (Ball 1964, David 1988, Page-Dumroese et al. 1990), and (3) the analysis of organic layers in ecological studies (Covington 1981).

The dry-ashing method involves the removal of organic matter by combustion of the sample at medium temperature (375 to 600°C) in a temperature-regulated muffle furnace. If necessary, samples are dried prior to ashing. The substance remaining after ignition is the ash and includes mineral impurities such as sand. Vessels suggested for ashing are porcelain, quartz, or platinum dishes (Williams and Vlamis 1961). Samples weights used vary from 0.25 to 2.0 g. The ash may be further dissolved in an acid solution for elemental analysis.

44.3.2 MATERIALS

1   Muffle furnace — controlled to ±5°C for ashing at 550°C.

2   High-form porcelain, 30-mL crucible.

3   Porcelain crucible cover or aluminum foil, heavy duty.

4   Desiccator cabinet or nonvacuum, desiccator with desiccant.

5   Analytical balance, spoons, etc.

6   Electric drying oven: regulated to a constant temperature of 105°C.

44.3.3 PROCEDURE

1   Weigh a 2-g sample of 2-mm oven-dried soil (105°C) into a tared high-form porcelain, 30-mL crucible. Determine the mass of covered high-form porcelain crucible. Remove the cover and place the crucible in a muffle furnace.

2   Gradually bring the temperature in the furnace to 375°C and maintain it for 1 h and then ash the sample either at 550°C for 16–20 h (Andrejko et al. 1983) or at 600°C for 6 h (Goldin 1987).

3   Remove the crucible from the furnace, cover, place it in a desiccator, cool, and weigh with 0.1-mg accuracy. Save the crucible and its contents for metal ion determination.

44.3.4 Calculation

Calculate the ash content as follows:

$$\text{Ash, g/100g} = [(a - c)/(b - c)] \times 100 \tag{44.2}$$

where a = final weight (g) of crucible and ash; b = weight (g) of crucible and sample; c = weight (g) of empty crucible.

The procedure described above can be used to determine the amount of organic matter as follows:

$$\% \text{ Organic matter} = 100 - \% \text{ mineral content (ash)} \tag{44.3}$$

44.3.5 COMMENTS

1   Using the above procedure for determining ash content it has been shown by Andrejko et al. (1983) that a temperature setting of 550°C is satisfactory for most purposes.

2   Dry ashing may overestimate the amount of organic matter in the soil. Positive errors are dependent on soil properties, such as the amount of carbonates and the amount and type of clay present in the mineral fraction of the soil (Goldin 1987).

3   The procedure outlined above measures the mass percentage of ash and organic matter in organic soil, including moss, humus, and reed-sedge types (Day et al. 1979).

4   Samples should be placed in the muffle furnace cold and the temperature allowed to rise slowly to avoid volatilization losses which are aggravated by violent deflagration.

5 Use high-form porcelain crucibles with covers or equivalent if ashes are retained for elemental analysis. These crucibles eliminate possible contamination of the ash by boron, which may volatilize from the furnace walls (Williams and Vlamis 1961).

44.4 TOTAL ELEMENTAL ANALYSIS

44.4.1 Major and Minor Elements (Other than Nitrogen and Carbon)

The determination of total elements in organic soils is common practice in characterization and pollution studies. Total elemental analysis may require a complete destruction of both organic and inorganic fractions of the soil matrix. The chemical procedure involved in the destruction of organic materials (peat, plants, sediments, soils) fall basically into two groups: (a) dry-ashing methods and (b) wet-ashing (or digestion) methods.

In the dry-ashing procedure, the organic material is ignited in a muffle furnace at medium temperature (550–660°C) to oxidize organic matter and the ions are extracted from the ash with either 1.5 M HCl (Ali et al. 1988) or 2 M HNO_3 (Day et al. 1979) solutions. This method of ignition is suitable for metal ion determinations.

The wet digestion involves complete dissolution of the organic material to convert elements to soluble forms by various digestion/oxidation procedures. This phase is then followed by determination of the liberated ions. The digestion procedure uses a variety of strong oxidizing and mineral acids, most frequently in combinations; for example HNO_3 + $HClO_4$ + HF (Desjardins 1978) and HNO_3 + $HClO_4$ (Mathur et al. 1980). In principle, a portion of the organic matter is oxidized by HNO_3 and the refractory remainder with $HClO_4$ (Van Lierop 1976). In order to reduce the danger of explosion during the digestion with $HClO_4$, the organic sample is pretreated with HNO_3 (Lim and Jackson 1982). Concentrated $HClO_4$ is a strong oxidizing agent for organic materials when heated to high temperatures (King 1932). Hydrofluoric acid is an excellent decomposition agent for siliceous materials present in the soil (Pritchard and Lee 1984). The acid digestion procedure is applicable for routine advisory soil analysis, but requires some specialized equipment such as a $HClO_4$ fumehood.

In most instances, metal in sample digests or extracts are determined independently by atomic absorption spectrometry and nonmetal ions by colorimetric methods. In some analytical soil laboratories, metal and nonmetal ions are determined simultaneously by inductively-coupled plasma emission spectrometry. The determination of phosphorus, boron, and sulfur can be done conveniently by autoanalyzer.

Since the analytical procedures given for mineral soils are also applicable to organic soils, the analyst may choose an appropriate method from previous chapters concerning mineral soils. The dry-ashing method is simpler, safer, and more economical than wet ashing (Ali et al. 1988). Wet ashing in concentrated acids is somewhat more accurate and rapid, but cumbersome and dangerous (Finck 1982). However, with proper equipment and procedures, a strong oxidizing reagent such as $HClO_4$ is no more hazardous than other strong acids (Day et al. 1979). According to Johnson and Ulrich (1959), wet ashing of organic material such as plant tissue is superior to dry ashing for the determination of elements except B and Cl. If the element content of the organic soil is low, it should be necessary to take a larger sample size (Desjardins 1978).

44.4.2 Total Nitrogen

The determination of nitrogen in organic soils is important as it is one of the primary nutrient elements necessary for plant growth. In organic soils the element occurs largely in the organic fraction, and its availability for plant growth is associated with the activity of microorganisms which synthetize and decompose the organic matter (Fitts and Hanway 1971). Organic soils are major contributors to the total N in soils (Stevenson 1982). The total nitrogen content of horticultural peats ranges from 0.6% in sphagnum moss peat to 3.5% in decomposed peat (Foth and Turk 1973). The range for N concentration of the cultivated layer of many organic soils from Canada is about 0.82 to 3.84% (MacLean et al. 1964, Mathur and Sanderson 1978). In general, very acid organic soils require substantial amounts of nitrogen for satisfactory plant growth, because of low soil-nitrogen content, poor microbial activity, and unfavorable carbon:nitrogen ratio (Lucas and Davis 1961).

The most-used methods for determination of total N in organic soils are semi-, micro-, and macro-Kjeldahl methods. These digestion techniques are similar to those for mineral soils. The analyst may determine total Kjeldahl N as outlined in Chapter 22, using an air-dried and ground (0.5 mm or finer) sample of appropriate size containing up to 10 mg of N for the macro-Kjeldahl method or up to 2 mg of N for the micro-Kjeldahl method.

44.5 SOLUBLE OR EXTRACTABLE ELEMENTS

Most organic soils are low in plant-available nitrogen, potassium, phosphorus, and some micronutrients (Donahue et al. 1983), mainly due to the lack of clay minerals and the high presence of metal-chelating humus in the soils (Lucas et al. 1975, Mathur and Farnham 1985, Mathur and Lévesque 1988). Several chemical methods have been proposed as indices of availability of the essential elements in soils. The extractants are generally dilute solutions of mineral or organic acids, simple salts, or organic and mineral complexing agents. The methods used for organic soil are essentially similar to those for mineral soils. However, it is recommended to increase the soil-extracting solution ratio in order to accommodate the high water-adsorptive capacity of organic soils (Bigger et al. 1953) and to eliminate the effects of neutralizing the extracting solution through reaction with limed organic soils (van Lierop et al. 1980). Prolonged extraction time and a wider soil:extractant ratio are useful for minimizing the effects of rewetting time variability of dried organic soils (van Lierop et al. 1980).

Chemical methods that have been used for extraction and estimation of plant-available elements in organic soils are outlined in Table 44.1. Lévesque and Mathur (1988) compared the effectiveness of eight extractants for Cu, Fe, Mn, and Zn in cultivated organic soils (histosols, peat, muck). These include 0.5 M HCl + 0.05 M AlCl$_3$, 1.0 M NH$_4$HCO$_3$ + 0.005 M diethylene triamine pentaacetic acid (DTPA), 0.005 M DTPA + 0.1 M triethanolamine (TEA), 1.0 M ammonium acetate (NH$_4$OAc) + 0.2 M ethylene diamine tetraacetic acid (EDTA) disodium salt, 0.1 M EDTA, 0.1 M HCl, 2.5% acetic acid (HOAc), and 1.0 M NH$_4$OAc. The overall results suggested that the various criteria for soil tests were best met by the extractant 0.5 M HCl + 0.05 M AlCl$_3$ for Cu, Fe, and Zn and by the extractant 0.1 M HCl for Mn (Mathur and Lévesque 1989).

Some chemical methods, especially those including chelating agents, remove organic matter in addition to inorganic material, which gives a dark brown-colored extract. Such organic matter must be removed from the soil extracts before the inorganic elements (P, B, etc.)

Table 44.1 Commonly Used Chemical Methods for Determination of Extractable Elements in Organic Soils

Chemical extractants	Soil-extracting solution ratio (w:v)	Extraction time	Extractable elements	Ref.
0.5 M HOAc; 0.5 M NaHCO$_3$; 0.03 M NH$_4$F + 0.1 M HCl; distilled water	1:20	30 min	P	MacLean et al. (1964)
0.135 M HCl; 0.018 HOAc	1:10	1 min	K	Bigger et al. (1953)
0.1 M (NH$_4$)$_2$SO$_4$	1:20	1 h	P, K, Mg	Stahlberg (1980)
Deionized water	1:12.5[a]	24 h	P	Anderson and Beverly (1985)
5 M HOAc	1:2.5[a]	24 h	K	Anderson and Beverly (1985)
1.0 M NH$_4$OAc (pH 7.0)	1:2.5[a]	1 h	Fe, Zn, Cu, Mn	Mathur and Lévesque (1989)
0.005 M DTPA + 0.1 M TEA + 0.01 M CaCl$_2$ (pH 7.3)	1:2.5[a]	1 h	Fe, Zn, Cu, Mn	Mathur and Lévesque (1989)
0.5 M HCl + 0.05 M AlCl$_3$	1:2.5[a]	1 h	Fe, Zn, Cu, Mn	Mathur and Lévesque (1989)
0.1 M HCl	1:2.5[a]	1 h	Fe, Zn, Cu, Mn	Mathur and Lévesque (1989)
Hot distilled water	1:2	5 min	B	Mathur and Rayment (1977)
0.05 M HCl + 0.0125 M H$_2$SO$_4$	>1:4	>5 min	P, K, Ca, Mg	van Lierop et al. (1980)

[a] Volume: volume ratio.

can be assessed by colorimetric methods. The removal of interfering organic matter can be done by treating a portion of the soil extract with activated charcoal, separating the charcoal from the extract by filtration, and analyzing an aliquot of the charcoal-treated extract for the inorganic element. Activated charcoal should be pretreated with 6 N HCl, followed by repeated washings with distilled water to remove P contained in some commercial activated charcoal. According to McGeehan et al. (1989), charcoal is effective in reducing extract color, but should be used with caution to avoid negative error due to B sorption by decolorizing charcoal during filtration.

Many commonly used methods for mineral soils that are not presented in Table 44.1 could be suitable for organic soils. The analyst may determine extractable elements as outlined in previous chapters concerning mineral soils. Many of these methods could be modified by increasing the soil-extracting solution ratio to adapt them to organic soils. If the nutrient content of the organic soil is low, it should be necessary to take a larger sample size.

Extractable Zn, Cu, Fe, and Mn in cultivated organic soils may be determined by individual soil test extractants (Lévesque and Mathur 1988) as follows:

1   Weigh a 10-mL, air-dried and ground (<2 mm) sample, of known moisture content in a 50-mL centrifuge vial with a cap.

2   Add 25 mL of the extractant and shake for 1 h.

3   Centrifuge and filter the supernatant through a Whatman® No. 541 filter. Interference from organic matter can be eliminated by treating the extract with HNO_3-$HClO_4$-HF (wet oxidation). Clear filtrates or extracts of the digests could be analyzed by atomic absorption spectroscopy to determine their Cu, Fe, Mn, and Zn contents. The extractant may include: 0.5 M HCl + 0.05 M $AlCl_3$, 0.005 M DTPA + 0.1 M TEA + 0.01 M $CaCl_2$ (pH 7.3), 0.1 M HCl or 1.0 M NH_4OAc (pH 7.0).

44.6 SOIL pH AND LIME REQUIREMENT

Critical soil acidity related to optimal crop yields in organic soils have usually been reported as measurements made in water, 0.01 M $CaCl_2$, or 1 N KCl as the suspending media (van Lierop et al. 1980, van Lierop 1981a). Soil organic pH values obtained in calcium chloride solution usually run about a 0.5 to 0.8 pH unit lower than measurements in water due to release of more hydrogen ions by cation exchange (Day et al. 1979). Methods using pastes of distilled water and air- or oven-dried peat may give pH values slightly lower than methods using fresh peat (Stanek 1973).

The pH of organic soils has been determined potentiometrically in soil suspensions consisting of: (1) 1:2 or 1:4 volumetric ratio of compacted field-moist soil (or samples with reconstituted bulk density) to water, 0.01 M $CaCl_2$, or 1 M KCl. Measure pH about 30 min after mixing; preparation of samples with reconstituted bulk densities is described by van Lierop (1981b) and is outlined in Chapter 43. (2) 3:50 weight to volume ratio of air-dried peat or equivalent amount of moist material to water, 0.01 M $CaCl_2$, or 1 M KCl. Measure pH about 30 min after mixing (ASTM D2976 1988). Soil pH determinations are described by van Lierop and Mackenzie (1977) and van Lierop (1981a). For very fibrous materials, such as sphagnum moss peat, additional water may be needed (Rund 1984).

**Table 44.2 Calibration to Determine Lime Requirement of the Surface
20 cm of Organic Soil Using Several Buffer Methods
(van Lierop 1983)**

Soil-buffer pH	SMP	Woodruff	Mehlich	Ba(OAc)$_2$	NH$_4$OAc
6.1	0.0[a]	0.0	0.0	0.0	0.0
6.0	0.0	1.5	0.0	0.0	0.0
5.9	1.1	3.9	1.1	0.0	0.0
5.8	2.2	6.3	2.8	0.0	0.0
5.7	3.4	8.6	4.5	0.0	0.0
5.6	4.6	11.0	6.3	0.0	0.0
5.5	5.7	13.3	8.0	0.0	0.0
5.4	6.9	15.7	9.8	0.0	0.0
5.3	8.0	18.1	11.5	0.0	1.5
5.2	9.2	20.4	13.3	2.8	5.4
5.1	10.3	22.8	15.0	5.6	9.3
5.0	11.5	25.1	16.7	8.4	13.1
4.9	12.7		18.5	11.2	17.0
4.8	13.8		20.2	14.0	21.0
4.7	15.0		22.0	16.7	24.7
4.6	16.1		23.7	19.5	28.6
4.5	17.3		25.5	22.3	
4.4	18.4		27.2	25.0	
4.3	19.6			27.8	
4.2	20.7				
4.1	21.9				
4.0	23.0				
3.9	24.2				

[a] Rates are expressed in meq of CaCO$_3$/100 mL soil and are equivalent
    to tons of CaCO$_3$/ha to a depth of 20 cm.

The soil suspension after pH measurement may be used for lime requirement (LR) determination as outlined in Chapter 14. The choice of buffer for determining the LR of organic soils is not very important (van Lierop 1983). However, it is simpler to use either the Ba(OAc)$_2$ or NH$_4$OAc buffer if a buffer has to be prepared for the determination of LR of both mineral and organic soils (van Lierop 1983). In addition, the pH should be measured in water. The analyst may select and record the amount of lime required to bring the organic soil to pH 5.4 (H$_2$O), based on the soil-buffer pH of Table 44.2.

44.7 CATION-EXCHANGE CAPACITY AND EXCHANGEABLE CATIONS

The cation exchange of organic soils is considered a result of the exchange or the replacement with other cations of the dissociable hydrogen ions in certain organic functional groups (Puustjärvi 1956). Carboxyl and phenolic groups are the predominant sites of cation exchange in the soil organic matter (Schnitzer and Skinner 1965, Schnitzer 1969). Other functional groups that may hold exchangeable cations are imide, enol, amine, and hydroxyl groups (Bohn et al. 1985). Cation-exchange phenomenon in organic soils is more complex than for mineral soils; therefore, the determination of the cation-exchange capacity (CEC) of organic soils is somewhat arbitrary (MacLean et al. 1964).

Exchangeable bases and exchange acidity of organic soils have usually been determined respectively by the NH_4OAc method buffered at pH 7.0 and the $BaCl_2$-TEA method buffered at pH 8.0. The two most common methods of determining CEC of organic soils are the NH_4OAc method buffered at pH 7 and the unbuffered HCl-$Ba(OAc)_2$ method. The barium acetate method may yield somewhat higher values than the ammonium acetate method (MacLean et al. 1964). All these methods are described by Osborne (1978) and Day et al. (1979).

It is also possible to determine the CEC by separate determinations, and subsequent summing up of exchangeable bases (as determined by the NH_4OAc method buffered at pH 7) and exchangeable or titratable acidity (as determined by the $BaCl_2$-TEA method buffered at pH 8).

In general, the barium acetate method (Osborne 1978) is the preferred method for CEC of organic soil materials. Other commonly used methods for mineral soils may also be suitable for organic soils. The analyst may determine exchangeable cations and CEC of organic soils as outlined in previous chapters concerning mineral soils. Since the CEC of organic soils is much higher than that of mineral soils, it should be preferable to take a lower sample size for the determination of CEC. The values of CEC may vary, depending on the analytical procedure used for its determination. CEC measurements of some organic soil materials by either the compulsion exchange method (Hendershot and Duquette 1986) or the more classical exchange method may not be possible to perform in Büchner funnels because the organic material may float.

REFERENCES

Ali, M. W., Zoltai, S. C., and Radford, F. G. 1988. A comparison of dry and wet ashing methods for the elemental analysis of peat. Can. J. Soil Sci. 68: 443–447.

Anderson, D. L. and Beverly, R. B. 1985. The effects of drying upon extractable phosphorus, potassium and bulk density of organic and mineral soils of the Everglades. Soil Sci. Soc. Am. J. 49: 362–366.

Andrejko, M. J., Fiene, F., and Cohen, A. D. 1983. Comparison of ashing techniques for determination of inorganic content of peats. Pages 5–20 in P. M. Jarret, Ed. Testing of peats and organic soils. ASTM STP 820. American Society for Testing and Materials, Philadelphia.

ASTM (American Society for Testing and Materials). 1988. Annual book of ATM standards. Volume 04.08. ASTM, Philadelphia.

Ball, D. F. 1964. Loss-on-ignition as an estimate of organic matter and organic carbon in non-calcareous soils. J. Soil Sci. 15: 84–92.

Bigger, T. C., Davis, J. F., and Lawton, K. 1953. The behaviour of applied phosphorus and potassium in organic soil as indicated by soil tests and the relationship between soil tests, green-tissue tests and crop yields. Soil Sci. Soc. Am. Proc. 17: 279–283.

Bohn, H. L., McNeal, B. L., and O'Connor, G. A. 1985. Soil chemistry. 2nd ed. Wiley-Interscience, New York.

Covington, W. W. 1981. Changes in forest floor organic matter and nutrient content following clearcutting in northern hardwoods. Ecology 61: 41–48.

David, M. B. 1988. Use of loss-on-ignition to assess soil organic carbon in forest soils. Commun. Soil Sci. Plant Anal. 19: 1593–1599.

Dawson, J. E. 1956. Organic soils. Adv. Agron. 8: 337–401.

Day, J. H., Rennie, P. J., Stanek, W., and Raymond, G. P. 1979. Peat testing manual.

Associate committee on geotechnical research. National Research Council of Canada. Technical Memorandum No. 125. Ottawa.

Desjardins, J. G. 1978. Determination of total major and minor elements (other than C, N, P and S). Pages 140–141 in J. A. McKeague, Ed. Manual on soil sampling and methods of analysis. 2nd ed. Can. Soc. Soil Sci., Ottawa.

Donahue, R. L., Miller, R. W., and Shickluna, J. C. 1983. Soils: an introduction to soils and plant growth. 5th ed. Prentice-Hall, Englewood Cliffs, NJ.

Farnham, R. S. and Finney, M. R. 1965. Classification and properties of organic soils. Adv. Agron. 17: 115–162.

Finck, A. 1982. Fertilizers and fertilization: introduction and practical guide to crop fertilization. Verlag Chemie. Florida.

Fitts, J. W. and Hanway, J. J. 1971. Prescribing soil and crop nutrient needs. Pages 57–79 in R. A. Olson, T. J. Army, J. J. Hanway, and V. J. Kilmer, Eds. Fertilizer technology and use. 2nd ed. American Society of Agronomy, Madison, WI.

Foth, H. D. and Turk, L. M. 1973. Fundamentals of soil science, 5th ed. Wiley Eastern Private Limited, New Delhi.

Fox, R. L. and Kamprath, E. J. 1971. Adsorption and leaching of P in acid organic soils and high organic matter sand. Soil Sci. Soc. Am. Proc. 35: 154–156.

Goldin, A. 1987. Reassessing the use of loss-on-ignition for estimating organic matter content in noncalcareous soils. Commun. Soil Sci. Plant Anal. 18: 1111–1116.

Hendershot, W. H. and Duquette, M. 1986. A simple barium chloride method for determining cation exchange capacity and exchangeable cations. Soil Sci. Soc. Am. J. 50: 605–608.

Ivarson, K. C. 1977. Changes in decomposition rate, microbial population and carbohydrate content of acid peat bog after liming and reclamation. Can. J. Soil Sci. 57: 129–137.

Johnson, C. M. and Ulrich, A. 1959. Analytical methods for use in plant analysis. Calif. Agr. Exp. Sta. Bull. 766.

King, E. J. 1932. The colorimetric determination of phosphorus. Bio. Chem. J. 26: 292–297.

Lévesque, M. and Mathur, S. P. 1988. Soil tests for copper, iron, manganese, and zinc in histosols. 3. A comparison of eight extractants for measuring active and reserve forms of the elements. Soil Sci. 145: 215–221.

Lim, C. H. and Jackson, M. L. 1982. Dissolution for total elemental analysis. Pages 1–12 in A. L. Page, R. H. Miller, and D. R. Keeney, Eds. Methods of soil analysis. 2nd ed. Agronomy No. 9. Part 2. American Society of Agronomy, Madison, WI.

Lucas, R. E. and Davis, J. F. 1961. Relationships between pH values of organic soils and availabilities of 12 plant nutrients. Soil Sci. 92: 172–182.

Lucas, R. E., Rieke, P. E., Shickuna, J. C., and Cole, A. 1975. Lime and fertilizer requirements for peats. Pages 40–70 in D. W. Robinson and J. G. D. Lamb, Eds. Peat in horticulture. Academic Press, London.

MacLean, A. J., Halstead, R. L., Mack, A. R., and Jasmin, J. J. 1964. Comparison of procedures for estimating exchange properties and availability of phosphorus and potassium in some eastern Canadian organic soils. Can. J. Soil. Sci. 44: 66–75.

Mathur, S. P. and Farnham, R. S. 1985. Geochemistry of humic substances in natural and cultivated peatlands. Pages 53–85 in G. Aiken, D. McKnight, R. Wershaw, and P. MacCarthy, Eds. Humic substances in soil, sediment and water. Wiley-Interscience, New York.

Mathur, S. P. and Lévesque, N. P. 1988. Soil tests for copper, iron, manganese, and zinc in histosols. 2. The distributions of soil iron and manganese in sequentially extractable forms. Soil Sci. 145: 102–110.

Mathur, S. P. and Lévesque, N. P. 1989. Soil tests for copper, iron, manganese, and zinc in histosols. 4. Selection on the basis of soil chemical data and uptakes by oats, carrots, onions, and lettuce. Soil Sci. 148: 424–432.

Mathur, S. P., MacDougall, J. I., and McGrath, M. 1980. Levels of acitvities of some carbohydrates, protease, lipase, and phosphatase in organic soils of differing copper content. Soil Sci. 129: 376–385.

Mathur, S. P. and Rayment, A. F. 1977. Influence of trace element fertilization on the decomposition rate and phosphatase activity of a mesic fibrisol. Can. J. Soil Sci. 57: 397–408.

Mathur, S. P. and Sanderson, R. B. 1978. Relationships between copper contents, rates of soil respiration and phosphatase activities of some histosols in an area of southwestern Quebec in the summer and the fall. Can. J. Soil Sci. 58: 125–134.

McGeehan, S. L., Topper, K., and Naylor, D. V. 1989. Sources of variation in hot water extraction and colorimetric determination of soil boron. Commun. Soil Sci. Plant Anal. 20: 1777–1786.

Mehlich, A. 1973. Uniformity of soil test results as influence by volume weight. Commun. Soil Sci. Plant Anal. 4: 475–486.

Orem, W. H. 1989. Dissolved organic matter in peats: chemical structure and origin. Page 30 in GEOC. Abstracts of papers. 197th ACS National Meeting. American Chemical Society, Dallas, TX.

Osborne, V. E. 1978. Cation exchange capacity and exchangeable cations. Pages 72–86 in J. A. McKeague, Ed. Manual on soil sampling and methods of analysis. 2nd ed. Can. Soc. Soil Sci., Ottawa.

Page-Dumroese, D. S., Loewenstein, H., Graham, R. T., and Harvey, A. E. 1990. Soil source, seed source, and organic-matter content effects on Douglas-fir seedling growth. Soil Sci. Soc. Am. J. 54: 229–233.

Parent, L. E., Mackenzie, A. F., and Perron, Y. 1985. Rate of P uptake by onions compared with rate of pyrophosphate hydrolysis in organic soils. Can. J. Soil Sci. 65: 1091–1095.

Pritchard, M. W. and Lee, J. 1984. Simultaneous determination of boron, phosphorus and sulphur

in some biological and soil materials by inductively-coupled plasma emission spectrometry. Anal. Chim. Acta 157: 313–326.

Puustjärvi, V. 1956. On the cation exchange capacity of peats and on the factors of influence upon its formation. Acta Agric. Scand. 6: 410–449.

Rayment, A. F. and Mathur, S. P. 1977. Observations on the subsidence of drained peat soils under grassland culture in Newfoundland. Proc. 17th Maskeg Res. Conf. Saskatoon. National Research Council of Canada. Ottawa.

Rund, R. C. 1984. 2. Fertilizers. Pages 8–37 in S. Williams, Ed. Official methods of analysis of the Association of Official Analytical Chemists. 14th ed. Association of Official Analytical Chemists, Arlington, VA.

Schnitzer, M. 1969. Reactions between fulvic acid, a soil humid compound and inorganic soil constituents. Soil Sci. Soc. Am. Proc. 33: 75–81.

Schnitzer, M. and Skinner, S. I. M. 1965. Organo-metallic interactions in soils. 4. Carboxyl and hydroxyl groups in organic matter and metal retention. Soil Sci. 99: 278–284.

Skoropanov, S. G. 1968. Reclamation and cultivation of peat-bog soils. Translated from Russian. Mary Edestein, Ed. Israël program for scientific translations.

Stahlberg, S. 1980. A new extraction method for estimation of plant-available P, K and Mg. Acta Agric. Scand. 30: 93–107.

Stanek, W. 1973. Comparisons of methods of pH determination for organic terrain surveys. Can. J. Soil Sci. 53: 177–183.

Stevenson, F. J. 1982. Distribution of nitrogen in soil. Pages 1–41 in F. J. Stevenson, Ed. Nitrogen in agricultural soils. Agronomy No. 22. American Society of Agronomy, Madison, WI.

van Lierop, W. 1976. Digestion procedures for simultaneous automated determination of NH_4, P, Ca, and Mg in plant material. Can. J. Soil Sci. 56: 425–432.

van Lierop, W. 1981a. Conversion of organic soil pH values measured in water, 0.01 M CaCl₂ or 1 N KCl. Can. J. Soil Sci. 61: 577–579.

van Lierop, W. 1981b. Laboratory determination of field bulk density for improving fertilizer recommendation of organic soils. Can. J. Soil Sci. 61: 475–482.

van Lierop, W. 1983. Lime requirement determination of acid organic soils using buffer-pH methods. Can. J. Soil Sci. 63: 411–423.

van Lierop, W. and MacKenzie, A. F. 1977. Soil pH and its application to organic soils. Can. J. Soil Sci. 57: 55–64.

van Lierop, W., Martel, Y., and Cescas, M. P. 1980. Optimal soil pH and sufficiency concentration of N, P and K for maximum alfalfa and onion yields on acid organic soil. Can. J. Soil Sci. 60: 107–117.

White, R. E. 1987. Introduction to the principles and practice of soil science. 2nd ed. Blackwell Scientific, California.

Williams, D. E. and Vlamis, J. 1961. Boron contamination in furnace dry ashing of plant material. Anal. Chem. 33: 967–968.

Chapter 45
Micromorphological Methodology for Organic Soils

C. A. Fox

Agriculture Canada
Ottawa, Ontario, Canada

L. E. Parent

Laval University
Sainte-Foy, Quebec, Canada

45.1 INTRODUCTION

The preparation of organic materials for microscopic evaluation requires that special care be taken to preserve the *in situ* field quality of the organic constituents. The methodology described in this chapter will apply to the preparation of samples obtained from forest (unsaturated environments) as well as peat (saturated environments) materials. Micromorphological methodology specific for inorganic materials is presented in Chapter 65.

45.2 SAMPLE PREPARATION

45.2.1 Field Sampling of Forest Materials

Murphy (1986) describes a procedure for removing *in situ* surface humus forms by carving a soil column, encasing the sample with plaster of Paris, and then excavating the column. However, obtaining undamaged, *in situ* samples of forest materials (also referred to as folic materials) is often difficult because of the numerous live roots of varying sizes, the fibrous nature of the accumulated residues, fallen twigs and branches, and sometimes, logs and roots from periodic tree-fall events. This is especially applicable to the soils of the Folisol great group (Agriculture Canada Expert Committee on Soil Survey 1987, Fox et al. 1987). For soils with thick accumulations of folic materials the following approach to micromorphological sampling is taken.

Soil Sampling and Methods of Analysis, M. R. Carter, Ed.,
Canadian Society of Soil Science. © 1993 Lewis Publishers.

45.2.1.1 MATERIALS

1   Soil sampling equipment, i.e., shovel, knife, pruning shears, pruning saw or dry-wall hand saws with short blades, hand trowel, scissors, plastic bags, masking tape, tape measure, permanent (waterproof) marking pens.

2   Sampling containers (Kubiena boxes); see Chapter 65 (Section 65.2.1, item 3). The edges of the sample container on one of the open ends can be sharpened to facilitate cutting into fibrous organic materials.

3   Modified Macaulay auger. See Chapter 42 (Sampling Organic Soils).

4   Portable, gas-operated 16-in chainsaw plus required safety equipment for head, eye, and leg protection.

5   Compass, topographic maps (or aerial photograph), field notebook.

45.2.1.2 PROCEDURE

1   For both thin surface horizons (usually less than 10 cm) composed of accumulated forest materials and organic horizons below the surface layer that have no wood fragments and only few small live roots, the procedures for locating the representative soil site and sampling are the same as for inorganic materials outlined in Section 65.2.2 (steps 1 to 15).

2   For thick accumulations of folic materials, having large live roots, abundant woody materials, or extreme accumulations of fibrous materials, the following procedure is recommended:

   a. At the sampling site, determine a representative location for the soil pit by using transect sampling schemes. Use the Macaulay auger to obtain several soil cores for assessing the overall spatial distribution of the organic materials with respect to the type of material and the depth of occurrence.

   b. Mark out an area 1 m² in size. Use the Macaulay auger to determine the exact depth to the mineral contact below the accumulated organic material. Note the exact location and depth of any incorporated stones or fragmental material. These steps are crucial for the safe operation of the chainsaw.

   c. Insert the chainsaw into the surface organic materials to a depth of about 10 cm. Cut around the perimeter of the designated area and then cut a grid pattern to divide the area into smaller, easily handled portions. Remove the loosened organic material with a shovel. Repeat at 10-cm depth increments until all live roots and fibrous and woody materials are passed through.

3   Describe the profile face. See Fox (1985a), Fox et al. (1987), and Klinka et al. (1981) for descriptive systems for forest soils, specifically folic materials.

4   Follow procedures described in Chapter 65, Section 65.2.2 (steps 3 to 15) for taking *in situ* soil samples.

45.2.2 Field Sampling of Peat Materials

The saturated environment of the peat deposit is the main obstacle that often prevents taking samples directly with containers. In order to have access to the soil face, a pumping or hand-bailing system must be readily available to remove the rapid influx of water, and even then, the sides of the soil pit may slump and be extremely unstable and dangerous. In addition, peat deposits frequently extend to depths beyond all feasibility of digging. Consequently, peat samplers that remove an intact core are used (see Chapter 42). Intact samples suitable for micromorphological studies can be obtained easily with the modified Macaulay auger.

45.2.2.1 MATERIALS

1   Macaulay auger with 1-m extension rods and a selected barrel size. A barrel having dimensions of 3.8 cm diameter and length of 50 cm was found most suitable for obtaining micromorphological samples.

2   Sample containers. Prepare half-cylinder containers from 3.8-cm (1.5-in.) diameter PVC plumbing pipe cut into 15-cm lengths. Other lengths can also be used, depending on the requirements of the study. Cut lids from 0.64-cm (1/4-in) plywood to fit the dimensions of the open end of the half-cylinders.

3   Knife, plastic bags, label tags and twist-ties, masking tape, white or yellow wax pencil to write on black plastic PVC pipe, permanent water-insoluble marking pen.

4   Sturdy container for transportation of samples, styrofoam packing beads.

45.2.2.2 PROCEDURE

1   With the Macaulay auger, first obtain from the required depth a core of peat material for field description. Describe the peat material according to a field descriptive system such as the Canada Wetland Registry forms (Tarnocai 1980). Take samples for chemical and physical analyses.

2   With the Macauley auger, adjacent (approximately 20 cm) to the auger hole in step 1, take a second core of peat material at the same depth. Open the auger so that the sampled core is positioned on the blade.

3   Place the PVC container on top of the extracted core positioned at the selected sampling depth. Use the wax pencil to record on the PVC container the sample number and the depth range, and indicate the direction to the peat surface with an arrow. With a sharp knife, cut through the extracted peat core at both ends of the sample container. To remove the sample from the blade of the auger, hold securely onto the PVC container, and then, while maintaining the auger in a horizontal position, invert the auger barrel. The portion of the extracted core not secured by the PVC container will fall away. Slowly slide the container off of the blade.

4   Place the core sample still in its container into a plastic bag. Over the open end of the PVC container, place the wooden lid on top of the plastic bag and secure with masking tape.

5   Record on the masking tape of the wrapped container the sample number, the organic layer, the vertical orientation to the surface, the depth range, and other specific data as required.

6   To transport to the laboratory, place the samples *horizontally* into a sturdy container packed preferably with styrofoam beads. Never position the containers in a vertical direction because movement and vibration from transportation results in severe compression of the saturated samples.

7   Store the samples (except for the sedimentary peat samples) in a freezer or cold room set below 0°C. Ensure that these samples are placed in a horizontal position to prevent ice lenses from segregating to one end of the sample and causing structural disruption. Store sedimentary peat samples in a cold room at 4–10°C; do not freeze, because freezing has a desiccating effect on the morphology. Prepare for resin impregnation (Section 45.4) as soon as possible.

45.3 MICROTOMIC THIN-SECTIONING (L. E. PARENT AND F. J. PAUZÉ)

Botanical microtechniques are particularly well suited for peat materials, even those containing quartz grains which can damage steel microtome knives and tear specimen sections if not properly treated. The traditional paraffin embedding medium was replaced by amorphous celloidin, to prevent disturbing the peat particle arrangements on slides when extracting the crystalline paraffin. Parent et al. (1980) modified the original celloidin method described by Sass (1958) and could obtain peat (unconsolidated material) and sedimentary peat or gyttja (consolidated material) thin sections of 15–20 μm.

45.3.1 MATERIALS AND REAGENTS

Caution: many of the reagents indicated below are hazardous and/or toxic. W.H.M.I.S. safety procedures must be used at all times.

1   Fine-mesh nylon cloth (as for nylon stockings) for unconsolidated peat; capsules with top and bottom grids (as in Canlab No. M7616-1) for consolidated peat.

2   Acrolein (CH_2=CHCHO), 10% solution in water. Acrolein commercial solution comes in a brown bottle which, upon arrival, should be wrapped with tin-foil and stored in a freezer for it is unstable, polymerizes under light, and has a pungent odor. A fumehood must be used whenever acrolein is diluted or used in any of the procedure steps.

3   Tetrahydrofuran (THF:tetramethylene oxide C_4H_8O). For safety, THF must be used in a fumehood at all times. Stock solutions can be kept at room temperature in brown bottles.

4   Chloroform (trichloromethane). Use only in a fumehood.

5   Hydrofluoric acid (HF, 50%)-ethanol (50%) solution (1:1). This solution need be prepared only if quartz grains must be dissolved from peat block specimens. *Caution:* nalgene, not glassware, should be used to prepare this solution, as the HF acid will dissolve glass. A fumehood must be used at all times when handling this chemical.

6   *Pure THF-ethanol (ETOH) 95% (1:1) solution* and *ethyl ether-ETOH 95% (1:1) solution*. The latter solution is highly volatile and flammable.

7   Celloidin: a form of nitrocellulose, also named parlodion or pyroxylin, made by Mallinkrodt Chemical Works, St. Louis, MO.

8   Celloidin solutions: 2, 4, and 6%, in pure THF-ETOH 95% (1:1) (made by adding 2, 4, or 6 g of celloidin chips to 100 mL of the THF-ETOH solution), prepared on a regular basis for short-term use.

9   Celloidin solutions: 6, 8, 10, and 12% in ethyl ether-ethanol 95% (1:1) solution, prepared on a regular basis for short-term use. *Waste celloidin can be salvaged* (for more details, cf. Sass 1958).

10   Glycerin [$C_3H_5(OH)_3$]-ETOH 95% solution (1:1).

11   Sliding microtome.

12   Steel microtome knife (two), and knife sharpener.

13   ETOH (95%)-chloroform solution (4:1). Use a fumehood; be aware of highly flammable nature.

14   Benzene. This chemical must be handled wearing proper protection and at all times used only under a fumehood. The purpose of this chemical is that it is to be used only in the final clearing of the sections and as a solvent of the glue (Permount, in this instance), to affix cover slips and prevent air bubbles. L. E. Parent (personal communication) suggests that benzene not

be replaced by acetone for a clearing solvent, as acetone will dissolve the hardened celloidin of the thin sections.

15   Permount® (Fisher Scientific).

16   Neck-free specimen bottles (125 mL) that can be corked and covered with a metal clamp to ensure that they remain corked when in the oven.

17   Ventilated fumehood: ventilated oven.

18   Scalpels, section lifters (spatulas), frosted microslides 7.6 × 2.5 cm (3 × 1 in), cover slips, balsam bottle, hair brushes with quill handle (camel hair brush type), forceps, needles, small staining dishes with cover, and weights consisting of rectangular lead blocks 20 × 15 × 15 mm (≈50 g to flatten sections at the end of the procedure).

19   Wood blocks or metal bases for mounting specimens. Wood blocks are safer, and their size should be approximately 20 × 20 × 25 mm, for cross sections, or preferably larger for longitudinal sections.

20   Binocular microscope, with polarizing filters, to distinguish mineral grains.

45.3.2 FIELD SAMPLING AND SPECIMEN PREPARATION

1   Collecting unconsolidated peat block samples (UPBS) in the field is a critical step. Each peat block is cut on lateral sides and at the base with a sharpened knife, over which is carefully slid a 1-L waxed cardboard carton (type used for milk), the top part of it having been previously cut off and one side opened. The back side and top of the block are cut, the block containing cardboard is carefully withdrawn from the soil, the open side of the cardboard is flipped back and secured with rubber bands. The top part is covered and secured with saran wrap foil. Placed on dry ice in a cooler, the blocks are transported to a laboratory freezer maintained at −10°C. Coherent peat (sedimentary peat) block samples (CPBS) can be collected in a similar manner, but are stored at 4°C; never allow these samples to freeze. After each sampling, blocks are identified as to the horizon of origin and orientation in the horizon.

2   Prior to slicing blocks to prepare the specimens, frozen UPBS are reduced by 20 mm on all faces, using a diamond saw, in order to remove disturbed material while sampling. Examination with the binocular microscope of the portions of the peat material that were removed should indicate whether or ot quartz grains/mineral fragments are present. If such grains are observed, make a note of the sample because further treatment will be necessary to remove the grains (Section 45.3.4, step 3 and Section 45.3.8).

3   Slices 25-mm thick are then cut all the way through a block with the diamond saw. Slices are identified as to their number and orientation in the block.

Each slice is then cut anew into specimens of 20 × 20 × 25 mm, which are again identified as to their position and orientation in slices.

4   Each UPBS specimen thus obtained is wrapped in a nylon cloth, the ends of which are secured by sewing, and an identification tag is attached to it.

5   Sedimentary peat material (CPBS) needs only to be cut into slices 5–7 mm thick. The specimens obtained are placed into metal or plastic capsules with top and bottom grids so as to assure rapid penetration and/or circulation of fluids.

45.3.3 UPBS AND CPBS SPECIMEN FIXATION AND DEHYDRATION

Caution: the following operations must be performed in a fumehood.

1   Each UPBS frozen specimen is placed in a 125-mL neck-free specimen bottle and submerged with four to five times the volume of the specimen with a 10% acrolein fixation solution. Specimen bottles are then corked and refrigerated for 2 d at 4°C.

2   Specimens of CPBS can also be fixed with acrolein solution, but are preferably submitted to dehydration only, as indicated in Section 45.3.4.

NOTE: from now on (A) any fluid to be added to a specimen bottle will have to equal four to five times the volume of the specimen; (B) only the supernatant will be removed, in order to keep specimens bathed at all times; (C) removing the supernatant or adding fresh fluid will be done by pipetting along the upper wall of bottles so as not to disturb the specimen fabric; (D) specimens will never be moved by the tag or otherwise laterally or upwards.

45.3.4 UPBS AND CPBS SPECIMEN DEHYDRATION AND CELLOIDIN INFILTRATION

Caution: the following operations must be performed in a fumehood or ventilated hood.

1   UPBS and/or CPBS specimens in fixing solution can now be dehydrated with pure THF *in four 1-h successive steps,* so as to reduce residual water to approximately less than 1%. Remaining water will be removed by pure THF and ethanol 95%, during the first steps of celloidin infiltration (THF or acrolein mix well with ethanol). At this stage, UPBS or CPBS specimens are infiltrated and covered with nearly pure THF. The 2, 4, and 6% celloidin solutions can now be used.

2   Remove THF supernatant from specimen bottles and add an equivalent amount of the 2% celloidin solution. Cork the bottles, place a metal clamp

over the cork and bottom of each bottle, and place them in a 50°C oven for 24 h. Repeat this 2% procedure, and then follow with the 4 and 6% solutions.

3   At this stage, if quartz grains were identified in the UPBS specimens, they have to be dissolved according to the procedure that is described in Section 45.3.8.

4   Switch now to the 6, 8, 10, and 12% celloidin solutions *prepared with ethyl ether-ethanol 95% (1:1) solution* and resume as in step 1 above, except for the repetition. This change from THF to ethyl ether as celloidin solvent aims at obtaining a translucent embedding medium at the end of the process, instead of a milky one generated by the THF-ETOH celloidin solutions.

5   After the 12% concentration step, specimen bottles are kept in the 50°C oven for a week and celloidin chips are added daily to thicken the medium to a concentration of 16–20% celloidin. Checking the proper consistency of the embedding medium is done by dipping a dried matchstick into the celloidin solution, lifting out a mass of celloidin, then immersing it in pure chloroform for 1 h. The hardened celloidin forms a clear, firm mass that can be sliced easily with a razor blade.

6   Then, scoop tagged specimens out of their bottle with a spoonlike spatula (never handle the specimen by the tag); with a large chunk of celloidin surrounding them, immerse them in chloroform for 3 d in order to remove celloidin solvents and harden the celloidin to the center of the specimens. The hardened specimens can be stored indefinitely thereafter by immersing them into a glycerin-ethanol 95% (1:1) solution.

45.3.5 BLOCKING AND MOUNTING SPECIMENS

A specimen of UPBS or CPBS can be damaged by compression in the microtome clamp and therefore needs to be mounted on a wood block *preferably* (or on a metal base supplied with a sliding microtome). Blocks must be prepared a few days in advance of specimen mounting.

Caution: use a fumehood and/or ventilated hood.

1   Prepare mounting blocks by drying them thoroughly in a 110°C oven, soak them in anhydrous methyl alcohol, then store in *waste 6% celloidin (ethyl ether-ETOH solution)*. One day before using, in a fumehood, soak the blocks in used chloroform and let dry. Just before mounting, scrape off with a scalpel excess celloidin on one pitted end and the four sides near the pitted end of blocks and wrap the scraped portion with a generous amount of 6% *"unheated" celloidin* (ethyl ether-ETOH solution) and affix a prepared specimen as indicated in step 2 below.

2   Withdraw specimens, one at a time, from the glycerin-alcohol storage fluid and trim, cutting away excess celloidin with a scalpel or a razor blade, but

leaving on the sides and ends about 3–5 mm of celloidin. Make sure that the cuttings leave all sides approximately parallel. Then, soak the specimens for 2–3 h in anhydrous ethyl alcohol to remove the residual water left from the storage fluid. This will also soften the celloidin somewhat, but just enough for adequate mounting of specimens on prepared wood blocks. Now, soak the block and mounted specimens for about 1 h in used chloroform to harden the celloidin prior to clamping the block rigidly into the sliding microtome clamp.

45.3.6 SPECIMEN SECTIONING

Caution: microtoming should be performed under a ventilated hood.

1   Once the specimen wood block is fixed into the microtome clamp, place the knife in the sliding carriage and lubricate the tracks with a fine film of light oil, and wipe the excess with a cloth, so as to obtain regular strokes. The operation of the sliding microtome should be carefully studied. Such modern microtomes have an automatic feed. Adjust the feed until the specimen almost touches the underside of the knife. Set the thickness gauge to 18 μm to begin. The knife should be oriented at a wide oblique angle in order that as much of the length of the blade as possible may be used for cutting purposes. The knife bending angle should be set at 30° or between 30 and 40° (depending on the type of knife used, it could be a little less than 30°). *Practice with the specimen will tell.* Make sure that there will be ample clearance between the knife carriage and the specimen carrier even after many sections have been cut.

2   Prior to sectioning, moisten the cutting knife area and the top of the specimen with ETOH 95%-chloroform solution (Section 45.3.1, item 13), using a camel hair brush, and repeat for each stroke. When the first celloidin sections are obtained, check for nicks in the material sectioned; if present, have this knife sharpened and proceed with a freshly sharpened one.

3   *In front of the microtoming area,* place two rows of a dozen small dishes with covers containing fresh ETOH 95%-chloroform (4:1) solution. The first row will receive the first 12 sections in seriation: the second row will serve for transfer of the same seriated 12 sections until ready to be mounted on microslides. The purpose of the second row is to allow more clearing of the sections, and at the same time, more sectioning of the specimen for the first row.

4   Float each embedded peat section on the cutting knife with ETOH 95% using a camel hair brush; drag them carefully with the brush over a section lifter, and place them *serially* in the small dishes of the first row. Sections can be left in this row or later, in the second one for 1 h. If at this stage sectioning has to be postponed for a longer period of time, return the block-mounted specimen to used storage fluid; never let it dry. Sections should be mounted or transferred serially into benzene-containing dishes. Solutions should be kept clear at all times.

45.3.7 MOUNTING SECTIONS ON MICROSLIDES

Caution: procedures in this section must be performed under a ventilated hood, except for steps 1 and 4.

1   Clean the "factory-precleaned" microslides, leaving no film or paper fibers over the mounting area, and write on the frosted end, with a lead pencil, the specimen and section numbers. Remove the pencil lead dust with a light brush.

2   Before mounting sections, those of the second row (refers to step 4, Section 45.3.6) are first flooded for 5 min in dishes containing chloroform-ETOH 95% (1:1) supplemented with a few drops of benzene and transferred one at a time (maintaining seriation) to a dish containing pure benzene as a final clearing agent and solvent for Permount® (glue). As soon as a section from the second row is transferred to benzene, the corresponding section in the first row is transferred to the second row.

3   Transfer a section on a slide, add enough Permount® (two to three drops or more) along one side of the section, bring one side of a cover slip of appropriate dimension in contact with the Permount® and let this glue run along and cover the section entirely as the cover slip is gradually laid down over the section; this procedure should prevent the presence of air bubbles over the section area. If tiny bubbles are present, gently press them out with a bent needle over the cover slip. If large ones are present, do not move the cover slip, mix a drop or two of Permount® with an equal amount of benzene, and bring this mixture with a pipette along the sides of the cover slip; it will usually penetrate and chase air bubbles. If totally unsuccessful, transfer the microslide into a staining dish containing benzene, maintain slides upright, and let stand. *Do not agitate:* wait until the slide, section, and cover slip set apart and then recuperate the section.

4   The above procedure completed, place the microslides for 24 h at room temperature on a flat plate or working bench covered with a light waxy cardboard. The next day, place a small lead cube carefully over each cover slip so as to flatten sections and chase excess Permount®.

5   After two or more days, scrape off excess Permount® with a scalpel, and *under a fumehood*, using surgical gloves and paper facial tissues (or a lint-free cotton cloth) moistened with benzene, clean the slides, and more so, the section areas. The slides are now ready for microscopic examination.

45.3.8 SPECIAL TREATMENT FOR UPBS SPECIMENS WITH QUARTZ GRAINS

It is to be noted that calcareous and siliceous bioliths do not interfere with UPBS specimen sectioning, but quartz grains (sand grains) do and must be dissolved. Referring to Section

45.3.4, step 2, celloidin infiltrations are first done *with 2, 4, and 6% celloidin-THF solutions.* After the last infiltration, the following procedure is to be followed.

Caution: HF acid is to be used in this section. HF fumes are very hazardous; make sure all safety precautions are used with regard to handling and disposal of material with HF. Avoid any possible contact of fumes with glass, especially with any microscope lens.

1   In a fumehood, scoop tagged specimens out of their bottle with a spoonlike spatula, so as to keep specimens surrounded with some celloidin, transfer to a large-volume jar containing pure chloroform and let stand for 3 d. The celloidin will harden and take on a milky appearance.

2   *Under a fumehood for HF acid,* place specimens in HF-ETOH (see Section 45.3.1, item 5) in containers with screw caps resistant to HF acid (nalgene type); make sure the cap is firmly screwed on at the end of the process. Ensure the front window of the fumehood is closed as required for maximum ventilation. The period of time required to dissolve the quartz grains is at least 1 week.

3   At the end of 1 week, wash the specimens in a continuous flow of tap water for a period of time equivalent to that of the HF treatment. At the end of this time period, check the specimen for HF residues with a razor blade, which will darken if exposed. If HF residues are still present, washing must continue until no trace is detected. For samples designated clear of HF residues, if these are observed to be similar with respect to quartz content, remove a large piece from one of the specimens and examine it under a binocular microscope equipped with polarizing filters. If no quartz is present, proceed to step 4 below. If quartz is still present, repeat step 2 above. If the quartz content varies considerably among the specimens, several of them will require checking under the microscope.

4   Finally, dehydrate specimens in pure THF for 2 h and resume the procedure at Section 45.3.4, step 4.

45.3.9 Comments

Celloidin-impregnated sections can be manipulated without dissolving the matrix. Staining of sections can be achieved specifically in stain baths. As a consequence of the amorphous nature of celloidin, the impregnated material keeps its original arrangement and is well suited for microscopical studies.

45.4 THIN SECTION PRODUCTION BY POLYESTER RESIN IMPREGNATION

Thin sections of organic soils provide a means to examine the spatial arrangement of the various organic constituents and pores. The methodology involves water replacement by acetone and the impregnation of the sample with polyester resin. The procedures used are similar to those outlined for mineral soils in Chapter 65. The procedure outlined below details specific steps to follow for organic soils, pointing out any differences or precautions needed.

45.4.1 MATERIALS AND REAGENTS

1   For preparation of samples for water removal by acetone replacement, refer to Chapter 65.3.2 for materials and reagents.

2   For materials and reagents for resin impregnation, see Chapter 65.4.1.

3   For materials and reagents for thin-section preparation, see Chapter 65.6.1.

45.4.2 PROCEDURE FOR POLYESTER RESIN IMPREGNATION

1   Remove the frozen core (Section 45.2.2, step 7) from the sample container. Cut the sample in half. Place one half onto the cotton gauze which has been placed on the stainless steel screening. Replace the other half of the sample into the sample container and rewrap; this portion is to be used in the event of poor polymerization. Alternatively, the sample can be processed as a replicate sample.

2   Remove the water from the organic samples by acetone replacement either by capillarity (see Chapter 65.3.2) or by vapor exchange (see Chapter 65.3.3). DO NOT AIR DRY any organic materials; severe volume loss and cracking will occur. DO NOT IMMERSE core samples of peat materials (Section 45.2.2) in acetone; these samples will float due to their low bulk density. Fill the acetone-exchange tray so that the samples are just sitting in the acetone (about 0.25 cm depth or less).

3   Monitor the acetone for percent of water content as outlined in Chapter 65.3.2 (steps 7 to 9).

4   Prepare the polyester resin solution according to Chapter 65.4.2 (step 1). The amount of catalyst and accelerator required to produce good polymerization must be predetermined with test samples for the particular polyester resin when used in context of organic materials. The quantities given in Chapter 65.4.2 (step 1) for the catalyst and accelerator were determined for *inorganic* samples, the specific polyester resin shipment as well as the influencing effect of local laboratory conditions at the Centre for Land and Biological Resources Research, Ottawa. Each new shipment must be tested to confirm that the predetermined amounts of catalyst and accelerator are still applicable.

5   Impregnate the sample with the resin solution as described in Chapter 65.4.2 (steps 2 to 11).

6   Prepare thin sections. See Chapter 65.6 for procedure.

45.5 MICROMORPHOLOGICAL DESCRIPTION

Several descriptive systems have been developed for characterizing the micromorphology of organic materials. These systems are reviewed in Bullock (1973), Lee (1983), and Fox (1985b). The main objective for any micromorphological description of a thin section is to provide a clear, concise, and easily understandable discussion of both the observed features and spatial relationships between those features. An approach for describing thin sections is outlined in Chapter 65.10.

REFERENCES

Agriculture Canada Expert Committee on Soil Survey. 1987. The Canadian system of soil classification. 2nd ed. Agriculture Canada Publ. 1646. 164 pp.

Bullock, P. 1973. The micromorphology of soil organic matter — a synthesis of recent research. Pages 49–66 in G. K. Rutherford, Ed. Soil microscopy. Proc. 4th International Working Meeting on Soil Micromorphology, Kingston, Ontario. 27–31 Aug. 1973. The Limestone Press, Kingston.

Fox, C. A. 1985a. Characteristics of the Folisolic soils of British Columbia. Pages 205–232 in J. A. Shields and D. J. Kroetsch, Eds. Expert Committee on Soil Survey, Proc. of the Sixth Annual Meeting, Guelph, Ont., November 26–30, 1984. Research Branch, Agriculture Canada, Ottawa, Ontario.

Fox, C. A. 1985b. Micromorphological characterization of Histosols. Pages 85–104 in L. A. Douglas and M. L. Thompson, Eds. Soil micromorphology and soil classification. Special Publ. No. 15. Soil Science Society of America, Madison, WI.

Fox, C. A., Trowbridge, R. L., and Tarnocai, C. 1987. Classification, macromorphology and chemical characteristics of Folisols from British Columbia. Can. J. Soil Sci. 67: 765–778.

Klinka, K., Green, R. N., Trowbridge, R. L., and Lowe, L. E. 1981. Taxonomic classification of humus forms in ecosystems of British Columbia, First Approximation. Land Management Report ISSN 0702-9861; No. 8. Information Services Branch, Ministry of Forests, Victoria, BC. 54 pp.

Lee, G. B. 1983. The micromorphology of peat. Pages 485–501 in P. Bullock and C. P. Murphy, Eds. Soil micromorphology. Vol. 2. Soil Genesis. Proc. 6th Int. Work. Meet. Soil Micromorphology, London. 17–21 Aug. 1981. A B Academic Publishers, Berkhamsted, Herts, U.K.

Murphy, C. P. 1986. Thin section preparation of soils and sediments. A B Academic Publishers, Berkhamsted, Herts, U.K. 149 pp.

Parent, L. E., Pauzé, F. J., and Bourbeau, G. A. 1980. Methode nouvelle de préparation de coupes minces des tourbes et des gyttja. Can. J. Soil Sci. 60: 487–496.

Sass, J. E. 1958. Botanical microtechnique. 3rd ed. Iowa State University Press, Ames, IA.

Tarnocai, C. 1980. Canadian Wetland Registry. Pages 9–38 in C. Rubec and F. Pollet, Eds. Proc. of a Workshop on Canadian Wetlands, Saskatoon, 1979. Ecological Land Classification Ser. 12. Lands Directorate, Environment Canada, Ottawa, Ont.

Chapter 46
Palynological Assessment
of Organic Materials

Pierre J. H. Richard

University of Montreal

Montreal, Quebec, Canada

46.1 INTRODUCTION

Pollen analysis is a micropaleontological technique and a biostratigraphical tool. As such, it relies heavily on the stratigraphical conformity of the material studied. When there is no disturbance of the stratigraphy, the method produces time-series statistics on the pollen representation of the vegetation, and leads to paleoecological reconstructions. In disturbed stratigraphies, it may nevertheless yield valuable evidence of the vegetation involved before and after the process. In soil science, the method is best suited for the study of organic soils where the material is accumulated in stratigraphical order. However, pollen analysis of mineral soil horizons also provides interesting insights on the history of the soil-forming environment and processes (Aaby 1983, Andersen 1979, 1986, Dimbleby 1957, 1961, Guillet 1972, Iversen 1969, among others). It can, of course, yield valuable information for surficial (humus) or buried organic layers of mineral soils.

The palynological study of organic materials is useful for soil scientists interested in the history of the underlying material in organic soils (Fibrisols, Mesisols, Humisols). Along with additional micropaleontological and paleontological techniques, the method provides unique information about the local succession of the vegetation covers that gave rise to the organic soils. The method allows the estimation of the net rate of accumulation of the different layers of the material at a given site, and the variation of the rate within a landscape unit of organic terrain. It has been successfully used to correlate the layers of different control sections within a given peatland unit and to establish the rate of erosion of organic soils through agricultural practice of differing duration (Lévesque et al. 1982). Through a careful study of the state of preservation of the pollen grains and spores along a profile, the method can yield useful information about the degree of oxidation to which the material has been exposed during its accumulation.

The basics of pollen analysis are masterly developed in the textbook by Faegri et al. (1989). Any pollen-analytical study of organic materials should be done along the guidelines and with the methods amply discussed in their landmark book, to which we refer the reader (see

Soil Sampling and Methods of Analysis, M. R. Carter, Ed.,

Canadian Society of Soil Science. © 1993 Lewis Publishers.

also Moore and Webb 1978). In addition, the *Handbook of Holocene Palæoecology and Palæohydrology* edited by Berglund (1986) shall provide the reader with a recent update on the various methods applied to peat (and lake sediments) for reconstructing past environments (see also Birks and Birks 1980).

46.2 FIELD SAMPLING

Pollen analysis should only be performed on uncontaminated samples taken from the organic soil profile, including the underlying material. The sampling conditions (weather) should be recorded in detail, along with the botanical composition of the local vegetation at the sampling site. The flora of the whole peatland and the surrounding regional, upland flora should also be recorded if the information is not available. The same should be done for the local, extralocal, and regional vegetation types (*sensu* Janssen 1973, 1984).

The choice of locality for taking samples for pollen analysis depends on the nature of the problem under study. It is assumed here that the emplacement of the control sections for the study of organic soils is determined by the soil scientist for pedological purposes, and that the palynological sampling is performed in the same profiles. The palynological study of the development of the entire peatland unit would require sampling strategies based on an adequate profiling and mapping of the sediment types within the peatland.

Whatever the location of the control sections, there is a need for the sampling of surface material (*Sphagnum* or other mosses producing compact moss polsters) to provide reference material (pollen spectra) connected with the present-day vegetation on the peatland. These surface samples should encompass the whole array of vegetation types that could have thrived at the sampling point (e.g., over the control section). They shall be needed for the interpretation of the sequence of pollen spectra from the section.

Pollen analysis is performed on the material extracted from small samples (1 cc) of the original peat material. The pollen concentration of such samples varies usually from some thousands to several hundred thousands. There is consequently no need to get large samples of peat for pollen and spores analysis alone. The field sampling may be made at intervals of 1, 2, 5, or 10 cm on an open face of the control section, the sample density being adapted to the local stratigraphy and the problem investigated. When working along ditches, peat horizons can be observed in great detail and the quantity of material sampled can be easily adjusted to analytical needs other than pollen analysis. Peat monoliths may be cut from the open faces and taken in containers. This method is interesting because subsampling of the peat in the laboratory can be performed under controlled conditions and can be made as detailed as necessary. In many cases, water will be a problem down the middle tier of the control section, and pumping may be required. Coring devices that yield undisturbed material from known depths could then be used. If a complete pollen record is needed for the site, the coring should be done on the sediments below the 160-cm depth until the inorganic geological substrate is reached. Peat might be underlain by lacustrine organic muds (e.g., gyttja, dy, and marl), in which case the modified piston sampler could be used to provide cores from these types of sediments. Those coring devices may be used for retrieving sediments from the entire column, including the first 160 cm. Coring devices are described in Chapter 42.

46.2.1 MATERIALS

1   Surface sampling: plastic bags, scissors, waterproof marking pens.

2   Sampling from the open face of the control section: shovel, knife, pruning shears (for roots), trowel, spatula, mounted needle, plastic bags, vials (5 cc), waterproof marking pens, measuring tape, masking tape, clean cloth, clean water.

3   Coring devices: Hiller sampler, Macauley/Russian sampler, Coûteaux sampler, modified piston sampler (see Chapter 42).

46.2.2 Procedures

46.2.2.1 *Surface Sampling*

A handful of moss polster is picked and the living part (2–5 upper cm) of it is cut and put in a properly labeled plastic bag. A number (three to five or more) of these samples should be picked from the same spot and mixed in the same bag, to ensure proper representation of the pollen grains or spores of all the plants growing within the vegetation type (to avoid casual overrepresentation of any single plant). The surface sampling should cover all the dominant vegetation types found on the peatland, on and around the control section. The samples should be kept cool (4°C) or air dried until processed in the laboratory, to avoid fungal or other microbiological activity.

46.2.2.2 SAMPLING FROM THE OPEN FACE OF THE CONTROL SECTION

1   Direct sampling for pollen-analytical purposes: vials can be forced in the material and labeled according to their depth. Care must be given to leave an outlet for the air to get out from the vials to ensure proper sampling of about 5 cc of peat. Gentle compaction should not be a serious problem for future processing of such a small volume. If more material is needed (e.g., for radiocarbon dating or for physical and chemical analyses), it can be carefully scraped horizontally from the face and put in a properly labeled plastic bag. There could be a need to sample layers thinner than the mouth of the vial or to sample millimetric layers. The necessary volume of the material could then be taken with a small spatula or a needle and transferred in the vial. The samples should be kept cool (4°C) and should maintain their original moisture content until processed to avoid fungal development or other microbiological activity and to allow significant volumetric pollen analysis. The tools should be thoroughly cleaned with water between each depth sampled to avoid contamination. Subsamples for certain chemical analyses should be taken with equipment rinsed with distilled water. Direct sampling should usually not be performed on rainy days nor during days where pollen shedding by plants is significant.

2   Sampling peat monoliths: see Chapter 42.

46.2.2.3 *Coring*

The description and operation of coring devices are presented in many publications (Aaby and Digerfeldt 1986, Coûteaux, 1962, Faegri et al. 1989, Larouche 1979, Wright 1980, Wright et al. 1965) and described in Chapter 42.

46.3 LABORATORY PREPARATION

The preparation of the surface samples (moss polsters) and the samples of peat (or sediment) for microscopical observation aims at extracting and concentrating the spores and pollen grains present. The laboratory technique for peats normally includes the following main steps (modified from Berglund and Ralska-Jasiewiczowa 1986):

1   Subsampling of a standard volume of material, and addition of a known amount of exotic pollen for later calculation of pollen concentration and pollen accumulation rate.

2   Dispersion of sediment (deflocculation) by stirring and by chemical treatment (KOH, NaOH).

3   Removal of coarse (>ca. 0.3 mm) organic or inorganic particles by sieving.

4   Chemical removal of fine extraneous matter, such as calcium carbonate (using HCl), humic acids (KOH or NaOH), pyrite and lignin (HNO_3), silica (HF).

5   Digestion of cellulose by acetolysis.

6   Staining (with neutral red, safranin, or basic fuchsine).

7   Embedding (glycerin or silicon oil).

8   Mounting on slides.

The technique for surface samples (moss polsters) does not involve step 1.

46.3.1 MATERIALS AND REAGENTS

1   Fumehood, dispensable plastic gloves, polypropylene rubber gloves, face mask, rubber apron, centrifuge, vibration stirrer, block heater, mini (1-cc) piston sampler (homemade), plastic bottles (reagent and distilled water dispensers), polypropylene conical test tubes (15 mL), test tube holders, scissors, small metal spatulas, porcelain sieves (mesh about 0.3 mm), Nitex micro sieve cloth (10 μm); porcelain dishes, glass covers, glass beakers, petri dishes, microscope slides and cover slips, nail polish.

2   Suspension or tablets of exotic (*Eucalyptus* or *Lycopodium*) pollen grains of known concentration.

3   KOH, NaOH, HCl, H_2SO_4; HNO_3, HF, at concentrations noted in the procedures. $(CH_3CO)_2O$ (acetic anhydride), CH_3COOH (glacial acetic acid), C_2H_5OH (ethyl alcohol), $(CH_3)_3COH$ (tertiary butyl alcohol), distilled water, glycerine (or silicon oil at viscosity ≥ 2000 cSt), neutral red or safranin.

46.3.2 Procedure

46.3.2.1 SURFACE SAMPLES

1   Cut the moss polsters into 1-cm pieces, thoroughly mix the material and put about 5 cc (compressed) in a 250-mL glass beaker. Add slowly a convenient, but minimal amount of concentrated H_2SO_4 to cover the material, stir, and leave overnight for digestion.

2   Add two parts of water and stir. Remove coarse particles by sieving in a porcelain sieve over a porcelain dish. Use gentle water jets from a squeeze bottle and use a minimum amount of water. The <0.3-mm suspension is then transferred progressively in a 15-mL conical test tube using repeated centrifugation (10 min at 1500 $\times$ *g*) and decantation of the water. Make sure that all the material is transferred in the tube.

3   Remove the >100-μm particles through sieving on a Nitex sieve over a 250-mL beaker, and concentrate the material in a conical test tube as described in step 2.

4   Remove the lignin by treating with an oxidizing solution (100 mL of 50% HNO_3 and four to five drops of pure HCl). Add several milliliters of the oxidizing solution to the residue in the test tube, stir, and leave for 1–2 min maximum. Add water to balance the tubes and centrifuge. Decant and rinse twice. Then proceed with step 3 of the procedure below.

46.3.2.2 PEAT AND MUD SAMPLES

1   Remove the outer layer to avoid contamination, subsample 1 cc of fresh sediment (e.g., with a mini-piston sampler), and put it in a 15-mL polypropylene conical test tube. Make sure a representative sample of the peat matrix is obtained, not a piece of wood. Add a known volume of the exotic pollen suspension (*Eucalyptus*) or in form of calibrated tablets (*Lycopodium*). Add 10 mL of 10% KOH. Stir thoroughly with the spatula first, then with a vibration stirrer. *Work under a fumehood for the remainder of the procedure.* Heat 20 min at 100°C on a block heater and stir frequently with the spatula or with disposable wooden sticks using one stick per sample per stir to avoid cross contamination.

2   Remove coarse particles by sieving in a porcelain sieve over a porcelain dish. Use gentle water jets from a squeeze bottle and use a minimum amount of water. The <0.3-mm suspension is then transferred back progressively into the 15-mL conical test tube using repeated centrifugation and decantation of the water, leaving fine sand, if any. Make sure that all the organic material is transferred in the tube. The material on the sieve should be kept for examination under a low-magnification microscope (some macrofossils might be identified at that step and the >0.3-mm inorganic fraction might be weighed). KOH must be washed completely out in the process in order not to react with HF.

3   Remove silica by slowly pouring 10 mL of 48% HF in the tube and stir with the spatula. Place the spatula in a beaker with water to stop the reaction of the metal with HF. Leave overnight. This stage may be accelerated by heating and stirring every 20 min for 1 to 1.5 h. Add 3 mL of water, centrifuge for 10 min, and decant into a HF waste plastic container.

4   Remove silica gels and carbonates by pouring 10 mL of 10% HCl in the tube. Stir, and heat 20 min at 100°C. Add distilled water, stir, centrifuge 10 min, and decant the supernatant.

5   Remove cellulose through acetolysis. Add 10 mL of glacial acetic acid, stir, and leave for 20 min to dehydrate the material. Equilibrate the tubes with glacial acetic acid, centrifuge for 10 min, and decant. Add 9 mL of pure acetic anhydride and 1 mL of sulfuric acid (98.08%), stir, heat 1–2 min on the block heater, add acetic acid to equilibrate the tubes, and centrifuge 10 min. Decant.

6   Embed in glycerin. Rinse by adding distilled water, stir, centrifuge, and decant. Add 10 mL of 30% glycerined water, stir, centrifuge for 30 min. Keep the tubes upside down at the end of the decantation operation, and leave the tubes upside down on a shelf covered with paper for 1 h to get rid of most of the supernatant.

7   Stain with neutral red (one to two drops; stir well with the spatula).

8   Mount in pure glycerin. Add a small amount (1–5 mL) of glycerine and stir thoroughly with the spatula. Transfer a drop of the material on a slide with the spatula and cover gently with an 18- × 18-mm coverslip. Seal with nail polish, putting a drop in two corners to keep the coverslip in place. The amount of glycerin added should be adjusted to the pollen concentration of the residue in order to get from 500 to 1000 pollen grains per slide. It is a good practice not to add too much of glycerin at first. If the pollen density on the slide is too great, dilute the residue by adding more glycerin and stir. The size of the drop is adjusted to the size of the coverslip, taking care that the material should not flow outside it.

46.3.3 COMMENTS

1 Most pollen grains and spores resist the above treatments, provided the material is not submitted too long to the oxidants. The removal of pyrite and lignin with HNO_3 (see Section 46.3, step 4) is not included in our standard procedure and should be performed only if necessary.

2 Steps 3 and 4 may at first appear unnecessary for peats, but our experience tells us that even very small amounts of siliceous matter in the sediments (windblown sands) lead to slides that are difficult to observe under the microscope.

3 In minerogenic sediments (silts and clays), an extra sieving step is included after step 2 with Nitex sieves to remove particles smaller than 10 μm (Cwynar et al. 1979).

4 Always stop the reactions with distilled water (or acetic acid after step 5 above) before centrifuging.

5 A different spatula should be used to stir each sample, and the same one used for and rinsed between the different steps for each sample to avoid intersample contamination and loss of pollen grains and spores. Disposable wood sticks can also be used.

6 Great care should be given to the decantation operations in order not to lose the material. Adding some alcohol to the tubes before centrifugation lowers somewhat the density of the liquid and ensures a proper settling of the material during centrifugation. The conical shape of the bottom of the tubes ensures an appropriate extrusion of the liquid during decantation. The decanting movement should be done regularly, rather rapidly, without hesitations or halts. We usually process 6, 12, or 18 samples simultaneously.

7 The previously cited references provide additional information on the procedures and should be consulted.

8 Embedding in silicon oil instead of glycerin involves different procedures for steps 6 and 7 (see Andersen 1960). Silicon oil gives more permanent slides that need no sealing, and the size of the material embedded does not change with time. However, the swelling of the pollen grains that occurs with time in the highly hydrophilous glycerin medium helps with the identification of the fine structures of the palynomorphs during microscopy. The measurements on pollen grains should, however, be made soon after mounting if the results are to be compared from slide to slide. The same holds for reference collections.

46.4 POLLEN STATISTICS

The slides hopefully contain only spores, pollen grains, and other microfossils resistant to the treatments. The organic or inorganic debris should be sparse. The pollen grains should contrast visually because of the staining. The ideal ratio of exotic pollen to the original pollen content is one. The slide is regularly scanned under a high-quality light microscope at a routine magnification of 400 (1000 needed for identification at genera or species levels). The pollen grains and spores are tallied along with the grains of the exotic spiker (*Eucalyptus* pollen, *Lycopodium* spores). The palynomorphs are identified with the aid of a reference collection and identification keys (McAndrews et al. 1973, Kapp 1969, Richard 1970). The tally is stopped when a sufficiently high Pollen Sum is reached, such as 300 or 500 pollen grains (see Faegri et al., 1989). The Pollen Sum usually includes the pollen grains of all the terrestrial vascular plant species. It thus *excludes* the pollen of the aquatic plants and the spores of ferns and fern allies, and of mosses. The percentages of the various pollen and spore taxa are usually calculated on the basis of the above-defined Pollen Sum. In the study of peats, this kind of Pollen Sum ensures that the spores of the locally growing mosses, fern-allies, or ferns, the sporulation of which is erratic, will not obscure the pollen representation of the wind-pollinated vascular plants. The question of what plants should be included in the Pollen Sum for the calculation of percentages is presented in Faegri et al. (1989); they very often correspond to the pollen of all the terrestrial vascular plants. The same set of species should be included in the Pollen Sum throughout the samples.

46.4.1 The Pollen Spectrum and the Pollen Diagram

The pollen spectrum of a sample is expressed by the percentages of each taxon (each pollen type or spore type) represented in that sample. The percentages of the taxa included in the Pollen Sum total 100%. The percentages of the taxa not included in the Pollen Sum are based on the Pollen Sum and added to the spectrum. The sequence of pollen spectra from the different depths of a control section of a core constitutes a pollen diagram. The vertical sequence of percentages for a given taxon throughout the control section or the core constitutes its pollen curve, or spore curve (Faegri et al. 1989).

The surface pollen spectrum reflects the recent vegetation at the sampling point. The pollen diagram represents change in the pollen representation of the vegetation through time. The individual pollen curves represent change in the pollen representation of individual taxa through time. The changes in the pollen representation of the vegetation (pollen spectra) or the individual taxa (pollen curves) may represent actual changes in the composition of the former or abundance of the latter. The interpretation of the pollen spectra in terms of local, extra-local, or regional vegetation is aided by the comparison with the surface pollen spectra from the moss polsters of the various vegetation types within the peatland.

For the soil scientist, the interpretation of the pollen diagram may answer questions about the ancestry and stability of the contemporaneous peat-forming vegetation, about the developmental history of the peatland system, and about the changes in its trophic regime through indicator plants. Radiocarbon and other dating methods (e.g., volcanic ash layers) ensure appropriate chronology. Once a well-dated standard pollen diagram is available for a mire, other pollen diagrams can be dated by means of the pollen stratigraphy.

46.4.2 The Pollen Concentration

The pollen concentration is the number of grains per volume unit of the original sediment (grains/cc). It is calculated by using the number of exotic pollen grains tallied along with the palynomorphs identified. Since we know the number of exotic pollen added to 1 cc of sediment at the beginning of the laboratory treatment of the sample, and the number of palynormorphs counted, the pollen concentration is proportional to the ratio between the exotic pollen added and the exotic pollen tallied. The standard total pollen concentration is that of the taxa included in the Pollem Sum, but the concentration of each taxa identified can be easily computed.

The pollen concentration is affected by the quantity of pollen deposited on the surface of the peat, quantity that varies with the pollen production, dispersal, and preservation of the different plant species, with the characteristics of the deposition site (lake, bog, fen) and with their proximity of the sampling point. The pollen concentration is highly affected by the degree of humification of the organic material, or the length of time covered by the 1 cc of the peat or sediment sample. For parts of the pollen diagram where the regional (outside the peatland) pollen sources are acceptably constant, the variations in the pollen concentration of the regional taxa reflect the variations in humification. For soil scientists, this allows an estimate of this parameter that could be compared to other (chemical or physical) methods. Various usage of the pollen concentration is illustrated by Aaby (1986), Aaby and Tauber (1975), and Middeldorp (1982, 1984), among others.

46.4.3 The Pollen Accumulation Rate

The pollen accumulation rate is the net number of pollen grains per cm^2 of sediment per year (radiometric year). It is based on the pollen concentration multiplied by the net accumulation rate (cm per year) of the sediment. The peat accumulation rate is obtained through radiometric measurements, usually radiocarbon, but also lead or cesium, for the topmost recent layers (Aaby 1986). The pollen accumulation rate curve of an individual taxon is more directly related to its varying abundance than its percentages curve, since the percentages depend on the abundance of the other taxa. In peats, the net accumulation rate of the sediment is usually too crudely estimated to allow usage of the pollen accumulation rate values. If the net accumulation rate of the sediments can be evaluated for closely spaced samples, then the pollen accumulation rates of the various taxa can be usefully calculated.

46.5 ASSESSMENT

Interpretation of pollen and spores statistics in terms of vegetation requires knowledge of all the natural processes involved in the palynological representation of plants, including pollen or spores production, dispersion, deposition, and preservation (see Birks and Birks 1980). Those differ for each taxon, and for various ecological contexts (depositional environments, structure and composition of plant communities and of landscape units). The magnitude of the source area of the pollen grains and spores has to be evaluated for each taxon, and the relationship between pollen abundance and plant abundance has to be assessed for differing combinations of plants and of various potential source areas (Delcourt et al. 1984, Janssen 1973). These are done through the study of current pollen and spores deposition from known vegetational patterns in given depositional environments. The reconstructions are ideally aided by the identification of the macrofossil remains of plants, along with pollen analysis, since these remains are of more local origin. The paleoecological interpretation of the data (pollen or macrofossil diagrams) depends on the knowledge of the present-day relationships between plants and their environment (uniformitarianism; see Rymer 1978), but the method may in turn reveal ecologically sound relationships that have no counterparts in the modern world, thus greatly enhancing our views on the potentialities of plants, and on the variety of the ecological combinations (Davis 1976).

REFERENCES

Aaby, B. 1983. Forest development, soil genesis and human activity illustrated by pollen and hypha analysis of two neighbouring podzols in Draved Forest, Denmark. Geological Survey of Denmark, Series II 114: 5–114.

Aaby, B. 1986. Palæoecological studies of mires. Pages 145–164., in B. Berglund, Ed. Handbook of Holocene palæoecology and palæohydrology. John Wiley & Sons, New York.

Aaby, B. and Digerfeldt, G. 1986. Sampling techniques for lakes and bogs. Pages 181–194, in B. Berglund, Ed. Handbook of Holocene palæoecology and palæohydrology. John Wiley & Sons, New York.

Aaby, B. and Tauber, H. 1975. Rates of peat formation in relation to degree of humification and local environment, as shown by studies of a raised bog in Denmark. Boreas 4: 1–17.

Andersen, S. T. 1960. Silicon oil as a mounting medium for pollen grains. Dan. Geol. Unders. IV Series 4: 1–24.

Andersen, S. T. 1979. Brown earth and podzol: soil genesis illuminated by microfossil analysis. Boreas 8: 59–73.

Andersen, S. T. 1986. Palæoecological studies of terrestrial soils. Pages 165–177 in B. Berglund, Ed. Handbook of Holocene palæoecology and palæohydrology. John Wiley & Sons, New York.

Berglung, B., Ed. 1986. Handbook of Holocene palæoecology and palæohydrology. John Wiley & Sons, New York.

Berglund, B. E. and Ralska-Jasiewiczowa, M. 1986. Pollen analysis and pollen diagrams. Pages 455–485, in B. E. Berglund, Ed. Handbook of Holocene palæoecology and palæohydrology. John Wiley & Sons, New York.

Birks, H. J. B. and Birks, H. H. 1980. Quaternary palaeoecology. Edward Arnold, London.

Coûteaux, M. 1962. Notes sur le prélèvement et la préparation de certains sédiments. Pollen Spores 4: 317–322.

Cwynar, C., Burden, E., and McAndrews, J. H. 1979. An inexpensive sieving method for concentrating pollen and spores from fine-grained sediments. Can. J. of Earth Sci. 16: 1116–1120.

Davis, M. B. 1976. Pleistocene biogeography of temperate deciduous forests. Geosci. Man 24, 13: 13–26.

Delcourt, P. A., Delcourt, H. R., and Webb, Th., III 1984. Atlas of mapped distributions of dominance and modern pollen percentages for important tree taxa of Eastern North America. AASP (American Association of Stratigraphic Palynologists), Contributions Series 14: 131 p.

Dimbleby, G. W. 1957. Pollen analysis of terrestrial soils. New Phytol. 56: 12–28.

Dimbleby, G. W. 1961. Soil pollen analysis. J. Soil Sci. 12: 1–11.

Faegri, K., Kaland, P. E., and Krzywinski, K. 1989. Textbook of pollen analysis. 4th ed. John Wiley & Sons, New York. 338 p.

Guillet, B. 1972. Relation entre l'histoire de la végétation et la podzolisation dans les Vosges. Thèse de doctorat, Université de Nancy 1, no. CNRS: A-0-7640, 112 p.

Iversen, J. 1969. Retrogressive development of a forest ecosystem demonstrated by pollen diagram of a fossil mor. Oikos (Suppl.) 12: 35–49.

Jannsen, C. R. 1973. Local and regional pollen deposition. Pages 31–42 in H. J. B. Birks and

R. G. West, Eds. Quaternary plant ecology. Blackwell Scientific, Oxford.

Jannsen, C. R. 1984. Modern pollen assemblages and vegetation in the Myrtle Lake Peatland, Minnesota. Ecol. Monogr. 54: 213–252.

Kapp, R. O. 1969. How to know pollen and spores. WM. C. Brown Co. Publishers, Dubuque, IA, 249 p.

Larouche, A. C. 1979. Histoire postglaciaire comparée de la végétation à Sainte-Foy et au mont des Eboulements, Québec, par l'analyse macrofossile et l'analyse pollinique. Mémoire de M.Sc., Faculté de foresterie et de géodésie, Université Laval, Québec. 117 p.

Lévesque, M. P., Mathur, S. P., and Richard, P. J. H. 1982. A study of physical and chemical changes in a cultivated organic soil based on palynological synchronization of subsurface layers. Nat. Can. 109: 181–187.

McAndrews, J. H., Berti, A. A., and Norris, G. 1973. Key to the Quaternary pollen and spores of the Great Lakes Region. Royal Ontario Museum, Life Sciences, Miscellaneous publications, 61 p.

Middeldorp, A. A. 1982. Pollen concentration as a basis for indirect dating and quantifying net organic and fungal production in a peat bog ecosystem. Rev. Palaeobot. Palynol. 37: 225–282.

Middeldorp, A. A. 1984. Functional Paleoecology of Raised Bogs — an Analysis by Means of Pollen Density Dating, in Connection with the Regional Forest History. Ph.D. thesis, University of Amsterdam, 124 p.

Moore, P. D. and Webb, J. A. 1978. An illustrated guide to pollen analysis. Hodder and Stoughton, Unibooks, 133 p.

Richard, P. J. H. 1970. Atlas pollinique des arbres et de quelques arbustes indigènes du Québec. Nat. Can. 97: 1–34; 97–161; 241–306.

Rymer, L. 1978. The use of uniformitarianism and analogy in palæoecology, particularly pollen analysis. Pages 245–258 in D. Walker, and J. C. Guppy, Eds. Biology and Quaternary environments. Australian Academy of Science.

Wright, H. E., Jr. 1980. Cores of soft lake sediment. Boreas 9: 107–114.

Wright, H. E., Jr., Livingstone, D. A., and Cushing, E. J. 1965. Coring devices for lake sediments. Pages 494–520 in B. Kummel and D. Raup, Eds. Handbook of paleontological techniques. W. H. Freeman, San Francisco.

Soil Physical Analyses

Chapter 47
Particle Size Distribution

B. H. Sheldrick and C. Wang
Agriculture Canada
Ottawa, Ontario, Canada

47.1 INTRODUCTION

Particle size analysis is the measurement of the proportions of the various sizes of primary soil particles as determined usually either by their capacities to pass through sieves of various mesh size or by their rates of settling in water. The proportions are usually represented by the relative weights of particles within stated size classes. The limits of these size classes differ in various commonly used systems of soil particle size classification (Figure 47.1). In this chapter the CSSC system is used. For engineering interpretations, the AASHO and the Unified systems are used.

Both pipette and hydrometer methods for particle size distribution analysis to be described in this chapter are based on the principle of sedimentation known as Stokes' Law. It describes small spherical particles of density PS and diameter D settling through a liquid of density PL and viscosity n at a rate of

$$v = D^2 g \, (PS - PL)/18n \tag{47.1}$$

Where g = acceleration of gravity. It is important to know that all sedimentation tables used for particle size distribution analysis, such as Table 47.1, were based on the following assumptions: (1) soil particles are spherical (most silicate clay particles, in fact, are platy); (2) PS = 2.65 or 2.60 Mg m^{-3} (density could vary from 2.0 to 3.2 Mg m^{-3}); and (3) temperature of H_2O is constant throughout sedimentation (it is important to use H_2O of room temperature and to operate in a temperature-controlled room). It is also assumed that the settling of soil particles in a cylinder is not influenced by other particles nor by the cylinder wall.

47.2 PIPETTE METHOD

The following pipette method (Soil Conservation Service 1984) was adapted from the U.S. Soil Survey, Lincoln, NE, and reduces the time required to do the analysis by replacing the numerous centrifuge washing steps with a filter candle system. The data determined using the filter candle system corresponds well with the data from the standard pipette method and centrifuge washing (Sheldrick and Wang 1987).

Soil Sampling and Methods of Analysis, M. R. Carter, Ed.,
Canadian Society of Soil Science. © 1993 Lewis Publishers.

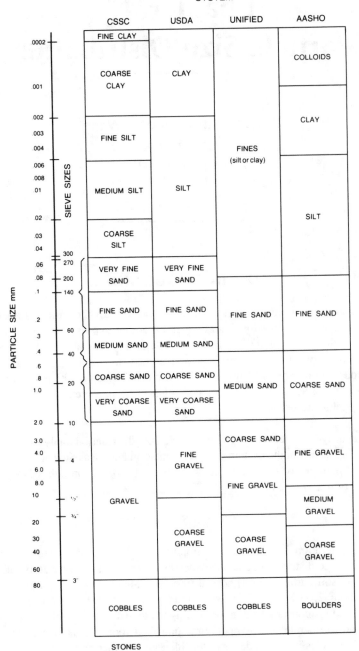

FIGURE 47.1. A comparison of particle size limits in four systems of particle size classification. (From McKeague, J. A., Ed., *Can. Soc. Soil Sci.*, Ottawa, Ontario. With permission.)

Table 47.1 Settling Depths for Specific Times and Temperatures
for Particle Size = 2 μm (McKeague 1978)

Temp (°C)	Time			
	4½ h Depth (cm)	5 h Depth (cm)	5½ h Depth (cm)	6½ h Depth (cm)
20.0	5.79	6.44	7.08	8.37
20.3	5.81	6.48	7.13	8.43
20.5	5.86	6.52	7.17	8.47
20.7	5.89	6.55	7.20	8.51
21.0	5.93	6.59	7.25	8.57
21.3	5.97	6.64	7.30	8.63
21.5	6.01	6.68	7.33	8.68
21.7	6.04	6.72	7.39	8.73
22.0	6.09	6.75	7.43	8.78
22.3	6.13	6.80	7.49	8.85
22.5	6.15	6.83	7.51	8.88
22.7	6.18	6.86	7.55	8.92
23.0	6.22	6.91	7.60	8.98
23.3	6.27	6.96	7.66	9.05
23.5	6.29	6.98	7.68	9.08
23.7	6.33	7.04	7.74	9.15
24.0	6.37	7.08	7.78	9.20
24.3	6.40	7.12	7.83	9.25
24.5	6.43	7.15	7.86	9.29
24.7	6.45	7.18	7.89	9.33
25.0	6.51	7.24	7.96	9.41
25.3	6.56	7.28	8.01	9.47
25.5	6.58	7.31	8.04	9.50
25.7	6.61	7.35	8.08	9.55
26.0	6.66	7.40	8.14	9.62
26.3	6.69	7.44	8.18	9.67
26.5	6.72	7.47	8.22	9.72
26.7	6.76	7.51	8.26	9.76
27.0	6.81	7.56	8.32	9.83
27.3	6.85	7.61	8.37	9.89
27.5	6.87	7.64	8.40	9.93
27.7	6.91	7.68	8.44	9.98
28.0	6.97	7.74	8.51	10.06
28.3	7.01	7.79	8.57	10.13
28.5	7.04	7.82	8.61	10.17
28.7	7.07	7.86	8.65	10.22
29.0	7.12	7.91	8.70	10.28
29.3	7.16	7.95	8.75	10.34
29.5	7.19	7.99	8.79	10.39
29.7	7.22	8.02	8.82	10.43
30.0	7.27	8.08	8.88	10.50

If centrifuge washing is preferred, the soil samples after pretreatments can be washed in 250-mL centrifuge bottles with approximately 50 mL H_2O and centrifuged for 10 min at $500 \times g$. Repeat the washing procedure three times and tests for salts as described in step 2 under Removal of Soluble Salts.

47.2.1 MATERIALS AND REAGENTS

1   Fleakers — 300 mL plus plastic caps.

2   Filter candles — FP-88-02 (available from Sciquip Ltd., 233 Millrage Court, Mississauga, Ontario).

3   Shakers

   a. End over end (40–60 rpm).

   b. Sieve shaker (500 oscillations per minute).

4   Cylinder — soil suspension (1205 mL) marked at 1000 mL.

5   Racks

   a. Custom-built metal frame to hold four motor-driven stirrers equipped with propeller-type stirrer and Teflon® guard.

   b. Shaw pipette rack modified to hold four 25-mL Lowry pipettes.

   c. Custom-built wood frame to support fleakers, filter candles, and vacuum system.

6   Styrofoam pipe insulating cover.

7   100-mL beakers or wide-mouth glass pill bottles.

8   Balance (0.1-mg sensitivity).

9   Sieves

   a. 300 mesh, 15 cm.

   b. Set of sieves, brass, 6.3 cm. U.S. series and Tyler screen scale equivalent designations as follows:

Opening (mm)	U.S. No.	Tyler mesh size
1.00	18	16
0.50	35	32
0.25	60	60
0.105	140	150
0.046	300	300

10   Hydrogen peroxide (30 or 50%).

11   Hydrochloric acid, 1 *M*.

12   Citrate-bicarbonate buffer. Prepare a 0.3-*M* solution of sodium citrate (88.4 g L^{-1}) and add 125 mL of 1 *M* sodium bicarbonate (84 g L^{-1}) to each liter of citrate solution.

13   Sodium hydrosulfite (dithionite).

14   Saturated sodium chloride solution.

15   Prepare a solution of sodium metaphosphate with enough sodium carbonate added to bring the pH to 10 (NaPO$_3$), 35.7 g L^{-1} + Na$_2$CO$_3$ 7.9 g L^{-1} is suitable.

47.2.2 PROCEDURE FOR PRETREATMENTS

47.2.2.1 REMOVAL OF CARBONATES

1   Weigh 10 g of 2-mm air-dried soil into a 300-mL fleaker (tared to 1 mg). If the sample appears to be sandy, weigh a larger sample (e.g., 30 g).

2   Add 50 mL of water, mix, and add 1 *M* HCl slowly until the pH falls between 3.5 and 4.0 and remains there for 10 min. Stronger HCl can be used to avoid having a large volume of solution in soils high in carbonate content. Soils requiring a large amount of HCl to adjust the pH are washed several times with water to remove excess acid by using the filter candle system.

47.2.2.2 REMOVAL OF ORGANIC MATTER

1   Add 10 mL of hydrogen peroxide (H$_2$O$_2$, 30 or 50%) to the fleakers, cover, and allow to stand. If a violent reaction occurs, repeat the cold H$_2$O$_2$ treatment until no more frothing occurs.

2   When frothing subsides, heat to about 90°C. Continue adding H$_2$O$_2$ and heating until most of the organic matter is destroyed (as observed by the color and the rate of reaction of the sample).

3   Rinse down the sides of the reaction vessel occasionally. Continue heating the sample for about 45 min after the final addition of hydrogen peroxide to remove excess hydrogen peroxide.

NOTE: 1. It may be necessary to transfer samples containing high amounts of organic matter (>5%) to large beakers (e.g., 1000-mL tall form). 2. If excessive frothing occurs, cool the container either with cold water or by the addition of methyl alcohol to avoid sample loss.

47.2.2.3 REMOVAL OF SOLUBLE SALTS

1   Place the fleakers in a rack and filter the remaining peroxide and water off from step 3 under Removal of Organic Matter using a filter candle system.

2   Add 150 mL water in a jet strong enough to stir the sample, and filter the suspension through the filter candle system. Five such washings and filterings are usually enough, except for soils containing much coarse gypsum. To test for salts, check with silver nitrate ($AgNO_3$) for Cl^- and barium chloride for SO_4^{2-}.

3   Remove soil adhering to the filter candle by applying back pressure gently and using a rubber-tipped finger as a policeman.

NOTE: If iron oxides are to be removed *DO NOT* complete step 4 at this time.

4   Place the sample in an oven overnight at 105°C, cool in a desiccator, and weigh to the nearest milligram. Use the weight of the oven-dried treated sample as the base weight for calculating percentages of the various fractions.

47.2.2.4 REMOVAL OF IRON OXIDES (OPTIONAL)

1   Add 150 mL of citrate-bicarbonate buffer to the samples in the fleakers. Stir and add 3 g of sodium hydrosulfite ($Na_2S_2O_4$) gradually, as some samples may froth.

2   Put the fleakers in a water bath at 80°C and stir intermittently for 20 min.

3   Remove the fleakers from the bath, place in the holding rack, and filter the suspension through the filter candle system. If a brownish color remains, repeat steps 1 to 3 inclusively. If the samples are completely gleyed (gray), proceed to step 4.

4   Wash five times with a jet of water strong enough to stir the sample and filter the suspension through the filter candle system.

5   To determine the oven dried weight repeat step 4 under Removal of Soluble Salts.

47.2.3 PROCEDURE FOR PARTICLE SIZE SEPARATION

47.2.3.1 DISPERSION OF SAMPLE

1   Add 10 mL of sodium metaphosphate dispersing agent to the fleakers containing the oven-dried treated samples. Make the volume to 200 mL with distilled water.

2   Stopper tightly and shake end-over-end (50–60 rpm) overnight.

47.2.3.2 SEPARATION OF SAND FRACTIONS

1   Pour the suspensions through a 300-mesh (47-μm) sieve into a sedimentation cylinder marked at 1 L. The 300-mesh sieve of about 14-cm diameter is placed in a large funnel held above the 1205-mL cylinder by a retort stand.

2   Wash the sand retained on the sieve thoroughly with a fine jet of water and collect the washings in the cylinder until the volume in the cylinder is about 950 mL. Remove the sieve and make the final volume to 1000 mL.

3   Transfer the sand to a 100-mL beaker and oven dry at 105°C. Weigh the sand and record the weight at this time if only total sand is being determined. Otherwise, proceed with sand fractionation.

4   Transfer the dried sand to a set of sieves (6-cm diameter) arranged as follows from top to bottom: 1, 0.5, and 0.25 mm (60 mesh); 0.105 mm (140 mesh); 0.047 mm (300 mesh) and pan. Pour the sand on the top sieve, put the cover in place, and shake the sieves on a sieve shaker. The time of shaking depends on the type of shaker and the volume of sand (usually 5 to 10 min is sufficient). Weigh each sand fraction and record the weight.

47.2.3.3 DETERMINATION OF CLAY (0–2 μm)

1   Before placing the cylinders in a sedimentation room (any vibration-free area equipped with Shaw pipette rack), stir the material in the sedimentation cylinders for 4 min with a motor-driven stirrer (8 min if suspension has stood for longer than 16 h).

2   Remove from stirrer and slip a length of sytrofoam pipe-insulating cover over the sedimentation cylinder. Stir the suspension for 30 s with a hand stirrer, using an up-and-down motion. Note the time at the completion of stirring.

3   Sample the 2-μm fraction after a predetermined settling time (usually 4.5 to 6.5 h), varying the depth according to the time and temperature (Table 47.1).

About 1 min before sedimentation is complete, lower the tip of a closed Lowy 25-mL pipette slowly into the suspension to the proper depth with a precalibrated Shaw pipette rack. Regulate the filling time of the pipette to about 12 s. Fill the pipette and empty it into a tared 90-mL wide-mouth bottle (or 100-mL beaker) and rinse the pipette into the bottle once.

4   Evaporate the water and dry in an oven at 105°C for at least 24 h. Cool in a desiccator containing phosphorus pentoxide (P_2O_5) or Drierite as a desiccant. Weight and record the weight.

47.2.3.4 DETERMINATION OF FINE CLAY (0.2 μm) (OPTIONAL)

1   Pour about 200 mL of suspension from the sedimentation cylinders into 250-mL centrifuge bottles. Shake the suspensions, and centrifuge at the appropriate speed for the time necessary to sediment particles coarser than 0.2 μm to a depth of 5 cm (54 min at 510 g at 25°C on an IEC® Model V centrifuge). The formula to use is based on Stokes' law:

$$t = \frac{63.0 \times 10^8 \, n \log R/S}{N^2 D^2 \Delta s} \tag{47.2}$$

where

n   = viscosity in poises at the existing temperature
R   = radius of rotation (cm) of the top of the sediment in the tube
S   = radius of rotation (cm) of the surface of the suspension in the tube
N   = revolutions per minute
D   = particle diameter in μm
Δs   = difference in specific gravity between the solvated particle and the
         suspending liquid (usually use $\Delta s = 1.65$)
t   = time in minutes
         (See Jackson [1956] for tables of centrifuging times and speeds.)

2   Withdraw a 25-mL aliquot from a depth of 5 cm. Discharge the sample into a tared weighing bottle or beaker, rinse the pipette, add the rinsing to the weighing bottle, dry at 105°C, cool in a desiccator, and weigh.

47.2.4 CALCULATIONS

1   A = weight (g) of pipetted fraction (2 or 0.2 μm)
     B = weight correction for dispersing agent (g)

NOTE: To determine the correction factor, add 10 mL of the sodium metaphosphate solution to a 1-L cylinder, make to volume, stir thoroughly, withdraw duplicate 25-mL samples, dry, and weigh (about 0.012 g).

$$K = \frac{1000}{\text{Volume of pipette (mL)}} \tag{47.3}$$

$$D = \frac{100}{\text{pretreated oven-dried total sample (g)}} \tag{47.4}$$

2   Sand fraction(s):

Percentage of sand fraction (s) = weight (g) of fraction on sieve times D   (47.5)

Pipetted fraction(s):

Percentage of pipetted fraction(s) = (A − B)KD   (47.6)

Silt fraction:

Percentage of silt = 100 − (0–2 μm clay + sand)   (47.7)

47.2.5 COMMENTS

1   Ultrasonic dispersion in water (Genrich and Brenner 1974) is commonly used for dispersion of clay. This specific procedure has not, however, been tested for particle size distribution in Canada.

2   Iron oxides should be removed from samples to permit the determination of phyllosilicate minerals by X-ray diffraction, but this is not necessary for most samples. However, if the interest is in iron oxides in the clay fraction, pretreatments with dithionite-citrate-bicarbonate must be avoided.

47.3 HYDROMETER METHOD

A hydrometer can be used to measure the density of a soil suspension after various times of settling and, hence, the particle size distribution. Such measurements can be made on suspensions prepared by any of the pretreatments outlined in Section 47.2.2. In reality, however, the hydrometer method is commonly used to estimate particle size distribution without any pretreatment, except dispersion with Calgon®. The hydrometer method outlined here is that of a simplified procedure of Day (1965).

47.3.1 MATERIALS AND REAGENTS

1   Standard hydrometer, ASTM No. 1. 152 H, with Bouyoucos scale in g/L.

2   Electric stirrer.

3   Plunger.

4   End-over-end shaker.

5   Cylinders with 1-L mark 36 ± 2 cm from the bottom of the inside.

6   Amyl alcohol.

7   Calgon® solution (50 g/L).

8   Constant-temperature room (if possible).

47.3.2 CALIBRATION OF HYDROMETER

1   Add 100 mL of Calgon® solution to the cylinder and make the volume to 1 L with distilled water. Mix thoroughly with plunger and let stand until the temperature is constant (between 20 and 25°C).

2   Lower the hydrometer into the solution carefully, and determine the scale reading R_L at the *upper edge of the meniscus* surrounding the stem.

47.3.3 HYDROMETER PROCEDURE

1   Weigh 40.0 g of soil (100 g of loamy sand or sand) into a 600-mL beaker, add 100 mL of Calgon® solution and about 300 mL distilled water, and allow the sample to soak overnight.

2   Weigh another sample of the same soil (about 10 g) for determination of oven-dried weight. Dry overnight at 105°C, cool, and weigh.

3   Transfer the Calgon®-treated sample to a dispersing cup and mix for 5 min with the electric mixer (milkshake machine), or transfer the suspension to shaker bottles and shake overnight on an end-over-end shaker.

4   Transfer the suspension to a cylinder and add distilled water to bring the volume to 1 L.

5   Allow time for the suspensions to equilibrate with room temperature between 20 and 25°C.

6   Insert the plunger and move it up and down to mix the contents thoroughly. Dislodge sediment with strong upward strokes of the plunger near the bottom and by spinning the plunger while the disk is just above the sediment. Finish stirring with two or three slow, smooth strokes. Record the time of completion of stirring. Add a drop of amyl alcohol if the surface of the suspension is covered with foam.

7   Lower the hydrometer carefully into the suspension and take readings after 40 s ($R_{40 s}$).

8   Remove the hydrometer carefully after the 40-s reading, rinse it, and wipe it dry.

9   Reinsert the hydrometer carefully and take reading at 7 h ($R_{7\,h}$).

47.3.4 Calculations

$$\text{Sand \%} = (R_{40\,s} - R_L) \times \frac{100}{\text{oven-dried soil Wt. in g}} \tag{47.8}$$

$$\text{Clay \%} = (R_{7\,h} - R_L) \times \frac{100}{\text{oven-dried soil Wt. in g}} \tag{47.9}$$

$$\text{Silt \%} = 100 - (\text{sand \% + clay \%}) \tag{47.10}$$

47.3.5 Comments

The simplified hydrometer method described in this chapter is not recommended for calcareous or saline soils or soils with more than 2% organic C. For the detailed hydrometer method please refer to Day (1965).

47.4 SIEVE ANALYSIS (MECHANICAL METHOD)

The grain-size analysis is used in the classification of soils for engineering purposes. The resulting grain-size distribution curves are used as part of the criteria for road embankment construction and for prediction of the susceptibility of a soil to frost action.

The grain-size analysis is an attempt to determine the relative proportions of the different grain sizes which make up a given soil mass.

47.4.1 MATERIALS AND REAGENTS

1   Sieves and pan (20-cm dia). Recommended ASTM sieve sizes No. 4 (4.76 mm), No. 10 (2.00 mm), No. 40 (0.42 mm), and a No. 200 (0.074 mm).

2   Sieve brush.

3   Glass beaker (500 mL).

4   Porcelain evaporation dish (20-cm dia).

5   Balance (capacity 1000 g; sensitivity 0.1 g).

6   Mortar and rubber-tipped pestal.

7   Drying oven (capable of 105°C).

8   Sieve shaker.

47.4.2 PROCEDURES

1   Thoroughly clean and weigh each sieve to be used to 0.1 g.

NOTE: Sieves should always be brushed clean from the bottom side. Particles which are forced through the sieve from the top enlarge the opening and reduce the life expectancy of the sieve. Particles which are tightly lodged in the mesh may be loosened by tapping the side wall of the sieve against the palm of the hand.

2   Select and weigh a representative sample of approximately 500 g, separate the soil into individual particles by crushing with either the fingers or a rubber-tipped pestle.

The size of sample which is considered to be representative is dependent upon the maximum size fragment present or be analyzed. The following representative sample guidelines are commonly employed:

   Particles up to 5 mm — 500 g
   Particles up to 20 mm — 5 kg
   Particles up to 75 mm — 20 kg

NOTE: For fine-grained soils which dry into hard aggregates, the most reliable and most easily duplicated method of performing the sieve analysis is to take a quantity of oven-dried soil, break the sample as fine as possible, wash on a No. 200 sieve, oven-dry and sieve the residue through a stack of sieves by shaking horizontally by hand or mechanically for at least 10 min.

3   The initial washing of the soil should be carefully conducted to avoid damaging the sieve or losing any soil by splashing the material out of the sieve. The soil is washed through the sieve using tap water until the water runs clear.

4   Using a wash bottle, carefully back wash the residue into a large porcelain evaporation dish, decant as much of the excess water as possible, ensuring that none of the sample is lost in the process, and oven-dry the remaining soil-water suspension for 16–24 h.

5   Remove the sample from the drying oven, place watch glass on top of the evaporation dish, and allow the dish and contents to cool to room temperature. Record the weight of the sample.

6   Pass the sample through the stack of sieves, using as an "absolute" minimum the following sieve sizes: #4, #10, #40, and #200. Since the object of this

exercise is to obtain a semilogarithmic curve, it is highly recommended that the following sieves be included in the sieve stack: #20, #60, #100, or #140.

7   Following the required 10 min of shaking, weigh each sieve and record the gross weight of the sieve plus the soil. Subtract the weight of the sieve as determined in step 1 and determine the amount of the total sample retained on each sieve as a percentage.

8   Sum the weight of the residue collected on each sieve and compare this to the sample weight recorded in step 5. A discrepancy of more than 2% by weight is considered unsatisfactory and the test should be repeated.

9   Compute the percentage passing each sieve by starting with 100% and subtracting the percent retained on each sieve as a cumulative procedure.

10 Plot a semilogarithmic grain-size distribution curve. If less than 10% of the total sample passes the #200 sieve, this completes the test; if more than 10% passes, then continue with a particle size distribution method.

11 From the grain-size distribution curve compute the coefficient of uniformity ($Cu = D_{60}/D_{10}$); where D refers to the effective diameter of the soil particles and the subscripts (10 and 60) denote the percent which is smaller. An indication of the spread or range of grain sizes is given by Cu, with a large Cu value indicating that the D_{60} and D_{10} sizes differ appreciably.

REFERENCES

Day, P. R. 1965. Particle fractionation and particle-size analysis. Pages 545–567 in Methods of soil analysis, C. A. Black, Ed. Agronomy No. 9, Part 1. American Society of Agronomy, Madison, WI.

Genrich, D. A. and Bremner, J. M. 1974. Isolation of soil particle-size fractions. Soil Sci. Soc. Am. Proc. 38: 222–225.

Jackson, M. L. 1956. Soil Chemical Analysis — Advanced Course. Published by the author, Department of Soils, University of Wisconsin, Madison.

McKeague, J. A., Ed. 1978. Manual on soil sampling and methods of analysis. 2nd ed. Can. Soc. Soil Sci. Ottawa, Ont.

Sheldrick, B. H. and Wang, C. 1987. Compilation of data for ECSS Reference Soil Samples. Land Resource Research Centre, Ottawa.

Soil Conservation Service, 1984. Soil survey laboratory methods and procedure for collecting soil samples. Soil Survey Investigations, Report No. 1 (Revised 1984), U.S. Department of Agriculture, Washington, D.C.

Chapter 48
Soil Shrinkage

B. P. Warkentin

Oregon State University
Corvallis, Oregon

48.1 INTRODUCTION

The change in volume of a soil with change in water content is a characteristic of interest to pedologists as well as to various users of soil. Volume change determines soil structure, aggregate shape, pore size distribution for infiltration and aeration, root damage due to cracking, and safety of buildings, roads, and dams on soil.

Volume change is determined by the nature of the solid surface, e.g., the presence of "swelling", minerals, and the arrangement or fabric of primary and secondary soil particles. Shrinkage and swelling involve soil particle rearrangement. Shrinkage is rapid, and is complete when water content decrease is complete. Swelling, however, is a slow process, occurring in the field over months and years. The increase in "field capacity" of a clay soil under continuously moist conditions is an example of slow swelling. Swelling of soils confined under buildings or roads continues for several years. Soils in the field can be categorized as (a) no volume change, e.g., sands; (b) swelling and shrinking within defined limits as water content changes, e.g., Vertisols; and (c) shrinkage on first drying with little volume increase on wetting, e.g., Andisols.

Shrinkage of a soil sample measured in the laboratory often does not correctly predict volume change in the field, because the stress distribution, especially at the boundaries, is very different in the two situations. Measured shrinkage is greatest for remolded samples, and least for measurements *in situ*. The measurement to be used is determined by the objectives of the study; some measurements must be made in the field. Changes in height of a soil may not be important in crop production, but changes in the other two directions lead to cracking which may be detrimental or beneficial. Uses of soil for bearing loads of buildings or roads are very sensitive to changes in height. Soil cracks resulting from shrinkage are large if the soil is cohesive, and small, but numerous where the soil has good aggregate structure with little cohesion between aggregates.

Shrinkage curves, volume as a function of decreasing water content, measured on samples in the laboratory show several distinct stages. At the wet end the decrease in volume (ΔV) is less than the decrease in water content (ΔW), then there is a stage of normal shrinkage with $\Delta V = \Delta W$, and finally residual shrinkage where ΔV again is less than ΔW. There is a limiting water content below which no further shrinkage takes place. The swelling curves are rarely measured, but show the same stages (Chang and Warkentin 1968).

There are four groups of measurements of shrinkage, depending upon the soils and the objectives of the study.

1 Aggregated soils, for characterization.

2 Cohesive soils, e.g., subsoils, for characterization, measuring isotropy of structure, etc.

3 Field measurements of height change and crack characteristics.

4 Maximum volume change, for engineering design purposes.

The method for aggregated soils will be described in detail after some general comments on the other measurements.

48.2 SHRINKAGE OF COHESIVE SOIL SAMPLES

Undisturbed samples in the size range from 3–10 cm on a side are cut from moist subsoils. Samples may be cylinders or cubes. The volume is measured as the samples are dried slowly to prevent cracking. The samples should be kept under high humidity or a saturated atmosphere to assure slow, uniform drying. Dimensions are measured with vernier calipers or from the projected image of the sample on a screen (e.g., Warkentin and Bozozuk 1961).

Measurements can be made on irregularly shaped samples by measuring the distance between pins inserted into the sample, or marks scored on the sample face. A traveling microscope is used to get precision for small distances.

48.3 MEASUREMENT OF VOLUME CHANGE ON UNDISTURBED SOILS IN THE FIELD

Changes in height of the soil surface are measured to determine engineering behavior of soils. Some measurements have been made for pedology information. A few methods exist for measuring cracks and hence estimating shrinkage in the horizontal axes.

Changes in vertical dimension are measured from rods placed on the soil surface, with surveying levels against a reference level, which is usually a post driven into the ground below the level of volume change. Sometimes the post is inside a sleeve to isolate the post from heaving of the soil where friction or cohesion could bind the post to the soil.

An automatic recording device has been described by Yaalon and Kalmer (1972), where a stand placed on the soil surface is hooked mechanically to a recorder to provide continuous height measurements on a chart. The greatest difficulty is assuring that the small area of soil on which the height measurement is made is representative of the soil to be described.

El Abedine and Robinson (1971) describe a systematic method for measuring width and depth, and for estimating volume, of soil cracks. The frequency and volume of cracks intersecting a straight line were determined.

48.4 SHRINKAGE OF REMOLDED SAMPLES

These methods involve remolding the soil sample at high water content (ASTM 1989). The methods measure the potential volume change due to the mineral surfaces without modification by soil particle arrangement and bonding. The methods are valid for comparing different soils and for screening soils. When a correlation is established for a group of similar soils between undisturbed and remolded volume change, the remolded values can be used to predict volume change in the field. The methods are used most widely in characterizing soils for engineering applications.

A sample of <2-mm sieved soil is mixed with water to about the liquid limit (the water content at which soil changes from plastic to liquid properties). The wet soil should be allowed to equilibrate several hours or overnight. The soil is then poured into special molds or extruded from a syringe as rods 6–10 cm long and 1 cm in diameter (Schafer and Singer 1976). The length of rods is measured wet and after drying, and linear extensibility is calculated over the range of water contents from wet to dry.

Shrinkage methods involving the use of mercury to measure volume of soil samples, either undisturbed or remolded, are no longer recommended because of the undesirability of handling mercury.

48.5 SHRINKAGE OF AGGREGATED SOILS

The method most commonly used for surface soils is measurement of bulk density at different soil water potentials on soil clods coated with Saran plastic (Grossman et al. 1968, Soil Conservation Service 1982).

Clods of about 150 cm³ are coated in the field by dipping in a solution of Saran plastic in methyl ethyl ketone. These coated clods are weighed in air and in water to determine mass and volume, to calculate bulk density. The clods are equilibrated near water saturation by removing the plastic from one side of the clod and smoothing the surface to make good contact with a ceramic plate. The clods are then oven dried to get dry bulk density.

The volume change, change in total porosity, change in void ratio, or coefficient of linear extensibility, are calculated from these measurements.

Soils with large amounts of coarse grains (gravel and larger) may show little volume change by these measurements even though they have actively swelling clays. Shrinkage can be expressed on the basis of <2-mm soil by measuring the proportion of >2-mm grains and correcting the measured volume of the clod.

48.5.1 MATERIALS AND REAGENTS

1   Methyl ethyl ketone.

2   Dow® Saran F310 resin. Dissolve the resin in the methyl ethyl ketone in the ratio of 1:5 resin:solvent. Ratios from 1:4 to 1:7 have been used. Stir, until dissolved, with a wooden stick or a nonsparking stirrer under an exhaust hood. This coating solution can be stored in 2- to 4-L metal cans with tight lids.

Caution: the solvent is flammable and the vapors form an explosive mixture with air.

Acetone can also be used as the solvent, but is less desirable because it penetrates the soil pores more readily.

3   Hair nets.

4   Balance.

5   Drying oven.

6   Tension table apparatus.

48.5.2 PROCEDURE

1   Sample collection: collect natural clods (in triplicate) of about 100 to 200 cm³ (fist-sized). Remove a piece of soil larger than the clod from the face of a sampling pit with a spade. Gently cut or break off protruding peaks and material sheared by the spade. Roots can be cut with side cutters. The procedure must be adjusted to meet the conditions of the soil in the field at the time of sampling.

2   The clod is placed in a hairnet and suspended from a clothesline string. Dry clods may be moistened with a fine-mist spray to prevent loose pieces falling off at the surface. The suspended clod is dipped in the container of the resin coating solution and immersed momentarily. These coated clods should be allowed to dry for 30 min or longer. Store the clods in waterproof plastic bags as soon as the coating dries, since the coating is permeable to water vapor. Transport to the laboratory in rigid containers.

3   Measuring wet and dry bulk density: in the laboratory, apply an additional coating of resin to prevent disruption of the sample on wetting. Remove a patch of Saran coating from a flat surface with a diamond saw. Place the exposed area in firm contact with a tension table equilibrated to 5-cm water tension. When clods have reached a constant weight (7 to 14 d), remove them and coat the exposed surface.

4   Weigh the sample first in air, and then immersed in water to determine volume by Archimedes' principle.

5   After determining moist volume, oven dry the clods at 105°C. Record the weight of the oven-dried clods suspended in air and in water. The difference is oven-dry volume. Calculate bulk density.

48.5.3 Calculations

Bulk density is the oven-dry weight of soil, divided by the wet or dry volume.

The shrinkage, or volume change from wet to dry, can be expressed as a volume decrease per unit weight of soil (m^3 Mg^{-1}), as a decrease in porosity per unit volume of soil (m^3 m^{-3}), or as an equivalent linear shrinkage.

$$\text{Shrinkage} = \frac{1}{D_b\ \text{wet}} - \frac{1}{D_b\ \text{dry}}; \ m^3Mg^{-1} \tag{48.1}$$

$$\text{Change in porosity} = \frac{D_b\ \text{dry} - D_b\ \text{wet}}{D_s} \tag{48.2}$$

$$\text{Coefficient of linear extensibility} = \left[\frac{D_b\ \text{dry}}{D_b\ \text{wet}}\right]^{1/3} - 1 \tag{48.3}$$

Where: D_b wet = bulk density at 5-cm water tension
$\quad\quad\quad D_b$ dry = bulk density of oven-dried soil
$\quad\quad\quad\quad D_s$ = density of soil solids

48.5.4 Corrections

The weight of the resin coating on the clod is not large enough to cause an error for routine measurement of bulk density, except for small samples. The resin coating can be weighed, it has a density of about 1.3 Mg m^{-3}. Appropriate corrections can be made.

Correction can also be made for the amount of coarse grains (larger than 2 mm) in the soil. The weight of >2-mm grains is determined after separation by wet sieving. The volume of >2-mm grains is obtained by direct measurement (displacement of water) or from calculation using the density of the minerals. This volume is subtracted from the total volume to express shrinkage on the basis of <2-mm soil fraction. This may be useful in comparing different soils.

REFERENCES

ASTM 1989. American Society for Testing and Materials, Annual Book of Standards, Vol. 04.08. Method D 3877-80, Standard testing methods for one-dimensional expansion, shrinkage and uplift pressure of soil-lime mixtures.

Chang, R. K. and Warkentin, B. P. 1968. Volume change of compacted clay soil aggregates. Soil Sci. 102: 106–111.

El Abedine, A. Z. and Robinson, G. H. 1971. A study on cracking in some vertisols of the Sudan. Geoderma 5: 229–248.

Grossman, R. B., Brasher, B. R., Franzmeier, D. P., and Walker, J. L. 1968. Linear extensibility as calculated from natural-clod bulk density measurements. Soil Sci. Soc. Am. Proc. 32: 570–573.

Schafer, W. M. and Singer, M. J. 1976. A new method of measuring shrink-swell potential using soil pastes. Soil Sci. Soc. Am. J. 40: 805–806.

Soil Conservation Service 1982. Procedures for collecting soil samples and methods of analysis for soil survey. Soil Survey Investigations Report No. 1, Revised. U.S. Department of Agriculture.

Warkentin, B. P. and Bozozuk, M. 1961. Shrinking and swelling properties of two Canadian clays. Pages 851–855 in Proc. 5th Int. Conf. Soil Mech. and Found. Eng. 3A149.

Yaalon, D. H. and Kalmer, D. 1972. Vertical movement in an undisturbed soil: continuous measurement of swelling and shrinkage with a sensitive apparatus. Geoderma 8: 231–240.

Chapter 49
Soil Consistency Limits

R. A. McBride
University of Guelph
Guelph, Ontario, Canada

49.1 INTRODUCTION

The (Atterberg) consistency limits of soils are used primarily in classifying cohesive soil materials for engineering purposes, and are strongly correlated to other fundamental soil properties (DeJong et al. 1990). They are also used widely in the estimation of other test indices useful for soil engineering interpretations, such as shear strength and bearing capacity, compressibility, swelling potential, and specific surface (reviewed in McBride 1989).

As gravimetric moisture contents, the shrinkage limit, lower plastic limit (w_P), and upper plastic (''liquid'') limit (w_L) test indices represent the three major points of transition in soil consistency among the solid, semisolid, plastic, and liquid states, respectively (Das 1982). The standard ASTM (1981) test procedures for w_L (ASTM D423-72) and w_P (ASTM D424-71) determination are somewhat subjective and arbitrary shear tests, and so are prone to significant operator and mechanical variability. Test reproducibility has not been improved with the introduction of the one-point w_L test method, which reduces the testing time (and cost) as compared to the normally iterative procedures.

Considerable effort has been directed at researching alternative and procedurally unified test methods, including desorption by pressure-plate extraction (reviewed in McBride 1989), consolidation of soil-water suspensions (McBride and Bober 1989, McBride and Baumgartner 1992), and drop-cone penetration. Some have shown very promising results, but only w_L determination by cone penetration has been standardized to date, and is the preferred method of the British Standards Institute (BS 1377:1975 Test 2[A]). It appears unlikely that a direct measure of w_P by the cone penetration method is possible (Harison 1988).

This chapter outlines only those test protocols that have been standardized and that are widely accepted and used. The primary sources for these procedures were BSI (1975), ASTM (1981), and Sheldrick (1984).

Soil Sampling and Methods of Analysis, M. R. Carter, Ed.,
Canadian Society of Soil Science. © 1993 Lewis Publishers.

49.2 UPPER PLASTIC ("LIQUID") LIMIT

The upper plastic ("liquid") limit of a cohesive soil is defined as the gravimetric moisture content (percent) corresponding to an arbitrary limit between the liquid and plastic states of consistence when in a remolded condition. In accordance with the standard ASTM (Casagrande) test procedure, it is the moisture content at which a pat of soil, cut by a standard-sized groove, will flow together for a distance of 13 mm under the impact of 25 blows in a standard ASTM liquid limit device. The undrained cohesion of soils in this consistency state is approximately 1.7 kPa (Wroth and Wood 1978).

49.2.1 CASAGRANDE METHOD
(ASTM D423-72; BS 1377:1975, TEST 2[B])

49.2.1.1 APPARATUS AND MATERIALS

1   ASTM liquid limit device with grooving tool.

2   Metal spatula (7–8 cm in length, 2 cm wide).

3   Evaporating dish (10–12 cm in diameter).

4   Moisture containers.

5   Balance (sensitive to 0.01 g).

6   Drying oven (105°C).

7   Air-dried soil sample of about 100 g and passing a No. 40 (425 μm) sieve (ASTM D421).

49.2.1.2 PROCEDURE

1   Place the soil sample in the evaporating dish and thoroughly mix with 15 to 20 mL of distilled water by alternately and repeatedly stirring, kneading, and chopping with a spatula. Make further additions of water in increments of 1 to 3 mL. Thoroughly mix each increment of water with the soil, as previously described, before adding another increment of water.

2   When sufficient water has been thoroughly mixed with the soil to produce a consistency that will require 30 to 35 drops of the cup to cause closure, place a portion of the mixture in the cup above the spot where the cup rests on the base, and squeeze it down and spread it into position with as few strokes of the spatula as possible. Care should be taken to prevent the entrapment of air bubbles within the soil mass. With the spatula, level the soil and at the same time trim it to a depth of 1 cm at the point of maximum thickness. Return the excess soil to the evaporating dish. Divide the soil in

the cup by firm strokes of the grooving tool along the diameter through the centerline of the cam follower so that a clean, sharp groove of the proper dimensions will be formed. To avoid tearing of the sides of the groove or slipping of the soil pat on the cup, up to six strokes, from front to back or from back to front counting as one stroke, are permitted. Each stroke should penetrate a little deeper until the last stroke from back to front scrapes the bottom of the cup clean. Make the groove with as few strokes as possible.

3   Lift and drop the cup by turning the crank at the rate of 2 rps, until the two halves of the soil pat come in contact at the bottom of the groove along a distance of about 13 mm. Record the number of drops (N) required to close the groove along a distance of 13 mm.

4   Remove a slice of soil approximately the width of the spatula, extending from edge to edge of the soil pat at right angles to the groove and including that portion of the groove in which the soil flowed together, and place in a suitable tared container. Weigh and record the mass. Oven dry the soil in the container to constant mass at 105°C and reweigh as soon as it has cooled, but before hygroscopic moisture can be absorbed. Record this mass. Record the loss in mass due to drying as the mass of water.

5   Transfer the soil remaining in the cup to the evaporating dish. Wash and dry the cup and grooving tool, and reattach the cup to the carriage in preparation for the next trial.

6   Repeat steps 2 to 5 for at least two additional trials, with the soil collected in the evaporating dish, to which sufficient water has been added to bring the soil to a more fluid condition. The object of this procedure is to obtain samples of such consistency that the number of drops required to close the groove will be above and below 25. The number of drops should be less than 35 and exceed 15. The test shall always proceed from the drier to the wetter condition of the soil.

49.2.1.3 CALCULATIONS

1   Calculate the gravimetric moisture content of the soil (w), expressed as a percentage of water in the sample on a dry-mass basis (%kg kg^{-1}), as follows:

$$w = (\text{mass of water/mass of oven-dry soil}) \times 100 \qquad (49.1)$$

2   Plot a "flow curve" representing the relationship between gravimetric moisture content and corresponding numbers of drops of the cup on a semilogarithmic graph with w as abscissae on the linear scale, and the number of drops (N) as ordinates on the logarithmic scale. The flow curve is a straight line drawn as nearly as possible through the three or more plotted points.

3   Take the moisture content corresponding to the intersection of the flow curve with the N = 25 ordinate as the upper plastic limit ("liquid limit") of the soil. Report the w_L test index to the nearest whole number.

49.2.1.4 Comments

Prior to testing, inspect the ASTM liquid limit device to determine that the device is in good working order, that the pin connecting the cup is not worn sufficiently to permit side play, that the screws connecting the cup to the hanger arm are tight, and that a groove has not been worn in the cup through long usage. Also ensure that the dimensions of the grooving tool are to specification (ASTM 1981).

By means of the gauge on the handle of the grooving tool and the adjustment plate, adjust the height to which the cup is lifted so that the point on the cup that comes in contact with the base is exactly 1 cm above the base. Secure the adjustment plate by tightening the screws. With the gauge still in place, check the adjustment by revolving the crank rapidly several times. If the adjustment is correct, a slight ringing sound will be heard when the cam strikes the cam follower. If the cup is raised off the gauge or no sound is heard, make further adjustments.

49.2.2 One-Point Casagrande Method (ASTM D423-72; BS 1377:1975 Test 2[C])

49.2.2.1 PROCEDURE

1   The requirements for the apparatus, the soil sample preparation, and the mechanical device adjustments are identical to those under Section 49.2.1.

2   Proceed in accordance with procedural steps 1 through 5 under Section 49.2.1, except that a moisture content sample shall be taken only for the accepted trial. The accepted trial requires between 20 and 30 drops of the cup to close the groove and at least two consistent consecutive closures are to be observed before taking the moisture content sample for calculation of the upper plastic limit. The test should always proceed from the drier to the wetter condition of the soil.

49.2.2.2 CALCULATIONS

1   Calculate the percent gravimetric moisture content (w) of the soil for the accepted trial as per Equation 49.1.

2   Determine the upper plastic limit using the following formula:

$$w_L = w(N/25)^{0.12} \qquad (49.2)$$

where: N = the number of drops of the cup required to close the groove at the test moisture content.

3   Report the w_L test index to the nearest whole number.

49.2.3 DROP-CONE PENETROMETER METHOD
(BS 1377:1975 TEST 2[A])

49.2.3.1 APPARATUS AND MATERIALS

1   A drop-cone penetrometer as used in bituminous material testing complying with the requirements of BS 4691.

2   A cone of stainless steel or duralumin approximately 35 mm long, with a smooth, polished surface and an angle of $30 \pm 1°$. The mass of the cone together with its sliding shaft is 80.00 ± 0.05 g.

3   A noncorrodible, air-tight container.

4   A metal cup approximately 5.5 cm in diameter and 4.0 cm deep with rim parallel to the flat base.

5   Metal spatula (7–8 cm in length, 2 cm wide).

6   Evaporating dish (10–12 cm in diameter).

7   Moisture containers.

8   Balance (sensitive to 0.01 g).

9   Drying oven (105°C).

10   Air-dried soil sample of about 200 g and passing a No. 40 (425 μm) sieve (ASTM D421).

49.2.3.2 PROCEDURE

1   A sample weighing at least 200 g is placed on the evaporating dish and mixed thoroughly with distilled water using the spatula until the mass becomes a thick, homogeneous paste. This paste is then allowed to stand in the air-tight container for about 24 h to allow the water to permeate throughout the soil mass.

2   The sample is then removed from the container and remixed for at least 10 min. If necessary, further water is added so that the first cone penetration reading is approximately 15 mm.

3   The remixed soil is pushed into the cup with a spatula, taking care not to trap air. The excess soil is removed to give a smooth surface. The cone is lowered so that it just touches the surface of the soil. When the cone is in the correct position, a slight movement of the cup will just mark the surface

of the soil and the reading of the dial gauge is noted to the nearest 0.1 mm. The cone is then released for a period of 5 ± 1 s. If the apparatus is not fitted with an automatic release and locking device, care should be taken not to jerk the apparatus during these operations. After the cone has been locked in position, the dial gauge is lowered to the new position of the cone shaft and the reading noted to the nearest 0.1 mm. The difference between the readings at the beginning and end of the test is recorded as the depth of cone penetration.

4   The cone is lifted out and cleaned carefully. A little more wet soil is added to the cup and the process repeated. If the difference between the first and second penetration readings is less than 0.5 mm, the average of the two penetrations is recorded. If the second penetration is more than 0.5 mm and less than 1 mm different from the first, a third test shall be carried out. If the overall range is then not more than 1 mm, a moisture content sample (about 10 g) is taken from the area penetrated by the cone and the moisture content determined. The average of the three penetrations is recorded. If the overall range is more than 1 mm, the soil shall be removed from the cup, remixed, and the test repeated until consistent results are obtained.

5   The operations described in steps 3 and 4 are to be repeated at least four times using the same sample to which further increments of distilled water have been added. The amount of water added shall be chosen so that a range of penetration values of approximately 15 to 25 mm is covered.

49.2.3.3 *Calculations*

The relationship between the gravimetric moisture content and the depth of cone penetration is plotted with the percentage moisture contents as abscissae and the cone penetrations as ordinates, both on linear scales. The best straight line fitting the plotted points is drawn through them. The moisture content corresponding to a cone penetration of 20 mm is taken as the upper plastic limit of the soil and is expressed to the nearest whole number. The method of obtaining the w_L shall be stated (i.e. using the cone penetrometer).

49.2.3.4 COMMENTS

1   The automatic release and locking penetrometers are preferred over the manual instruments if operator variability is to be minimized.

2   The 20-mm penetration depth standard should be used with caution, as many studies outside of Britain (including Canada) have documented both under- and overestimation of the Casagrande w_L using the cone penetration method (cf. McBride and Baumgartner 1992).

49.3 LOWER PLASTIC LIMIT

The lower plastic limit of a cohesive soil is defined as the gravimetric moisture content (percent) corresponding to an arbitrary limit between the plastic and semisolid states of consistence when in a remolded condition. In accordance with the standard ASTM test procedure, it is the moisture content at which a soil will just begin to crumble when rolled into a thread approximately 3.2 mm in diameter. The undrained cohesion of soils in this consistency state is approximately 170 kPa (Wroth and Wood 1978).

49.3.1 ATTERBERG METHOD
(ASTM D424-71; BS 1377:1975, TEST 3)

49.3.1.1 APPARATUS AND MATERIALS

1   Evaporating dish (10–12 cm in diameter).

2   Metal spatula (7–8 cm in length, 2 cm wide).

3   Surface for rolling (e.g., a ground-glass plate).

4   Moisture containers.

5   Balance (sensitive to 0.01 g).

6   Drying oven (105°C).

7   Air-dried soil sample of about 15 g and passing a No. 40 (425 μm) sieve (ASTM D421).

49.3.1.2 PROCEDURE

1   If the lower plastic limit only is required, take about 15 g of air-dried soil, place in an evaporating dish, and thoroughly mix with distilled water until the mass becomes plastic enough to be easily shaped into a ball. Take a portion of this ball weighing about 8 g for the test sample.

2   If both the upper and lower plastic limits are required, take a test sample weighing about 8 g from the thoroughly wet and mixed portion of the soil prepared in accordance with ASTM D423 (see Section 49.2.1). Take the sample at any stage of the mixing process at which the mass becomes plastic enough to be easily shaped into a ball without sticking to the fingers excessively when squeezed. If the sample is taken before completion of the upper plastic limit test, set it aside and allow to season in air until the w_L test has been completed. If the sample is taken after completion of the w_L test, and is still too dry to permit rolling to a 3.2-mm thread, add more water.

3.   Squeeze and form the 8-g test sample taken in accordance with steps 1 or
      2 into an ellipsoidal-shaped mass. Roll this mass between the fingers and
      the ground-glass plate lying on a smooth horizontal surface with just suf-
      ficient pressure to roll the mass into a thread of uniform diameter throughout
      its length. The rate of rolling shall be between 80 and 90 strokes/min, count-
      ing a stroke as one complete motion of the hand forward and back to the
      starting position again.

4   When the diameter of the thread becomes about 3.2 mm, break the thread
      into six or eight pieces. Squeeze the pieces together between the thumbs
      and fingers of both hands into a uniform mass roughly ellipsoidal in shape,
      and reroll. Continue this alternate rolling to a thread 3.2 mm in diameter,
      gathering together, kneading, and rerolling, until the thread crumbles under
      the pressure required for rolling and the soil can no longer be rolled into
      a thread. The crumbling may occur when the thread has a diameter greater
      than 3.2 mm. This shall be considered a satisfactory endpoint, provided the
      soil has been previously rolled into a thread 3.2 mm in diameter. The crum-
      bling will manifest itself differently with the various types of soil. Some soils
      fall apart in numerous small aggregations of particles; others may form an
      outside tubular layer that starts splitting at both ends. The splitting pro-
      gresses toward the middle, and finally, the thread falls apart in many small
      platy particles. Heavy clay soils require much pressure to deform the thread,
      particularly as they approach the lower plastic limit, and finally, the thread
      breaks into a series of barrel-shaped segments each about 6.4 to 9.5 mm in
      length. At no time shall the operator attempt to produce failure at exactly
      3.2 mm diameter by allowing the thread to reach 3.2 mm, then reducing the
      rate of rolling or the hand pressure, or both, and continuing the rolling
      without further deformation until the thread falls apart. It is permissible,
      however, to reduce the total amount of deformation for marginally plastic
      soils by making the initial diameter of the ellipsoidal-shaped mass nearer to
      the required 3.2 mm final diameter.

5   Gather the portions of the crumbled soil together and place in a suitable
      tared container. Weigh the container and soil and record the mass. Oven
      dry the soil in the container to constant mass at 105°C and weigh. Record
      this mass. Record the loss in mass as the mass of water.

49.3.1.3 CALCULATIONS

1   Calculate the percent gravimetric moisture content of the remolded soil as
      per Equation 49.1. Report this value as the w_P test index to the nearest whole
      number.

2   Calculate the plasticity index of a soil as the difference between its upper
      and lower plastic limits, as follows:

$$\text{Plasticity index} = w_L - w_P \tag{49.3}$$

3   Report the difference calculated in Equation 49.3 as the plasticity index, except under the following conditions:

    a. When the w_L or w_P test indices cannot be determined, report the plasticity index as NP (nonplastic).

    b. When the soil is extremely sandy, the w_P test is to be performed before the w_L test. If the w_P cannot be determined, report the plasticity index as NP.

    c. When the w_P is greater than or equal to the w_L, report the plasticity index as NP.

REFERENCES

American Society for Testing and Materials. 1981. 1981 Annual Book of A.S.T.M. Standards, Part 19. American Society for Testing and Materials, Philadelphia, PA.

British Standards Institution. 1975. Methods of test for soils for civil engineering purposes. British Standard 1377:1975. British Standards Institution, London.

Das, B. M. 1982. Soil mechanics laboratory manual. Engineering Press, San Jose, CA. 250 p.

DeJong, E., Acton, D. F., and Stonehouse, H. B. 1990. Estimating the Atterberg limits of southern Saskatchewan soils from texture and carbon contents. Can. J. Soil Sci. 70: 543–554.

Harison, J. A. 1988. Using the BS cone penetrometer for the determination of the plastic limit of soils. Geotechnique 38: 433–438.

McBride, R. A. 1989. A re-examination of alternative test procedures for soil consistency limit determination. II. A simulated desorption procedure. Soil Sci. Soc. Am. J. 53: 184–191.

McBride, R. A. and Bober, M. L. 1989. A re-examination of alternative test procedures for soil consistency limit determination. I. A compression-based procedure. Soil Sci. Soc. Am. J. 53: 178–183.

McBride, R. A. and Baumgartner, N. 1992. A simple slurry consolidometer designed for the estimation of the consistency limits of soils. J. Terramechanics 29(2): 223–238.

Sheldrick, B. H., 1984. Analytical methods manual 1984. L.R.R.I. Contribution No. 84-30. Land Resource Research Institute, Research Branch, Agriculture Canada, Ottawa, Ont.

Wroth, C. P. and Wood, D. M. 1978. The correlation of index properties with some basic engineering properties of soils. Can. Geotech. J. 15: 137–145.

Chapter 50
Density and Compressibility

J. L. B. Culley

Agriculture Canada
Ottawa, Ontario, Canada

50.1 INTRODUCTION

Soil bulk density (D_b) has been defined by Blake and Hartge (1986a) as the ratio of the mass of oven-dried solids (M_g) to the bulk volume of the solids plus pore space at some specified soil water content, usually that at sampling. D_b is an extremely useful parameter, as it is required to calculate porosity when particle density is known, to convert weights to volumes, and to estimate weights of soil volumes too large to weigh. It is also required to convert weight-based determinations to a volume basis which are often of more interest. For example, the volumetric content of water in a soil layer is obtained by multiplying the gravimetric water content by the product of the bulk density and the volume of the layer.

Particle density (D_p) of soil refers to the density of the soil particles; no account is taken of the volume of voids between the particles. Its measurement is important in determinations of soil porosity; its value for a soil is determined by both the type and relative abundance of mineral and organic constituents. Methods to accomplish measurement of D_p have been described and discussed by Blake and Hartge (1986b).

In the soil mechanics literature, reference is made to both consolidation and compression tests of saturated and unsaturated soils. Consolidation tests are undertaken to estimate the amount of settlement that may be expected from a soil under a given load. Compression for this purpose is usually achieved by uniaxial (vertical) loading of confined cores. There is a temporal component to such investigations, as a long equilibrium time can be required for fine-textured soils. Bradford and Gupta (1986) have presented the principles and procedures for estimating soil compressibility under these conditions.

The unconfined compression test is undertaken to determine the strength of either undisturbed or remolded cohesive soil. The test is a more rapid and economic means of obtaining the shear strength of cohesive soil than the standard triaxial compression test, the performance of which is extremely involved and expensive. According to Bowles (1970), shear strength data from the unconfined test are reasonably reliable; however, the use of this test to obtain the modulus of elasticity will, in general, be very unreliable. The American Society for Testing and Materials (1966) has published a standard procedure (ASTM D2166-66) on which this procedure is based. Unconfined compression tests have also been conducted on individual soil aggregates (Braunack and Dexter 1989). The method described is for the measurement of the unconfined compressive strength of cylindrical soil samples.

50.2 BULK DENSITY

D_b is a dynamic soil property. The D_b of coarse-textured and other soils which do not shrink on desiccation depends on the nature and degree of packing of the soil particles. For soils which shrink and swell, it depends on the soil water content as well as the nature and packing of the particles. The rate of change in D_b with water content depends on soil structure and water content. Greater rates of change are associated with intermediate water contents (the plant-available range) than at the wet and dry extremes. Despite this complex behavior, D_b is often used to characterize soil structure.

The D_b of soils can be determined by both *in situ* and destructive sampling techniques. *In situ* determination involves the measurement of the attenuation of gamma rays through a fixed soil length; with proper calibration, D_b bulk density can be computed if volumetric water content is determined simultaneously. Core, clod, and excavation are the most common methods of determining D_b destructively. Methods for clod and core techniques as well as a discussion of gamma ray attenuation are presented below. Excavation methods, which involve removal of a volume of soil followed by determination of the volume of the excavation using water, sand, foam, or linear measures, have been presented by Blake and Hartge (1986a) and will not be discussed.

D_b is considered to exhibit relatively low variability spatially (Warrick and Neilson 1980). Generally, coefficients of variability for measurements of D_b for a given profile horizon of a soil series do not exceed about 10%. Thus, about four samples should be sufficient to estimate the mean density to within 10% of the true value, 95% of the time for a uniform soil type.

50.2.1 CORE METHOD

50.2.1.1 MATERIALS AND SUPPLIES

1   Double-cylinder core sampler either hand operated, such as the modified Uhland, or hydraulically driven (see Blake and Hartge 1986a).

2   Clean, dry, and uniform cylinder with a known diameter, d (mm) and height, h (mm); the volume, V (cm³) is calculated as:

$$V = \pi d^2 h / 4000 \tag{50.1}$$

3   Sharp and rigid knife or spatula.

4   Balance sensitive to 0.01 g.

5   Drying oven capable of 105°C, preferably equipped with a circulating fan.

6   Plastic bags and corrosion-resistant weighing tins large enough to hold soil sample and the cylindrical core.

7   Disks to protect ends of cores.

8   Masking tape.

9   Glass beads (260 μm), used to measure core volume not occupied by soil.

50.2.1.2 PROCEDURE

1   Identify and weigh cylindrical cores; record weight as W1 (g).

2   Identify both tins and tops and record weight as W2 (g).

3   Prepare a smooth "undisturbed" vertical or horizontal surface at the sampling depth.

4   Drive or press sampler into the soil sufficiently to fill the inner core without inducing compression. Do not rock the sampler. In frictional or dense soils, careful excavation above the bottom to minimize soil-metal adhesion may help in obtaining a representative core. In soils exhibiting soil-metal adhesion, an application of mineral oil to the core may also be beneficial. Use of oil may affect wetting and drying within the core if it is to be used for water desorption characterization.

5   After careful removal of the undisturbed soil core, examine it for signs of shattering or compression. Trim the ends of acceptable cores flush with the end of the cylinder.

6   For cores which completely fill the cylinders, and if only density is to be determined, the contents of the cylinder can be pushed out into the pre-weighed weighing tin, which is then closed and weighed (W3). Proceed to step 10.

7   Otherwise, cover the ends of the core with disks and place in plastic bag which is then sealed with tape to prevent water loss.

8   If the soil does not completely fill the cylinder, proceed to Section 50.2.1.4 on partially filled cores.

9   At the laboratory, remove disks from core and place the full core into the tin, weigh, and record the weight as W4.

10 Place samples in an oven set to 105°C. Drying time varies with core size and oven type. Cores of about 350 cm^3 usually require about 72 h of drying in ovens equipped with circulating fans. Smaller cores require less time. After drying and cooling in a desiccator, record the weight of tin plus dry soil plus core cylinder as W5 (g).

50.2.1.3 *Calculations*

For soils retained in the cylinder, calculate the bulk density, D_b as follows:

$$D_b = (W5 - W2 - W1)/V \text{ (g cm}^{-3} \text{ or Mg m}^{-3}) \tag{50.2}$$

and volumetric water content, Θ,

$$\Theta = [(W4 - W5)/(W5 - W2 - W1)]D_b \text{ (g cm}^{-3} \text{ or Mg m}^{-3}) \tag{50.3}$$

For the case when cyclinders were emptied into tins at sampling,

$$D_b = (W5 - W2)/V \tag{50.4}$$

$$\Theta = [(W3 - W5)/(W5 - W2)]D_b \tag{50.5}$$

The wet density (D_{bw}), used in soil mechanics, and which is useful for making comparisons between samples of soil which exhibit volume changes on drying, is

$$D_{bw} = D_b + \theta \text{ (g cm}^{-3}) \tag{50.6}$$

50.2.1.4 DENSITY OF PARTIALLY FILLED CORES

1 Tare a graduated cylinder of volume Vg1 and then fill with glass beads; the weight of the beads is recorded as Wg1.

2 Weigh (W3) the soil core, place a disk under one end, and put on a tray.

3 Pour glass beads onto the soil and level to the top of the cylinder with a spatula.

4 Place a disk over the top of the cylinder, invert the core, and fill the other end with beads. Transfer the core to a preweighed tin (W2) and dry at 105°C.

5 Return excess glass beads from the tray to the cylinder and record their volume and weight as Vg2 and Wg2, respectively.

6 After drying the sample, cool in a desiccator and weigh (W5).

7 Calculate soil volume, Vs, as:

$$Vs = V - (Wg1 - Wg2)/C \tag{50.7}$$

where C is the packing density of the glass beads, which each analyst should determine for himself. Beads having a nominal diameter of 260 μm pack to a density of 1.499 g/cm³. Alternatively, the volume of beads can be calculated as Vg1 − Vg2.

8   Calculate D_b as:

$$D_b = (W5 - (Wg1 - Wg2) - W2 - W1)/Vs \qquad (50.8)$$

and

$$\Theta = [(W4 - W1 - (W5 - (Wg1 - Wg2) - W1 - W2))/Vs]D_b \qquad (50.9)$$

50.2.1.5 *Correction for Coarse Fragments*

For certain applications the D_b of the fine fraction, defined as those particles less than 2 mm in diameter, is of interest. This density is obtained by sieving the oven-dried soil through a 2-mm sieve. The material retained on the sieve is washed, dried, and its weight recorded as W6. The volume of this fraction, Vc, can be determined by measuring the displacement of water in a graduated cylinder when the fragments are added. D_b is then:

$$D_b = (W5 - W2 - W1 - W6)/(V - Vc) \qquad (50.10)$$

50.2.1.6 *Comments*

In addition to double cylinders, other tools have been used to obtain soil samples for density determinations. For example, McCauley peat augers, which remove a one-half cylinder, are often used in peat soils (see Chapter 42). Erbach (1987) has discussed various tools for obtaining soil cores. Raper and Erbach (1987) observed that densities obtained by augering cores were significantly lower and more accurate than those obtained by pushing cores; application of a Teflon® coating on sampler tips did not materially improve results. Cameron et al. (1990) have reported a novel lysimeter casing which included an internal cutting ring. The apparatus produces a small annular gap between the wall and the soil which could be sealed with petroleum jelly to prevent edge flow in leaching experiments as well as reducing soil-metal adhesion.

50.2.2 CLOD METHOD

Blake and Hartge (1986a) have presented details of the method. The procedure involves the application of Archimedes' Principle to determine clod volume. The method is time consuming. Only quite stable (and cohesive) aggregates should be used, and, as Blake and Hartge (1986a) point out, it is not very precise. Shrinkage of clods from field-moist to oven dryness can be measured by adapting the method as described in McKeague (1978). In this procedure the whole aggregate, rather than just a subsample, is oven dried at 105°C. The clod method assumes that about 10% of saran initially present is lost on drying; this assumption does not strongly affect results. Shrinkage over other water content ranges of interest can be determined by equilibrating clods on pressure plates or tension tanks before coating with saran.

50.2.3 NUCLEAR METHODS

50.2.3.1 *Instrumentation*

Gamma radiation attenuation can be used to determine soil D_b. Gardner (1986) has discussed the principles of gamma ray attenuation as well as the sources of error associated with its

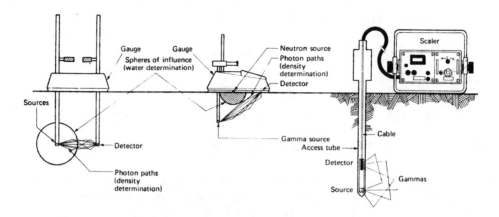

FIGURE 50.1. Schematic representations of commercially available nuclear instruments for measuring soil density.

measurement. Briefly, attenuation of gamma rays by air is negligible, but it is exponentially related to wet density (D_{bw}). The mass attenuation coefficient of dry soil is little affected by typical variations in the relative abundance of the elements which constitute soils. However, the mass attenuation coefficient of water exceeds that for typical soil material by 5–10%. Thus, changes in water content strongly affect soil bulk attenuation, and so D_b can be determined only if Θ is also determined. Neutron moderation is usually used for determination of Θ. γ-Rays at two energies can be used, but the precision capability is very limited.

Reliable commercial instruments are now available to measure gamma ray attenuation and neutron moderation. Two U.S. manufacturers, Troxler and CPN, supply a range of instruments for different applications. The best current instruments are equipped with both gamma ray (Cs^{137}) and neutron (Am^{241}/Be) sources and detectors. Inclusion of the neutron source and detector permit the independent determination of water contents. There are two different modes, backscatter and transmission, by which gamma ray attenuation is measured with these instruments. With backscatter, both the source and the detector, usually a Geiger-Mueller (GM) tube, are located within the instrument, thus permitting a nondestructive test. This configuration works well for near-surface layers and is widely used in road construction applications. Neutron source and detector are also located within the instrument, which similarly restricts the soil volume within which the fast neutrons are thermalized.

Attenuation by transmission requires that the source and/or the detector are lowered down a pre-augered role. Currently, there are two types of single-probe instruments as well as dual-probe transmission instruments (Figure 50.1).

One single-probe design, which is commonly used in geotechnical applications, has the Cs^{137} located within a stainless steel probe of up to 0.3 m length. The neutron source as well as both detectors are located on or near the base plate of the instrument, which is placed in complete contact with the soil surface. This instrument is well suited to near-surface measurements. Another single-probe design incorporates sensors and detectors into the probe; this geometry permits measurements ranging in depth from about 0.25 to more than 10 m. Thus it is not, in general, suitable for near-surface agronomic applications. Dual probes usually contain both sources and the neutron detector in one probe, and the GM tube for gamma ray detection in the other probe. Older instruments often contained no neutron-attenuation capability.

From this brief description it should be clear that unless care is taken, measurements of soil water contents may not be representative of the zone where density is being evaluated.

50.2.3.1 COMMENTS ON USE OF NUCLEAR GAUGES

Manufacturers of commercial available equipment provide detailed instructions which should be read and followed.

1 Manufacturers supply detailed instructions on the safe use of these potentially hazardous instruments. If used carefully within these guidelines, there is minimal risk of radiation exposure to the operator.

2 Experience (e.g., Rawitz et al. 1982) indicates that factory calibrations may not be applicable to all soils. Thus, it is critical to assess the relationships provided, particularly those for soil water determinations. It appears that careful verification under relevant field situations using the core method may be the most appropriate way to accomplish the check for density.

3 If the soil profile to be characterized is thought to exhibit a significant hydraulic gradient, the profile should be thoroughly wetted and allowed to drain under gravity before the measurements are made. Bruce and Luxmoore (1986) have described a procedure for wetting field soil. This minimizes gradients in water content.

4 If the soil profile is thought to contain highly variable soil layers, it is advisable to determine the gravimetric water contents of these layers; Gardner (1986) has presented the method.

5 Follow the manufacturer's instructions for determining standard and measured counts. In general, longer counting times are required for thermal neutrons than for gamma rays.

50.2.3.2 CALCULATIONS AND COMMENTS

1. Data analysis depends on the design of the instrument. Of course, transmission-type probes measure properties between source and detector. For single probes, the overall layer properties are measured. Usually it is individual layers which are of interest. Layer values can be obtained by depth-weighting the overall depth averages. Estimates of dry density when no measurement of volumetric water content is available or when soil horizons are widely variable can be obtained for soil layers according to the equation:

$$D_b = D_{bw}/(1 + w) \qquad (50.11)$$

where w is the gravimetric (g/g) soil water content.

2   Comparisons of single- and double-probe instruments have been undertaken (Culley and McGovern 1990). If care is taken, single probes can provide as satisfactory results as double probes, which are considerably more cumbersome.

50.3 PARTICLE DENSITY

The total volume of pores within a soil sample can be measured directly or calculated if both the soil D_b and D_p have been measured. The total porosity of a soil sample can be calculated as:

$$St = (1 - D_b/D_p) \qquad (50.12)$$

where St is the fractional total pore space. D_p is relatively invariant with time, and so changes in porosity are due to D_b, which depends strongly on previous stresses (both mechanically applied and environmental).

Commonly, a soil D_p of 2.65 Mg/m^3 is assumed. This value corresponds to the specific gravity of quartz. Many silicate and nonsilicate minerals, such as feldspars, granites, limestone, micas, and kaolinite, exhibit specific gravities between about 2.3 and 3.0. The specific gravity of hematite, goethite, and other iron-containing minerals frequently exceeds 3.3, while organic matter has a specific gravity of about 1.4. The D_p for a soil consisting of three constituents x_1, x_2, and x_3 (fractions expressed by weight) with specific gravities of D_{p1}, D_{p2}, and D_{p3} can be calculated as follows:

$$\frac{1}{D_p} = x_1/D_{p1} + x_2/D_{p2} + x_3/D_{p3} \qquad (50.13)$$

50.3.1 PROCEDURE

Soil D_p is measured by application of Archimedes' Principle and involves the use of volume displacement of dry soil by a liquid of known density in a pycnometer. Complete details of the method may be found in Blake and Hartge (1986b).

50.4 UNCONFINED COMPRESSION

50.4.1 Definition

The unconfined compressive strength is the total load per unit area at which a cylindrical or prismatic soil specimen fails. The strength is taken as the maximum load per unit attained or that at 20% axial strain, whichever occurs first during the test.

50.4.2 MATERIALS AND METHODS

1   Samples should have a minimum diameter of 33 mm; samples less than 71 mm in diameter should have no particle with a diameter greater than 10%

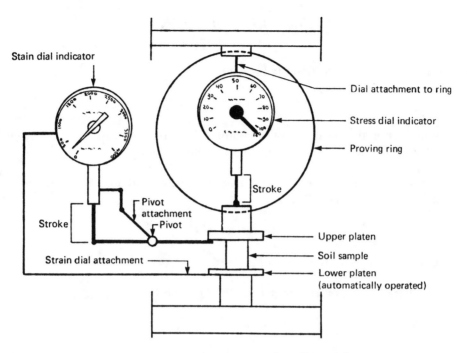

Stain dial indicator

Dial attachment to ring

Stress dial indicator

Proving ring

Stroke

Pivot
attachment
Pivot

Stroke

Strain dial attachment

Upper platen

Soil sample

Lower platen
(automatically operated)

FIGURE 50.2. Components of a unconfined compression apparatus. (Adapted from Rogoski et al., Soil Sci. Soc. Am. Proc. 32: 720–724, 1968.)

of that of the specimen. For thicker samples, individual particles should be less than one sixth of the sample diameter. Height-to-diameter ratio should range between 2 and 3.

2   Undisturbed samples of uniform cross section with ends perpendicular to the longitudinal axis should be carved so as to prevent specimen drying. Application of thin plastic coatings can help to prevent crumbling.

3   Record the weight of the sample (g).

4   Measure the height (L_o) and diameter (d) of the soil specimen to the nearest 0.25 mm using vernier callipers.

5   Place the sample in the loading device (Figure 50.2) so that it is centered on the bottom platen. Adjust the upper platen such that it is just in contact with the soil sample. Zero the deformation (strain) indicator.

6   Apply the load so as to produce axial strain at a rate of 0.5 to 2% per minute. Record load, P (kg of force), and deformation every 30 s until load decreases with increasing strain or 20% strain is reached. The complete test should not require more than 10 min to complete.

7.   Place a sample of the failed sample in a preweighed tin, weigh the tin plus the moist soil, dry at 105°C, cool, and weigh the tin plus the dry soil.

50.4.3 CALCULATIONS

1   Calculate total strain (ϵ) at failure as:

$$\epsilon = \Delta L / L_o \qquad (50.14)$$

where ΔL is the total change in sample length as measured by the deformation indicator and L_o is the original sample length.

2   Calculate initial cross-sectional area ($A_o = \pi d^2/4$) and adjusted area, $A(mm^2)$ at each applied load, P, as:

$$A = A_o/(1 - \epsilon) \qquad (50.15)$$

where ϵ is the strain at load P.

3   Calculate the pressure (stress), $\sigma(kPa)$, at each load P (kg) as:

$$\sigma = (P/A)E \qquad (50.16)$$

where E equals 9810 ($kN\ mm^2\ kg^{-1}\ m^{-2}$)

4   Plot on graph paper the applied pressure (stress) vs. strain. Unconfined shear strength is determined either by the maximum pressure or 20% strain, whichever comes first.

5   Cohesion is calculated at one half the shear strength, which assumes that the angle of internal friction is zero, a reasonable assumption for wet clay.

REFERENCES

American Society for Testing and Materials. 1966. Unconfined compressive strength of cohesive soil. D 2166. In ASTM (1989) 04, 08: 253–257.

Blake, G. R. and Hartge, K. H. 1986a. Bulk density. Pages 363–375 in A. Klute, Ed. Methods of soil analysis. Part 1, Agronomy No. 9, American Society of Agronomy, Madison, WI.

Blake, G. R. and Hartge, K. H. 1986b. Particle density. Pages 377–382 in A. Klute, Ed. Methods of soil analysis. Part 1. Agronomy No. 9. American Society of Agronomy, Madison, WI.

Bowles, J. E. 1970. Engineering properties of soils and their measurement. McGraw-Hill, New York.

Bradford, J. M. and Gupta, S. C. 1986. Compressibility. Pages 479–492 in A. Klute, Ed. Methods of soil analysis. Part 1. Agronomy No. 9. American Society of Agronomy, Madison, WI.

Braunack, M. V. and Dexter, A. R. 1989. Soil aggregation in the seedbed: a review. I. Properties of aggregates and beds of aggregates. Soil Tillage Res. 14: 259–279.

Bruce, R. R. and Luxmoore, R. J. 1986. Water retention: field methods. Pages 663–686 in A. Klute, Ed. Methods of soil analysis. Part 1. Agronomy No. 9. American Society of Agronomy, Madison, WI.

Cameron, K. C., Harrison, D. F., Smith, N. P., and McKay, C. D. A. 1990. A method to prevent edge flow in undisturbed soil cores and lysimeters. Aust. J. Soil Res. 28: 879–886.

Culley, J. L. B. and McGovern, M. A. 1990. Single and dual probe nuclear instruments for determining water contents and bulk densities of a clay loam soil. Soil Tillage Res. 16: 245–256.

Erbach, D. C. 1987. Measurement of soil bulk density and moisture. Trans ASAE 30: 922–931.

Gardner, W. H. 1986. Water content. Pages 493–544 in A. Klute, Ed. Methods of soil analysis. Part 1. Agronomy No. 9. American Society of Agronomy, Madison, WI.

McKeague, J. A., Ed. 1978. Manual on soil sampling and methods of analysis. Canadian Society of Soil Science, Ottawa, Ont.

Raper, R. L. and Erbach, D. C. 1987. Bulk density measurement variability with core samplers. Trans. ASAE 30: 878–881.

Rawitz, E., Etkin, H., and Hazan, A. 1982. Calibration and field testing of a two-probe gamma gauge. Soil Sci. Soc. Am. J. 46: 461–465.

Warrick, A. W. and Neilson, D. R. 1980. Spatial variability of soil physical properties in the field. In D. Hillel, Ed. Applications of soil physics. Academic Press, New York.

Chapter 51
Soil Water Content

G. C. Topp

Agriculture Canada
Ottawa, Ontario, Canada

51.1 INTRODUCTION

The determination of water content is among the most commonly performed kinds of soil analysis. Soil water content affects so much the behavior and use of soil that practically every type of soil study requires measurement of water content. Traditionally, water content has been expressed as the ratio of the mass of water present in the sample to the mass of the sample after it has been dried at 105°C to a constant mass. Alternatively, the volume of water present in a unit volume of soil may be used as a measure of water content. Thus, water content as usually used in soil studies is a dimensionless ratio of two masses or two volumes, and it is important to specify which is being used. The values are sometimes expressed as percentages resulting from multiplying the dimensionless ratios by 100.

Although water content is a very commonly measured soil parameter, it is not an independent variable in the physical sense. For example, the water content of soil does not indicate how much water is available for plant growth unless we know how tightly the water is held by the soil. The amount of plant-available water in a soil is dependent on the soil water potential. Because the water potential is much more difficult to determine, the water content has often been used as the indicator of the state of water in soil. The water content of soil is a dynamic and changing parameter whose measurement may be needed on a time basis. The geographical or spatial pattern of water content in the field may vary considerably and depend on numerous factors such as texture, structure, position in the landscape, drainage, cropping practice, and so on. The spatial characterization of soil water content over a period of time requires a large number of determinations.

The amount of water in a soil affects directly the growth of crops, microbes, and insects. The strength of soil, which determines root penetration, and the energy requirements for tillage are dependent on water content. Although a variety of techniques are available for measurement of soil water content, they are not all equivalent nor applicable in all situations. Thus, choosing a suitable method or methods and following a well-defined procedure are important for obtaining reproducible and reliable data.

51.2 GRAVIMETRIC METHOD WITH OVEN DRYING

This method involves weighing a moist sample, oven drying it at 105°C, reweighing, and calculating the mass of water lost as a percentage of the mass of the dried soil. The method is apparently straightforward and it is commonly thought to yield absolute results. In fact, this is not so for several reasons. Water is retained by the components of the soil at a wide range of energy levels and there is no magic time at which the soil reaches a "dry" state when maintained at 105°C. Soil samples continue to decrease in mass slowly at 105°C for many days (Gardner 1986). Many soil samples contain organic materials, some of which are volatile at 105°C, so some of the decrease in mass may be due to volatilization of components other than water. In addition, there is the problem of temperature control. The drying ovens in common use in most soil laboratories may maintain the temperature in the range of 100–110°C if they are carefully adjusted. Repeated checking is essential. Temperatures within the oven vary, depending on the location in the oven chamber. The actual temperature of the soil sample is not measured.

In spite of these imperfections, however, the oven-drying method is a commonly used, convenient method to obtain an estimate of soil water content.

51.2.1 APPARATUS AND MATERIALS

1   Soil-sampling tools. Augers, shovels, or sampling tubes may be used, depending on soil conditions and the depth from which samples are to be taken.

2   Tape measure for determining the depth limits of each sample.

3   Rust-resistant, numbered drying tins with tight-fitting, numbered lids. The tins should be 5 cm in diameter and 5 cm high for many applications.

4   A drying oven with a temperature-control device that will maintain a temperature between 100 and 110°C. Forced-air circulating ovens will dry samples more rapidly, but convective ovens are also suitable.

5   Desiccators, each containing active desiccant.

6   A balance, accurate to the nearest 0.01 g.

51.2.2 PROCEDURE

1   Obtain soil samples at the desired depths. It is important to place each sample immediately into an impermeable container and to close it tightly to prevent evaporation of the water. Samples may be placed directly into tared drying tins and enclosed by putting on the tight-fitting lids, or transported to the laboratory in sealed plastic bags from which the moist soil is subsequently transferred to the drying tins.

2　　Weigh the moist sample to the nearest 0.01 g as soon as possible after the sample is placed in the tared drying tin.

3　　Remove the lid from the tin, place it on the bottom of the tin, and place the open tin of moist soil in the oven at the preset temperature of 105 ± 5°C.

4　　Dry the samples until loss of weight with time is very slight, for example, <0.1% in 6 h. Usually, for loose samples with a thickness of a few cm, 24 to 48 h is an adequate drying time in a convective oven. It is important to establish the required drying period for a set of samples. This may be done by monitoring the rate of change of weight of a few representative samples.

5　　Remove the samples from the oven, replace the lid firmly on each drying tin, and place the tins in a desiccator.

6　　After the samples have cooled to ambient room temperature, reweigh the samples to a precision of 0.01 g.

7　　Calculate the water content (mass basis) as a percentage of the mass of dry soil as follows:

$$100 \left[\frac{(\text{mass of moist soil } + \text{ tin}) - (\text{mass of dry soil } + \text{ tin})}{\text{mass of dry soil}} \right] \qquad (51.1)$$

The water content on a volume basis, by percentage, may be calculated by multiplying the above result from Equation 51.1 by the bulk density of the soil.

51.2.3 COMMENTS

1　　For samples of organic soils, drying at 105°C may result in mass losses arising from oxidation and volatilization of organic components (Gardner 1986). There is, however, no magic temperature at which water can be removed without incurring any loss of organic substances. Most of the organic matter in soils is humified, relatively stable material. For general purposes, it is probably best to dry all soil samples at 105°C for measurement of water content.

2　　The gravimetric method has the advantage of simplicity in principle and in the equipment requirements. It has the disadvantage that measurements are not made *in situ* and sampling is destructive of the site. Also, the conversion of water content to a volume basis requires the bulk density as additional information. It is possible to get the volumetric water content directly from the determination of the gravimetric loss of water during drying, provided the soil sampling is done with a sampler of known total volume. The mass change on drying divided by the *in situ* volume of the soil sample gives the volumetric water content directly. Inherent in this approach is the assumption that the density of water is constant at 1 Mg m^{-3}. The mass of the dry soil is available from this measurement sequence and thus the bulk density of the soil also.

51.3 TIME-DOMAIN REFLECTOMETRY

Time-domain reflectometry (TDR) makes use of the unique electrical properties of the water molecule to determine the water content of soil. At radio frequencies, the dielectric constant of water is about 80. Most of the other solid components of soil have dielectric constants in the range 2 to 7, and that of air is effectively 1. Thus, a measure of the dielectric constant of soil is a good measure of the water content of the soil. The TDR technique measures the velocity of propagation of a high-frequency signal. The velocity of propagation is decreased in higher dielectric materials according to

$$v = \frac{c}{\sqrt{K'}} \tag{51.2}$$

where v is the velocity of propagation in the soil, c is the propagation velocity of electrical signals and light in free space (i.e., vacuum) (c = 3×10^8 m $\cdot$ s^{-1}), and K$'$ is the dielectric constant of the soil. TDR, applied to soil water content determinations, is essentially cable radar in which a very fast rise-time voltage pulse is propagated down and reflected back from the end of a transmission line or wave guide in the soil. By determining the travel time t of the step pulse traveling in the transmission line or wave guide of length L, one can get the velocity as L/t. Equation 51.2 can be rearranged to give the apparent dielectric constant as

$$K_a = \left(\frac{ct}{2L}\right)^2 \tag{51.3}$$

The relationship in Equation 51.2 is approximate so that in Equation 51.3 we use K$_a$, the apparent dielectric constant, instead of K$'$. Topp et al. (1980) showed for a variety of soils that the relationship between volumetric water content (θ) and dielectric constant (K$_a$) is essentially independent of soil texture, porosity, and salt content. They proposed a third-degree polynomial for conversion of K$_a$ values to volumetric water content.

$$\theta - 5.03 \times 10^{-2} + 2.92 \times 10^{-2}K_a - 5.5 \times 10^{-4}K_a^2 + 4.3 \times 10^{-6}K_a^3 \tag{51.4}$$

This relationship has been tested by others (Dalton 1992, Zegelin et al. 1992) and found to be satisfactory for most mineral soils and conditions. Recent work by Herkelrath et al. (1991) showed that Equation 51.4 cannot be used for soils having high organic content.

51.3.1 APPARATUS AND MATERIALS

1   The instrument for TDR measurements (Figure 51.1) includes (a) a pulser, giving short rise-time pulses having rise times of the order of <10^{-10} s; (b) a sampling receiver for detecting voltage levels on the output cable after each pulse is sent; (c) a precise timing system which synchronizes all the

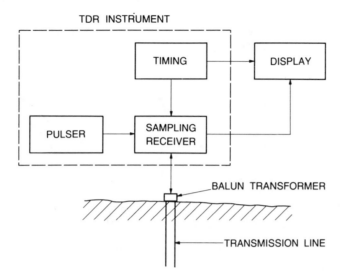

TDR INSTRUMENT

FIGURE 51.1. A block diagram of a TDR instrument and its display unit.

timing for the pulser, the receiver, and the data display; and (d) the display which shows the time and voltage magnitude (usually as reflection coefficient). The step function (voltage pulse) travels past the impedance-matching transformer (balun), and on into the transmission line in the soil that is being measured. The receiver monitors the voltages on the line, detecting the transmitted pulse and any reflections from the transmission line.

The receiver employs an electronic sampling technique to measure the high-frequency signals and produces an audio frequency facsimile as its output for display and analysis. This audio frequency facsimile is generated using a similar principle to that used where strobe lights are used to create a "slow"-motion facsimile, such as in timing automobile ignition.

2 Soil probes act, in the electrical sense, as transmission lines or wave guides for the TDR pulses in the soil. These probes may have either of two configurations (Figure 51.2). The balanced-pair (two-pronged) transmission line requires two prongs or rods in the soil. The simulated coaxial transmission line uses at least three prongs or rods in the soil. A variety of handles or connectors have been used to connect the TDR instrument to the rods in the soil (Dalton 1992, Zegelin et al. 1992, Topp et al. 1984). Dalton (1992) and Zegelin et al. (1992) give a thorough discussion of criteria for construction and installation of TDR probes for both laboratory and field use.

3 Tools for insertion or installation of soil probes.

4 Cables for connecting TDR instrument and soil probes. Cable combinations between the TDR instrument and soil probes is determined by the type of probes used. If one is using a balanced-pair (two-pronged) transmission line, the analysis and interpretation of the TDR waveform is less ambiguous if an impedance-matching transformer (balun) is used. This balun must be located between the 50Ω unbalanced cable from the instrument and the balanced-pair transmission line. Azac model TP 103 is most commonly used

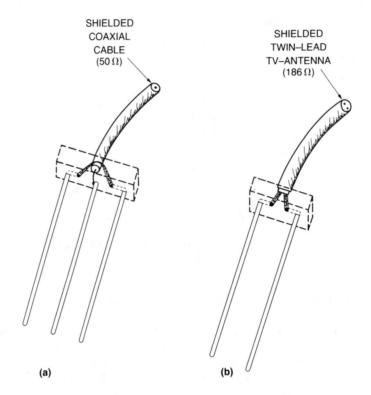

SHIELDED
COAXIAL
CABLE
(50 Ω)

SHIELDED
TWIN–LEAD
TV–ANTENNA
(186 Ω)

(a)   (b)

FIGURE 51.2. TDR transmission line used as soil probes. (a) Simulated coaxial (unbalanced); (b) twin lead (balanced pair).

with 50Ω unbalanced to 200Ω balanced, which is reasonably well matched to 186Ω shielded TV antenna (Belden 9090 and Columbia® 5720 are examples). The two conductors in the shielded TV antenna cable are attached to the two rods in the soil (Figure 51.2). The three or more pronged probes simulate the coaxial line and require no impedance-matching transformer. Thus, the center conductor of the coaxial cable attaches to the center rod and the shield of the cable attaches to the outer two or more rods (Figure 51.2).

5   Switching or multiplexing devices, shown schematically in Figure 51.3 and described by Baker and Allmaras (1990), Heimovaara and Bouten (1990), Herkelrath et al. (1991).

51.3.2 PROCEDURE

1   Install probes of appropriate length and separation in the soil to meet the requirements of the desired measurement. See Dalton (1992) and Zegelin et al. (1992) for limiting criteria.

2   Connect the probes to the multiplexer, then to the TDR instrument or directly to the TDR instrument if no multiplexer is being used (Figures 51.1 and 51.3).

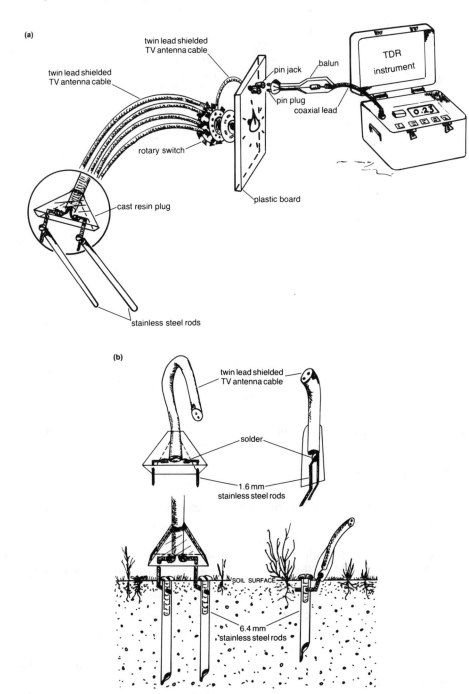

FIGURE 51.3. (a) Details, given schematically, of connections, components, and switching (multiplexing) used for long-term, in-field installations; (b) the connection between the soil TDR line and the shielded TV antenna lead.

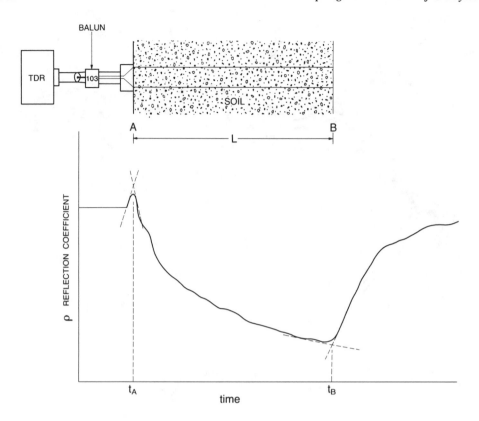

FIGURE 51.4. A typical TDR trace from a twin-lead soil probe as portrayed at the top. The time interval $t_B - t_A$ represents the time of travel of the signal for one return "trip" from A to B and back to A. This time is used in Equation 51.3 to calculate Ka.

3   Initiate TDR signal transmission and measure the travel time, t, as shown in Figure 51.4.

4   Calculate the apparent dielectric constant, Ka, as given by Equation 51.3 and substitute Ka into Equation 51.4 to get the volumetric water content θ.

51.3.3 COMMENTS

1   The empirical relationship between Ka and θ proposed by Topp et al. (1980) has been shown to be largely independent of the texture and porosity of the soil. Soil texture and organic matter content affect the Ka vs. θ relationship slightly so that greater precision of water content may be possible by calibration for a specific soil. Relative or comparative measurements of change of water content generally do not require an improved calibration curve.

2   Currently, considerable effort is being aimed at automatic interpretation of the TDR traces to give water contents directly (Baker and Allmaras 1990, Heimovaara and Bouten 1990, Herkelrath et al. 1991). Automatic interpre-

tations are made by commercial instruments such as in TRASE by Soilmoisture Equipment and IRAMS by CPN and by Campbell Scientific using a Tektronix® cable tester.

3　There is considerable flexibility in design of the probes used in the soil, which allows the measurement of water content to be "tailored" to the experimental or monitoring requirements. The spacing and geometry of the probes can be altered to change the sample size and its orientation (Kachanoski et al. 1990). The simulated coaxial (three or more pronged) probes (Zegelin et al. 1989) offer considerable promise and utilize simpler cables and connections. The choice of soil probe type and orientation depend on the intended use of the resulting data. Topp (1987) discusses the relative merits and associated accuracy limitations of vertically and horizontally oriented probes in the field.

4　It appears that there will be developments and improvements of TDR applications to soil water measurements. These are occurring both in TDR instrumentation and in soil probe design. Currently, the accuracy of measurement of water content is about ± 0.01 m^3 m^{-3} and the precision or repeatability is about ± 0.005 m^3 m^{-3}, under ideal conditions.

5　From the magnitude of the reflected TDR signal it is possible, in theory at least, to calculate the electrical conductivity of the soil (Dalton et al. 1984, Dalton 1987, 1992, Topp et al. 1988, Zegelin et al. 1989). As a result of multiple reflections of the signal in the soil and cables, it is not possible to interpret the TDR signals unequivocally for electrical conductivity. As a consequence, one must use an empirical approach and determine a coefficient for the particular combination of probes and cables being used. Kachanoski et al. (1992) have shown that TDR offers a convenient and straightforward way of measuring solute transport parameters from chemical tracer experiments.

6　The dielectric constant of soil can be measured in ways other than by TDR. Thomas (1966) demonstrated the potential for using capacitance measurement at radio frequencies for determining the soil dielectric constant and thus its water content. Dean et al. (1987) and Bell et al. (1987) have recently reported an improved capacitance technique for use in access tubes in the field, similar to neutron access tubes.

51.4 NEUTRON THERMALIZATION

High-energy neutrons released into soil from a radioactive source are scattered and slowed to the thermal energy level. Hydrogen nuclei, being near the same size and mass as neutrons, serve particularly well in the thermalization process. Although H is a component of organic matter (approximately 5%), the vast majority of the H in soils occurs as a component of water. Thus, a measure of the quantity of thermalized neutrons in the vicinity of a neutron source provides a reasonable basis of estimating the water content of soils. When a fast neutron source is placed in moist soil, the high-energy or fast neutrons are immediately thermalized by a series of collisions with nuclei, such as those of hydrogen, forming a cloud of thermal neutrons. The density of this cloud depends on the rate of emission of fast neutrons, the concentration of hydrogen nuclei (related to water content), and the rate of

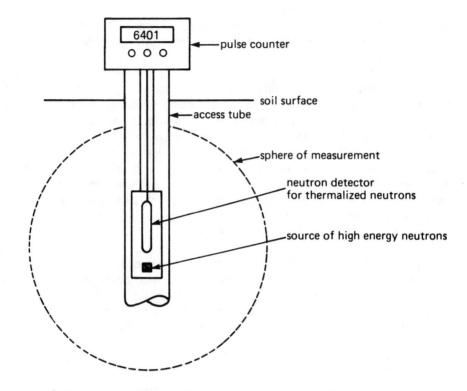

FIGURE 51.5. A schematic representation of the measurement of soil water content using neutron thermalization.

capture of neutrons by absorbing nuclei in the surrounding soil. The concentration of hydrogen nuclei (water content) determines how far the neutrons must travel before undergoing enough collisions to be slowed to the level of thermal energy. Thus, the density of thermal neutrons can be calibrated against the water content on a volume basis.

The presence of a cloud of neutrons around the neutron source restricts the use of this technique to the field or very large bulk samples. The radius of the sphere of neutrons varies with water content from about 0.15 m for saturated soil to over 0.60 m for dry soil. With this large and variable sphere of measurement this method offers only limited spatial resolution of water content. The resolution of changes in water content with time can attain a precision of 0.01 m^3 m^{-3} (Gardner 1986).

51.4.1 APPARATUS AND MATERIALS

1   Neutron moisture depth probe and meter (Figure 51.5). The depth probe or down-hole portion consists of the fast neutron source (usually americium-241/beryllium) and a detector for thermalized neutrons. Protective shielding composed of lead (for gamma rays) and polyethylene or paraffin (for neutron absorption) is part of the box which houses the meter. The depth probe is stored within the shield when not involved in measurement. The shield also serves as a reference material for standardizing the instrument. The meter assembly includes a scaler for registering the counts generated by the ther-

mal neutron detector, a built-in computer, a display, and a keyboard. The entire units are commercially available from several sources to meet a variety of requirements, including combination with gamma ray density units using back scatter of gamma rays.

2   Thin-walled aluminum, steel, or plastic access tubing. Tubing diameter needs to be consistent with the probe size to optimize the performance of the instrument. Manufacturers recommend the sizes appropriate for their instruments. The lengths of access tubing will be dictated by the data requirements of the project.

3   Soil auger for installing access tubing. A slightly smaller diameter than the tubing assures a tight fit of the tubing to the augered hole.

4   Radioisotope license, radioisotope leak test kits, and radiation monitoring instruments as required by the license.

5   Calibration curves or parametric data for setting the calibration in the computer of the instrument, if applicable.

51.4.2 PROCEDURE

1   Install the access tubes in augered holes. As air gaps or other spaces in the vicinity of the access tube have a large impact on the neutron count, much effort is made to prevent their occurrence during augering. It is advisable, therefore, to drill the access hole through the access tubing. If a high water table is probable at the site, a plug should be placed in the bottom of the tube to prevent the entry of water at times of high water table. If installation takes place into a water table situation, special installation procedures must be adopted to remove both water and loosened soil from inside the tube prior to installation of the plug in the bottom. About 0.10 m of tubing is left above the soil surface to allow the placement of the instrument for measurements. Between measurements the top of the access tube is covered with an empty can or stopper to keep out water and debris.

2   Take a reference measurement with the probe in its shield. Place the probe unit on one of the access tubes and select an appropriate counting time as specified by the manufacturer. Initiate the counting process. Several standard counts or reference measurements, taken initially, will indicate the stability of the instrument and repeatability of the measurements. The reference count can be used to remove the effects of instrument drift from each subsequent measurement until another reference count is made. If one assumes that the drift is monotonic, the amount of drift between reference measurements can be proportioned over the time interval between readings on the reference. The appropriate portion of the drift can then be assigned or removed from the individual measurement for its correction.

3   Make measurements by lowering the probe into the desired access tube to the selected depth. Initiate the counting of thermalized neutrons. Make one

or more counts at each selected depth in the soil access tube. As the zone of measurement is spherical and >0.15 m in radius, depth increments for measurement should be no <0.15 m. The approach to the soil surface should be no closer than 0.15 m.

4   Convert count ratios (count in soil/count in standard shield) to volumetric water content from the calibration curve or from the internal computation of the instrument.

51.4.3 COMMENTS

1   The manufacturer's calibration curve is often adequate for many practical applications. Site-specific calibration may be required where soil has unusual neutron-absorption properties or if the diameter or material of the access tubing is different from the manufacturer's recommendations. The large and varying zone of influence of the thermalized neutrons limits the depth resolution to greater than 0.15 m and results in low reliability in layered soil and in soils where soil water gradients are steep.

2   In recent years, concern has increased in regard to the radiation hazard associated with using radioactive sources. In addition, regulations concerning the use of radioactive devices have increased. Although we admit that there is always some risk associated with the use of radioactive devices, we are of the opinion that their use according to the manufacturer's instructions and in compliance with the radioactive source license constitutes a safe procedure. The neutron-moderation method for the measurement of soil water content is considered to be useful and reliable.

51.5 GAMMA RAY ATTENUATION

When gamma rays pass through soil, some of them are absorbed by the soil particles and the water in the path of the rays. The degree to which a beam of monoenergetic gamma rays is attenuated or reduced in intensity in passing through a soil column depends on the constituent elements and the density of the soil. If the soil solids do not move, i.e., if the soil bulk density remains constant, then changes in attenuation can be attributed to changes in water content.

The attenuation of gamma rays makes use of a radioactive source of gamma rays on one side of the measured soil sample and the radiation detector on the opposite side. The gamma rays pass through soil, water, air, and the container walls. The attenuation equation, neglecting air, is

$$I = I_0 \exp[-(\mu_s \rho_b + \mu_w \theta_v)d_s - (\mu_c \rho_c)2d_c] \tag{51.5}$$

where I is the intensity of the gamma ray beam measured at the detector, I_0 is the intensity of gamma rays measured by the detector with nothing in the beam, μ_s is the attenuation coefficient of the soil solids, ρ_b is the bulk density of the soil, μ_w is the attenuation coefficient of water, θ_v is the volumetric water content, d_s is the thickness of the soil in the direction of the beam of gamma rays, μ_c is the attenuation coefficient of the container material, and d_c is the wall thickness of the container ($2d_c$ represents the total path length of the gamma ray beam in the container material).

For dry soil the equation becomes

$$I_d = I_0 \exp[-(\mu_s\rho_b)d_s - (\mu_c\rho_c)2d_c] \tag{51.6}$$

where I_d is the gamma ray intensity measured for dry soil. Division of Equation 51.5 by Equation 51.6 gives

$$\frac{I}{I_d} = \exp[(-\mu_w\theta_w)d_s] \tag{51.7}$$

θ_v will result from Equation 51.7, by calculation, for each measured value of I when values of I_d, μ_w, and d_s have been determined for the soil and experimental arrangement. Although μ_w can be obtained from published tables, it is preferable to measure the absorption coefficient using the same physical arrangement as that used during the measurements.

51.5.1 APPARATUS AND MATERIALS

1   Source of gamma radiation. Cesium-137, which emits gamma rays of energy 0.662 MeV and has a 30-year half-life, and [241]Am at 0.060 MeV and 470 year half-life, are both well suited for water content measurement. The size of the source depends on the use to be made of the measurement system. The reader is referred to Gardner (1986) for a discussion of source selection and to Chang et al. (1990) for a description of a recently reported complete system.

2   Radiation shielding and beam collimation. The radioactive source is retained within lead shielding to prevent irradiation of the operator and to achieve beam collimation. The optimum measurement relies on a collimated beam of gamma rays (Figure 51.6) streaming out from the source and trained on the surface of the soil being measured. The aim of collimation is to obtain a beam of gamma rays that passes through the soil and reaches the radiation detector as approximately parallel rays. Gardner (1986) discusses extensively the collimation requirements.

3   Radioisotope license, radioisotope leak test kits, and radiation monitoring instruments as required by the license. The radioisotope license will specify the minimum shielding requirement for the chosen source strength.

4   Gamma-sensitive detector probe. The most commonly used probe consists of a scintillation crystal, a photomultiplier tube, and a preamplifier. A thallium-activated sodium iodide crystal (NaI-Th), 2.5 cm thick in the direction of the gamma ray beam, provides a good, general-purpose scintillator from which light pulses are captured by a 5-cm phototube and preamplifier. A collimator slit is usually placed in front of the scintillation crystal and lead shielding surrounds the phototube assembly (Figure 51.6).

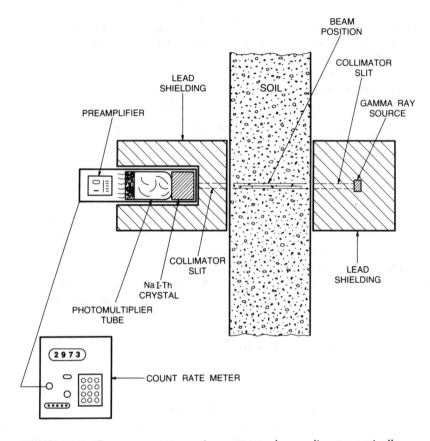

FIGURE 51.6. Gamma ray-attenuation system, shown diagrammatically.

5   Soil container. The soil container shape and size will depend on the intent of the experiment. Gamma ray measurements are usually made at a number of positions within the soil and are usually achieved by a mechanical scanning device, as is described below. The optimum thickness of soil in the direction of the beam of gamma rays depends on the energy of the incident gamma rays (Singh and Chandra 1977, Gardner 1986, Mudahar and Sahota 1985). For ^{137}Cs, soil thickness may range over 10 to 35 cm, and for ^{241}Am from 2 to 8 cm. The walls of the soil container should be as thin as practical and of low-density material so as to absorb a minimum of the gamma rays. The walls must, however, hold the soil at a uniform thickness.

6   Mechanism for scanning the soil container. Initially, the scanning of soil was one dimensional along a soil column (Gurr 1962, Davidson et al. 1963, De Swart and Groenevelt 1971, Goit et al. 1978). More recently, the scanning is in two dimensions and automatically controlled (Chang et al. 1990, Hopmans and Dane 1986). The choice of mechanism for scanning will depend on the requirements of the experiment.

7   Scaler or ratemeter. A number of different scalers and ratemeters are available. The main requirements include an amplifier-discriminator which allows selecting energy levels so that only primary or direct gamma rays are counted. The time resolution should be at least 1 μs or less. The desirable complete

system includes a programmable automatic controller for timing of the scanning of the soil container and a data acquisition system. The data manipulation and calculation possibilities give output as water content values with time and at selected positions over the soil container (Chang et al. 1990, Hopmans and Dane 1986).

51.5.2 PROCEDURE

1   Determine the attenuation coefficients of water, soil, and container walls. This is achieved by measuring the gamma ray counts with the soil container (box or cylinder) empty, then filled with dry soil packed to a known bulk density, and finally with the container filled with water only. As shown by Equation 51.7 it is sufficient, in practice, to measure I_d which combines soil and container in one measurement. For blocks or columns of soil taken with minimal disturbance it will be necessary to measure the attenuation coefficient of soil and container at known water contents different from dry. Usually, this will be achieved at the end of an experiment. In addition, the attenuation must be measured at each measurement location because the bulk density is generally variable.

2   Make measurements of I at the number of locations and times as dictated by the experiment.

3   Convert I values to volumetric water content from calculation using Equation 51.7.

51.5.3 COMMENTS

1   The use of a single-source gamma ray system for measurement of water content depends upon the assumption that bulk density remains constant when water content changes. In most soils, the bulk density changes as the soil water content changes. It is possible to measure both water content and bulk density simultaneously by using two sources of gamma rays, having gamma rays at different energy levels (Gardner 1986, Nofziger and Swartzendruber 1974, Gardner et al. 1972). Thus, the dual-gamma technique improves the precision of water content measurement as compared to that when a constant bulk density is assumed (Gardner 1986).

2   The use of gamma rays requires the operator(s) of the equipment to comply with the radioisotope license. There is, however, some inevitable exposure to gamma rays.

3   The achievable precision for measuring water content with gamma rays depends on a number of soil factors, the source strength, the counting time, and so on. Gardner (1986) has given detailed error analysis using a single gamma source. His error analysis contains useful and usable information

for designing gamma ray systems, selecting source strengths and their energy levels, and choosing soil thickness.

REFERENCES

Baker, J. M. and Allmaras, R. R. 1990. System for automating and multiplexing soil moisture measurement by time domain reflectometry. Soil Sci. Soc. Am. J. 54: 1–6.

Bell, J. P., Dean, T. J., and Hodnett, M. G. 1987. Soil moisture measurement by an improved capacitance technique. Part II. Field techniques, evaluation and calibration. J. Hydrol. 93: 79–90.

Chang, C., Pang, D. W., and Sommerfeldt, T. G. 1990. Automated laboratory gamma-ray soil density measurement system. Can. Agric. Eng. 33: 027–030.

Dalton, F. N. 1987. Measurement of soil water content and electrical conductivity using time-domain reflectometry. Proceedings: International Conference on Measurement of Soil and Plant Water Status. Logan, UT. 1: 95–98.

Dalton, F. N. 1992. Development of time-domain reflectometry for measuring soil water content and bulk soil electrical conductivity. In G. C. Topp et al., Eds. Advances in measurement of soil physical properties: bringing theory into practice. Spec. Publ. No. 30, Soil Science Society of America, Madison, WI. 143–167.

Dalton, F. N., Herkelrath, W. N., Rawlins, D. S., and Rhoades, J. D. 1984. Time domain reflectometry: simultaneous measurement of soil-water content and electrical conductivity with a single probe. Science 224: 989–990.

Davidson, J. M., Biggar, J. W., and Nielsen, D. R. 1963. Gamma-radiation attenuation for measuring bulk density and transient water flow in porous materials. J. Geophys. Res. 68: 4777–4783.

Dean, T. J., Bell, J. P., and Baty, A. J. B. 1987. Soil moisture measurement by an improved capacitance technique. Part I. Sensor design and performance. J. Hydrol. 93: 67–78.

De Swart, J. G. and Groenevelt, P. H. 1971. Column scanning with 60 keV gamma radiation. Soil Sci. 112: 419–424.

Gardner, W. H. 1986. Water content. Pages 493–544 in A. Klute, Ed. Methods of soil analysis. Part 1. Physical and mineralogical methods. Agronomy No. 9, 2nd ed. American Society of Agronomy, Soil Science Society of America, Madison, WI.

Gardner, W. H., Campbell, G. S., and Calissendorff, C. 1972. Systematic and random errors in dual gamma energy soil bulk density and water content measurements. Soil Sci. Soc. Am. Proc. 36: 393–398.

Goit, J. B., Groenevelt, P. H., Kay, B. D., and Loch, J. P. 1978. The applicability of dual gamma scanning to freezing soils and the problem of stratification. Soil Sci. Soc. Am. J. 42: 858–863.

Gurr, C. G. 1962. Use of gamma rays in measuring water content and permeability in unsaturated columns of soil. Soil Sci. 94: 224–229.

Heimovaara, T. J. and Bouten, W. 1990. A computer-controlled 36-channel time domain reflectometry system for monitoring soil-water contents. Water Resour. Res. 26: 2311–2316.

Herkelrath, W. N., Hamburg, S. P., and Murphy, F. 1991. Automatic, real-time monitoring of soil moisture in a remote field area with time domain reflectometry. Water Resour. Res. 27: 857–864.

Hopmans, J. W. and Dane, J. H. 1986. Calibration of a dual-energy gamma radiation system for multiple point measurements in soil. Water Resour. Res. 22: 1109–1114.

Kachanoski, R. G., Van Wesenbeeck, I. J., Von Bertoldi, P., Ward, A., and Hamlen, C. 1990. Measurement of soil-water content during axial-symmetric water flow. Soil Sci. Soc. Am. J. 54: 645–649.

Kachanoski, R. G., Pringle, E., and Ward, A. 1992. Measurement of field solute travel times using time domain reflectometry. Soil Sci. Soc. Am. J. 56: 47–52.

Mudahar, G. S. and Sahota, H. S. 1985. Optimal thickness of soil between source and detector for different gamma-ray energies. J. Hydrol. 80: 265–269.

Nofziger, D. L. and Swartzendruber, D. 1974. Material content of binary physical mixtures as measured with dual-energy beam of gamma rays. J. Appl. Phys. 45: 5443–5449.

Singh, B. P. and Chandra, S. 1977. Evaluation of the optimal thickness of soil between source and detector in the gamma-ray attenuation method. J. Hydrol. 32: 189–191.

Thomas, A. M. 1966. In situ measurement of moisture in soil and similar substances by fringe capacitance. J. Sci. Instrum. 43: 21–27.

Topp, G. C. 1987. The application of time-domain reflectometry (TDR) to soil water content measurement. Proceedings: International Conference on Measurement of Soil and Plant Water Status. Logan, UT. 1: 85–93.

Topp, G. C., Davis, J. L., and Annan, A. P. 1980. Electromagnetic determination of soil water content: measurement in coaxial transmission lines. Water Resour. Res. 16: 574–582.

Topp, G. C., Davis, J. L., Bailey, W. G., and Zebchuk, W. D. 1984. The measurement of soil water content using a portable TDR hand probe. Can. J. Soil Sci. 64: 313–321.

Topp, G. C., Yanuka, M., Zebchuk, W. D., and Zegelin, S. J. 1988. Determination of electrical conductivity using time domain reflectometry. Water Resour. Res. 24: 945–952.

Zegelin, S. J., White, I., and Jenkins, D. R. 1989. Improved field probes for soil-water content and electrical conductivity measurement using time domain reflectometry. Water Resour. Res. 25: 2367–2376.

Zegelin, S. J., White, I., and Russell, G. F. 1992. A critique of the time domain reflectometry technique for determining field soil-water content. G. C. Topp et al., Eds. in Advances in measurement of soil physical properties: bringing theory into practice. Spec. Publ. No. 30, Soil Science Society of America, Madison, WI. 187–208.

Chapter 52
Soil Water Potential

N. J. Livingston

University of Victoria
Victoria, British Columbia, Canada

52.1 INTRODUCTION

There are numerous techniques to measure soil water potential. Generally they involve (1) the measurement of some property of the soil that changes with changes in water potential or, more commonly, (2) the measurement of some property of the water in a reference medium that is in thermodynamic equilibration of the liquid or gas phase of the water in the soil. In the latter case, a critical assumption must be made at some stage during the measurement process that a thermodynamic equilibration has been reached between the sensor and the surrounding medium. Significant measurement errors can and frequently do occur if such an equilibration has not been reached. Good measurement technique and instrument design, therefore, should primarily focus on ensuring that there is good hydraulic and thermal contact and exchange between the sensor and the surrounding soil.

It should be appreciated that there can be enormous spatial variability in soil water potential and that a representative field average might require a very large number of measurements. It is likely that inhomogeneities in soil water, in many cases, will exceed errors inherent in particular measurement techniques.

Soil water potential can be determined indirectly by recourse to measurements of soil water content and soil water release or soil moisture characteristic curves that relate volumetric or gravimetric water content to soil water potential. The methodology used to generate water release curves is discussed in Chapter 53. Alternatively, an average root zone water potential can be estimated from measurements of predawn xylem potential made with a pressure chamber. This chapter will focus on direct measurements of soil water potential. It will not provide detailed descriptions of the many instruments and techniques available. Such descriptions can be found in Richards (1965), Brown and Van Haveren (1972), and Hanks and Brown (1987). Instead, only a brief overview of some of the more important techniques will be provided. However, practical considerations will be emphasized.

52.1.1 Definitions

The measurement of water potential is widely accepted as fundamental to quantifying both the water status in various media and the energetics of water movement in the soil-plant-atmospheric continuum.

Soil Sampling and Methods of Analysis, M. R. Carter, Ed.,
Canadian Society of Soil Science. © 1993 Lewis Publishers.

Water potential (ψ_w) is classically defined as the amount of work to transport, isothermally and reversibly, an infinitesimal quantity of water from a pool of pure water, at a specified pressure and elevation, to the system under condition. For most purposes it is more convenient to consider the amount of work required to remove water from the system. This is, in effect, equal to all the various forces holding on to the water. In many studies it is desirable to consider the separate effects of these forces. Unfortunately, this is difficult because some of these forces are not mutually independent.

Another commonly used term that is frequently confused with water potential is the chemical potential of water (μ_w). Again, this term indicates the capacity of water to do work and is equal to its partial molal Gibbs free energy in units of energy per mole. Potentials are referenced to a standard state (μ_w^*) that by convention is set to zero. In this standard state water is pure and free from any external forces. Water potential is simply the chemical potential divided by the partial molal volume of water (V_w; m^3 mol^{-1}) so that:

$$\psi_w = (\mu_w - \mu_w^*)/V_w = J\ mol^{-1}/m^3 mol^{-1} \tag{52.1}$$

52.1.2 Units

Water potential can also be described in terms of energy per unit mass of water ($J\ kg^{-1}$) or energy per unit volume of water ($J\ m^{-3}$). Since $J = N\ m$, water potential may be expressed in terms of force per unit area or pressure which has units of pascals or megapascals (1 MPa $= 10^6$ Pa $= 10$ bars). The use of energy per unit volume can cause some confusion because of the tendency to think of water potentials, expressed in pressure terms, as being pressures that can be used to calculate rates of flow. This is incorrect because not all the components of water potential are equally effective in causing flow. For example, a semipermeable membrane is required before flow can be induced by gradients in osmotic potential. In soil physics, water potential is often equated to the height of a column of pure water exerting pressure on a surface, where 100 kPa or 1 bar is equivalent to the pressure exerted by a column of water 1000 cm high. This notation is known as hydraulic head.

52.1.3 Components of Water Potential

The total water potential (ψ_t) of a system is the sum of the component water potentials so that:

$$\psi_w = \psi_t = \psi_m + \psi_\pi + \psi_p + \psi_g \tag{52.2}$$

where ψ is the potential energy per unit mass, volume, or weight of water, and the subscripts, m, π, p, and g are for matric, osmotic, pressure, and gravitational components, respectively. In soil, the gravitational potential energy per unit mass of soil is the gravitational constant multiplied by the vertical distance to the reference position, and the pressure potential, if present, is the gravitational constant multiplied by the distance from the point of measurement to the free water surface above it. Therefore, the only instrument required to measure these two potentials is a ruler. In some cases a piezometer tube is used to determine the level of the free water surface.

The term matric potential is often used to describe the potential arising from the attraction of the soil matrix for water, that is, the forces binding water to structural elements or colloids. Campbell (1987) reviews the various means of measuring matric potential and the type of sensors employed. It is strongly recommended that readers refer to Passioura (1980) for a more detailed discussion of the meaning of matric potential.

The osmotic potential in soil arises from dissolved solutes and lowered activity of water attributable to interaction with charged surfaces. A reasonable estimate of osmotic potential can be derived from measurements of electrical conductivity corrected for water content (Gupta and Hanks 1972). A more reliable measure can be obtained by extracting soil solution from the soil and measuring the potential directly in a thermocouple psychrometer (see Section 52.2.3). Campbell (1987) describes an instrument that combines a pressure chamber and thermocouple psychrometer that provides the most reliable measurements.

52.2 MEASUREMENT DEVICES

52.2.1 Resistance Blocks

The use of resistance blocks is a relatively inexpensive means of providing a continuous measure of the matric potential of soil as an analog of total soil water potential. Typically, resistance blocks are composed of a matrix of porous material, for example, gypsum, fiberglass or nylon. Each block contains two embedded electrodes. When the matrix comes to equilibration with the water content of the surrounding soil its electrical conductivity (or resistance) is determined by passing a fixed alternating current (AC) voltage (typically about 0.5 V) through the block. The voltage can be induced and measured manually, or, alternatively, automated measurements can be made by a number of commercially available data loggers, for example, the Campbell Scientific 21X (Campbell Scientific Canada Inc., Edmonton, Alberta).

The electrical conductivity of a dry block (such as gypsum) is zero. However, when permeated with water containing electrolytes, the conductivity of the block should equal that of the surrounding soil. The conductivity of the block increases with increasing water content, which is a function of how well the block competes with the soil for water. Commercial gypsum blocks permit a small amount of calcium sulfate (a weak electrolyte) to go into solution with the permeating water, thereby creating a relatively stable conducting medium. Blocks do not work well in saline soils or in soils where saline irrigation water is applied. There is often significant variability between sensors, so each block should be individually calibrated to develop the logarithmic curve that relates soil water potential to conductivity or resistance. Calibration should be conducted on the same soil in which field or laboratory measurements are to be taken. Preferably, blocks should be calibrated on a pressure plate apparatus (see Chapter 53). In doing so, conductance, water content, and water potential can be measured simultaneously. Blocks display considerable hysteresis, so there could be significant differences between calibrations performed during wetting and drying cycles. Blocks can degrade over time and sometimes become subject to peculiar drainage patterns in which water flows directly into cracks in the block. It is imperative, therefore, that blocks be recalibrated at regular intervals (approximately 3 months). Resistance blocks work best in soils drier than -0.05 MPa, so it is recommended that alternative devices (i.e., tensiometers) be used in wet soils. Additionally, resistance blocks react relatively slowly to changes in soil water and therefore should not be used to follow wetting fronts. It has been reported that in some highly porous soils blocks do not equilibrate well with the soil, so they remain wet as the surrounding soil dries. In such cases measurement errors can be very large. Carlson and El Salam (1987) report that significant measurement errors arise, particularly in wet soils, if blocks are subjected to large variations in temperature.

Considerable care should be taken when installing blocks that there is the best possible hydraulic contact between the sensor and surrounding soil. It is recommended that before installation blocks be soaked for at least 24 h in a slurry made up of the field or laboratory soil and distilled water. Precautions, similar to those outlined in Chapter 64, should be made to minimize soil disturbance during sensor installation and data collection. Soil compaction or damage to vegetation can lead to dramatic changes in soil water balance. Provided resistance blocks have been carefully calibrated and properly installed, they can provide a continuous and automated measure of soil water potential (and volumetric water content, if the appropriate soil water release curve has been determined) in dry soils. Errors are quite variable and range from ± 0.1 to 0.5 MPa.

52.2.2 Tensiometers

Tensiometers have been widely and successfully used for many years to measure soil water availability in soils where soil water potentials range from 0 to -0.08 MPa. They are relatively inexpensive, simple, and easy to install and are particularly suited to studies where large numbers of measurements may be required.

Typically, tensiometers consist of a porous ceramic cup cemented to a PVC tube. The tops of the tubes are sealed with rubber or septum stoppers (a detailed list of the materials required and the procedures necessary to construct tensiometers is available from Soil Measurement Systems, 7344 North Oracle Road, Suite 170, Tucson, AZ 85704).

The porous cup allows water and solutes to pass freely between an internal solution and the soil, but prevents the passage of air. The hydrostatic pressure within the tensiometer cup equilibrates with the soil water potential, which is measured with a manometer, a pressure gauge, or a pressure transducer (Tensicorder, Soil Measurement Systems). Tensiometers are insensitive to soil solution osmotic potential and therefore cannot be used to measure water potential in saline soils.

Tensiometers have a restricted measurement range because tensions in the internal water column lead to the formation of air bubbles that allow most of the contained water to flow out, thereby permitting air entry. Once air enters the tensiometer tube, measurements are no longer valid. An additional problem is caused by the movement of water out of the tensiometer into adjacent soil, leading to an increase in soil water potential. This is a function of the hydraulic conductivity of both the tensiometer cup and the surrounding soil. The maximum depth to which a conventional tensiometer can be usefully inserted is about 4 m. This is because the absolute pressure (P) inside the tensiometer at a height (h) above the tensiometer cup is given by:

$$P = A - \psi - h \tag{52.3}$$

where A is atmospheric pressure. In this example all terms are in units of hydraulic head (m). For a tensiometer with a cup 3 m below ground and about 0.5 m of the instrument above ground, and assuming an atmospheric pressure of 10 m, the lowest pressure in the system will be about 6.5 m H_2O (0.065 MPa). For deeper installation depths this limit is reduced, that is, the range available for the matric potential component decreases linearly with depth. It is recommended that for measurements at depth, mercury manometers not be used and that, alternatively, short column tensiometers that use pressure transducers be installed at the relevant depth.

Installation of tensiometers is generally straightforward, but care should be taken in rocky or stony soils because tensiometer cups are quite fragile. A simple installation procedure is to bore a hole with an auger with the same outside diameter as the tensiometer tubing. Soil at the bottom of the hole should be removed, sieved to remove any large stones, mixed with distilled water, and then repacked. Alternatively, an aluminum access tube can first be installed and then the tensiometer inserted into the access tube with its lower end (the porous cup) protruding about 0.1–0.2 m below the access tube into the soil.

Above-ground tubing should be kept to a minimum and should be shielded to avoid radiative heating. Readings are best taken early in the morning to minimize temperature effects. It is useful to incorporate a short section of clear plastic tubing at the upper end of the tensiometer if pressure transducers are to be used (see Marthaler et al. [1983] for a full description of field-installed tensiometers equipped with pressure gauges). Small-diameter (3 mm O.D.) nylon tubing should be used to connect the water (deaerated) column to mercury manometers.

Tensiometers can provide accurate and reliable measurements of soil water potential in moist soils. Sensitivity can be as high as 0.1 kPa. Field-installed tensiometers with mercury manometers, while being inexpensive to build and providing high sensitivity, require careful maintenance. Tensiometers equipped with pressure gauges also require some maintenance and their cost is high if large numbers are used. Tensiometers equipped with pressure transducers provide an excellent alternative. Portable transducers can be used to monitor large numbers of tensiometers manually, or an automated system can be used whereby a single transducer is used to monitor several tensiometers through the use of a switching valve and an automated recording system.

52.2.3 Thermocouple Psychrometers

Thermocouple psychrometers ideally should provide an excellent compliment to tensiometers in that they operate in a range of water potentials with an upper limit of approximately -0.1 MPa. There are a number of systems that allow automated or manual *in situ* measurements of water potential with soil psychrometers (Wescor, Inc., Logan, UT). Alternatively, there are systems in which soil samples are placed in sample chambers and the water potential is determined after an approximately 15-min equilibration time (Decagon Devices, Inc., Pullman, WA).

Total water potential is determined by measurement of relative vapor pressure of air that is in equilibration with the soil pores. The relation between water potential and relative vapor pressure is described by the equation:

$$\psi_t = (RT/V_w) \ln (e/e_s) \tag{52.4}$$

where R is the universal gas constant (J mol^{-1} °K^{-1}), T is the absolute temperature (°K), e is the vapor pressure of the air (Pa), and e_s is the saturation vapor pressure (Pa) at the air temperature. The ratio e/e_s, the relative humidity, is dimensionless.

Rawlins and Dalton (1967), Lang (1968), and Weibe et al. (1971) have described devices that can be used to measure soil water potential *in situ*. The later device was commercially produced (Wescor, Inc., Logan, UT). A soil psychrometer usually consists of a small ceramic cup (about 1 cm wide and long) that contains a single thermocouple (50–100 μm in diameter). The ceramic cup allows water vapor to diffuse between the soil and the inside of the cup until a vapor equilibration is established. The sensing junction of the thermocouple is constructed of very fine welded chromal and constantan wires. The other junctions (the reference junction) are connected to massive (>0.40 mm diameter) copper wires. The open end of the porous cup is sealed with a Teflon® plug. In some psychrometers (Szietz 1975) another thermocouple is embedded in this region to provide a measure of the absolute temperature.

54.2.3.1 *Psychrometric Mode of Operation*

When an appropriate current is applied to the sensing junction in a thermally equilibrated system, the junction cools by the Peltier effect. With continued cooling the junction temperature falls below the dew point so that a drop of water forms. The maximum cooling is about 5°C below the ambient temperature. When the soil water potential is less than 0 MPa, then the relative humidity inside a porous cup, in equilibration with the soil, will be less than 100% and water will evaporate from the condensed drop and cool the junction to a temperature that can be related to the relative humidity in the cup. This decline in temperature can be measured with a voltmeter that has microvolt or nanovolt sensitivity. Typically, the output voltage is about 0.5 μV MPa^{-1}. The temperature of the psychrometer is also measured so that the soil water potential can be calculated from Equation 52.4.

Calibrations between psychrometer output and water potential can be obtained by the immersion of the psychrometer chamber in a range of salt solutions with varying potentials. The water potentials of these solutions, at varying temperatures, can be calculated using the equations and coefficients provided by Lang (1967). These calibrations should be carried out in a well-controlled water bath. Excellent temperature control is critical.

Very small temperature depressions are generated at the wet sensing junction during measurements, so any temperature gradients between the sensing and reference junction or between the evaporating surface and the sensing junction will lead to large errors. For example, a 0.001°C temperature difference corresponds to a 10-kPa (0.1-bar) error. It is therefore imperative that there are no temperature gradients across the sensor or leads. This tends to preclude the use of soil psychrometers near the soil surface where there are extreme gradients in temperature.

There have been numerous attempts to design thermocouple psychrometers that measure soil water potential in the presence of temperature gradients. Campbell (1979) describes a psychrometer constructed from high-thermal conductivity materials to minimize temperature gradients within it. To minimize additional errors due to condensation, the psychrometer is symmetrical with respect to ceramic windows that allow vapor exchange between the chamber and the soil. Measurement errors due to temperature gradients were almost a third of those from the best of previous designs.

54.2.3.2 *Dew Point Mode of Operation*

Many psychrometers can be used in a dew point mode, whereby there are continuous feedback dew point measurements (Neumann and Thurtell 1972). Peltier cooling is again used to condense a water droplet on the sensing junction. However, the cooling current is continuously adjusted so that there is no net gain or loss of water vapor. In unsaturated conditions, the dew point temperature will be below ambient, and this temperature difference will be measured as a differential voltage between the reference and sensing junctions when no current is flowing.

Dew point hygrometers, while still sensitive to thermal gradients, are much less sensitive than psychrometers to changes in ambient temperature and therefore require no temperature correction. In addition they provide a relatively large signal with a sensitivity of approximately 0.75 μV MPa^{-1}, about 50% greater than that provided by psychrometers. Additionally, since there is no net movement of water from the thermocouple junction to the chamber at the dew point, the signal is stable for a long period and the vapor equilibrium in the chamber is not disturbed.

54.2.3.3 *Installation and Maintenance*

Whether using dew point hygrometers or psychrometers, it is very important to ensure that there are no leaks or contamination of thermocouples. Deposits of hydroscopic materials on chamber surfaces dramatically delay vapor equilibration, so it is critical to keep all the internal surfaces of the devices very clean. Furthermore, it is imperative that there is a uniform rate of evaporation from the thermocouple junction and a uniform concentration of water vapor throughout the chamber. Salt deposits on the junction change the rate of evaporation and therefore invalidate the calibration. Ceramic cups that surround the thermocouple generally have a pore diameter of 3 μm, while stainless steel screens have pores approximately ten times as large. These shields protect a thermocouple from soil particles, but allow the movement of dissolved salts which can then precipitate out on the inner walls of the chamber or on the thermocouple itself.

Psychrometers fabricated from stainless steel are more resistant to corrosion, but are generally more expensive than those made from brass. Corrosion resistance of brass psychrometers can be improved quite considerably by chrome or nickel plating. In saline soils psychrometers should be checked regularly for corrosion of the fine thermocouple wires. Soil psychrometers can often be cleaned, without disassembly, by running water over the units for several hours. Most commercial units are relatively easy to take apart for more thorough cleaning. Thermocouples and their mounts can be cleaned by immersing them in steam or solvents, such as reagent-grade acetone or 10% ammonium hydroxide solution. Parts should be thoroughly rinsed in distilled or deionized water. Wescor strongly recommends that the water used for rinsing have a resistance of at least 1 MΩ cm^{-3}. Psychrometers should be dried by blowing with clean air.

Savage et al. (1987) recommended that the screen cage covers used in commercially available soil psychrometers be removed and soaked in a 10:1 mixture of water and hydrochloric acid to remove any traces of rust. This is particularly important if the devices were calibrated in salt solutions. The screens should then be soaked in acetone to reduce the possibility of fungal growth.

To minimize thermal gradients, psychrometers should be installed with the axis of the sensor parallel to the soil surface. Extreme caution should be taken if sensors are placed near the soil surface where they can experience large temperature gradients. Measurements at depths shallower than 15 to 30 cm should not be attempted with *in situ* methods. To reduce heat conduction along lead wires, at least two loops of wire (about 4 cm long) should be wrapped up behind the psychrometer sensing head and buried at the same depth. Rundel and Jarrell (1989) recommend that if continuous data records are required, extra psychrometers should be installed to replace those that fail or are removed for calibration checks. Installation of psychrometers in pots often leads to difficulties because of large temperature gradients which are difficult to control.

As with all types of sensors, psychrometers should be installed with the minimum amount of soil disturbance. Despite the potential disadvantages of using psychrometers in nonisothermal soils, they do offer the advantage of being relatively easy to install and providing repeated measurements, if properly maintained and calibrated, over a wide range of water potentials, at a point source without redisturbing the soil.

52.2.4 Filter Paper Method

Greacen et al. (1987) have provided an extensive evaluation of the filter paper method for measuring soil matric potential. In this technique, filter paper is wrapped around a plastic cone which is then mounted on a wooden dowel of appropriate length which is pushed into the soil. The filter paper is left for about 6 d to equilibrate and then withdrawn and weighed. The soil matric potential is determined by reference to a water release curve derived in the laboratory for the specific type of filter paper used.

The filter paper method, when used properly, offers a simple and inexpensive method for measuring soil matric potential over the range -0.05 to 0.1 MPa. It is only useful, because of the slow equilibration times, in static or slowly changing conditions.

52.3 CONCLUSIONS

The selection of the most appropriate device or technique for a particular experimental design depends primarily on the researcher having a clear understanding of the measurement and data requirements. If precise measurements, confined to wet soils within the 0 to -0.08 MPa range, are required, then tensiometers are highly recommended. If there is a need for measurements to be made over a greater range, then the tensiometers should be supplemented with thermocouple psychrometers. Great care should be taken with the latter devices, since very large errors can occur if they are incorrectly calibrated or poorly maintained.

REFERENCES

Brown, R. W. and Van Haveren, B. P. 1972. Psychrometry in water relations research. Proceedings of the symposium on thermocouple psychrometers. Utah Agricultural Experimental Station, Logan, UT.

Campbell, G. S. 1979. Improved thermocouple psychrometers for measurement of soil water potential in a temperature gradient. J. Phys. E. 12: 739–743.

Campbell, G. S. 1987. Soil water potential measurement. Pages 115–119 in Proceedings of International Conference on Measurement of Soil and Plant Water Status, Vol. 1. Logan, UT, July 1987.

Carlson, T. N. and El Salam, J. 1987. Measurement of soil moisture using gypsum blocks. Pages

193–200 in Proceedings of International Conference on Measurement of Soil and Plant Water Status, Vol. 1. Logan, UT, July 1987.

Hanks, R. J. and Brown, R. W., Eds. 1987. Pages 115–119 in Proceedings of International Conference on Measurement of Soil and Plant Water Status, Vol. 1. Logan, UT, July 1987.

Greacen, E. L., Walker, G. R., and Cook, P. G. 1987. Evaluation of the filter paper method for measuring soil water suction. Pages 137–144 in Proceedings of International Conference on Measurement of Soil and Plant Water Status, Vol. 1. Logan, UT, July 1987.

Gupta, S. C. and Hanks, R. J. 1972. Influence of water content of electrical conductivity of the soil. Soil Sci. Soc. Am. Proc. 36: 855–857.

Lang, A. R. G. 1967. Osmotic coefficients and water potentials of sodium chloride solutions from 0 to 40°C. Aust. J. Chem. 20: 2017–2023.

Lang, A. R. G. 1968. Psychrometric measurement of soil water potential in situ under cotton plants. Soil Sci. 106: 460–464.

Neumann, H. H. and Thurtell, G. W. 1972. Pages 103–121 in R. W. Brown and B. P. Haveren, Eds. Psychrometry in water relations research. Utah Agricultural Experimental Station, Logan, UT.

Marthaler, H. P., Vogelsanger, W., Richard, F., and Wierenga, P. J. 1983. A pressure transducer for field tensiometers. Soil Sci. Soc. Am. J. 47: 624–627.

Passioura, J. B. 1980. The transport of water from soil to shoot in wheat seedlings. J. Exp. Bot. 31: 1161–1169.

Rawlins, S. L. and Dalton, F. N. 1967. Psychrometric measurements of soil water potential without precise temperature control. Soil Sci. Soc. Am. Proc. 31: 297–300.

Richards, L. A. 1965. Physical conditions of water in soil. Pages 128–152 in C. A. Black, Ed. Methods of soil analysis. Agronomy No. 9, American Society of Agronomy, Madison, WI.

Rundel, P. W. and Jarrell, W. R. 1989. Water in the environment. In R. W. Pearcy, J. Ehleringer, H. A. Mooney, and P. W. Rundel, Eds. Plant physiological ecology. Chapman and Hall, New York.

Savage, M. J., Ritchie, J. T., and Khuvutlu, I. N. 1987. Soil hygrometers for obtaining water potential. Pages 119–124 in Proceedings of International Conference on Measurement of Soil and Plant Water Status, Vol. 1. Logan, UT, July 1987.

Szietz, G. 1975. Instruments and their exposure. Pages 229–272 in J. L. Monteith, Ed. Vegetation and the atmosphere, Vol. 1. Academic Press, London.

Weibe, H. H., Campbell, G. S., Gardner, W. H., Rawlins, S., Cary, J. W., and Brown, R. W. 1971. Measurement of plant and water status. Utah Agricultural Experiment Bulletin 484. Utah State University, Logan.

Chapter 53
Soil Water Desorption Curves

G. C. Topp and Y. T. Galganov

Agriculture Canada

Ottawa, Ontario, Canada

B. C. Ball

The Scottish Agricultural College

Edinburgh, Scotland, U.K.

M. R. Carter

Agriculture Canada

Charlottetown, Prince Edward Island, Canada

53.1 INTRODUCTION

Soil water desorption curves are useful directly and indirectly as indicators of other soil behavior traits, such as drainage, aeration, infiltration, and rooting patterns. Desorption curves are the relationship between soil water content and soil water potential (matric potential) for soil during a drying phase. The independent measurement of each of these variables is discussed in two earlier chapters (Chapters 51 and 52). The ideal arrangement for obtaining the soil water desorption curves would be to have simultaneous *in situ* measurements along a vertical profile of both the water content and potential under conditions of a steady decrease of both variables. Several methods have been described by Bruce and Luxmoore (1986). A number of factors limit this approach, such as methods to make continuous and simultaneous measurements over wide ranges, particularly for the soil water potential. The system depends on favorable weather. For example, rainfall infiltration reverses the direction of change from drying to wetting. As the relationship between water content and matric potential is not reversible and exhibits hysteresis, it is often difficult to interpret *in situ* measurements unless very carefully controlled boundary conditions are applied. As a result, soil water desorption curves are most generally measured in the laboratory using well-controlled initial and boundary conditions.

The desorption curves are sometimes called soil water retention curves, indicating that the measurement determines how much water is retained by the soil at each successively lower matric potential. The measurement of soil water desorption curves and the reliability of the resulting data are very important for adequate characterization of a number of other soil physical attributes. The information can be used to indicate the soil pore size distribution

Soil Sampling and Methods of Analysis, M. R. Carter, Ed.,

Canadian Society of Soil Science. © 1993 Lewis Publishers.

interpreted in relation to aeration and water availability for plants, for water flow and infiltration in soil, and frost susceptibility (see Chapters 54 and 72).

The soil water desorption curves have been most widely used in the prediction of the flow of water and other fluids in soil, and many attempts have been made to define adequate and yet useful mathematical functions for the desorption curve (Brooks and Corey 1964, Clapp and Hornberger 1978, Van Genuchten 1980). Chapters 58 describes how one of these approaches is used for the estimation of unsaturated hydraulic conductivity. Parameters which describe the shape of the desorption curve have been shown to be useful for describing the influence of tillage and compaction on soil structure (O'Sullivan and Ball 1992).

The nature of the desorption curves obtained depends upon the soil conditions during measurement. In order that the obtained data represent the *in situ* condition, the desorption curve measurement must be carried out on soil in its natural or "undisturbed" condition. This is most important when desorbing to potentials down to -300 kPa. At lower potentials, desorption is more dependent on soil microstructure, which is more related to the textural matrix. Transporting a sample to the laboratory always results in some disturbance, but this should be minimized as much as possible. Klute (1986) and Reeve and Carter (1991) give extensive treatments for measuring desorption curves in both disturbed and undisturbed soils.

Here, first, we indicate the important considerations for assuring minimal disturbances of the soil within the soil cores used in the measurements. We then describe the procedures for measurement of the desorption curves of soil cores utilizing a combination of tension tables and pressure plates.

53.2 SOIL CORE SAMPLING

Sampling with metal cylinders for "undisturbed" soil cores is dependent on soil conditions and core size. Cores should be obtained at or near field capacity to avoid soil shattering (dry soil) or compaction (wet soil) during sampling. Stony soils are obviously less well suited for this technique. Soil cores should be large enough to sample a representative volume of soil. Such a volume should contain at least 20 peds and will increase as soil texture become finer and soil structure becomes coarser (Bouma 1983). For most soils, core internal diameter and length should exceed 7.5 cm to minimize the effects of edge disturbance during sampling. Reducing the length of the core to 5 cm where soil conditions permit gives more rapid and uniform equilibration during the measurements to be described. Various core samplers are available to obtain vertical or horizontal cores at different soil depths. Guidance for the construction of sampling equipment and for collection and preparation of minimally disturbed samples is given by McIntyre (1974). Excavated cores should be trimmed flush with the end of the cylinder, capped, and wrapped in plastic or placed in a plastic bag to prevent evaporation and to provide protection during transport. Soil cores should be stored at 0 to 4°C to reduce faunal activity.

53.3 TENSION TABLES

In this method the soil core is brought into good hydraulic contact with the tension medium and the water potential is lowered (tension increased) to some selected value. Time is allowed for water to drain from the soil in response to the imposed gradient in matric potential until an equilibrium is established. The water remaining in the soil at equilibrium represents one point on the soil water desorption curve at the set potential. This general process is repeated for each desired matric potential.

53.3.1 APPARATUS AND MATERIALS

1   The tension table consists of a saturated tension medium and a means of changing and controlling the matric potential of the water saturating the medium (Figure 53.1). This follows the same principles used by Stakman et al. (1969). The tension medium is chosen to have an air entry potential lower than that to be used during measurement and a high saturated hydraulic conductivity (Topp and Zebchuk 1979). Early studies used fine sand as the tension medium, and more recent procedures have used silica flour (diameter 10–50 μm for −20 kPa), glass beads (diameter 30 μm for −10 kPa), and aluminum oxide powder (diameter 9 μm for −50 kPa). Beneath the tension medium there are a retaining mesh and drainage system which allow easy removal of water extracted from the soil cores. The top of the tension medium may be protected by a cloth (nylon mesh) cover to prevent the tension medium from adhering to the cores. Two equivalent approaches have been used successfully for the container and drainage system. A description of the design and construction of a rectangular perspex tension table is provided by Ball and Hunter (1988). They have used glass microfiber paper as their retaining mesh over drainage channels in the perspex. The approach of Topp and Zebchuk (1979) has been modified by the use of a cylindrical polyethylene tray (0.6 m diam × 0.15 m wall) which can hold up to 24 cores (76 mm diam × 76 mm high) (Figure 53.1[a]). The drainage has been achieved by coarse-mesh woven screen and the retaining mesh was 6 μm nylon mesh. In order to deter algal growth, laboratory windows adjacent to the tables should be darkened. The preparation of these or equivalent tension tables is relatively time consuming.

2   The apparatus for control of matric potential and extraction of the water uses a suction, or tension, applied to the water saturating the tension medium. This may be achieved by a hanging water column (Figure 53.1[a]) or by regulated vacuum (Figure 53.1[b]). Achieving a potential of −20 kPa by a hanging water column requires a laboratory with a suitably high ceiling. A trap for fines from the tension medium is recommended as shown in Figure 53.1(b).

3   Soil core cylinders of appropriate dimensions are required. The bottom end of the filled soil core is covered with a nylon cloth (mesh 53 μm) which is held in place with a rubber elastic band. The upper end of the core is covered with a disk to prevent losses or gains of soil and water and to prevent contamination during weighing, etc.

4   A balance for weighing the soil cores should have a range of 0–2 kg and a sensitivity of 0.1 g.

5   Temperature-controlled room at 22 ± 2°C.

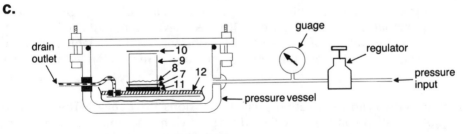

a.

1. tray
2. drainage mesh
3. cementing seal
4. retaining mesh
5. tension medium

6. cloth cover
7. rubber band
8. nylon cloth
9. core cylinder
10. cover disc

b.

c.

11. hydraulic contact slurry
12. ceramic plate

FIGURE 53.1. Schematic diagrams of the essential components for the determination of desorption curves on soil cores.

53.3.2 PROCEDURE

1   Saturate the soil cores within their cylinders at room pressure using deaired, temperature-equilibrated water. Water is deaired by boiling and cooling or by applying a vacuum of 65 kPa for 45 min in a vacuum desiccator. Water used for saturation should have similar solute composition to the resident native soil water to prevent dispersion of soil aggregates. In general, tap water contains sufficient salts to meet this requirement. To saturate, place the cores into trays and add water until the level is 1 cm below the top of the core and leave until the core is fully saturated (about 1 d). Increasing the depth of water around the cores in equal increments over a 3- to 4-d period is the preferred method, particularly for fine-textured soils.

2   Weigh the saturated core (preferably under water, in which case the cores require support in a cradle) and place on the tension table with the hanging water column set at -0.1 kPa (-1 cm) relative to the surface of the tension medium. Make sure hydraulic contact is achieved between the bottom of the core and the tension medium, by a slight twist of the core to deform the tension medium to the shape of the soil surface. Allow 6 h for equilibration and weigh the core. Note: as a result of gravity, the matric potential in a 7.6-cm core equilibrated as given here will vary from -0.1 kPa at the base to -0.86 kPa at the upper surface. We generally quote the potential in reference to the midcore position, or -0.48 kPa in this case.

3   Reestablish hydraulic contact and lower the hanging water column to provide the next desired matric potential. (Note: a 10-cm head of water applies 1 kPa matric potential.) The time required for equilibration depends on the target matric potential, the height of the soil core, the adequacy of the hydraulic contact of the soil with the tension medium, and, to a limited extent, on the tension medium. Guidelines for equilibration times are given in Table 53.1 as derived from the findings of Ball and Hunter (1988) and Topp and Zebchuk (1979). Record the weight of the equilibrated cores.

4   Return individual cores to the same place and orientation on the tension medium and ensure hydraulic contact is established. Marks on the cores can help to achieve this. Topp and Zebchuk (1979) reestablished hydraulic contact by raising the hanging water column to zero and giving a slight twist to each core. Ball and Hunter (1988) sprayed water beneath each core before replacing each core in the same place. Set the matric potential to the new desired value.

5   Repeat the above procedure to give a stepwise decrease in matric potential to -50 kPa. The values chosen for matric potential will depend on the intended use of the data (see Section 53.6). Some often-used values are 0, -0.5, -1, -5, -10, and -33 kPa.

6   Retain the cores at 0–4°C for additional measurements on the pressure plate.

Table 53.1 Equilibration Times for 76 mm-High Cores

Potential (kPa)	Hydraulic head (m)	Approx. equilibration times in days			
		Glass beads	Aluminum oxide	Silica flour	Pressure plate
-0.5	-0.05	0.25	—	—	—
-1	-0.1	1	—	4	—
-5	-0.5	3	—	8	—
-10	-1.0	4	—	10	—
-33	-3.3	—	10	14	—
-100	-10.0	—	—	—	12–15
-400	-40.0	—	—	—	15–20
-1500	-150.0	—	—	—	20–25

53.4 PRESSURE PLATES

Pressure plates are used to provide information on soil water desorption for matric potentials from -50 to -1500 kPa. Drying of the soil is achieved by applying a sequence of pressures inside the container surrounding the soil which is in hydraulic contact with a ceramic plate with the opposite side open to the atmosphere. The pressure applied within the container decreases the matric potential of the water in the plate, resulting in movement of the water from the soil through the plate.

53.4.1 APPARATUS AND MATERIALS

1   A container which will withstand the required applied pressures. Large ($\approx$10-L) pressure cookers have been modified for use at pressures up to 100 kPa. More sturdy containers have been manufactured specifically for the higher pressures. Pressure vessels may require testing and the fitting of blow-off devices. In the interest of safety of the operator(s), check local regulations before using any vessel.

2   Porous ceramic plates with the required bubbling pressures are available in three grades for measurements to matric potentials of -100, -500, and -1500 kPa, for example, from Soilmoisture Equipment Corp., California.

3   A source of air pressure, such as a compressor or a tank of compressed air or nitrogen.

4   Pressure regulators should be nonbleeding to minimize the requirement for gas from the pressure source. This is particularly important when using tanks of compressed air or nitrogen.

5   Hydraulic contact between the soil core and the ceramic plate is achieved by a layer of water-saturated slurry of fine particulate material, such as kaolin.

6   A balance for weighing the soil cores should have a range of 0–2 kg and a sensitivity of 0.1 g.

7   Temperature-controlled room at 22 $\pm$ 2°C.

8 A drying oven with a temperature-control device that will maintain a temperature between 100 and 110°C. Forced-air circulating ovens will dry samples more rapidly, but convective ovens are also suitable.

9 Desiccators, each containing active desiccant.

10 Rust-resistant, numbered drying tins with tight-fitting, numbered lids. The tins should be >8 cm in diameter and height to accommodate 7.6-cm cores.

53.4.2 PROCEDURE

1 Saturate the ceramic plates by immersion in deaired, temperature-equilibrated tap water overnight and remove excess free water from the plate before placing it into position inside the pressure vessel. A wire or string cradle is useful for locating or removing plates from pressure vessels. Connect the plate outlet to the feed-through of the pressure vessel.

2 Spread a layer of hydraulic contact medium having 1–3 mm thickness on the cloth-covered bottom of the cores from Section 53.3.2, above. Immediately place each core upright on the saturated ceramic plate.

3 Close the pressure vessel, apply pressure gradually to the desired level, and allow drainage to take place until equilibration has been achieved, as indicated by no water draining from the vessel outlet or when weight loss from the cores has ceased. Establishing when soil cores have reached equilibrium is difficult because the rate of water loss is very slow and approaches zero asymptotically, particularly as the potential approaches -1500 kPa. An alternative indirect method is offered in Section 53.6, given below. Guidelines for equilibration times are included in Table 53.1.

4 Remove the core from the plate and remove from the core all adhering hydraulic contact medium. Record the weight of the soil core.

5 Repeat the above steps 2, 3, and 4 at the selected pressures to give the required soil matric potentials. Subsampling of the soil core may be advisable to achieve shorter equilibration times for target potentials lower than -300 kPa. Subsamples can be taken by packing smaller samples (28 mm diam $\times$ 20 mm length) from soil removed from the main core.

6 After the final equilibration, remove the nylon cloth, rubber band, and cover disk from the soil core. Be sure the cloth, band, and disk are free of soil by removing any adhering soil and make sure the removed soil is retained with the soil core. Dry the cloth, band, and disk, after which their total weight should be recorded. At this point it may be advisable to measure the volume difference between that of the soil at the final measured matric potential and the total volume of the soil core cylinder — the initial volume. This allows correction for volume change resulting from shrinkage (Sheldrick 1984).

7   Place soil and cylinder in drying tins and oven dry at 105°C until weight loss ceases (72 h for 7.6-cm length cores in a forced-air oven).

8   Remove the samples from the oven, replace the lid on each drying tin, and place the tins in a desiccator.

9   After the samples have cooled to ambient room temperature, reweigh the samples to a precision of 0.1 g. If the core was subsampled in step 5, ensure that all the dry soil is collected for weighing to get the total mass of the soil from the core.

10 Remove the core cylinder and clean it of all adhering soil and record the weight of the cylinder.

53.5 CALCULATIONS

1   Obtain the mass of dry soil by subtracting the weighed values of cylinder and the drying tin.

2   Obtain the mass of wet soil at each measured matric potential by subtracting the weights of the cylinder, the nylon cloth, the rubber band, and the cover disk for each core.

3   Calculate the volumetric water content ($\theta[\psi]$) at each matric potential (ψ) as:

$$\theta(\psi) = 100\left[\frac{M(\psi) - M_d}{V_s \rho_w}\right] \tag{53.1}$$

where $M(\psi)$ and M_d are the masses of soil (and water) measured at $-\psi$ and oven dried, respectively, V_s is the volume of the soil (taken as the cylinder volume if the soil filled the cylinder and shrinkage was minimal), and ρ_w is the density of water.

4   The saturated water content ($\theta[0]$) is a particular case of $\theta(\psi)$ and is used to convert water content on a volume basis to degree of saturation ($S[\psi]$) as:

$$S(\psi) = \frac{\theta(\psi)}{\theta(0)} \tag{53.2}$$

5. Soil water desorption data are often presented as graphical relationships as in Figure 53.2.

53.6 COMMENTS

1   The choice of equilibration tensions and pressures is best made in relation to the intended use of the data. Where least data are required, e.g., for soil

Soil Water Desorption Curves
Ap Horizons

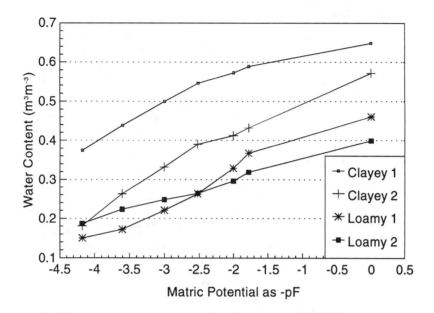

FIGURE 53.2. Soil water desorption curves of clayey and loamy soils.

survey purposes, then values at saturation, "field capacity" (-5 to -30 kPa), and the lower limit of water readily available to plants (-200 to -1500 kPa) would suffice. Where the influence of experiment treatments on soil structure and drainage is required, assessment of several size fractions of macroporosity may be needed and equilibration at a greater number of high matric potentials will be required, such as 0, -0.5, -1, -5, -10, and -33 kPa.

2 Most measurements with pressure plates have been described primarily for use with sieved samples (Klute 1986, McKeague 1978, Sheldrick 1984, Reeve and Carter 1991). The increased interest in soil structure and strength has increased the need for undisturbed soil at lower matric potentials (B. D. Kay 1992, personal communication). We have given this procedure for measurement using soil cores to indicate the requirements for measuring soil water desorption curves on undisturbed soil. Soil cores also provide relatively large volumes for other measurements, such as penetrability measurements.

3 The length of the samples is the critical factor contributing to longer equilibration times, but the hydraulic contact between the soil and the plate or tension medium is also very important. We have found by recent studies on 76 × 76 mm intact soil cores that the rate of approach to equilibrium has an initial quite rapid portion lasting between 24 and 48 h, followed by a much slower drainage rate of longer duration (Figure 53.3). The latter drainage rate follows a curve of the form:

$$\text{Core mass (g)} = a - b* \ln [\text{Time (h)}]; \text{Time} > 48 \text{ h} \qquad (53.3)$$

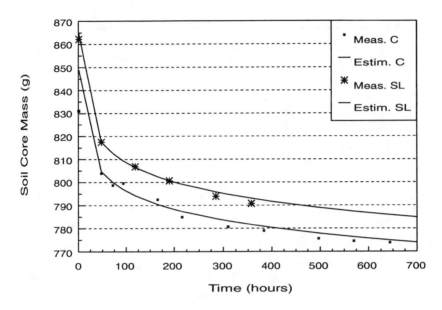

Pressure Plate Desorption
To -1500 kPa

FIGURE 53.3. Rate of weight loss for cores of a sandy loam (SL) and a clay (C) soil equilibrating to −1500 kPa. The data points are the measured weights and lines are from fitting Equation 53.3 to those data points between approximately 48 and 200 h. The last measured point was selected as the end of equilibration using a rate of weight loss criterion, such as ≤1 g per 100 h for the clayey soil.

where the y-intercept, a, and the slope, b, can be determined from a series of weights of the sample or the weight of the water volume outflow after the initial 48 h of drainage. A minimum of three weights from within the range of 48 to 200 h are required to establish a reasonable estimate of equilibration time and/or weight. The length of time is a function of the soil texture and structure as well as the pressure within the pressure vessel. Figure 53.3 gives representative curves for two soils draining to equilibration at −1500 kPa. The use of the range from 48 to 200 h gave good estimation of the drainage path for up to 600 h or 25 d. This method provides an estimate of the equilibration time based on the performance of the soil during the early experimental measurement. Employing this approach when only the weight (i.e., the water content) is required, one is able to shorten the experimental time by a factor of 2 or 3. If one is using the soil for other measurements which depend on having the soil at the target potential, the method can be used to predict the required equilibration time.

4   These desorption methods, which depend on achieving equilibrium, are very time consuming. It is feasible to make dynamic measurements of water content and matric potential simultaneously while the soil is drying (Malicki et al. 1992). Research is required, however, to develop the necessary combination of techniques and requirements beyond the range of tensiometers.

Although soil water desorption data are widely used, little effort has gone into methodology improvements.

REFERENCES

Ball, B. C. and Hunter, R. 1988. The determination of water release characteristics of soil cores at low suction. Geoderma 43: 195–212.

Bouma, J. 1983. Use of soil survey data to select measurement techniques for hydraulic conductivity. Agric. Water Manag. 6: 177–190.

Bruce, R. R. and Luxmoore, R. J. 1986. Water retention: field methods. Pages 663–683 in A. Klute, Ed. Methods of Soil Analysis. Part I. Physical and mineralogical methods. Agronomy No. 9, 2nd ed. American Society of Agronomy, Madison, WI.

Brooks, R. H. and Corey, A. T. 1964. Hydraulic properties of porous media. Hydrology Paper, Colorado State University, Fort Collins. 3: 1–27.

Clapp, R. B. and Hornberger, G. M. 1978. Empirical equations for some hydraulic properties. Water Resour. Res. 14: 601–604.

Klute, A. 1986. Water retention: laboratory methods. Pages 635–662 in A. Klute, Ed. Methods of soil analysis. Part I. Physical and Mineralogical Methods. Agronomy No. 9, 2nd ed. American Society of Agronomy, Madison, WI.

Malicki, M. A., Plagge, R., Renger, M., and Walczak, R. T. 1992. Application of time-domain reflectometry (TDR) soil moisture miniprobe for the determination of unsaturated soil water characteristics from undisturbed soil cores. Irrig. Sci. 13: 65–72.

McIntyre, D. S. 1974. Soil sampling techniques for physical measurements. Pages 12–20 in J. Loveday, Ed. Methods for analysis of irrigated soils. Tech. Commun. No. 54, Commonwealth Agriculture Bureau, Wilke and Company, Clayton, Australia.

McKeague, J. A., Ed. 1978. Manual on soil sampling and methods of analysis. 2nd ed. Canadian Society of Soil Science, Ottawa.

O'Sullivan, M. F. and Ball, B. C. 1992. The shape of the water release characteristic as affected by tillage, compaction and soil type. Soil Tillage Res. in press.

Reeve, M. J. and Carter, A. D. 1991. Water release characteristic. Pages 111–160 in K. A. Smith and C. E. Mullins, Eds. in "Soil Analysis: Physical Methods" Marcel Dekker, New York.

Sheldrick, B. S., Ed. 1984. Analytical methods manual 1984, Land Resource Research Institute. LRRI Report No. 84-30. Pages 35/1–36/2. Agriculture Canada, Ottawa.

Stakman, W. P., Valk, G. A., and van der Harst, G. G. 1969. Determination of soil moisture retention curves. I. Sand-box apparatus — range pF 0 to 2.7. Institute for Land and Water Management Research, Wageningen, The Netherlands.

Topp, G. C. and Zebchuk, W. D. 1979. The determination of soil-water desorption curves for soil cores. Can. J. Soil Sci. 59: 19–26.

Van Genuchten, M. Th. 1980. A closed-form equation for predicting the hydraulic conductivity of unsaturated soils. Soil Sci. Soc. Am. 44: 1072–1081.

Chapter 54
Soil Porosity

M. R. Carter
Agriculture Canada
Charlottetown, Prince Edward Island, Canada

B. C. Ball
The Scottish Agricultural College
Edinburgh, Scotland, U.K.

54.1 INTRODUCTION

Soil porosity and pore size distribution are useful parameters to assess the physical condition and structure of the soil. In many cases pore size distribution is considered the best indicator of the soil physical condition (Greenland 1979). Associated porosity factors, such as macropore volume, pore continuity, and air-filled pore space, are also important guides to characterize soil structure.

Total soil porosity (St) is easily measured and commonly used, but limited in application. It reflects the character of soil as a three-phase system of solids, liquids, and gases. In moist soil, total porosity can be partitioned into air-filled porosity (Fa) and water-filled porosity (relative saturation).

Of greater interest is the distribution of the total soil pore space into different categories of pore diameter, especially the larger soil pores which are associated with the transfer and movement of water and air. Calculation of pore size distribution has proven useful for predicting water infiltration rates, water availability to plants, water storage capacity in the soil and soil aeration status (Cary and Hayden 1973), and classification of soil structure in the field (Thomasson 1978, McKeague et al. 1986). Measurements of pore size distribution in soil provide descriptive information about the soil pore system, rather than absolute measurements. The techniques are limited by the basic assumptions of the capillary model which represents soil pores as parallel tubes of varying radii (Ball 1981a), and by the confusion in regard to terminology for classifying the various pore sizes (Danielson and Sutherland 1986). Pore sizes are thus expressed as equivalent pore diameters (epd).

Micromorphological methods can be used to characterize the actual morphology of soil pores, and are especially useful for studying the shape and continuity of pores above 100 μm in diameter. Details on micromorphology are given in Chapter 65. Morphological techniques are also required to describe large pores and cracks (greater than 2 mm diameter or width) such as worm holes and pores between large structural units.

Here we describe soil physical methods to measure total porosity and its air- and water-filled fractions and pore size distribution from measurements of water desorption using soil cores. The latter is suitable for determination of equivalent pore sizes below 150 μm (Bouma 1991). Indices of pore continuity from measurements of gas diffusion or air permeability are also discussed.

54.2 TOTAL POROSITY, MACROPOROSITY, AIR-FILLED POROSITY, AND RELATIVE SATURATION

The measurements of soil porosity described in this chapter require sampling of a known volume and weight of soil. This is achieved by taking cores of soil from the field or packing cores with sieved, moist soil.

54.2.1 Soil Cores

Sampling with metal cylinders for ''undisturbed'' soil cores is dependent on soil conditions and core size. Cores should be obtained at or near field capacity to avoid soil shattering or compaction during sampling. Stony soils are obviously not suited for this technique. Small soil cores are generally unsuitable, since they fail to sample a representative volume of soil. For most soils, core internal diameter and length should exceed 7.5 cm to reduce edge effects. Various core samplers are available to obtain vertical or horizontal cores at different soil depths. Lubrication of the cores with oil (e.g., sunflower oil; Blackwell et al. 1990) can improve core insertion into the soil. Coating the inner surface of the core with petroleum jelly can improve the soil-metal contact, which is important for pore continuity estimates. Cameron et al. (1990) discuss ways to improve the seal between the soil and the metal core to reduce edge flow for measurements for pore continuity. For soils with significant amounts of clay, the incidence of smearing during soil core preparation can be ameliorated using the epoxy resin ''peel'' method of Koppi and Geering (1986) to provide unsmeared surfaces for pore continuity studies. Guidance for the construction of sampling equipment and for collection and preparation of minimally disturbed samples is given by McIntyre (1974). Excavated cores should be trimmed flush with the end of the cylinder and wrapped in plastic to prevent evaporation and to provide protection during transport. Soil cores should be stored at 0 to 4°C to reduce macrofauna activity.

54.2.2 Total Porosity

Total soil porosity (St) is a measure of the volume percentage pore space and is usually derived from measurements of soil dry bulk density (D_b) and soil particle density (D_p). In mineral soils, total porosity can vary from 20 to 70%. Usually, the value of D_p in mineral soils is assumed to be 2.65 Mg m^{-3}, which is the value for quartz. Both the submersion and pycnometer methods, as outlined by Blake and Hartge (1986), provide greater accuracy for the determination of D_p. D_p measured on <2-mm soil samples may not be representative of undisturbed soils which may contain stones and organic material. Procedures to obtain D_b, such as core, clod, and nuclear methods, are given in Chapter 50.

54.2.2.1 Calculation

Since D_p represents the ratio of the mass of dry soil to the combined volume of solids and pores, the ratio D_b/D_p will give the volume fraction occupied by the solids. The fraction occupied by the pore space is therefore calculated as follows:

$$St = 1 - (D_b/D_p) \tag{54.1}$$

Results for St can be expressed as a volume fraction or as a percentage. In this chapter we will use percentages for presentations of porosity indices.

54.2.3 Macroporosity

Macroporosity is a useful index to gauge soil response to different management and tillage systems (Carter 1988). The volume of macropores is also approximately equal to the Fa porosity of the soil at field capacity (air capacity). Generally, macropore volume should exceed 10% of the soil volume to maintain optimum soil aeration. However, this can be modified by the degree of pore continuity.

Macroporosity (pores with an epd >50 μm) can be determined using soil cores on a tension table as outlined in Chapter 53 using a potential of -6 kPa and Equation 54.7.

54.2.4 Air-Filled Porosity

Fa porosity is a measure of the fraction of the soil bulk volume occupied by air. Soil matric potential information should be given when reporting Fa porosity. Assuming that the total pore space is filled with water at saturation, the air-filled pore space at -6 kPa matric potential would equal the volume of macropores.

Fa porosity, as a percentage of total volume, can be calculated from the tension table procedure for any specific matric potential using the following equation:

$$Fa = (W_s - W_p)/(V_t)0.01 \tag{54.2}$$

where W_s is the weight of the saturated core; W_p is the weight of the core at specific potential; and V_t is the volume of the core. The difference method based on D_b and soil water content can also be used for determining the Fa (volume basis) of a core at any gravimetric water content by calculating total porosity (St) using Equation 54.1 and the following equation:

$$Fa = St - (water content, \%)(D_b) \tag{54.3}$$

Ball and Hunter (1988) describe some of the limitations and errors involved in the determination of Fa porosity.

54.2.5 Relative Saturation

Relative saturation (S), sometimes called "water-filled pore space", expresses the volume of water in the soil relative to the total volume of pores. Thus, it ranges from zero in a dry soil to 100% in saturated conditions. A S over 65–70% can indicate that the soil may become anaerobic (Linn and Doran 1984) and has proved a useful index for porosity studies in wet soils (Carter 1988). S is calculated in soil cores as follows:

$$S = \text{Weight of water (g)}/(V_t\text{-[Dry weight of soil}/D_p]) \tag{54.4}$$

The S index can be calculated from Fa and St (i.e., $S = St - Fa$).

54.3 PORE SIZE DISTRIBUTION

54.3.1 Pore Size Classification

Attempts have been made to classify pores in regard to function rather than just size alone. Greenland (1979) describes various terminologies for epd and proposed a scheme to describe the functional properties of pore size groups as follows: fissures (>500 μm), transmission pores (500–50 μm), storage pores (50–0.5 μm), and residual pores (<0.5 μm). Emphasis has also been placed on large pores (termed "macropores"), above 50 μm in diameter, which are associated with hydraulic conductivity (Germann and Beven 1981). Generally, the range of pore size of interest will be dependent on the purpose of the measurement (Ball and Hunter 1988). For example, soil survey purposes may require pore size ranges associated with field capacity (50 μm) and the lower limit of readily available water for plants (3 to 1.5 μm). Thomasson (1978) used the soil volume of pores greater than 60 μm diameter (termed "air capacity") and the volume occupied by pores of between 60 and 0.2 μm (termed "available water") as a classification of soil structural condition. In contrast, soil structure or tillage studies require information about macroporosity (>50 μm). Studies on infiltration and preferential path flow of water require information on pores >1 mm epd (Luxmoore et al. 1990).

54.3.2 Pore Size Distribution <1 mm epd

Water desorption, using soil cores, has wide application for determination of pore size distribution. The method is suitable for undisturbed cores containing relatively small structural units (<2 cm diameter) and for remolded cores. This method is based on determination of the water desorption curve (moisture characteristic). The curve relates soil moisture content to the energy state of the soil water, the water potential, expressed usually in terms of matric potential. Based on the assumptions of the capillary model, the volume of water removed from the soil in response to a change in matric potential corresponds to the change in energy status of the soil water and the volume of pores above a specific diameter. Mathematically, this is expressed by the Kelvin equation

$$d = 4\gamma \cos \alpha (pgh)^{-1} \tag{54.5}$$

where d is the diameter (m) of the largest pores which remain full of water after a negative potential (h) in meters of water is applied; γ is the surface tension of water (72.75 mJ m^{-2} at 20°C); α is the contact angle of the water held in the pore (taken to be zero); p is the density of water (0.998 Mg m^{-3} at 20°C); and g is the acceleration due to gravity (9.8 m s^{-2}). The equation indicates that pores which can maintain a water meniscus against the combined force of pgh have an upper limiting radius. Therefore, applying a negative potential of -1 m of water (equivalent to -0.1 bar or -10 kPa matric potential) will give a diameter of 30 μm at 20°C for the largest pores full of water.

Use of the above procedure, however, requires that the soil is progressively drained by decreasing matric potential; otherwise, soil hysteresis influences water content at a given potential. The pore size distribution must also remain stable during determination of the water desorption curve.

In fine-textured soils (clay content above 30%) shrinkage can change the pore size distribution over time, while in very sandy soils (sand content above 80%) low water adhesion causes gravitational loss of water (Olson 1985). Under these conditions, other techniques to determine pore size distribution, such as mercury intrusion and nitrogen sorption, have been developed (Lawrence 1977). Both of these methods are based on the cylindrical pore model for pore size distribution. Danielson and Sutherland (1986) provide details on the mercury intrusion method. The water desorption method described below is thus best suited for medium-textured soils and for both sandy and fine-textured soils. However, the method is applicable in clay soils at relatively high matric potentials (0 to -1 kPa). Generally, the water desorption method is mainly used at high potentials (-1 to -100 kPa) where Equation 54.5 is most applicable.

Methods to determine water desorption between 0 and -20 kPa matric potential use tension table water extraction (Topp and Zebchuk 1979, Ball and Hunter 1988). Methods to determine water desorption at lower than -20 kPa potential require the use of pressure plate extraction (20–500 kPa) and pressure membrane extraction (100–1500 kPa). These three extraction methods and procedures are described in Chapter 53.

54.3.2.1 Calculations

To calculate the water desorption curve, plot volumetric moisture content (θ_v) vs. matric potential. Determine volumetric soil moisture content (%) at each matric potential as follows:

$$\theta_v = (W_p - W_2)/(V_t)0.01 \qquad (54.6)$$

where W_p is the weight of soil plus core at a specific matric potential, and W_2 is the weight of dry soil plus core. Determination of pore size distribution is based on the information given in Equation 54.5 where the diameter of pores remaining full of water was 30 μm after equilibration at a potential of -1 m (-10 kPa). Thus, the diameter of the smallest pore drained at a specific potential would be as follows:

$$D(\mu m) = 300/(kPa) \qquad (54.7)$$

Calculation of the volume of water removed between two specific pore size diameters would equal the volume of pore space for that pore size range.

54.3.2.2 COMMENTS

1   The pore size diameters are approximate and therefore designated as epd. In addition, the contact angle (α) in Equation 54.5 is assumed to be zero, although this may not be always correct. Further, the method provides no information of pore continuity or shape. Also, for soil cores with a relatively large volume of macropores (epd >50 μm) it is difficult to obtain complete

saturation by capillary wetting. Under these circumstances use of vacuum wetting (Ball and Hunter 1988) or estimation of total porosity from D_b using Equation 54.1 is advised.

2   Use of undisturbed soil cores will prove a problem on recently cultivated soils due to loose cores. Generally, some degree of compaction by weathering must occur before intact cores can be obtained.

3   During the saturation procedure, formaldehyde solution (4% w/v) can be sprayed on the surface of the core to remove or suppress the activity of earthworms and other macrofauna (Ball and Hunter 1988).

54.3.3 Pore Size Distribution >1 mm epd

The size and continuity of pores >1 mm epd are particularly important for environmental modeling of infiltration of water and solute solution. Douglas (1986) describes morphological procedures for assessing the dimensions of large macropores. Procedure 1 is recommended for soils which do not shrink measurably on drying. Procedure 2 is recommended for soils which remain stable during saturation.

54.3.3.1 *Procedure 1*

Oven-dry core sample. Record the number and diameter of channel-type pores of diameter ≥ 1 mm judged to have openings at both ends of the sample. Measure pore diameter using a set of 1- to 8-mm rigid aluminum rod gauges.

Douglas (1986) also proposed a method of identification of functional macropores, i.e., those responsible for flow-through samples.

54.3.3.2 *Procedure 2*

Saturate core sample. Pour solution of rhodamine-B dye through the samples. After drainage, extrude the cores from their retaining rings for a distance of 1 cm and carefully pick off the extruded 1 cm of soil. The continuous, stained pores thus exposed are traced onto acetate sheets, counted, and measured.

An alternative to staining or direct measurement techniques is X-ray computed tomography (Anderson et al. 1990). In this technique, undisturbed cores 8×8 cm are scanned and pore diameters are given as output. Another technique is image analysis of soil porosity using difference imagery of stereo photographs (Grevers and de Jong 1990).

54.4 PORE CONTINUITY INDICES

The continuity of soil macropores is particularly important in determining the conductivity of the soil for gas movement, water infiltration, and root exploration. Indices of continuity relate to the ability of the soil porosity to conduct fluid. Gas is commonly the fluid used, since it does not change the soil structure when it diffuses or flows through the soil. Use of gas movement allows measurements to be repeated at different soil moisture contents. This allows assessment of the change of pore continuity with Fa porosity.

In order to measure pore continuity, measure gas relative diffusivity (D/Do) and/or air permeability (Ka) (in units of μm^2) and Fa porosity (volume percentage) in soil cores in the laboratory. Techniques for measuring Ka are given in Chapter 60. In addition, Ball and Smith (1990) present an extensive review of suitable techniques.

In terms of gas relative diffusivity, pore continuity Cd is

$$Cd = (D/Do)/(Fa \times 0.01) \tag{54.8}$$

This factor, Cd, derived by Ball (1981b), ranges from 0 for completely blocked pores to 1 for straight tubular pores aligned with the direction of diffusing gas.

In terms of air permeability, pore continuity Ck is either:

$$Ck = Ka/(Fa \times 0.01) \tag{54.9}$$

as derived by Groenevelt et al. (1984), or O:

$$O = Ka/(Fa \times 0.01) \tag{54.10}$$

as derived by Blackwell et al. (1990). O is termed macropore organization because it is applicable to soil samples drained at field capacity (i.e., when the macropores only are conducting). Other indices of pore continuity and pore tortuosity were reviewed by Ball et al. (1988).

REFERENCES

Anderson, S. H., Peyton, R. L., and Gantzer, C. J. 1990. Evaluation of constructed and natural soil macropores using X-ray computed tomography. Geoderma 46: 13–29.

Ball, B. C. 1981a. Modelling of soil pores as tubes using gas permeabilities, gas diffusivities and water release. J. Soil Sci. 32: 465–481.

Ball, B. C. 1981b. Pore characteristics of soils from two cultivation experiments as shown by gas diffusivities and permeabilities and air-filled porosities. J. Soil Sci. 32: 483–498.

Ball, B. C. and Hunter, R. 1988. The determination of water release characteristics of soil cores at low suctions. Geoderma 43: 195–212.

Ball, B. C. and Smith, K. A. 1991. Gas movement. Pages 511–549 in K. A. Smith and C. M. Mullins, Eds. Soil analysis: physical methods. Marcel Dekker, New York.

Ball, B. C., O'Sullivan, M. F., and Hunter, R. 1988. Gas diffusion, fluid flow and derived pore continuity indices in relation to vehicle traffic and tillage. J. Soil Sci. 39: 327–339.

Blackwell, P. S., Ringrose-Voase, A. J., Jayawardane, N. S., Olsson, K. A., McKenzie, D. C., and Mason, W. K. 1990. The use of air-filled porosity and intrinsic permeability to characterize structure of macropore space and saturated hydraulic conductivity of clay soils. J. Soil Sci. 41: 215–228.

Blake, G. R. and Hartge, K. H. 1986. Particle density. Pages 377–382 in A. Klute, Ed. Methods of soil analysis. Part 1. Agronomy No. 9, 2nd ed. American Society of Agronomy, Madison, WI.

Bouma, J. 1991. Influence of soil macroporosity on environmental quality. Adv. Agron. 46: 1–37.

Cameron, K. C., Harrison, D. F., Smith, N. P., and McLay, C. D. A. 1990. A method to prevent edge-flow in undisturbed soil cores and lysimeters. Aust. J. Soil Res. 28: 879–886.

Carter, M. R. 1988. Temporal variability of soil macroporosity in a fine sandy loam under mouldboard ploughing and direct drilling. Soil Tillage Res. 12: 37–51.

Cary, J. W. and Hayden, C. W. 1973. An index for soil pore size distribution. Geoderma 9: 249–256.

Danielson, R. E. and Sutherland, P. L. 1986. Soil porosity. Pages 443–461 in A. Klute, Ed. Methods of soil analysis. Part 1. Agronomy No. 9, 2nd ed. American Society of Agronomy, Madison, WI.

Douglas, J. T. 1986. Macroporosity and permeability of some cores from England and France. Geoderma 37: 221–231.

Germann, P. and Beven, K. 1981. Water flow in soil macropores. III. A statistical approach. J. Soil Sci. 32: 31–39.

Greeniand, D. J. 1979. Structural organization of soils and crop production. Pages 47–56 in R. Lal and D. J. Greenland, Eds. Soil physical properties and crop production in the tropics. John Wiley & Sons, New York.

Grevers, M. C. J. and De Jong, E. 1990. The characterization of soil macroporosity of a clay soil under ten grasses using image analysis. Can. J. Soil Sci. 70: 93–103.

Groenevelt, P. H., Kay, B. D., and Grant, C. D. 1984. Physical assessment of soil with respect to rooting potential. Geoderma 34: 101–114.

Koppi, A. J. and Geering, H. R. 1986. The preparation of unsmeared soil surfaces and an improved apparatus for infiltration measurements. J. Soil Sci. 37: 177–181.

Lawrence, G. P. 1977. Measurement of pore sizes in fine-textured soils: a review of existing techniques. J. Soil Sci. 28: 527–540.

Linn, D. M. and Doran, J. W. 1984. Effect of water-filled pore space on carbon dioxide and nitrous oxide production in tilled and non-tilled soils. Soil Sci. Soc. Am. J. 48: 1267–1272.

Luxmoore, R. J., Jardine, P. M., Wilson, G. V., Jones, J. R., and Zelazny, L. W. 1990. Physical and chemical controls of preferred path flow through a forested hillslope. Geoderma 46: 139–154.

McIntyre, D. S. 1974. Soil sampling techniques for physical measurements. Pages 12–20 in J. Loveday, Ed. Methods for analysis of irrigated soils. Tech. Commun. No. 54, Commonwealth Agricultural Bureau, Wilke and Company, Clayton, Australia.

McKeague, J. A., Wang, C., and Coen, G. M. 1986. Describing and interpreting the macrostructure of mineral soils — a preliminary report. LRRI Contribution No. 84-50 Research Branch, Agriculture Canada, Ottawa. 47 pp.

Olson, K. R. 1985. Characterization of pore size distributions within soils by mercury intrusion and water-release methods. Soil Sci. 139: 400–404.

Thomasson, A. J. 1978. Towards an objective classification of soil structure. J. Soil Sci. 29: 38–46.

Topp, G. C. and Zebchuk, W. 1979. The determination of soil-water desorption curves for soil cores. Can. J. Soil Sci. 59: 19–26.

Chapter 55
Saturated Hydraulic Conductivity: Laboratory Measurement

W. D. Reynolds

Agriculture Canada
Ottawa, Ontario, Canada

55.1 INTRODUCTION

Saturated hydraulic conductivity (K_s) is a measure of the "ease" or "ability" of a saturated porous medium to transmit water. This parameter is defined by Darcy's law, which, for one-dimensional vertical flow, may be written in the form

$$q = K_s i \tag{55.1}$$

where q is the water flux through the porous medium (volume of water per unit cross-sectional area of flow per unit time), and i is the hydraulic head gradient (dimensionless) in the porous medium. As indicated by Equation 55.1, the dimensions of K_s are the same as those for q (i.e., volume of water per unit cross-sectional area of flow per unit time); however, these dimensions are usually simplified arbitrarily to length per unit time so that K_s may be expressed in the more convenient (but physically incorrect) units of velocity (i.e., $m\ s^{-1}$, $cm\ h^{-1}$, $m\ d^{-1}$, etc.). For this reason, K_s is sometimes referred to (incorrectly) as a "velocity". The K_s value is a constant for rigid, homogeneous porous media.

The primary factors determining K_s include the texture (grain size distribution) and structure (cracks, worm holes, root channels, etc.) of the porous medium, the temperature of the water, the ionic (salt) concentration in the water, and the presence of air bubbles in the porous medium.

The K_s value tends to increase greatly with coarser texture and increasing structure, due to an increase in the number and size of large, highly water-conductive pores. Consequently, coarse-textured and/or structured porous media tend to have larger K_s values than fine-textured and/or structureless porous media. In addition, the texture and structure can interact in such a way that a fine-textured, but structured, material (i.e., a structured clay soil) can have a larger K_s than a coarse-textured, but structureless, material (i.e., a structureless sand soil). An important implication of this texture/structure interaction is that the antecedent structure must be preserved by the measuring technique in order for the K_s measurement to be representative of the porous medium in its "natural" condition.

Soil Sampling and Methods of Analysis, M. R. Carter, Ed.,
Canadian Society of Soil Science. © 1993 Lewis Publishers.

Hydraulic conductivity is directly proportional to water viscosity, and viscosity is inversely proportional to temperature. Consequently, the measured value of K_s will tend to increase with the temperature of the water used. An increase in water temperature from 10 to 25°C will result in a 45% increase in K_s, all other factors remaining equal. Precise measurements and comparisons of K_s values should therefore always be referenced to a specific water temperature.

The ionic (dissolved salt) concentration of the water can affect K_s through flocculation or dispersion of clay within the porous medium. The measured K_s will increase if clay is flocculated and decrease if clay is dispersed. In extreme cases, clay dispersion can reduce K_s to virtually zero. The water used for measuring the K_s of a natural porous medium should therefore be either "native" water extracted from the porous medium or a laboratory "approximation" which has the same major ion chemistry as the native water. Local municipal tap water is often an adequate approximation to native water, although this should always be checked. Distilled or deionized water should never be used for measuring the K_s of a natural porous medium.

Bubbles of air entrapped within the porous medium physically block pores and can thereby cause the measured K_s to be less than when the material is completely saturated. In relatively structureless materials, entrapped air can reduce the measured K_s by more than a factor of 2. Air bubbles occur due to entrapment of resident air by the infiltrating water when the porous medium is being wetted up from a dry or partially dry state. They can also "grow *in situ*" within the pores due to exsolution of air from the influent water. These problems are minimized (but not eliminated) by slow wetting of the porous medium with water that has been deaired and equilibrated to the temperature of the porous medium.

Further information concerning the theoretical basis and other aspects of K_s can be obtained from Bouwer (1978), Koorevaar et al. (1983), and other standard texts.

55.2 CONSTANT-HEAD SOIL CORE METHOD (ADAPTED FROM ELRICK ET AL. 1981; SEE ALSO KLUTE AND DIRKSEN 1986)

This method determines the K_s of cylindrical, 7.62-cm (3-in) I.D. by 7.62-cm (3-inch) long cores of soil or other porous medium. The soil is first wetted to saturation and then water is flowed through the soil at a steady rate under a constant hydraulic head gradient.

55.2.1 APPARATUS AND PROCEDURE

1   Collect samples in 7.62-cm internal diameter by 7.62-cm long by 0.3-cm wall thickness aluminum cylinders. If the K_s measurements are intended to represent field conditions, then the samples should be "undisturbed" so that the antecedent soil structure remains intact to as great an extent as possible (see Section 55.1). Criteria and procedures for obtaining minimally disturbed samples and for minimizing preferential flow along the cylinder walls can be found in McIntyre (1974), Rogers and Carter (1986), Amoozegar (1988), and Cameron et al. (1990). Trim the sample flush with the cylinder at both ends and remove any material (soil) adhering to the outside of the cylinder. Cover the top end of the core with 270-mesh (53-μm pore size) woven nylon

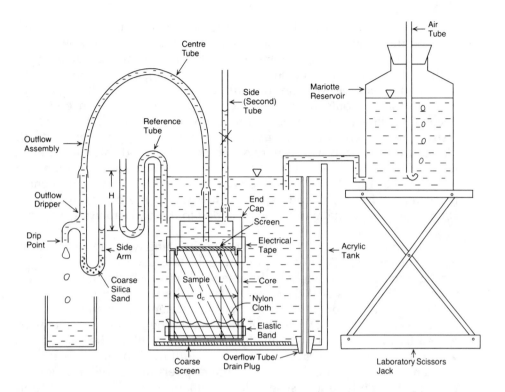

FIGURE 55.1. Schematic of the apparatus for the constant head soil core method. (Adapted from Elrick, D. E., Sheard, R. W., and Baumgartner, N. Proc. IV Int Turfgrass Research Conf., Guelph, 189, 1981.)

cloth ("Nitex" nylon bolting cloth), held in place with a stout elastic band (Figure 55.1). Press the acrylic end cap onto the bottom end of the core (see also comment 2, Section 55.2.3) and seal by wrapping the joint with heavy-duty, vinyl plastic, electrical tape (Figure 55.1). A durable, water-tight seal is made by stretching the tape tightly over the joint for three overlapping wraps, and then adding an additional three overlapping wraps of tape without stretching. The final three wraps of nonstretched tape prevents the tape from gradually peeling back over time. During the wrapping operation, the sample should be placed in a plastic petri dish in order to prevent tearing of the nylon cloth through frictional wear against the bench top. It may also be possible to seal the end cap to the cylinder by brushing melted (100°C) paraffin wax over the joint.

2   Place the prepared cores in an empty acrylic tank which has a coarse mesh (2-mm openings), woven stainless steel screen in the bottom (Figure 55.1) to allow unrestricted wetting of the core through the nylon cloth. The tank should be approximately 30 cm high so that a wide range of hydraulic head gradients can be applied to the cores (discussed further below). A combined overflow tube-drain plug (Figure 55.1) prevents overfilling and facilitates drainage of the tank at the end of the measurements.

3   Saturate the cores slowly over a 4-d period of adding deaired, temperature-equilibrated water (either native water or a laboratory approximation). Sub-

merge an additional one third of the length of the core for each of the first 3 d, and then add water to the full tank level (top of the overflow tube — Figure 55.1) on the fourth day. It may be advisable to add algicide/fungicide to the water to prevent excessive growth of microorganisms both within the cores and within the water in the tank. During the saturation process, the end cap tubes should be held vertical to prevent their open ends from becoming immersed in the water. Immersion of the ends of the tubes may restrict air flow and thereby impede the saturation of the cores and the filling of the end cap with water. At the end of the 4-d saturating period, the acrylic end caps should be filled with water, except perhaps for a small entrapped air bubble. This bubble is of no concern as long as the end of the central tube in the end cap is submerged. If the central tube is not submerged, additional water may be added to the end cap using a syringe.

4   Fill a mariotte reservoir (25-L carboy) with the same type of water that was used to fill the tank. Connect its outflow to the tank using 0.8-cm I.D. Tygon® tubing, and adjust the height of the reservoir so that it maintains the desired water level in the tank (Figure 55.1). Placing the reservoir on a laboratory scissors jack allows for convenient height adjustment. The mariotte reservoir maintains a constant water level in the tank during the measurements (required by the theory) by supplying water to the tank at the same rate at which it is being withdrawn through the outflow drippers. It is not recommended that water collected from the drippers be poured back into the tank, as this tends to add suspended sediment (e.g., silt and clay) to the tank water, which may then plug the nylon cloth.

5   Using a syringe, fill the water outflow assembly with water and clamp. Fill the central tube of the end cap tube with water by submerging it in the tank. Connect the outflow assembly and central end cap tube, making sure that there are no air bubbles within the tubing. Clamp the second end cap tube. Fill the reference tube with water by submerging it in the tank, and set it up as indicated in Figure 55.1.

6   Arrange the water outflow assembly such that the drip point of the outflow dripper is 1–2 cm below the water level in the tank. Unclamp the outflow assembly slowly and carefully so that air is not sucked into the tubing. Adjust the elevation of the outflow dripper so that a drop falls no faster than one every 0.25–0.5 s and no slower than every 1–2 min. If the minimum possible dripper elevation still produces less than one drop every 1–2 min, then the falling head soil core method should be used (Section 55.3).

7   Determine the flow rate through the core either by measuring the volume of water collected in a set period of time (a 1-min time interval is convenient) or by measuring the time required to collect a set volume of water (a 20-mL volume is convenient). It is recommended that about four flow rate determinations be made and the results averaged to reduce the effects of variability. Concurrent with the flow rate measurements, the difference in hydraulic head between the top and bottom of the core is determined by measuring the difference in elevation (H) between the water level in the reference tube and the water level in the side arm of the outflow dripper (Figure 55.1).

55.2.2 Calculations

Calculate K_s using

$$K_s = \frac{4QL}{\pi t H d_c^2} \qquad (55.2)$$

where Q (mL) is the volume of water collected during time interval, t (s), L (cm) is the length of sample in the core, H (cm) is the difference in elevation between the water level in the reference tube and the water level in the side arm of the outflow dripper, and d_c (cm) is the inside diameter of the core. This yields K_s in units of cm s^{-1}.

55.2.3 COMMENTS

1 The shorter side tube in the acrylic end cap allows sufficient air escape during the saturation process to ensure submergence of the center tube and thus a continuous hydraulic connection between the outflow dripper and the sample. The woven stainless steel retaining screen (0.1-cm openings) in the end cap serves to reduce possible lifting of the sample during the measurements. The end cap also features a small inside lip that projects into the sample. This serves to center the cap on the core, to provide a better seal with the core, and to reduce "short-circuit" flow along the inside wall of the core. It is not recommended that filter paper be used on either end of the core, due to its potential for gradual plugging with suspended sediment during the flow measurements.

2 Because water flows upward in this constant head method, the end cap should be placed on the bottom of the core (i.e., the core is "upside down" in the tank) so that water flows from the top of the sample to the bottom.

3 The outflow dripper is designed to allow accurate determination of the hydraulic head difference across the core, which is particularly important when the head difference is small (i.e., 1–2 cm), as used on highly permeable samples. The side arm accounts for the pressure required for drop formation, which can cause the level of true zero pressure (i.e., the water level in the side arm) to be as much as 0.6 cm above the drip point. The coarse silica sand (16-mesh size) in the dripper elbow damps out the pressure pulses of drop formation and drip, which could otherwise induce a 0.5- to 1.0-cm oscillation of the side arm water level. The outflow dripper, side arm, and reference tube should all be the same internal diameter in order to cancel out capillary rise effects (a capillary rise of 0.6 cm is possible for 0.5-cm-diameter tubing).

4 Advantages of the tank method include:

a. Up to 36 cores can be saturated at a time in a 120-cm long by 30-cm wide tank, and one operator can measure the K_s of several cores simultaneously by using several outflow assemblies.

b. The hydraulic head difference (H) can be conveniently adjusted from approximately 0.5–25.0 cm (assuming that the tank is approximately 30 cm high), thus allowing a wide range of K_s (about 3×10^{-1} cm s^{-1} to 5×10^{-8} cm s^{-1}) to be measured rapidly.

c. The use of 270-mesh nylon cloth allows the cores to be moved directly to a tension table for measurement of the water desorption relationship (Chapter 53).

55.3 FALLING HEAD SOIL CORE METHOD
(KLUTE AND DIRKSEN 1986)

This method determines the K_s of cylindrical, 7.62-cm (3-inch) I.D. by 7.62-cm (3-inch) long cores of soil or other porous medium. The soil is first wetted to saturation and then water is flowed through the sample under a falling head of water. The apparatus and procedures described below are designed to complement the constant head soil core method (Section 55.2).

55.3.1 APPARATUS AND PROCEDURE

Collect and prepare the samples as indicated for the constant head soil core method (Section 55.2.1), except for the following changes:

1 Reverse the ends of the core to which the end cap and nylon cloth are attached, i.e., the end cap is attached to the top end of the core, and the nylon cloth to the bottom end of the core. This causes water to flow from the top of the sample to the bottom (see also comment 1, Section 55.3.3).

2 At the end of the 4-d saturation period, a syringe (or some alternate means) is used to fill the end cap completely with water. A small air bubble in the end cap is not a problem; however, an air-free end cap is preferable (see also comment 2, Section 55.3.3).

3 The mariotte reservoir, water outflow assembly, and reference tube used in the constant head method are not required for the falling head method.

4 Fill the central tube of the end cap with water by submerging it in the tank. Connect the central tube to a falling head standpipe which has previously been filled with water from the tank (Figure 55.2). A particularly convenient falling head standpipe consists of a 50-mL capacity by 1-cm^2 cross-sectional area glass burette which is fitted with a stopcock and a scale that increases upward, rather than downward. The 1-cm^2 cross-sectional area combined with the upward-reading scale means that the 1-mL graduations on the burette are 1 cm apart and correspond with height above an arbitrary datum. Fill the second end cap tube with water by submerging it in the tank, and then clamp. Make sure that the water level in the tank is at the maximum allowed by the overflow tube. Adjust the height of the standpipe so that the zero point of its scale corresponds with the water level in the tank. With

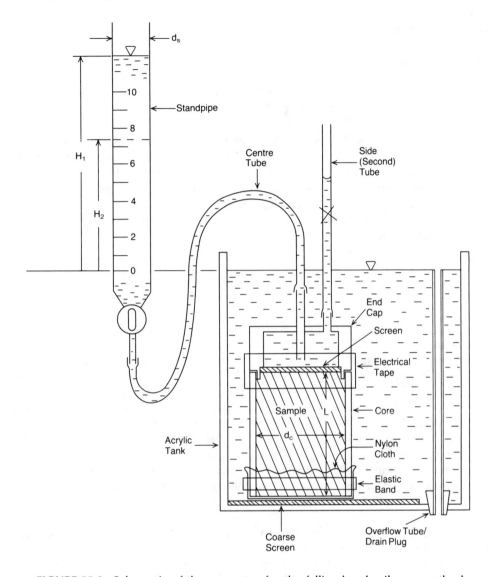

FIGURE 55.2. Schematic of the apparatus for the falling head soil core method.

this arrangement, the constant water level in the tank serves as a convenient datum and the standpipe scale gives directly the height of the standpipe water level (H) above the datum.

5   Open the standpipe stopcock and measure the time for the water level to fall from H_1 to H_2, H_3, H_4, etc. (Figure 55.2). The initial water level in the standpipe should be somewhat above the level chosen for H_1 so that initial, short-term transient effects are not included in the measurements. Such effects are due primarily to the sudden, hydrostatic pressure-induced expansion of the end cap tubing and/or compression of air inside the end cap, when the standpipe stopcock is opened. Successive measurement of about five H levels (i.e., H_1, H_2, H_3, H_4, H_5) is recommended, starting at $H_1 = 50$–70 cm (see Section 55.3.2 and comment 2, Section 55.3.3).

55.3.2 Calculations

Calculate K_s using

$$K_{si} = \left(\frac{d_s}{d_c}\right)^2 \frac{L}{t} \ln\left(\frac{H_i}{H_{i+1}}\right); \quad i = 1,2,3,4,5 \tag{55.3}$$

where d_s (cm) is the inside diameter of the standpipe, d_c (cm) is the inside diameter of the core, L (cm) is the length of sample in the core, H (cm) is the height of water in the standpipe relative to the datum, and t (s) is the time required for the water level in the standpipe to fall from H_i (cm) to H_{i+1} (cm). This yields K_s in units of cm s^{-1}. Successive measurement of five H levels allows four individual K_s values to be determined, which may then be averaged to obtain a more representative mean K_s value. The standpipe may also be refilled to obtain additional K_s measurements.

55.3.3 COMMENTS

1   Ideally, the end cap and nylon cloth should always be placed so that water flows from the top of the sample to the bottom. For the falling head method, this requires that the end cap be attached to the top of the core, since water flows downward through the sample. For the constant head method (Section 55.2), however, the end cap should be placed on the bottom of the core (i.e., the sample is "upside down" in the tank), since water flows upward through the sample in this method. Under most circumstances, however, the direction of water flow through the sample has no significant effect on the K_s calculation. Consequently, core orientation is usually of little concern, and more importantly, the end cap does not need to be moved to the opposite end of the core when switching from the constant head method to the falling head method or vice versa.

2   As the water level in the standpipe falls during a measurement, any air entrapped within the end cap will expand in response to the declining hydrostatic pressure. This can introduce a significant error into the K_s calculation (Equation 55.3) if the volume of entrapped air and/or the (H_i/H_{i+1}) ratio are large. It is therefore recommended that the volume of air in the end cap and associated tubing be kept to a practical minimum and that the (H_i/H_{i+1}) ratio be no greater than about 1.1.

3   The range of K_s that can be measured practically using the above described falling head method is about 10^{-3} cm s^{-1} to 10^{-7} cm s^{-1}. This range can be extended somewhat by adjusting the diameter of the standpipe. Increase the standpipe diameter for a larger maximum K_s and decrease the diameter for a smaller minimum K_s.

4   As mentioned earlier, the falling head method described here is designed to complement the constant head method described in Section 55.2. When flow is found to be too slow for the constant head method, the falling head method can be set up quickly with a minimum of effort. The two methods therefore constitute a versatile "package" which allows a wide range of K_s to be measured rapidly (approximately 10^{-1} cm s^{-1} to 10^{-7} cm s^{-1}).

55.4 OTHER METHODS FOR LABORATORY MEASUREMENT OF K_s

Other methods for measuring K_s in the laboratory include the clod method and the undistributed soil block method. These methods are useful in soils having characteristics that preclude the collection of good, undisturbed (intact) core samples for use in the constant head or falling head soil core methods (Sections 55.2 and 55.3, respectively).

In essence, the clod method involves the collection of an intact, naturally occurring soil clod (approximately 10–15 cm diameter), trimming the clod to a cylindrical or cuboid shape, encasing the sides of the clod in paraffin wax and installing end caps, saturating the clod in water, and determining K_s using the standard constant or falling head test procedures. An important practical disadvantage of this method is the difficulty in trimming the clod to a cylindrical or cuboid shape. Perhaps more importantly, there is also a concern that the K_s value obtained is not representative of the whole soil because intact clods usually do not contain the full range of the soil macrostructure. Details of the clod method may be found in Sheldrick (1984).

The undisturbed soil block method involves carving a cuboid block of soil out of the wall of a sampling pit. The block can range in size from 10–30 cm on a side (or more), and its vertical and horizontal cross sections can be either square or rectangular. The walls of the block are encased in paraffin wax, resin, or gypsum. A screen is placed on the base of the block to prevent slumping of the soil, and the block is then set on a funnel which drains into a graduated cylinder. A small, constant head of water (≈ 1–3 cm) is ponded on the upper surface of the block and the flow rate measured until it becomes constant (i.e., the rate of water flow in the top of the block equals the rate of flow out the bottom). The K_s value is determined using the standard constant or falling head test procedures. Important advantages of this method are that the size of the blocks can be adjusted so that the full range of soil macrostructure is adequately sampled; smearing and compaction of clayey and low-strength soils can be minimized; and the anisotropy in K_s (i.e., its x, y, and z components) can be measured on the same soil block by alternately rotating the block and flowing water through the side walls (after sealing off the block ends and opening up the appropriate sides). Disadvantages of the method are that it is slow and labor intensive, the blocks are difficult to saturate completely with water, and it is difficult to measure a wide range in K_s values due to the size of the blocks. Details on the undisturbed soil block method may be found in Bouma and Dekker (1981) and references contained therein. Additional methods for laboratory measurement of K_s may be found in Klute and Dirksen (1986) and Youngs (1991).

REFERENCES

Amoozegar, A. 1988. Preparing soil cores collected by a sampling probe for laboratory analysis of soil hydraulic properties. Soil Sci. Soc. Am. J. 52: 1814–1816.

Bouma, J. and Dekker, L. W. 1981. A method for measuring the vertical and horizontal K_{sat} of clay soils with macropores. Soil Sci. Soc. Am. J. 45: 662–663.

Bouwer, H. 1978. Groundwater hydrology. McGraw-Hill, Toronto. 480 pp.

Cameron, K. C., Harrison, D. F., Smith, N. P., and McLay, C. D. A. 1990. A method to prevent edgeflow in undisturbed soil cores and lysimeters. Aust. J. Soil Res. 28: 879–886.

Elrick, D. E., Sheard, R. W., and Baumgartner, N. 1981. A simple procedure for determining the hydraulic conductivity and water retention of putting green soil mixtures. Pages 189–200 in Proc. IV Int. Turfgrass Research Conf., Guelph, Ont.

Klute, A. and Dirksen, C. 1986. Hydraulic conductivity and diffusivity: laboratory methods.

Pages 687–734 in A. Klute, Ed. Methods of soil analysis. Part 1. Physical and mineralogical methods. 2nd ed. Agronomy No. 9, American Society of Agronomy, Madison, WI.

Koorevaar, P., Menelik, G., and Dirksen, C. 1983. Elements of soil physics. Elsevier, New York. 228 pp.

McIntyre, D. S. 1974. Procuring undisturbed cores for soil physical measurements. Pages 154–165 in J. Loveday, Ed. Methods for analysis of irrigated soils. Commonwealth Agricultural Bureaux, Wilke and Company, Clayton, Australia.

Rogers, J. S. and Carter, C. E. 1986. Soil core sampling for hydraulic conductivity and bulk density. Soil Sci. Soc. Am. J. 51: 1393–1394.

Sheldrick, B. H. 1984. Analytical methods manual 1984. Land Resource Research Institute, Agriculture Canada, Ottawa, Pages 38/1–38/2.

Youngs, E. G. 1991. Hydraulic conductivity of saturated soils. Pages 170–179 in K. A. Smith and C. E. Mullins, Eds. Soil analysis: physical methods. Marcel Dekker, New York.

Chapter 56
Saturated Hydraulic Conductivity: Field Measurement

W. D. Reynolds

Agriculture Canada
Ottawa, Ontario, Canada

56.1 INTRODUCTION

It is well established that field or "*in situ*" measurements of saturated hydraulic conductivity (K_s) are essential for accurate determination of water movement in the field. As a consequence, field measurements of K_s are critically important in the design and monitoring of irrigation and drainage systems, manure impoundments, septic tanks, canals and reservoirs, sanitary landfills, and many other agricultural, industrial, and environmental installations (Bouwer 1978). It has also been established that field measurements of K_s are useful for predicting rainfall infiltration and runoff, and for characterizing changes in soil macrostructure as a result of changing land management practices (e.g., Hillel 1980, Gregorich et al. 1992).

When K_s is measured in the unsaturated zone (i.e., above the water table), it is often referred to as the "field-saturated" hydraulic conductivity, K_{fs} (Reynolds et al. 1983). This is in recognition of the fact that air bubbles are usually entrapped in porous media when it is "saturated" by downward-infiltrating water, particularly when the infiltration occurs under ponded conditions. The water content of a porous medium at "field saturation" is consequently lower than at complete or true saturation (for a brief comparison of "field saturation" and "unsaturation", see Chapter 59, Section 59.1). Depending on the amount of entrapped air, K_{fs} can be a factor of 2 or more below the truly saturated (i.e., no entrapped air) hydraulic conductivity, K_s (Bouwer 1978, Reynolds and Elrick 1987). For many unsaturated zone applications, K_{rs} is considered more appropriate than K_s because most natural and man-made infiltration processes result in significant air entrapment within the porous medium.

When K_{fs} is measured in the unsaturated zone, an important companion parameter, matric flux potential (ϕ_m), can also be determined. The ϕ_m parameter (which is related to sorptivity — see Chapter 57; Reynolds and Elrick 1987) is an indicator of the "capillarity" of an unsaturated porous medium; i.e., its ability to absorb water. Measures of both K_{rs} and ϕ_m (or sorptivity) are required for field prediction of ponded infiltration into unsaturated porous media (e.g., Swartzendruber and Hogarth 1991). Generally speaking, the magnitude of ϕ_m

Soil Sampling and Methods of Analysis, M. R. Carter, Ed.,
Canadian Society of Soil Science. © 1993 Lewis Publishers.

depends on the texture and structure of the porous medium and on the amount of water already present within the porous medium at the time of measurement. Coarse-textured and/or structured and/or wet porous media tend to have lower capillarity than fine-textured and/or structureless and/or dry porous media. Porous media which are already saturated or field saturated at the start of the measurement have zero capillarity (i.e., $\phi_m = 0$).

56.2 CONSTANT HEAD WELL PERMEAMETER METHOD (REYNOLDS AND ELRICK 1985, 1986, ELRICK ET AL. 1989)

This method provides *in situ* determinations in the unsaturated (vadose) zone of field-saturated hydraulic conductivity (K_{fs}) and matric flux potential (ϕ_m). It involves the measurement of the steady-state infiltration rate (recharge) required to maintain a steady depth of water in an uncased, cylindrical auger hole which terminates above the water table.

This method is also capable of giving estimates of the Green-Ampt wetting front potential for ponded infiltration, ponded and zero-ponded sorptivity, and the flow-weighted mean pore radius for "field-saturated" flow. Procedures for obtaining these parameters are given in Elrick and Reynolds (1992).

Several constant head well permeameter methods exist which differ to varying degrees in theory, procedure, and apparatus (e.g., Zanger 1953, Philip 1985, Reynolds and Elrick 1986, Stephens et al. 1987, Amoozegar 1989, Banton et al. 1991). The constant head well permeameter technique described below is known as the "Guelph Permeameter" method (Reynolds and Elrick 1986, Elrick et al. 1989), which is commercially available from Soilmoisture Equipment Corporation, Santa Barbara, CA.

56.2.1 APPARATUS AND PROCEDURE

1   Using a screw-type or bucket auger, excavate a hole or "well" to the desired depth (Figure 56.1). The well should be cylindrical and have a reasonably flat bottom (a flashlight works well for checking this), as these features are required by the well permeameter theory. The bottom of the well should be at least 20 cm above the water table to avoid interference caused by "mounding" of the water table up into the well. Auger-induced smearing and compaction of the well surfaces in fine-textured materials (which can result in artificially low K_{fs} and ϕ_m values) should be minimized within the measurement zone (Figure 56.2) by not augering when the material is very wet (above field capacity), by using a very sharp auger, by applying very little downward pressure on the auger, and by taking only small bites with the auger before emptying it out. The "two-finger/two-turn" rule for augering within the measurement zone seems to work reasonably well; i.e., once the top of the measurement zone is reached, use only two fingers on each hand to apply downward pressure on the auger (i.e., the weight of the auger provides most of the downward pressure), and make only two complete turns of the auger before emptying it out. If inspection of the well reveals smearing within the measurement zone (a smear layer generally appears as a smooth, "polished" surface under the light of a flashlight),

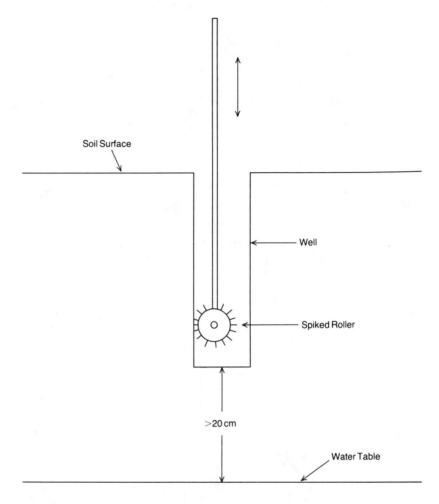

FIGURE 56.1. Schematic of the well and a desmearing apparatus for the constant head well permeameter method. (Adapted from Reynolds, W. D. and Elrick, D. E., Ground Water Monit. Rev. 6, 84, 1986.)

steps should be taken to remove it. A small, spiked roller mounted on a handle (Figure 56.1) seems to work reasonably well for this purpose. When the roller is run up and down the well wall several times the smear layer is broken up and plucked off by the sharpened, paddle-shaped spikes (see also comment 1, Section 56.2.3).

2 Stand the empty well permeameter in the well and support it with a large tripod (Figure 56.2) that clamps solidly to the permeameter reservoir (a tripod with 1.2-m adjustable legs works well). This gives the permeameter good stability against wind, plus a means for carrying the weight of the permeameter (when full of water) so that the water outlet tip (Figure 56.2) does not sink into the base of the well during a measurement. In unstable and/or low-permeability materials, it is advisable to backfill around the permeameter to the top of the measurement zone using pea gravel or coarse sand (Figure 56.2). The backfill material, which must have a much greater per-

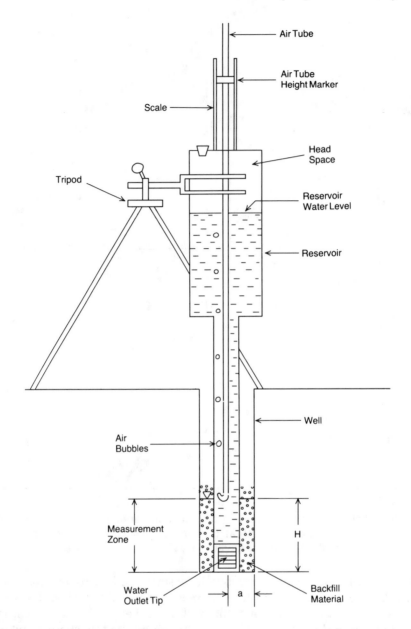

FIGURE 56.2. Schematic of a Guelph permeameter for use in the constant head well permeameter method.

meability than the material being tested, serves to prevent collapse of the well during a measurement and to produce a faster, more uniform bubbling of the permeameter air tube. It also helps to reduce well siltation (discussed further below).

3   Close the water outlet of the permeameter by pushing the air tube down into the outlet tip, and then fill the permeameter with water (Figure 56.2). Use water at ambient temperature to minimize the accumulation within the reservoir of bubbles of degassed air, which can obscure the permeameter

scale. Do not use distilled or deionized water, as this may encourage clay dispersion and subsequent siltation of the well surface during the measurement. In porous media that are particularly susceptible to siltation (primarily materials with high silt content), use of native water and/or addition of flocculant to the water may be advisable. Fill the permeameter reservoir to the top, leaving no air space. This minimizes overfilling of the well when flow is started (see also comment 2, Section 56.2.3).

4   Lift the air tube out of the outlet tip to establish and maintain the desired depth of water in the well (H) (Figure 56.2). The air tube should be raised slowly in order to prevent a sudden rush of water, which can erode the well (especially if backfill material has not been used), cause well siltation by stirring silt and clay into suspension, and cause excessive air entrapment within the porous medium. The desired depth of ponding in the well (H) is obtained by setting the air tube height marker at the appropriate point on the scale at the top of the permeameter. The permeameter is operating properly when air bubbles rise regularly up through the permeameter and into the reservoir. Further comments on this and related points are given in Section 56.2.3.

5   The rate of water flow out of the permeameter and into the porous medium is measured by monitoring the rate of fall of the water level in the permeameter reservoir (using the reservoir measurement scale and a stopwatch). When the rate of fall becomes constant (R), the flow rate out of the permeameter and into the porous medium has reached steady state. the K_{fs} and ϕ_m can now be calculated as indicated in Section 56.2.2.

56.2.2 Calculations

Calculate K_{fs} and ϕ_m using

$$K_{fs} = CAR/[2\pi H^2 + C\pi a^2 + (2\pi H/\alpha^*)] \tag{56.1}$$

$$\phi_m = CAR/[(2\pi H^2 + C\pi a^2)\alpha^* + 2\pi H] \tag{56.2}$$

where C is a dimensionless shape factor obtained from Figure 56.3, A (cm²) is the cross-sectional area (cell constant) of the permeameter reservoir, R (cm s⁻¹) is the steady rate of fall of the water level in the permeameter reservoir, H (cm) is the steady depth of water in the well (set by the height of the air tube), a (cm) is the radius of the well, and α^* (cm⁻¹) is a soil texture/structure parameter obtained from Table 56.1 (see also comment 4, Section 56.2.3). This yields K_{fs} in units of cm s⁻¹ and ϕ_m in units of cm² s⁻¹.

56.2.3 COMMENTS

1   Because water flows out of the well and into the soil in the constant head well permeameter method, any significant smearing, compaction, or siltation in the measurement zone can result in K_{fs} and ϕ_m values (Equation 56.1

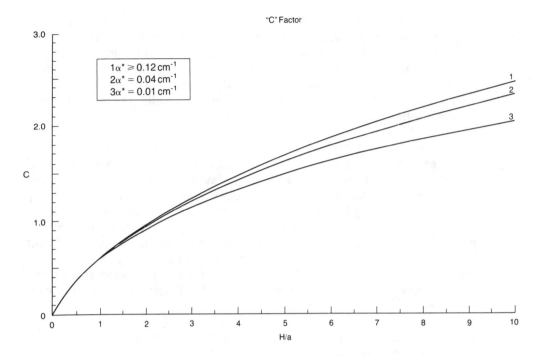

"C" Factor

$1\alpha^* \geq 0.12 \, \text{cm}^{-1}$
$2\alpha^* = 0.04 \, \text{cm}^{-1}$
$3\alpha^* = 0.01 \, \text{cm}^{-1}$

H/a

FIGURE 56.3. Shape factors (C) for use in the constant head well permeameter method. H is the depth of water in the well and a is the well radius. (Adapted from Reynolds, W. D. and Elrick, D. E., Soil Sci. 144, 282, 1987.)

and 56.2) which are much too low. In wet, structured silty clay soils, for example, the K_{fs} and ϕ_m values are often reduced by more than an order of magnitude if "normal" augering techniques are used, rather than the "two-finger/two-turn" rule. Proper and careful augering of the well is therefore essential in materials that are susceptible to smearing, compaction, and siltation. Desmearing techniques other than the spiked roller include the use of a cylindrical brush or quick-setting resin (Koppi and Geering 1986). It appears so far, however, that none of these desmearing procedures are completely effective under all conditions.

2 If there is no permeameter response (i.e., no bubbling) after the air tube has been raised to the desired H level, the well may be overfilled with water (i.e., the water level in the well is above the base of the air tube). To remedy this, remove water from the well (a syringe and Tygon® tubing often work well for this) until bubbling starts, which indicates that the set H level has been reached. Alternatively, one can simply allow the water level to fall under natural drainage, and the permeameter will start itself when the H level is eventually reached. A slowly bubbling permeameter means that the porous medium is of low permeability and/or the well is smeared/compacted/silted up. In low-permeability materials where the permeameter bubbles slowly, it is advisable to shade the permeameter reservoir from direct, hot sun in order to minimize solar heating of the head space (Figure 56.2) above the water surface. Thermal expansion of the air in the head space can prevent bubbling. In addition, extreme solar heating of the water in the permeameter will cause a significant reduction in the water viscosity. This will introduce

3   The time required for a permeameter to reach steady flow is determined primarily by the permeability of the material being tested, ranging from about 5–30 min in highly permeable materials ($K_{fs} \geq 10^{-4}$ cm s^{-1}) to as much as several hours in low-permeability materials ($K_{fs} < 10^{-5}$ cm s^{-1}). The range of K_{fs} that can be measured practically with a well permeameter is on the order of 10^{-2} cm s^{-1} to 10^{-6} cm s^{-1}, although this range has been extended somewhat with careful use and some equipment modifications (see Elrick and Reynolds 1992).

errors in the K_{fs} and ϕ_m calculations if not corrected (see Section 56.1 and Chapter 55, Section 55.1). A permeameter that will not bubble may also have an air (vacuum) leak in the reservoir. A permeameter "troubleshooting" section is included in the Soilmoisture Equipment Corp. user's manual.

4   The α^* value (Table 56.1) can usually be determined adequately by simple visual estimation of the appropriate soil texture/structure category for the measurement site. The texture/structure categories are broad enough that one should not be in error (using visual estimation) by more than one category, which generally introduces an error of no more than a factor of 2 or 3 into the K_{fs} and ϕ_m calculations. This level of accuracy is usually adequate for most practical applications. In addition, the sensitivity of K_{fs} and ϕ_m to the choice of α^* can be reduced further by adjusting the H level. The sensitivity of K_{fs} to the choice of α^* decreases as H increases, while the sensitivity of ϕ_m to α^* decreases as H decreases. Consequently, if one is interested primarily in K_{fs}, H should be as large as possible; and if the interest is primarily in ϕ_m, H should be as small as possible.

5   Other procedures for collecting and analyzing constant head well permeameter data may be found in Amoozegar and Warrick (1986) and Stephens et al. (1987). Further discussion of the method may be found in Elrick and Reynolds (1992).

56.3 CONSTANT HEAD PRESSURE (SINGLE-RING) INFILTROMETER METHOD (REYNOLDS AND ELRICK 1990)

This method provides *in situ* determinations in the unsaturated (vadose) zone of field-saturated hydraulic conductivity (K_{fs}) and matric flux potential (ϕ_m). It involves the measurement of the steady-state infiltration rate (recharge) required to maintain a steady depth of water (or constant water pressure) within a single ring inserted a small distance into the porous medium (soil). The definitions of K_{fs} and ϕ_m are given in Section 56.1.

The most important difference between the pressure infiltrometer method (Reynolds and Elrick 1990) and the traditional single-ring infiltrometer method (Bouwer 1986) is the theoretical treatment of water flow out of the ring and into the unsaturated soil. In the traditional single-ring infiltrometer approach, K_{fs} is determined assuming one-dimensional (1-D) vertical flow through and below the ring. It is well established, however, that the flow beneath the ring is not 1-D, but diverges laterally due to the capillarity of the unsaturated soil and the hydrostatic pressure of the ponded water in the ring (Bouwer 1986). This unaccounted for divergence results in an overestimate of K_{fs} when using the 1-D analysis. Attempts to prevent the flow divergence by adding a concentric outer guard ring (i.e., the "double-ring" infiltrometer) have been shown to be unsuccessful (Swartzendruber and Olson 1961a, b), and Bouwer (1986) has concluded that the traditional 1-D analysis can be accurate only for shallow ponding in impractically large rings. The pressure infiltrometer, on the other hand, takes flow divergence into account by using a three-dimensional flow analysis. This results in more realistic K_{fs} values and in estimates of ϕ_m, as well.

The pressure infiltrometer is also capable of giving estimates of the Green-Ampt wetting front potential for ponded infiltration, ponded and zero-ponded sorptivity, and the flow-weighted mean pore radius for "field-saturated" flow. Procedures for obtaining these parameters are given in Elrick and Reynolds (1992) and references contained therein.

The apparatus and procedures described below are designed to complement the constant head well permeameter method (Section 56.2), in that it allows measurements to be made at the soil surface (rather than below the soil surface), and the ring is designed to attach directly to the "Guelph permeameter" reservoir. For these reasons, the method is sometimes referred to as the "Guelph pressure infiltrometer" method.

56.3.1 APPARATUS AND PROCEDURE

1 Using a block of wood and a hammer, a hydraulic ram, or a driving apparatus (Figure 56.4), insert a 10- to 20-cm diameter ring into the soil to a depth of 3–5 cm. The ring should be thin-walled ($\approx$0.3 cm wall thickness) and beveled to a sharp cutting edge at the base (approximately a 20° bevel on the outside surface) in order to minimize resistance and soil compaction/shattering during the insertion process. It is advisable to construct the ring from stainless steel to prevent rusting. A stainless steel flange (1.6 cm wide by 1.0 cm thick) containing four equally spaced, threaded holes (0.5 cm diameter) should be welded to the top of the ring to facilitate inserting the ring into the soil, and to facilitate sealing the ring to the end-cap arrangement (Figure 56.4). The ring should be held as straight (vertical) as possible during the insertion process. If the ring tilts during insertion, no attempt should be made to straighten it. A slight tilt to the ring can be accommodated; however, if the tilt is significant the ring should be removed and restarted at a new location (see also comment 2, Section 56.3.3). Preparation of the soil surface (i.e., scraping, leveling, etc.) is neither required nor recommended, as this may alter the hydraulic properties of the soil.

2 After the ring has been inserted, the contact between the inside surface of the ring and the soil should be tamped lightly (using the blunt end of a pencil or similar implement) to minimize the possibility of "short-circuit" flow along the inside wall of the ring (Figure 56.4).

3 Attach the end cap to the mariotte reservoir and seal the end cap/reservoir arrangement to the ring using the four wing bolts, which screw into the threaded holes in the flange (Figure 56.5). Support the reservoir using a large tripod (same design as recommended for the constant head well permeameter — Section 56.2.1) to provide stability against wind and to carry the weight of the reservoir and water (Figure 56.5).

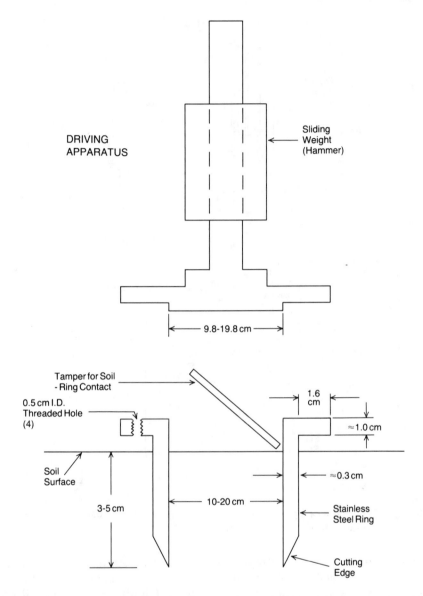

FIGURE 56.4. Schematic of a driving apparatus, ring, and soil tamper for use in the constant head pressure infiltrometer method.

4   Close the water outlet of the end cap by firmly pushing the air tube down into the outlet port, and then fill the reservoir to the top with water (Figure 56.5). The water should meet the same specifications as given in step 3 of Section 56.2.1 (constant head well permeameter).

5   Lift the air tube out of the outlet port to fill completely with water the head space between the soil surface and the end cap. This action also establishes and maintains the desired pressure head of water (H) on the soil surface inside the ring. The air tube should be raised slowly in order to prevent a sudden rush of water, which can erode the infiltration (soil) surface, cause siltation of the infiltration surface by stirring silt and clay into suspension,

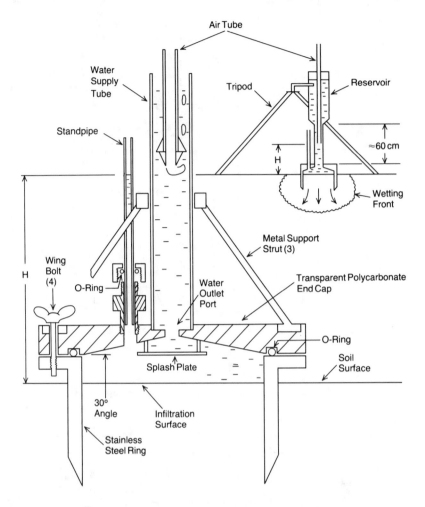

FIGURE 56.5. Schematic of a pressure infiltrometer apparatus for use in the constant head pressure infiltrometer method.

and cause excessive air entrapment within the porous medium. The pressure head of water acting on the infiltration surface (H) is determined by measuring the height of water in the standpipe (Figure 56.5). The infiltrometer is operating properly when air bubbles rise regularly up into the reservoir from the base of the air tube. There should be no water leaks or air leaks in the end cap or at the seal between the end cap and the ring.

6   The rate of water flow out of the infiltrometer and into the porous medium is measured by monitoring the rate of fall of the water level in the reservoir (using the reservoir measurement scale and a stopwatch). When the rate of fall becomes constant (R), the flow rate out of the infiltrometer and into the porous medium has reached steady state. The K_{fs} and ϕ_m values are calculated as indicated in Section 56.3.2.

56.3.2 Calculations

The K_{fs} and ϕ_m values can be calculated using single- or multiple-head procedures.

Table 56.1 Soil Texture/Structure Categories for Site Estimation of $\alpha*$ in the Constant Head Well Permeameter and Constant Head Pressure Infiltrometer Methods

Soil texture/structure category	$\alpha*$ (cm^{-1})
Compacted, structureless, clayey materials such as landfill caps and liners, lacustrine, or marine sediments, etc.	0.01
Soils which are both fine textured (clayey) and unstructured	0.04
Most structured soils from clays through loams; also includes unstructured medium and fine sands; the first choice for most soils	0.12
Coarse and gravelly sands; may also include some highly structured soils with large cracks and macropores	0.36

Adapted from Elrick, D. E., Reynolds, W. D., and Tan, K. A. *Ground Water Monit. Rev.*, 9, 184, 1989.

For the single-head approach,

$$K_{fs} = \alpha*GAR_1/[a(\alpha*H_1 + 1) + G\alpha*\pi a^2] \tag{56.3}$$

$$\phi_m = GAR_1/[a(\alpha*H_1 + 1) + G\alpha*\pi a^2] \tag{56.4}$$

where $\alpha*$ (cm^{-1}) is a soil texture/structure parameter obtained from Table 56.1 (see also comment 4, Section 56.2.3), A (cm^2) is the cross-sectional area (cell constant) of the infiltrometer reservoir, R_1 (cm s^{-1}) is the steady rate of fall of the water level in the infiltrometer reservoir, a (cm) is the inside radius of the ring, H_1 (cm) is the steady pressure head on the infiltration surface (set by the height of the air tube and determined by measuring the height of water in the standpipe), and G is a dimensionless shape factor obtained using

$$G = 0.316 \, (d/a) + 0.184 \tag{56.5}$$

where d (cm) is the depth of ring insertion into the soil. Equation (56.5) applies for 5 cm $\leq$ a $\leq$ 10 cm, 3 cm $\leq$ d $\leq$ 5 cm, and 5 cm $\leq$ H$_1$ $\leq$ 25 cm. Recent theoretical work suggests, however, that the upper limit of H$_1$ can often be extended to well over 1 m without significant reduction in accuracy. The units of K$_{fs}$ and ϕ_m are cm s^{-1} and cm^2 s^{-1}, respectively.

For the multiple-head approach,

$$K_{fs} = (G/a)[\Delta(AR)/\Delta H] \tag{56.6}$$

$$\phi_m = (G/a)(I - \pi a^2 K_{fs}) \tag{56.7}$$

where $\Delta(AR)/\Delta H$ and I are the slope and intercept, respectively, of the linear least squares regression line through a plot of (AR) vs. H, and the other parameters are as defined in Equation 56.3 to Equation 56.5. Equations 56.6 and 56.7 apply when two or more heads (H values) are sequentially set in ascending order (i.e., H$_1$ set first, H$_1$ < H$_2$ < H$_3$ ...) and without intervening drainage to obtain the corresponding (AR) values (i.e., A$_1$R$_1$, A$_2$R$_2$, A$_3$R$_3$, ...).

Comments on the relative merits of the single- and multiple-head calculations are given in Section 56.3.3, comment 6.

56.3.3 COMMENTS

1　In order to obtain values of K_{fs} and ϕ_m which are representative of the soil, care must be taken to minimize smearing and siltation of the infiltration surface, compaction or shattering of the soil during ring insertion, and "short-circuit" flow along the contact between the soil and the inside surface of the ring wall. Procedures similar to those discussed in Section 56.2.1 (constant head well permeameter) may be used to minimize smearing and siltation. Compaction and shattering, which occur primarily in clay-rich soils, can usually be minimized by avoiding the soil when it is excessively wet (compaction prone) and when it is excessively dry (shattering prone). Short-circuit flow can be minimized by lightly tamping the contact between the soil and the ring wall (Figure 56.4) and by careful insertion of the ring (see also comment 2 below).

　　Other factors that affect the accuracy of pressure infiltrometer measurements include macrostructure collapse during insertion of the ring and cut-off of nonvertical macropores by the ring wall. There appears to be little that can be done to avoid these problems.

2　No attempt should be made to straighten a ring if it tilts during the insertion process. This is because straightening will tend to compact the soil inside the ring, as well as open up a gap between the ring and the soil on one side. A ring with excessive tilt (say, greater than 10°) should be removed and started again at a new location. Such a degree of tilt will cause non-vertical flow through the soil, as well as impart an impractical lean to the infiltrometer reservoir. These situations may have an adverse effect on the representativeness of the K_{fs} and ϕ_m determinations.

3　If ponding occurs around the outside of the ring during a measurement, critical assumptions in the theory of the method have been violated and the measurement should be abandoned. The appearance of a wetting front, however, is both admissible and even expected when the depth of ring insertion is small (Figure 56.5).

4　A slowly bubbling infiltrometer means that the soil is of low permeability and/or the infiltration surface is smeared/compacted/silted up. A rapidly bubbling infiltrometer means that the soil is of high permeability and/or the soil has been shattered during ring insertion and/or short-circuit flow is occurring. An infiltrometer that is bubbling slowly is susceptible to solar heating effects, and the procedures for minimizing these adverse effects (as well as for general "troubleshooting") are similar to those given in Section 56.2.3 for the constant head well permeameter.

5　The time required for the infiltrometer to reach steady flow tends to increase as the permeability and water content of the material decrease, and as the ring radius and depth of insertion increases. Steady flow is often attained within 5 min using small rings (10-cm diameter) in highly permeable soils, and within 60–120 min using large rings (20-cm diameter) in low-permeability materials. The range of K_{fs} that can be measured practically with a pressure infiltrometer is on the order of 5×10^{-2} cm s^{-1} to 1×10^{-7} cm s^{-1}.

6 The primary advantages of the multiple-head analysis for calculating K_{fs} and ϕ_m (Equations 56.6 and 56.7) are that independent measures of K_{fs} and ϕ_m are obtained, and an α^* value does not have to be estimated. An important limitation, however, is that soil heterogeneity in the form of layering, cracks, worm holes, root channels, etc., can result in unrealistic and invalid (i.e., negative) K_{fs} and ϕ_m values. When a measurement produces either a negative K_{fs} or a negative ϕ_m, both K_{fs} and ϕ_m should be recalculated using the single-head analysis (Equations 56.3 and 56.4). Fortunately, the pressure infiltrometer method is not highly susceptible to soil heterogeneity, and "success" rates for multiple-head measurements (i.e., both K_{fs} and ϕ_m positive) are often greater than 90%.

The primary advantages of the single-head analysis for calculating K_{fs} and ϕ_m (Equations 56.3 and 56.4) are the increased speed of setting only one head (H_1) and the avoidance of negative K_{fs} and ϕ_m values. A limitation, however, is the necessity for estimating α^*, which, if done incorrectly, may produce K_{fs} and ϕ_m values of reduced accuracy. Notwithstanding this, recent studies suggest that the single-head analysis yields K_{fs} and ϕ_m values which are usually accurate to within a factor of 2 when α^* is site estimated and selected from the categories in Table 56.1 (see also comment 3, Section 56.2.3). This level of accuracy is sufficient for most practical applications. In addition, the sensitivity of K_{fs} and ϕ_m to the choice of α^* can be further reduced by choosing the H_1 value as per the recommendations for the constant head well permeameter method (comment 3, Section 56.2.3).

7 Once a pressure infiltrometer measurement has been completed, the infiltrometer ring and its contained soil provide a convenient intact soil core for laboratory measurement of K_s (Chapter 55), the soil water desorption relationship (Chapter 53), bulk density (Chapter 50), and porosity (Chapter 54).

8 Additional information concerning the pressure infiltrometer method may be found in Elrick and Reynolds (1992).

56.4 OTHER METHODS FOR MEASURING K_{fs} IN THE FIELD

There are many other methods for field measurement of K_{fs} both above and below the water table. The most popular methods for application above the water table include the air-entry permeameter method, the shallow-well pump-in method, the inversed auger hole method, and the single- and double-ring infiltrometer methods. The auger hole and piezometer methods are the most popular for use below the water table. These more traditional methods are described in detail in several established manuals (e.g., Bouwer and Jackson 1974, Kessler and Oosterbaan 1974, U.S. Bureau of Reclamation 1978, Bouwer 1986, Amoozegar and Warrick 1986, Hendrickx, 1990, Youngs, 1991) and therefore will not be discussed here.

REFERENCES

Amoozegar, A. 1989. A compact constant-head permeameter for measuring saturated hydraulic conductivity of the vadose zone. Soil Sci. Soc. Am. J. 53: 1356–1361.

Amoozegar, A., and Warrick, A. W. 1986. Hydraulic conductivity of saturated soils: field methods. Pages 735–770 in A. Klute, Ed. Methods of soil analysis. Part 1. Physical and mineralogical methods. 2nd ed. Agronomy No. 9. American Society of Agronomy, Madison, WI.

Banton, O., Côté, D., and Trudelle, M. 1991. Determination in the field of saturated hydraulic conductivity using the Côté constant head infiltrometer: theory and mathematical approximations. Can. J. Soil Sci. 71: 119–126.

Bouwer, H. 1978. Pages 123–124 in Groundwater hydrology. McGraw-Hill, Toronto.

Bouwer, H. 1986. Intake rate: cylinder infiltrometer. Pages 825–844 in A. Klute, Ed. Methods of soil analysis. Part 1. Physical and mineralogical methods. 2nd ed. Agronomy No. 9. American Society of Agronomy, Madison, WI.

Bouwer, H. and Jackson, R. D. 1974. Determining soil properties. Pages 611–672 in J. van Schilfgaarde, Ed. Drainage for agriculture, Agronomy No. 17. American Society of Agronomy, Madison, WI.

Elrick, D. E. and Reynolds, W. D. 1992. Infiltration from constant head well permeameters and infiltrometers. Pages 1–24. in G. C. Topp, W. D. Reynolds, and R. E. Green, Eds. Advances in measurement of soil physical properties: bringing theory into practice. SSSA Spec. Publ. 30. Soil Science Society of America, Madison, WI.

Elrick, D. E., Reynolds, W. D., and Tan, K. A. 1989. Hydraulic conductivity measurements in the unsaturated zone using improved well analyses. Ground Water Monit. Rev. 9: 184–193.

Gregorich, E. G., Reynolds, W. D., McGovern, M. A., Culley, J. L. B., and Curnoe, W. E. 1992. Changes in some physical properties with depth in a structurally degraded soil under no-tillage. Soil Tillage Res. (in press).

Hendrickx, J. M. H. 1990. Determination of hydraulic soil properties. Pages 67–76 in M. G. Anderson and T. P. Burt, Eds. Process studies in hillslope hydrology. John Wiley & Sons, Toronto.

Hillel, D. 1980. Pages 5–72 in Applications of soil physics. Academic Press, Toronto.

Kessler J. and Oosterbaan, R. J. 1974. Determining hydraulic conductivity of soils. Pages 253–296 in Drainage principles and applications, Vol. 3, Publ. 16. International Institute for Land Reclamation and Improvement, Wageningen, The Netherlands.

Koppi, A. J. and Geering, H. R. 1986. The preparation of unsmeared soil surfaces and an improved apparatus for infiltration measurements. J. Soil Sci. 37: 177–181.

Philip, J. R. 1985. Approximate analysis of the borehole permeameter in unsaturated soil. Water Resour. Res. 21: 1025–1033.

Reynolds, W. D. and Elrick, D. E. 1985. *In situ* measurement of field saturated hydraulic conductivity, sorptivity, and the alpha-parameter using the Guelph permeameter. Soil Sci. 140: 292–302.

Reynolds, W. D. and Elrick, D. E. 1986. A method for simultaneous in-situ measurement in the vadose zone of field-saturated hydraulic conductivity sorptivity, and the conductivity-pressure head relationship. Ground Water Monit. Rev. 6: 84–95.

Reynolds, W. D. and Elrick, D. E. 1987. A laboratory and numerical assessment of the Guelph permeameter method. Soil Sci. 144: 282–299.

Reynolds, W. D. and Elrick, D. E. 1990. Ponded infiltration from a single ring. I. Analysis of steady flow. Soil Sci. Soc. Am. J. 54: 1233–1241.

Reynolds, W. D., Elrick, D. E., and Topp, G. C. 1983. A reexamination of the constant head well permeameter method for measuring saturated hydraulic conductivity above the water table. Soil Sci. 136: 250–268.

Stephens, D. B., Lambert, K., and Watson, D. 1987. Regression models for hydraulic conductivity and field test of the borehole permeameter. Water Resour. Res. 23: 2207–2214.

Swartzendruber, D. and Hogarth, W. L. 1991. Water infiltration into soil in response to ponded-water head. Soil Sci. Soc. Am. J. 55: 1511–1515.

Swartzendruber, D. and Olson, T. C. 1961a. Sand-model study of buffer effects in the double-ring infiltrometer. Soil Sci. Soc. Am. Proc. 25: 5–8.

Swartzendruber, D. and Olson, T. C. 1961b. Model study of the double-ring infiltrometer as affected by depth of wetting and particle size. Soil Sci. Soc. Am. Proc. 92: 219–225.

U.S. Bureau of Reclamation. 1978. Drainage manual. U.S. Department of the Interior, Superintendent of Documents, U.S. Government Printing Office, Washington, D.C. Pages 74–97.

Youngs, E. G. 1991. Hydraulic conductivity of saturated soils. Pages 180–199 in K. A. Smith and C. E. Mullins, Eds. Soil analysis: physical methods. Marcel Dekker, New York.

Zanger, C. N. 1953. Theory and problems of water percolation. Pages 69–71 in U.S. Department of the Interior, Bureau of Reclamation, Eng. Monogr. No. 8, Denver, CO.

Chapter 57
Unsaturated Hydraulic Conductivity and Sorptivity: Laboratory Measurement

F. J. Cook

CSIRO, Canberra, ACT, Australia

G. P. Lilley and R. A. Nunns

Landcare Research New Zealand

Lower Hutt, New Zealand

57.1 INTRODUCTION

Measurements of hydraulic conductivity and sorptivity (Philip 1957) are necessary to make use of soil physics (Clothier and Scotter 1982, White and Perroux 1987, 1989, Clothier et al. 1988, White and Broadbridge 1988) in applications such as design of dripper irrigation (Cook et al. 1986), sprinkler irrigation (Cook 1988), and drainage (Scotter et al. 1990). Many methods exist for the measurement of sorptivity (S) and unsaturated hydraulic conductivity (K) in the laboratory (Clothier and Wooding 1983, Klute and Dirksen 1986, Samani and Willardson 1987, White 1989). However, it is often more pertinent to obtain a measure of K and S for undisturbed field cores. The method described here measures, on undisturbed cores exhumed from the field, S and K at matric potentials in the range 0 to −1 kPa (Ward and Hayes 1991). This is a modification of the method of Clothier and White (1981). Although Clothier and White (1981) originally used their method to make measurements in the field, the method described here can be used either in the laboratory or in the field.

The method uses a tension permeameter as the means of supplying water to the soil. White et al. (1992) have described the history and rationale for the development of tension permeameters (Figure 57.1). The design of tension permeameters is described in Perroux and White (1988).

57.1.1 Calculation of Sorptivity

At early times in the infiltration of water into an initial dry soil, the process is dominated by the capillary forces (Philip 1957). Philip (1957) showed:

$$\lim_{t \to 0} I \to St^{1/2} \tag{57.1}$$

Soil Sampling and Methods of Analysis, M. R. Carter, Ed.,

Canadian Society of Soil Science. © 1993 Lewis Publishers.

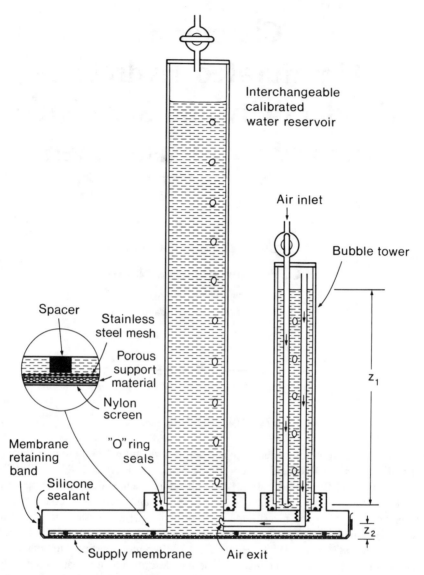

FIGURE 57.1. Tension permeameter. (Adapted from Perroux, K. M. and White, I., *Soil Sci. Soc. Am.*, 1205, 1988.)

where I is the cumulative infiltration (m), and t is time (s). Thus, for early times if I is plotted against $t^{1/2}$ then the slope of the straight line is S. Sand is used as a contact medium between the permeameter and the soil. This has a much greater S than the soil, so the S for the soil can easily be distinguished from the sand.

57.1.2 Calculation of Unsaturated Hydraulic Conductivity

In the original method of Clothier and White (1981) water was supplied to the top of the exhumed core by the tension permeameter, while the bottom was open to the atmosphere. The steady-state flux of water passing through this core was q, while the top of the core was at a matric potential determined by the permeameter, ψ^0, and the bottom was at zero

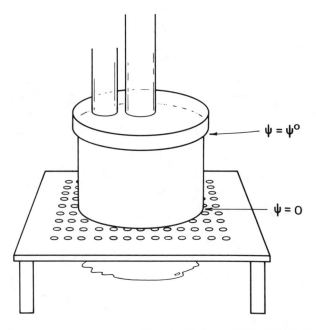

FIGURE 57.2. Experimental apparatus as used by Clothier and White (1981). (From Clothier, B. E. and White, I., *Soil Sci. Soc. Am. J.*, 45, 241, 1981. With permission.)

matric potential (Figure 57.2). Clothier and White (1981) then calculated the hydraulic conductivity, K, assumed to be at the mean matric potential of $\psi^0/2$, using:

$$k(\psi^0/2) = \frac{qL}{(\psi^0 + L)} \tag{57.2}$$

where L is the length of the core.

The nonlinear dependence of hydraulic conductivity on potential results in a nonlinear potential gradient within the core. The nonlinearity of potential gradient makes Equation 57.2 an incorrect approximation (Cook 1991) and it overestimates the value of $K(\psi^0/2)$. Cook (1991) provided a method for calculating K if the saturated hydraulic conductivity is also measured on the same core. However, one way to overcome the problem associated with the method of Clothier and White (1981) is to apply the same matric potential, ψ^0, to the bottom of the core (Figure 57.3). Calculation of K is thus simplified to:

$$k(\psi^0) = q \tag{57.3}$$

57.2 METHODS

57.2.1 CORE SAMPLING

The taking of the soil core so that disturbance to the structure and fabric is minimized is an important step in any laboratory measurements of hydraulic conductivity. The method outlined here is that which we have found to be satisfactory. The radius of tension permeameters used is approximately 52 mm, hence the core liners used have an outside radius of approx-

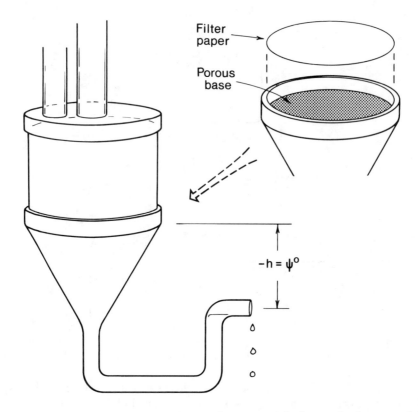

FIGURE 57.3. Büchner funnel, as used to apply a potential of ψ^0 to the bottom of the soil core upon which a tension permeameter maintains ψ^0 at the top.

imately 50 mm. The length of the cores is here 75 mm, but other lengths could be used. The material used for the core liner can be PVC, aluminum, or stainless steel. We have found stainless steel to be the most satisfactory because of its rigidity for thin wall thicknesses, and a sharp cutting edge can easily be maintained. Although initially expensive, the liners have a long life.

The soil cores are taken using the following procedure:

1   A small amount of light grease is smeared on the inside bottom 20 mm of the core liner. This lubricates the core and can prevent preferential flow down the side of the core if saturated hydraulic conductivity measurements are subsequently made.

2   The core liner is placed on the surface of the soil from which the core sample is to be taken. Subsequently, this may be the exposed surface of a soil horizon at some depth. The core is then pressed lightly into the soil to some small depth. A soil pedestal about 20 mm in height and of slightly larger radius than the liner is then carefully carved with a sharp knife (Figure 57.4a).

3   The core liner is carefully pushed downwards about 15–20 mm until that portion of the pedestal of soil created is enclosed within the liner (Figure 57.4b).

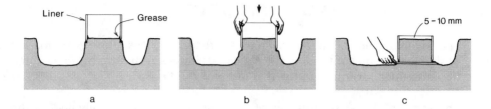

FIGURE 57.4. Taking a soil core sample: (a) carving a soil pedestal; (b) pushing the core liner down; and (c) soil core sample removal.

4   Steps 2 and 3 are repeated until the soil is approximately 5 mm below the top of the liner (Figure 57.4c).

5   The soil in the liner is subsequently cut off carefully below the bottom of the liner using a large spatula or paint scraper (Figure 57.4c). The excess soil from the bottom of the core liner is then trimmed smooth with the bottom of the core liner using a sharp knife. Snap-off blade knives are useful for this task. Any excess grease on the top rim of the soil can be carefully removed with a spatula.

6   Finally, the core is placed in a plastic bag and sealed, or wrapped in self-adhesive plastic film. This package is then placed in a foam-lined carrying case for transport to the laboratory.

57.2.2 MEASURING SORPTIVITY

57.2.2.1 MATERIALS

1   A uniform fine-grade sand. This is used to provide good contact between the soil in the core and the membrane on the bottom of the permeameter. Details on the required properties of this sand can be found in Perroux and White (1988).

2   A soil core, obtained by the method described in Section 57.2.1.

3   A tension permeameter.

4   A stand or table as shown in Figure 57.2.

5.   A stopwatch.

6.   A sheet on which to record the results.

57.2.2.2 PROCEDURE

1   The weight of the core and its volume are recorded. This requires estimation of the volume by subtracting the space left between the top of the soil and the core liner, from the internal volume of the liner.

2   The sand is placed on the top of the soil in the core and leveled. If the volume of sand is measured, then this volume can be used as an estimate of space between the top of the soil and the top of the core liner.

3   The tension permeameter is filled by placing the base in a container of water, opening the valve on the top of the standpipe, and sucking water into the permeameter until the reservoir standpipe is filled. The tension permeameter is set to the required potential,

$$\psi^0 = z_2 - z_1 \qquad (57.4)$$

(where z_1 and z_2 and defined in Figure 57.1) by adjusting the water level in the bubble tower (z_1). Care needs to be taken to remove all the air bubbles out of the base of the permeameter before using it.

4   Sorptivity is measured by initially placing the core on a surface with the base open to the atmosphere (Figure 57.2). The water level in the permeameter is recorded. The permeameter is placed on the top of the core with a slight twisting motion to ensure good contact, and a stopwatch is started. The time for every 10-mm drop in the reservoir is recorded until water starts to drip from the bottom of the core.

5   The cumulative infiltration rate, I, is calculated by:

$$I = \frac{(h_t - h_o) r_s^2}{r_c^2} \qquad (57.5)$$

where h_t and h_o are the levels in the permeameter reservoir (m), at time t (s), and at the start, respectively; r_s is the internal radius of the permeameter reservoir (m) and r_c is the internal radius of the liner (m). I is plotted against $t^{1/2}$ and the straight line portion of the graph associated with infiltration into the soil used to calculate S by regression (Figure 57.5).

6   The permeameter is carefully slid off the core. The core is then transferred to the Buchner funnel for measurement of hydraulic conductivity.

57.2.3 MEASURING UNSATURATED HYDRAULIC CONDUCTIVITY

57.2.3.1 MATERIALS

1   Soil core from sorptivity measurements (Section 57.2.2).

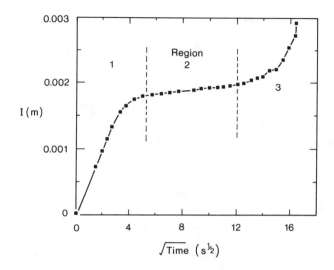

FIGURE 57.5. I vs. time$^{1/2}$. Region 1 of the graph is due to sorption by the sand; region 2 is due to sorption by the soil, and the data in this region can be used to calculate S; region 3 is where the flux is tending towards steady state.

2   Buchner funnel apparatus (Figures 57.3 and 57.6).

3   "Fast-flow" filter paper.

4   Tension permeameter.

5   Stopwatch.

6   Sheet for recording results.

57.2.3.2 PROCEDURE

1   The Buchner funnel needs to be set up. This is done by placing the funnel in a rack or stand (Figure 57.6) and placing a "fast-flow" filter paper in the Buchner funnel (Figure 57.3). Water is forced by means of a squeeze bottle through the tubing, funnel, and filter paper (Figure 57.6). After removing the squeeze bottle the tubing is lowered to give the same potential:

$$\psi^0 = -h \qquad (57.6)$$

where h is defined in Figure 57.3. The core is then placed on the filter paper in the Buchner funnel.

2   The tension permeameter is filled by the method described in step 2 (Section 57.2.2) and the potential checked to see if it is the same as the Buchner funnel. If the potential on the tension permeameter needs adjusting, then the procedure is described in step 3 (Section 57.2.2). The initial level of water in the reservoir of the permeameter is recorded.

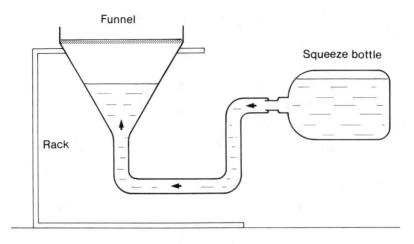

FIGURE 57.6. Setting up the Büchner funnel for use.

3   The permeameter is then placed on the top of the core with a slight twisting motion in order to ensure good contact. A stopwatch is started. Periodic measurements of the water level in the reservoir are made (Table 57.1). From these data the flux density can be determined. The onset of the steady-state regime can be observed. For the example given in Table 57.1, steady-state was reached in 20 min. The time taken to reach steady state can vary from minutes to hours, depending on the soil. The interval between readings will also depend on the soil. Hence, it is probably better to record the time of every 10- or 20-mm drop in the reservoir level, rather than the level change at set time intervals.

4   Calculation of the hydraulic conductivity can be carried out using:

$$k(\psi^0) = \frac{C(h_2 - h_1)r_s^2}{r_c^2(t_2 - t_1)} \tag{57.7}$$

where h_1 and h_2 are the levels in the reservoir at times t_2 and t_1, respectively, during steady-state flow. C is a conversion factor to change K into the SI units of m s^{-1}. An example using the data from Table 57.1 is given below.

Given,

$$\begin{array}{ll} r_s = 9.5 \text{ mm} & r_c = 49 \text{ mm} \\ t_1 = 20 \text{ min} & h_1 = 72 \text{ mm} \\ t_2 = 70 \text{ min} & h_2 = 93 \text{ mm} \\ C = 1/(60 \times 1000) \end{array}$$

then,

$$k(\psi^0) = \frac{(93 - 72) \times 9.5^2}{60000 \times 49^2 \times (70 - 50)}$$

$$= 1.36 \times 10^{-6} \text{ m s}^{-1}$$

Table 57.1 Data from an Example of the Measurement of K with a Tension Permeameter

Time (min)	Height in standpipe (mm)	Height difference time (mm/min)	Comments
0	10	—	Sand wetting
5	50	8	
10	65	1	
15	69	0.8	
20	72	0.6	Steady-state begins
30	76	0.4	
40	80	0.4	
50	85	0.5	
60	89	0.4	
70	93	0.4	

5   Once steady state has been obtained the permeameter can be removed. The sand is quickly removed from the top of the core and the core weight recorded, so that the water content associated with ψ^0 can be calculated.

6   However, if K at other potentials is to be measured, we must place a fresh layer of sand on the core and return to step 2 (Section 57.2.3) and repeat the procedure for a new value of ψ^0. To avoid hysteresis the sequence of potentials should monotonically increase or decrease.

7   After the last measurement on the core, the whole core is oven dried and the weight recorded.

8   The bulk density and water contents associated with the ψ^0 values are calculated.

57.3 COMMENTS

1   The method described here can be used to make measurements of hydraulic conductivity and sorptivity in the range of potential from 0 through −1 kPa. This range can be extended to −4 kPa potential by the use of special membranes (R. Sides, personal communication).

2   The advantage of this laboratory method is that a large number of cores can be taken quickly in the field and the measurements of sorptivity and hydraulic conductivity can be made later in the laboratory. However, it is preferable to measure sorptivity and hydraulic conductivity *in situ*, as little soil disturbance occurs. The tension permeameter can be used to make *in situ* measurements) (Smettem and Clothier 1989, Ankeny et al. 1991, Reynolds and Elrick 1991, White et al. 1992).

REFERENCES

Ankeny, M. D., Ahmed, M., Kaspar, T. C., and Horton, R. 1991. Simple field method for determining hydraulic conductivity. Soil Sci. Soc. Am. J. 55: 467–470.

Clothier, B. E. and White, I. 1981. Measurement of sorptivity and soil water diffusivity in the field. Soil Sci. Soc. Am. J. 45: 241–245.

Clothier, B. E. and Scotter, D. R. 1982. Constant-flux infiltration from a hemispherical cavity. Soil Sci. Soc. Am. J. 46: 696–700.

Clothier, B. E., Sauer, T. J., and Green, S. R. 1988. The movement of ammonium nitrate into unsaturated soil during unsteady absorption. Soil Sci. Soc. Am. J. 52: 340–345.

Clothier, B. E. and Wooding, R. A. 1983. The soil water diffusivity near saturation. Soil Sci. Soc. Am. J. 47: 636–640.

Cook, F. J. 1988. Design criteria for droplet irrigation of effluent at Waingawa Freezing Works. N.Z. Soil Bureau Contract Report 88/13.

Cook, F. J., Beecroft, F. G., Joe, E. N., and Balks, M. R. 1986. Flow of water in a gravelly sandy loam from a point source. Surface soil management, proceedings, N.Z. Soc. Soil Sci. and Aust. Soc. Soil Sci. Inc., Joint Conference, 20 Nov–23 Nov., 1986, Rotorua, New Zealand.

Cook, F. J. 1991. Calculation of hydraulic conductivity from suction permeameter measurements. Soil Sci. 152, 321–325.

Klute, A. and Dirksen, C. 1986. Hydraulic conductivity and diffusivity: laboratory methods. In A. Klute, Ed. Methods of soil analysis. Part 1. Physical and mineralogical methods. 2nd ed. American Society of Agronomy, Soil Science Society of America, Madison, WI.

Perroux, K. M. and White, I. 1988. Designs for disc permeameters. Soil Sci. Soc. Am. 52: 1205–1215.

Philip, J. R. 1957. The theory of infiltration. 4. Sorptivity and algebraic infiltration equations. Soil Sci. 84: 257–264.

Reynolds, W. D. and Elrick, D. E. 1991. Determination of hydraulic conductivity using a tension permeameter. Soil Sci. Soc. Am. J. 55: 633–639.

Samani, Z. and Willardson, L. S. 1987. Simple laboratory measurement of unsaturated hydraulic conductivity. J. Irr. Drainage Eng. 113: 405–412.

Scotter, D. R., Heng, K. L., Horne, D. J., and White, R. E. 1990. A simplified analysis of soil water flow to a tile drain. J. Soil Sci. 41: 189–198.

Smettem, K. R. J. and Clothier, B. E. 1989. Measuring unsaturated sorptivity and hydraulic conductivity using multiple disc permeameters. J. Soil Sci. 40: 563–568.

Ward, R. D. and Hayes, M. J. 1991. Measurement of near-saturated hydraulic conductivity of undisturbed cores by use of the permeameter tube. DSIR Land Resources Technical Record 20. 6p.

White, I. 1989. Simple laboratory measurement of unsaturated hydraulic conductivity. Discussion. J. Irr. Drainage Eng. 115: 515–518.

White, I. and Perroux, K. M. 1987. Use of sorptivity to determine field soil hydraulic properties. Soil Sci. Soc. Am. J. 51: 1093–1101.

White, I. and Broadbridge, P. 1988. Constant rate rainfall infiltration: a versatile linear model. 2. Applications of solutions. Water Resour. Res. 24: 155–162.

White, I. and Perroux, K. M. 1989. Estimation of unsaturated hydraulic conductivity from field sorptivity measurements. Soil Sci. Soc. Am. J. 53: 324–329.

White, I., Sully, M. J., and Perroux, K. M. 1992. Measurement of surface-soil hydraulic properties: disc permeameters, tension infiltrometers, and other techniques. In Advances in Measurement of Soil Physical Properties: Bringing Theory into Practice. Soil Sci. Soc. Am. Special Publication No. 30, Madison, WI.

Chapter 58
Unsaturated Hydraulic Conductivity: Estimation from Desorption Curves

R. de Jong

Agriculture Canada

Ottawa, Ontario, Canada

58.1 INTRODUCTION

Soil water flow processes such as infiltration, redistribution, evapotranspiration, and drainage or capillary rise involve flow in unsaturated soil. In order to describe the flow quantitatively, reliable estimates of the parameters entering the flow equations are necessary. In addition to the soil water retention curve (or soil water desorption curve), the hydraulic conductivity as a function of soil water content (θ) or pressure head (ψ) is the most important function governing the movement of water and solutes in the soil.

Although field and laboratory techniques for measuring the unsaturated hydraulic conductivity have been developed (see, e.g., Klute 1986), the methods remain time consuming and costly. As a result, many methods have been developed to estimate the unsaturated hydraulic conductivity from empirical equations (e.g., Clapp and Hornberger 1978, Bloemen 1980, Rawls and Brakensiek 1985). Alternatively, several investigators have proposed theoretical pore size distribution models that predict the hydraulic conductivity function from more easily measured soil water retention data. Tabular data are produced from the methods of Childs and Collis-George (1950) and Millington and Quirk (1961), whereas the Brooks and Corey (1964) and Campbell (1974) solutions are in closed form and the methods of Burdine (1953) and Mualem (1976) are in integral form. Van Genuchten (1980) combined an empirical S-shaped curve for the soil water retention function with the pore size distribution theory of Mualem (1976) to derive a closed-form analytical expression for the unsaturated hydraulic conductivity. Except for the saturated hydraulic conductivity, K_s, the resulting conductivity function contains parameters that can be estimated from measured soil water retention data.

Soil Sampling and Methods of Analysis, M. R. Carter, Ed.,
Canadian Society of Soil Science. © 1993 Lewis Publishers.

58.2 PROCEDURE

1   Soil water retention data (on a volumetric basis) are fitted to the following empirical equation (van Genuchten 1980):

$$\theta = \theta_r + \frac{\theta_s - \theta_r}{[1 + |\alpha\psi|^n]^m} \quad (\theta_r \leq \theta \leq \theta_s) \tag{58.1}$$

where θ_r and θ_s refer to the residual and saturated volumetric water contents $[L^3\, L^{-3}]$, respectively, ψ [L] is the soil water pressure head (matric potential); and $\alpha\ [L^{-1}]$, n [dimensionless] and m [dimensionless] are parameters which determine the shape of the $\theta(\psi)$ curve. The residual water content, θ_r, is the water content value where the gradient, $d\theta/d\psi$, becomes zero, which in theory occurs only as ψ approaches $-\infty$. In practice, however, θ_r is the water content at some large, but finite, negative value of ψ. The parameter n determines the rate at which the S-shaped retention curve turns toward the ordinate for large negative values of ψ, thus reflecting the steepness of the curve, while α equals approximately the inverse of the pressure head at the inflection point where $d\theta/d\psi$ has its maximum value. The procedure for fitting the data is described below.

2   Assuming m = 1 − (1/n), van Genuchten (1980) combined Equation 58.1 with the following theoretical pore size distribution model which was derived by Mualem (1976):

$$K(\psi) = K_s S^L \left[\int_0^s \frac{1}{\psi(x)}\, dx \Big/ \int_0^1 \frac{1}{\psi(x)}\, dx \right]^2 \tag{58.2}$$

where L [dimensionless] is an unknown parameter, X is a dummy variable, and S is relative saturation ($0 \leq S \leq 1$) defined as:

$$S = \frac{\theta - \theta_r}{\theta_s - \theta_r} \tag{58.3}$$

Combining Equations 58.2 and 58.3 leads to (van Genuchten 1980):

$$K(S) = K_s S^L [1 - (1 - S^{1/m})^m]^2 \tag{58.4}$$

or, in terms of soil water pressure head:

$$K(\psi) = K_s \frac{[(1 + |\alpha\psi|^n)^m - |\alpha\psi|^{n-1}]^2}{[1 + |\alpha\psi|^n]^{m(L+2)}} \tag{58.5}$$

Equations 58.1 and 58.5 are the analytical functions describing the $\theta(\psi)$ and $K(\psi)$ relationships, respectively. Although L is presumably a soil-specific parameter, Mualem (1976) concluded from an analysis of 45 soil hydraulic data sets that L is on average about 0.5. Wösten and van Genuchten (1988) show how the parameters θ_r, α, n, and L affect the shapes of the calculated $\theta(\psi)$ and $K(\psi)$ functions. The computationally more complicated case when

Table 58.1
Example of Input File

Manotick, sandy loam, depth 0–20 cm					
13					
	0.050		0.45	0.010	3.00	36.0
	1		0	1	1	
0.	0.4531				
5.	0.4459				
10.	0.4440				
20.	0.4431				
40.	0.4403				
60.	0.4333				
80.	0.4221				
100.	0.4102				
150.	0.3956				
225.	0.3723				
300.	0.3476				
500.	0.3126				
15000.	0.2015				

the parameters n and m in Equation 58.1 are independent is discussed by van Genuchten and Nielsen (1985).

3   A nonlinear least-squares optimization program, called RETC.FOR (van Genuchten 1986, unpublished), determines simultaneously some or all of the unknown coefficients (θ_r, θ_s, α, n, m, L, and/or K_s) from measured soil water retention and hydraulic conductivity data. A shortened, and more restrictive version of the program, RETC3.FOR, estimates only the independent parameters, θ_r, θ_s, α, and n, from the measured retention data. It assumes that m $= 1 - (1/n)$, L = 0.5, and K_s have been measured independently. Associated input and output files, RETC3.IN and RETC3.OUT, are listed in Tables 58.1 and 58.2, respectively.

4   Input data (for RETC3.IN)

Line 1:   TITLE (Format A60)
          Use any label you desire.
Line 2:   NWC (Free format)
          Number of measured soil water retention data pairs. If NWC = 0, the hydraulic properties (θ, ψ, and K) are calculated using the initial estimates as the correct coefficient values: lines 4, 5, . . . are then not needed.
Line 3:   (B(I), I = 1,5) (Free format)
          Initial estimates for the unknown coefficients, to be entered in the following sequence: θ_r, θ_s, α, n, K_s. Routinely, rerun the program with different initial estimates (see, e.g., Wösten and van Genuchten 1988) to make sure that the program converges to the same final parameter values.
Line 4:   (Index (I), I = 1,4) (Free format)
          Indices for the coefficients, B(I), indicating whether the Ith coefficient is an unknown and to be fitted to the data (INDEX (I) = 0). That is, INDEX (I) = 0 indicates that coefficient B(I) is known and its value specified. The index for K_s (INDEX (5)) is set internally to 0.

Table 58.2 Example of Output File

Analysis of soil hydraulic properties
Manotick, sandy loam, depth 0–20 cm
  Mualem-based restriction, M = 1 − 1/N
  Analysis of retention data

Initial values of the coefficients

No.	Name	Initial value	Index
1	WCR	.0500	1
2	WCS	.4500	0
3	ALPHA	.0100	1
4	N	3.0000	1
5	M	.6667	0
6	EXPO	.5000	0
7	CONDS	36.0000	0

Nonlinear least-squares analysis: final results

Variable	Value	S.E. Coeff.	T-Value	95% Confidence limits Lower	Upper
WCR	.18018	.00598	30.12	.1669	.1935
ALPHA	.00629	.00029	21.52	.0056	.0069
N	1.56133	0.4317	36.17	1.4652	1.6575

WCR	.18018
WCS	.45000
ALPHA	.00629
N	1.56133
M	.35952
EXPO	.50000
CONDS	36.00000

Observed and fitted data

No.	P	Log-P	WC-OBS	WC-Fit	WC-DIF
1	.1000E − 04	− 5.0000	.4531	.4500	.0031
2	.5000E + 01	.6990	.4459	.4496	− .0037
3	.1000E + 02	1.0000	.4440	.4487	− .0047
4	.2000E + 02	1.3010	.4431	.4463	− .0032
5	.4000E + 02	1.6021	.4403	.4396	.0007
6	.6000E + 02	1.7782	.4333	.4315	.0018
7	.8000E + 02	1.9031	.4221	.4229	− .0008
8	.1000E + 03	2.0000	.4102	.4142	− .0040
9	.1500E + 03	2.1761	.3956	.3938	.0018
10	.2250E + 03	2.3522	.3723	.3684	.0039
11	.3000E + 03	2.4771	.3476	.3488	− .0012
12	.5000E + 03	2.6990	.3126	.3143	− .0017
13	.1500E + 05	4.1761	.2015	.2012	.0003

Table 58.2 Example of Output File (continued)

Analysis of soil hydraulic properties
Manotick, sandy loam, depth 0–20 cm
  Mualem-based restriction, M = 1 − 1/N
  Analysis of retention data

R Squared for regression of observed vs. fitted values = .99853

Soil hydraulic properties

WC	P	LOGP	COND	LOGK
.1900	.5817E + 05	4.765	.8765E − 08	− 8.057
.2000	.1664E + 05	4.221	.6201E − 06	− 6.208
.2100	.8028E + 04	3.905	.7388E − 05	− 5.131
.2200	.4787E + 04	3.680	.4274E − 04	− 4.369
.2300	.3203E + 04	3.506	.1667E − 03	− 3.778
.2400	.2303E + 04	3.362	.5074E − 03	− 3.295
.2500	.1739E + 04	3.240	.1302E − 02	− 2.885
.2600	.1361E + 04	3.134	.2952E − 02	− 2.530
.2700	.1093E + 04	3.039	.6091E − 02	− 2.215
.2800	.8960E + 03	2.952	.1168E − 01	− 1.933
.2900	.7460E + 03	2.873	.2110E − 01	− 1.676
.3000	.6286E + 03	2.798	.3636E − 01	− 1.439
.3100	.5347E + 03	2.728	.6021E − 01	− 1.220
.3200	.4581E + 03	2.661	.9649E − 01	− 1.016
.3300	.3945E + 03	2.596	.1504E + 00	− .823
.3400	.3408E + 03	2.532	.2293E + 00	− .640
.3500	.2948E + 03	2.470	.3428E + 00	− .465
.3600	.2549E + 03	2.406	.5048E + 00	− .297
.3700	.2198E + 03	2.342	.7344E + 00	− .134
.3800	.1885E + 03	2.275	.1059E + 01	.025
.3900	.1602E + 03	2.205	.1520E + 01	.182
.4000	.1342E + 03	2.128	.2179E + 01	.338
.4100	.1099E + 03	2.041	.3139E + 01	.497
.4200	.8663E + 02	1.938	.4587E + 01	.662
.4300	.6350E + 02	1.803	.6911E + 01	.840
.4400	.3882E + 02	1.589	.1122E + 02	1.050
.4500	.0000E + 00		.3600E + 02	1.556

Line 5:   X (I), Y (I) (Free format) (NWC lines)
                Starting with line 5, the NWC observations of pressure head and volumetric water content are read in. Use one line per observation. Pressure heads are read in as positive values, for the sake of convenience.

58.2 COMMENTS

1   The input data file, RETC3.IN, in Table 58.1 is for a Manotick sandy loam soil. The water retention curve consists of 13 (NWC = 13) measured data pairs, i.e., absolute pressure head (cm of water) vs. volumetric water content (cm^3 cm^{-3}). The initial estimates for θ_r, θ_s, α, and n are, respectively, 0.05 cm^3 cm^{-3}, 0.45 cm^3 cm^{-3}, 0.10 cm^{-1}, and 3.00. K_s was measured to be 36.0 cm d^{-1}. Since INDEX(1), INDEX(3), and INDEX(4) are set to 1, the program

will calculate the parameters θ_r, α, and n, but, because INDEX(2) = 0, the water retention curve will be fitted through the initial estimate of θ_s, i.e., 0.45.

2   The output file, RETC3.OUT, is listed in Table 58.2. The initial values of the coefficients are printed, along with their indices, using the following notation:

WCR	= θ_r (cm³ cm⁻³)	N	= n	[dimensionless]
WCS	= θ_s (cm³ cm⁻³)	M	= 1 − 1/N	[dimensionless]
ALPHA	= α (cm⁻¹)	EXPO	= L (= 0.5)	[dimensionless]
CONDS	= K_s (cm d⁻¹)			

The results of the regression analysis are presented next, along with appropriate statistics for each calculated coefficient. Depending on the measured data, the program may or may not automatically adjust the θ_r value to 0.00 to achieve an adequate model-data fit to the data. Observed data and fitted water contents are printed, along with the differences between observed and fitted water contents. A correlation coefficient, R^2, is printed for observed vs. fitted water contents. Finally, the absolute value of pressure head, the common log of the pressure head, the hydraulic conductivity, and the common log of the hydraulic conductivity are printed for volumetric water contents ranging between θ_r and θ_s.

3   Important advantages of this method (and similar approaches) are that the program uses $\theta(\psi)$ data which are relatively easily obtained and which are routinely collected in many soil surveys, and that the program can provide estimates of K(ψ) for a wide range of ψ ($-15{,}000$ cm $\leq \psi \leq 0$). However, disadvantages include the limited accuracy with which the empirical $\theta(\psi)$ function can fit the data, especially for soils with bimodal pore size distributions and the many simplifying assumptions in the theoretical model for estimating K(ψ).

4   A copy of the program, RETC3.FOR, can be obtained upon receipt of a blank diskette and post-paid mailer. Send your requests to Dr. R. de Jong, CLBRR/ LRD, Agriculture Canada, Central Experimental Farm, Ottawa, Ont., K1A 0C6.

58.3 ACKNOWLEDGMENTS

The author is indebted to Dr. M. Th. van Genuchten for providing the original computer program RETC.

REFERENCES

Bloemen, G. W. 1980. Calculation of hydraulic conductivities of soils from texture and organic matter content. Z. Pflanzenernaehr. Bodenk. 143: 581–615.

Brooks, R. H. and Corey, A. T. 1964. Hydraulic properties of porous media. Hydrology Paper, Colorado State University, Fort Collins 3: 1–27.

Burdine, N. T. 1953. Relative permeability calculation from size distribution data. Am. Soc. Agric. Eng. 10: 400–404.

Campbell, G. S. 1974. A simple method for determining unsaturated conductivity from moisture retention data. Soil Sci. 117: 311–314.

Childs, E. C. and Collis-George, N. 1950. The permeability of porous materials. Proc. R. Soc. London 201A: 392–405.

Clapp, R. B. and Hornberger, G. M. 1978. Empirical equations for some hydraulic properties. Water Resour. Res. 14: 601–604.

Klute, A., Ed. 1986. Methods of soil analysis. Part 1. Physical and mineralogical methods. 2nd ed. Agronomy No. 9. American Society of Agronomy, Madison, WI.

Millington, R. J. and Quirk, J. P. 1961. Permeability of porous solids. Trans. Faraday Soc. 57: 1200–1206.

Mualem, Y. 1976. A new model for predicting the hydraulic conductivity of unsaturated porous media. Water Resourc. Res. 12: 513–522.

Rawls, W. J. and Brakensiek, D. L. 1985. Prediction of soil water properties for hydrologic modelling. Pages 293–299 in Proc. Symp. on Watershed Management, American Society of Civil Engineers, New York.

van Genuchten, M. Th. 1980. A closed-form equation for predicting the hydraulic conductivity of unsaturated soils. Soil Sci. Soc. Am. Proc. 44: 892–898.

van Genuchten, M. Th. and Nielsen, D. R. 1985. On describing and predicting the hydraulic properties of unsaturated soils. Ann. Geophys. 3: 615–628.

Wösten, J. H. M. and van Genuchten, M. Th. 1988. Using texture and other soil properties to predict the unsaturated soil hydraulic functions. Soil Sci. Soc. Am. J. 52: 1762–1770.

Chapter 59
Unsaturated Hydraulic Conductivity:
Field Measurement

W. D. Reynolds

Agriculture Canada
Ottawa, Ontario, Canada

59.1 INTRODUCTION

Water movement in the field above the water table occurs primarily as unsaturated flow, and is consequently controlled to a large extent by the unsaturated hydraulic conductivity of the porous medium. Field, or "*in situ*", measurements of unsaturated hydraulic conductivity [K(ψ)] are therefore critically important for characterizing many aspects of unsaturated water flow, such as: rainfall infiltration and runoff; aquifer recharge; migration of nutrients, pesticides, and contaminants through the soil profile; and design and monitoring of irrigation and drainage systems (Hillel 1980). Field measurements of K(ψ) are also useful for characterizing soil macrostructure and its changes as a result of changing land management practices (White et al. 1992).

The K(ψ) parameter is usually a highly nonlinear function of negative pore water pressure head, ψ (i.e., tension). This relationship is shown schematically in Figure 59.1 for a range of soil textures and structures. Note also in Figure 59.1 that for the special case of zero pore water pressure head ($\psi = 0$), the K(ψ) value is equal to the field-saturated hydraulic conductivity, K_{fs} (Chapter 56).

The $\psi = 0$ point in Figure 59.1 also serves to separate the porous medium conditions of "unsaturation" and "field saturation". For unsaturation, ψ is negative and the porous medium contains an air phase which is more or less continuous and at atmospheric pressure. For field saturation, ψ is positive and the air phase tends to exist as discrete bubbles which are above atmospheric pressure. Due to the presence of the air phase, the unsaturated and field-saturated water contents of a porous medium are less than that for complete or "true" saturation. The K(ψ) parameter consequently applies to unsaturated flow, K_{fs} applies to field-saturated flow, and K_s (Chapter 55) applies to completely saturated flow.

Soil Sampling and Methods of Analysis, M. R. Carter, Ed.,
Canadian Society of Soil Science. © 1993 Lewis Publishers.

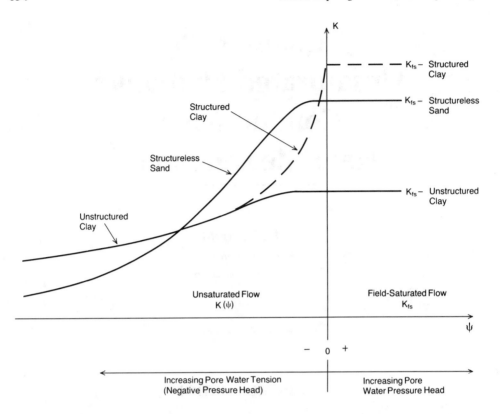

FIGURE 59.1. Schematic of interactions between field-saturated hydraulic conductivity (K_{fs}), near-saturated hydraulic conductivity [$K(\psi)$], soil texture, and soil structure. The ψ value represents positive and negative pore water pressure head.

The primary factors affecting K_{fs} and $K(\psi)$ are the texture (grain size distribution) and structure (cracks, worm holes, root channels, etc.) of the porous medium. The K_{fs} value tends to increase greatly with coarser texture and increasing structure, due to an increase in the number and size of large, highly water-conductive pores. At the same time, the low tension region of the $K(\psi)$ relationship usually becomes steeper (i.e., K vs. ψ has a larger slope), because these same large pores also tend to desaturate and become nonwater conducting at lower pore water tensions. Consequently, coarse-textured and/or structured porous media tend to have larger K_{fs} values and steeper $K(\psi)$ relationships (in the low-tension region) than fine textured and/or structureless porous media. In addition, the texture and structure can interact in such a way that a fine-textured, but structured, material (i.e., a structured clay soil) can have both a larger K_{fs} and a steeper $K(\psi)$ (in the low-tension region) than a coarse-textured, but structureless, material (i.e., a structureless sand soil). A schematic illustration of this behavior is given by the "structured clay" and "structureless sand" curves in Figure 59.1. An important implication of this texture-structure interaction is that the antecedent structure must be preserved in order for K_{fs} and $K(\psi)$ measurements to be representative of the porous medium in its "natural" field condition.

Other important factors affecting K_{fs} and $K(\psi)$ include the temperature and ionic (salt) concentration of the water used for making the measurements. These aspects, which apply for K_s as well, are discussed in Chapter 55, Section 55.1.

59.2 TENSION INFILTROMETER METHOD
(REYNOLDS AND ELRICK 1991)

This method provides *in situ* determination in the unsaturated (vadose) zone of field-saturated hydraulic conductivity, K_{fs}, and near-saturated hydraulic conductivity, $K(\psi)$. It involves the measurement of the steady-state infiltration rate (recharge) into the porous medium (soil) through a porous plate or membrane on which a constant negative water pressure (tension) is applied. "Near-saturated" $K(\psi)$ refers to $K(\psi)$ measurements made over the pressure head range, $-20 \text{ cm} \le \psi \le 0$, where water contents are nearly as high as those for saturation or field saturation.

The tension infiltrometer is also capable of giving estimates of the matric and tension flux potentials, the zero-ponded and tension sorptivities, and the flow-weighted mean pore radii for field-saturated and near-saturated flow. These parameters are useful for characterizing soil structure. Procedures for obtaining these parameters are given in Smettem and Clothier (1989), Elrick and Reynolds (1992), and White et al. (1992).

The tension infiltrometer method is designed to complement the constant head well permeameter and pressure infiltrometer methods (Chapter 56) in that it provides measurements of near-saturated (as opposed to field-saturated) soil hydraulic properties, and the infiltrometer disk is designed to attach directly to the "Guelph permeameter" reservoir. As a result, this method is occasionally referred to as the "Guelph tension infiltrometer" method.

59.2.1 APPARATUS AND PROCEDURE

1   For a measurement at the soil surface, choose a relatively level site which is at least as large as the retaining ring for the contact material (Figure 59.2). Remove from the measurement site all debris which is both loose and "large" (e.g., corn husks, corn cobs, etc., which are more than about 0.4 cm high), as it may cause poor contact between the soil and the contact material. Avoid measurement sites which contain large "attached" debris, i.e., partially exposed roots, partially buried corn cobs, large clods, etc., which are greater than about 0.4 cm high. The disruption of the soil caused by removal of "attached" debris may change the hydraulic properties of the infiltration surface. Avoid vacuuming, blowing off, or sweeping the site (to remove loose material which is present naturally), as this may also change the effective hydraulic properties of the infiltration surface. For a subsurface measurement, the site should be excavated using procedures that minimize smearing and compaction, as this can alter hydraulic properties as well.

2   Gently press a 10.7- to 21.4-cm diameter by 5.0- to 12.0-cm long, sharpened "retaining ring" into the infiltration surface (i.e., measurement site) to a depth of about 0.5 cm (Figure 59.2). The retaining ring, which is conveniently constructed from PVC sewer pipe, should be just large enough to allow the chosen size of infiltrometer (10.0–20.0 cm diameter) to fit inside (Figure 59.2).

3   Lay a circle of 270-mesh nylon bolting cloth (53-μm pore size, "Nitex" brand, precut to the same inside diameter as the retaining ring) on the infiltration

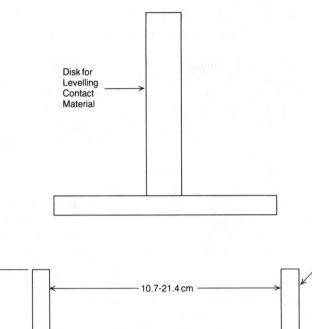

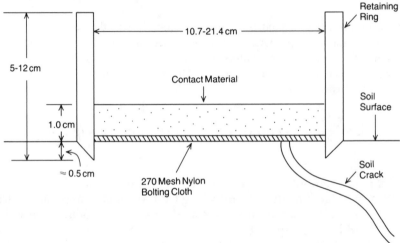

FIGURE 59.2. Schematic of the leveling disk, retaining ring, contact material, and nylon bolting cloth for use in the tension infiltrometer method.

surface inside the ring (see also comment 1, Section 59.2.3). Pour the contact material (air-dried borosilicate glass spheres, 53- to 105-μm size range) on top of the nylon cloth to an average depth of 1 cm (a 1-cm reference line on the inside of the retaining ring helps achieve this). Level and smooth the contact material using a wooden disk or similar device (Figure 59.2).

4   Attach the presaturated infiltrometer disk-supply tube assembly (see comment 2, Section 59.2.3 for details concerning the construction and saturation of the infiltrometer disk) to the constant head well permeameter reservoir and stand upright with the infiltrometer disk submerged in a flat-bottomed pail below a few centimeters of water. Support the infiltrometer using a large tripod (same design as recommended for the constant head well permeameter — Chapter 56). Close the water outlet of the infiltrometer disk by firmly pushing the base of the air tube down into the outlet port, and then fill the reservoir to the top with water. The base of the reservoir air

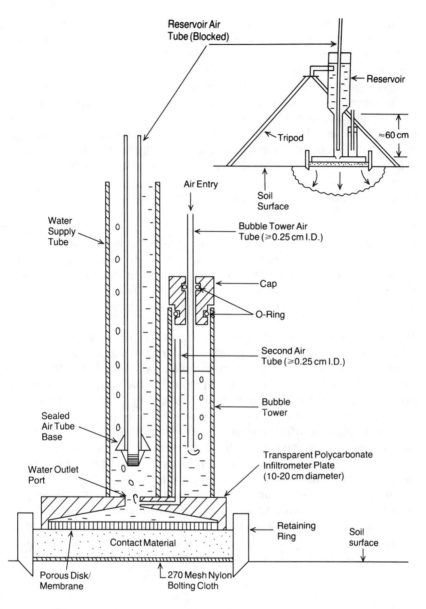

FIGURE 59.3. Schematic of a tension infiltrometer apparatus for use in the tension infiltrometer method.

tube should be sealed (clear silicon caulking works well for this) so that the infiltrometer bubble tower provides the only source of air (Figure 59.3). Remove the cap-air tube arrangement from the bubble tower, establish the correct water level in the bubble tower by adding or removing water, and then replace the cap (see also comment 2, Section 59.2.3). Set the bubble tower air tube at the highest tension to be measured. Open the water outlet of the infiltrometer disk by lifting the base of the reservoir air tube out of the water outlet port (as indicated in Figure 59.3). The water used in the infiltrometer should meet the same specifications as given in Chapter 56, step 3 of Section 56.2.1 (constant head well permeameter method).

5   Lift the infiltrometer out of the bucket and tilt slightly to remove water from
     the upper surface of the infiltrometer plate. Carefully lower the infiltrometer
     plate onto the contact material, using a slight twist to ensure contact. Use
     the tripod to hold the infiltrometer in place and to carry the weight of the
     reservoir and water. When initial contact with the contact material is made,
     air bubbles should rise rapidly in the bubble tower and up into the reservoir,
     indicating rapid saturation of the contact material. Air bubbles should not
     appear from any location other than the bubble tower air tube and the
     connection between the bubble tower and the supply tube (Figure 59.3 —
     see also comment 2, Section 59.2.3). The infiltrometer is operating properly
     when air bubbles rise regularly up into the reservoir from the connection
     between the bubble tower and the supply tube.

     Measurements should be limited to soils which are at field capacity or drier
     (i.e., matric potential ≤ -100 cm), otherwise there may be insufficient cap-
     illarity in the soil to draw water out of the infiltrometer against the tension
     applied by the bubble tower. Also, the tension infiltrometer theory requires
     that the soil be sufficiently dry to have a negligibly small antecedent $K(\psi)$
     value relative to the minimum $K(\psi)$ being measured.

6   The rate of water flow out of the infiltrometer and into the porous medium
     is measured by monitoring the rate of fall of the water level in the reservoir
     (using the reservoir measurement scale and a stopwatch). When the rate of
     fall becomes constant, the flow rate out of the infiltrometer and into the
     porous medium has reached steady state (see also comment 3, Section
     59.2.3). The method requires that at least two tensions (ψ_1, ψ_2) are set se-
     quentially on the infiltrometer disk (by adjusting the elevation of the air
     tube in the bubble tower) and the corresponding steady flow rates (Q_1, Q_2)
     measured. The tensions, which are actually negative pressure heads, should
     be set in descending order of magnitude (i.e., $|\psi_1| > |\psi_2| > |\psi_3|$, . . .), with
     the first and largest tension ($|\psi_1|$) no greater than about 20 cm (i.e., $\psi_1 =
     -20$ cm), and the last and smallest tension ($|\psi_f|$) at or near zero (i.e., $\psi_f \approx
     0$). It is recommended that three to five tensions be used to provide adequate
     definition of the near-saturated $K(\psi)$ relationship (see also comment 4, Sec-
     tion 59.2.3). The K_{fs} and $K(\psi)$ values are calculated as indicated in Section
     59.2.2 below.

59.2.2 CALCULATIONS

1   As mentioned above, a sequence of steady flow rates (Q_1, Q_2, Q_3, . . .) are
     measured by setting a sequence of tensions (ψ_1, ψ_2, ψ_3, . . .) on the infil-
     trometer disk. The Q values are obtained using the relationship

$$Q_x = AR_x; . . x = 1,2,3 . . . \tag{59.1}$$

     where A (cm^2) is the cross-sectional area (cell constant) of the infiltrometer
     reservoir, and R_x (cm s^{-1}) is the steady rate of fall of the water level in the
     infiltrometer reservoir with the tension in the bubble tower set at ψ_x (cm).

2   The semilog slope, $S_{x,y}$, between adjacent (ψ, Q) data pairs is calculated using

$$S_{x,y} = \frac{\ln (Q_x/Q_y)}{(\psi_x - \psi_y)} \tag{59.2}$$

where $x = 1,2,3, \ldots , y = x + 1$.

3   The value of $K^*_{x,y}$ is calculated from the semilog intercept for each set of adjacent (ψ, Q) data pairs using

$$K^*_{x,y} = \frac{G_d S_{x,y} Q_x}{a(1 + G_d S_{x,y} \pi a)(Q_x/Q_y)^p} \tag{59.3}$$

where a (cm) is the radius of the retaining ring for the contact material, G_d is a dimensionless shape factor equal to 0.237, and $P = \psi_x / (\psi_x - \psi_y)$.

4   For $\psi_2, \psi_3, \ldots , \psi_{f-1}$, the values of $K(\psi)$ are calculated using

$$K(\psi_x) = (K_1 + K_2)/2 \tag{59.4}$$

where $K_1 = K^*_{x-1,x} \exp(S_{x-1,x} \psi_x)$
       $K_2 = K^*_{x,x+1} \exp(S_{x,x+1} \psi_x)$
For ψ_1, $K(\psi_1) = K_2$. For ψ_f, $K(\psi_f) = K_1$ and $K_{fs} = K^*_{f-1,f}$.

5   Due to their complexity, Equations 59.1 to 59.4 are best solved using a computer program. A Fortran program for this purpose is available from W.D. Reynolds, Centre for Land and Biological Resources Research, Research Branch, Agriculture Canada, 960 Carling Ave., Ottawa, Ont., K1A 0C6. Further discussion of the $K(\psi)$ calculations is given in comments 4 and 5, Section 59.2.3.

59.2.3 COMMENTS

1   The primary purpose of the nylon bolting cloth is to prevent the contact material from infilling cracks, worm holes, and other macropores present in the infiltration surface (Figure 59.2). Significant infilling of this macrostructure could change the hydraulic properties of the soil, as well as increase greatly the amount of contact material required. The nylon cloth also facilitates reclamation of the contact material from the infiltration surface for subsequent reuse at other sites (after air drying and sieving to remove entrained debris).

The purpose of the contact material is to provide a uniform hydraulic link between the uneven infiltration surface and the flat, rigid porous disk of the infiltrometer. The required hydraulic properties of the contact material are that it has a large enough water entry value to spontaneously saturate (from an air-dry state) under the maximum tension set on the infiltrometer (ψ_1), and that it have a K_{fs} value larger than the $K(\psi_x)$ of the soil being measured. The contact material mentioned above (i.e., the 53- to 105-μm borosilicate glass spheres) has a water entry value of about 30 cm and a K_{fs} value of 10^{-2} cm s^{-1}. This should be adequate for infiltrometer tensions of 20 cm and less and for use on most agricultural soils. Other contact materials can be used if the above specifications are not adequate.

2   The tension infiltrometer consists essentially of a porous disk or membrane sealed over a cone-shaped cavity in the underside of a transparent plastic plate (Figure 59.3). Water is supplied to the infiltrometer via a water-supply tube. Tension is set on the porous disk/membrane via the bubble tower.

The seal between the porous disk/membrane and the plate (usually made using epoxy glue) must be air tight or it will leak air when tension is applied. The porous disk/membrane must be hydrophilic (water wettable) and have a saturated air entry tension greater than the maximum tension set by the bubble tower. If the air entry tension is exceeded, air will leak through the plate/membrane and appear as a stream of bubbles. Porous plates may be constructed of porous metal (stainless steel, brass), sintered glass, or porous plastic that has been treated to make them water wettable. Membranes are usually made of fine-mesh (270 mesh or smaller) nylon cloth or metal sieve screen. The wettability of the disk/membrane can often be enhanced by first cleaning it with a solvent (e.g., ethanol, dithionite-citrate solution, dilute bleach solution) to remove hydrophobic oils, oxides, organic matter, etc., and then treating it with a wetting agent (e.g., a solution of isopropyl alcohol, distilled water, and commercial surfactant). Porous disks/membranes should have an air entry value of at least 20 cm and a K_{fs} of 10^{-1} to 10^{-3} cm s^{-1}. The porous disk or membrane is saturated by submerging the entire infiltrometer plate in a water bath containing several centimeters of deaired, temperature-equilibrated water, and then drawing the water up through the disk/membrane (by sucking through a one-way valve inserted into the top of the water-supply tube) until the cone-shaped cavity and supply tube are full (Figure 59.4). The bubble tower should also be filled to the proper level with water (discussed below) and the air tube set at the maximum tension. Leave the infiltrometer standing in the water bath for a couple of days, periodically jarring the plate against the wall of the water bath to remove air that may be entrapped in the disk/membrane. Test the infiltrometer by removing it from the water bath and setting it on several layers of dry paper towel. Air bubbles should rise rapidly from the air tube and up into the supply tube as water flows out of the infiltrometer and into the paper towel. If air bubbles rise from the disk/membrane itself, then it is either not completely saturated, it contains a hole or large pore, or there is a break in the seal between the disk/membrane and the plate. Repeating the above procedures should stop the leak if incomplete saturation is the problem. The leak will persist if it is due to a hole in the disk/membrane or to a break in the seal. Such holes and breaks can usually be plugged using a couple of drops of epoxy glue. To maintain saturation during transport, place the infiltrometer in a flat-bottomed pail with the plate submerged below several centimeters of water. For long-term storage, the infiltrometer should be air dried to prevent algal growth in the disk/membrane, etc.

The bubble tower must be calibrated to produce the correct tensions on the infiltration surface. This can be accomplished using a tension table-

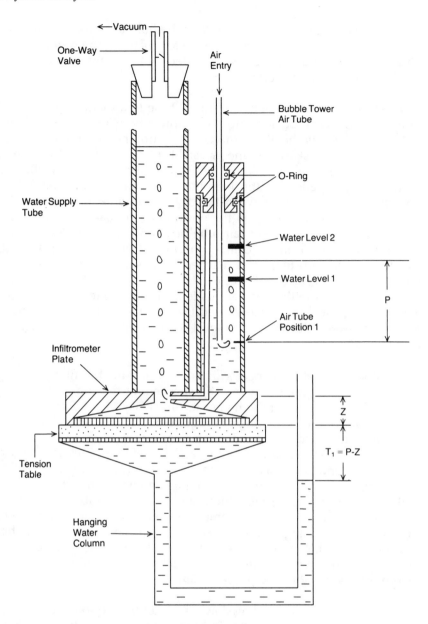

FIGURE 59.4. Calibration of a tension infiltrometer using a vacuum source and a tension table-hanging water column arrangement.

hanging water column arrangement (Figure 59.4). Adjust the water level in the bubble tower so that it is about 4–5 cm below the top of the second tube when the cap and air tube are removed (the "second tube" is identified in Figure 59.3). Mark the position of the water level on the side of the bubble tower using a waterproof marker (water level 1 in Figure 59.4). Install the cap and push the air tube down until it touches the bottom of the bubble tower. Mark the position of this water level as well (water level 2 in Figure 59.4). This second water level should be at least 2 cm below the top of the "second tube" in the tower (Figure 59.3). The first water level mark (water level 1) provides a reference for refilling the bubble tower, and the second mark (water level 2) provides a reference for checking the water level to see if adjustment is required. Set the saturated infiltrometer on a tension table which is set at zero tension.

Connect the one-way valve in the supply tube to a vacuum source to draw air through the air tube and into the supply tube (Figure 59.4). Using trial-and-error, adjust the height of the air tube until the maximum desired tension (ψ_1) on the disk/membrane is established as indicated by the water elevation in the hanging column (i.e., $\psi_1 = T_1$ in Figure 59.4). Be sure to allow in the calibration for the 1-cm thickness of contact material that provides the hydraulic link between the disk/membrane and the soil (Step 3, Section 59.2.1). Mark and label the position of the base of the air tube on the side of the bubble tower (air tube position 1 in Figure 59.4), and then raise the air tube to find and mark the next desired tension. Repeat this procedure until zero tension is reached. A tension sequence of $\psi_1 = -15$ cm, $\psi_2 = -10$ cm, $\psi_3 = -5$ cm, $\psi_4 = -3$ cm, $\psi_5 = -1$ cm, $\psi_6 = 0$ cm is recommended. Note that the various tensions will generally not be equally spaced along the bubble tower because the level of water in the tower changes (between the level 1 and level 2 limits) with changing height of the air tube.

3   Steady flow is usually attained within 15–60 min for the first (and largest) tension set on the infiltrometer. The succeeding tensions usually require less equilibration time. At the higher tensions (say 15, 10, 5 cm), the steady flow rates are often low enough to require use of the "small barrel" in the reservoir in order to obtain adequate measurement accuracy. When the small barrel is used, it will require periodic refilling. This is best accomplished at the end of a measurement for a given tension. Adjust the reservoir valve to allow slow flow of water from the large barrel into the small barrel. At the same time, slowly raise the bubble tower air tube to establish the next tension. This technique accomplishes refilling, while at the same time ensuring that the change from one tension to the next is monotonic, thereby preventing possible hysteresis effects. Details on switching between the barrels in the reservoir are given in the Soilmoisture Equipment Corp. procedure manual for the Guelph permeameter.

An infiltrometer that is bubbling slowly is susceptible to solar heating effects, and the procedures for minimizing these adverse effects (as well as for general "trouble shooting") are similar to those given in Chapter 56 (constant head well permeameter method, Section 56.2.3).

4   The K(ψ) values are determined by fitting an exponential curve segment between adjacent (ψ,Q) data pairs (Section 59.2.2). Since K(ψ) relations are generally exponential over only small ranges, then the accuracy of the K(ψ) values tends to increase with the number of tensions (and thereby exponential curve segments) used within a particular tension range. For the common tension range of -15 cm $\leq \psi \leq 0$, it is recommended that a minimum of four tensions be used (e.g., $\psi_1 = -15$ cm, $\psi_2 = -10$ cm, $\psi_3 = -5$ cm, $\psi_4 = 0$ cm).

The range of $K(\psi)$ that can be measured conveniently with a tension infiltrometer appears to be on the order of 10^{-2} to 10^{-6} cm s^{-1}.

The tension infiltrometer may not give accurate estimates of K_{fs} if the soil is significantly more permeable than the porous disk/membrane and contact material. This is because the disk/membrane and/or the contact material can act as restricting layers in this situation.

Equations 59.1 to 59.4 will occasionally yield unrealistic and invalid (i.e., negative) $K(\psi)$ values. This occurs when soil heterogeneity, a strong vertical water content gradient, or nonattainment of steady flow causes one or more steady flow rates (Q) to decrease with decreasing tension (ψ), rather than to increase with decreasing tension as required by the theory. When this occurs, the aberrant data should either be discarded or the Q values adjusted in a defensible way to produce realistic $K(\psi)$ values.

Other factors affecting the accuracy of tension infiltrometer measurements include macrostructure collapse under the infiltrometer during the infiltration measurement, and inadequate or changing hydraulic contact between the infiltrometer and the infiltration surface (soil). Macrostructure collapse is caused by the weight of the infiltrometer combined with a decline in soil strength as the soil wets up. Inadequate or changing hydraulic contact is caused by wind-induced vibration of the infiltrometer, and by decreasing weight of the infiltrometer as the water empties out of the reservoir. Both of these problems can be reduced by supporting the infiltrometer with a large tripod (as recommended above) which clamps solidly to the reservoir.

5   Additional information concerning the tension infiltrometer method, including alternative infiltrometer designs and approaches for determining near-saturated $K(\psi)$, may be found in Ankeny (1992), Elrick and Reynolds (1992), Smettem and Clothier (1989), and White et al. (1992).

59.3 OTHER METHODS FOR DETERMINING $K(\psi)$ IN THE FIELD

Other useful methods for measuring $K(\psi)$ in the field include the instantaneous profile method and the unit gradient drainage method. Important advantages of these techniques are that a large volume of soil is sampled (on the order of 10–200 m^3); $K(\psi)$ and $\theta(\psi)$ can be measured simultaneously at several depths in the soil profile; and $K(\psi)$ and $\theta(\psi)$ can often be measured over a reasonably large range of tensions (say, -500 cm $\leq \psi \leq 0$). Important limitations, however, are that the methods require considerable time (up to several weeks) and large quantities of water (up to thousands of liters); they require several trained field staff; and they require considerable specialized and expensive equipment (e.g., neutron probe or time-domain reflectometry cable tester for measuring water content, several tensiometers of various lengths for measuring matric potential, high-volume water delivery system). The instantaneous profile and unit gradient drainage methods are described in detail in Hillel (1980) and Green et al. (1986), and will therefore not be discussed here.

Additional methods for field measurement of $K(\psi)$ may be found in Dirksen (1991).

REFERENCES

Ankeny, M. D. 1992. Methods and theory for unconfined infiltration measurements. Pages 123–141 in G. C. Topp, W. D. Reynolds, and R. E. Green, Eds. Advances in measurement of soil physical properties: bringing theory into practice. SSSA Spec. Publ. 30. Soil Science Society of America, Madison, WI.

Dirksen, C. 1991. Unsaturated hydraulic conductivity. Pages 209–269 in K. A. Smith and C. E. Mullins, Eds. Soil analysis: physical methods. Marcel Dekker, New York.

Elrick, D. E. and Reynolds, W. D. 1992. Infiltration from constant head well permeameters and infiltrometers. Pages 1–24 in G. C. Topp, W. D. Reynolds, and R. E. Green, Eds. Advances in measurement of soil physical properties: bringing theory into practice. SSSA Spec. Publ. 30. Soil Science Society of America, Madison, WI.

Green, R. E., Ahuja, L. R., and Chong, S. K. 1986. Hydraulic conductivity, diffusivity, and sorptivity of unsaturated soils: field methods. Pages 771–798 in A. Klute, Ed. Methods of soil analysis. Part 1. Physical and mineralogical methods. 2nd ed. Agronomy No. 9, American Society of Agronomy, Madison, WI.

Hillel, D. 1980. Fundamentals of soil physics. Academic Press, Toronto.

Reynolds, W. D. and Elrick, D. E. 1991. Determination of hydraulic conductivity using a tension infiltrometer. Soil Sci. Soc. Am. J. 55: 633–639.

Smettem, K. R. J. and Clothier, B. E. 1989. Measuring unsaturated sorptivity and hydraulic conductivity using multiple disc permeameters. J. Soil Sci. 40: 563–568.

White, I., Sully, M. J., and Perroux, K. M. 1992. Measurement of surface-soil hydraulic properties: disc permeameters, tension infiltrometers and other techniques. Pages 69–103 in G. C. Topp, W. D. Reynolds, and R. E. Green, Eds. Advances in measurement of soil physical properties: bringing theory into practice. SSSA Spec. Publ. 30. Soil Science Society of America, Madison, WI.

Chapter 60
Air Permeability

C. D. Grant
University of Adelaide
Glen Osmond, S.A., Australia

P. H. Groenevelt
University of Guelph
Guelph, Ontario, Canada

60.1 INTRODUCTION

Air permeability of soils (and other porous materials) is the coefficient, k_a, governing convective transmission of air through the soil under an applied total pressure gradient. The theory of air flow through soil is based on Darcy's law, which states that the velocity of fluid flowing through a porous column is directly proportional to the pressure difference and inversely proportional to the length of the column. Furthermore, Poiseuille's equation dictates that flow of air through a single pore varies as the fourth power of the pore radius, and so large pores and wide cracks make the greatest contribution to air permeability in soils. The air permeability coefficient, k_a, has units of m^2 and is also known as the intrinsic permeability to air (Reeve 1953). It can be derived from Darcy's law (for laminar flow of liquids) using simple assumptions about isothermal, nonturbulent flow of a viscous gas (Kirkham 1946). It has been used since at least the early part of the 20th century for describing, defining, and characterizing the structural arrangement and continuity of the pore space (e.g., Green and Ampt 1911, Buehrer 1932). Air permeability is very sensitive to differences in soil structure (Corey 1986) and has been widely used to characterize changes in structure that result from different soil management practices (e.g., Ball 1981, Groenevelt et al. 1984, Ball et al. 1988). The theory of air permeability in isotropic and anisotropic media has been investigated thoroughly (e.g., Maasland and Kirkham 1955, Corey 1986).

Measurement of air permeability in the field (as opposed to in the laboratory) is a desirable technical feature (cf. Green and Fordham 1975), but the surface layers of the soil are structurally anisotropic and so variability is large (and anormally distributed; McIntyre and Tanner 1959) and the interpretation of field data is difficult (Janse and Bolt 1960). In this regard, the field measurement of air permeability is not yet as advanced (and may never be) as that of hydraulic conductivity. The use of isolated field samples (cores) taken into the laboratory for measurement of k_a is therefore generally preferred (Janse and Bolt 1960) despite its limitations (Blackwell et al. 1990). However, certain simple precautions need to be observed. For example, a very serious error (overestimate) in k_a can result if the sample boundaries are not perfectly sealed during measurement (Corey 1986).

Soil Sampling and Methods of Analysis, M. R. Carter, Ed.,
Canadian Society of Soil Science. © 1993 Lewis Publishers.

Air permeability of soils can be measured in the laboratory in a number of ways (Corey 1986), but two principal methods will be discussed here, namely the constant pressure gradient method (with a measured flux of air), and the constant flux method (with a measured pressure). The choice of method is usually a matter of convenience, but each method has advantages.

60.2 CONSTANT PRESSURE GRADIENT METHOD

Numerous variations of the constant pressure gradient method have been used for measuring air permeability (e.g., Grover 1955, Ball et al. 1981, Groenevelt et al. 1984). The method consists of exposing a soil core (in a cylindrical ring) to a large volume of air having constant pressure (above atmospheric), and measuring the volume of air that passes through the soil core with time. The intrinsic permeability to air, k_a (m^2), is calculated for small air pressures (<0.2 m H_2O) from:

$$\frac{Q}{A} = k_a \frac{\rho_w g}{\eta} \frac{\Delta h}{L}$$ (60.1)

where Q is the volume of air measured at the high-pressure (inlet) side of the soil core, passing into the soil core per unit time (m^3 s^{-1}), A is the cross-sectional area (orthogonal to direction of air flow) of the soil core (m^2), L is the length of the soil core (m), and Δh is the air pressure difference (head of water, m) over the soil core between the air inlet side, i, and the air outlet side, a, at atmospheric pressure; $\Delta h = h_i - h_a$; $h_i = P_i/\rho_w g$ and $h_a = P_a/\rho_w g$; ρ_w is the density of water (kg m^{-3}), g is the gravitational constant (m s^{-1}) and η is the viscosity of air (kg m^{-1} s^{-1}).

60.2.1 MATERIALS

Specifications for this apparatus may be found in Grover (1955) and in Tanner and Wengel (1957). Only the basics are given here.

1   Large-volume air tank. This can be made of stainless steel or stiff plastic (or nalgene) containers.

2   Float can; made of same materials.

3   Guide rod (and calibration device). This ensures that the float can will sink evenly into the water reservoir in the air tank, and also serves as a convenient measuring stick for determining the volume of air flowing out of the air tank and into the soil core.

4   Annular weights. These may be required for increasing the pressure in the air tank, and can be fitted over the guide rod in the center of the tank.

5   Pressure gauge. A simple water manometer will suffice.

6   Soil sample. A soil core (at a specified water content or water potential) collected in a rigid cylindrical sample holder is required.

60.2.2 PROCEDURES

1   The soil core sample is connected to a holder, which is fitted with a rubber ring to ensure a complete seal with no leaks. The outflow side of the soil core is left open to the atmosphere.

2   The float can is allowed to fall freely under its own weight (or with the addition of annular weights for higher pressures) in order to build up some air pressure head, which is then maintained throughout the experiment.

3   The guide rod is monitored with time and two readings are selected, which enables the calculation of the air flux under constant pressure across the sample.

4   Calculation of the intrinsic permeability is done using Equation 60.1.

60.2.3 COMMENTS

1   This method is best used for highly permeable samples and for samples held at water potentials close to zero, because the air pressure gradient (small) across the moist soil sample can be controlled to prevent alteration of the liquid phase and to avoid turbulent flow of air. Pressure heads should be kept as small as possible, and certainly less than 0.20 m.

2   Air leaks should be checked for by allowing the float can to fall, and then sealing off the outlet for the air. If the float can stops sinking, then the system is not leaking.

3   An alternative to the float can method that avoids the necessity for calibration is the direct measurement of air flow using flow meters (see Green and Fordham 1975).

60.3 CONSTANT FLUX METHOD

The constant flux method, as used by Groenevelt and Lemoine (1987) and Blackwell et al. (1990), consists of imposing a constant air flux and then measuring the resulting pressure difference across a soil core sample. Methods for imposing air fluxes are easily constructed and depend only upon the resources at hand in the laboratory. A small cylinder and flow meter supplying compressed air, for example, would suffice. Alternatively, Figure 60.1 shows the use of a syringe (variable sizes) whose air volume, V, is delivered by a slowly advancing, motor-driven piston. Any other apparatus that can supply a nonvariable and easily measured air flux is adequate.

Steady-state air flow (i.e., constant air flux and pressure gradient across soil sample) may or may not occur, and when it does, the pressure gradient can become quite large ($\Delta h > 0.2$

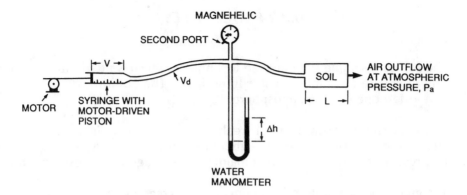

FIGURE 60.1. Instrumentation for air permeability measurements. V is the volume in the syringe and V_d is the "dead" volume.

m H_2O). For this reason, the intrinsic air permeability, k_a, must be calculated from the more correct form of Equation 60.1 derived from the ideal gas law after Kirkham (1946):

$$\frac{Q}{A} = K\frac{\Delta h}{L}\left[1 - \frac{\Delta h}{2h_1}\right]$$

(60.2)

where $K = \rho_w\, g\, k_a/\eta\ (\text{m s}^{-1})$

If Δh does not go above 0.2 m H_2O, however, the term in parentheses approaches unity and may be ignored, so that Equation 60.2 reduces to Equation 60.1.

60.3.1 MATERIALS

The apparatus used is shown in Figure 60.1.

1 Syringes: they can be relatively large, having volume, V, anywhere from 35 to 140 cm³; the inner walls of the syringe can be greased with vaseline to prevent air leakage; Becton Dickinson® disposable syringes were used by Groenevelt and Lemoine (1987).

2 Motor-driven piston: this is used to push the air out of the syringe and needs to have a number of speed settings so that a range of air supply rates can be produced. A Saga® Instruments model 355 pump was found by Groenevelt and Lemoine (1987) to be adequate for this purpose.

3 Pressure gauge: either a simple water manometer or a magnehelic pressure gauge (Dwyer® Instruments) is suitable. The conditions for the use of each are outlined below.

4 Soil sample: same as for the constant pressure gradient method discussed above.

5 Relatively stiff, small-inside diameter connection tubing, which minimizes the "dead" volume, V_d.

60.3.2 PROCEDURES

1   The soil sample is connected to a holder, which is fitted with a rubber O-ring to ensure a complete seal with no leaks. The outflow side of the soil core is left open to the atmosphere.

2   The motor-driven piston in the syringe is set to move at some constant velocity and builds up a pressure difference across the attached soil core, and this is monitored using the pressure gauge. The actual flux chosen will depend upon how permeable the soil sample is, and some initial adjustments are often needed prior to data collection. When the pressure difference becomes constant, the mass flux of air through the soil core has also become constant.

3   When steady-state flow is established and both the flux and the pressure difference across the soil core are constant, the intrinsic permeability to air can be calculated from Equation 60.2.

4   If steady-state air flow conditions are not reached, the intrinsic permeability to air must be calculated from observations taken during the transient state. In this instance, the magnehelic pressure gauge is not suitable because it "bleeds" air until steady-state conditions are reached, so the water manometer has to be used. The procedures and mathematical tools for calculating the permeability in this case are outlined in detail along with examples in Groenevelt and Lemoine (1987).

60.3.3 COMMENTS

1   The principal advantages of the constant flux method are that the apparatus can be quite simple, and measurements can be made on samples of very low permeability using either steady-state or transient-state flow conditions.

2   The apparatus must first be checked for air leaks so that equipment errors are not introduced. This can be done by sealing off the air outlet end of the apparatus and allowing the pressure to build up, then shutting off the pump and monitoring the pressure; it should stay constant.

REFERENCES

Ball, B. C. 1981. Pore characteristics of soils from two cultivation experiments as shown by gas diffusivities and permeabilities and air-filled porosities. J. Soil Sci. 32: 483–498.

Ball, B. C., Harris, W., and Burford, J. R. 1981. A laboratory method to measure gas diffusion and flow in soil and other porous materials. J. Soil Sci. 32: 323–333.

Ball, B. C., O'Sullivan, M. F., and Hunter, R. 1988. Gas diffusion, fluid flow and derived pore continuity indices in relation to vehicle traffic and tillage. J. Soil Sci. 39: 327–339.

Blackwell, P. S., Ringrose-Voase, A. J., Jayawardane, N. S., Olsson, K. A., McKenzie, D. C., and Mason, W. K. 1990. The use of air-filled porosity and intrinsic permeability to air to characterize structure of macropore space and saturated hydraulic conductivity of clay soils. J. Soil Sci. 41: 215–228.

Buehrer, T. F. 1932. The movement of gases through the soil as a criterion of soil structure. Arizona Agricultural Experiment Station Tech. Bull. 39: 57 pp.

Corey, A. T. 1986. Air permeability. Pages 1121–1136 in A. Klute, Ed. Methods of soil analysis. Part 1. Physical and mineralogical methods. 2nd ed. Monograph No. 9, American Society of Agronomy, Madison, WI.

Green, R. D. and Fordham, S. J. 1975. A field method for determining air permeability in soil. Pages 273–288 in Soil physical conditions and crop production. Tech. Bull. 29. Ministry of Agriculture Fish. Food, London.

Green, W. H. and Ampt, G. A. 1911. Studies on soil physics. Part 1. The flow of air and water through soils. J. Agric. Sci. 4: 1–24.

Groenevelt, P. H., Kay, B. D., and Grant, C. D. 1984. Physical assessment of a soil with respect to rooting potential. Geoderma 34: 101–114.

Groenevelt, P. H. and Lemoine, G. G. 1987. On the measurement of air permeability. Neth. J. Agric. Sci. 35: 385–394.

Grover, B. L. 1955. Simplified air permeameters for soil in place. Soil Sci. Soc. Am. Proc. 19: 414–418.

Janse, A. R. P. and Bolt, G. H. 1960. The determination of the air-permeability of soils. Neth. J. Agric. Sci. 8: 124–131.

Kirkham, D. 1946. Field method for determination of air permeability of soil in its undisturbed state. Soil Sci. Soc. Am. Proc. 11: 93–99.

Maasland, M. and Kirkham, D. 1955. Theory and measurement of anisotropic air permeability in soil. Soil Sci. Soc. Am. Proc. 19: 395–400.

McIntyre, D. S. and Tanner, C. B. 1959. Anormally distributed soil physical measurements and nonparametric statistics. Soil Sci. 88: 133–137.

Reeve, R. C. 1953. A method for determining the stability of soil structure based upon air and water permeability measurements. Soil Sci. Soc. Am. Proc. 17: 324–329.

Tanner, C. B. and Wengel, R. W. 1957. An air permeameter for field and laboratory use. Soil Sci. Soc. Am. Proc. 21: 663–664.

Chapter 61
Aggregate Stability to Water

D. A. Angers

Agriculture Canada

Sainte-Foy, Quebec, Canada

G. R. Mehuys

McGill University

Sainte Anne de Bellevue, Quebec, Canada

61.1 INTRODUCTION

An aggregate is a group of primary particles that cohere to each other more strongly than to surrounding soil particles (Kemper and Rosenau 1986). Aggregate stability can be defined as the resistance of the bonds within the aggregates to external forces of impact, shearing, abrasion, or disruption arising from the escape of entrapped compressed air (slaking). Therefore, soil aggregate stability largely determines the susceptibility of soil to water erosion, crust formation, hardsetting, and compaction.

Stability measurements can be made at the scale of the whole soil or large aggregates (>250 μm) or at the scale of the clay and silt-size particles. At the large scale, aggregates or whole soils are exposed to disruptive forces, usually by wet sieving. The proportion of aggregates remaining on one or several sieves represents the stable aggregates. At the scale of clay or silt-size particles, the methods generally consist of characterizing the suspension created as a result of exposing the aggregates to disruptive forces either by turbidimetry or densitometry (e.g., pipetting).

In this chapter, we describe methods for determining the stability of large aggregates (wet-aggregate stability, WAS), the size distribution of water-stable aggregates (WSA), and turbidity.

61.2 WET-AGGREGATE STABILITY

Aggregate stability is determined by measuring the proportion of aggregates of a given size (usually 1 to 2 mm) which do not break down into units smaller than a preselected size (usually 250 μm) under the influence of disruptive forces.

For most soils, a correction for the presence of primary particles (sand or coarse fragments) in the stable aggregate fraction has to be made. This is achieved by dispersing the stable aggregates, generally with sodium hexametaphosphate, resieving the dispersed aggregates, and substracting the mass of coarse primary particles from the previously obtained mass of stable aggregates.

61.2.1 MATERIALS, EQUIPMENT, AND REAGENTS

1 A wet-sieving apparatus similar to that described by Bourget and Kemp (1957) or Kemper and Rosenau (1986). Sieves with openings of 250 μm.

2 Erlenmeyer flasks, 250-mL capacity.

3 Sodium hexametaphosphate (0.5% w/v).

61.2.2 PROCEDURE

1 Weigh 10 g of 1- to 2-mm aggregates (w1). Either air-dried or field-moist aggregates can be used, depending on the objective of the study (see Section 61.5).

2 Spread the aggregates on a 250-μm sieve and place the sieve on the wet-sieving apparatus.

3 Lower the sieve to the water surface and allow the aggregates to wet by capillarity for 10 min. Other wetting procedures can be used (see Section 61.5).

4 Adjust the height of the sieve so that all the aggregates remain immersed in water on the upstroke of the machine. Start the motor and allow the sieve to be raised and lowered 3.7 cm, 29 times/min for 10 min. Other specifications can be used.

5 Remove the sieve and wash the stable aggregates into a tared 250-mL Erlenmeyer flask.

6 Dry the aggregates at 105°C and weigh (w2).

7 Add approximately 50 mL of 0.5% Na-hexametaphosphate to the flask and shake for 45 min.

8 Wash the dispersed aggregates on a 250-μm sieve.

9 Collect the primary particles remaining on the sieve into the same 250-mL Erlenmeyer. Dry the primary particles at 105°C and weigh (w3).

10 Weigh a subsample of 1- to 2-mm aggregates and measure gravimetric water content (wc), expressed as $g\ g^{-1}$, using the method given in Chapter 51 (Section 51.2).

61.2.3 Calculations

Percent WAS can be calculated from:

$$\%WAS = 100 \ (w2 - w3)/((w1/(1 + wc)) - w3) \qquad (61.1)$$

61.2.4 Comments

Some published methods specify that the height of the sieve be adjusted so that the aggregates are just covered with water on the downstroke of the wet-sieving apparatus. This procedure introduces additional disruption to the aggregates, i.e., a lapping motion of the water on the aggregates.

61.3 SIZE DISTRIBUTION OF WATER-STABLE AGGREGATES

In this method, the entire soil fraction is considered. The size distribution of the aggregates is measured after sieving in water. As in Section 61.2, a correction for coarse primary particles has to be made for most soils.

61.3.1 MATERIALS, EQUIPMENT, AND REAGENTS

1   A wet-sieving apparatus similar to that described by Bourget and Kemp (1957) or Kemper and Rosenau (1986). A nest of sieves with openings of 4.5, 2.0, 1.0, 0.5, and 0.25 mm.

2   Erlenmeyer flasks, 250-mL capacity.

3   Sodium hexametaphosphate (0.5% w/v).

61.3.2 PROCEDURE

1   Weigh 40 g of soil (w1) that passes an 8-mm sieve (see Section 61.3.4). Either air-dried or field-moist soil can be used, depending on the objective of the study (see Section 61.5).

2   Spread the soil evenly over the top of a nest of sieves with openings of 4.75, 2.0, 1.0, 0.5, and 0.25 mm.

3   Place the sieves on the wet-sieving apparatus.

4   Lower the sieves into the water until the top sieve is level with the water surface. Allow the soil to wet by capillarity for 10 min. Other wetting procedures can be used (see Section 61.5).

5   Adjust the height of the nest of sieves so that the aggregates on the top sieve remain immersed in water on the upstroke of the machine. Start the

motor and allow the sieves to rise and lower 3.7 cm, 29 times/min for 10 min. Other specifications can be used.

6   Raise the sieves and wash the stable aggregates on each sieve into separate tared 250-mL Erlenmeyer flasks with distilled water.

7   Dry each fraction of aggregates at 105°C and weigh ($w2_i$).

8   Add approximately 50 mL of 0.5% Na-hexametaphosphate to each flask and shake for 45 min.

9   Return each fraction of dispersed aggregates onto the specific sieve size used in step 6. Collect the primary particles remaining on each sieve into the corresponding tared Erlenmeyer. Dry at 105°C and weigh ($w3_i$).

10   Weigh a subsample of soil and determine its gravimetric wc, expressed in $g\ g^{-1}$.

61.3.3 Calculations

The proportion of WSA in each of the size fractions (WSA_i) can be calculated from:

$$WSA_i = (w2_i - w3_i)/((w1/(1 + wc)) - w3_i) \tag{61.2}$$

where $i = 1, 2, 3, \ldots, n$ and corresponds to each size fraction.

Several indices can be calculated if it is desired to express the size distribution with a single parameter. The most widely used parameter is the mean weight diameter (MWD):

$$MWD = \sum_{i=1}^{n} x_i\ WSA_i \tag{61.3}$$

where $i = 1, 2, 3, \ldots, n$ and corresponds to each fraction collected, including the one that passes the finest sieve; x_i is the mean diameter of each size fraction (i.e., mean intersieve size); and WSA_i is as defined in Equation 61.2.

Because the size distribution of soil aggregates is approximately log-normal rather than normal (Gardner 1956), the geometric mean diameter (GMD) (Kemper and Rosenau 1986) is also used. Baldock and Kay (1987) used a power function to describe the size distribution and proposed that the regression constant be used as an index of aggregate size distribution. Perfect and Kay (1991) have recently proposed that fractal theory can be used to characterize soil aggregate size distribution. Attempts have been made to measure agriculturally valuable aggregates by assigning a weight value to each aggregate size range (Dobrzanski et al. 1975, MacRae and Mehuys 1987). Assuming that aggregates between 1 and 5 mm are desirable, weight values of 3, 8.5, 9.5, 4, and 0 were assigned to the aggregate fractions 8–4.75, 4.75–2, 2–1, 1–0.25, and <0.25 mm (McRae and Mehuys 1987).

61.3.4 COMMENTS

1   The largest source of error in sieving work is in sample preparation. Samples, whether air dried or field moist, should be disturbed as little as possible. A representative subsample must be taken for the analysis to be reproducible. The following procedure can be used. Create a cone with the soil to be analyzed. Divide the cone into quarters with a large spatula. Take two sub-samples of 40 g, one from each of two quarters. One subsample is used for aggregate size distribution, the other to determine the water content of the sample.

2   Aggregates passing an 8-mm sieve are commonly used for aggregate size analysis, but aggregates passing a 6-mm sieve can also be used. In this latter case, the 4.75-mm sieve can be omitted from the nest of sieves, thus reducing the number of manipulations slightly.

61.4 A COMBINED METHOD FOR WET-AGGREGATE STABILITY AND TURBIDITY

Pojasok and Kay (1990) have proposed a method which combines the measurement of the stability of large aggregates (1–2 mm) and turbidity, to characterize two different scales of structural units. The method is briefly described in this chapter. The reader is referred to the original paper for more details.

Methods for measuring only turbidity are described, among others, by Williams et al. (1966) and Molope et al. (1985).

61.4.1 PROCEDURE

1   Field-moist aggregates are wetted under tension (-0.1 kPa) and shaken end-over-end in water for 10 min.

2   The material is poured through a 250-μm sieve. The aggregates left on the sieve are the WSA and are analyzed further as in Section 61.2.

3   The percentage of light transmission of the filtrate is measured at a wave-length of 620 nm at a depth calculated from Stokes' law. If desired, the amount of suspended clay particles can be determined using a calibration curve relating the percentage of suspended clay to the percentage of light transmission.

61.4.2 COMMENTS

1   This method offers the advantage of stability measurements on structural units of different scales and can be easily adapted for routine analysis of a large number of samples with minimum equipment and space.

2   Measurements on whole soils or aggregates of different sizes can also be made.

61.5 GENERAL COMMENTS

The measurement of the stability of aggregates of a given size fraction (e.g., 1 to 2 mm) usually requires less time than a measurement made on the whole soil using a nest of sieves. More information is obtained, however, when the whole soil fraction is considered. For example, management effects such as tillage, cropping, or organic amendments are often detected only in a specific-size fraction. Angers and Mehuys (1988) found that cropping had a large effect on the amount of WSA in the 2- to 6-mm size fraction, whereas no effect was found in the 1- to 2-mm fraction. Nevertheless, in other studies both parameters have been found to be correlated (Kemper and Rosenau 1986). Correlations have also been found between turbidity and the stability of large aggregates, although the scales of the measurement differ considerably (Williams et al. 1966, Molope et al. 1985). Molope et al. (1985) also found turbidity to be sensitive to management effects.

Two major factors control the stability of soil aggregates in water: the initial wc of the aggregates and the wetting procedure. When aggregates approaching air dryness are immersed directly in water, slaking of the aggregates can occur. In some studies, this may be desirable if slaking is of concern: for example, in irrigation studies or if differences among stable soils are to be determined (Angers et al. 1992). If slaking is to be avoided, air-dried aggregates should be wetted under vacuum, under tension, or using a fine spray or mist of water. A complete discussion on this subject can be found in Kemper and Rosenau (1986). Measurements can also be made on field-moist aggregates. Under this condition, wetting by capillarity or even by direct immersion can be used with minimal slaking.

Also of great importance is the field sampling procedure, especially if measurements are to be made on field-moist aggregates. The soil is sampled with a shovel and gently crumbled by hand to pass an 8- or 6-mm sieve. The soil is kept in a rigid-wall plastic container at 4°C in order to minimize microbial activity and water loss and to avoid compression of the aggregates during storage.

Several other methods have been proposed for the determination of aggregate stability. Most are variations of the methods described in this chapter. Variations include the use of chemical pretreatments prior to wet sieving to characterize bonding mechanisms. For example, aggregates can be treated with sodium periodate to oxidize carbohydrates or with sodium pyrophosphate to break cation bridges (e.g., Baldock and Kay 1987).

REFERENCES

Angers, D. A. and Mehuys, G. R. 1988. Effects of cropping on macro-aggregation of a marine clay soil. Can. J. Soil Sci. 68: 723–732.

Angers, D. A., Pesant, A., and Vigneux, J. 1992. Early cropping-induced changes in soil aggregation, organic matter, and microbial biomass. Soil Sci. Soc. Am. J. 56: 115–119.

Baldock, J. A. and Kay, B. D. 1987. Influence of cropping history and chemical treatments on the water-stable aggregation of a silt loam soil. Can. J. Soil Sci. 67: 501–511.

Bourget, S. J. and Kemp, J. G. 1957. Wet sieving apparatus for stability analysis of soil aggregates. Can. J. Soil Sci. 37: 60.

Dobrzanski, B., Witkowska, B., and Walczak, R. 1975. Soil-aggregation and water-stability index. Polish J. Soil Sci. 8: 3–8.

Gardner, W. R. 1956. Representation of soil aggregate-size distribution by a logarithmic-normal distribution. Soil Sci. Soc. Am. Proc. 20: 151–153.

Kemper, W. D. and Rosenau, R. C. 1986. Aggregate stability and size distribution. Pages 425–442 in A. Klute, Ed. Methods of soil analysis. Part 1. 2nd ed. American Society of Agronomy, Madison, WI.

MacRae, R. J. and Mehuys, G. R. 1987. Effects of green manuring in rotation with corn on physical properties of two Québec soils. Biol. Agric. Hortic. 4: 257–270.

Molope, M. B., Page, E. R., and Grieve, I. C. 1985. A comparison of soil aggregate stability tests using soils with contrasting cultivation histories. Commun. Soil Sci. Plant Anal. 16: 315–322.

Perfect, E. and Kay, B. D. 1991. Fractal theory applied to soil aggregation. Soil Sci. Soc. Am. J. 55: 1552–1558.

Pojasok, T. and Kay, B. D. 1990. Assessment of a combination of wet sieving and turbidimetry to characterize the structural stability of moist aggregates. Can. J. Soil Sci. 70: 33–42.

Williams, B. G., Greenland, D. J., Lindstrom, G. R., and Quirk, J. P. 1966. Techniques for the determination of the stability of soil aggregates. Soil Sci. 101: 157–163.

Chapter 62
Dry Aggregate Distribution

W. M. White
University of Calgary
Calgary, Alberta, Canada

62.1 INTRODUCTION

The literature describes a variety of methods for determining the distribution and stability of dry soil aggregates, primarily for assessing potential erodibility by wind action or as a measure of seedbed tilth. Kemper and Rosenau (1986) discuss a number of these — mean weight diameter (MWD), geometric mean diameter, weighted mean diameter, log standard deviation — and observe that either of the first two are preferred. The correlation between these techniques is approximately 0.90, so the choice may be somewhat arbitrary; however, they go on to note that " . . . the mean weight diameter (MWD) is easier to calculate and for most individuals to visualize". All involve some sort of sieve analysis, the results of which are then expressed via one of these indices. Since MWD is the traditional approach used in wind erosion studies and the closest to a standard measurement presently in use, it will be outlined in some detail.

62.2 DRY-SIEVE METHOD

The most frequently used technique is the rotary sieving method first advanced by Chepil and Bisal (1943) and later modified by Chepil (1952). This device is preferred over either hand sieving or a RO-TAP® nested sieve shaker or similar unit, as it is more consistent than the former and is believed to cause less aggregate comminution than the latter. Rotary sieves of the Chepil design are not too common, so the technique may be applied to whichever sieving method is available, bearing in mind the previously noted caveats. The number of sieve screens and their mesh sizes will depend on the goals of a particular study, but the end result of the operation will be a series of particle size class weight values from which the MWD calculation may be made.

62.2.1 EQUIPMENT

1   Suitable equipment for collecting samples in the field such as shovels, trowels, etc.

Soil Sampling and Methods of Analysis, M. R. Carter, Ed.,
Canadian Society of Soil Science. © 1993 Lewis Publishers.

2   Open trays or paper bags into which the samples may be transferred and stored with a minimum amount of disturbance.

3   Rotary sieve, RO-TAP®, or some other shaking device that will accommodate the required number of standard 8-in inner diameter sieves. *Note* that a rotary sieve requires its own specific kind of cylindrical sieve inserts that are not compatible with the typical round, flat-bottomed sieves most commonly used.

4   A balance capable of measuring to two decimal places.

5   A dust mask is advisable.

62.2.2 PROCEDURE

1   Carefully collect soil samples from the field site, making sure to handle the material as little as possible during transfer to minimize aggregate breakage.

2   Remove samples to the sieving facility, again taking care to minimize bouncing, crushing, etc., that might cause further aggregate disruption.

3   Each sample should be sieved in a separate operation — as gently as possible place the soil into the hopper/top sieve and turn the machine on (see comments below for sample prepration before sieving is initiated).

4   For the rotary machine, continue sieving until no more material remains on the screens. Other systems, such as standard sieves on a RO-TAP®, present a problem as their is no guide as to how long to continue the sieving operation making this a personal judgment and, thereby, introducing a degree of subjectivity that may affect the consistency and replicability of the results.

5   Weigh the amount of material in each collecting pan or retained on each sieve. At this stage the MWD calculations may be made.

62.2.3 COMMENTS

1   Some debate arises over just how to treat the samples before sieving — usually they are left in the trays/bags and allowed to air dry for at least 7 d before proceeding; however, there is some conjecture as to whether or not this might cause the formation and/or strengthening of aggregates, thus creating a departure from the true field status. Since air drying appears to be the general practice, it is recommended that this step be followed.

2   If an aggregate stability assessment is to be made, the materials should be gently replaced in the hopper/top sieve and steps 4 and 5 repeated (i.e., another dry-sieving operation performed) to yield a second set of weight measurements from which a second MWD calculation may be made.

3   It should be observed that the collection, transfer, and sieving operations by their very nature are bound to result in some degree of mechanical breakdown of the aggregates but, if care is taken at each step, this may be reduced to a minimum and thus have a relatively minor impact on the results of the MWD computations.

62.2.4 Mean Weight Diameter Calculation

The MWD was originally proposed as an index for describing the dry-aggregate distribution of a soil by van Bavel (1949). His technique entailed the plotting of many data points on a graph, then calculating the area under the resulting curve as the sum of a number of smaller areas. However, while this technique is quite accurate, it may be a cumbersome procedure. Youker and McGuinness (1956) suggested an alternative method that uses the summation equation:

$$MWD = \sum_{i=1}^{n} XiWi \qquad (62.1)$$

where:
Xi = mean diameter of size fraction/size class midpoint;
Wi = proportion of total sample retained on sieve.

Youker and McGuinness (1956) state that using Equation 62.1 would produce a slight overestimation of the MWD value. This is also noted by Kemper and Rosenau (1986), and they advocated an adjustment to compensate for this in the following manner. The raw summation figure would be substituted into a regression equation of the form:

$$Y = 0.876X - 0.079 \qquad (62.2)$$

where:
Y = adjusted MWD;
X = MWD figure from summation Equation 62.1.

The MWD calculation given in Table 62.1 is based on a sample taken from a loam/sandy loam soil and sieved on a Chepil-style rotary machine.

Substitution of the unadjusted MWD value into the regression Equation 62.2:

$Y = 0.876X - 0.079$
$Y = 0.876 \, (1.435) - 0.079$
$Y = 1.257 - 0.079 = 1.178$

Therefore, the standardized MWD for this soil is 1.178, which may be compared to the index for other soils or, perhaps, the same soil at a different time during the year to examine spatial or temporal variations in the dry-aggregate distribution. Sillanpaa and Webber (1961) provide an example of the application of this procedure in a Canadian context.

Table 62.1 Sample MWD Calculation

Sieve class mm diameter	Midpoint (A)	% Retained on sieve (B)	A × B/100	Cumulative product
>38.00	50.00[a]	0.080	0.040	0.040
37.79–12.70	25.35	3.450	0.875	0.915
12.69–6.40	9.55	2.280	0.218	1.133
6.39–2.00	4.20	4.280	0.180	1.313
1.99–0.83	1.415	2.260	0.032	1.345
0.82–0.47	0.65	11.140	0.072	1.417
<0.47	0.235	76.510	0.018	1.435
		Σ = 100%	Σ = 1.435	

[a] The upper limit of this size class is determined by measuring the diameter of the largest aggregate before sieving is initiated. In this example it was 62.0 mm.

REFERENCES

Chepil, W. S. 1952. Improved rotary sieve for measuring state and stability of dry soil structure. Soil Sci. Soc. Am. Proc. 16: 113–117.

Chepil, W. S. and Bisal, F. 1943. A rotary sieve method for determining the size distribution of soil clods. Soil Sci. 56: 95–100.

Kemper, W. D. and Rosenau, R. C. 1986. Pages 425–442 in A. Klute and A. L. Page, Eds. Aggregate stability and size distribution. Methods of soil analysis. Part 1. Physical and mineralogical methods. Agronomy No. 9, 2nd ed. Soil Science

Society of America, American Society of Agronomy, Madison, WI.

Sillanpaa, M. and Webber, L. R. 1961. The effect of freezing-thawing and wetting-drying cycles on soil aggregation. Can. J. Soil Sci. 41: 182–187.

van Bavel, C. M. H. 1949. Mean weight diameter of soil aggregates as a statistical index of aggregation. Soil Sci. Soc. Am. Proc. 14: 20–23.

Youker, R. E. and McGuinness, J. L. 1956. A short method of obtaining mean weight diameter values of aggregate analyses of soils. Soil Sci. 83: 291–294.

Chapter 63
Soil Air

R. E. Farrell, J. A. Elliott, and E. de Jong

University of Saskatchewan

Saskatoon, Saskatchewan, Canada

63.1 OVERVIEW

The composition of the soil atmosphere depends upon the balance between the rates of production and the use of various gases in the soil, and the rate of exchange between the soil air and the air above the soil surface. Biological processes (soil respiration) normally release CO_2 and use O_2, but can also result in the release of other gases. Chemical processes can also release gases, e.g., the volatilization of ammonia from fertilizers or the release of radon gas from the mineral fraction of the soil. The soil water plays key roles in controlling the composition of the soil atmosphere: it affects the rate of the biological processes, is a major factor controlling the rate of exchange, and provides temporary storage of soluble gases.

The soil atmosphere reflects the nature of the soil respiration process (aerobic or anaerobic), and when combined with suitable transport coefficients can provide estimates of the rates with which these processes occur. For example, de Jong et al. (1974) used the CO_2 distribution in the soil to calculate CO_2 fluxes when the diffusion constant of the soil was known; Rolston (1978) and Colbourn et al. (1984) did the same for estimating N_2O fluxes in anaerobic soils. In these studies, steady-state conditions were assumed, and only exchange through gaseous diffusion was taken into account. A complete description of changes in the soil atmosphere would have to take into account diffusion, mass flow, and solution and dissolution of various gases in the liquid phase.

In some methods of analysis, the gas of interest is reduced or absorbed by the detector. When used *in situ* these methods may yield erroneous results, as the total amount of the gas absorbed reflects not only the concentration in the gaseous phase, but also the rapidity with which it can be replaced by diffusion from the surrounding soil mass. Thus, the initial rate of absorption might reflect the existing concentration, but the rate of absorption over longer periods might reflect diffusive flow to the sampling chamber when the diffusion constant of the soil is small.

Soil Sampling and Methods of Analysis, M. R. Carter, Ed.,
Canadian Society of Soil Science. © 1993 Lewis Publishers.

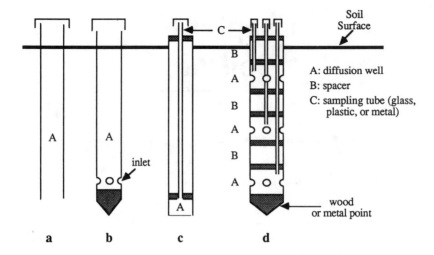

FIGURE 63.1. Air well designs.

63.2 SAMPLING OF THE GASEOUS PHASE

In some studies, "grab" samples are taken by inserting a probe to the desired depth in the soil and withdrawing a soil air sample under vacuum. Depending on the care used in inserting the probe, the size of the sample, and the rapidity of withdrawal, serious contamination can occur from air leaking around the probe or preferential withdrawal of air from the larger pores. Nowadays, grab samples are generally avoided and, instead, air samples are withdrawn from air — or diffusion — wells.

Figure 63.1 shows some designs of air wells. The air inside the well equilibrates with the soil air at the inlet through diffusion. The time required for equilibration depends on the diffusion constant of the soil, the cross section of the inlets, and the length of the well. Figures 63.1a and 63.1b shw the simplest well designs consisting of a piece of pipe closed at the top. The air in the space at the bottom of the well shown in Figure 63.1c would equilibrate quickly. Although long equilibration times would be required for the air in the withdrawal tube, the air in it can be purged and discarded prior to taking a sample for analysis. Figure 63.1d shows a design for a multiple-well tube which has the advantage of a similar geometry for each sampling depth and of reducing the number of tubes that have to be inserted in a plot.

63.2.1 Sampling Procedures

63.2.1.1 *Materials*

Metal and plastic tubing have been used in the construction of the wells and of the gas-withdrawal tubes. In our experience, translucent plastic tubing used for the withdrawal tube turns brittle with time when exposed to sunlight. Occasionally, rodents attack this type of tubing. Serum caps or similar are used to seal the tops of the wells or the gas-withdrawal tubes.

63.2.1.2 *Installation*

Gas wells should be sealed tightly against the surrounding soil. This can be achieved by punching a hole of a slightly smaller diameter than the well and forcing the well into it.

Pushing the well into the soil is facilitated by a pointed tip on the wells, as illustrated in Figures 63.1b and 63.1d. Soil pushed into a tube of the design illustrated in Figure 63.1a can be augered out; this is not possible for tubes of the design illustrated in Figure 63.1c, but the size of the unlined cavity is not critical. If the wells are installed in a wet soil, shrinkage cracks may develop around them in clay soils, and installation in a dry soil is recommended.

63.2.1.3 Equilibration

Van Bavel (1954) has suggested that up to $1^{1}/_2$ h are needed for 15-cm-long wells, as shown in Figure 63.1a, to equilibrate to within 2% of their final value, and up to 8 h for wells that are 45 cm long. These calculated equilibration times were for wells inserted in a soil with 15% air space. Longer equilibration times will be needed in soils with lower air space or for wells with a smaller inlet. In practice, the concentration in the well will always lag the concentration in the soil air. The shorter the well, the closer the concentration in the air well will be to the concentration in the soil air at the inlet.

63.2.1.4 Sample Collection

Samples can be withdrawn from the gas-sampling wells using a variety of procedures. Evacuated test tubes (Vacutainers®) have been used by de Jong et al. (1974) and Amundson and Smith (1988) when samples needed to be stored for a relatively long time prior to analysis. Vacutainers® can be reused, and covering the tops of the Vacutainers® with a thin layer of silicone helps to prevent gas leakage. Where the samples can be analyzed quickly, plastic syringes may be used. The needles are left on the syringe and are inserted in rubber stoppers to prevent leakage. Samples in plastic syringes should be analyzed within 8 h, whereas samples in Vacutainers® can be stored for several days, especially if they are under a slight positive pressure (Amundson and Smith 1988).

63.3 ANALYSIS OF THE GASEOUS PHASE

Early measurements of soil air composition often relied on absorption techniques, e.g., CO_2 in an alkali and O_2 in pyrogallol. Large volumes of soil air, for example, 50 cm³ or more, were taken to the laboratory and O_2 and CO_2 determined by absorption in gas burettes. These techniques will not be discussed because of problems associated with the sampling (see above) and with analysis when CO_2 is present in low concentrations.

63.3.1 Laboratory Analysis by Gas Chromatography

Usually, small gas samples withdrawn by hypodermic syringes or into Vacutainers® are taken to the laboratory for analysis using gas chromatography. A wide range of methods, detectors, and column packings are currently used in soil atmosphere research. The most comprehensive technique for gas chromatographic analysis of the soil atmosphere was proposed by Blackmer and Bremner (1977). The method allows rapid and precise determination of N_2, O_2, Ar, CO_2, CH_4, and N_2O. Neon, H_2, CO, and NO are also separated and C_2H_4 and C_2H_2 form a composite peak. A Tracor® Model 150G gas chromatograph fitted with an ultrasonic detector and a dual-phase meter is used in the analysis. The gases are separated on two stainless steel columns of 3.2 mm O.D. and 2.1 mm I.D. packed with 50- to 80-mesh Porapak® Q. Gases pass through the first column, which is 4.3 m long and maintained at 45°C, into the A side of the detector and then enter the second column, which is 7.6 m

**Table 63.1 Methods and Detectors Used for Components of the Soil Atmosphere
Commonly Measured Using Gas Chromatography**

Gas	Detector	Detection limit ($cm^3 m^{-3}$)	Ref.
O_2	Thermal conductivity	75	Hall and Dowdell (1981)
	Ultrasonic	1	Blackmer and Bremner (1977)
CO_2	Thermal conducitivity	20	Smith and Dowdell (1973)
	Ultrasonic	1	Blackmer and Bremner (1977)
	Helium ionization	1	Mitchell (1973)
N_2O	Ultrasonic	1	Blackmer and Bremner (1977)
	Electron Capture	0.1	Mosier and Mack (1980)
CH_4	Ultrasonic	1	Blackmer and Bremner (1977)
	Flame ionization	0.02	Smith and Dowdell (1973)
C_2H_4	Flame ionization	0.02	Smith and Dowdell (1973)
	Flame ionization	0.005	Wood (1980)
	Taguishi gas sensor	10	Holfeld et al. (1979)
S gases	Flame photometric	0.2	Banwart and Bremner (1974)
	Flame photometric		de Souza (1984)

long and submerged in a dry ice-methanol bath, and exit into side B of the detector. Helium is used as the carrier gas and flow is maintained at 50 mL min^{-1} by regulating the gas supply at a pressure of 4.2 kg cm^{-2} and the back pressure regulator on side B at 2.1 kg cm^{-2}. It must be noted, however, that operational procedures may vary for different gas chromatographic systems. The system described above can analyze samples from a flask (see Blackmer and Bremner [1977] for details) or from gas-tight syringes and can be modified to include another column. Most soil air samples can be analyzed in 6.5 min, but if large amounts of C_2H_4, C_2H_2, H_2, or Ne are present the interval between samples must be increased.

Many other methods are available and should be selected according to the gas(es) of interest, the sensitivity required, and the cost or availability of a detector. For example, although the method of Bremner and Blackmer (1977) gives precise determinations of O_2 and CO_2, an alternative method using a thermal conductivity detector would be adequate (Anderson 1982) if these were the only gases of interest. In studies which require the analysis of N_2O in air, an electron capture detector is usually chosen over the ultrasonic detector because it is more sensitive to N_2O. Smith and Arah (1991) give an overview of the principles of gas chromatography, column selection, and detector capability for gas chromatographic analysis of the soil atmosphere. Other useful reviews have been prepared by Bremner and Blackmer (1982) and Smith (1977). References for some alternative gas chromatographic methods are provided in Table 63.1.

Accurate sample injection is important. If the sample volume varies from the standard, errors will result. In some instances, we have found it useful to use N_2 as an internal standard for the volume of the sample, or have used the sum of N_2, O_2, Ar, and CO_2 as a correction factor. In well-aerated soils, N_2 should be close to 78%, and N_2, O_2, Ar, and CO_2 should account for approximately 99% (assuming that water vapor is included) of the total sample.

63.3.2 *In Situ* Analysis

Van Bavel (1965) describes the use of a paramagnetic oxygen analyzer and a portable carbon dioxide analyzer, based on a thermal conductivity detector, for measuring O_2 and CO_2, respectively, in the field. Soil air is pumped from a diffusion well, through the analyzer, and returned to the well. Although the accuracy of these O_2 and CO_2 measurements was adequate, it was much less than that achievable with gas chromatography. Polarographic O_2 analyzers are also available commercially.

63.4 ANALYSIS OF THE GAS-LIQUID INTERFACE

Gas chromatography is not suitable for *in situ* monitoring of the dynamics of the soil atmosphere or the determination of gases dissolved in the soil solution. Electrochemical sensors, on the other hand, can measure the partial pressure of gases in both air and solution and, hence, are adaptable to these situations. Indeed, electrodes have been described for the *in situ* continuous monitoring of O_2 (Willey and Tanner 1963) and CO_2 (Jensen et al. 1965). Likewise, a commercial O_2 analyzer that employs an amperometric O_2 sensor has been used to measure the diffusivity of O_2 in field soils (Rolston 1986). In this section, we discuss the platinum (Pt) microelectrode as a tool for measuring the oxygen diffusion rate (ODR) and O_2 concentration in the gaseous and liquid phases of the soil. Detailed reviews of the theory, validity, and methodology of the Pt microelectrode methods can be found in the literature (McIntyre 1971, Phene 1986).

63.4.1 ODR Measurements

63.4.1.1 *Principles and Apparatus*

The usefulness of ODR measurements is based on the premise that a reasonable assessment of the soil aeration status can be obtained by measuring the diffusion of O_2 through the liquid phase of the soil to a reducing surface which approximates a plant root. The basic ODR measurement system (Figure 63.2) consists of a Pt microelectrode (the cathode), a nonpolarizable reference electrode (the anode; usually a Ag,AgCl electrode with a KCl-saturated agar salt bridge), a power supply and its associated electrical circuit (to apply an electrical potential between the cathode and anode), and a milliammeter (to measure the output current). When the electrical potential of the Pt microelectrode is lowered sufficiently with respect to the reference electrode (a potential of -0.65 V is usually recommended for purposes of standardization), the O_2 at the microelectrode surface is electrolytically reduced until the concentration of O_2 at the surface is zero. At this time, the rates of reduction and diffusion of O_2 are equal and independent of the applied voltage, and the resulting electrical current is proportional to the flux of O_2 at the surface of the Pt microelectrode. The ODR can be calculated from the equation:

$$ODR = C(i/A) \tag{63.1}$$

where C is a constant (0.00497 μg μA^{-1} min^{-1}), i is the output current in μA, and A is the surface area of the microelectrode in cm^2.

63.4.1.2 PROCEDURES

1   Construction of Pt microelectrodes: Pt microelectrodes can be constructed by spotwelding 1.25-cm lengths of 20- or 22-) gauge platinum wire to 15-cm lengths of 18-gauge insulated copper wire. Disposable, 1-mL plastic pipette tips are inserted over the Cu-Pt wires to function as molds and, leaving 1.0

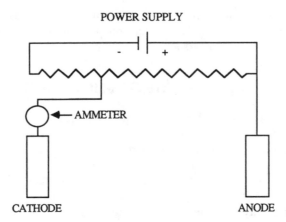

POWER SUPPLY

← AMMETER

CATHODE

ANODE

FIGURE 63.2. Basic components of the ODR measurement system.

cm of the Pt wire exposed, filled with epoxy, which is then left to harden (Farrell et al. 1991). The electrodes are cleaned in detergent and hydrogen peroxide, then rinsed with distilled water.

2   The Pt microelectrodes are inserted into the soil, taking care not to damage the tip.

3   The reference electrode is connected to the soil by means of the salt bridge. If dry, the soil near the salt bridge may be moistened with some distilled water to ensure that there is good contact between the reference electrode and the soil.

4   Measurement of the electrical current: apply a potential of -0.65 V between the electrodes and wait for a steady-state current to be achieved (usually about 5–10 min). Measure the steady-state current and calculate the ODR using Equation 63.1.

63.4.1.3 COMMENTS

The major factors affecting ODR measurements can be grouped into two categories: (1) electrochemical factors (e.g., choice of the applied potential, establishment of the steady-state current, installation of the electrodes, and poisoning of the Pt electrodes) and (2) soil factors (e.g., moisture and salt content, O_2 concentration, and temperature). These factors have been discussed in detail by McIntyre (1971). Erroneous ODR measurements often can be traced to one of three causes: (1) selection of an inappropriate applied potential, (2) poisoning of the Pt electrode, or (3) a change in the moisture or salt content of the soil.

1   Applied potential: under ideal conditions, reduction of O_2 at the cathode begins at an applied potential of about -0.2 V (the decomposition potential of oxygen). Increasing (negatively) the applied voltage further produces an increase in the current until the *limiting potential* (the potential at which the rate of reduction is controlled by the rate at which O_2 can diffuse to the surface of the cathode) is reached. At this stage, the plot of current vs.

applied voltage (i.e., a polarogram) forms a plateau and no further increase in current is observed until the applied voltage reaches the discharge potential of the hydrogen ion (Armstrong and Wright 1976). Although ODR measurements usually employ an applied potential of -0.65 V (Phene 1986), Armstrong and Wright (1976) and Blackwell (1983) recommended that the applied potential be derived from a current-voltage plot of the limiting potential *(in situ)* and that this be the first step in any ODR measurement. They further recommended foregoing the ODR measurement if the plateau of the current-voltage plot is absent. Blackwell (1983) reported that the Armstrong and Wright method yields ODR values which often are an order of magnitude lower than those measured with the standard method (i.e., applied potential $= -0.65$ V). Hence, considerable thought should be given to the choice of an appropriate applied voltage. Because of the expense and inconvenience involved in making multiple determinations of the limiting potential, it has been suggested that a single measurement obtained at each sampling depth is adequate (Armstrong and Wright 1976).

2   Electrode poisoning: poisoning can be defined as any chemical or physical change in the Pt electrode that prevents the electrochemical reduction of O_2 (Devitt et al. 1989). Poisoning is not usually a factor if the electrodes are removed from the soil after each measurement. If the electrodes are left in place for extended periods, however, poisoning may become a factor. Electrode poisoning can result from the movement of colloidal material to the electrode surface or the precipitation of carbonates or mixed carbonate-aluminosilicates on the surface of the electrode (McIntyre 1971, Devitt et al., 1989). The effects of poisoning can be minimized by removing and cleaning the electrodes every 4 to 8 weeks.

3   Corrections for soil moisture and salts: changes in the moisture or salt content of the soil will be reflected by changes in the electrical resistance of the soil. This, in turn, will affect the "true" potential between the Pt and reference electrodes. Thus, because the output current is dependent on the "true" potential between the electrodes as well as the flux of O_2 to the Pt electrode, the "true" ODR will depend partly on the soil resistance. Methods of correcting ODR for changes in the soil resistance have been described by Kristensen (1966) and Callebaut et al. (1980).

Despite these problems and some reservations about the theoretical validity of using the Pt microelectrode to measure soil O_2 (McIntyre 1971), it is generally agreed that until a better method is developed, ODR measurements obtained with the Pt microelectrode can provide valuable information regarding soil aeration. Moreover, the basic ODR measurement system can be adapted to include a data acquisition system, thus allowing rapid sequential measurements of up to 72 electrodes (Phene et al. 1976).

63.4.2 O_2 Concentration Measurements

63.4.2.1 *Principles and Apparatus*

The O_2 electrode is an amperometric sensor that consists of a Pt (or Au) cathode and an Ag,AgCl anode (electrically connected to the cathode by an electrolyte, e.g., KCl), which

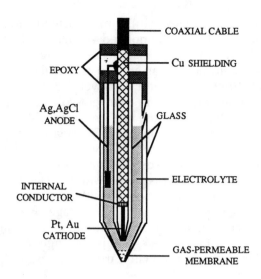

COAXIAL CABLE

EPOXY

Cu SHIELDING

Ag,AgCl
ANODE

GLASS

INTERNAL
CONDUCTOR

ELECTROLYTE

Pt, Au
CATHODE

GAS-PERMEABLE
MEMBRANE

FIGURE 63.3. O$_2$ microelectrode.

contact the soil through a gas-permeable membrane (Figure 63.3). Theoretical and operational considerations have been reviewed by Davies (1962). Briefly, when an appropriate potential is applied between the cathode and anode, the O$_2$ that diffuses across the gas-permeable membrane is reduced at the cathode, the O$_2$ concentration at the surface of the cathode is zero, and the output current is proportional to the concentration of O$_2$ reduced at the cathode. Willey and Tanner (1963) modified the basic O$_2$ electrode by incorporating a thermistor to compensate for temperature effects. Willey (1974) modified the electrode further by adding a second gas-permeable membrane, separated from the first by a piece of nylon mesh, to prevent water from condensing on the membrane adjacent to the Pt cathode. In this way, even though water may condense on the outer membrane, a low-impedance path for the diffusion of O$_2$ to the cathode exists between the two membranes. Probes such as these are suitable for the determination of O$_2$ concentrations in both the gaseous and solution phases of the soil.

63.4.2.2 Procedures

Oxygen probes incorporating features of the electrodes described above are available commercially. Activation, calibration, and use of the electrodes should be performed according to the manufacturer's instructions. In general, the electrodes will require recalibration on a monthly basis, although more frequent recalibrations will be necessary for studies requiring increased precision. Likewise, to ensure reliable performance it is advisable that the internal electrolyte be changed on a monthly basis.

63.4.2.3 Comments

The continued development of O$_2$ electrodes has led to some novel experimental designs. For example, in an attempt to establish the presence of anaerobic microsites within individual soil aggregates and determine the intra-aggregate O$_2$ diffusion coefficient, Sexstone et al. (1985) employed an O$_2$ microelectrode for the direct measurement of the O$_2$ concentration profiles within individual soil aggregates. Although such an experiment is not without its difficulties, it would have been virtually impossible with any other analytical technique. Furthermore, it demonstrates the enormous potential presented by the development of other gas-sensing probes.

63.4.3 Miscellaneous Gas-Sensing Probes

Gas-sensing probes have been developed for a number of gases, e.g., CO_2, NH_3, SO_2, and CH_4. To date, however, investigations of the soil atmosphere have employed only the potentiometric CO_2 probe (Jensen et al. 1965). Nevertheless, the continuing development of chemical sensors and biosensors (Janata 1990) may be expected to yield new opportunities for quantifying the various components of the soil atmosphere and studying its dynamics *in situ*.

REFERENCES

Amundson, R. G. and Smith, V. S. 1988. Annual cycles of physical and biological properties in an uncultivated and an irrigated soil in the San Joaquin Valley of California. Agric. Ecosyst. Environ. 20: 195–208.

Anderson, J. P. E. 1982. Soil respiration. Pages 831–871 in A. L. Page, Ed. Methods of soil analysis. Part 2. Chemical and microbiological properties. Agronomy No. 9, 2nd ed. American Society of Agronomy, Madison, WI.

Armstrong, W. and Wright, E. J. 1976. A polarographic assembly for multiple sampling of soil oxygen flux in the field. J. Appl. Ecol. 13: 849–856.

Banwart, W. L. and Bremner, J. M. 1974. Gas chromatographic identification of sulfur gases in soil atmospheres. Soil Biol. Biochem. 6: 113–115.

Blackmer, A. M. and Bremner, J. M. 1977. Gas chromatographic analysis of soil atmospheres. Soil Sci. Soc. Am. J. 41: 908–912.

Blackwell, P. S. 1983. Measurements of aeration in waterlogged soils: some improvements of techniques and their application to experiments using lysimeters. J. Soil Sci. 34: 271–285.

Bremner, J. M. and Blackmer, A. M. 1982. Composition of soil atmospheres. Pages 873–901 in A. L. Page, Ed. Methods of soil analysis. Part 2. Chemical and microbiological properties. Agronomy No. 9, 2nd ed. American Society of Agronomy, Madison, WI.

Callebaut, F., Balcaen, M., Gabriels, D., and De Boodt, M. 1980. Data acquisition system for field determination of redox potential, oxygen dif-

fusion rate and soil electrical resistance. Meded. Fac. Landbouwwet. Rijksuniv. 45: 15–24.

Colbourn, P., Harper, I. W., and Iqbal, M. M. 1984. Denitrification losses from ^{15}N-labelled calcium nitrate fertilizer in a clay soil in the field. J. Soil Sci. 35: 539–547.

Davies, P. W. 1962. The oxygen cathode. Pages 137–179 in W. L. Nastuk, Ed. Physical techniques in biological research. Academic Press, New York.

de Jong, E., Schappert, H. J. V., and MacDonald, K. B. 1974. Carbon dioxide evolution from virgin and cultivated soil as affected by management practices and climate. Can. J. Soil Sci. 54: 299–307.

de Souza, T. L. C. 1984. Supelpak-S: the GC separating column for sulphur gases. J. Chromatogr. Sci. 22: 470–472.

Devitt, D. A., Stolzy, L. H., Miller, W. W., Campana, J. E., and Sternberg, P. 1989. Influence of salinity, leaching fraction, and soil type on oxygen diffusion rate measurements and electrode "poisoning". Soil Sci. 148: 327–335.

Farrell, R. E., Swerhone, G. D. W., and van Kessel, C. 1991. Construction and evaluation of a reference electrode assembly for use in monitoring *in situ* soil redox potentials. Commun. Soil Sci. Plant Anal. 22: 1059–1068.

Hall, K. C. and Dowdell, R. J. 1981. An isothermal gas chromatographic method for the simultaneous estimation of oxygen, nitrous oxide and carbon dioxide content of gases in the soil. J. Chromatogr. Sci. 19: 107–111.

Holfeld, H. S., Mallard, C. S., and LaRue, T. A., 1979. Portable gas chromatograph. Plant Soil 52: 595–598.

Janata, J. 1990. Chemical sensors. Anal. Chem. 62: 33R–44R.

Jensen, C. R., Van Gundy, S. D., and Stolzy, L. H. 1965. Recording CO_2 in soil-root systems with a potentiometric membrane electrode. Soil Sci. Soc. Am. Proc. 29: 631–633.

Kristensen, K. J. 1966. Factors affecting measurements of oxygen diffusion rate (ODR) with bare platinum microelectrodes. Agron. J. 56: 295–301.

McIntyre, D. S. 1971. The platinum microelectrode method for soil aeration measurement. Adv. Agron. 22: 235–283.

Mitchell, M. J. 1973. An improved method for microrespirometry using gas chromatography. Soil Biol. Biochem. 5: 271–274.

Mosier A. R. and Mack, L. 1980. Gas chromatographic system for precise, rapid analysis of nitrous oxide. Soil Sci. Soc. Am. J. 44: 1121–1123.

Phene, C. J. 1986. Oxygen electrode measurement. Pages 1137–1159 in A. L. Page, Ed. Methods of soil analysis. Part 1. Physical and mineralogical methods. Agronomy No. 9, 2nd ed. American Society of Agronomy, Madison, WI.

Phene, C. J., Campbell, R. B., and Doty, C. W. 1976. Characterization of soil aeration in situ with automated oxygen diffusion measurements. Soil Sci. 122: 271–281.

Rolston, D. E. 1978. Application of gaseous-diffusion theory to measurement of denitrification. Pages 309–335 in D. R. Nielsen and J. G. MacDonald, Eds. Nitrogen in the environment, Vol. 1. Academic Press, New York.

Rolston, D. E. 1986. Gas diffusivity. Pages 1089–1102 in A. L. Page, Ed. Methods of soil analysis. Part 1. Physical and mineralogical methods. Agronomy No. 9, 2nd ed. American Society of Agronomy, Madison, WI.

Sexstone, A. J., Revsbech, N. P., Parkin, T. B., and Tiedje, J. M. 1985. Direct measurement of oxygen profiles and denitrification rates in soil aggregates. Soil Sci. Soc. Am. J. 49: 645–651.

Smith, K. A. 1977. Gas chromatographic analysis of the soil atmosphere. Pages 197–229 in J. C. Giddings et al., Eds. Advances in chromatography. Vol 15. Marcel Dekker, New York.

Smith, K. A. and Arah, J. R. M. 1991. Gas chromatographic analysis of the soil atmosphere. Pages 505–546 in K. A. Smith, Ed. Soil analysis: modern instrumental techniques. 2nd ed. Marcel Dekker, New York.

Smith, K. A. and Dowdell, R. J. 1973. Gas chromatographic analysis of the soil atmosphere: automatic analysis of gas samples for O, N, Ar, CO, NO and C-C hydrocarbons. J. Chromatogr. Sci. 11: 655–658.

van Bavel, C. H. M. 1954. Simple diffusion well for measuring soil specific diffusion impedance and soil air composition. Soil Sci. Soc. Am. Proc. 18: 229–234.

van Bavel, C. H. M. 1965. Composition of soil atmosphere. Pages 315–346 in C. A. Black et al., Eds. Methods of soil analysis. Part 1. Physical and mineralogical properties, including statistics of measurement and sampling. Agronomy No. 9. American Society of Agronomy, Madison, WI.

Willey, C. R. 1974. Elimination of errors caused by condensation of water on membrane-covered oxygen sensors. Soil Sci. 117: 343–346.

Willey, C. R. and Tanner, C. B. 1963. Membrane covered electrode for measurement of oxygen concentration in soil. Soil Sci. Soc. Am. Proc. 27: 511–515.

Wood, M. J. 1980. An application of gas chromatography to measure concentrations of ethane, propane, and ethylene in interstitial soil gases. J. Chromatogr. Sci. 18: 307–310.

Chapter 64
Soil Temperature

N. J. Livingston

University of Victoria

Victoria, British Columbia, Canada

64.1 INTRODUCTION

There are numerous sensors or transducers that can provide measurements of soil temperature. The ideal temperature transducer converts variations of temperature into another easily measured quantity, for example, voltage. Typically, the measurement must be made at one location and the information used elsewhere, sometimes several hundred feet away. The selection of the most appropriate transducer for a given experimental design should include consideration of instrument accuracy, sensitivity, stability, range, reliability, maintenance, response time, ruggedness, size, ease of construction and of installation, compatibility with available display and recording devices, and cost.

There are a large number of texts in which there are extensive reviews of a range of temperature sensors (Tanner 1963, Fritchen and Gay 1979, Unwin 1980, Woodward and Sheehy 1983, Stathers and Spittlehouse 1990, Omega Engineering 1991). Accordingly, this chapter will not focus on detailed descriptions of instruments, but on general principles.

64.2 MEASUREMENT PRINCIPLES AND PROCEDURES

64.2.1 Range and Sensitivity

Soil temperatures can range from approximately −40 to 60°C. The greatest extremes in temperature occur at the surface and decrease rapidly with increasing depth. The absolute temperature and its variation with time and depth will be greatly influenced by surface cover and by the thermal properties and water content of the soil.

The sensitivity required depends primarily upon the purpose of the measurement, particularly the accuracy required and where the measurements are made (e.g., at the soil surface or at greater depth), the degree of spatial variability, and the number of measurement sensors. Perrier (1971) has suggested that a sensitivity of ±1°C is sufficient for the analysis of plant growth and development. A greater sensitivity of ±0.1°C is needed for determination of heat fluxes. Tanner (1963) is especially critical of scientists who insist upon very high sensitivity without being able to physically or biologically justify or understand this requirement.

Soil Sampling and Methods of Analysis, M. R. Carter, Ed.,
Canadian Society of Soil Science. © 1993 Lewis Publishers.

64.2.2 Sampling Requirements

There are no hard and fast rules for determining how many soil temperature samples are required to obtain a given accuracy. The number of samples depends on the heterogeneity of both the soil and the surface cover. For example, enormous variability in surface temperature can be caused by differential shading of plants, especially in the spring before a complete crop cover has been established. Tanner (1963) suggests that the simplest procedure to determine sampling size for a particular accuracy is to place several thermometers at a given depth in a random spacing on a line or in a circle and measure the variability. The number of samples required will decrease rapidly with depth because of the exponential decrease in the amplitude of periodic temperature fluctuations. Averages of temperatures within a soil layer can be determined by connecting thermocouples in parallel or series within that layer (see Section 64.3.4). This is a useful technique for providing estimates of soil heat storage. Additionally, replicates of measurements at a particular depth or series of depths can be obtained using the same technique. Suomi (1975) used 5-cm-long resistance thermometers installed vertically to obtain average layer temperatures. In addition, he attached 12 of these resistance elements in series to obtain a spatial average.

64.2.3 Sampling Rates

Soil temperatures and temperature gradients are often recorded on multichannel recorders that sample data at fixed intervals. Researchers need to select the optimum sampling rate that allows them to reconstruct the sensor output signal without being overwhelmed with redundant data. The output from the temperature sensor can be considered to be a fluctuating signal with a particular frequency distribution. Shannon's sampling theorem (discussed in Tanner 1963 and Fritchen and Gay 1979) predicts that for a sensor with an output signal f as a function of time (t), the sampling rate should be at least twice as high as the highest frequency of interest (i.e., 2f). This assumes that the sensor does not affect the recorded signal. However, because there is a finite lag before a sensor responds to changing conditions, it is important to consider the response time of the sensor itself when determining the optimum sampling frequency.

The response time of an instrument for a given change in environmental conditions is often given in terms of its time constant (τ). The time constant is defined as the time for a sensor, subjected to a step change in conditions, to reach 63.2% of the total change. In general, the time constant of a sensor is related to its physical construction, size and thermal characteristics, and to the environmental conditions in which it is placed. For example, a freely exposed mercury-in-glass thermometer in air with a wind speed of 0.1 m s^{-1} might have a time constant of 360 s. However, for a 100-fold increase in wind speed the time constant might drop to 40 s (Woodward and Sheehy 1983). The time constant of a soil surface, by comparison, is approximately 100 s (Weigland and Swanson 1973).

Increasing the time constant of a sensor by, for example, encasing it in a waterproof epoxy sheath, is a useful way of smoothing out unwanted high frequencies when only time averages are required.

Environmental conditions seldom, if ever, change in a stepwise manner. However, it is reasonable to assume that soil temperature fluctuates sinusoidally. The amplitude of the measured sine wave will be influenced by the time constant of the sensor used. Following the theory of Panofsky and Brier (1958) and Fuchs (1971), it can be shown that in the simplest case the sampling interval (ΔT) is

$$\Delta T = \alpha \pi \tau \tag{64.1}$$

where α, the coefficient of attenuation, is

$$\alpha = (1 + 4\pi^2 f^2 \tau^2)^{-0.5} \tag{64.2}$$

It can be seen that if the time constant of a sensor is taken into account when determining the appropriate sampling interval, the calculated sampling interval is considerably smaller (i.e., has a greater frequency) than that predicted by the Shannon theorem.

64.2.4 Installation

The measurement location should be chosen to best represent the area of interest. Physical properties of the surface (reflectivity, the presence of woody debris, vegetation, organic matter, etc.) should be considered because they have a significant influence on the surface energy balance and therefore on surface and subsurface temperatures.

There should be minimal soil disturbance during installation. Soil profiles and surface conditions should be carefully reconstructed, and soils repacked to the original bulk density. Failure to do so will lead to significant changes in the thermal and hydraulic regime of the soil.

Fritschen and Gay (1979) describe a modification of a technique given by Portman (1957) for inserting temperature sensors in undisturbed soil. A triangular pit is dug to a desired depth. A board with predrilled holes at selected increments is then pushed up to a smoothed side of the pit with its top flush with the surface. Brass or copper tubes (for example, 0.2 cm diameter by 15 cm long) with pointed and sealed ends, and each housing an appropriate temperature sensor, are pushed through the holes in the reference board into the soil. The hole is then back filled.

Stathers and Spittlehouse (1990) list a number of precautions and procedures that should be adopted when installing sensors:

1 Approach the measurement site from one direction (preferably the downhill side) to minimize disturbance. Avoid walking in the area or install removable boardwalks if there is significant vegetation. Flag the site boundaries.

2 Soil removed from a pit should be stored on a tarpaulin to facilitate reconstruction of the soil profile once the temperature sensors have been installed.

3 Depth and other appropriate information should be put on waterproof labels attached to sensor leads prior to installation.

4 Use the surface as the reference point to determine lower profile depths. Starting from the bottom of the profile and working towards the surface, repeat the following procedure at each measurement depth:

a. Drill a 100- to 150-mm-deep hole with a diameter equal to that of the sensor, horizontally into the vertical face of the profile.

b. Push the sensor into the hole, ensuring that there is good thermal contact between the sensor and the surrounding soil. Measurement depth should correspond to the center of the sensor.

c. Bury at least 20 cm of lead wire at the same depth as the sensor to minimize errors caused by heat conduction toward the sensor.

d. Refill the soil pit to the depth of the next sensor, taking care to reconstruct the soil profile.

5 Cover the sensor leads or put them in a protective box to prevent animal damage.

64.2.5 Sensor Calibration and Preparation

Temperature sensors should always be individually calibrated over the range of interest before field installation. Never rely on the factory calibration. Ideally, sensors should be calibrated in a stirred water bath against a standard — for example a good-quality platinum resistance thermometer or NBS mercury-in-glass thermometer.

An ice bath made up of 50% distilled water provides a convenient reference temperature. Provided the bath is well mixed, temperatures are reproducible to 0.01°C. However, if too little ice is used and the mixture is not well mixed errors can exceed 1°C. Tap water provides another convenient and reproducible reference, provided that the water has been circulated for at least 5 min.

Most temperature sensors are relatively stable and should not require frequent recalibration. Those sensors that are placed in the field for long periods (greater than one or two field seasons) should be recalibrated at the beginning and end of the experiment. If sensors are to be in the field over winter they should, before installation, be put through a complete freeze-thawing cycle in the laboratory.

It is highly recommended that sensors be encased in robust, waterproof material before installation. This will greatly increase sensor reliability and longevity, but will significantly increase the response time of the device. Potted sensors are commercially available (for example, thermistor probes from Campbell Scientific Canada, Edmonton, Alberta) but are easily made in the laboratory. A simple procedure is to place the sensor (after the appropriate leads have been attached) at the center of a section of a drinking straw (approximately 5 cm long by 1 cm in diameter). Clamp the bottom end of the straw and then fill it with an industrial-grade epoxy resin or other suitable potting compound. Once the compound has cured (approximately 24 h), cut away the straw.

Cable integrity should always be carefully checked (using a resistance meter) before field installation. Well-insulated, good-quality cable should always be used. Heat-shrink tubing should always be placed over exposed cable-sensor connections.

There is a huge range of recording devices or data loggers suitable for recording the output of field-installed sensors (see, for example, Pearcy 1989). Whatever the device selected, it should be carefully checked in the laboratory. A complete measurement cycle (data collection, data transfer and analysis) should be performed before installation in the field.

64.3 TEMPERATURE TRANSDUCERS

Instruments to measure soil temperature can be classified into four general categories based on their operating principles: expansion, chemical, radiative, and electrical.

64.3.1 Expansion

Thermal expansion instruments make use of the fact that the dimensions of all substances, whether solids, liquids, or gases, change with temperature. Instruments operating on this physical principle include liquid-in-glass thermometers, bimetallic strips (often used in thermographs), and the pressure thermometer. Generally, these type of sensors are slow to respond, are relatively large, and must be read manually. As Tanner (1963) points out, a major source of error in expansion-type devices is that the sensors have greatly different thermal properties from that of soil. Differences in heat capacity and thermal conductivity may greatly affect the soil temperature at the sensor location, particularly if the sensor volume is quite large. This problem is particularly acute near the soil surface. Mercury-in-glass thermometers have an additional problem caused by the conduction of heat down the stem to the bulb. Mercury-in-glass thermometers are commonly used in meteorological stations that do not have automated data collection systems. While they can provide quite precise measurements and are very stable, they are extremely fragile and are not recommended for field use. These and other expansion devices are discussed in detail by Morris (1988).

64.3.2 Chemical

There are a number of chemical sensors available. These include heat-sensitive films or crystals that change color with temperature, solutions that undergo chemical or electrolytic changes with temperature, and compounds that have differing melting points. All these sensors have the distinct disadvantage that they do not provide any electrical output. They are generally inaccurate, have low resolution, and are not recommended. Electrical sensors are generally the most appropriate type of sensors for soil temperature measurements, particularly when automated recording is required.

64.3.3 Radiative

All objects (at temperatures above 0 K) emit radiation. The radiative flux is proportional to the fourth power of the temperature of the emitting surface (Stefan-Boltzmann Law). Terrestrial objects emit long-wave or infrared radiation as distinct from solar or short-wave radiation.

Most infrared thermometers have filters that exclude radiation outside the wave band 8–13 μm. They either employ thermogenerative detectors, such as thermopiles, or photogenerative detectors, such as cadmium selenide, that produce a current that varies linearly with incident quantum flux. Commercially available infrared thermometers usually have viewing angles that range from 2 to 30°.

While infrared thermometers have the great advantage that they can provide noncontact measurements of surface temperature, and therefore are the preferred instrument for measuring soil surface temperatures, there are a number of sources of measurement error. The output of the sensing element is usually temperature dependent, so it will change with changes in ambient temperature (additional measurement errors result from temperature gradients within the instrument because the sensing element cannot distinguish between radiation received from the detector housing and that emitted from the target).

An infrared sensor positioned above a soil surface will not only receive long-wave radiation emitted by the surface, but will receive reflected atmospheric long-wave radiation. If this latter source of radiation is neglected, surface temperatures are considerably overestimated. In addition, more errors will arise if the emissivity of the emitting surface is not known. Small differences in surface emissivity result in large systematic errors in measured temperatures. The degree of such errors varies with such factors as viewing angle, soil moisture, and cloudiness, so it is difficult if not impossible to apply corrections. Finally, infrared thermometers require frequent recalibration because of their lack of stability.

Graham et al. (1989) describe a calibration procedure and mathematical treatment for a small, inexpensive, thermopile-type infrared sensor (Dexter Research model 2M-HS, Dexter, MI). This allows accurate ($\pm 0.2°C$) and continuous measurement of surface temperatures in the field. However, because of the complexity of the equations required to calculate target temperatures, on-line measurements require a computer-based data acquisition system.

64.3.4 Electrical

Electrical sensors are generally the most appropriate type of sensors for soil temperature measurement, particularly when automated recording is required.

64.3.4.1 *Thermocouples*

Thermocouples are inexpensive, respond relatively rapidly to changes in temperature, are easy to construct, are stable, and generate electrical signals that can be amplified and transmitted over great distances without introducing measurement errors. They rely on the principle that when two different metals are joined at both ends, and two junctions are exposed to different temperatures, a small voltage is created and a current flows round the circuit — this is known as the Seeback effect. The voltage is a function of the type of metals and is directly proportional to the temperature difference. Once this proportionality is established for a given pair of metals, the user can convert voltage readings into temperature differences. If one of the junctions is kept at a known reference temperature while the other is exposed to the medium whose temperature is to be measured, the latter measurement can be calculated directly from the measured voltage. The temperature of the measuring junction is simply the sum of the temperature difference between the measuring and reference junction and the reference junction temperature.

Many data loggers (for example, the 21X, CR7, and CR10 from Campbell Scientific, Canada) have a built in reference junction. A platinum resistance thermometer is typically mounted in the center of the terminal strip designed to minimize temperature gradients between connectors. The reference junction is established when the two leads from the measurement thermocouple are connected to the terminal panel.

An external reference box is often used either when the data logger or measuring device lacks a reference junction, or to reduce the cost of the thermocouple wire when temperature

Table 64.1 Seeback Coefficients for Four Commonly Used Thermocouples

Type	Metals	Seeback coefficient $(\mu V°C^{-1})$
E	Chromel/constantan	58.5
J	Iron/constantan	50.5
K	Chromel/alumel	39.4
T	Copper/constantan	38.7

measurements are made some distance from the measuring device. Small battery- or AC-powered electronic reference junctions are available. For laboratory work, reference junctions can be placed in melting ice (0°C) or in constant-temperature ovens.

No type of thermocouple has a precisely linear relationship between voltage (E, mV) and temperature (T, °C), but the relationship is well described by the second-order polynomial:

$$E = a + bT + cT^2 \qquad (64.3)$$

where a, b, and c are coefficients that depend on the type of thermocouple used. For example, for copper/constantan thermocouples (type T thermocouples), the coeffcients are 0.09 μV, 38.7 $\mu V°C^{-1}$, and 0.41 $\mu V°C^{-2}$, respectively. The term bT, where b is the Seebeck coefficient, dominates Equation 64.3, so for most purposes this term can be used alone. For example, the assumption of a linear relationship between E and T for a copper/constantan thermocouple over the range -20 to 50°C leads to an error range of about -1 to 0.5°C. The error range for iron/constantan thermocouples (type J thermocouples) is about half that of copper/constantan thermocouples. Table 64.1 gives the Seeback coefficients for a range of thermocouples.

Tables of all the coefficients (a, b, and c) for a wide range of thermocouples can be found in an excellent and comprehensive manual on thermocouples published by the American Society for Testing and Materials (1974).

Chromel/alumel thermocouples (type K) are less resistant to oxidation than type T or J thermocouples and operate over a wide temperature range (-250 to 1260°C). They are well suited for studies on the effects of burns on soil properties. Iron/constantan thermocouples are used extensively in industry because of their relative low cost and high Seebeck coefficient. However, because of a lack of uniformity and standardization of J-type materials, there can be considerable deviation between the actual performance and that predicted from standard tables.

When a number of thermocouples (n) are joined in series to form a thermopile, measurement sensitivity increases by a factor n. This is useful for amplifying small signals and lessens the requirements of the measurement device. Differential measurements are made by connecting two thermocouples in series and placing them in different environments. The output of the two thermocouples will indicate the algebraic sum of the two temperatures. When large differential output voltages are required, two thermopiles may be connected in series to measure the differential output. When thermocouples are connected in parallel, the output voltage is not increased, but a mean temperature is obtained. This is particularly useful when determining the mean temperature of a profile or layer. Another advantage to this arrangement is that only one channel of a data acquisition device is needed to obtain the average soil temperature reading.

Particular care should be taken when using thermocouples to measure surface temperatures. The thermocouple must be in good thermal contact with the surface and must be small enough that the sensor itself does not alter the radiation balance of the surface. Fritchen and Gay (1979) describe the complex conduction errors that can arise if thermocouples are not properly installed. It is recommended that thermocouples be inserted a few millimeters below the surface to minimize radiation errors.

Thermocouples are delicate instruments that should be treated carefully. Oxidation can lead to changes in thermoelectric behavior, so it is recommended that thermocouples be enclosed in a protective sheath (see Section 64.2.5), even though this will lead to an increased time constant. As mentioned previously, this can be an advantage if rapid fluctuations in soil temperature are not needed and only integrated time averages are required.

Thermocouples are constructed by either soldering or welding. Soft soldered junctions are perfectly adequate for most field situations. The softer thermocouple wire should be twisted around the other wire and the twisted area cleaned with a grease solvent and washed with trichloroethylene. A low-flux solder should be used to construct the junction. Silver solder provides mechanically stronger junctions than those fashioned with soft solder. Brazing occurs at 550 to 650°C. Wires should first be cleaned and the junction coated with the appropriate flux. The junction should be heated until the flux becomes liquid and then the brazing rods applied to the heated area.

Welded junctions are more reliable and robust than soldered junctions, but require specialized equipment. They are required if thermocouples are exposed to high temperatures (that occur, for example, in burn experiments). There are numerous welding techniques, all requiring specialized equipment. These include percussion welding (LeMay 1958), resistance welding (Stover 1960), and arc welding (Lopushinsky 1971). Thermocouple wire is available in a wide range of diameters. The time constant of the thermocouple increases significantly with increasing wire gauge. For most soil applications a 24-AWG wire (0.51 mm diameter) is recommended. The same thermocouple gauge need not be maintained all the way from the measuring junction to the measurement device. For example, fine wire (44 AWG, 0.05 mm diameter) might be used for soil surface measurements and then connected to more robust 24-AWG wire. Connections between thermocouple wire should be made with thermocouple connectors (available from all good supply houses) which maintain continuity within the lines.

Thermocouples are excellent measurement devices and are highly recommended for soil temperature measurement.

64.3.4.2 *Resistance Thermometers*

Both conductor and semiconductor devices are available for measurements of soil temperature. In conductor devices the resistance of the metal increases with increasing temperature. In most cases the relationship is nonlinear and so is inconvenient for measurement purposes. However, platinum resistance thermometers are linear to within $\pm 0.4\%$ over the temperature range between -200 and $40°C$. Platinum thermometers are extremely stable, but are expensive. Cheaper, but less accurate alternatives are copper and nickel. However, such resistance thermometers are susceptible to corrosion and oxidation and should always be protected from bending in an epoxy or glass sheaf because the resistance of the element changes with strain. An additional problem is that the length of the leads connecting the sensor to the measurement device can become significant if the resistance of the leads is large relative to that of the sensor.

In semiconductor devices the resistance decreases exponentially with temperature at a rate of between -3 and $-5\%°C^{-1}$. Thermistors are beads of semiconductor material prepared from oxides of the iron group of metals, such as chromium, cobalt, iron, manganese, and nickel. The major advantage of thermistors is that they are inexpensive, stable, have large temperature coefficients and therefore sensitivity, and are available in a wide range of sizes and shapes. Because of their high resistance, lead resistance is negligible.

Thermistors do have a number of disadvantages in that they require some type of direct current (DC) excitation and linearization circuitry to provide a lienar relationship between voltage and temperature. Examples of such circuits are given by Fritchen and Gay (1979). Without linearization circuitry, thermistors, unlike thermocouples, cannot be connected in series to measure an arithmetic mean of the temperature at several locations.

There can be significant variability between thermistors so that each should be individually calibrated before field use. Care should be taken to avoid excessive heating, which can cause thermistor instability, when soldering cable to thermistor leads.

Thermistors can be encased in flexible epoxies or RTV silicones to make thermal probes. As with all temperature sensors, care should be taken that there is good thermal contact between the probe and the surrounding soil.

Silicon and germanium diodes are a useful and cheap alternative to thermistors (see, for example, Black and McNaughton 1971). More recently, semiconductor thermometers have been constructed as an integrated circuit. Such devices include National Semiconductor® LX5600 and Analog® Devices AD590. These are factory calibrated, linear, and require very few external components. They can provide an accuracy of $\pm 0.25°C$ over a range of -20 to $50°C$, but are subject to drift.

64.3.4.3 *Quartz Thermometers*

The quartz thermometer employs the principle that the resonant frequency of a material such as quartz is a function of temperature, so that temperature changes can be translated into frequency changes. The temperature sensing element consists of a quartz crystal enclosed within a probe. The crystal is connected electrically so as to form the resonant element within an electronic oscillator. Measurement of the oscillator frequency allows the measured temperature to be calculated. Commercially available thermometers have a typical resolution of $0.001°C$ over the range -40 to $250°C$, and a nonlinearity of $\pm 0.02°C$ from 0 to $100°C$. However, because of their expense and complexity they have a specialized usage and are not recommended for routine soil temperature measurements.

64.4 CONCLUSIONS

The selection of the most appropriate type of temperature sensor and the arrangement and number of sensors used for a particular experimental design depends on many factors, not the least of which is the type of measuring or recording device available. It is imperative that, before the selection and installation of measurement devices, the researcher clearly understands and justifies the measurement and data requirements.

That given, it is recommended that thermocouples be used whenever possible. Despite the fact that they provide a relatively small output signal, they are simple to construct, inexpensive, are highly reliable, and provide considerable measurement flexibility.

REFERENCES

American Society for Testing and Materials. 1974. Manual on the use of thermocouples in temperature measurement. Special technical publication 470A. American Society for Testing and Materials, Philadelphia, PA.

Black, T. A. and McNaughton, K. G. 1971. Psychrometric apparatus for Bowen-ratio determination over forests. Boundary-Layer Meteorol. 2: 246–254.

Fritchen, L. J. and Gay, L. W. 1979. Environmental instrumentation. Springer-Verlag, New York.

Fuchs, M. 1971. Data logging and scanning rate considerations in micrometeorological experiments. Agric. Meteorol. 9: 285–286.

Graham, M. E. D., Thurtell, G. W., and Kidd, G. E. 1989. Calibration of a small infrared sensor for measuring leaf temperature in the field: non-steady state conditions. Agric. For. Meteorol. 295–305.

Le May, J. 1958. More accurate thermocouples with percussion welding. J. Instrum. Soc. Am. 5: 42–45.

Lopushinsky, W. 1971. An improved welding jig for peltier thermocouple psychrometers. Soil. Sci. Soc. Am. Proc. 35: 149–150.

Morris, A. S. 1988. Principles of measurement and instrumentation. Prentice-Hall, New York.

Omega Engineering Inc. 1991. Temperature measurement handbook. Omega Press, Stanford, CN.

Panofsky, H. A. and Brier, G. W. 1958. Some applications of statistics to meteorology. College of Mineral Industries, Pennsylvania.

Pearcy, R. W. 1989. Field data acquisition. In R. W. Pearcy, J. Ehleringer, H. A. Mooney, and P. W. Rundel, Eds. Plant physiological ecology. Chapman and Hall, New York.

Perrier, A. 1971. Plant photosynthetic production — manual of methods. In Z. S. Sestak, J. Catsky, P. J. Jarvis and W. Junk, Eds. The Hague.

Portman, D. J. 1957. Soil thermocouples. Exploring the atmosphere's first mile. Pergamon Press, New York.

Stathers, R. J. and Spittlehouse, D. L. 1990. Forest soil temperature manual. FRDA report no 130. B.C. Ministry of Forests, Research Branch.

Stover, C. M. 1960. Method of butt welding small thermocouples 0.001 to 0.01 inch in diameter. Rev. Sci. Instrum. 31: 605–608.

Suomi, V. E. 1957. Soil temperature integrators. Exploring the atmosphere's first mile. Pergamon Press, New York.

Tanner, C. B. 1963. Basic instrumentation and measurements for plant environment and micrometeorology. Department of Soils, Bull. 6, University of Wisconsin, Madison.

Unwin, D. M. 1980. Microclimate measurements for ecologists. Academic Press, New York.

Weigland, C. L. and Swanson, W. A. 1973. Time constants for thermal equilibration of leaf canopy, and soil surfaces with changes in insulation. Agron. J. 65: 722–724.

Woodward, F. I. and Sheehy, J. E. 1983. Principles and measurements in environmental biology. Butterworths, London.

Chapter 65
Micromorphological Methodology
for Inorganic Soils

C. A. Fox, R. K. Guertin
Agriculture Canada
Ottawa, Ontario, Canada

E. Dickson, S. Sweeney, and R. Protz
University of Guelph
Guelph, Ontario, Canada

A. R. Mermut
University of Saskatchewan
Saskatoon, Saskatchewan, Canada

65.1 INTRODUCTION

Micromorphological concepts and procedures provide unique methodology for evaluating the spatial arrangement of soil constituents (mineral, organic, and biological components) and the associated soil pores; that is, by confirming whether the soil constituents are distributed uniformly or as heterogenous materials; determining how the soil constituent arrangement changes across boundaries; and most important, assessing where soil processes have affected the soil and to what degree of intensity. Thin sections and polished blocks provide the means for examining the spatial arrangement and interrelationships. No other chemical and physical analyses of the soil provide this kind of spatial information and data.

Some examples of the application of micromorphological procedures to soil research are the following: in tillage research, micromorphological studies have provided data on the unique spatial arrangement and characteristics of the soil pores that result from applying different tillage implements and management systems (Pagliai et al. 1989, Singh et al. 1991). In soil genesis and classification research, micromorphology has been used extensively to examine the arrangement of the soil constituents within soil horizons, provide evidence for movement of fine materials, as well as quantify the chemical composition of selected features.

The main instruments used are the stereomicroscope, the polarizing light microscope, and the scanning electron microscope. These instruments facilitate descriptive assessments to be made at magnifications ranging from less than $10\times$ to more than $10,000\times$. Recently,

Soil Sampling and Methods of Analysis, M. R. Carter, Ed.,
Canadian Society of Soil Science. © 1993 Lewis Publishers.

because of technological advances, image analysis is being used more frequently in conjunction with these instruments to obtain quantitative data about the observed soil constituents and pores. Microprobe analysis (energy-dispersive X-ray analyzer, EDXRA) has also been used for determining the elemental composition of particular soil mineral constituents.

It is essential that for any micromorphological assessment of the soil that the continuity of scale be maintained (Protz et al. 1987, Koppi and McBratney 1991) between both the actual field description at the time of sampling and the chosen magnifications at which observations and measurements are being made; otherwise, any interpretations about the spatial arrangement of the soil constituents in terms of the soil itself cannot be made.

It cannot be emphasized enough that for all micromorphological research applications it is crucial that the integrity of the samples be maintained during both sample collection and preparation. This ensures that observations and quantitative analyses can be related directly to the soil and that the interpretations and inferences resulting from the evaluation of the arrangement of the soil components and pore distributions are real and representative.

Murphy (1986), FitzPatrick (1984), and Bouma (1969) provide valuable references for the preparation of samples for micromorphological studies. The following chapter will present selected methodology used primarily by Canadian micromorphologists for sampling and preparation of inorganic materials. Methodology specific for organic soils is presented in Chapter 45.

65.2 FIELD PROCEDURES FOR MICROMORPHOLOGICAL SAMPLES

Because of the recognized importance of sampling to any micromorphological study, the procedures for obtaining good-quality samples for most soils and sediments will be presented in some detail. Refer to Murphy (1986) for the methodology for sampling problem soils such as the following: very stony, gravelly, or fragile materials; very coherent or cemented materials; loose, unconsolidated sands and granular surface materials; and surface organic layers (humus forms). See Chapter 45 for sampling procedures for organic soils.

65.2.1 MATERIALS

1   Topographic map and/or aerial photograph, compass, field notebook.

2   Soil sampling equipment, i.e., shovel, knife, pruning shears, dry-wall hand saws with short blades, scissors, plasterer's trowel, hand gardening tools (hand trowel, weed uprooter), plastic bags, plastic wrap, masking tape, permanent (waterproof) marking pens, tape measure, hammer, piece of hardwood (approximately 10 × 8 × 5 cm).

3   Sample containers (often referred to as Kubiena boxes). The size of the container will be dependent on research requirements (i.e., whether blocks for image analyses or thin sections are needed) and the laboratory facilities available for preparing the samples. The container must be constructed of a rust-free material such as galvanized tin or aluminum, be strong enough to support the sample during sampling, transportation, and laboratory preparation (suggest 20 to 22 gauge), and open on both faces. Suggested con-

tainer dimensions for preparing blocks for image analyses are 7.5 cm (l) $\times$ 6.5 cm (w) $\times$ 5 cm or 3 cm depth. For preparing 2.5 cm $\times$ 4.5 cm thin sections, the minimum container size need only be 6.5 cm (l) $\times$ 4.0 cm (w) $\times$ 3.0 cm in depth. Prepare lids (from 0.64 cm ($^1/_4$") plywood) to cover the top and bottom faces of the container.

4   Hydraulic jacking system (Wires and Sheldrick 1987).

65.2.2 PROCEDURE

1   Site selection. With the available soil maps, select a representative site. On location, select, by using transect sampling schemes, a representative pedon that is characteristic of the landscape and the spatial variability of the soil properties. Depending on the objectives of the study, the full range of the observed soil properties can be included in the sampling or only specific attributes can be emphasized.

2   Open the soil pit approximately 1 m^2. Describe the soil profile using a soil data descriptive system such as the CanSIS Soil Data File — Detailed Form (Day 1983) or Canada Wetland Registry form (Tarnocai 1980).

3   With the knife and hand trowels, prepare a smooth, level soil face. Remove any protruding roots with the pruning shears, scissors, or saw.

4   Orient the container (either vertically or horizontally) on the prepared soil face. Mark an arrow on the container to indicate the vertical direction to the soil surface. Place the container at approximately the midpoint of the soil horizon. Avoid taking samples which span two horizons unless such samples are required for a specific purpose or the horizons are very thin. In such cases, mark on the side of the container the boundary between the horizons.

5   Mark the sample number and/or the site number on the container. For soils with a preferred structural orientation, other directional information should be marked on the container, such as indicating the open end of the container which represents the front or soil face and marking the compass direction to which the soil face is exposed.

6   The following procedures are used to obtain samples:

     a. *Hand pressure.* The container is pushed slowly into the soil using only hand pressure. This procedure is most suitable for soils that offer little resistance to hand pressure, for example, unsaturated sands and silts as well as moist clays.

     b. *Knife.* Cut into the soil around the sides of the container and with hand pressure ease it into the soil. Alternatively, remove a large clod of the soil (noting the orientation), cut it slightly larger than the dimensions of the container, then carefully ease the container onto this cut block of soil. This procedure is often used for very dense soils, fibrous organic layers, or situations (often wet soils) where the pressure from hand pushing would result in compaction.

c. *Hydraulic system.* With the hydraulic jacking system (Wires and Sheldrick 1987), the container can be pushed slowly and continuously into the soil with minimum disturbance. This method is extremely useful for sampling very dense clays that are stone free. Release the pressure very slowly to reduce possible rebounding of the soil.

d. *Hammer.* Place the piece of hardwood on top of the container, then, with the hammer pound the container into the soil. This method very often results in shattering of the structure, especially if the soil is dry and compact. Consequently, it is not a preferred method and should be avoided. It is used sometimes as a method of last resort.

e. *Clods.* For extremely dense materials, cemented layers, or very stony soils, taking samples with containers may be difficult or impossible. For these situtions, clods (minimum 5 cm diameter) should be removed from the soil. Note the orientation of the clods.

7 Sketch (draw to scale) the soil profile with the sample locations noted. Murphy (1986) recommends photographing the soil profile with the containers still in place.

8 Remove the sample container from the soil. With a knife, cut away the surrounding soil on the sides of the container, cut into the soil beneath the container, and make a V-cut into the soil behind the container. Carefully lift the container from the soil with a hand trowel, garden weed uprooter, or shovel.

9 With a knife, remove the excess soil until level with the open ends of the container. If the sample does not fill the container completely (usually occurs at the corners or if surface samples were taken), fill the remainder of the space with nonabsorbent material to prevent structural disturbance during transportation. Do not use loose soil or surface vegetation fragments, as this can lead to erroneous interpretations in assuming later that this material is part of the sample. If faunal activity (i.e., earthworms, fungal growth) is expected to cause severe disturbance to the sample, Murphy (1986) suggests applying a drop of fungicide or Formalin to the surface.

10 Place the sample (do not remove the metal container) into a plastic bag and wrap tightly around the container to prevent moisture loss. For clods, carefully wrap the sample in plastic wrap or bags and secure with masking tape. Moisture loss is a particular hazard for organic materials and fine-textured samples, as the resultant shrinkage cracks and volume loss can severely disrupt the morphology.

11 Cover both open ends of the plastic wrapped container with the precut plywood lids. Use masking tape to secure the lids.

12 Record the following minimum information about each sample on the masking tape and in a field notebook:

   a. Profile site number.

   b. Horizon designation.

   c. Soil depth.

   d. Orientation information, especially the direction to the soil surface.

13 Take samples for determining the chemical and physical properties of the soil in close proximity to the micromorphology samples so that the data can be directly related.

14 Transport the samples in heavy cardbord boxes, 5-gallon steel cans, or wooden crates. Layer the bottom and fill any empty spaces with styrofoam packing material to prevent movement during transportation.

15 Store the samples in a cold room (about 4°C) to inhibit faunal activity and fungal growth. Do not freeze mineral soil samples, as ice lens formation can produce marked changes to the morphology that were not present at the time of sampling. Prolonged storage will lead to moisture loss with subsequent disruption of the morphology. Consequently, process the samples as soon as possible after sampling.

65.3 REMOVAL OF WATER FROM MICROMORPHOLOGICAL SAMPLES

In order to prepare thin sections and/or polished blocks, polyester resins are used to consolidate the soil samples so that the required cutting and grinding can be achieved. Because polyester resins are immiscible with water, all water must be removed from the sample to ensure that polymerization can proceed and result in a resin-impregnated sample of high quality. The main methods used to remove the water are the following: air or oven drying; acetone exchange by immersion, and acetone exchange by vapor replacement.

65.3.1 Water Replacement by Air/Oven Drying (Modified from Murphy 1986)

Air drying is recommended for coarse-textured sands and silts and soils that are thixotropic. These soils are unstable if immersed in acetone. Air drying is also used for preparing samples for studies which emphasize only the chemical composition of specific soil constituents or are limited only to the identification and occurrence of specific features. Air drying is not recommended for studies where the objective is to quantify the spatial arrangement of the soil pores and soil constituents as they existed at the time of field sampling; acetone exchange is used for preparing the samples for such studies.

Although the preparation time when compared to acetone exchange is considerably less, air or oven drying of the samples often results in artifact cracks and considerable volume decrease due to moisture loss. This is especially critical for fine-textured soils with swelling clays, organic soils, and mineral soils with a high organic matter content. Consequently, for these soils, air drying is not recommended.

65.3.1.1 MATERIALS

1   Weighing balance (minimum 3 decimals).

2   Fumehood.

3   Ventilated oven.

4   Desiccator or enclosed cabinet with desiccant.

65.3.1.2 PROCEDURE

1   Record the initial weight of the sample.

2   Place the container (or clod) onto a tray in a well-ventilated area such as a fumehood for several days (>1 week) until samples appear extremely dry.

3   Transfer to an oven set at 40°C maximum.

4   Weigh the samples approximately every 24 h until a minimum constant weight is obtained and maintained for at least three consecutive weighings.

5   Transfer the samples to the enclosed cabinet to maintain the constant weight until processing by resin impregnation.

65.3.2 Water Replacement by Immersion in Acetone

Acetone replacement of the water in the sample is the preferred method and used most frequently by micromorphologists (Murphy 1985, 1986, Moran et al. 1989b) for maintaining the integrity of the pore morphology at the time of field sampling. It is not recommended for samples that may flow or slump when immersed in a liquid; such samples should be prepared by air drying (see Section 65.3.1). See Chapter 45 for procedures specific for organic soils.

65.3.2.1 MATERIALS AND REAGENTS

1   Acetone, technical grade.

2   Fumehood.

3   Polypropylene tray (tank) that is acetone resistant, has a secure lid, and is fitted with a wire rack on the bottom. A glass desiccator with ceramic plate can also be used.

4   Magnetic stirring bar and magnetic stirrer with wide, flat base.

5 Stainless steel screening and pieces of cotton gauze cut to dimensions 1–2 cm larger than the size of sample containers.

6 Balance (4 decimals).

65.3.2.2 PROCEDURE

1 **In the fumehood,** position and secure the polypropylene tank on the magnetic stirrer. Place the magnetic stirring bar in the bottom of the tank.

2 Place the cotton gauze onto the stainless steel screen, then position either the soil sample (the container is not removed) or the stable clod onto the cotton.

3 Place the sample with its screen onto the wire rack in the tank. Avoid the area directly above the magnetic stirring bar in the event of disturbance from the motion of the acetone.

4 Record the sample locations in the tank, noting the sample number, field orientation, and any other data about the sample. This is crucial as all markings made on the container itself at the time of sampling will be removed by the acetone. Label flags marked with pencil can also be inserted with a stainless steel straight pin into a corner of the container.

5 Add acetone (100%) very slowly into the tank, allowing capillary rise in the samples. Continue to add the acetone until the level is 0.5–1.0 cm from the top of the sample container. For some samples, where stability may be a problem, do not immerse the sample, but maintain the acetone at a depth of less than 0.5 cm from the bottom of the container.

6 Agitate the acetone very slowly by rotating the magnetic stirring bar.

7 Soak the samples for a minimum time period of 3 d. At the end of the selected soaking period, before removing the acetone from the tank, determine the water content of the acetone. A simple procedure is to determine the specific gravity of water in acetone by weighing (to 4 decimal places) 10-mL subsamples of the acetone solution and comparing this weight to a standard curve of weights for 10 mL of 0.5, 1.0, 1.5, 2.0, 3.0, 5.0, and 10.0% distilled water in 100% acetone. Additional procedures are suggested in Murphy (1985) and Moran et al. (1989b).

8 If water is present, remove the acetone from the tank and replace with 100% acetone. Continue replacements and testing (step 7) until the amount of water is less than 2% (preferably less than 1%). Most resins (must be determined by testing) can only tolerate less than 2% water.

9 When the percentage of the water content of the acetone-water solution is less than 2%, replace the acetone at least two additional times, continuing to test each time for water. This is to ensure that the water content remains

less than 2% (preferably less than 1%). The samples are now ready for impregnation with polyester resins (see Section 65.4 or 65.5).

65.3.3 Water Replacement by Acetone Vapor (After FitzPatrick 1984)

Water replacement by acetone vapor is a very slow procedure and requires careful monitoring. It works on the principle of differences in vapor pressure between the acetone and the water; the sample takes up the acetone faster than the water can evaporate, and eventually the water drips from the sample. FitzPatrick (1984) recommends that the samples should not be thicker than 2 cm. This method is often considered for use with organic materials and may be suitable for some mineral materials that would be unstable if completely immersed in acetone.

65.3.3.1 Materials and Reagents

Same as for Section 65.3.2. It is best to use a glass desiccator (as shown in Murphy 1986) so that the condition of the sample can be continuously observed without having to remove the lid.

65.3.3.2 PROCEDURE

1   In a fumehood, add 100% acetone to the bottom of the glass desiccator to within 5 cm of the bottom of the ceramic plate or rack. A magnetic stirring bar can be added to circulate the acetone. FitzPatrick (1984) suggests successive concentrations of acetone in water ranging from 4 to 100% to avoid oversaturation of the sample with water which occurs when 100% acetone is used from the start. For weakly cohesive samples, oversaturation can severely affect stability. For such samples successive concentrations are recommended.

2   Prepare samples as stated in Section 65.3.2.2 (step 2).

3   Insert a wire rack or other structure onto the cearmic plate in order to elevate the samples to facilitate free dripping of the water from the sample. Place the wire screen with sample onto the rack. Record the location of the sample as directed in Section 65.3.2.2 (step 4). Replace the desiccator cover.

4   Monitor the water content of the 100% acetone as described in Section 65.3.2.2 (step 7). Replace the acetone as in Section 65.3.2.2 (steps 8 and 9). If using successive concentrations, start monitoring for the percentage of water after reaching 100% acetone.

5   Observe periodically the surface condition of the sample, especially when the water stops dripping from the sample. At this stage, the sample can become excessively dried. Organic soils are particularly susceptible with damage occurring to the tissues.

65.3.3 Comment

To reduce the amount of acetone required, recycling procedures can be used either by reusing the acetone or removing the water with desiccants. For example, the acetone used for the last three exchanges (if the percentage of water content is about 2%) can be saved and used for the initial exchanges of the next batch of samples. The use of anhydrous calcium chloride to remove water from the acetone as suggested by FitzPatrick (1984) will render the acetone unsuitable for recycling, as the calcium chloride dissolves in the presence of water. Moran et al. (1989b) describe a closed system for continuously recycling the acetone during the exchange procedure by using a polyacrylamide superabsorbent material and zeolite to remove the water.

65.4 SAMPLE IMPREGNATION USING THE POLYESTER RESIN METHOD (Modified from Sheldrick 1984)

65.4.1 MATERIALS AND REAGENTS

1 Fumehoods. Access to a vacuum line (22–30 cm Hg) that can be adjusted.

2 Solvent-resistant plastic containers with lids. The height of these containers should be more than twice the height of the sample container.

3 Impregnation apparatus. Consists of a 20- to 25-cm diameter desiccator with a Teflon® O-ring on the cover. The cover lid is modified to hold a 500-mL reservoir with a lead to the vacuum source and a Teflon® stopcock fitted to allow air entry after evacuation. See Murphy (1986), FitzPatrick (1984), or Sheldrick (1984) for typical designs.

4 Acetone, technical grade.

5 Polyester resin, umpromoted (base resin 349-8015, available from Guards-man® Products, Cornwall, Ontario). Note: an unpromoted resin must be used to have control over the rate of polymerization by being able to adjust the amount of the catalyst and accelerator added to the resin.

6 Catalyst: 60% methyl ethyl ketone peroxide in dimethyl phthalate (Lupersol® DDM, available from Lucidol Div. of Penwalt Inc.).

7 Accelerator: 6% cobalt napthanate.

8 Uvitex OB fluorescent dye (Ciba-Geigy®). Note: use Uvitex OB only when the samples are to be evaluated with image analysis techniques. Do not add if the natural fluorescence of features (i.e., organic tissues) is to be examined or if other staining methods will be used.

9 Magnetic stirring bar and magnetic stirrer, automatic pipette.

10   A 4000-mL glass beaker.

11   Ventilated oven to fumehood.

65.4.2 PROCEDURE

1   In a fumehood, prepare the resin solution as follows:

   a. Into a 4000-mL beaker, add 2000 mL acetone. For all mixing, use a mag-
      netic stirring bar. Add (optional) 2.0 g of Uvitex OB and mix well.

   b. Add 2000 mL polyester resin, stirring constantly.

   c. With a clean automatic pipette, add 2.0 mL (see note below) of accelerator
      (item 7 above) and mix extremely well. Clean the pipette.

   d. With the clean pipette, add 4.0 mL (see note below) of catalyst (item 6
      above) and mix the solution extremely well.

Note: DO NOT MIX THE CATALYST AND ACCELERATOR TOGETHER *AT ANY
TIME* AS AN EXPLOSION MAY RESULT. The quantity of catalyst and accelerator added
above is predetermined by testing each new shipment of resin for the suitable combination
of catalyst and accelerator required to slow the polymerization process sufficiently to achieve
high-quality impregnation of the sample.

2   Insert the prepared sample (Section 65.3; chapter 45.4.2, step 4) carefully
      into the plastic container; do not invert the sample at any time. Record the
      orientation and sample number on the container with pencil. When the
      container is removed from the impregnation apparatus, permanent marking
      pens can be used.

3   Place the sample container into the impregnation apparatus. Make sure the
      desiccator cover is well sealed to maintain a constant vacuum.

4   Fill the reservoir with the resin solution prepared in step 1 above.

5   Adjust the vacuum line to remove slowly the air from both the desiccator
      and the resin reservoir. Reduce the vacuum if the resin solution boils vig-
      orously.

6   Add the resin *very slowly* along the inside of the plastic container. Do not
      drip the resin onto the sample. The resin will enter the sample by capillary
      flow. Continue to add resin until the volume added is a little more than
      twice the height of the sample. *Note:* overfilling with resin solution will
      eliminate the need to add resin periodically during the polymerization pro-
      cess to maintain complete immersion of the sample. Once the acetone,
      present in a 1:1 ratio with the resin, completely volatilizes from the resin,
      the final level of the resin will be just above the sample.

7   Slowly release the vacuum from the impregnation apparatus. Remove the sample container and cover with the container lid to prevent rapid volatilization of the acetone and allow for a soaking period.

8   Place the sample in the fumehood. Keep covered for 7 to 10 d, then remove the lid to allow the acetone to volatilize.

9   As the polymerization process takes place, the resin consistency will change from a liquid to a viscous gel to a hard plastic. When it is very hard to the touch, place the samples into the ventilated oven at a temperature of 30°C and increment 10°C daily to a maximum of 70°C. After 3 d, shut the oven off and let the samples cool to room temperature in the oven before removing them.

10   Label the resin-impregnated sample by both scratching on the surface and marking with a permanent marking pen or red wax pencil the laboratory number and the orientation direction to the soil surface.

11   Remove the sample from the container. If the sample is sticky on the bottom, invert the sample in the container and return it to the oven set at 50°C, incremented by 10°C to 70°C, and then leave it for a minimum of 2 d. The sample is now ready for thin section or polished-block preparation.

65.5 POLYMERIZATION WITH GAMMA RADIATION (McCARRICK AND PROTZ 1978, SHIPITALO AND PROTZ 1987)

65.5.1 3-HYDROXY-BUTYL METHYLMETHACRYLATE (3-HBMA) SYNTHESIS

65.5.1.1 MATERIALS AND REAGENTS

1   Sodium metal stored in paraffin oil (prevent water contact).

2   Sodium sulfate (Na_2SO_4), anhydrous, granular, 1-kg quantity.

3   Sodium chloride (NaCl), 2.26-kg quantity.

4   Calcium chloride ($CaCl_2$), anhydrous, 4–20 mesh size (granular form).

5   1,3-Butanediol (TERRCHEM Laboratories Ltd., 80 Galaxy Drive, Rexdale, Ontario M9W 4Y8).

6   Methyl methacrylate (Rohm and Haas® Canada Inc., 2 Manse Road, West Hill, Ontario M1E 3T9).

7   Uvitex OB, fluorescent dye (Ciba-Geigy®).

8   Drierite moisture sorbent, anhydrous $CaSO_4$, 8 mesh, 1 lb.

9　　Two large area/based magnetic stirrers; two 74-mm magnetic stirring bars; a long magnetic stirring bar retriever rod.

10　　Several amber 4-L bottles for storage of the resin.

11　　Drying column assembly with #10 and #9 rubber stoppers.

12　　Tray and shallow glass dishes.

13　　Forceps, 15 cm long, and narrow, thin-bladed spatula.

14　　Glass wool, Kimwipes tissues, and paper towels.

15　　Volumetric flasks, 2 and 1 L.

16　　Graduated, 3- to 4-cm-diameter, 1-L burette.

17　　Laboratory balance.

18　　Four 6-L Erlenmeyer flasks and #10 rubber stoppers.

19　　Two 2-L or one 4-L Erlenmeyer flasks.

20　　Separatory funnels, one 2 L and one 1 L.

21　　Cabinet desiccator.

65.5.1.2 PROCEDURE

CAUTION: conduct all resin synthesis steps in a fumehood (unless otherwise stated) to prevent gas explosion or fire. The fumehood must have a very strong draft, not be connected to any other fumehood, and have a straight-line venting system directly to the roof.

NOTE: the following sequence of steps is suggested:
　　Day 1: steps 1 and 2, also 10.
　　Day 2: steps 3 to 7.
　　Day 3: steps 8, 9, 11 to 16.
　　Day 4: steps 17, 18.
　　Day 5: steps 19, 20.

1　　Oven dry 50 mL (measured in a beaker) $CaCl_2$ and 1 kg Na_2SO_4 at 100–110°C for 24 h. Cool in cabinet desiccator.

2　　Prepare all glassware beforehand and set aside. Do not use soap to clean glassware; rinse with water only and air dry.

3　　Prepare drying tube assembly in fumehood as follows:

　　a. Insert the top of a #10 stopper (with a centered, precut 5- to 7-mm hole) into the bottom of the drying tube assembly cylinder.

b. Place a lightly packed glass wool plug on top of the stopper.

c. Add a 5-mm layer of Drierite moisture sorbent and cover with a lightly packed glass wool plug.

d. Add 50 mL of anhydrous $CaCl_2$ (14–20 mesh) and cover with a lightly packed glass wool plug.

e. Close cylinder with a #9 stopper (centered, 5-mm precut hole).

4   Place a 6-L Erlenmeyer flask containing a stirring bar onto the magnetic stirrer. Pour 1.5 L of 1,3-butanediol into the flask.

5   Line a glass weighing dish with kimwipes. Place it onto the balance. Use several large kimwipes to cover a working area on the lab bench. Use forceps to remove a piece of sodium metal from the paraffin and pat it dry with kimwipes. Use proper safety protocols for handling the sodium; use forceps and avoid contact with water. With a spatula, cut small slivers of sodium to total 7.5 g.

6   Add the metal slivers to the 6-L Erlenmeyer flask containing 1,3-butanediol (step 4). Start the magnetic stirrer. A vigorous reaction may result — care must be taken to avoid an explosion. Small sodium metal slivers may completely dissolve in 3 to 6 h. Larger ones may require 10 h.

7   Cap the reaction flask with the drying tube assembly.

8   When the sodium metal has completely dissolved, add exactly 2.7 L of methyl methacrylate monomer to the 6-L flask and continue stirring until the solution is homogeneous.

9   Add another 1.5 L of 1,3-butanediol to the 6-L flask. Stir 3-HBMA precursor solution for 2 h and then stopper the flask. If the solution is to be left at this stage, displace the air with dry nitrogen gas.

10   Prepare two brine solutions for purifying (steps 11 to 14) the precursor 3-HBMA solution:

a. Prepare a 17% NaCl brine solution. Dissolve 680 g NaCl in 4 L of deionized water in a 4-L Erlenmeyer flask. With the magnetic stirrer, mix until dissolved.

b. Prepare a 21% saturated NaCl solution. Dissolve 840 g NaCl in 4 L of hot (70°C) deionized water in a 4-L Erlenmeyer flask. Mix as in (a).

11   Transfer 1.4 L of 3-HBMA precursor solution to a 2-L separatory funnel mounted on a stand. Attach a 1-L separatory funnel to the stand above the 2-L funnel. Add 680 mL of 17% NaCl solution. Wash the precursor solution. Remove the 1-L funnel.

12 Stopper the 2-L separatory funnel, remove it from the stand, and shake carefully by hand once. Carefully open the stopcock to release pressure from generated gases. Repeat this step two or three times. Place the separatory funnel back on the stand to allow water and organic phases to separate (occurs within 10 min). The upper layer is 3-HBMA. The lower layer contains water, NaCl, and organic phase remnants.

13 Drain off the lower aqueous layer, but retain some in the funnel until final washing. Handle waste product according to WHMIS safety standards.

14 Place the 1-L separatory funnel above the 2-L funnel to wash precursor 3-HBMA with 400 mL of 21% NaCl solution. Remove the funnels from the stand, stopper the 2-L funnel, de-gas, and remove the aqueous layer as in steps 12 and 13. Repeat 21% NaCl solution wash (step 14) two times.

15 Transfer the final washed 3-HBMA (step 14) to an oven-dried 6-L Erlenmeyer flask through the top of the funnel to avoid contamination.

16 Add 250 mL granular anhydrous Na_2SO_4 to the 3-HBMA in the 6-L flask. Place the flask on a magnetic stirrer for 16 h.

17 Filter dry the washed 3-HBMA. Use a 3- to 4-cm-diameter, 1-L graduated burette with a lightly packed glass wool bottom plug covered by a layer of anhydrous Na_2SO_4 (about 100 mL) and another thin, lightly packed layer of glass wool. Mount the burette on a stand over an oven-dried 6-L Erlenmeyer flask. Open the stopcock. Slowly pour the contents of the step 16 flask through the filter burette.

18 Displace the atmosphere above 3-HBMA with nitrogen. Stopper the flask securely. Let stand approximately 16 h.

19 Repeat steps 17 and 18 if 3-HBMA has not remained clear. Store clear 3-HBMA in oven-dried 4-L amber glass bottles.

20 Uvitex OB fluorescent dye can be added to the 3-HBMA if desired. Mix 0.8 g of dye per 1 L of 3-HBMA with a magnetic stirrer for 2 h. Store in oven-dried, 4-L amber glass bottles.

65.5.2 SAMPLE PREPARATION AND POLYMERIZATION WITH 3-HBMA RESIN

65.5.2.1 MATERIALS AND REAGENTS

1 3-HBMA resin.

2 Uvitex-OB (Ciba-Geigy®), an ultraviolet fluorescent dye.

3   Item 1 (fumehoods) and item 3 (impregnation apparatus) as specified in Section 65.4.1.

4   Waxed aluminum foil, clothes iron, corrugated cardboard.

5   Wooden forms for foil boxes (dimensions slightly larger than sampling containers).

6   2.0-L plastic jars, 5-kg clear plastic bags, twist ties.

7   Pressure chamber, 4 bar (Soil Moisture Equipment Co., Hoskin Scientific)

8   Acetone (technical grade) and paper towels for cleaning.

9   Gamma cell (Cobalt 60 source).

65.5.2.2 PROCEDURE

1   Cut the 2.0-L plastic jars across their diameters at approximately 10 cm below their lids; keep the upper portion (see step 6). The height of the lower portion of the jar is determined by the physical space available in the impregnation apparatus, the pressure chamber, and the Gamma cell. Place two 5-kg plastic bags inside one another to double line the inside of the plastic jar (as you would for a garbage can).

2   Construct leakproof foil boxes by folding sheets of waxed aluminum foil around wooden forms. The height of the foil boxes should exceed the sample container heights by about 2 cm. With the clothes iron, heat the aluminum foil to melt the wax along the seams of the boxes. Test the foil boxes for leaks with water, correct as required, and thoroughly dry. Cut pieces of corrugated cardboard to fit the bottoms of the foil boxes to allow better resin access to the base of the sample.

3   Remove the water from the soil according to the methodology outlined in Section 65.3. When complete, carefully transfer the sample to the foil box and label as suggested in Section 65.4.2, step 2.

4   Place the foil boxes into the plastic bag-lined jars (step 1). Gently pull the plastic bags away from the top edges of the jar where they overlapped. Wedge pieces of cardboard into the space between the sides of the jar and the outermost side of the plastic bag to support the sample and reduce the amount of resin required. Overlap the ends of the plastic bags down over the top edge of the jar once the wedging has been completed.

5   Prepare the impregnation apparatus for adding 3-HBMA resin under vacuum according to procedures in Section 65.4.2 (steps 2–5). Maintain a vacuum of 25 to 30 cm Hg. Add the resin into the space between the foil box and the sample container; use a slow drip rate over 2 to 3 h. Do not let the resin drip directly onto the soil surface. Impregnation is completed when the soil

surface is covered by approximately 0.5 cm of ponded resin. *Caution:* Whenever the sample requires handling (steps 6–9), keep the sample container level to prevent resin drainage from the soil pores in the event of the surface being exposed to air.

6   Turn off the resin drip and very slowly release the vacuum from the impregnation apparatus. Remove the plastic jar and cover with the upper portion of the jar retained in step 1 above. Transfer to a fumehood and let stand for 1 week.

7   Because 3-HBMA will volatilize, each day periodicaly inspect the sample to ensure the level of resin is still 0.5 cm above the surface. Replenish, if required, by slowly adding new 3-HBMA at the edge of the sample. DO NOT ALLOW ANY EXPOSURE OF THE SOIL SURFACE TO AIR.

8   When the ponded resin level has stabilized, place the sample into the pressure chamber (without the lid and with the plastic bags pulled down over the upper edges of the jar). Operate the pressure chamber according to safety manual instructions. Increase the pressure slowly to 480 kPa and maintain for approximately 0.5 h. Release the pressure slowly. Remove the samples from the pressure chamber. Pull the plastic bags up over the sample and close with a twist tie. Place the lid on top of the container to protect the samples from foreign material and to minimize the release of resin vapors during transport to the Gamma Cell Facility.

9   Resin hardening is initiated by exposure to gamma radiation. At the Gamma Cell site: remove the jar lid. Just before the sample is placed into the Gamma Cell chamber, open the plastic bags and roll the ends down over the jar sides. Close the chamber and lower it into the gamma radiation source. Expose each sample to 15-min incremental exposures of radiation for a total of 1.5 h. The sample will heat up and emit an unpleasant vapor during this process. At the end of each 15-min period in the gamma cell, remove the sample from the cell using "hot mits". DO NOT BREATHE IN THE VAPORS. Alternate between two samples to allow for cooling and venting of vapors of one while the other is being exposed to radiation.

10   After the gamma radiation exposures are completed, the surface of the resin will be hardened. This is readily observable. In a fumehood, remove the sample from the plastic jar and bags. Wipe off any liquid or gel-like material from the sides of the sample block with a paper towel. Place the block onto a labeled paper towel to "cure" for approximately 1 week or until the outer surface feels completely dry to the touch and vapors are no longer present.

11   Use a chisel to remove the metal Kubiena box from around the hardened block. Excess plastic on the block edges can be trimmed off with a rock saw.

12   Label the hardened block as indicated in Section 65.4.2, step 10. It is now ready for further processing.

65.6 THIN SECTION PREPARATION FOR MICROMORPHOLOGICAL DESCRIPTION

A brief description of the procedures for hand production of thin sections is presented below. For more specific details (i.e., the grinding stages), one should refer to the following: FitzPatrick (1984) and Murphy (1986).

65.6.1 MATERIALS AND REAGENTS

1   Rock saws: large (46-cm) and small (18- or 25-cm) diamond blades. Observe all WHMIS safety precautions when using the saws.

2   Lapping fluid: for the large rock saw, Multicool, a water-soluble synthetic oil (Mantek®, Brampton, Ontario). For the small rock saw, use water.

3   Vacuum resectioning holder (Copeland 1965).

4   Slide holder (Cochrane and King 1957).

5   Diamond disks to fit grinding wheels: sizes 165, 70, 45, 30, and 15 μm.

6   Carbimet adhesive-backed disks: grit sizes 120, 240, 320, 400, and 600 (available from Buehler).

7   Galvanized metal trays lined with aluminum foil.

8   Fumehoods and ventilated oven to fumehood.

9   Resin solution (see Section 65.4.2, step 1).

10   Pipette with attached bulb.

11   Aluminum foil, straight pins, styrofoam sheet (45 × 30 × 3 cm).

12   Hot plate.

13   Epoxy: Araldite® 502 resin with hardener HY956, ratio 5:1. Prepare immediately before use (Ciba-Geigy®). Other low viscosity epoxy resins are also available.

14   Petrographic slides: size 27 × 46 mm.

15   Cover glasses: size 22 × 40 mm, No. 1.5 thickness.

16   Canada balsam, neutral (Buehler).

17   Alcohol flame.

18   Diamond pencil.

19   Polarizing microscope.

20   Chloroform.

21   Ultrasonic cleaner.

65.6.2 PROCEDURE

1   Determine whether vertical or horizontal orientations are required for the thin section. Orient the hardened block (obtained from Section 65.4.2, step 11; Section 65.5.2, step 12; or Chapter 45.4.2, step 4) accordingly in the large rock saw and cut a few slabs about 1 cm thick.

2   Place the slabs on paper towels to absorb the excess oil and remove the last traces with tap water. Dry with compressed air. With a red wax pencil, mark the orientation (to the soil surface) and the sample number on the slab.

3   Place the slabs into the galvanized trays. In a fumehood, puddle the surface of the slabs with the resin solution by applying small amounts with the pipette until all pores become filled. Leave in the fumehood overnight. The next day, place in a ventilated oven at 50°C and leave for 24 h or until the resin is hardened. Remove from the oven and allow to cool.

4   With the small rock saw, cut a slab into smaller sections (or chips) which correspond to the size dimensions of the petrographic slide. Mark the orientation on the side of the chip with a red wax pencil. Place the chips into a paper cup labeled with the sample number.

5   If part of the chip was coated on the surface with plastic during puddling (step 3 above), grind off this excess plastic with 120-grit Carbimet disks.

6   Grind each chip successively on 70-, 45-, 30-, and 15-μm diamond laps to obtain a smooth, flat surface. Rinse the chip with water and dry it immediately with compressed air.

7   From the chips prepared in step 6 above, select three representative chips for final thin-section production. Store the chips not chosen. Place the selected chips on a piece of aluminum foil about 1 cm larger than the chip itself. Mark on the foil the sample number.

8   Select those chips which have a large proportion of plastic in relation to soil material, and place into a metal tray. Place the tray into a ventilated oven at 50°C for 24 h. Remove from the oven and coat the chips with epoxy (Section 65.6.1, item 13). Select those chips having a large proportion of soil material in relation to plastic, and in a fumehood, place the chips directly onto the hot plate set at approximately 80°C and coat with epoxy (Section 65.6.1, item 13).

9	Using a hinged action (such as closing a book cover), place a clean, heated (on the hot plate) petrographic slide on the epoxy-covered chip and gently squeeze out any air bubbles immediately.

10	Place the mounted chip carefully onto the piece of styrofoam sheet. Hold the glass slide in place with straight pins. Maintain in place until the epoxy is completely cured (approximately 24 h).

11	With the diamond pencil, label the end of the glass slide (on the underside of the slide, i.e., not the chip side) with the sample number and orientation.

12	Place the mounted chip in the vacuum resectioning holder (Copeland 1965). With the small saw, cut off the excess soil material so that approximately 0.5 mm remains.

13	Insert the mounted chip into the slide holder (Cochrane and King 1957). Grind the chip on successive diamond laps (70, 45, 30, and 15 μm), taking utmost care to obtain a flat surface. Adjust the grinding accordingly. At the final grinding stage with the 15-μm lap, quartz grains must appear white to grey; the section is now about 30–35 μm thick. It is now ready for final mounting for thin sections (steps 14–15 below) or for preparation as polished sections (Section 65.7).

14	Clean the section with water in the ultrasonic cleaner, rinse, and dry immediately with compressed air.

15	Place the glass side of the slide directly onto the hot plate set at 120°C and apply to the surface of the prepared section neutral Canada balsam. Pass a clean cover glass through an alcohol flame to ensure it is moisture free. Using a hinged action, place the cover glass onto the thin section and remove immediately any air bubbles by using gentle pressure. Leave the section on the hot plate for 30 min.

16	In the fume hood, when the thin section has cooled, remove with chloroform any excess Canada balsam surrounding the cover glass. Do not leave the slide in contact with the chloroform for longer than 1 minute as the cover glass can be removed and the section damaged. Rinse in clean, unused chloroform to remove any remaining traces of balsam.

17	Label the top of the slide with the sample number and the orientation using a permanent marking pen. It is now ready for microscopic study.

65.7 POLISHED SECTIONS FOR SCANNING ELECTRON MICROSCOPE/ENERGY DISPERSIVE X-RAY ANALYZER (SEM/EDXRA)

65.7.1 MATERIALS

1	Thin-section production materials and reagents as listed in Section 65.6.1.

2   Nylon or Texmet polishing cloth.

3   Plate glass.

4   Diamond paste (6 μm and 1 μm, Buehler).

5   Aluminum oxide (9 μm and 0.3 μm, Buehler).

6   Low-speed polisher/grinder with Petro-thin polishing attachment (Buehler).

7   Lapping oil (Buehler).

65.7.2 PROCEDURE

1   Following the procedures for thin-section production in Section 65.6.2 (steps 1–13), grind the section to about 35-μm thickness.

2   In a fumehood, clean the section with the ultrasonic cleaner filled with water, rinse, and dry immediately with compressed air.

3   Cover the back of the glass slide with masking tape to prevent scratching of the glass while inserted in the polishing holder. Wet the masking tape with water and insert into the polishing holder.

4   On the plate glass, make a paste with 9-μm aluminum oxide and distilled water. By hand, and using a straight, one-direction movement, grind the surface of the section smooth and flat in order to condition it to the holder.

5   Clean the holder and section using the ultrasonic cleaner filled with water. Using lapping oil, insert the section in the polishing holder.

6   Proceed to grind the section with 6-μm diamond paste and 2–3 mL of lapping oil on the polishing cloth as per directions supplied with the polisher. Check the section every 10–15 min for the degree of polish. This step is completed when the mineral grains have a smooth, unpitted surface with sharp, well-defined boundaries.

7   In a fumehood clean the holder and section as in an ultrasonic cleaner using acetone as the organic solvent for the lapping oil. Inspect the masking tape for wear and reapply if needed. Insert the section into the polishing holder using lapping oil.

8   Proceed to grind with the 1-μm diamond paste and 2–3 mL lapping oil on the polishing cloth as per directions of the polisher. Check the sections every 10–15 min. This polishing stage is completed when the mineral grains have a very smooth surface with very sharp boundaries.

9   Thoroughly clean the section and holder as in step 7, but use distilled water to reinsert the section into the holder. As per directions of the polisher, polish the section using a distilled water/0.3-μm aluminum oxide slurry pre-

pared on the cloth. This stage should only be carried out for less than 30 s in order to prevent the appearance of relief on the polished surface. The polishing is now complete. Clean the section thoroughly as in step 2.

10 Examine the polished section under the polarizing microscope. Identify and photograph the features to be analyzed.

11 Prepare the section as required for the specific SEM being used.

12 Following directions for the operation of the SEM and EDXRA system, locate the features of interest. Examine with the SEM several areas at increasing scales of magnifications from low to extremely high. Select features or microregions for elemental analysis with the EDXRA.

65.7.3 Comment

For polishers that hold more than one sample at a time, grind samples with similar hardness together; for example, all sandy or all clayey materials, to avoid differences in the rate of removal of the soil material.

Each polishing step is completed on a different polishing wheel which is fitted with its own cloth for each diamond paste size used, thus avoiding contamination. Do not clean the cloth when the polishing is completed in order to establish on the cloth a loading of a particular diamond paste size.

65.8 POLISHED BLOCKS FOR IMAGE ANALYSIS

With the advances in image analysis technology, quantitative data about the morphology of specific features can be obtained (Ringrose-Voase 1991). Most applications of image analysis concepts and techniques have been directed towards obtaining data about soil pores (Ismail 1975, Murphy et al. 1977, Ringrose-Voase 1987, 1991, Moran et al. 1989). Samples that are prepared for image analysis must be representative of the soil at the time of sampling and be of excellent quality with respect to polymerization. Quantitative data can be obtained about soil features observed from polished blocks, thin or polished sections, or the SEM. See Section 65.6 for preparing thin sections and Section 65.7 for polished sections. The following describes a procedure for preparing polished blocks.

65.8.1 MATERIALS

1 Materials and reagents for resin impregnation of samples (see Section 65.4 or 65.5).

2 Rock saws: large (46-cm) and small (18-cm) diamond blades.

3 Lapping oil: for large saw, Multicool (synthetic, water-soluble oil available from Mantek®, Brampton, Ontario).

4   Carbimet disks: grit sizes 60 and 120 (Buehler).

5   Silicon carbide: 400 and 600 grit.

6   Aluminum oxide: 9-μm size.

7   Ultrasonic cleaner.

8   Resin solution (see Section 65.4.2, step 1).

9   Galvanized metal tray lined with aluminum foil.

65.8.2 PROCEDURE

1   Impregnate the sample with resin according to the chosen method as outlined in either Section 65.4 or 65.6.

2   With the large rock saw, cut away the excess plastic surrounding the sample container. Wipe the block with tissues and clean in water to remove any oil.

3   With a hammer and chisel, remove the impregnated sample from the container. Remove any remaining oil by cleaning with water.

4   Orient the sample according to whether a vertical or a horizontal cross section is required. Cut a slab 1- to 2-cm thick using the small rock saw (lubricant is water). Mark, with red wax pencil, the sample number on the slab. For vertically oriented samples, indicate the direction to the soil surface.

5   Place the slab into the galvanized tray and reimpregnate the surface with the resin solution as described in Section 65.6.2, step 3.

6   Remove any plastic that may cover the surface of the slab by grinding with 60- and 120-grit Carbimet disks. If the surface is well impregnated, go to step 7. If the surface is not well impregnated, holes of varying sizes will still be visible, indicating that the pores were incompletely filled with resin. In such cases, repeat steps 5 and 6. If the entire sample has resulted in poor impregnation, that is, the soil material is soft, loose, breaks apart easily, or the center portion is not impregnated, try as a last resort to reimpregnate the entire slab. If this is unsuccessful in maintaining the structure intact, discard the sample.

7   Mark a cross-hatch pattern over the slab surface with a felt marking pen (water-proof). Grind the surface with 400- and 600-grit silicon carbide and finish with 9-μm aluminum oxide. If low areas still exist after being ground, the cross-hatch markings will still be present. Adjust the grinding accordingly. Thoroughly clean the block after each grinding in the ultrasonic cleaner filled with water.

8   The sample is now ready for photography under ultraviolet illumination. Use high-contrast, black and white Kodak® T-Max® 100 print film (the pores will appear black in the negative). Orient the ultraviolet lamps at a 45° angle for best resolution. Include in each photograph a scale in both the x and y directions.

9   Refer to the image analysis system manual for operating directions and specific requirements for image input.

65.9 STAINING TECHNIQUES AS APPLIED TO IMAGE ANALYSIS

Staining techniques have been used to enhance and identify features on micromorphological thin sections. Bouma (1969) and FitzPatrick (1984) review methodology for staining of specific minerals in both thin sections and polished blocks. The stained areas can be used to advantage with the image analyzer, as the staining provides sufficient contrast to adequately differentiate the features. Bui and Mermut (1989) developed a procedure for analyzing carbonate-stained regions, which is outlined in Section 65.9.1 (see also Chapter 69, Section 69.3). This approach can be adapted for other stains, but may require adjustments of the lighting and filters.

65.9.1 QUANTIFICATION OF SOIL CALCIUM CARBONATES ON THIN SECTION OR POLISHED BLOCK (AFTER BUI AND MERMUT, 1989)

65.9.1.1 MATERIALS AND REAGENTS

1   HCl, hydrochloric acid, technical grade (used in step 1 below, either as a concentrated acid fume or as 1% HCl solution).

2   Alizarin red-S (0.1 g dissolved in 100 mL 0.2% cold hydrochloric acid, commercial grade).

3   Fumehood.

4   Light table with an opaque white plexiglas cover and illuminated with different types of tungsten filament light sources as follows:

     a. Transmitted light from four 75-W flood lights (usual light source of the light table) having a color temperature of about 2820 K.

     b. Reflected light from a Schott® KL 1500 lamp with tungsten halogen reflector 150-W bulb operated at 3000 K.

     c. Reflected light from a desk lamp with a 60-W "soft white" bulb with a color temperature of about 2750 K.

5   Color Wratten filters (green Wratten filter no. 16; orange Wratten filter no. 60).

6   Image-analyzing system with video camera (telemacro lens) interfaced to a computer monitor and appropriate software programs to interact with the image to provide quantitative data about the features (i.e., Kontron-IBAS system).

65.9.1.2 PROCEDURE

1   In a fumehood, etch the surface of the thin section or polished block by either exposing to concentrated HCl acid fumes or immersing in 1% HCl for 5 min. Wash thoroughly in water to remove the acid. Etched carbonate on thin sections and polished blocks will appear three dimensional.

2   Immerse the samples in the alizarin red-S solution for about 2–3 min until the calcite appears a deep red-brown color. The time will vary with the composition, porosity, and grain size of the sample. Rinse the sample quickly in water and dry immediately. FitzPatrick (1984) advises to not overwash, as the stain is soluble in water, dry quickly, and do not touch the stained surface, as it is a precipitate and can be easily removed.

3   Place the stained section (or block) on the light table. Illuminate the sample with the different tungsten filament light sources to obtain in conjunction with the video camera the best projection of the surface onto the computer monitor of the image analyzer. In order to enhance the contrast between the stained portions and the surrounding soil materials, it may be necessary to place selected color Wratten filters between the sample and the telemacro lens of the video camera. For their system setup, Bui and Mermut (1989) used Wratten filters no. 16 and 60.

4   Refer to the operating manual of the image analysis system for programming procedures to obtain specific quantitative data about the stained regions.

65.10 MICROMORPHOLOGICAL DESCRIPTION

The main goal for micromorphological descriptions is to produce a clear, concise, and easily understandable discussion. Descriptive systems are available for characterizing the observed features and arrangements. Some of the major contributions are the following: Kubiena (1938) presented the basic concepts and principles for describing the soil morphology. Brewer (1964, 1976) developed a descriptive system based on the observed morphology and not on the genetic interpretation of the processes involved. Brewer and Pawluk (1975) expanded the terminology to include unique morphologies observed in permafrost soils. Stoops and Jongerius (1975) approached the classification of soil morphology from a concept of the variation in distribution of the coarse and fine materials. Specific systems were developed for characterizing the organic matter distributions within and overlying the mineral soil, such as those by Barratt (1968/1969), Babel (1965), and Bal (1973), as well as for peat materials and organic layers (Fox 1984, 1985). Bullock et al. (1985) developed an internationally accepted system which emphasizes the description of the morphology, uses a minimum of terminology and jargon, and is based on the Stoops and Jongerius (1975) concepts.

At present, micromorphologists seem to approach the description of soil features in the following ways: (1) adopting exclusively the terminology of Brewer (1964, revised 1976) and Brewer and Pawluk (1975); (2) referring to selected terminology from several descriptive systems; (3) referencing exclusively the descriptive system (Bullock et al. 1985) compiled by the International Working Group on Soil Micromorphology. Comparing the various descriptive systems to say that one should be used exclusively for all soil descriptions may be too restrictive. Each system has particular merits, and as such the application of the conceptual framework and selected terminology of a specific system may be more appropriate for a particular study. However, it is highly recommended, especially for research publications, that the equivalent terminology according to the *Handbook For Soil Thin Section Description* (Bullock et al. 1985) be referenced and included for comparative purposes.

65.10.1 MATERIALS

1   Stereomicroscope; polarizing light microscope with attachments for both fluorescent and incident light capabilities.

2   Microscope camera attachments: Polaroid® and single-lens reflex camera for 35-mm slides.

65.10.2 PROCEDURE

1   With both the stereomicroscope and the light microscope, examine several fields of view at various magnifications to delineate regions of similar morphology and characterize specific features using transmitted, polarized, incident, and fluorescent light observations as required.

2   Record the interrelationships among the various soil features and the spatial distribution patterns within these regions according to the selected descriptive systems.

3   Record quantitative data about the occurrence and areal distribution of the observed features. These data can be obtained at different levels of accuracy, depending on the procedure chosen; for example, in order of increasing reliability, (1) visual estimates, (2) point counting, or (3) image analysis techniques can be used.

4   Photograph selected features and/or distribution patterns for presentation or further analysis using image analysis techniques.

REFERENCES

Babel, U. 1975. Micromorphology of soil organic matter. Pages 69–473 in J. E. Gieseking, Ed. Soil components. Vol. 1. Organic components. Springer-Verlag, New York.

Bal, L. 1973. Micromorphological analysis of soils: Lower levels in the organization of organic materials. Soil Survey Paper No. 6. Soil Survey Institute, Wageningen, The Netherlands. 174 pp.

Barratt, B. C. 1968/1969. A revised classification and nomenclature of microscopic soil materials with particular reference to organic components. Geoderma 2: 257–271.

Bouma, A. H. 1969. Methods for the study of sedimentary structures. Wiley-Interscience (John Wiley & Sons), New York. 458 pp.

Brewer, R. 1964. Fabric and mineral analysis of soils. John Wiley & Sons, New York. 470 pp.

Brewer, R. 1976. Fabric and mineral analysis of soils. Robert E. Krieger, Huntington, NY.

Brewer, R. and Pawluk, S. 1975. Investigations of some soils developed in hummocks of the Canadian Sub-Arctic and Southern Arctic regions. 1. Morphology and micromorphology. Can. J. Soil Sci. 55: 301–319.

Bui, E. N. and Mermut, A. R. 1989. Quantification of soil calcium carbonates by staining and image analysis. Can. J. Soil Sci. 69: 677–682.

Bullock, P., Fedoroff, N., Jongerius, A., Stoops, G., and Tursina, T. 1985. Handbook for soil thin section description. Waine Res. Pub., Wolverhampton, England. 152 pp.

Cochrane, M. and King, A. G. 1957. Two new types of holders used in grinding thin sections. Am. Mineral. 42: 422–425.

Copeland, D. A. 1965. A simple apparatus for trimming thin sections. Am. Mineral. 50: 1128–1130.

Day, J. H., Ed. 1983. The Canada Soil Information System (CanSIS) manual for describing soils in the field 1982 (Revised). Expert Committee on Soil Survey. Agriculture Canada, Research Branch, Land Resource Research Institute, Ottawa. LRRI Contribution No. 82-52.

FitzPatrick, E. A. 1984. Micromorphology of soils. Chapman and Hall, London. 433 pp.

Fox, C. A. 1984. A morphometric system for describing the micromorphology of organic soils and organic layers. Can. J. Soil Sci. 64: 495–503.

Fox, C. A. 1985. A morphometric system for describing the micromorphology of organic soils and organic layers: further quantitative and qualitative characterization. Can. J. Soil Sci. 65: 695–706.

Ismail, S. N. A. 1975. Micromorphometric soil-porosity characterisation by means of electro-optical image analysis (Quantimet 720). Soil Survey Paper No. 9. Netherlands Soil Survey Institute, Wageningen, The Netherlands.

Kubiena, W. L. 1938. Micropedology. Collegiate Press, Ames, IA.

Koppi, A. J. and McBratney, A. B. 1991. A basis for soil mesomorphological analysis. J. Soil Sci. 42: 139–146.

McCarrick, T. and Protz, R. 1978. The effects of soil composition upon three acrylic soil impregnating resins. Commun. Soil Sci. Plant Anal. 9: 955–962.

Moran, C. J., McBratney, A. B., and Koppi, A. J., 1989a. A rapid method for analysis of soil macropore structure. I. Specimen preparation and digital binary image production. Soil Sci. Soc. Am. J. 53: 921–928.

Moran, C. J., McBratney, A. B., Ringrose-Voase, A. J., and Chartres, C. J. 1989b. A method for the dehydration and impregnation of clay soil. J. Soil Sci. 40: 569–575.

Murphy, C. P. 1985. Faster methods of liquid phase acetone replacement of water from soils and sediments prior to resin impregnation. Geoderma 35: 39–45.

Murphy, C. P. 1986. Thin section preparation of soils and sediments. A B Academic Pub. Berkhamsted, Herts, England. 149 pp.

Murphy, C. P., Bullock, P., and Biswell, K. J. 1977. The measurement and characterization of voids in soil thin sections by image analysis. Part I. Principles and techniques. J. Soil Sci. 28: 498–508.

Pagliai, M., Pezzarossa, B., Mazzoncini, M., and Bonari, E. 1989. Effects of tillage on porosity and microstructure of a loam soil. Soil Technol. 2: 345–358.

Protz, R., Shipitalo, M. J., Mermut, A. R., and Fox, C. A. 1987. Image analysis of soils — present and future. Geoderma 40: 115–125.

Ringrose-Voase, A. J. 1987. A scheme for the quantitative description of soil structure by image analysis. J. Soil Sci. 38: 343–356.

Ringrose-Voase, A. J. 1991. Micromorphology of soil structure: description, quantification, application. Aust. J. Soil Res. 29: 777–813.

Singh, P., Kanwar, R. S., and Thompson, M. L. 1991. Measurement and characterization of macropores by using AUTOCAD and automatic image analysis. J. Environ. Qual. 20: 289–294.

Sheldrick, B. H., Ed. 1984. Analytical methods manual, LRRC 84-30. Land Resource Research Centre, Agriculture Canada, Ottawa.

Shipitalo, M. J. and Protz, R. 1987. Comparison of morphology and porosity of a soil under conventional and zero tillage. Can. J. Soil Sci. 67: 445–456.

Stoops, G. and Jongerius, A. 1975. Proposals for a micromorphological classification in soil materials. 1. A classification of the related distributions of coarse and fine particles. Geoderma 13: 189–200.

Tarnocai, C. 1980. Canadian Wetland Registry. Pages 9–38 in C. Rubec and F. Pollet, Eds. Proc. Workshop on Canadian Wetlands. Saskatoon. 1979. Environment Canada. Ecological Land Classification Ser. 12.

Wires, K. C. and Sheldrick, B. H. 1987. A soil sampling procedure for micromorphological studies. Can. J. Soil Sci. 67: 693–695.

Soil Mineralogical Analyses

Chapter 66
Soil Separation for
Mineralogical Analysis

C. R. de Kimpe

Agriculture Canada
Ottawa, Ontario, Canada

66.1 INTRODUCTION

Mineralogical analysis is undertaken to determine the nature and proportions of the mineral components in soils. Minerals have a major influence on the physical, chemical, and biological behavior of a soil. The mineral assemblage also provides information on the potential fertility level and the transformations occurring during pedogenesis.

Ideally, soil mineralogical analysis would be performed without any pretreatment in order to provide the most accurate information about the nature of soil components. However, the nature and composition of the soil material make a correct analysis very difficult under this condition. For example, many soils contain aggregates that cannot be disrupted by water dispersion only; therefore, the material that is collected in the finer fraction would not be representative of the actual assemblage. With different cations being generally adsorbed on the exchange sites, dispersion may also be difficult to achieve. Stirring the sample in water will not break down the organometallic complexes present in the soil.

The diversity of the minerals and the wide range of particle sizes, from less than 1 μm to 2 mm, preclude a single global analysis by the most general technique used, X-ray diffraction. Separation of the soil into the major size fractions is a prerequisite for mineralogical analysis. It will provide a rough sorting between phyllosilicates present mostly in the fine fractions and other silicates present mainly in the coarser fractions. These minerals have distinct requirements for their analysis and identification.

As pretreatments may affect the mineralogical composition, it is preferable to keep them to a minimum. Four methods will be successively described, with an increasing pretreatment severity from the first to the fourth. A method would be selected according to the objectives of the study.

Obviously, water dispersion will be less time consuming than any other separation procedure. Yet, it may provide enough information for some quick tests and may therefore be preferred to time-consuming techniques.

Soil Sampling and Methods of Analysis, M. R. Carter, Ed.,
Canadian Society of Soil Science. © 1993 Lewis Publishers.

66.2 WATER-DISPERSIBLE CLAY

This is the simplest treatment that can be applied to a soil in order to collect clay- and silt-size material. The method will not provide a quantitative recovery of the fine materials, but it will provide a sample large enough to obtain some information about the clay components that are present in the soil. Identification procedures are given in Chapter 67.

66.2.1 MATERIALS AND REAGENTS

1   Shaker bottles and end-over-end shaker.

2   20-cm diameter, 300-mesh sieve.

3   25-cm diameter funnel and stand.

4   Low-speed centrifuge, about 2000 rpm.

5   1-L tall-form beakers.

6   1 M $MgCl_2$ solution.

66.2.2 PROCEDURE

1   Weigh about 20 g of soil into shaker bottles and half fill with distilled water. Stopper the bottle.

2   Shake overnight on an end-over-end shaker.

3   Sieve the suspension on a 300-mesh sieve placed on the funnel. Collect the filtrate in a 1-L tall-form beaker.

4   Rinse the sand with distilled water.

5   Fill the beaker to about 800 mL. Stir thoroughly.

6   Let the beaker stand 7 h at 25°C.

7   Siphon the upper 10 cm containing the clay.

8   Repeat steps 5 to 7 until the supernatant is clear.

9   Flocculate the clay with 1 M $MgCl_2$ and discard the supernatant. Buffered Mg acetate may be used to avoid precipitation of $Mg(OH)_2$ in some systems at high pH.

10   Using a centrifuge, wash the clay three times with 1 M $MgCl_2$ to make sure it is well Mg-saturated.

11 Wash the clay at least three times with distilled water to remove the excess salt. Test the supernatant to make sure it is Cl⁻ free (use AgNO$_3$).

12 Freeze dry the clay and let the silt air dry.

66.3 DISPERSION OF THE SOIL IN WATER AND SONIFICATION

Shaking the soil in water may not be sufficient to break the aggregates and to release much of the clay-sized material. Sonification is an effective way to break aggregates (Genrich and Bremner 1974). However, it will not provide complete dispersion unless chemical treatments are also applied (see further, Sections 66.4 and 66.5).

66.3.1 MATERIALS AND REAGENTS

1 1-L tall-form beakers.

2 20-cm diameter, 300-mesh sieve.

3 25-cm diameter funnel and stand.

4 Sonifier or ultrasonic dip-type probe.

5 Low-speed centrifuge.

6 1 *M* MgCl$_2$ solution.

66.3.2 PROCEDURE

1 Weigh about 20 g of soil into a 1-L beaker.

2 Add 250 mL of distilled water.

3 Place the beaker in the sonifier.

4 Sonify for 30 s.

5 Sieve the soil on a 300-mesh sieve and rinse the sand.

6 Collect the silt + clay suspension in a 1-L tall-form beaker.

7 To collect more fine material, return the sand to the original beaker and repeat steps 1 to 5.

8 To separate the clay from the silt, proceed as indicated under Section 66.2.2, steps 5 to 12.

66.3.2.1 COMMENTS

1   A dip-type probe may be used for the sonification. Read the probe specifications. A larger volume of water may be required to plunge the probe deeply enough and a longer time of vibration may also be required. If this is the case, it is recommended to use a cooling jacket around the beaker to prevent excess heating that can damage the clay-size components.

2   There may be a problem in determining and maintaining a constant amount of ultrasonic energy to the sample, especially when extended treatments are necessary. It is essential to follow the recommendations provided with the different types of equipment.

66.4 DISPERSION IN WATER AT pH 9 AFTER REMOVAL OF ORGANIC MATTER

This method will provide a better dispersion of the clay-size material through the decomposition of organometallic complexes. It will also preserve the crystalline Fe and Al hydroxides and oxyhydroxides.

66.4.1 MATERIALS AND REAGENTS

1   1-L tall-form beakers.

2   20-cm diameter, 300-mesh sieve.

3   25-cm diameter funnel and stand.

4   Hot plate or sand bath.

5   Low-speed centrifuge, about 2000 rpm.

6   Sonifier.

7   Hydrogen peroxide, concentrated.

8   1 M $MgCl_2$ solution.

66.4.2 PROCEDURE

1   Weigh about 20 g of soil into a 1-L beaker.

2   Moisten the soil with distilled water.

3   Add 5 mL of concentrated H_2O_2. Gentle heating may be necessary to initiate the reaction.

4   When the reaction is complete, heat gently on a hot plate until close to dryness.

5   Repeat steps 3 and 4 as required to achieve complete oxidation of organic matter. This is demonstrated by the lighter color of the sample.

6   Transfer the soil to a 250-mL centrifuge bottle.

7   After centrifuging, discard the supernatant and wash three times with distilled water, discarding each time the supernatant after centrifugation.

8   Transfer the soil back to the 1-L beaker.

9   Add 250 mL of H_2O.

10   Adjust pH to 9.0 with M NaOH.

11   Sonify the suspension for 30 s.

12   Proceed as in Section 66.3.2, steps 5 to 8. Make sure that the pH of the suspension remains close to 9.0.

66.4.3 COMMENTS

1   The presence of carbonates, as in calcareous soils or sometimes in the deeper horizons of other soils, may prevent the dispersion of clay-size material. If this is the case, it will be necessary to first proceed with carbonate dissolution using diluted HCl before organic matter oxidation.

2   Na hypochlorite has been proposed as an organic matter oxidizing agent to substitute for hydrogen peroxide (Lavkulich and Wiens 1970). Using this reagent will bring the pH of the suspension close to 9.0.

66.5 DISPERSION IN WATER AFTER ORGANIC MATTER AND IRON OXIDES REMOVAL

This procedure will provide complete dispersion of the clay minerals through decomposition of all organometallic complexes and dissolution of the sesquioxides which cement and bind mineral particles (Mehra and Jackson 1960). However, this method will prevent the identification of all secondary crystalline oxyhydroxides that were possibly present in the soil.

When separation and recovery of the various fractions are performed quantitatively, it will be possible to establish a particle-size distribution on an organic matter- and sesquioxide-free basis.

66.5.1 MATERIALS AND REAGENTS

1   1-L tall-form beakers.

2   20-cm diameter, 300-mesh sieve.

3   25-cm diameter funnel and stand.

4   Controlled-temperature water bath.

5   Hot plate.

6   Low-speed centrifuge, about 2000 rpm.

7   Hydrogen peroxide.

8   Citrate-bicarbonate solution.

9   Na dithionite.

10   1 M MgCl$_2$.

66.5.2 PROCEDURE

1   Weigh 20 g of soil into a 1-L beaker.

2   Moisten the soil with distilled water.

3   Proceed with organic matter oxidation as described above in Section 66.4.2, steps 3 to 7.

The next step is the dissolution of sesquioxides.

8   Add 150 mL of citrate-bicarbonate solution.

9   Put the centrifuge bottle in a water bath at 80°C and check that the temperature in the bottle reaches at least 75°C.

10   Add 1 g of Na dithionite and stir thoroughly.

11   Let reaction take place for 5 min, then repeat step 10 twice. The soil will take a greyish color, indicating complete reduction of all iron oxides.

12   When the reduction is complete, add 10 mL of saturated NaCl, mix, centrifuge, and pour off the supernatant.

13   Using the centrifuge, wash the soil several times until the supernatant is Cl$^-$ free.

14 Sieve the suspension on a 300-mesh sieve and proceed with the sand, silt, and clay separation as in Section 66.2.2, steps 3 to 12.

66.6.3 COMMENTS

1 Sonification is not necessary with this latter procedure, as all aggregates have been decomposed, leaving only individual mineral particles.

2 In the case of soils containing carbonates, it is necessary to proceed with the dissolution of carbonates with HCl before organic matter oxidation.

REFERENCES

Genrich, D. A. and Bremner, J. M. 1974. Isolation of soil particle-size fractions. Soil Sci. Soc. Am. Proc. 38: 222–225.

Lavkulich, L. M. and Wiens, J. H. 1970. Comparison of organic matter destruction by hydrogen peroxide and sodium hypochlorite and its effect on selected mineral constituents. Soil Sci. Soc. Am. Proc. 34: 755–758.

Mehra, O. P. and Jackson, M. L. 1960. Iron oxide removal from soils and clays by a dithionite-citrate system buffered with sodium bicarbonate. Clay. Clay Miner. 7: 317–327.

Chapter 67
Clay and Silt Analysis

C. R. de Kimpe
Agriculture Canada
Ottawa, Ontario, Canada

67.1 INTRODUCTION

X-Ray diffraction (XRD) is the most generalized method of identifying minerals in the soil clay- and silt-sized fractions. This analysis is often complemented by some chemical methods (see Chapter 70). It is worth mentioning that other techniques, such as infrared spectroscopy, differential thermal and thermogravimetric analyses, Mossbauer spectroscopy, single crystal electron diffraction, and nuclear magnetic resonance have applications in the identification of clay minerals. However, these methods have specific applications and will therefore not be reviewed in this chapter.

Clay minerals are phyllosilicates and maximum information will be obtained from XRD patterns of preferentially oriented specimens showing specific variations of d001-spacings, i.e., specific spacings of basal reflections (Brindley and Brown 1980). This behavior depends to a large extent on the nature of the cations present on the exchange sites. Therefore, a prerequisite for adequate identification is to proceed with cation exchange to ensure homoionic clay specimens.

The silt fraction is usually analyzed in the same manner as the clay fraction. However, the silt fraction separated from sandy soils may sometimes contain a large percentage of coarser (50–20 μm) particles. Whenever this is the case, mild grinding of the specimen (this can be done under acetone) may be sufficient to reduce the particle size, in order to increase the number of particles with a preferred orientation, and therefore the peak intensities.

Nonbasal as well as basal reflections may be useful for mineral identification. For example, the former are used to differentiate dioctahedral and trioctahedral minerals. There is thus some usefulness in recording the XRD pattern for randomly oriented as well as for oriented clay sample preparations.

67.2 SAMPLE PREPARATION

67.2.1 MATERIALS AND REAGENTS

1   Low-speed centrifuge, around 2000 rpm.

2   Vortex shaker.

3   1 M $MgCl_2$.

4   1 M KCl.

5   Glycerol solution: 20 mL glycerol L^{-1} in H_2O or 2% glycerol in ethanol.

67.2.2 PROCEDURE

1   Weigh 30 mg of clay in a 10-mL centrifuge tube. If the clay has been completely Mg-saturated during the clay separation (see Chapter 66), proceed to step 6. If not, proceed as follows:

2   Add 7 mL of 1 M $MgCl_2$ solution. Stir thoroughly.

3   Centrifuge and discard the supernatant.

4   Repeat steps 2 and 3 twice to ensure full saturation of the exchange sites with Mg. Shaking the suspension for a few hours or overnight may ensure more complete exchange.

5   After exchange is complete, wash at least three times with H_2O, using the centrifuge, and discard the supernatant. Wash until the water is Cl^- free. Washing with alcohol may also be necessary when clay does not flocculate.

6   Add 1 mL of distilled H_2O and disperse thoroughly. Sonification may help.

7   Spread the suspension on a clean glass slide, 25 × 30 mm.

8   Let air dry on a horizontal surface to ensure even distribution of the particles.

A second slide should be prepared with K-saturated clay. Proceed as under Section 67.2.2, steps 1 to 8, but using M KCl solution.

67.2.3 COMMENTS

1   Some mineralogists prefer to substitute a ceramic plate for the glass slide. The ceramic plate is fitted on a suction pump or in a centrifuge tube with

a special adapter. The clay is then poured on the plate and the water evacuated with the appropriate technique. Film density will depend on the concentration of the clay suspension.

2   Others prefer the paste method. After the last washing and decantation, a small amount of clay is removed with a microspatula and smeared on a glass slide so as to provide a smooth surface.

67.3 RECORDING THE XRD PATTERN

Most information will be found in the 2–35° 2θ portion of the spectra. Recording specifications vary from instrument to instrument and it is therefore impossible to make specific recommendations. For a diffractometer equipped with a rotating goniometer, a scanning speed of 1° 2θ/min is generally adequate. For other instruments, a setting would be selected in order to provide well-separated peaks with strong intensity for the major peaks.

67.4 TREATMENTS

Shrinking and swelling properties, as well as decomposition or transformation temperatures are two factors which are taken into account for mineral identification. Shrinking/swelling properties depend very much on the cation present on the exchange sites.

67.4.1 Mg SLIDE

1   Record the XRD pattern without treatment. Swelling minerals and chlorite have a 1.4-nm spacing.

2   Record the XRD pattern after glycerol solvation. Smectite will have expanded to 1.8 nm.

Two methods have been proposed that provide good swelling of smectite:

1   If enough sample is available, suspend 30 mg of Mg-saturated clay in 1 mL of a 20-mL glycerol L^{-1} in H_2O solution. Stir thoroughly and spread on a glass slide (25 × 30 mm). Let air dry. Record the XRD pattern.

2   Proceed with solvation of the initial slide. Add dropwise, at several places on the slide, 0.4 mL of a 2% glycerol in ethanol solution. This technique provides quick penetration of the glycerol, and after evaporation of the ethanol, smectite expansion to 1.8 nm is easily observed. This method may be preferred when clay material is scarce (Miles and De Kimpe 1985).

Note: some clay mineralogists prefer Ca^{2+} rather than Mg^{2+} on the exchange sites; however, Mg^{2+} is the most common cation used.

67.4.2 K SLIDE

1   Record the XRD pattern at 25°C. Under K saturation, vermiculite readily collapses to 1.0 nm.

2   Heat the slide at 300°C for 1 h. Record the XRD pattern again. After this treatment, Fe and Al oxyhydroxides are decomposed and transformed to oxides. New peaks may therefore appear, but they are generally not intense. Smectite shows collapse to about 1.0 nm.

3   Heat the slide at 550°C for 1 h. Record the XRD pattern again. After this treatment, the following is normally observed:

   a. Vermiculite is fully collapsed to 1.0 nm.

   b. Smectite is collapsed close to 1.0 nm.

   c. Kaolinite peaks have disappeared after transformation of the mineral to metakaolin.

   d. Chlorite shows an inversion of intensities of the two major peaks at 1.4 and 0.7 nm.

67.4.3 COMMENTS

1   Heating the K-specimen slide first to 100°C may provide additional information on collapse, especially of vermiculite and vermiculitic clay.

2   Oriented specimens may peel off the slide on heating. Binders, such as PVP, have been successfully used to address this situation.

67.4.4 Relative Humidity Treatments

Ideally XRD analysis should be done at controlled relative humidity. Additional information on expanding minerals and on mixed layer minerals with expanding components is obtained by running the samples at room temperature and the following relative humidities: 0, 50, and 100%. Clays that collapse on heating to 300°C may reexpand if the sample is left in a humid room. Relative humidity conditions are achieved by setting the slide in a dessicator over appropriate mixtures of sulfuric acid and water.

67.5 CRITERIA FOR IDENTIFICATION

Criteria for identification of clay minerals by XRD are summarized below. Further details can be found in the references cited (Brown 1961; Harward and Brindley 1965, Sayegh et al. 1965, Harward et al. 1969, Brindley and Brown 1980).

67.5.1 APPROXIMATE 001 SPACINGS (nm) OBTAINED WITH THE CHARACTERIZATION TREATMENTS

1   Smectite

Mg	— saturation,	54% R.H.	1.53 nm
Mg	— "	and glycerol	1.74
Mg	— "	and eth. glycol	1.69
K	— " ,	54% R.H.	1.21–1.24
K	— "	105°C; dry air	1.03
K	— "	300°C; dry air	0.99–1.04
K	— "	300°C; glycerol	1.77
K	— "	500°C; dry air	0.98–1.02

2   Beidellite has essentially the same behavior as montmorillonite. See also comment 1 below.

3   Vermiculite

Mg	— saturation	54% R. H. at 25°C	1.43 nm
Mg	— "	Glycerol	1.41
Mg	— "	Eth. Glycol	1.41
K	— "	54% R.H.	1.04
K	— "	105°C; dry air	1.03
K	— "	300°C; glycerol	1.04
K	— "	500°C; dry air	1.0

See also comments 2 and 3 below.

4   Chlorite

a. Chlorite minerals have a 001 spacing at 1.4 nm, which does not vary with glycerol solvation or K-saturation and heating.

b. Most chlorites are trioctahedral chlorites. At room temperature the 002 spacing is normally more intense than the 001 spacing, but after heating at 550°C, the contrary is observed.

c. In dioctahedral chlorite (Brydon et al. 1961, Sudo and Sato 1966) the 001 spacing is more intense than the 002 spacing at room temperature. See also comment 4 below.

d. In the presence of kaolinite, there may be some confusion between the various lines, but the 0.473-nm reflection is typical for chlorites.

5   Micaceous minerals

a. They are recognized by the presence of a 0.99- to 1.04-nm spacing regardless of saturating cation, humidity, or solvation.

b. In the case of illite, K-saturation may contribute to a better organization of the layers and result in sharper reflections.

6   Kaolin minerals

Kaolinite is characterized by the presence of a 0.72- to 0.73-nm reflection regardless of saturation cation, humidity, or solvation. Well-crystallized kaolinite has a 001 spacing of 0.715 nm with sharp peaks. Some natural halloysites, prior to drying at 110°C, have a higher water content than kaolinite and show a strong reflection at 1.01 nm that collapses irreversibly to about 0.72 nm on drying. Reflections are also broader than for kaolinite. This is due to the tubular morphology of the mineral. Electron-optical observations may be a useful complementary tool in this particular situation. In the presence of chlorite and chloritized intergrades, possible confusion may arise between the second-order chlorite line and the first order of kaolinite. However, since kaolinite is transformed into metakaolin below 550°C, heat treatment at this temperature will indicate whether or not kaolinite is present. With poorly crystallized chlorites, heat treatments at temperatures lower than that for the dehydration of kaolinite may decompose these chlorites. Treatment with HCl is also useful in differentiating chlorites from kaolinites due to the more rapid dissolution of chlorites (Grim 1953, Ross 1967, 1969).

7   Interstratified clay minerals

a. Many examples of interstratified minerals have been described. They can be divided into two groups, regularly and randomly interstratified minerals. The two groups differ in the degree of regularity in the stacking of the successive layers. The presence of a long spacing (generally larger than 2.0 nm) mineral indicates that the mixed-layer mineral involved has some tendency towards regularity. Regularity can be evaluated from the degree of rationality based on observed basal spacings. In soils, many cases show the occurrence of mixed-layered minerals which have no specific long-spacing reflections. The randomly interstratified minerals are therefore discussed here.

b. Weaver (1956) documented the mixed-layer mica-montmorillonite minerals. Various spacings can be observed which are not necessarily either the sum or the arithmetic mean of the sum of the original spacings. When interstratification involves swelling minerals, the treatments (solvation or heat) will influence one of the constituents and thus modify the spacings of the interstratified minerals.

c. Most complexes involve mica (illite), smectite, vermiculite, and chlorite, either as binary or ternary mixtures. As an example we will consider the case of randomly interstratified clay minerals commonly found in soil clays, which belong to a binary system. The identification of the two layer components may be made from a cross examination of results from diagnostic tests. Some examples for cross examination are tabulated below (Kodama and Brydon 1968). When two spacings are given, they indicate the range within which a peak may be expected.

An approximate estimation of the mixing ratio of the two component layers may be made by an inspection of the positions of basal reflections, since diffraction theory indicates that the basal reflections of an interstratification migrate according to interference phenomena between layers of two similar types. One should, however, realize that the shifts of peak positions are not always linearly proportional to the mixing ratios. The direct Fourier transform method proposed by MacEwan may be applied not only to the binary system, but also to tertiary or more complicated systems. The method requires that structure factors of component layers be similar. With this so-called "situation method" (Sato et al. 1965), successful applications can provide useful information concerning types of component layers and modes of layer stacking sequence. If types of component layers are known and a large electronic computer is available, calculations for theoretical diffraction patterns of many types can be made according to a complete XRD theory (Brown 1961). Thus answers may be sought by a comparison between calculated and experimental diffraction patterns (Reynolds and Howes 1970, Sato et al. 1965).

Table 67.1 Identification of Two Layer Clay Components

Type of interstratification	Glycerol	Humid 100% RH	Air dry 50% RH	Dry 0% RH	550°C
Mica-vermiculite	X	X	1.0–1.4	1.0–1.15	1.0
Mica-chlorite	X	X	1.0–1.4	X	1.0–1.4
Mica-smectite	1.0 or 1.8	Expansion	1.0–1.55	1.0–1.15	1.0
Vermiculite-chlorite	X	X	1.4	1.15–1.4	1.0–1.4
Vermiculite-smectite	1.4–1.8	Expansion	1.4–1.54	1.15	1.0
Chlorite-smectite	1.4–1.8	Expansion	1.4–1.54	1.15–1.4	1.0–1.4

X indicates that the basal spacing is unchanged from air dry.

8 Other minerals

a. **Quartz** is a common constituent of soil specimens which gives a strong diffraction pattern with sharp reflections. The strongest reflection at 0.344 nm can be seen with less than 1% of quartz. As this line is confounded with the mica reflection at 0.333 nm, the presence of quartz should also be ascertained by the presence of the 0.426-nm reflection.

b. Feldspars may be common in the clay fraction and constitute sometimes a major part of some samples. All feldspars appear to have two or three strong reflections in the 0.328–0.316 nm region.

c. Amphiboles are often present in soils. Typical reflections occur in the 0.85- to 0.84-nm area, and they persist through the 700°C heat treatment.

d. Among the iron oxides, four may be found in soil clays. When present, they may give sharp reflections, the most intense of which are indicated here for the various forms:

goethite:	0.418 — 0.269 — 0.245 nm
hematite:	0.269 — 0.251 — 0.169
magnetite:	0.253 — 0.148 — 0.297
lepidocrocite:	0.627 — 0.329 — 0.247

See Norrish and Taylor (1962) for further information.

e. Gypsum has the most intense reflection at 0.756 nm, and less intense ones at 0.306 and 0.427 nm.

f. Among the carbonates, the most frequent ones are calcite and dolomite (see Chapter 69).

g. Short-range ordered minerals like imogolite and ferrihydrite, and also "noncrystalline" minerals such as allophane, opaline silica, and hisingerite may be present. Techniques such as infrared spectroscopy, electron microscopy, and differential XRD are useful for their characterization.

67.5.2 COMMENTS

1 An additional test for beidellite is based on the irreversible collapse of Li-saturated smectite when heated to 200–300°C.

2 The Li-saturated specimen is heated overnight at 200–300°C. It is then saturated with glycerol. A 001 spacing of 1.77 nm will indicate beidellite and a 0.95-nm spacing will indicate montmorillonite.

3 Low-charge Ca vermiculite may expand with ethylene glycol.

4 With K-saturation and 50% R.H., a line at 1.4 nm may still be present with vermiculite, particularly for low-charge vermiculites, derived from nonmicaceous minerals (Weaver 1958).

5 In soil, mainly under strong weathering conditions, smectite and vermiculite may form chloritic intergrades with aluminum hydroxide between the layers. The properties will be intermediate between chlorite and vermiculite or smectite. The predominant feature is the resistance to complete collapse upon K-saturation and heating, especially when the amount and stability of hydroxy interlayers increase. Samples toward the chlorite end of the spectrum (complete and well-organized hydroxy layers) may also exhibit resistance to expansion.

67.6 SEMIQUANTITATIVE ESTIMATION OF CLAY MINERALS

Quantitative analysis of the components in a clay mineral mixture may be desirable and useful in geochemical and pedological studies. The principle of the following method is similar to that of Johns et al. (1954) and of Sudo et al. (1961). The intensities of basal reflections are directly compared to the intensity ratios established from artificial mixtures of standard clay minerals. See also a more detailed treatment of quantitative clay mineral analysis by Kodama et al. (1977).

67.6.1 PROCEDURE

1   Determine analytical conditions for instruments:

    a. Radiation, filter, operational high-voltage, and X-ray tube current.

    b. Slit system, scanning speed of the goniometer.

    c. Electronic instrument for detector system.

    d. Electronic instrument for recording system.

2   Establish the intensity ratio of basal reflections of clay mineral components:

    a. Choose standard clay minerals with the same particle size range (probably in many cases, <2μm).

    b. Prepare quartz powder for sample dilution and calcite or fluorite powder for an internal standard (powder size should be the same as that of clay minerals).

    c. Prepare binary mixtures of clay minerals with equal weights of each mineral.

    d. Dilute the prepared binary mixtures with quartz powder in varius proportions (for example, 10 mg kaolinite and 10 mg mica and 5 mg quartz equals 25 mg, etc.).

    e. Add internal standard (add 5 mg to 25-mg mixture and make up 30 mg, for example).

    f. Prepare oriented aggregates on glass slides (37.5 × 25 mm) using a fixed amount of mixture (usually 20, 30, or 40 mg).

    g. Collect intensity data by measuring diffraction peak heights or areas for the standard, and normalize by the internal standard.

    h. Construct a calibration curve for each binary system to obtain intensity ratio (for example, S [smectite 1.5 nm]: M [mica 1.0 nm]: K [kaolinite 0.7 nm] equals 22.6:1.0:1.2) after comparison of results for various binary systems.

    i. If standard minerals with various crystallinities are available, mean intensity ratios may be obtained.

3   Collect intensity data for unknown samples.

    a. Carry out X-ray analysis of the samples under the conditions used for the standards.

b. Divide the intensity of basal reflections attributable to a component by the corresponding intensity ratio [for example, I (kaolinite)/1.2].

c. Estimate "relative amounts" of clay minerals in the system from the corrected intensity data (B).

d. In the case where a basal reflection belongs to more than one component, apply appropriate diagnostic tests and try to isolate the basal reflection of a single component. Standards should be subjected to the same diagnostic tests and subsequently analyzed.

67.6.2 Comments

Other semiquantitative methods (Klug and Alexander 1954) are the spike method and the direct method from mass absorption coefficient.

67.7 ADDITIONAL TREATMENTS

The behavior of soil clay minerals is not always as clear as that of reference minerals. A major problem, especially in soils with an acid reaction, is the presence of interlayer material, most commonly Al hydroxypolymers, that may prevent expansion of smectite and the collapse of vermiculite and smectite. The presence of interlayer material is often indicated by a collapse between 1.05 and 1.1 nm after heating at 550°C.

1   If more precision about the nature of clay minerals is required, interlayer material can be dissolved as follows (Tamura 1958).

a. Weigh 100 mg of clay into a 25-mL centrifuge tube.

b. Add 20 mL of 0.5 M Na citrate (pH 7.3) solution.

c. Heat in a water bath at 100°C for 1 h.

d. Centrifuge and discard the supernatant or collect it for Al analysis.

e. Repeat steps 2–4 twice.

f. Wash with water and freeze dry the clay.

g. Prepare Mg and K slides as described in Section 67.2 and proceed again with the XRD recording as indicated.

2   To differentiate between smectitic and vermiculitic clay minerals, it is recommended to heat a K-specimen slide at 300°C for 1 h, then solvate with glycerol. A smectitic clay would expand, whereas a vermiculitic clay would not.

3   It is sometimes desirable to remove chlorite to improve other mineral identification. Most chlorites are highly susceptible to acid dissolution, using M HCl and gentle heating.

4   Other treatments can also be considered that will assist in the proper identification of clay minerals (see Chapter 70):

a. Treatment with NaOH will enrich the proportion of iron oxide minerals.

b. Treatment with HCl will help to dissolve chlorite and biotite.

c. Treatment with Tiron will help remove amorphous materials.

d. Treatment with acid oxalate will dissolve imogolite and noncrystalline forms of Fe.

REFERENCES

Brindley, G. W. and Brown, G., Eds. 1980. Crystal structures of clay minerals and their X-ray identification. Mineralogical Society, London.

Brown, G. 1961. The X-ray identification and crystal structures of clay minerals. Mineralogical Society, London.

Brydon, J. E., Clark, J. S., and Osborne, V. 1961. Dioctahedral chlorite. Can. Mineral. 6: 595–609.

Grim, R. E. 1953. Clay mineralogy. McGraw-Hill, Toronto.

Harward, M. E. and Brindley, G. W. 1965. Swelling properties of synthetic smectites in relation to lattice substitution. Clays Clay Miner. 13th Conf. 2099–222.

Harward, M. E., Carstea, D. D., and Sayegh, A. H. 1969. Properties of vermiculite and smectites: expansion and collapse. Clays Clay Miner. 16: 437–447.

Johns, W. D., Grim, R. E., and Bradley, W. F. 1954. Quantitative estimations of clay minerals by diffraction methods. J. Sed. Petrol. 24: 424–251.

Klug, H. P. and Alexander, L. E. 1954. X-Ray diffraction procedures, John Wiley & Sons, New York.

Kodama, H. and Brydon, J. E. 1968. A study of clay minerals in podzol soils in New Brunswick. Clay Miner. 77: 295–309.

Kodama, H., Scott, G. C., and Miles, N. M. 1977. X-Ray quantitative analysis of minerals in soils. Soil Research Institute, Ottawa. 49 pp.

Miles, N. M. and De Kimpe, C. R. 1985. Application of glycerol/ethanol solutions for solvation of smectites dried on glass slides. Can. J. Soil Sci. 65: 229–232.

Norrish, K. and Taylor, R. M. 1962. Quantitative analysis by X-ray diffraction. Clay Miner. Bull. 5: 98–102.

Reynolds, R. C., Jr. and Howes, J. 1970. The nature of interlayering in mixed-layer illite-montmorillonite. Clays Clay Miner. 18: 25–36.

Ross, G. J. 1967. Kinetics of acid dissolution of an orthochlorite mineral. Can. J. Chem. 45: 3031–3034.

Ross, G. J. 1969. Acid dissolution of chlorites: release of magnesium, iron and aluminum and mode of acid attack. Clays Clay Miner. 17: 347–354.

Sato, M., Oinuma, K., and Kobayashi, K. 1965. Interstratified mineral of illite and montmorillonite. Nature 208: 179–180.

Sayegh, A. H., Harward, M. E., and Knox, E. G. 1965. Humidity and temperature interaction with respect to K-saturated expanding clay minerals. Am. Miner. 50: 490–495.

Sudo, T. and Sato, M. 1966. Dioctahedral chlorite. Proc. Int. Clay Conf. 1966. Jerusalem, pp. 33–39.

Sudo, T., Oinuma, K., and Kobayashi, K. 1961. Mineralogical problems concerning rapid clay mineral analysis of sedimentary rocks. Acta Univ. Carol. Geol. Suppl. 1: 189–219.

Tamura, T. 1958. Identification of clay minerals from acid soils. J. Soil Sci. 9: 141–147.

Weaver, C. E. 1956. Mixed layer clays in sedimentary rocks. Am. Mineral. 41: 202–221.

Weaver, C. E. 1958. Potassium fixation by expandable clay minerals. Am. Mineral. 43: 839–861.

Chapter 68
Sand Analysis

C. R. de Kimpe

Agriculture Canada
Ottawa, Ontario, Canada

68.1 INTRODUCTION

Sand mineralogy is determined for a number of reasons. The most common reasons for sand analysis are to assess the degree of weathering of soil layers and to assess the degree of uniformity of successive layers in a soil profile.

Sand analysis may also provide useful information about the status of nutrients readily or potentially available to plants in the soil. This may be particularly important in sandy soils with a low percentage of clay, and in a more general way in all acidic soils undergoing podzolization. Soil acidification generally results in the dissolution of ferromagnesian minerals and the release of various cations to the soil solution.

Sand analysis is a complementary analysis that is not commonly performed. This chapter is intended to provide only limited information about the available techniques, whereas textbooks on mineralogy will give more details about the different techniques that may be applied.

Techniques used for sand analysis differ from those used for clay-size particles. Furthermore, a number of analyses may be necessary to obtain an adequate identification. Most soils contain a large percentage of quartz and feldspars, and the abundance of these minerals makes difficult the determination of other primary minerals. Further separation within the sand fraction is therefore recommended.

The sand fraction covers the particle-size range from 2000 to 50 μm. Coarser grains are often aggregates of several minerals, sometimes covering a wide range of specific gravity, and therefore this may prevent a complete and appropriate separation. On the other hand, the very fine sand, <100 μm, is not convenient for several analyses, as the grains are too small to allow proper handling and identification on a single-grain basis. It is generally considered that the analysis of the 250- to 100-μm size fraction will provide the best information, as most of the grains in that size fraction are monomineralic.

Soil Sampling and Methods of Analysis, M. R. Carter, Ed.,
Canadian Society of Soil Science. © 1993 Lewis Publishers.

68.2 GRINDING AND SIEVING

68.2.1 MATERIALS

1   Sieves with openings of 250 and 100 μm.

2   Agate or corundum mortar, or wiggle bug.

68.2.2 PROCEDURE

1   Pass the sand fraction through a nest of sieves and save the 250- to 100-μm material for further analyses. Discard the very fine sand.

2   In an agate or corundum mortar, grind the grains, preferentially by hammering, in order to break the coarse grains. Grinding can also be done in a wiggle bug.

3   Repeat steps 1 and 2 in order to obtain a representative sample.

68.3 SEPARATION ACCORDING TO SPECIFIC GRAVITY

Separation of heavy minerals (sp. gr. >2.8) is carried out with one or several high-density liquids:

Bromoform	2.89 g/cc
S-Tetrabromoethane	2.96
Methylene iodide	3.30
Clerici solution	4.20
Molten lead salts	5.80

S-Tetrabromoethane is the most frequently used liquid.

68.3.1 MATERIALS

1   Separatory funnel.

2   Low-speed centrifuge, about 2000 rpm with adapter for separatory funnel.

3   Acetone.

4   Whatman® filter paper No. 42.

68.3.2 PROCEDURE

1　Pour 100 mL heavy liquid into a 250-mL separatory funnel.

2　Spread 10 g of sand (250- to 100-μm size fraction) on the liquid. Stopper the funnel.

3　Shake vigorously to wet all grains.

4　Let funnel stand to allow sedimentation to occur, or better, use a centrifuge to perform the separation (there is a special adapter for the separatory funnels).

5　Collect the heavy minerals in a beaker.

6　Repeat the separation.

7　Filter the solution on a Whatman® filter paper No. 42 to recover as much solution as possible (expensive material).

8　Wash the grains with acetone and let air dry.

Further separation may then be performed with other liquids if desired.

68.4 SEPARATION ACCORDING TO MAGNETIC SUSCEPTIBILITY

A further separation, especially useful for the dark, heavy minerals, is performed with the Frantz® isodynamic electromagnetic separator, a technique based on the magnetic susceptibility of the minerals (Parfenoff et al. 1970).

With this instrument, slope and current are adjusted in order to collect successively various fractions.

68.5 IDENTIFICATION OF THE MINERALS

Methods for identification are based on specific properties of the minerals:

1　Crystallographic properties. Single-crystal XRD examination using a camera.

2　Optical properties. Under the microscope, examination of the cleavage, fractures, color, refractive index, etc.

3　Physical properties. Cleavage, color, density, etc.

4　Chemical properties. Total chemical analysis, etc.

The light fraction can be analyzed microscopically and by X-ray diffraction (XRD). For the latter method, it is necessary to grind the material to a finer size more appropriate to the technique. Primary minerals do not require preferential orientation for the identification, and it is even better to ensure random orientation to obtain a maximum number of useful reflections. This is achieved in powder diffraction patterns.

The sand grains can also be examined under the microscope. A polished thin section is prepared by embedding enough sand grains in a resin on a glass slide, and the section is then polished to the desired thickness, suitable for microscopic examination.

A staining method (Reeder and McAllister 1957) was described for the quantitative determination of feldspars.

68.5.1 MATERIALS

1   Sodium cobaltinitrite solution. Dissolve 1 g sodium cobaltinitrite reagent in 4 mL distilled H_2O.

2   Hematein solution. Dissolve 0.05 g hematein in 100 mL of 95% ethanol.

3   Buffer solution. Dissolve 20 g sodium acetate in 100 mL of distilled H_2O, then add 6 mL glacial acetic acid and dilute the whole to 200 mL. The solution is approximately 0.5 N in acidity, and is buffered at pH 4.8.

4   48% hydrofluoric acid (HF).

5   Lead crucibles.

68.5.2 PROCEDURE

1   Minerals of specific gravity fraction <2.70 g/cc (quartz and feldspar group) are treated by direct contact with 48% HF in a lead crucible for 2 min.

2   Immediately dilute the acid with water and siphon off the supernatant liquid.

3   Spread the sample over the bottom of the crucible, one layer deep.

4   Add the sodium cobaltinitrite solution and allow to remain in contact for 1 min.

5   Wash the sample free of the solution by siphoning off the liquid.

6   Add ten drops of hematein solution, followed by five drops of buffer solution. Mix well by swirling the crucible for 2–3 min.

7   The solution is allowed to remain in contact for 5 min.

8   Wash the sample free of the solution with 95% ethanol and finally wash twice with acetone and air dry.

Table 68.1 Effect of Two Staining Reagents on Various Minerals

Mineral	Sodium cobaltinitrite	Hematein
Albite	None	Light purple
Andesine	None	Medium purple
Anorthite	None	Deep purple
Calcite	None	None
Chlorite	None	Greenish purple
Gypsum	None	None
Labradorite	None	Medium to deep purple
Microcline	Yellow	None
Oligoclase	None	Light to medium purple
Orthoclase	Yellow	None
Quartz	None	None

Minerals are then ready for microscope examination and grain counts. Table 68.1 shows the effect of differential staining of various minerals.

68.5.3 COMMENTS

1 Minerals of known species should be carried through the staining procedure to check the reagents and to act as standards.

2 Heavy minerals are most commonly ferromagnesian opaque minerals. The preferred method is therefore XRD analysis. Sample preparation again requires grinding to the size of very fine silt in order to produce enough grains with adequately oriented planes. Examination is performed on randomly oriented specimens. For primary minerals, it is necessary to record the XRD pattern over a wider range, up to 70° 2θ, in order to obtain a substantial number of reflections for each mineral.

3 For identification purpose, some information has been collected by Brindley and Brown (1980). However, it may be necessary to consult the ASTM file for more detail.

4 To assist in the identification of minerals from the sand fraction, total chemical analysis of the separates may be useful (see Chapter 70 on total chemical analysis).

REFERENCES

Brindley, G. W. and Brown, G., Eds. 1980. Crystal structures of clay minerals and their X-ray identification. Mineralogical Society, London.

Parfenoff, A., Pomerol, C., and Toureng, J. 1970. Pages 69–86 *in* Les minéraux en grains. Masson et Cie.

Reeder, S. W. and McAllister, A. L. 1957. A staining method for the quantitative determination of feldspars in rocks and sands from soils. Can. J. Soil. Sci. 37: 57–59.

Chapter 69
Identification and Measurement of Carbonate Minerals

R. J. St. Arnaud and A. R. Mermut

University of Saskatchewan

Saskatoon, Saskatchewan, Canada

Tee Boon Goh

University of Manitoba

Winnipeg, Manitoba, Canada

69.1 INTRODUCTION

A variety of methods can be used for the determination of calcite, dolomite, and magnesian calcite in soils. For more specialized research no one method is completely satisfactory and a combination of chemical, mineralogical and microscopic techniques are recommended. X-Ray diffraction (XRD) in combination with chemical methods (see Chapter 20) is useful for both qualitative and quantitative work. The microscopic examination of stained carbonate minerals permits *in situ* observation of carbonate minerals for studies in soil genesis.

69.2 X-RAY DIFFRACTION METHODS

Quantitative XRD techniques such as those proposed by Tennant and Berger (1957) and Weber and Smith (1961) provide a simple and reliable method for the detection and recognition of carbonate minerals in soils. Characteristics such as peak amplitude or area under the characteristic XRD peaks (Table 69.1) can provide a quantitative or semiquantitative measure of amounts present.

Such methods must be used with caution, since factors such as particle size, crystallinity, and chemical composition of the carbonate minerals and the nature of the soil matrix markedly affect the results obtained (Figure 69.1).

One of the most troublesome aspects of quantifying carbonate mineralogy by XRD lies in the selection of minerals to be used as standards for the comparisons required, since minerals from different sources can have widely different diffraction intensities. It is evident that estimates of calcite and dolomite using different, randomly selected standards (Figure 69.1) would yield widely divergent values. An accurate evaluation of the carbonate mineralogy

Soil Sampling and Methods of Analysis, M. R. Carter, Ed.,
Canadian Society of Soil Science. © 1993 Lewis Publishers.

Table 69.1 Diagnostic X-Ray Spacings for Common Ca, Mg, and Fe Carbonates Found in Soils. (Compiled from Doner and Lynn 1989.)

Mineral	Formula	X-ray spacings $d = nm\ (I/I_1)$
Calcite	$CaCO_3$	0.304 (100), 0.229 (18), 0.210 (18)
Aragonite	$CaCO_3$	0.340 (100), 0.198 (65), 0.327 (52)
Dolomite	$CaMg(CO_3)_2$	0.289 (100), 0.219 (30), 0.179 (30)
Mg-calcite	$Ca_{1-x}Mg_xCO_3$	0.304 to 0.297
Siderite	$FeCO_3$	0.279 (100), 0.173 (35), 0.174 (30)

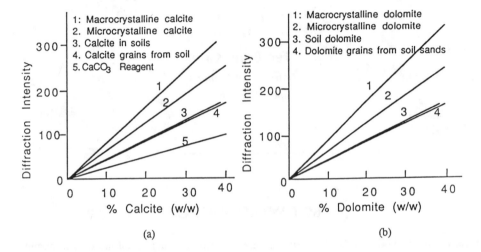

FIGURE 69.1. Relationship between XRD intensities (counts s^{-1}) and percentage calcite (a) and dolomite (b) using standard minerals indicated. The "soil" lines indicate this relationship using calcite and dolomite values determined for the total soil or soil separates by the citrate buffer method. Matrix for the standards consisted of carbonate-free soil (Bm horizon) developed from the same parent material. (Unpublished data, St. Arnaud.)

would require use of standards which have similar XRD intensities to the same minerals present in the soil being studied. More accurate results require techniques utilizing internal or external standards (Brindley and Brown 1980, Kodama et al. 1977) which meet this requirement or compensate for absorption caused by other components in the soil or material being analyzed.

A comparison of the results by the XRD method described below and the citrate buffer method described earlier (Section 20.5) can be observed in Figure 69.2.

The procedures outlined below provide a semiquantitative evaluation of the contents of calcite and dolomite in soils and also allow detection of magnesian calcites (St. Arnaud and Herbillon 1973) and estimation of their Mg contents by XRD technique.

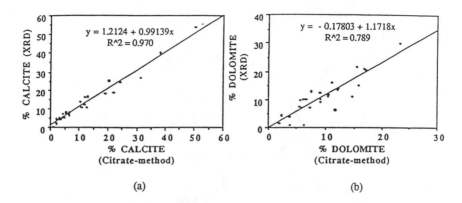

FIGURE 69.2. Comparison of calcite and dolomite content in soils as determined by two methods: (a) peak areas from X-ray diffractograms using soil carbonate minerals as standards and (b) citrate buffer extraction analysis. (Unpublished data, St. Arnaud.)

69.2.1 Sample Preparation

Soil samples, fractions, and standard minerals should be uniformly ground to obtain reproducible diffraction intensities. The samples should initially be crushed to pass a 300-mesh sieve, then ground in a motorized agate pestle and mortar for an additional 45 min.

69.2.2 Carbonate Mineral Identification

A preliminary XRD scan at a scanning rate of $1°$ min^{-1} using a spinner or well-type mount for the powdered sample is made to establish the presence of component carbonate minerals in the samples. Such identification is particularly important when the citrate-buffer method is used to quantify calcite and dolomite. Diagnostic peaks (Table 69.1) are used for identification. In the case of calcite and dolomite, a second, slower scan ($0.125°$ min^{-1}) over the range to include the distinctive peaks at 0.304 and 0.289 nm will verify the presence of calcite and dolomite, respectively, and provide suitable diffractograms for measuring peak heights and peak areas. A skewing or displacement of the calcite peak towards lower spacings will denote the presence of magnesian calcites (Figure 69.3); this feature may be completely missed unless scanned at the slower rate.

69.2.3 Quantification

As mentioned above, in order to obtain as accurate results as possible, it is desirable to obtain reference minerals as similar as possible to those contained in the soils to be analyzed. This could entail obtaining diffraction patterns on a series of samples of the desired minerals and selecting the one which most closely approaches the intensity of the same mineral from the bulk soil sample itself. Using staining techniques such as the alizarine red method (Section 69.3), it is possible to isolate and separate by hand, grains of calcite and dolomite from the sand or gravel fraction of the soil studied; these can then serve as standards. After the grains are selected by staining, they should be ground and their purity established by XRD before combining them as the mineral standards.

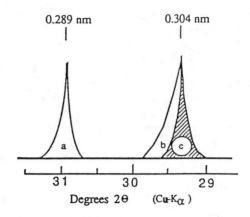

FIGURE 69.3. Typical diffractogram used to identify dolomite (a), magnesian calcite (b), and pure calcite (c) in soils or soil materials.

69.2.4 Matrix for Standard Minerals

Mineral standards should be mixed with material which has similar X-ray absorption properties to those of the samples. Soil material from a noncalcareous horizon from the profile studied generally proves satisfactory; alternatively, when dealing with calcareous materials, removal of all carbonates with an acid treatment will provide a suitable matrix material. Variable amounts of the calcite and dolomite standards mixed with the matrix selected can serve as a basis for the quantification of these two minerals by measurement and comparison of peak areas or intensities (peak heights or counts on the peaks). A transparency such as that depicted in Figure 69.4 based on the relationship established by Goldsmith and Graf (1958) can be used to estimate the % $MgCO_3$ in the calcite from its diffractograms; % $MgCO_3$ should be read at the one-third height of the curve.

69.3 STAINING OF THIN SECTIONS OR POLISHED BLOCKS

Initial quantification of carbonates in thin sections was first achieved by point counting (Chayes 1949). Recently, Protz et al. (1987) suggested that carbonates could be studied by image analysis.

Staining methods have been used by sedimentary petrologists to aid identification of carbonate mineral species in thin sections (Friedman 1959). Recently, Bui and Mermut (1989) attempted to combine staining with video image analysis to identify and quantify carbonates (see Chapter 65, Section 65.9.1). With this technique, for example, it is now possible to determine the size, shape, and distribution of individual soil carbonates by image analysis.

The staining procedure suggested by Friedman (1959) is given in Figure 69.5. This has proven to be successful for thin sections and polished blocks. The calcite is stained deep red with alizarine red S dye and can be easily recognized and separated from aragonite, dolomite, magnesite, high-magnesian calcite, anhydrite, and gypsum.

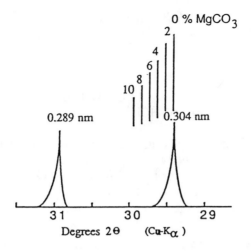

FIGURE 69.4. Transparent overlay used to estimate the % $MgCO_3$ in magnesian calcites. For skewed curves such as shown in Figure 69.3, use the midpoint of the curve at one-third height.

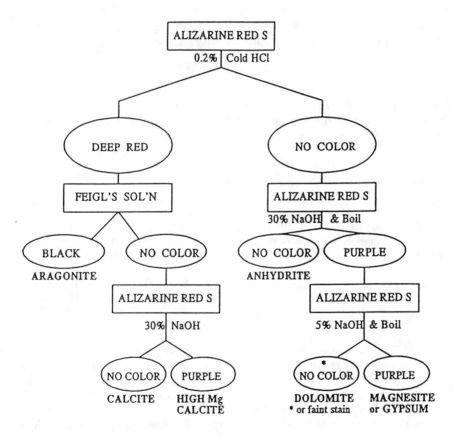

FIGURE 69.5. Staining procedure using alizarine red S and Feigl's solution. (From Friedman, G. M., *J. Sediment. Petrol.*, 29, 87, 1959. With permission.)

69.3.1 REAGENTS

1   10% (v/v) HCl.

2   Alizarine red S staining solution: 1 L of distilled water containing 1 g alizarine red S and 2 mL concentrated HCl.

3   Feigl's solution: 1 g of Ag_2SO_4 is added to a solution of 11.8 g $MnSO_4 \cdot 7H_2O$ dissolved in 100 mL distilled water and boiled. After cooling, the suspension is filtered, and one or two drops of 2 mol L^{-1} NaOH solution is added. The precipitate is filtered off after 1–2 h.

69.3.2 Procedure

The surface of the uncovered thin section or block is first etched by immersing the surface upwards in 10% HCl about a minute (see Chapter 65 for thin section preparation). The etched surface is then washed with distilled water and immersed in a warm (40°C) alizarine red S solution in a petri dish for 4 min, occasionally swirling the dish. The stained surface is then washed with distilled water. The slide, after drying, is ready for petrographic examination or for quantitative determination by image analysis (Chapter 65).

69.3.3 Comments

Katz and Friedman (1965) suggest a further treatment of the stained surface by acetate peel. The analysis should be performed immediately, as the stain may fade with time. For permanent storage, thin sections are covered with a cover glass.

REFERENCES

Brindley, G. W. and Brown, G. 1980. Crystal structures of clay minerals and their x-ray identification. Mineralogical Society, London.

Bui, E. N. and Mermut, A. R. 1989. Quantification of soil calcium carbonates by staining and image analysis. Can. J. Soil Sci. 69: 677–682.

Chayes, F. 1949. A simple point counter for thin section analysis. Am. Mineral. 34: 1–11.

Doner, H. E. and Lynn, W. C. 1989. Carbonate, halide, sulfate and sulfide minerals. Pages 279–330 in J. B. Dixon and S. B. Weed, Eds. Minerals in soil environments, 2nd. ed. Soil Science Society of America, Madison, WI.

Friedman, G. M. 1959. Identification of carbonate minerals by staining methods. J. Sediment. Petrol. 29: 87–97.

Goldsmith, J. R. and Graf, D. L. 1958. Relation between lattice constants and composition of the Ca-Mg carbonates. Am. Mineral. 43: 84–101.

Katz, A. and Friedman, G. M. 1965. The preparation of stained acetate peels for the study of carbonate rocks. J. Sediment. Petrol. 35: 248–249.

Kodama, H., Scott, G. C., and Miles, N. M. 1977. X-Ray quantitative analysis of minerals in soils. 49 pp. Soil Research Institute, Research Branch, Agriculture Canada, Ottawa.

Protz, R., Shipitalo, M. J., Mermut, A. R., and Fox, C. A. 1987. Image analysis of soils, present and future. Geoderma 40: 115–125.

St. Arnaud, R. J. and Herbillon, A. J. 1973. Occurrence and genesis of secondary magnesium-bearing calcites in soils. Geoderma 9: 279–298.

Tennant, C. B. and Berger, R. W. 1957. X-Ray determination of dolomite-calcite ratio of carbonate rock. Am. Mineral. 42: 23–29.

Weber, J. N. and Smith, F. G. 1961. Rapid determination of calcite-dolomite ratio in sedimentary rocks. J. Sediment. Petrol. 31: 130–132.

Chapter 70
Chemical Methods in
Mineralogical Analysis

G. J. Ross

Agriculture Canada
Ottawa, Ontario, Canada

70.1 INTRODUCTION

Chemical methods are useful in support of instrumental techniques in the mineralogical analysis of soils. Noncrystalline material is commonly intimately associated with more crystalline minerals. Thus, selective dissolution methods are needed to remove the noncrystalline material and clean up the sample for mineralogical analysis by X-ray diffraction or other physical methods. Furthermore, chemical analysis of dissolution extracts and comparisons of results for X-ray diffraction and other analyses obtained before and after dissolution treatments are helpful in determining the nature and amounts of noncrystalline material in a sample. Chemical analysis for characteristic properties of certain minerals (e.g., K-fixation capacity of vermiculite) as well as total and partial elemental analysis also assist in the determination of mineralogical composition.

70.2 TIRON METHOD FOR REMOVAL AND CHARACTERIZATION OF INORGANIC SOIL COMPONENTS
(Biermans and Baert 1977; Kodama and Ross 1991)

The Tiron method is an alkaline and complexing method that dissolves clay-size, poorly crystalline and noncrystalline components effectively without significantly attacking associated crystalline minerals (Kodama and Ross 1991). They found that Tiron was as effective as acid oxalate in the dissolution of allophane, imogolite, and poorly crystalline hydrous iron oxides. In contrast to acid oxalate, Tiron removed opaline silica and dissolved virtually no magnetite. A simple treatment with Tiron effectively cleaned up clay samples for X-ray diffraction of crystalline minerals with amounts of extracted Si, Al, and Fe indicating the quantities of poorly crystalline and noncrystalline aluminosilicates as well as hydrous oxides of Fe, Al, and Si.

70.2.1 REAGENTS

1   Dissolve 31.42 g of Tiron, 4,5-dihydroxy-1,3-benzene-disulfonic acid (diso-dium salt), $C_6H_4Na_2O_8S_2$ (Sigma® No. D-7389) in approximately 800 mL of H_2O in a plastic beaker.

2   Prepare 100 mL Na_2CO_3 solution containing 5.3 g Na_2CO_3 in a plastic beaker and add this solution to the Tiron solution while stirring. The pH rises to about 10.2 and the solution turns greenish.

3   Add 4 *M* NaOH in 0.1- to 0.2-mL increments until the pH is 10.5.

4   Add H_2O to a total volume of 1 L in a polypropylene flask and store in the refrigerator. The Tiron solution can be stored up to 3 months without significant loss of extraction capacity.

5   Before using the solution, check the pH and adjust to 10.5, if required.

70.2.2 PROCEDURE

1   Weigh 30 mg of sample in a 50-mL polypropylene centrifuge tube. In general, the sample will have been pretreated for organic matter removal.

2   Add 30 mL of the Tiron solution to the tube and weigh. Larger samples may be used in the same solid-to-solution ratio in larger containers.

3   Cover the tube loosely with a polypropylene cap and place it in a water bath at 80°C for 1 h. Stir the contents every 10 to 12 min.

4   After extraction cool the tube in a cold-water bath, weigh, and replace lost weight by adding H_2O.

5   Mix the contents and centrifuge for 10 min at 4000 × *g*.

6   Siphon off most of the supernatant (25 mL) for analysis of extracted Si, Al, and Fe by atomic absorption spectrophotometry (AAS). Use the same Tiron matrix in sample and standard solutions.

7   For cation saturation of the residue for X-ray diffraction analysis, follow the procedure in Chapter 67.

Samples extracted with alkaline solutions, such as Tiron, may be affected by hydrolysis reactions during subsequent cation saturation. This can be avoided by using slightly acid (pH ~6.5) buffered solutions. For example, for Mg saturation a half-and-half $MgCl_2$ and $Mg(OAc)_2$ solution could be used.

70.2.3 CALCULATIONS

1
$$\text{g kg}^{-1} \text{ Fe, Al, Si} = \frac{\mu g/mL \text{ in final sol'n} \times \text{extractant (mL)} \times \text{dil.}}{\text{sample wt (mg)}} \quad (70.1)$$

2 For example, for 0.030 g of sample, 30 mL of extractant, $5\times$ dilution and 8 μg/mL determined:

$$\text{g kg}^{-1} \text{ Fe in sample} = \frac{8 \times 30 \times 5}{30} = 40$$

To obtain a rough estimate of the amounts of poorly crystalline and noncrystalline material extracted, multiply each element by the factor required for its oxide form or for the form in which it is thought to be present, and sum the resulting amounts.

70.3 SODIUM HYDROXIDE METHOD FOR REMOVAL AND CHARACTERIZATION OF INORGANIC SOIL COMPONENTS (Hashimoto and Jackson 1960; Jackson et al. 1986)

In this method the sample is boiled for 2.5 min in 0.5 M NaOH (1 mg sample/mL solution) to dissolve poorly crystalline and noncrystalline aluminosilicate components. In addition to removing the noncrystalline aluminosilicates, however, the boiling alkaline solution also attacks relatively crystalline aluminosilicate minerals, particularly smectite (Kodama and Ross 1991). The much milder, but equally effective Tiron method is therefore preferred in most mineralogical studies. The NaOH method is described in detail by Jackson et al. (1986).

70.4 WEIGHT-LOSS METHOD FOR ESTIMATING AMOUNTS OF NONCRYSTALLINE INORGANIC SOIL COMPONENTS (Hodges and Zelazny 1980; Jackson et al. 1986)

As was shown in Section 70.2.3, a rough estimate of the amount of noncrystalline material in a sample can be obtained by converting the Al, Fe, and Si extracted by Tiron (or other dissolution methods, such as oxalate) to their respective oxides and adding the oxides. This calculation omits the water contents of the oxides because noncrystalline materials may have widely variable water contents. A less laborious method that provides an acceptable estimate of noncrystalline material is to determine the weight loss of the sample after the selective dissolution treatment and assign this weight loss to noncrystalline material. If required, the extracted elements can also be determined in this method.

70.4.1 PROCEDURE

1 For the removal of noncrystalline material by the oxalate method or the Tiron method, follow the respective procedures in Sections 25.3 and 70.2.

2 Before the dissolution treatment dry the sample (preferably initially freeze dried) in the tube at 110°C overnight.

3   Cool the tube and contents in a P_2O_5 vacuum desiccator and weigh to the nearest 0.01 mg. Alternatively, apply steps 2 and 3 to another sample of the same material to correct the water content.

4   After the dissolution treatment add 25 mL of 0.5 M $(NH_4)_2CO_3$ to the residues in the tube, centrifuge, and discard the supernatant. Repeat this step twice and then wash once with 25 mL H_2O.

5   Dry the residues in the tube overnight at 110°C to volatilize excess $(NH_4)_2CO_3$ as NH_3, CO_2, and H_2O.

6   Cool the residues in the tube in a P_2O_5 vacuum desiccator and weigh to the nearest 0.01 mg.

70.4.2 CALCULATIONS

1   g kg^{-1} of noncrystalline material $=$

$$\frac{(\text{110°C sample wt. before diss. tr.} - \text{110°C sample wt. after diss. tr.})\ 1000}{\text{110°C sample wt. before diss. treatment}} \qquad (70.2)$$

2   If the extracted Fe, Al, and Si were determined and converted to their respective oxides, the amount of water associated with the noncrystalline material can be calculated as follows:

g kg^{-1} H_2O of noncrystalline material $=$

$$\frac{(\text{g } kg^{-1} \text{ noncrystalline material} - \text{g } kg^{-1} \text{ oxide sum}) \times 1000}{\text{g } kg^{-1} \text{ noncrystalline material}} \qquad (70.3)$$

70.4.3 Comments

Because the results are calculated from relatively small differences in sample weights, extra care should be taken in the weighing and centrifugation steps. Duplicate or triplicate samples may be analyzed to assure precision.

70.5 OTHER METHODS FOR REMOVAL AND CHARACTERIZATION OF INORGANIC SOIL COMPONENTS (Jackson et al. 1986)

As was indicated in Chapter 25, the dithionite, oxalate, and hydroxylamine methods are useful as selective dissolution procedures in the mineralogical analysis of clays or whole soil samples. The amounts of crystalline iron oxides, commonly goethite, lepidocrocite, and hematite, in samples can be estimated from the difference in the amounts of Fe extracted by the dithionite and oxalate methods (Schwertmann 1959). Assuming that the very poorly crystalline iron oxides are mainly ferrihydrite, the amount of ferrihydrite can be estimated from the amount of Fe extracted ($\times 1.7$) by oxalate (Parfitt and Childs 1988). The poorly crystalline aluminosilicates, allophane and imogolite, are also extracted by oxalate and amounts of these minerals may be estimated by this method (Parfitt and Wilson 1985). In contrast to oxalate, hydroxylamine does not dissolve magnetite. The hydroxylamine method (or the Tiron method) is therefore preferred for samples that may contain this iron oxide mineral. Tiron effectively extracts opaline silica, whereas oxalate dissolves little or none of this material. The amount of opaline silica may therefore be estimated from the difference in Si extracted by these two methods (Kodama and Ross 1991).

70.6 METHOD FOR VERMICULITE QUANTIFICATION BASED ON POTASSIUM FIXATION
(Alexiades and Jackson 1965; Jackson et al. 1986)

Vermiculite fixes K against replacement by NH_4, and this property is used to estimate the amounts of vermiculite in clays and soils. Despite the overestimation of amounts of actual vermiculite present in some cases (Ross and Kodama 1987), the method is useful and has gained wide acceptance. The amount of vermiculite is calculated from the difference between Ca exchange capacity (CaEC) and K exchange capacity (K/EC) after heating the K-saturated sample at 110°C. Because the calculations are based on this difference, the precision of the results is limited ($\sim \pm 10\%$).

70.6.1 REAGENTS

1 Prepare a 0.25-*M* calcium chloride ($CaCl_2$) solution (27.8 g/L) and a 0.005-*M* $CaCl_2$ solution (dilute 20 mL of 0.25 *M* $CaCl_2$ to 1 L). Determine the actual Ca concentration in the diluted solution by atomic absorption.

2 Prepare a 0.25-*M* magnesium chloride ($MgCl_2$) solution (23.8 g/L)

3 Prepare a 0.50-*M* potassium chloride (KCl) solution (37.3 g/L). Also make up a 0.02-*M* KCl solution in 80% (v/v) methanol (1.49 g/L). The actual K concentration of this solution should be known.

4 Prepare a 0.50-*M* ammonium chloride (NH_4Cl) solution (26.7 g/L).

70.6.2 PROCEDURE

1 Weigh 100 mg of sample (200 mg of low-CaEC material) and transfer to a weighed (to the nearest 0.1 mg) 15-mL centrifuge tube. In most cases the sample will have been pretreated to remove organic matter. Samples containing free carbonates and considerable amounts of amorphous materials and iron oxides should also have been pretreated to remove these components.

2 Add 5 mL of 0.25 *M* $CaCl_2$ to the sample. Mix thoroughly with an ultrasonifier or a vortex mixer. Centrifuge at 510 × *g* or higher for 10 min, decant, and discard the supernatant.

3   Do step 2 three times using 0.25 M $CaCl_2$. Then do step 2 five times using 0.005 M $CaCl_2$. Wipe the tube externally and weigh the tube with the sample and entrained liquid to the nearest mg.

4   Add 5 mL of 0.25 M $MgCl_2$ to the sample. Mix thoroughly and centrifuge as before. Do step 4 five times. Collect the supernatant in a 50-mL flask and fill the flasks to 50 mL with H_2O. Measure Ca in the flask by AAS.

5   Add 5 mL of 0.50 M KCl to the sample, mix well, centrifuge, and discard the supernatant. Do this five times using 0.50 M KCl, and then five times using 0.02 M KCl in 80% methanol. Wipe the tube externally and weigh the tube with the sample and entrained KCl to the nearest mg.

6   Dry the tube at 110°C overnight and weigh to the nearest 0.1 mg.

7   Add 10 mL of 0.50 M NH_4Cl. Mix thoroughly, centrifuge, and collect the supernatant in a 50-mL flask. Do this five times and fill the flask to 50 mL with 0.50 M NH_4Cl. Measure K in the flask by AAS.

8   For determination of Ca and K follow standard atomic absorption procedures. Take into account the lower density of the entrained 80% methanol solution, as compared to H_2O, when converting weight to volume or vice versa.

70.6.3 CALCULATIONS

1   The CaEC and the K/EC after drying at 110°C are calculated and the difference ($\times 1000$) is divided by the CEC of vermiculite (154) to obtain the amount of vermiculite in g kg^{-1}. The entrained cations are subtracted in these calculations.

2   For example, for 0.100 g of sample, 50 mL of solution and 24 µg Ca/mL determined:

$$\text{CaEC (total)} = \frac{24 \text{ µg/mL} \times 50 \text{ mL}}{0.100 \text{ g}} = \frac{1200 \text{ µg}}{0.100 \text{ g}}$$

$$= \frac{1200 \text{ mg}}{100 \text{ g}} - \frac{60 \text{ meq}}{100 \text{ g}} = 60 \text{ cmol}_c \text{ kg}^{-1}*$$

If 0.300 g (or mL) of entrained $CaCl_2$ solution was present that contained 0.200 mg Ca/mL, the CaEC (total) is reduced by:

$$\frac{0.200 \text{ mg/mL} \times 0.300 \text{ mL}}{0.100 \text{ g}} = \frac{0.060 \text{ mg}}{0.100 \text{ g}} = \frac{600 \text{ mg}}{1000 \text{ g}} = 3 \text{ cmol}_c \text{ kg}^{-1}$$

CaEC = 60 − 3 = 57 cmol$_c$ kg^{-1}

* The units for CEC of meq 100 g^{-1} are equivalent to the SI units of cmol$_c$ kg^{-1}.

3   For 32 μg K/mL determined:

$$K/EC \text{ (total)} = \frac{32 \text{ μg/mL} \times 50 \text{ mL}}{0.100 \text{ g}} = \frac{1600 \text{ μg}}{0.100 \text{ g}} = \frac{16000 \text{ mg}}{1000 \text{ g}} = 41 \text{ cmol}_c \text{ kg}^{-1}$$

If 0.280 g of entrained KCl in 80% methanol (specific gravity 0.83) was present that contained 0.710 mg K/mL, the K/EC (total) is reduced by:

$$\frac{0.710 \text{ mg/mL} \times 0.280 \text{ g}}{0.100 \text{ g} \times 0.83 \text{ g/mL}} = \frac{0.240 \text{ mg}}{0.100 \text{ g}} = \frac{2400 \text{ mg}}{1000 \text{ g}} = 6 \text{ cmol}_c \text{ kg}^{-1}$$

$$K/EC = 41 - 6 = 35 \text{ cmol}_c \text{ kg}^{-1}$$

4    $$\text{g kg}^{-1} \text{ of vermiculite} = \frac{(CaEC - K/EC) \, 1000}{154} = \frac{(57 - 35) \, 1000}{154} = \frac{22000}{154} = 143$$

5   According to Jackson et al. (1986), the amount of smectite also may be calculated from the K/EC, from which 5 cmol kg^{-1} is subtracted for the K/EC of external surfaces. Thus:

$$\text{g kg}^{-1} \text{ smectite} = \frac{(K/EC - 5) \, 1000}{105}$$

in which 105 represents the interlayer CEC for various smectites. For the sample used in these calculations the smectite content is

$$\frac{(35 - 5) \, 1000}{105} = 286 \text{ g kg}^{-1}$$

70.7 METHOD FOR VERMICULITE QUANTIFICATION BASED ON RUBIDIUM FIXATION (Ross et al. 1989b)

As was noted in the previous section, the precision of the vermiculite contents obtained by the K fixation method is limited because the results are based on differences between cation exchange capacities. This method was therefore modified to measure the relatively small changes of vermiculite content in soils that may result, for example, from K and NH_4 fertilization (Ross et al. 1985, 1989a) and mica weathering (Protz et al. 1988). In the modified procedure, the exchangeable Rb of a Rb-saturated and heated sample is replaced by NH_4 and the remaining fixed Rb is determined directly by dissolving the sample. Compared to the K fixation method, the Rb fixation method gave similar, but much more precise results, as indicated by its 4.5 times smaller variance (Ross et al. 1989b).

70.7.1 REAGENTS

1   Prepare a 0.50-*M* rubidium chloride (RbCl) solution (60.46 g/L).

2   Prepare a 0.50-*M* ammonium chloride (NH_4Cl) solution (26.7 g/L).

70.7.2 PROCEDURE

The procedure is similar to the K fixation procedure (Section 70.6.2), except that the CaEC determination is omitted (steps 2, 3, and 4) and the remaining steps are modified.

1   Weigh 100 mg of sample (200 mg of low-CEC material) and transfer to a weighed (to the nearest 0.1 mg) 15-mL centrifuge tube. If necessary, the sample will have been pretreated to remove organic matter, free carbonates, amorphous materials, and iron oxides.

2   Add 5 mL of 0.50 M RbCl to the sample. Mix thoroughly with an ultrasonifier or a vortex mixer. Centrifuge at 510 $\times$ g or higher for 10 min decant, and discard the supernatant. Do step 2 three times.

3   Add 10 mL 80% methanol, mix thoroughly, centrifuge, and discard the supernatant. One washing is sufficient, since complete removal of excess salt is not necessary.

4   Dry the tube at 110°C overnight and weigh to the nearest 0.1 mg.

5   Add 10 mL of 0.50 M NH$_4$Cl. Mix thoroughly, centrifuge, and discard the supernatant. Do step 5 three times.

6   For sample decomposition, transfer the sample to a 100-mL Teflon® beaker, using a minimum of water.

7   Dry the sample and add 10 mL concentrated HNO$_3$, cover, boil gently for 30 min on a hot plate at 100 to 150°C, and cool

8   **In a perchloric acid fumehood,** add 10 mL concentrated HClO$_4$, cover, boil gently for 30 min on a hot plate at 230°C, cool, and remove covers.

9   Add 10 mL concentrated HF, heat in a perchloric acid fumehood for 1 h at 80°C, and gradually increase heat to intense white fumes and take nearly to dryness at 250°C, but do not dry completely.

10 Cool and wash down walls of the beaker with 25 mL 1 M HCl or 1 M HNO$_3$.

11 Cover and bring to a boil to dissolve the residue, cool, and make up to 50 mL in a volumetric flask.

12 For determination of Rb follow standard AAS procedures. At this time other constituent cations of the dissolved sample can also be determined to quantify other minerals. For example, amounts of mica may be estimated from the K content of the sample, taking 100 g kg^{-1} as the K$_2$O content of mica (Jackson 1964). Other decomposition procedures that may be used instead of the procedure given here in steps 6 to 11 have been described by Lim and Jackson (1982).

70.7.3 CALCULATIONS

1   The amount of vermiculite in g kg^{-1} is calculated by dividing the cmol$_c$ kg^{-1} of Rb ($\times$1000) in the sample by the CEC of vermiculite (154). For example, for 0.100 g of sample, 50 mL of solution and 34 µg Rb/mL determined:

$$\frac{34 \ \mu g/mL \times 50 \ mL}{0.100 \ g} = \frac{1700 \ \mu g}{0.100 \ g} = \frac{17000 \ mg}{1000 \ g} = 20 \ cmol \ kg^{-1} \ of \ Rb \ in \ the \ sample$$

$$g \ kg^{-1} \ of \ vermiculite = \frac{20 \times 1000}{154} = 130$$

70.7.4 Comments

If required, the native Rb content of the sample may be differentiated from its fixed Rb content by determining the Rb content of a duplicate, untreated sample and subtracting it from the Rb-treated sample. The native Rb contents of the soil clays analyzed by Ross et al. (1989b) were less than 0.50 cmol$_c$ kg^{-1}.

70.8 TOTAL AND PARTIAL ELEMENTAL ANALYSIS

Total elemental analysis of a largely monomineralic sample is sometimes useful to determine the chemical structure of the mineral in the sample. For soil clays, which are rarely monomineralic, partial elemental analysis is more appropriate. For example, the Rb fixation procedure, mentioned above, described the estimation of amounts of vermiculite and mica in a sample from its Rb and K contents. Saturation of a sample with Rb or Sr, which are generally not significant constituent cations of clay minerals, and subsequent analysis for Rb and Sr of the dissolved sample provide much more precise CEC results than those obtained by conventional CEC procedures. Dissolution and isolation procedures for the semiquantitative analysis of nonphyllosilicate minerals, such as quartz, feldspars, rutile, and anatase, have been described by Jackson et al. (1986).

REFERENCES

Alexiades, C. A. and Jackson, M. L. 1965. Quantitative determination of vermiculite in soils. Soil Sci. Am. Proc. 29: 522–527.

Biermans, V. and Baert, L. 1977. Selective extraction of the amorphous Al, Fe and Si oxides using an alkaline Tiron solution. Clay Miner. 12: 127–134.

Hashimoto, L. and Jackson, M. L. 1960. Rapid dissolution of allophane and kaolinite-halloysite after dehydration. Clays Clay Miner. 7: 102–113.

Hodges, S. C. and Zelazny, L. W. 1980. Determination of noncrystalline soil components by weight difference after selective dissolution. Clays Clay Miner. 28: 35–42.

Jackson, M. L. 1964. Soil clay mineralogical analysis. Pages 245–294. in C. I. Rich and G. W. Kunze, Eds. Soil clay mineralogy. University of North Carolina Press, Chapel Hill, NC.

Jackson, M. L., Lim, C. H., and Zelazny, L. W. 1986. Oxides, hydroxides, and aluminosilicates. Pages 101–150 in A. Klute, Ed. Methods of soils analysis. Part 2. 2nd ed. Agronomy No. 9, American Society of Agronomy, Madison, WI.

Kodama, H. and Ross, G. J. 1991. Tiron dissolution method used to remove and characterize inorganic components in soils. Soil Sci. Soc. Am. J. 55: 1180–1187.

Lim, C. H. and Jackson, M. L. 1982. Dissolution for total elemental analysis. Pages 1–12 in A. L. Page et al., Eds. Methods of soil analysis. Part 2. 2nd ed. Agronomy No. 9, American Society of Agronomy, Madison, WI.

Parfitt, R. L. and Childs, C. W. 1988. Estimation of forms of Fe and Al: a review and analysis of contrasting soils by dissolution and Mössbauer methods. Aust. J. Soil Res. 26: 121–144.

Parfitt, R. L. and Wilson, A. D. 1985. Estimation of allophane and halloysite in three sequences of volcanic soils, N.Z. Catena Suppl. 7: 1–8.

Protz, R., Ross, G. J., Shipitalo, M. J., and Terasmae, J. 1988. Podzolic soil development in the southern James Bay Lowlands, Ontario. Can. J. Soil Sci. 68: 287–305.

Ross, G. J. and Kodama, H. 1987. Layer charge characteristics of expandable clays from soils. Pages 355–370 in J. B. Dixon, Ed. New developments in soil mineralogy. Trans. Commun. on Soil Miner., 13th Congr. Int. Soc. Soil Sci., Hamburg, Germany. 13–20 Aug. 1986. Int. Soil Refer. and Information Ctr., Wageningen, The Netherlands.

Ross, G. J., Cline, R. A., and Gamble, D. S. 1989a. Potassium exchange and fixation in some southern Ontario soils. Can. J. Soil Sci. 69: 649–661.

Ross, G. J., Phillips, P. A., and Culley, J. L. B. 1985. Transformation of vermiculite to pedogenic mica by fixation of potassium and ammonium in a six-year field manure application experiment. Can. J. Soil Sci. 65: 599–603.

Ross, G. J., Schuppli, P. A., and Wang, C. 1989b. Quantitative determination of vermiculite by a rubidium fixation method. Soil Sci. Soc. Am. J. 53: 1588–1589.

Schwertmann, U. 1959. Die fractionierte Extraction der freien Eisenoxide in Böden, ihre mineralogischen Formen und ihre Entstehungsweisen. Z. Pflanzenernaehr. Dueng. Bodenkd. 84: 194–204.

Analysis of Frozen Soils

Chapter 71
Sampling Frozen Soils

Charles Tarnocai

Agriculture Canada
Ottawa, Ontario, Canada

71.1 INTRODUCTION

Considerable work has been carried out on perennially frozen soils in northern Canada during the last 20 years. As a result of this work, it became evident that the perennially frozen part of the soil, the permafrost, is a very important component of the soil environment. The examination and sampling of these soils must therefore incude not only the active layer (the layer which annually freezes and thaws), but also the perennially frozen layer. In order to do this, several types of equipment are used for both frozen mineral and organic soils. This section provides descriptions and illustrations of the sampling equipment which has been used in the pedological investigation of frozen soils. Some of this equipment is used for excavating a trench in the frozen soil to expose the soil profile for description and sampling. The trenching is usually carried out to depths of 1 or $1\frac{1}{2}$ m. Coring equipment is used to sample frozen soils to greater depths, usually between 1 and 5 m. Core samplers are used to recover a core for soil characterization and for collection of ice content and bulk density samples.

71.2 EQUIPMENT USED FOR CORING

71.2.1 Modified Hoffer Probe

The modified Hoffer probe (Figure 71.1) is used mainly for coring frozen peat and stone-free, frozen, fine-textured material. This probe was modified and described by Brown (1968). It was later further modified as described by Zoltai (1978) to eliminate structural weaknesses.

The probe consists of a serrated steel coring bit, five 1-m-long extension rods, and the T-handle. The bit and extension rods are made of heavy-gauge, seamless, steel tubing. The cutting edge of the bit is serrated and hardened. The dimensions of this probe are shown in Figure 71.2.

With this tool it is possible not only to determine the depth of the peat deposit and the thickness of the various peat layers, but also to obtain samples for physical and chemical analysis and for ice content and bulk density determinations. With the addition of the extension rods, the probe can penetrate to a maximum depth of 5 m; operating the auger at greater depths is very difficult.

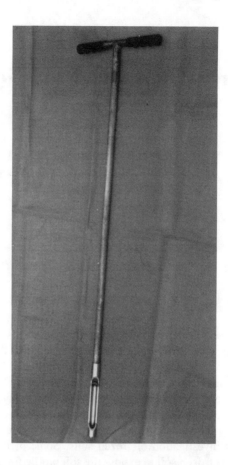

FIGURE 71.1. Modified Hoffer probe with one extension rod.

This probe is usually operated by one or two people. If two people operate the probe, one person usually holds the T-handle while the other holds the rod. A sample is obtained by ramming the probe down and turning it. This causes the coring bit to cut out a core sample 2 cm in diameter and 14 cm in length. The coring bit usually fills up within three or four ramming and turning motions and is pulled up to retrieve the sample. This ramming movement causes a slight amount of shattering and compaction, and this has to be accounted for when ice content and bulk density are determined.

The modified Hoffer probe is an excellent sampling tool because of its light weight and reliability. It is most successfully used for sampling fine-textured mineral and peat materials in areas where soil temperatures are approximately 0 to −4°C. Veillette and Nixon (1980) ran some trials with the probe on Somerset Island, N.W.T. (74°N) and found that in colder (temperatures of −5°C and lower) and harder permafrost the rate of penetration was very slow.

71.2.2 Modified CRREL Auger

The standard CRREL auger was designed by the U.S. Army Corps of Engineers to core snow, ice, and fine-grained mineral and organic materials. This auger consists of a stainless steel core barrel with an outside diameter of 11.3 cm and a length of 92 cm (Figure 71.3),

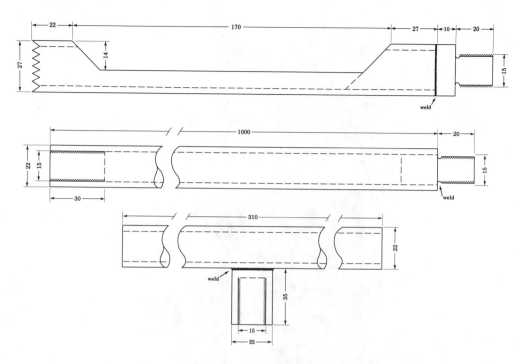

FIGURE 71.2. Dimensions of the bit, extension rod, and T-handle for the modified Hoffer probe. All measurements are in millimeters.

FIGURE 71.3. Standard USA-CRREL ice coring auger. Photograph courtesy of the Geological Survey of Canada (GSC 203205-G).

providing a 7.6-cm-diameter core. The barrel has a welded, double-helix flight configuration with a 20-cm pitch. The stainless steel cutting shoe fastened to the bottom of the barrel has removable cutting inserts. A removable head couples the barrel to the extension rods and permits the removal of the core through the top of the barrel. The T-handle couples directly to the head when starting a hole and to the 1-m aluminum extension rods fitted with stainless steel pin-type couplings when penetrating to greater depths.

The modified CRREL augers currently being used have a shorter barrel length and are usually driven by a gasoline-powered motor. The version used by the Geological Survey of Canada (Veillette and Nixon 1980) is equipped with barrels of three different diameters, producing cores with diameters of 3.8, 5.1, and 7.6 cm (Figure 71.4). All core barrels, heads, and

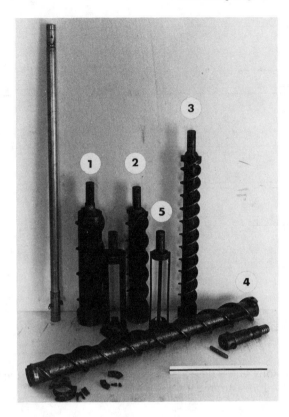

FIGURE 71.4. Modified CRREL core barrels: (1) length 41 cm, core diameter 7.6 cm; (2) length 41 cm, core diameter 5.1 cm; (3) length 61 cm, core diameter 3.8 cm; (4) length 92 cm, core diameter 5.2 cm; and (5) core catcher for 5.1-cm-diameter cores. Photograph courtesy of the Geological Survey of Canada (GSC 203205-A).

shoes for the modified CRREL auger are made of milled steel and are built in machine shops. The cutting inserts have milled steel shanks and tungsten carbide tips.

Prior to coring, a pit is dug to the frost table. Coring is started by chipping a small depression in the frost table to accept the shoe of the core barrel of the modified CRREL auger. The barrel is rotated slowly until the cutters are firmly engaged. When the barrrel is filled, it is removed from the hole and the head is disengaged to remove the core. A common problem occurs when the core remains solidly attached to the bottom of the borehole after the auger is withdrawn. In this case, the core must be removed by the core catcher. The core catcher consists of two metal rings held together by straight metal bars. Spring steel blades attached to the bottom ring are designed to grab the core. The core catcher, attached to the extension rods, is lowered over the core and a quick sideways movement or twisting motion usually breaks the core. The freed core is held inside the core catcher and can then be removed.

During drilling the barrel must be constantly rotated. If rotation is stopped, the barrel can freeze to the wall of the hole. Removing a frozen-in barrel from the hole is very difficult, and in some cases impossible.

Soil temperature affects the rate of penetration, although it is not as critical a factor as for other samplers. At temperatures between -1 and $0°C$, frozen clays with a large amount of unfrozen water will liquefy rapidly during coring, making them difficult or impossible to core.

This auger is an excellent sampling tool and test drill. It provides an undisturbed core and has a penetration range of 2 to 5 m. Its limitations are that any stony material damages the cutting bits and the core barrel can freeze in the drill hole.

71.2.3 Acker Portable Auger

The Acker portable auger must be operated by two people and is used for coring frozen peat and fine-textured, relatively stone-free, mineral materials. It consists of a split tube sampler, extension rods, a drop-hammer (weighing approximately 63 kg), an aluminum tripod, and a winch powered by a gasoline motor. The winch, whose drum has an outside diameter of 11.5 cm, is driven by a roller chain and powered by a 5-hp gasoline engine. The winch raises the drop-hammer; loosening the tension of the rope causes the hammer to drop on the top of the drill rod, driving it downward. The split tube sampler consists of a barrel that divides in half lengthwise to expose the entire sample.

With this auger it is possible not only to determine the depth of the frozen peat deposit and fine-textured mineral material, but also to obtain samples for physical and chemical analysis, bulk density, and ice content determination.

A common problem occurring with this auger is that the hammering action damages the extension rod joints and the core barrel. The heavy weight of the hammer (requiring a helicopter to move it) and the long time required to set up the auger make it a very expensive sampling method. In addition, a great deal of shattering and compaction of the frozen soil material takes place during the hammering process.

71.3 EQUIPMENT USED FOR TRENCHING

71.3.1 Electric Chain Saw

The chain saw is used only in perennially or seasonally frozen peat. It can be used to expose a vertical profile for the examination of frozen peat materials and for obtaining samples for physical and chemical analysis, ice content determination, and thin sections. The excavation is carried out by using the chain saw to make vertical cuts (Figure 71.5) in an approximately 30×30-cm grid pattern. These blocks, which are still attached at the base, are cut away by using a Kango® electric hammer or a Pico® gasoline-powered hammer, as shown in Figure 71.6.

No modification of the chain saw is required for this work and the chain saw is operated in the normal way. It is recommended that an electric chain saw be used instead of a gasoline type, since the fumes generated in the soil pit by the gasoline engine may result in serious headaches and discomfort to the operators.

71.3.2 Pick Axe

The pick axe is used mainly for limited excavation in frozen mineral materials. It is used to remove blocks of frozen material for ice content determination and to obtain frozen bulk samples for physical and chemical analysis.

FIGURE 71.5. Cutting of frozen peat with an electric chain saw.

FIGURE 71.6. Using an electric hammer to remove blocks cut by the electric chain saw.

71.4 EQUIPMENT USED FOR CORING AND TRENCHING

71.4.1 Kango® Rotary Electric Hammer

The Kango® rotary electric hammer was commonly used during the 1970s for excavating and coring frozen mineral soils. With this tool, it is possible not only to expose a trench in frozen soil, but also to penetrate high-ice content materials.

The 960-W rotary electric hammer is powered by a 1500-W portable AC gasoline generator (e.g., Honda®). The total weight of the generator is approximately 48 kg, and of the hammer and attachments approximately 30 kg.

One of the attachment tools used with the hammer is the spade, which is generally used for cutting and digging relatively stone-free frozen soils and pure ice. The chisel attachment is used in stony soils. The electric hammer, with the coring attachments, has both rotary and hammering movements. Two types of corers are used, one for stone-free materials and the other, which has carbide teeth, for stony materials. The former corer produces a core 50 mm in diameter and 60 mm in length, while the latter produces a core 30 mm in diameter and 50 mm in length.

The Kango® rotary electric hammer is a reliable tool both for trenching and coring. This hammer is useful for coring only to shallow depths and samples for bulk density and ice content are usually taken from the soil trench. One of the main drawbacks of this sampling tool is its heavy weight, which makes it difficult to transport. A helicopter or other transportation is needed to move it to the site.

71.4.2 Pico® Gasoline-Powered Hammer

The Pico® hammer is powered by a single-cylinder, air-cooled, two-stroke gasoline motor. The machine operates on the opposed-piston principle — both the motor piston and the hammer piston work in the same cylinder. The motor piston is connected via the connecting rod to the crankshaft unit and flywheel. The machine is equipped with a diaphragm carburetor, which permits work in any position.

The Pico® hammer has all the advantages of the Kango® hammer, plus its light weight (10 kg) and relatively short length (57 cm) make it easy to transport. The hammer comes with a metal frame pack which makes it possible for one person to carry it (Figure 71.7).

The Pico® hammer is equipped with all of the attachments described for the Kango® hammer and, in addition, a coring attachment which permits coring to a depth of approximately 2 m was designed and used by the author. This coring attachment (Figure 71.8) has a core barrel, extension rods, and a coupling device which connects it to the Pico® hammer. The core barrel is 40 cm long and is equipped with a changeable, serrated cutting edge and a removable end piece which connects it to the rod. The Pico® hammer is put in the rotary mode for coring. After the core barrel is filled, the end piece of the core barrel is removed with a pipe wrench and the frozen core is recovered.

The Pico® hammer is a very reliable sampling tool. Its light weight makes it easy to move and it is recommended as the best all-purpose equipment for sampling frozen soil. Care should be taken, however, to keep the exhaust fumes away from the operator. In confined areas (e.g., deep pits) serious health problems can occur if fumes are allowed to accumulate.

71.5 COLLECTION AND HANDLING OF FROZEN SOIL SAMPLES

Sampling and description of permafrost soils, especially those affected by cryoturbation, require special techniques. For these soils a soil unit (pedon), described in Agriculture Canada Expert Committee on Soil Survey (1987), is used as the basis for sampling and soil description. A trench is dug in order to expose a cross section of the pedon (for Cryosolic soils this is commonly the full cycle of the patterned ground). In the frozen portion of mineral soils this trench is excavated to a depth of approximately 1 m using a Pico® or Kango® hammer. In Organic Cryosols a chain saw can be used to cut the trench to a depth of about 1 m and samples can be obtained using one of the coring tools. Soil horizons

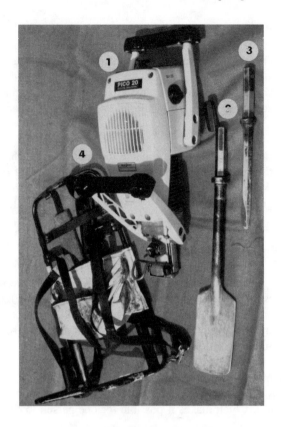

FIGURE 71.7A

FIGURE 71.7B

C

FIGURE 71.7. The Pico® gasoline-powered hammer with accessories. A: (1) Pico® hammer, (2) spade, (3) chisel, and (4) metal frame pack. B: (1) Pico® hammer with the spade and chisel packed on the metal frame pack and (2) the coring attachment. C: Pico® hammer with the spade attached.

exposed on the trench wall are then identified and, with the aid of a 10 × 10-cm-square grid, the horizon boundaries and other macro features are sketched on graph paper.

71.5.1 Collection of Samples

Frozen and unfrozen bulk samples are collected from each horizon for chemical and physical analysis. Unfrozen samples are collected in the conventional way (Sheldrick 1984). Frozen samples are collected with the aid of a power hammer or corer. Because of the ice contained in these samples, special containers and handling are required. If the samples contain moderate amounts of ice, they are usually placed in double, heavy-duty plastic bags. Both of these plastic bags should be sealed separately using waterproof ties or seals in order to avoid leakage. It is recommended that the high-ice content samples be placed in plastic jars having waterproof caps. The water in these thawed soil samples quickly separates from the solids.

FIGURE 71.8. The Pico® gasoline-powered hammer with the coring attachment: (1) 1-m-long extension rod; (2) core barrel; and (3) coupling device.

It is important that this water not be discarded, since this would result in the loss of nutrients and fine clay.

Frozen samples to be used for bulk density and ice content determinations are collected using a corer. After being measured (length and diameter) the cores are placed in watertight plastic jars. Undisturbed thin section samples are collected from low- and medium-ice content materials. While still frozen these samples are placed in Kubiana boxes. Before being sealed for shipping one side of the box is left open for a few hours in a cool and well-ventilated area to allow the samples to thaw and the excess moisture to evaporate. High-ice content samples cannot be collected by this method, since the large amount of thawed water quickly destroys the structure.

71.5.2 Shipping and Preparation

Although there is generally little possibility of drying the samples in the field, if facilities are available this is strongly recommended. The usual procedure, however, is for the various containers of bulk soil, thin section, and core samples to be placed in 5-gal metal sample pails for shipping. These sample containers should not be shipped in large canvas or plastic bags, since this method usually results in the rupture of the individual sample containers during shipping. The samples should be shipped by air to the laboratory as soon as possible. After the containers arrive they should be opened to allow the samples to dry. If the wet samples remain in closed containers for a long period, reduction could occur with resultant transferral of components such as nitrates and iron oxides.

REFERENCES

Agriculture Canada Expert Committee on Soil Survey. 1987. The Canadian System of Soil Clas-sification. 2nd ed. Research Branch, Agriculture Canada, Ottawa. Publication 1646. 164 p.

Brown, R. J. E. 1968. Permafrost investigation in northern Ontario and northeastern Manitoba. National Research Council Canada, Division of Building Research, Tech. Paper No. 291, 72 p.

Sheldrick, B. H. 1984. Analytical methods manual. Land Resource Research Institute, Research Branch, Agriculture Canada, Ottawa. L.R.R.I. Contribution No. 84-30.

Veillette, J. J. and Nixon, F. M. 1980. Portable drilling equipment for shallow permafrost sampling. Geological Survey Paper 79-21, Geological Survey of Canada, Ottawa. 35 p.

Zoltai, S. C. 1978. A portable sampler for perennially frozen stone-free soils. Can. J. Soil Sci. 58: 521–523.

Chapter 72
Hydrological Properties of
Frozen Soil

E. Perfect and B. D. Kay

University of Guelph
Guelph, Ontario, Canada

72.1 INTRODUCTION

This chapter reviews methods of measuring the hydrological properties of frozen soils. The methods are divided into two groups, depending on whether they require static or dynamic conditions. Each method is briefly described and evaluated as to its accuracy, ease of use, and degree of peer acceptance. Readers are referred to the original references for detailed information on instrumentation and protocol.

72.2 METHODS REQUIRING STATIC CONDITIONS

Water can occur as gas (water vapor), liquid (unfrozen water), and/or solid (ice) in frozen soils. Theoretically, the range of subzero temperatures over which all three phases coexist is relatively narrow (Figure 72.1). Freezing of unsaturated soil causes instabilities, with ice bodies either evanescing or spontaneously filling pores (Miller 1973). Thus, individual pores tend to contain air and unfrozen water, or ice and unfrozen water, rather than all three phases.

72.2.1 Vapor Content

The vapor content can be expressed in terms of vapor density (mass of water vapor per unit volume of air) or specific humidity (mass of water vapor per unit mass of air). The vapor density (ρ_v) can be calculated from the absolute temperature (T) and potential (P) of the second phase using the following relations:

$$\rho_v = (\rho_l)_o \exp(P_l V_l / RT) = (\rho_i)_o \exp(P_i V_i / RT) \tag{72.1}$$

where $(\rho_l)_o$ is the vapor density at $P_l = 0$, $(\rho_i)_o$ is the vapor density at $P_i = 0$, V is the partial specific volume, R is the gas constant, and the subscripts v, l, and i refer to the vapor, unfrozen water, and ice phases, respectively.

Soil Sampling and Methods of Analysis, M. R. Carter, Ed.,
Canadian Society of Soil Science. © 1993 Lewis Publishers.

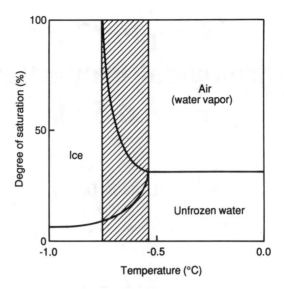

FIGURE 72.1. Phase composition of a two-dimensional pore as a function of temperature when the liquid water potential is held constant. The shaded area represents the range over which air, unfrozen water, and ice are all present. (Adapted from Miller, 1973.

72.2.2 Unfrozen Water Content

The energy status and quantity of water remaining in the liquid phase are crucial to the hydrology of frozen soils. Changes in the energy status of unfrozen water can be calculated using the Clapeyron equation:

$$dP_1 = H_f dT/V_1 T_o + V_i dP_i/V_1 + d\Pi \qquad (72.2)$$

where Π is the osmotic pressure associated with leachable solutes, H_f is the partial specific latent heat of fusion, and T_o is the freezing point of pure water in the bulk state.

Practical problems arise when one attempts to measure the energy status of unfrozen water; ordinary tensiometers freeze up and fail to record at subzero temperatures. McKim et al. (1976) designed a tensiometer that could operate below 0°C due to the presence of ethylene glycol. A semipermeable membrane prevented diffusion of the ethylene glycol into the frozen soil. McGAw et al. (1983) used this tensiometer to measure P_1 during transient freezing. The measured values of P_1 appear to be much smaller than those predicted by integration of Equation 72.2, assuming P_i and Π are both zero. The lack of coincidence between observed and predicted values of P_1 could be due to errors in the above assumptions, application of Equation 72.2 to transient conditions (Loch 1978), or the reduction in observed values of P_1 due to leakage of ethylene glycol.

The amount of liquid water present in frozen soil can be expressed gravimetrically (W_1) or volumetrically (θ_1). Whichever quantity is used, the unfrozen water content is a function of the total potential, $P_1 - \Pi$, which is temperature dependent when P_i remains constant. As the temperature is lowered below 0°C the liquid content initially remains constant due to freezing point depression and supercooling effects (Figure 72.1). Methods of measuring the freezing point depression and extent of supercooling of soil water are described by Anderson (1968).

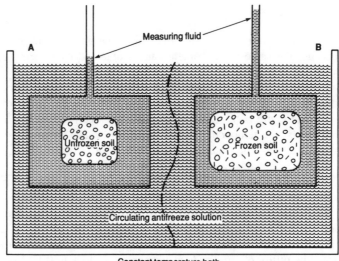

Measuring fluid

A

B

Unfrozen soil

Frozen soil

Circulating antifreeze solution

Constant temperature bath

FIGURE 72.2. Schematic diagram illustrating the principle of dilatometry: (A) cell before freezing, (B) cell after freezing. The actual volume change upon freezing will be $\leq \theta_w \times 9\%$.

Once ice is nucleated, the unfrozen water content decreases in a nonlinear fashion with decreasing temperature (Figure 72.1). This relation has been measured using acoustics (Deschartes et al. 1988), differential scanning calorimetry (Brown and Payne 1990), dilatometry (Williams 1976), pulsed nuclear magnetic resonance (Tice et al. 1978), time-domain reflectometry (Patterson and Smith 1981), and X-ray diffraction (Anderson and Hoekstra 1965). It can also be inferred from heat capacity data (Anderson 1966) and soil water retention determinations (Koopmans and Miller 1966). The recommended methods are dilatometry, inference from soil water retention data (similitude), and time-domain reflectometry.

72.2.2.1 *Dilatometry*

The unfrozen water content can be determined by measuring the expansion associated with the liquid-ice phase change (dilatometry). This method was developed by Bouyoucos (1917), and it continues to be used to this day (Black and Miller 1990). The soil sample, sealed in an impermeable latex membrane, is set within a liquid-filled chamber (Figure 72.2). The freezing point of the liquid must be below the minimum temperature of interest. The chamber is then placed in a constant-temperature bath and subjected to various temperature steps. Once nucleation is initiated the changing proportions of ice and unfrozen water can be calculated from the volume of liquid displaced from the chamber during each temperature step.

Because measurement errors are cumulative with cooling, dilatometry is restricted to temperatures within one or two degrees below 0°C. The soil must be fully saturated prior to freezing. The unfrozen pore water and measuring liquid are assumed to be incompressible. It is also assumed that the density change for soil water is the same as the 9% density change for pure water. Despite these constraints, the method is quite accurate. Williams (1976) reports that unfrozen water contents measured with a dilatometer compare favorably with those determined by similitude. The method is relatively direct. Furthermore, it requires only a modest investment in equipment compared to the other methods available.

72.2.2.2 Similitude

It is possible to predict the unfrozen water-temperature relationship from soil water retention data. This is because the same water content can be obtained by freezing or thawing air-free soil, as by drying or wetting ice-free soil (Miller 1965). First determine the soil water retention curve for the unfrozen soil (see Chapter 53). Values for the matric potential can then be converted into equivalent subzero temperatures using the integrated form of Equation 72.2, assuming Π and P_i are both zero. For clay soils this conversion is made directly. In the case of sandy soils, the temperature is multiplied by 2.20, an empirically determined ratio of the air-water to ice-water interfacial surface tensions.

The theory behind similitude has been described by Miller (1980). The samples must meet the general requirements of similar media and similar states. Consequently, it is assumed that soil in the frozen state contains no air, is at the same bulk density, has the same moisture distribution, and has followed a similar hysteretic pathway as soil in the unfrozen state. Despite these restrictions, the experiments of Koopmans and Miller (1966) and Black and Tice (1989) prove the utility of similitude at temperatures close to 0°C. In order to predict θ_l at lower temperatures it is necessary to extend the moisture retention curve to very negative potentials, requiring a pressure membrane apparatus. The similitude method is relatively simple and inexpensive when compared to more direct methods, which are often difficult to implement and require expensive equipment.

72.2.2.3 Time-Domain Reflectometry

Time-domain reflectometry (TDR) is based upon the velocity of propagation and reflection of an electromagnetic pulse in a dielectric medium. A cable tester, balun, and transmission lines are needed for the measurement (Chapter 51). The transmission lines should be installed prior to freezing. A step voltage or signal is propagated along the transmission lines, which act as a wave guide. The signal is reflected from the beginning and end of the transmission lines and returns back to the cable tester. It is the travel time, t, between these reflections that is determined (Figure 72.3). This value is related to the apparent dielectric constant of the medium, K_a, by Equation 51.3.

Liquid water has a dielectric constant >80, compared to values <6 for air, ice, and the soil matrix. Thus, K_a will be governed by the water content. Topp et al. (1980) published an empirical relationship between K_a and volumetric water content that is widely used for ice-free soils (Equation 52.4). Patterson and Smith (1981) suggested this equation could be used as a first approximation for the volumetric unfrozen water content (θ_l) of frozen soils. However, ice causes partial orientation of water dipoles through the Workman-Reynolds effect. Consequently, it is to be expected that the relationship between K_a and θ_l will depart from the Topp et al. (1980) equation as ice is substituted for air. In light of this concern, Smith and Tice (1988) developed an empirical relation between K_a and θ_l by independently measuring the unfrozen water content using pulsed nuclear magnetic resonance. Their equation can be written as:

$$\theta_l = -0.1458 + 0.0387K_a - 8.5 \times 10^{-4}(K_a)^2 + 9.9 \times 10^{-6}(K_a)^3 \quad (72.3)$$

with a standard error in θ_l of 1.55%. Recently, Oliphant (1985) and van Loon et al. (1991) have presented theoretical $K_a(\theta_l)$ functions. Their analyses confirm that the dielectric constant of unfrozen water is less than that for bulk water at the same temperature.

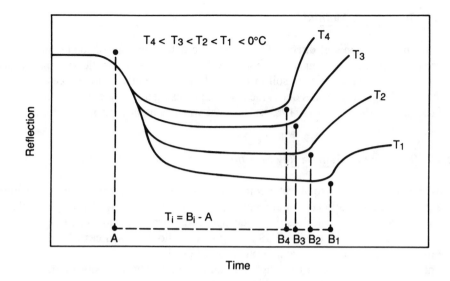

FIGURE 72.3. Idealized TDR traces for soil at different subzero temperatures (T_i), showing how the signal travel times (t_i) are determined. (Adapted from Oliphant, 1985.)

TDR is the preferred method for measuring the unfrozen water content of frozen soils. It is nearly independent of soil type, bulk density, and temperature. The cable tester is portable and can be easily interfaced to a microcomputer for automated data collection. It is fast (a single reading takes approximately 1 s) and accurate, and can be used for *in situ* measurements in the field or laboratory. Furthermore, the method offers the possibility of obtaining a simultaneous estimate of soil bulk electrical conductivity under frozen conditions (van Loon et al. 1990).

72.2.3 Ice Content

In the absence of air, ice content (θ_i) is normally determined as the difference between the total water content (θ_w) and the unfrozen water content (θ_l) on a volumetric basis:

$$\theta_i = \theta_w - \theta_l \tag{72.4}$$

Changes in the energy status of the ice can be determined, relative to the total potential of the liquid phase, using Equation 72.2.

72.2.4 Total Water Content

Measurements of the total water content of frozen soils may be either destructive (i.e., sectioning and gravimetric determination of water content) or nondestructive. The most popular nondestructive methods are dual energy gamma ray scanning and neutron thermalization. The neutron thermalization method is not recommended because of its large sampling volume (Hayhoe and Bailey 1985) and potential health hazards associated with frequent probe use (Gee et al. 1976).

72.2.4.1 *Gravimetric Method*

Efforts to sample frozen soil directly for gravimetric determinations of total water content are extremely destructive. Goit and Sheppard (1976) describe a durable and inexpensive corer for sampling shallow frozen soil in the field. More recently, some success has been achieved by cutting entire blocks out of the frozen zone using a concrete-cutting saw (Suzuki and Mackenzie 1979, Kay et al. 1985). The total water content is then determined gravimetrically (Chapter 51).

72.2.4.2 *Dual Energy Gamma Ray Scanning*

It is possible to measure the total water content of frozen soil nondestructively using gamma ray scanning. The attenuation of gamma rays by soil water is unaffected by the phase change from liquid to ice. Since bulk density can change during freezing, dual energy gamma ray scanning (DEGS) should be used. DEGS employs two radioactive sources: Cs-137 and Am-241, emitting 662 and 60 keV gamma photons, respectively. The radiation is collimated into a narrow beam that passes through the soil column. The intensity of this beam is measured by a scintillation detector, containing a sodium iodide thallium-activated crystal. The attenuation of the Cs-137 and Am-241 photons is related to the total water density, ρ_w (= mass of ice and unfrozen water per unit volume bulk) by the following equation:

$$\rho_w = \{(\mu_s)_{Am} \ln(I_o/I)_{Cs} - (\mu_s)_{Cs} \ln(I_o/I)_{Am}\}/x_{eff}\{(\mu_s)_{Am}(\mu_w)_{Cs} - (\mu_w)_{Am}(\mu_s)_{Cs}\} \quad (72.5)$$

where I is the measured count rate, I_o is the count rate through the empty soil column, μ_s and μ_w are mass attenuation coefficients for soil and water, respectively, and x_{eff} is the effective column thickness. The theory and instrumentation required for DEGS of frozen soils are described in detail by Goit et al. (1978) and Ayorinde (1983). Further information regarding experimental protocol can be found in Loch and Kay (1978). DEGS offers considerable promise for measuring total water contents in the laboratory. Data can be collected nondestructively over relatively short distances. However, theoretical and experimental refinements are required to make measurements in frozen soils containing discrete ice lenses (Goit et al. 1978).

72.3 METHODS REQUIRING DYNAMIC CONDITIONS

It is well known that water can move as a gas or liquid in unfrozen soils. In frozen soils, the solid phase (ice) can also move, due to melting and refreezing of water molecules (regelation) (Miller et al. 1975). Thus, transport of water in the solid phase must be added to whatever flow might be taking place in the liquid and/or vapor phases:

$$(j_w)_l = j_v + j_l + j_i \quad (72.6)$$

where $(j_w)_l$ is the total flux of water, specified in terms of the liquid phase, and J_v, j_l, and j_i are the fluxes of vapor, unfrozen water, and ice, respectively.

Water transport is described relative to the mineral matrix. Assuming there is no gradient of air pressure within the frozen soil, the total flux of water is given by (Perfect et al. 1991):

$$(j_w)_l = KdP_l/dx - L_{wT}(dT/T_o)/dx - L_{\Pi}d\Pi/dx - L_{w\epsilon}d\epsilon/dx \quad (72.7)$$

where x is distance, ϵ is the electrical potential, K is the direct transport coefficient or hydraulic conductivity, and L_{wn} is a coupled transport coefficient relating J_w to a gradient in the property n. Thus, water may move directly in response to a gradient in P_l, or indirectly through coupling to gradients in T, Π or ϵ. Coupled transport can be related to K through the use of reflection coefficients:

$$(j_w)_l = -K\{dP_l/dx + \sigma_T(dT/T_o)/dx + \sigma_\Pi d\Pi/dx + \sigma_\epsilon d\epsilon/dx\} \qquad (72.8)$$

where σ_n is the reflection coefficient ($\equiv L_{wn}/K$) for the property n. Equation 72.8 offers the possibility of determining K using methods that induce either direct or coupled transport.

72.3.1 Vapor Flux

72.3.1.1 *Direct Transport*

Nakano et al. (1984a) describe a direct method for measuring vapor diffusion using di-ethylphthalate (DEP) as a tracer. DEP was chosen because of its low sorptivity on clay minerals and negligible volatility. Furthermore, DEP is detectable at very low concentrations using electron capture gas chromatography. Thus, the amount required is so small (100 $\mu g \cdot cm^{-3}$) that no measurable changes occur in freezing point depression.

The experimental procedure involves two soil columns. One is prepared uniformly moist with the DEP solution, while the other has a negligible water content. The water content of the moist column must be lower than the equilibrium unfrozen water content at the temperature of interest, so that no ice forms. At time t = 0, the two columns are connected and placed in a constant-temperature bath. Over time, water is transported from the wet soil into the dry soil. At the end of the experiment, the two columns are sectioned, and total (gravimetric) water contents (W) and DEP concentrations are determined as a function of distance from the original connection.

Assuming $(dT/T_o)/dx = 0$, $d\Pi/dx = 0$, and $d\epsilon/dx = 0$, Equation 72.8 can be rewritten as:

$$(j_w)_l = -K\{(dP_l/dW)(dW/dx)\} = -DdW/dx \qquad (72.9)$$

where D is the soil-water diffusivity ($\equiv KdP_l/dW$). In the absence of ice, D is given by the sum of the diffusivities of the vapor, D_v, and liquid, D_l, phases:

$$D = D_v + D_l \qquad (72.10)$$

Nakano et al. (1984a) present theoretical equations for evaluating D and D_l from the measured W and DEP concentration vs. distance functions. The vapor diffusivity is then calculated by difference from Equation 72.10. Nakano et al. (1984b) modified the above method to measure vapor fluxes in frozen soil containing ice. Pore ice is allowed to form in the moist column, and a 2-mm air gap is introduced between the columns to interrupt water transport in the liquid and solid phases. However, this procedure may overestimate D_v (Smith and Burn 1987).

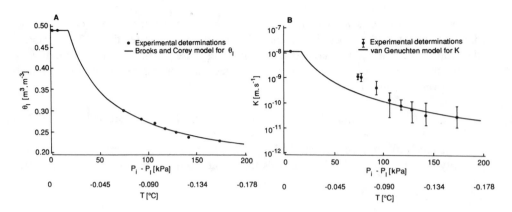

FIGURE 72.4. Prediction of the frozen hydraulic conductivity function (B) from the soil water retention curve (A) for a silt soil. Parameters in the Brooks and Corey and van Genuchten models were saturated water content = 0.49, residual water content = 0.02, ice-entry value = 17.06 kPa, α = 0.35, and β = 2 + (28/15)α. (Adapted from Black and Miller 1990.)

72.3.2 Unfrozen Water Flux

72.3.2.1 *Direct Transport*

In the absence of ice, liquid fluxes can be determined using the method described above (Section 72.3.1). Nakano et al. (1984b) also applied this method to evaluate the redistribution of unfrozen water in soil containing pore ice. The accompanying theoretical analysis (Nakano et al. 1984c) completely ignores the contribution of j_i to the total water flux. Thus, their equation for the soil-water diffusivity cannot be used to derive a valid transport coefficient. Despite these problems, the use of tracers to measure the flux of unfrozen water in frozen soils merits further investigation.

It is also possible to predict K in Equation 72.8 from the water retention curve (Chapter 53). Although this procedure was developed for ice-free soils, it can be applied in the presence of ice by invoking similitude (see Section 72.2.2.2). Several expressions are available for modeling the water retention curve (Black and Miller 1990). The resulting curve fit parameters are then used to estimate the hydraulic conductivity function (Figure 72.4). It is implicit that $j_w = j_l$. If segregated ice is present, the predicted flux may have to be adjusted with an empirical "impedance factor" (Taylor and Luthin 1978). Despite this weakness, the above approach should prove useful. The soil water retention curve is easily measured, whereas experimental determination of the frozen hydraulic function is difficult and time consuming.

72.3.3 Ice Flux

The flux of ice (regelation) can be measured using an "ice sandwich" apparatus. Two versions of this method are available, depending on which driving force is employed.

72.3.3.1 *Direct Transport*

The Darcian ice sandwich was introduced by Miller (1970). Porous phase barriers are used to separate an ice body from two reservoirs containing supercooled water (Figure 72.5).

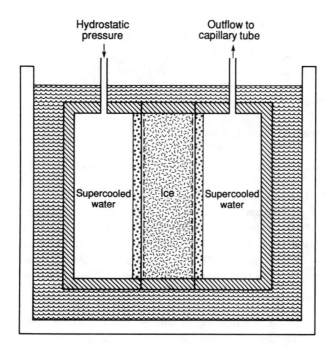

FIGURE 72.5. An "ice-sandwich" apparatus in cross section. From the central sample container outward, the components are phase barriers (semipermeable membranes), porous support plates, reservoirs, cell walls, and the refrigerated bath. (Adapted from Wood and Williams 1985.)

The cell is immersed in a refrigerated bath and hydrostatic pressure is applied to one end. A capillary tube connected to the other reservoir is used to measure the steady outflow of water. Because of the slow flow rates, evaporation from the meniscus may result in significant errors. Losses due to evaporation can be prevented by the introduction of oil between the meniscus and the open end of the capillary tube. The increased possibility of ice nucleation in the reservoirs with decreasing temperature means the "ice sandwich" method is restricted to temperatures close to 0°C.

Horiguchi and Miller (1980) and Wood and Williams (1985) studied the temperature and pressure dependence of regelation in the ice sandwich. They found that transport conforms to Darcy's law. Consequently, Equation 72.8 can be used to calculate the direct transport coefficient (K), assuming $(dT/T_o)/dx = 0$, $d\Pi/dx = 0$, and $d\epsilon/dx = 0$.

72.3.3.2 *Coupled Transport*

Equation 72.8 provides the basis for another technique that has been used to measure transport in the ice phase, namely the "osmotic ice sandwich" (Wood and Williams 1985). This method is similar to the Darcian ice sandwich described above. However, instead of applying a hydrostatic pressure, different concentrations of solutes are added to the end reservoirs, inducing an osmotic potential gradient, $d\Pi/dx$. The solute used is lactose monohydrate, since its relatively large molecules can be restrained from entering the ice chamber by semipermeable membranes.

Assuming perfect semipermeability, the reflection coefficient, σ_Π, in Equation 72.8 should equal one. However, the measured transport coefficients are about an order of magnitude lower than K at the same temperature (Wood and Williams 1985), indicating $\sigma_\Pi \ll 1$. The departure of σ_Π from one may be due to application of the ideal gas law (used to convert $d\Pi$ to dP_1) to concentrated solutions, failure to maintain $d\Pi/dx$ by continual flushing of reservoirs, or gradual passage of lactose molecules through the semipermeable membranes.

72.3.4 Total Water Flux

72.3.4.1 *Direct Transport*

The total water flux can be determined under steady-state conditions using a frozen permeameter similar in design to the Darcian ice sandwich, except the apparatus contains frozen saturated soil instead of pure ice. The soil sample is either prefrozen and cut to shape (Burt and Williams 1976) or prepared unfrozen, in which case ice is nucleated within the permeameter by brief exposure to a jet of decompressing CO_2 (Horiguchi and Miller 1980). Whichever procedure is used, attention must be paid to the distribution and morphology of the ice produced during freezing; ice segregation should be avoided. This is achieved through careful control of the freezing rate and envelope pressure (Black and Miller 1990).

Two varieties of frozen permeameter are in use, neither of which is entirely satisfactory. One employs solutes to prevent freezing in the end reservoirs (Burt and Williams 1976), while the other uses supercooled water (Horiguchi and Miller 1980).

The method of Burt and Williams (1976) employs equal concentrations of lactose monohydrate in the end reservoirs to reduce the freezing point of the bulk solution to that of the unfrozen soil water at the temperature of interest. Lactose molecules are restricted from entering the frozen soil by semipermeable membranes. However, some lactose molecules still enter the soil, causing a reduction in the total potential (Equation 72.2) and melting of ice, especially at the inflow end. Another drawback with this method is the progressive concentration and dilution of solutions within the inflow and outflow reservoirs, respectively. The resulting osmotic potential gradient may induce coupled transport of water in the opposite direction to that predicted by the gradient in matric potential.

The method of Horiguchi and Miller (1980) requires maintenance of the end reservoirs in a supercooled state. This protocol is acceptable provided the sample is carefully deaired and consolidated prior to freezing, to ensure mechanical equilibrium between the supercooled water and unfrozen soil water at subzero temperatures. The use of supercooled water restricts this method to a narrow temperature range, immediately below 0°C. Furthermore, elaborate precautions must be taken to minimize vibrations that might induce ice nucleation in the end reservoirs.

72.3.4.2 *Coupled Transport*

Equation 72.8 indicates that coupled transport of water can occur in response to a temperature gradient (thermoosmosis). Perfect and Williams (1980) designed on apparatus for measuring thermoosmosis under steady-state conditions. The driving force for water migration is a linear temperature gradient induced by thermoelectric cooling plates (Figure 72.6). The soil sample is prefrozen in a saturated state. After assembly, the apparatus is insulated and placed in a low-temperature incubator to reduce radial heat exchange. Lactose is used to prevent freezing in the end reservoirs, while semipermeable membranes retard melting of the frozen sample. Because of the temperature gradient, different concentrations of lactose solution must be used in the inflow and outflow reservoirs. Thus, data analysis is complicated due

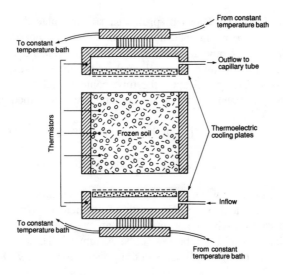

FIGURE 72.6. Exploded view of a thermoosmotic permeameter in cross section. From the central sample container outward, the components are semipermeable membranes, porous support plates, reservoirs, Peltier modules, and heat sinks. (Adapted from Perfect and Williams 1980.)

to the presence of an osmotic potential gradient in addition to the temperature gradient. Future users of this method should experiment with supercooled water or bentonite slurry in the reservoirs instead of lactose solutions.

Recently, a transient thermoosmotic method has been introduced for estimating K from measurements of the apparent thermal conductivity using the thermal probe method (van Loon et al. 1988, Fukuda and Jingsheng 1989). The apparent thermal conductivity, τ^*, is calculated as follows:

$$\tau^* = (Q/4\pi\Delta T) \ln (t_2/t_1) \tag{72.11}$$

where Q is the heat applied to the probe and ΔT is the temperature increase over the time interval t_1 to t_2. Kay et al. (1981) derived the following expression for τ^*, taking into account the coupled transport of heat due to movement of water in the vapor and liquid phases:

$$\tau^* = \tau + (H_f/V_l)^2 (AK/T_o) \tag{72.12}$$

where τ is the thermal conductivity at $T \ll 0°C$ and A is a constant accounting for dimensional consistency ($= 1.02 \times 10^{-4}$ m³·N⁻¹). Convective transport of heat by ice is assumed to be negligible. Substituting Equation 72.11 into Equation 72.12 yields the following expression for the frozen hydraulic conductivity:

$$K = (T_o/A)(V_l/H_f)^2 \{(Q/4\pi\Delta T) \ln (t_2/t_1) - \tau\} \tag{72.13}$$

The thermal probe method for measuring the apparent thermal conductivity of frozen soil is described in detail in Chapter 73.

Another transient thermoosmotic method involves back-calculating the flux of water from the change in water content over time in a closed system subjected to a linear temperature gradient (Oliphant et al. 1983). Tubes of prefrozen soil are placed within an aluminum block, the ends of which are maintained at different subzero temperatures by circulating antifreeze solutions from two temperature-controlled baths (Oliphant et al. 1983, Xu Xiaozu et al. 1985). The diameter of the tubes is small compared to their length, so that the thermal profile of the block is superimposed upon the tube contents. After a period of time, the tubes are removed and sectioned and the water contents are determined gravimetrically.

The mean water flux for segment n, j_n, can be calculated using the following expression:

$$j_n = j_{n-1} + \{(W_{2n} - W_{1n})\rho_b \Delta x\}/(t_2 - t_1) \tag{72.14}$$

where W_{2n} and W_{1n} are the total water contents of segment n at times t_2 and t_1, respectively, ρ_b is the dry bulk density, and Δx is the length of segment n. Water is assumed to be uniformly distributed within each tube at time t_1, so an average value can be used for W_{1n}. A mean flux is obtained for all the tube segments. Assuming $d\Pi/dx = 0$ and $d\epsilon/dx = 0$, this flux can be related to the hydraulic conductivity using Equation 72.8.

To determine K using thermoosmotic methods it is necessary to know the value of the reflection coefficient, σ_T in Equation 72.8. It can be shown that estimates of K obtained by substituting Equation 72.2 into Darcy's law, under the condition that $dP_i = 0$, will underestimate the true hydraulic conductivity by a factor of $1 + \sigma_T V_i/H_f$.

To measure σ_T, the following procedure is suggested (P. H. Groenevelt, personal communication). A closed system containing frozen saturated soil is required. A temperature gradient is then imposed, with the temperatures at both ends below 0°C. Once steady state is reached, unfrozen water should flow towards the cold end, inducing a matric potential gradient. The Darcian backflow in response to this gradient should be equal and opposite to the thermoosmotic flow. Thus, a steady circulation of water will be established, such that the macroscopic flux of water, $(j_w)_1$, is equal to zero. If the resulting matric potential gradient is measurable (see Section 72.2.2), σ_T will be given by:

$$(\sigma_T)_{j=0} = L_{wT}/K = -V_1 T_o dP_1/H_f dT \tag{72.15}$$

Equation 72.15 can then be used in conjunction with Equation 72.8 to estimate K, assuming $d\Pi/dx = 0$ and $d\epsilon/dx = 0$.

Because hydraulic conductivity changes with temperature, some form of averaging procedure, similar to the one suggested by Loch and Kay (1978), should be applied when dT/dx is used as the driving force. The overall hydraulic conductivity is then calculated from the hydraulic conductivity function integrated over the temperature range of interest:

$$\overline{K} = (T_2 - T_1)/\{\int_{T_1}^{T_2} 1/K(T)\, dT\} \tag{72.16}$$

where $\overline{K}$ is the overall hydraulic conductivity, T_1 and T_2 are the boundary temperatures, and $K(T)$ is the function relating hydraulic conductivity to temperature.

72.4 SUMMARY

The most important hydrological properties to measure under frozen static conditions are the unfrozen and total water contents. Preferred methods for measuring the unfrozen water content are dilatometry, similitude, and TDR. Preferred methods for measuring the total water content are destructive gravimetric sampling and nondestructive DEGS. No satisfactory method exists for measuring the energy status of unfrozen water.

The most important hydrological parameter to measure under frozen dynamic conditions is the total water flux. For direct transport the total water flux is expressed as a hydraulic conductivity. The preferred method of measuring the hydraulic conductivity is a frozen permeameter with supercooled water in the end reservoirs. The hydraulic conductivity can also be predicted from the water retention curve, assuming the contribution of ice and vapor to the total water flux is negligible. For coupled transport the thermoosmotic transport coefficient can be measured using the apparatus developed by Perfect and Williams (1980), with supercooled water or bentonite slurry in the end reservoirs. A method is suggested for measuring the reflection coefficient that relates thermoosmotic flow to hydraulic conductivity.

REFERENCES

Anderson, D. M. 1966. Phase composition of frozen montmorillonite-water mixtures from heat capacity measurements. Soil Sci. Soc. Am. Proc. 30: 670–675.

Anderson, D. M. 1968. Undercooling, freezing point depression and ice nucleation of soil water. Israel J. Chem. 6: 349–355.

Anderson, D. M. and Hoekstra, P. 1965. Migration of inter-lamellar water during freezing of Wyoming bentonite. Soil Sci. Soc. Am. Proc. 29: 498–504.

Ayorinde, O. A. 1983. Application of dual-energy gamma-ray technique for nondestructive soil moisture and density measurement during freezing. J. Energy Resour. Techn. 105: 38–42.

Black, P. B. and Tice, A. R. 1989. Comparison of soil freezing curve and soil water curve data for Windsor sandy loam. Water Resour. Res. 25: 2205–2210.

Black, P. B. and Miller, R. D. 1990. Hydraulic conductivity and unfrozen water content of air-free frozen silt. Water Resour. Res. 26: 323–329.

Bouyoucos, G. J. 1917. Classification and measurement of the different forms of water in the soil by means of the dilatometer method. Michigan Agricultural College Exp. Stn., Techn. Bull. No. 36. 48 pp.

Brown, S. C. and Payne, D. 1990. Frost action in clay soils. I. A temperature-step and equilibrate

differential scanning calorimeter technique for unfrozen water content determinations below 0°C. J. Soil Sci. 41: 535–546.

Burt, T. P. and Williams, P. J. 1976. Hydraulic conductivity in frozen soils. Earth Surface Processes 1: 349–360.

Deschartes, M. H., Cohen-Tenoudji, F., Aguirre-Puente, J., and Khastou, B. 1988. Acoustics and unfrozen water content determination, pages 324–328. Permafrost: Fifth International Conference Proceedings. Vol. 1. Tapir Publishers, Trondheim. Norway.

Fukuda, M. and Jingsheng, Z. 1989. Hydraulic conductivity measurements of partially frozen soil by needle probe method. Pages 251–265 in H. Rathmayer, Ed. Frost in Geotechnical Engineering. VTT offsetpaino, Espoo, Finland.

Gee, G. W., Stiver, J. F., and Borchert, H. R. 1976. Radiation hazard from Americium-Beryllium neutron moisture probes. Soil Sci. Soc. Am. J. 40: 492–494.

Goit, J. B., Groenevelt, P. H., Kay, B. D., and Loch, J. P. G. 1978. The applicability of dual gamma scanning to freezing soils and the problem of stratification. Soil Sci. Soc. Am. J. 42: 858–863.

Goit, J. B. and Sheppard, M. I. 1976. A durable and inexpensive sampler for shallow frozen soil. Can. J. Soil Sci. 56: 525–526.

Hayhoe, H. N. and Bailey, W. G. 1985. Monitoring changes in total and unfrozen water content in seasonally frozen soil using time domain reflectometry and neutron moderation techniques. Water Resour. Res. 21: 1077–1084.

Horiguchi, K. and Miller, R. D. 1980. Experimental studies with frozen soil in an "ice sandwich" permeameter. Cold Reg. Sci. Techn. 3: 177–183.

Kay, B. D., Fukuda, M., Izuta, H., and Sheppard, M. I. 1981. The importance of water migration in the measurement of the thermal conductivity of unsaturated frozen soils. Cold Reg. Sci. Techn. 5: 95–106.

Kay, B. D., Grant, C. D., and Groenevelt, P. H. 1985. Significance of ground freezing on soil bulk density under zero tillage. Soil Sci. Soc. Am. J. 49: 973–978.

Koopmans, R. W. R. and Miller, R. D. 1966. Soil freezing and soil water characteristic curves. Soil Sci. Soc. Am. Proc. 30: 680–685.

Loch, J. P. G. 1978. Thermodynamic equilibrium between ice and water in porous media. Soil Sci. 126: 77–80.

Loch, J. P. G. and Kay, B. D. 1978. Water redistribution in partially frozen, saturated silt under several temperature gradients and overburden loads. Soil Sci. Soc. Am. J. 42: 400–406.

McKim, H., Berg, R. L., McGaw, R., Atkins, R., and Ingersoll, J. W., 1976. Development of a remote reading tensiometer transducer system for use in subfreezing temperatures. Second Conference on Soil Water Problems in Cold Regions, Proceedings. Edmonton, Alberta, Canada.

McGaw, R., Berg, R. L., and Ingersoll, J. W., 1983. An investigation of transient processes in an advancing zone of freezing. Pages 821–825 in Permafrost: Fourth International Conference, Proceedings. National Academy Press, Washington, D.C.

Miller, R. D. 1965. Phase equilibria and soil freezing. Pages 193–197 in Permafrost: Second International Conference, Proceedings. National Academy of Sciences, Washington, D.C.

Miller, R. D. 1970. Ice sandwich: functional semipermeable membrane. Science 169: 584–585.

Miller, R. D. 1973. Soil freezing in relation to pore water pressure and temperature. Pages 344–352 in Permafrost: Second International Conference, Proceedings. National Academy Press, Washington, D.C.

Miller, R. D. 1980. Freezing phenomena in soils. Pages 254–299 in D. Hillel, Ed. Applications of soil physics. Academic Press, New York.

Miller, R. D., Loch, J. P. G., and Bresler, E. 1975. Transport of water and heat in a frozen permeameter. Soil Sci. Soc. Am. Proc. 39: 1029–1036.

Nakano, Y., Tice, A. R., and Jenkins, T. F. 1984a. Transport of water in frozen soil. V. Method for measuring the vapor diffusivity when ice is absent. Adv. Water Resour. 8: 172–179.

Nakano, Y., Tice, A. R., and Oliphant, J. L. 1984b. Transport of water in frozen soil. III. Experiments on the effects of ice content. Adv. Water Resour. 7: 28–34.

Nakano, Y., Tice, A. R., and Oliphant, J. L. 1984c. Transport of water in frozen soil. IV. Analysis of experimental results on the effects of ice content. Adv. Water Resour. 7: 58–66.

Oliphant, J. L. 1985. A model for dielectric constants of frozen soils. Pages 46–57 in D. M. Anderson and P. J. Williams, Eds. Freezing and thawing of soil-water systems. Technical Council on Cold Regions Engineering Monograph, A.S.C.E., New York.

Oliphant, J. L., Tice, A. R., and Nakano, Y. 1983. Water migration due to a temperature gradient in frozen soil. Pages 951–956 Permafrost: Fourth International Conference, Proceedings. National Academy Press, Washington, D.C.

Patterson, D. E. and Smith, M. W. 1981. The measurement of unfrozen water content by time domain reflectometry: results from laboratory tests. Can. Geotech. J. 18: 131–144.

Perfect, E. and Williams, P. J. 1980. Thermally induced water migration in frozen soils. Cold Reg. Sci. Techn. 3: 101–109.

Perfect, E., Groenevelt, P. H., and Kay, B. D. 1991. Transport phenomena in frozen porous media. Pages 243–270 in J. Bear and M. Y. Corapcioglu, Eds. Transport processes in porous media, NATO/ASI Series E, Vol. 202, Kluwer Academy Publ., Dordrecht, The Netherlands.

Smith, M. W. and Burn, C. R. 1987. Outward flux of vapor from frozen soils at Mayo, Yukon, Canada: results and interpretation. Cold Reg. Sci. Techn. 13: 143–152.

Smith, M. W. and Tice, A. R. 1988. Measurement of the unfrozen water content of soils: a comparison of NMR and TDR methods. Pages 473–477 in Permafrost: Fifth International Conference Proceedings. Vol. 1. Tapir Publishers, Trondheim, Norway.

Suzuki, M. and Mackenzie, D. N. 1979. A new method for sampling overwintering plants in frozen soil. Can. J. Plant Sci. 59: 549–550.

Taylor, G. S. and Luthin, J. N. 1978. A model for coupled heat and moisture transfer during soil freezing. Can. Geotech. J. 15: 548–555.

Tice, A. R., Burrous, C. M., and Anderson, D. M. 1978. Determination of unfrozen water in frozen soil by pulsed nuclear magnetic resonance. Pages 149–155 in Permafrost: Third International Conference, Proceedings. Vol. 1. N.R.C., Ottawa, Canada.

Topp, G. C., Davis, J. L., and Annan, A. P. 1980. Electromagnetic determination of soil water content: measurements in coaxial transmission lines. Water Resour. Res. 16: 574–582.

van Loon, W. K. P., Van Haneghem, I. A., and Boshoven, H. P. A. 1988. Thermal and hydraulic conductivity of unsaturated frozen sands. Pages 81–90 in R. H. Jones and J. T. Holden, Eds. Ground freezing 88. Balkema, Rotterdam, The Netherlands.

van Loon, W. K. P., Perfect, E., Groenevelt, P. H., and Kay, B. D. 1990. Application of time domain reflectometry to measure solute redistribution during soil freezing. Pages 186–194 in K. R. Cooley, Ed. Proc. Int. Symp. Frozen Soil Impacts on Agricultural, Range, and Forest Lands, CRREL Special Report 90-1.

van Loon, W. K. P., Perfect, E., Groenevelt, P. H., and Kay, B. D. 1991. Application of dispersion theory to time-domain reflectometry in soils. Transport in Porous Media 6: 391–406.

Williams, P. J. 1976. Volume change in frozen soils. Pages 233–246 in Laurits Bjerrum Memorial Volume, Oslo, Norway.

Wood, J. A. and Williams, P. J. 1985. Further experimental investigation of regelation flow with an ice sandwich permeameter. Pages 85–94 in D. M. Anderson and P. J. Williams, Eds. Freezing and thawing of soil-water systems. Technical Council on Cold Regions Engineering Monograph. A.S.C.E., New York.

Xu Xiaozu, Oliphant, J. L., and Tice, A. R. 1985. Experimental study on factors affecting water migration in frozen morin clay. Pages 123–128 in Fourth International Symposium on Ground Freezing. Vol. 1. Sapporo, Japan.

Chapter 73
Thermal Properties of Frozen Soils

W. K. P. van Loon, I. A. van Haneghem
Wageningen Agricultural University
Wageningen, The Netherlands

E. Perfect, B. D. Kay
University of Guelph
Guelph, Ontario, Canada

73.1 INTRODUCTION

The temperature field in frozen soil is governed by the heat balance equation which describes the storage and flows of heat. The heat balance equation is in its turn influenced by the thermal properties of the soil. Two independent thermal properties are to be considered: heat capacity and thermal conductivity.

The heat capacity reflects the ability of the soil to store heat. Therefore, it depends upon the amount of heat that can be stored in the individual components (i.e., mineral particles, liquid water, ice, and air).

The flow or transport of heat in frozen soil can be by conduction, convection, and/or diffusion. Radiation is generally neglected, although it plays a significant role in the flow of heat from the soil surface into the atmosphere. Thermal conductivity reflects the ability of the soil to transmit heat by conduction. It depends upon the thermal conductivities of individual components, as well as on their configuration, which controls the flow of energy between individual components. Convective transport of heat arises from the movement of individual components, particularly water in the liquid phase. In addition, the movement of water vapor may result in the transport of heat by diffusion. However, calculations by Hoekstra (1966) suggest that transport of vapor in frozen soils is of negligible magnitude.

In frozen soils, a temperature gradient generally causes both heat and water transport (Perfect et al. 1991). Because of this coupling, the heat balance equation must be solved simultaneously with the mass balance equation (Chapter 72). In the case of frost heave the equations for stress partition must also be solved simultaneously (O'Neill and Miller 1985). These coupled equations can be solved analytically only for some very special one-dimensional situations. In general, the equations are solved numerically using finite difference or finite element methods. Such numerical solutions are beyond the scope of this chapter. The reader is referred to Lewis and Sze (1988) and Shen and Ladanyi (1989) as recent examples.

Soil Sampling and Methods of Analysis, M. R. Carter, Ed.,
Canadian Society of Soil Science. © 1993 Lewis Publishers.

Table 73.1 Apparent Volumetric Heat Capacities for Freezing Soils (Williams 1973)

Soil type	Total water content (% dry wt.)	$\underline{C}$ at $-0.6°C$ $(MJm^{-3}K^{-1})$	ΔT max (°C)	ΔT min (°C)	$\underline{C}$ $(MJm^{-3}K^{-1})$
Sat. peat TD1	420.7	37	$-.03$	$-.32$	244
Sat. peat 16	574.2	17	$-.03$	$-.32$	252
Peat with sand 8	85.5	47	$-.12$	$-.32$	46
Organic clay TC4	35.9	42	$-.12$	$-.32$	34
Organic clay ND2	60.0	21	$-.04$	$-.32$	101
Cl. illite clay KNB	35.0	25	$-.1$	$-.32$	84
Montmorillonite WCB	48.0	66	$-.2$	$-.32$	25
Clay silt TD2	43.5	38	$-.06$	$-.32$	71
Org. silty clay TC3	34.8	—	$-.04$	$-.32$	63
Sandy gravel 10	7.5	1.7	$-.00$	$-.32$	34
Ottawa sand 20–30	15.0	1.7	$-.00$	$-.32$	71

Note:The normal heat capacity varies from 2 $MJm^{-3}K^{-1}$ (sand and gravel) to 4 $MJm^{-3}K^{-1}$ (peat).

73.2 HEAT CAPACITY

73.2.1 Principles

The measurement of specific heat c_p is based upon the temperature change ΔT in a sample with mass m, due to a given heat input ΔQ:

$$c_p \ (J\ kg^{-1}K^{-1}) = \Delta Q \Delta T^{-1} m^{-1} \qquad (73.1)$$

From this property and specific mass ρ, the volumetric heat capacity C can be calculated:

$$C \ (J\ m^{-3}K^{-1}) = \rho c_p \qquad (73.2)$$

As long as no phase change occurs, the volumetric heat capacity of a soil can be calculated from the heat capacities and volume fractions θ of its constituents, the solid matrix s, liquid water w, ice i, and air a:

$$C \ (J\ m^{-3}K^{-1}) = \theta_s C_s + \theta_w C_w + \theta_i C_i + \theta_a C_a \qquad (73.3)$$

However, at temperatures between -2 and $0°C$ the measured heat capacity is strongly influenced by the latent heat L of melting ice. When soil is heated at an initial temperature just below $0°C$, some ice will melt in order to reach a new thermodynamic equilibrium between water and ice in the soil. The measured temperature rise in the sample will be less than in the situation without melting ice. This effect can be incorporated in an apparent heat capacity $\underline{C}$, which is in fact the measured value:

$$\underline{C} \ (J\ m^{-3}K^{-1}) = C + \rho_w(\partial\theta_w/\partial T) \qquad (73.4)$$

Due to this latent heat effect, the measured or apparent heat capacity $\underline{C}$ can be more than twice as large as C (Table 73.1).

Heat capacity measurements can be performed with an adiabatic calorimeter or with a differential scanning calorimeter (DSC).

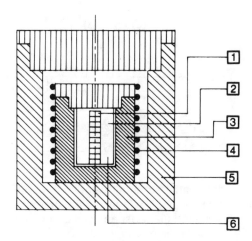

FIGURE 73.1. Principle of adiabatic calorimetry, with (1) electrical heating element, (2) removable sample container, (3) fixed insulation with removable lid, (4) thermal guard, (5) insulation with removable lid, and (6) thermocouples. Note: the entire setup should be put into a climate case or room. (Adapted from Johanson, O. and Frivik, P. E., Second Int. Symp. Ground Freezing Proc., Norwegian Inst. of Technol., Tronheim, Norway.)

73.2.2 Adiabatic Calorimetry

A well-known device for measuring (apparent) heat capacity is the adiabatic calorimeter (Mellor 1979, Johanson and Frivik 1980). This is a thermally insulated container, into which a known quantity of soil is placed (see Figure 73.1). The adiabatic calorimeter is preferably located in a temperature-controlled chamber to minimize heat (or cold) losses to the environment. For measurements in frozen soils the controlled ambient temperature must be set at values below 0°C.

Because of the supercooling problem, the measurement should start at low temperatures and the heat capacity be determined during a heating run. The temperature rise must take place slowly enough to ensure a uniform temperature in the soil sample. The temperature rise is recorded continuously as a function of the (electrical) heat input. The temperature response of the system must be recorded with great precision ($\pm 0.01°C$).

73.2.3 Differential Scanning Calorimetry

A method to determine the apparent heat capacity, which is commercially available, is DSC (Figure 73.2). The scanning calorimeter works by changing the temperature of a sample and a reference simultaneously and at a constant rate (Oliphant and Tice 1982). The sample and reference temperatures are measured separately and undergo exactly the same temperature course. To maintain a constant temperature rate in the sample a variable amount of heat must be supplied due to the changing apparent heat capacity.

The output of a DSC is shown in Figure 73.3. A typical scan goes from about -35 to $+15°C$ at a rate of approximately 10°C/min. To obtain an accurate measurement three scans are required:

1 Scan using an empty sample pan, resulting in the base line B(T).

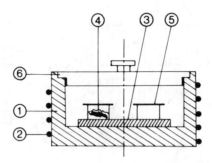

FIGURE 73.2. Principle of differential scanning calorimetry, with (1) metal furnace, (2) electrical heating coil, (3) DSC sensor with thermocouples, (4) sample pan, (5) reference pan, and (6) measuring cell cover. Note: the furnace must be insulated from the ambient temperature.

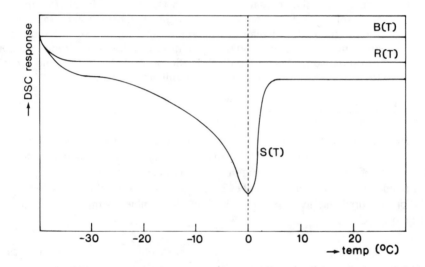

FIGURE 73.3. DSC response as a function of temperature at a scanning rate of 10°C per minute, with B(T) the base line, R(T) the reference material, and S(T) the sample measurement (Oliphant and Tice 1982). Note: the amount of heat supplied is negatively displayed along the vertical axis.

2   Scan using the sample pan filled with material of known heat capacity $C_R(T)$, resulting in the reference signal R(T).

3   Scan using the sample pan filled with the soil of interest, resulting in the signal S(T).

At each temperature T, the apparent heat capacity $\underline{C}(T)$ of the soil sample can be calculated from the following equation:

$$\underline{C}(T) = \frac{S(T) - B(T)}{R(T) - B(T)} \times \frac{V_R}{V_s} \times C_R(T) \tag{73.5}$$

where V_S and V_R are the volumes of the soil and reference samples, respectively. When both sample pans are entirely filled, the quotient V_R/V_S equals one.

73.2.4 COMMENTS ON CALORIMETRY IN FROZEN SOILS

1 Soil samples are normally stored in the unfrozen state. To determine their thermal properties, the samples need to be frozen. A common problem in isothermal freezing of soil samples is supercooling: the pore water remains unfrozen at temperatures below 0°C. To ensure ice nucleation the sample should be frozen to approximately $-10°C$ for sands and to $-20°C$ for clays. Other techniques can also be used to ensure ice nucleation: the seeding of ice by contacting a very cold needle ($-30°C$), or knocking the sample. Thermally induced redistribution of soil water during freezing (e.g., the development of ice lenses) is another complicating factor. Rapid freezing and, if the soil matrix is incompressible, an overburden (pressure) can be used to restrict such redistribution.

2 Smith and Riseborough (1985) have demonstrated the importance of accurate heat capacity values for thermal calculations. In fine-textured soils the latent heat effect has a significant influence over a larger temperature range than in coarse-textured soils. However, the difference between C and $\underline{C}$ is much greater in coarse-textured soils than in fine-textured soils as T approaches zero. In general, an independent estimate of the $\theta_w(T)$ relation is of great help in heat capacity measurements (Equation 73.4). For measuring methods to determine the unfrozen water content see Chapter 72.

3 Adiabatic calorimeters are relatively simple devices and can be constructed at low cost. Furthermore, the method can be applied to a wide range of soil materials, because sample volumes are not limiting. However, in the case of coarse materials large samples are required, resulting in long equilibration times (e.g., a sample of 10 cm³ needs a time of approximately 30 min to equilibrate). This a $\underline{C}(T)$ curve may take several days to complete.

4 In modern DSC the B(T) and R(T) curves are usually included in the computer program for the calculation of the apparent heat capacity (Figure 73.3). However, for accurate measurements it is better to determine these curves separately and compute C with the use of Equation 73.5. A disadvantage of DSC is the small sample size. Because the change in temperature with time is rapid the sample must be small in order to have a uniform temperature. The sample pan cannot hold more than approximately 0.05 cm³, which makes the method unsuitable for sands and gravels.

73.3 THERMAL CONDUCTIVITY

73.3.1 Principles

Thermal conductivity λ (W m^{-1}K^{-1}) is the direct transport coefficient between a temperature gradient and a heat flux q_T and is defined by Fourier's law:

$$q_r \ (\text{W m}^{-2}) = -\lambda \cdot \text{grad (T)} \tag{73.6}$$

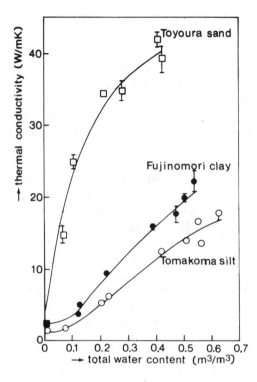

FIGURE 73.4. Thermal conductivities of sand, silt, and clay soils as a function of total water content at T = −10°C. (From Fukuda, M. and Jingsheng, Z., Frost in Geotech. Eng. 1, VTT Symp. 94, Espoo, Finland. With permission.)

Inaba (1983) and Sawada and Ohno (1985) provide evidence of the variation in thermal conductivity with temperature. Since the thermal conductivity of ice is larger than that of water, any variation in the thermal conductivity of frozen soil with subzero temperature must indicate a variation in ice content. The influence of total water content on λ can be seen in Figure 73.4 or in Yershov et al. (1988).

In contrast to the heat capacity, thermal conductivity is not only a simple function of the thermal conductivities and volume fractions of the soil constituents, but also depends upon their mutual configuration. Heat flows from one constituent to another (series conduction), as well as parallel to the different constituents (parallel conduction).

Many semitheoretical and empirical models have been developed to deal with this complex series — parallel behavior (Farouki 1986). De Vries (1963) used the analogy with electrical conductivity to derive a model for the thermal conductivity of unfrozen soil. Penner (1970) applied this model to frozen soils. More recently, Gori (1983) has developed a semitheoretical model specifically for frozen soils.

Johanson and Frivik (1980) have extended the empirical model of Kersten (1963), which gives the same accuracy as the above-mentioned theoretical models. Because of its relatively simple form and convenience, this model is prsented below. For frozen soils, the thermal conductivity λ is given by (Johanson and Frivik 1980):

$$\lambda\ (\mathrm{W\ m^{-1}K^{-1}}) = \lambda_{\mathrm{dry}} + (\lambda_{\mathrm{sat}} - \lambda_{\mathrm{dry}})\ \theta_i \phi^{-1} \tag{73.7}$$

where λ_{sat} is the thermal conductivity of the ice-saturated soil, λ_{dry} is the thermal conductivity of the dry soil, θ_i the volumetric ice content, and ϕ is the pore volume. Thus, the thermal conductivity must be measured in both extreme cases (dry and ice-saturated soil) and these measured values are then used in Equation 73.7.

The thermal conductivity is influenced by melting of small amounts of ice at temperatures close to 0°C. Furthermore, coupled transport of latent heat L in soils occurs as a result of water migration. Horiguchi and Miller (1980) were the first to demonstrate experimentally the transport of heat due to a water flux. The significance of convective transport of latent heat Q_L due to liquid water flow during the measurement of thermal conductivity has been demonstrated by Kay et al. (1981):

$$Q_L \; (W \; m^{-2}) \; = \; \rho_w L \cdot v_w \tag{73.8}$$

The liquid water flux v_w can be expressed with Darcy's law:

$$V_w \; (m \; s^{-1}) \; = \; -k \cdot A \cdot grad(p) \tag{73.9}$$

with A a constant accounting for unit consistancy ($A = 1.02 \cdot 10^{-4} m^3 \; N^{-1}$), k is the hydraulic conductivity, and grad(p) the pressure gradient.

Kay et al. (1981) reworked the pressure gradient term with use of the Clapeyron equation (see Chapter 72 or Van Loon 1991) and incorporated it with Darcy's law (Equation 73.9) into a theoretical expression for the apparent thermal conductivity $\underline{\lambda}$. Like the apparent heat capacity, the apparent thermal conductivity $\underline{\lambda}$ is also larger than λ:

$$\underline{\lambda} \; (W \; m^{-1}K^{-1}) \; = \; \lambda \; + \; L^2\rho_w^2 A \cdot k \cdot T_0^{-1} \tag{73.10}$$

where T_0 is the freezing temperature ($T_0 \approx 273$ K). Measurements of the apparent thermal conductivity have been made by Fukuda and Jingsheng (1989).

Thermal conductivity can be measured by steady state or transient methods.

73.3.2 Guarded Hot Plate Method

The most important steady state method for measuring the thermal conductivity of frozen soil is the guarded hot plate test (Farouki 1986). Two identical soil samples are placed above and below a flat plate which is the main heating unit (Figure 73.5).

A guard heater surrounding the main heater eliminates horizontal heat losses, ensuring unidirectional vertical heat flow from the main heater plate (Figure 73.5). This is achieved by keeping the two temperatures, as measured with the thermocouples next to the two heaters, at the same level. Liquid-cooled flat heat exchangers are placed on the outer surfaces of the soil sample. When the temperature gradient over the soil sample is measured together with the dissipated heat of the central heating plate, the thermal conductivity can be calculated using Fourier's law (Equation 73.6).

73.3.3 Nonsteady-State Probe Method

A typical device used as a nonsteady-state probe to measure thermal conductivity is shown in Figure 73.6. The length of the probe is about 200 mm and its diameter is 2 mm. The

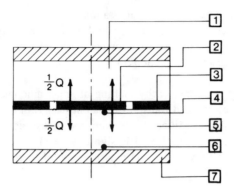

FIGURE 73.5. Principle of the guarded hot plate method for measuring thermal conductivity, with (1) upper soil sample, (2) main heating plate, (3) guard heater, (4) thermocouples near heating plates, (5) lower soil sample, (6) thermocouple near cooling plate, and (7) cooling plate.

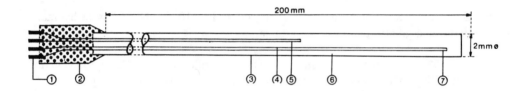

FIGURE 73.6. Cross section of a needle-shaped thermal conductivity probe, with (1) contacting wires, (2) polyvinyl chloride protector, (3) metallic tube, (4) hot junction of thermocouple, (5) heating wire, (6) insulating material, and (7) cold junction of the thermocouple. (From Van Loon, W. K. P. et al., *Int. J. Heat Mass Transfer*, 32, 1473, 1989. With permission.)

most important parts of the needle-shaped probe are a double-folded constantan heating wire and a constantan-manganine thermocouple, both with a diameter of 0.1 mm, carefully fitted into a stainless steel cylinder (Figure 73.6).

The "hot" junction of the thermocouple is placed very close to the heating wire; the "cold" junction is placed at the end of the probe and is considered to stay at its original temperature (Van Loon et al. 1989). To fix the position of the wires in the cylinder and to prevent electrical contact, the remaining space is filled with a silicon rubber compound.

The different components of the measuring system are shown in Figure 73.7. The probe is inserted into the soil sample. At time t = 0 a constant electrical heating current is switched on. The temperature response of the probe is then determined once per second. The measured response is inversely proportional to the thermal conductivity. The maximum measuring time is approximately 5 min and the sampling volume is a small cylinder around the probe. This gives the advantage that the thermal conductivity can be measured *in situ*.

The change in temperature of the probe with time T(t) is often used to calculate the thermal conductivity by assuming a perfect line source as described by Carslaw and Jaeger (1959). However, in this model the representation of the probe is highly simplified. More accurate

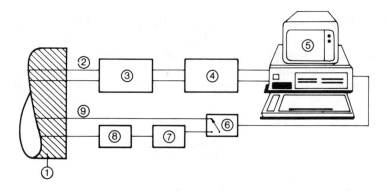

FIGURE 73.7. Schematic diagram of the experimental set up for apparent thermal conductivity measurement, using the nonsteady-state probe method, with (1) end of probe, (2) connecting thermocouple wires, (3) amplifier, (4) analog digital converter (ADC), (5) microcomputer, (6) relay, (7) stable currrent source, (8) current meter, and (9) connecting heating wires.

results can be obtained from the theoretical approach of Bruijn et al. (1983) and references therein. A general representation of their model can be written as:

$$T(t) \ (K) = A \cdot \ln(t) + B + D \ t^{-1} + E \cdot \ln(t) \ t^{-1} \tag{73.11}$$

With A, B, D, and E functions of the apparent thermal conductivity $\underline{\lambda}$, the apparent heat capacity $\underline{C}$, and the probe properties. To make the calculation of A, B, D, and E easier, it is assumed that these four coefficients are independent of one another. They can then be obtained by performing multiple regression analysis. For $\underline{\lambda}$ measurements only the A coefficient in Equation 73.11 is of interest:

$$A \ (K) = Q \ (4\pi\underline{\lambda})^{-1} \tag{73.12}$$

where Q is the dissipated heat per unit of length. Under well-defined conditions, it is also possible to obtain simultaneously $\underline{\lambda}$ and $\underline{C}$ in an isotropic and homogeneous medium (Van Loon et al. 1989).

73.3.4 COMMENTS

1   In the guarded hot plate method, relatively large temperature differences are necessary. These large temperature differences can introduce large measuring errors close to 0°C. A problem for measurements in frozen soils is the convective heat transport due to water movement during the measurement itself. In addition, evaporation at the "hot" side and condensation at the "cold" side (the heat pipe effect) might introduce large errors in unsaturated conditions. Apart from moisture migration the method is time consuming and only suitable for laboratory use.

2   In the transient or nonsteady-state probe method, measuring errors due to convection are reduced, because both the measuring time and the temperature difference are relatively small. The majority of results obtained with

the nonsteady-state probe method are for disturbed soils in the laboratory. Slusarchuk and Watson (1975) developed a stronger thermal probe applicable to field conditions. Van Loon (1991) discussed the complex influence of a macroscopic temperature gradient on the measured values of thermal conductivity. Careful interpretation is required, especially close to the freezing front.

73.4 THERMAL DIFFUSIVITY

73.4.1 Principles

The apparent thermal diffusivity $\underline{a}$ is usually obtained from separate measurements of apparent thermal conductivity $\underline{\lambda}$, and apparent heat capacity $\underline{C}$ by using its definition:

$$\underline{a}\ (m^2 s^{-1}) = \underline{\lambda}/\underline{C} \tag{73.13}$$

The calculated thermal diffusivity is an apparent thermal property, like the measured thermal conductivity and heat capacity. Thus, it will be influenced by the thermodynamic ice-water equilibrium in the soil. Thermal diffusivity is a consequence of the well-known Fourier equation for nonsteady-state diffusion (conduction):

$$\partial T/\partial t\ (K\,s^{-1}) = \underline{a} \cdot div\,(grad\,(T)) \tag{73.14}$$

When the equations for apparent heat capacity and thermal conductivity (73.4 and 73.10, respectively) are inserted into Equation 73.14 a more general heat balance equation can be obtained (Kay et al. 1981, Van Loon 1991):

$$C\frac{\partial T}{\partial t} + L\rho_w\frac{\partial \theta_w}{\partial t}\ (W\,m^{-3}) = div\{\lambda\,grad\,(T) + L\rho_w v_w\} \tag{73.15}$$

The convection of sensible heat has been neglected with respect to the convection of latent heat. The left-hand side of Equation 73.15 accounts for an increase in temperature of the medium and for the latent heat to melt ice. The right-hand side consists of the conduction term and the convection term of latent heat.

The apparent thermal diffusivity can be obtained from analytical solutions of Equation 73.14 for well-defined boundary conditions. Hoekstra et al. (1973) used sinusoidal boundary conditions, while Jackson and Taylor (1986) used a stepwise change. The latter is worked out in some detail. When at the surface of a semi-infinite soil sample, at initial temperature T_0, a different temperature T_1 is suddenly applied, the solution of the temperature $T(x,t)$ at distance x from the surface, and at time t, can be written as:

$$T(x,t) = T_1 + (T_0 - T_1)erf(y) \tag{73.16}$$

where y is a dimensionless position-time coordinate ($y = x(4at)^{-1/2}$) and erf, the error function or probability integral, is given by

$$erf(y) = 2\pi^{-1/2}\int_0^y exp(-u^2)du \tag{73.17}$$

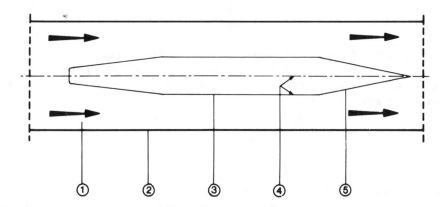

FIGURE 73.8. Schematic representation of one-dimensional cylindrical setup to measure thermal diffusivity, with (1) liquid flow at temperature T_1, (2) glass tube, (3) stainless steel sample holder, (4) thermocouple positions in liquid, sample surface, and sample center, and (5) soil sample. (From Mellor, J. D., Bulletin Int. Inst. of Refrigeration, Paris. With permission.)

where the dummy variable u is integrated from 0 to y. When the temperature is measured at two places in the soil sample at different times, the thermal diffusivity $\underline{a}$ can be calculated using Equation 73.16. Under field conditions, McGaw et al. (1978) have applied a finite difference procedure for estimating the thermal diffusivity from measured soil temperatures as a function of time.

73.4.2 Comparative Method

The apparent thermal diffusivity can be measured directly using the one-dimensional cylindrical method, which is shown in Figure 73.8. The apparatus consists of a cylindrical metal tube containing the soil sample and two thermocouples: one located at the center and the other on the inner wall. The sample container is heated or cooled by a liquid having a temperature T_1, flowing within a long glass tube. The temperature of the liquid is measured with a third thermocouple. The center temperature T at time t, in the cylindrical-shaped sample of homogeneous and isotropic soil, is given by a summation of Bessel functions (Mellor 1979). For a given configuration, these Bessel functions only depend on time t, position x, and apparent thermal diffusivity $\underline{a}$.

Bessel functions can be avoided using the comparative method. For a reference medium with a known diffusivity a_0, the half-time t_0 is determined. The half-time is the time interval in which the center temperature reaches a value half way between the initial and applied boundary temperatures:

$$T(t_0) = \tfrac{1}{2}(T_0 + T_1) \tag{73.18}$$

The half-time t_1 of the sample of interest is then determined. At a fixed position in the cylinder the temperature is only a function of $\underline{a}\cdot t$. Consequently, the thermal diffusivity of the unknown soil sample can be obtained from:

$$\underline{a} = a_0 \cdot (t_0/t_1) \tag{73.19}$$

73.4.3 Comments

Analogous to thermal conductivity measurements, thermal diffusivity measurements are hard to obtain for temperatures close to 0°C (remember $\underline{a} = \lambda/\underline{C}$). Small quantities of ice melt during the measurement, causing latent heat effects. It is therefore required to make the temperature changes as small as possible.

REFERENCES

Bruijn, P. J., Haneghem, I. A., van, and Schenk, J. 1983. Pages 359–366 in An improved nonsteady state probe method for measurements in granular materials. Part 1. Theory. High Temp. High Press. 15.

Carslaw, H. S. and Jaeger, J. C. 1959. Conduction of heat in solids. Clarendon, Oxford.

Farouki, O. T. 1986. Thermal properties of soils. Series on rock and soil mechanics, Vol. 11. Trans Tech Publications, Clausthal-Zellerfield, FRG.

Fukuda, M. and Jingsheng, Z. 1989. Hydraulic conductivity measurements of partially frozen soil by needle probe methods. Pages 251–266 in Frost in Geotech. Eng. 1, VTT Symp. 94, Espoo, Finland.

Gori, F. 1983. A theoretical model for predicting the effective thermal conductivity of unsaturated frozen soils. Pages 363–368 in Permafrost: Fourth Int. Conf. Proc. National Academy Press, Washington, D.C.

Horiguchi, K. and Miller, R. D. 1980. Experimental studies with frozen soil in an 'ice sandwich' permeameter, Cold Reg. Sci. Techn. 3: 177–183.

Hoekstra, P. 1966. Moisture movement in soils under temperature gradients with cold side temperature below freezing. Water Resour. Res. 2: 241–250.

Hoekstra, P., Delaney, A., and Atkins, R. 1973. Measuring the thermal properties of cylindrical specimens by the use of sinusoidal temperature waves. CRELL Technical Report 244, Hanover, NH.

Inaba, H. 1983. Experimental study on thermal properties of frozen soils. Cold Reg. Sci. Techn. 8: 181–187.

Jackson, R. D. and Taylor, S. A. 1986. Thermal conductivity and diffusivity. Pages 941–944 in A. Klute, Ed. Methods of soil analysis. Part I. Physical and mineralogical properties. American Society of Agronomy, Madison, WI.

Johanson, O. and Frivik, P. E. 1980. Thermal properties of soils and rock materials. Pages 427–453 Second Int. Symp. Ground Freezing Proc., Norwegian Inst. of Technol., Trondheim, Norway.

Kay, B. D., Fukuda, M., Izuta, H., and Sheppard, M. I. 1981. The importance of water migration in the measurement of the thermal conductivity of unsaturated soils. Cold Reg. Sci. Technol. 5: 95–106.

Kersten, M. S. 1963. Thermal properties of frozen ground. Pages 301–305 in Proc. First Int. Permafrost Conf., NAS-NRC, Washington, D.C.

Lewis, R. W. and Sze, W. K. 1988. A finite element simulation of frost heave in soils. Pages 73–80 in Fifth Int. Symp. Ground Freezing Proc., Balkema, Rotterdam, The Netherlands.

Loon, W. K. P., van, Haneghem, I. A., van, and Schenk, J. 1989. A new model for the nonsteady state probe method to measure thermal properties of porous materials. Int. J. Heat Mass Transfer 32: 1473–1481.

Loon, W. K. P. van 1991. Heat and Mass Transfer in Frozen Porous Media. Ph.D. thesis, Agricultural University, Wageningen, The Netherlands.

McGaw, R. W., Outcalt, S. I., and Ng, E. 1978. Thermal properties and regime of wet tundra soils at Barrow, Alaska. Pages 47–53 in Proc. Third Int. Permafrost Conf., NRC Ottawa, Canada.

Mellor, J. D. 1979. Thermophysical properties of foodstuffs. 3. Measurements. pages 1–14 in Bulletin Int. Inst. of Refrigeration, Paris.

O'Neill, K. and Miller, R. D. 1985. Exploration of a rigid ice model of frost heave. Water Resour. Res. 21: 281–296.

Oliphant, J. L. and Tice, A. R. 1982. Comparison of unfrozen water content mesured by DSC and NMR. Pages 115–121 in Third Int. Symp. on Ground Freezing Proc. CRREL Special Report 82-16, Hanover, NH.

Penner, E. 1970. Thermal conductivity of frozen soils. Can. J. Earth Sci. 7: 982–987.

Perfect, E., Groenevelt, P. H., and Kay, B. D. 1991. Transport phenomena in frozen porous media. Pages 243–270 in J. Bear and M. Y. Corapcioglu, Eds. Transport processes in porous media, NATO/ASI series E, Vol. 202. Kluwer Academic Press, Dordtrecht, The Netherlands.

Sawada, S. and Ohno, T. 1985. Laboratory studies on thermal conductivity of clay, silt and sand, in frozen and unfrozen states. Pages 53–58 in Fourth Int. Symp. Ground Freezing, Vol. 2. Hokkaido University Press, Sapporo, Japan.

Slusarchuk, W. A. and Watson, G. H. 1975. Thermal conductivity of some ice-rich permafrost soils. Can. Geotech. J. 12: 413–424.

Smith, M. and Riseborough, D. W. 1985. The sensitivity of thermal predictions to assumptions in soil properties. Pages 17–23 in Third Int. Symp. Ground Freezing Proc. Balkema, Rotterdam, The Netherlands.

Shen, M. and Ladanyi, B. 1989. Numerical solutions for freezing and thawing of soils using boundary conforming curvilinear coordinate systems. Pages 391–400 in Frost in Geotech. Eng. 1, VTT Symp. 94, Espoo, Finland.

Vries, D. A. de 1963. Thermal properties of soils. In W. R. van Wijk, Ed. Physics of plant environment. North-Holland, Amsterdam.

Williams, P. J. 1973. Determination of heat capacity of freezing soils. Pages 153–163 in Symp. on Frost Action on Roads. Proc.

Yershov, E. D., Komarov, I. A., Smirnova, N. N., Motenko, R. G., and Barkovskaya, Ye. N. 1988. Thermal characteristics of fine grained soils. Pages 135–140 in Fifth Int. Symp. Ground Freezing Proc. Balkema, Rotterdam, The Netherlands.

Chapter 74
Frost Heave Potential

J.-M. Konrad

Laval University

Sainte-Foy, Québec, Canada

74.1 INTRODUCTION

The problem of frost action damage is widespread; it occurs in all temperate zones wherever there is seasonal soil freezing as well in the active layer of more northerly permafrost regions. In addition, long-term freezing can occur beneath chilled pipelines, cryogenic storage tanks, and year-round ice surfaces. The problem arises because most saturated fine-grained soils under low stresses exhibit an affinity for water during freezing, leading to ice segregation and frost heaving.

There is general agreement that the water supply, soil type, thermal conditions, and over-burden pressure are the major factors governing the rate of frost heaving (Taber 1929, Beskow 1935, Penner 1968, 1972, Kaplar 1970). Moreover, these four main factors are all strongly interrelated, and it is therefore not easy to separate their effect on frost heaving.

From a phenomenological point of view, the mechanics of frost heave can be regarded as a problem of impeded drainage to an ice-water interface at the segregation front (Konrad and Morgenstern 1980, 1984). Substantial suctions are generated at this interface, but the reduced permeability of the frozen fringe (see Figures 74.1 and 74.2) impedes the flow of water to the ice lens. Thus, it appears that the frost heave potential (or frost susceptibility) of freezing soils is related to the characteristics of the frozen fringe that, in turn, depend essentially on the unfrozen water content-temperature-pressure relationship for the soil under study. It is well known that this relationship depends mainly on specific surface, grain size and gradation, amount of fines, void ratio (or density), quantity and type of exchangeable ions, solute concentration, and freeze-thaw cycles (Anderson and Morgenstern 1973, Tice et al. 1982, Konrad 1990).

This chapter does not attempt to review all the available frost-susceptibility criteria (this review has been done by Chamberlain 1981), but rather gives the basis for using the "Segregation Potential" (SP) concept which allows the estimation of frost heave potential for actual field conditions. After a brief review of the mechanics of frost heave, the chapter describes a simple laboratory frost heave test referred to as step freezing, which provides the basic frost heave parameter of a given soil. A procedure to obtain the frost heave potential at a given site is then outlined.

Soil Sampling and Methods of Analysis, M. R. Carter, Ed.,

Canadian Society of Soil Science. © 1993 Lewis Publishers.

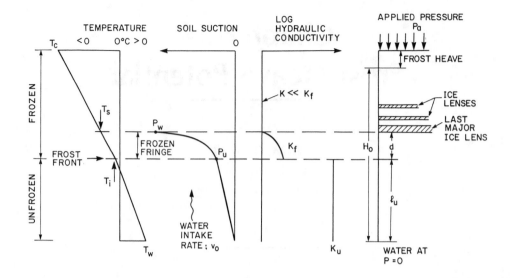

FIGURE 74.1. Mechanics of moisture transfer in freezing soils.

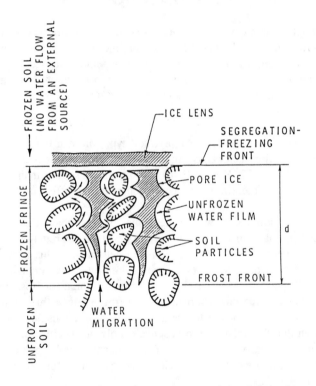

FIGURE 74.2. Schematic view of the frozen fringe.

74.2 SEGREGATION POTENTIAL

In conjunction with the development of a mechanistic frost heave theory, Konrad and Morgenstern (1980, 1984) proposed a new parameter to characterize freezing soils for transient thermal states. It was referred to as the SP and was defined as:

$$SP = v(\text{Grad } T_f) - 1 \qquad (74.1)$$

where v is the water migration rate within the unfrozen soil and Grad T_f the temperature gradient in the frozen fringe. It is noted that SP has a physical meaning only for transient freezing, i.e., when the frost front advances into unfrozen soil to seek thermal equilibrium. For a stationary frost front, frost heave is directly computed from net heat extraction rate as discussed below. It is also important to stress that Equation 74.1 assumes that the unfrozen soil is incompressible.

During transient freezing, SP of a given soil is not constant, but depends upon the overburden pressure P_e, the rate of cooling of the frozen fringe dT_f/dt, and the suction at the frost front P_u. The constitutive relationship (SP, dT_f/dt, P_u, P_e) is, however, unique for a given soil at a given void ratio, structure, and pore fluid chemistry. The rate of cooling of the frozen fringe is defined as the change of temperature per unit time at the frost front and can be calculated as:

$$dT_f/dt = \text{Grad } T_f(dX/dt) \tag{74.2}$$

where dX/dt is the rate of frost front advance.

The suction at the frost front is a function of the hydraulic conductivity of the unfrozen soil K_u and the length of flow path l_u. It can be calculated using Darcy's law:

$$P_u = v l_u/K_u + P_o \tag{74.3}$$

where P_o is the pressure at the base of the unfrozen soil.

74.3 FROST HEAVE POTENTIAL FOR FIELD CONDITIONS

In fairly incompressible soils (sands, silts, and overconsolidated clays), total heave rate under field conditions can be calculated from:

$$dh/dt = 1.09 \, v + 0.09 \, n \, dX/dt \tag{74.4}$$

where n is the porosity of the unfrozen soil, reduced to account for the percentage of *in situ* pore water that will not freeze. Using the SP concept and substituting Equation 74.1 in Equation 74.4 for v gives:

$$dh/dt = 1.09 \, SP(dTf/dt, Pu, Pe) \text{Grad} \, T_f + 0.09 \, n \, dX/dt \tag{74.5}$$

To predict successfully the frost heave in the field, it is imperative to determine the actual value of SP, and, in turn, the values of the rate of cooling, the suction at the frost front, and the overburden pressure at the warmest ice lens. It is noted that the value of Grad Tf and dX/dt can be obtained from a heat flow simulation, using the actual thermal and geometrical conditions in the field.

At thermal steady state, total heave is simply related to the net heat extraction rate (Konrad 1987b):

$$dh/dt = 1.09/L \, (k_f \text{Grad} \, T_f - k_u \text{Grad} \, T_u) \tag{74.6}$$

where L is the latent heat of fusion of pure water, k_f and k_u the thermal conductivities of frozen and unfrozen soil, respectively, and Grad T_u and Grad T_f the temperature gradient in the unfrozen soil near the frost front and in the frozen soil, respectively.

In the field, frost penetration rates are usually between 0.5 and 0.01 cm/d. The actual temperature gradients in the frozen fringe range from approximately 0.1 to 0.01°C/cm (Konrad 1987a). Consequently, the rate of cooling for most field conditions ranges down from 0.05 to 0.0001°C/d, which is extremely small.

The overburden pressure at the freezing front is calculated from

$$Pe = \gamma_f X(t) - u_o \qquad (74.7)$$

where γ_f is the unit weight of the frozen soil of thickness $X(t)$ and u_o is the hydrostatic pressure acting on the frost front.

74.4 SEGREGATION POTENTIAL FROM STEP-FREEZING TEST

The prediction of frost heave requires the determination of SP from laboratory freezing tests on representative samples and with representative values of dT_f/dt, P_u, and P_e.

Ideally, freezing tests on natural soils should be conducted on undisturbed samples. If this cannot be done, it is recommended to reconsolidate the soil to its *in situ* water content from a slurry with an initial water content of about 1.5 times the liquid limit. On the other hand, when the soil layers were caused by compaction, it is recommended to compact the soil into the freezing cell at the same void ratio. Freezing should be conducted for several freeze-thaw cycles.

In a step-freezing test, the sample is subjected to constant temperature boundary conditions throughout the freezing period. Initially, the sample is at a constant temperature above the freezing point and one end is then maintained at a constant temperature below 0°C. This procedure results in a period of decelerating frost penetration followed by a period of quasistationary frost front during which a major final ice lens can grow. Figure 74.3 shows in a schematic way the variations of heave, rate of cooling, suction at the frost front, and SP with time in a step-freezing test.

During transient freezing, the rate of cooling of the frozen fringe steadily decreases from extremely high values (>60°C/d) to a value of 0°C/d. Since the rate of cooling in the field is extremely small, it is proposed to determine the value of SP from step-freezing tests near thermal steady state.

74.5 A SIMPLIFIED METHOD OF FROST HEAVE DETERMINATION

To obtain the actual soil freezing characteristics requires a substantial number of tests, which might be a limitation on the practical use of the frost heave approach based on the SP. These problems, however, can be overcome by seeking an upper-bound value to frost heave and by considering the following simplifications inherent in field freezing conditions.

As discussed above, the rate of cooling in the field is usually extremely small. It is therefore argued that SP (field) may be approximated by the SP obtained near thermal steady state, i.e., at Point S in Figure 74.3b.

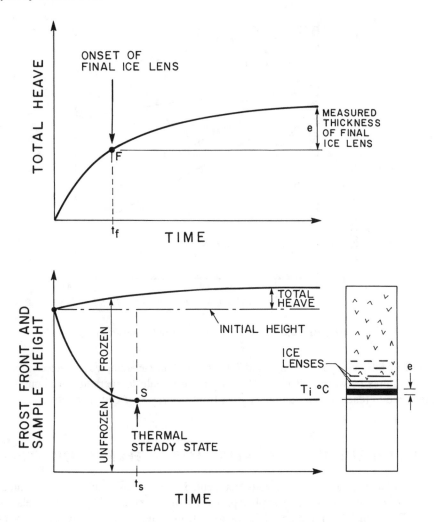

FIGURE 74.3. Typical characteristics of a step-freezing test.

The SP decreases with increasing suctions at the frost front (Konrad and Morgenstern 1981). Therefore, an upper-bound frost heave can be computed for the case where Pu is very small. In a laboratory test this would be achieved by minimizing the flow length in the unfrozen soil, i.e., choosing warm-plate temperatures in the range of 0 to 1°C.

In the field, for frost depths up to 10 m, effective overburden pressures vary between 0 and 100 kPa. Konrad and Morgenstern (1982a) have shown that the relationship between SP and applied pressure near thermal steady state can be expressed as:

$$SP = SP_o \exp(-aP_o) \tag{74.8}$$

where SP_o is the SP for $P_o = 0$ and a is a constant for a given soil.

In summary, it appears that only a limited number of well-controlled step-freezing tests may be required to characterize adequately the SP of homogeneous soil in the field over a wide range of overburden pressures. In practice, three step-freezing tests at three different applied pressures covering the expected range in the field suffice to define the field frost heave

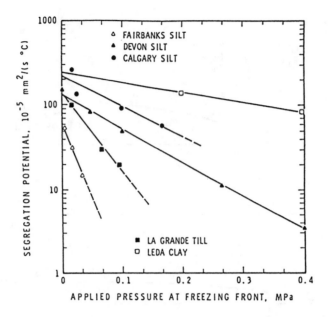

FIGURE 74.4. Summary of data for SP parameter.

characteristics. By using 10-cm-high and 7- to 10-cm-diameter samples subjected to subfreezing temperatures between -3 to $-6°C$, and a warm-plate temperature of about 1°C, the time required to reach a quasistationary frost front ($dTf/dt = 0°C/d$) is relatively short (less than 2 d).

74.6 SYNTHESIS OF SP VALUES FOR DIFFERENT SOIL TYPES

The results of a literature survey on well-documented cases of frost heave data are summarized in Figure 74.4. The SP for various applied pressures can be expressed by a simple relationship given in Equation 74.8 for many soils ranging from sandy silts to clays, thereby giving strong support for the general validity of this relationship to characterize any freezing soil.

Figure 74.4 also reveals the importance of including the overburden pressure in any frost heave potential evaluation. In general, SP of soils with low, unfrozen water contents (silty sands) will be very sensitive to small changes in overburden pressures; this is not the case for soils with higher unfrozen water contents, such as clayey silts and clays.

74.7 EXPERIMENTAL PROCEDURE

74.7.1 MATERIALS AND EQUIPMENT

It is suggested that the frost heave cell display the following key features (see Figure 74.5):

1 A sample container with a 10-cm I.D. and a wall thickness of about 10 to 30 mm. It is recommended that the PVC container be split vertically in two halves and is Teflon®-lined to reduce wall friction. It should also facilitate examination of the sample after freezing.

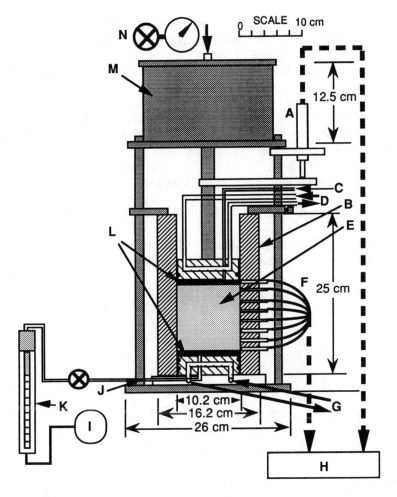

FIGURE 74.5. Schematic view of a freezing cell and set-up.

LEGEND:

A: DCDT	F: thermistors	J: water outlet (bottom)
B: PVC freezing cell	G: cooling fluid (bottom)	K: volume change indicator
C: water outlet (top)	H: data acquisition system -	L: porous plates
D: cooling fluid (top)	I: mercury manometer back	M: bellofram loading system
E: soil sample	pressure system	N: air pressure gauge

2 Insulation around the container is important and should be at least 20 cm thick.

3 Soil sample temperature is measured by an array of thermistors installed flush with the cell wall at regular spacing of about 10 to 15 mm. The accuracy of the thermistors should be better than ±0.05°C.

4 Heave is measured to within ±0.001 mm by a direct current displacement transducer (DCDT).

5 The water is supplied to the test sample with a Mariotte device accurate to ±0.1 mL.

6   Load may be applied to the sample with a bellofram air pressure system which provides up to 700 kPa loading or by a dead load.

7   Porous plates should be available at each heat exchanger in order to conduct hydraulic conductivity tests prior to freezing.

8   Heat exchangers are usually controlled with temperature baths used to impose the temperature boundary conditions during freezing. Alternatively, thermoelectric cooling plates can be used.

9   An independent bath at −10°C should be used to nucleate ice in the sample.

10   Data should be collected every hour by an automated acquisition system.

It is also recommended to use an environmental chamber maintained at +0.5 to 2°C to minimize heat transfer to the sample during freezing.

74.7.2 METHOD OF ANALYSIS

A step-freezing test gives the following basic data:

a. Temperature vs. time at various elevations in the sample.

b Total heave vs. time.

c. Volume of water intake vs. time.

The following data are then inferred from the basic data:

1   Position of frost front is derived from data "a" using a parabolic fit (usually the frost front is at 0°C, but can be any value of temperature if freezing point depression is different from 0°C owing to solutes, pressure, and pore size effects)

2   Rate of frost penetration is calculated from 1.

3   Temperature gradient in the frozen fringe is obtained from data "a" and data "1".

4   Rate of cooling is calculated from data "2" and data "3".

5   Rate of water intake is calculated from data "c". It is also possible to derive the rate of water intake from data "a" as shown by Konrad (1987a).

6   Suction is calculated from data "5" and data "1" (knowledge of the hydraulic conductivity of the unfrozen soil is required).

FROST HEAVE PREDICTION USING THE SP CONCEPT

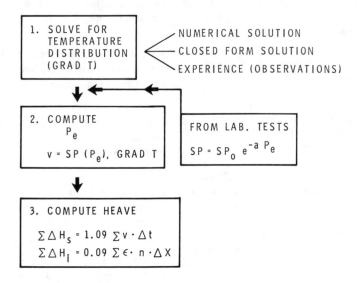

FIGURE 74.6. Flow chart for predicting frost heave in the field.

7 SP is calculated from data "5" and data "3" for a value of dT_f/dt close to the field value, i.e., 0°C/d. In general, the actual SP range is found between points F and S in Figure 74.3.

74.8 FROST HEAVE CALCULATION FOR FIELD CONDITIONS

Figure 74.6 outlines the procedure yielding frost heave vs. time for given thermal and geometrical conditions for a homogeneous site. Details of calculations are given in Konrad and Morgenstern (1984). In short, the temperature distribution in the field is obtained either from a numerical approach, from a closed form solution, or directly from observations on an instrumented site. The temperature gradient in the frozen soil adjacent to the frost front is then calculated as a function of time. Knowing the relationship between SP and overburden pressure obtained from laboratory step-freezing tests, the rate of water migration to the freezing front is readily computed using Equation 74.1. Integration over time provides the segregational heave which is added to the *in situ* heave to yield total heave with time.

For layered soils, it is necessary to estimate SP in each layer and use the SP corresponding to the soil layer in which the frozen fringe is located. If more than one soil layer is within the frozen fringe, an average value for SP is recommended.

REFERENCES

Anderson, D. M. and Morgenstern, N. R. 1973. Physics, chemistry and mechanics of frozen ground. Pages 257–288 in Proc. of the 2nd. Int. Conf. on Permafrost.

Beskow, G. 1935. Soil freezing and frost heaving with special application to roads and railroads

(translated by J. Osterberg). Northwestern University Tech. Inst., 1947.

Chamberlain, E. J. 1981. Frost susceptibility. Review of index tests. Monograph 81-2. U.S. CRREL. 110 pp.

Chamberlain, E. J., Gaskin, P., Esch, D., and Berg, R. L. 1982. Identification and classification of frost susceptible soils. ASCE Spring Convention, Las Vegas, Nevada, April 26–30.

Kaplar, C. W. 1970. Phenomenon and mechanism of frost heaving. Pages 1–13 in Highway Research Record No. 304.

Konrad, J.-M. 1987a. Procedure for determining the segregation potential of freezing soils. Geotech. Test. J. 51–38.

Konrad, J.-M. 1987b. The influence of heat extraction rate in freezing soils. Cold Reg. Sci. Technol. June issue.

Konrad, J.-M. 1990. Unfrozen water as a function of void ratio in a clayey silt. Cold Reg. Sci. Technol. 18: 49–55.

Konrad, J.-M. and Morgenstern, N. R. 1980. A mechanistic theory of ice lens formation in fine grained soils. Can. Geotech. J. 17: 473–486.

Konrad, J.-M. and Morgenstern, N. R. 1981. The segregation potential of a freezing soil. Can. Geotech. J. 18: 482–491.

Konrad, J.-M. and Morgenstern, N. R. 1982a. Effects of applied pressure on freezing soils. Can. Geotech. J. 19: 494–505.

Konrad, J.-M. and Morgenstern, N. R. 1982b. Prediction of frost heave in the laboratory during transient freezing. Can. Geotech. J. 19: 250–259.

Konrad, J.-M. Morgenstern, N. R. 1984. Frost heave prediction of chilled pipelines buried in unfrozen soils. Can. Geotech. J. 21: 100–115.

Miller, R. D. 1972. Freezing and heaving of saturated and unsaturated soils. Pages 1–11 in Highway Res. Rec. 393.

Penner, E. 1968. Particle size as a basis for predicting frost action in soils. Soils Found. 8: 21–28.

Penner, E. 1972. Influence of freezing rate on frost heaving. Pages 56–64 in Highway Res. Rec. 393.

Taber, S. 1929. Frost heaving. J. Geol. 37: 428–461.

Tice, A. R., Oliphant, J. L., Nakano, Y., and Jenkins, F. 1982. Relationships between the ice and unfrozen phases in frozen soil as determined by pulsed nuclear magnetic resonance and physical desorption data. U.S. CRREL Report 82-15.

Chapter 75
Depth of Frost Penetration

Fons J. Schellekens and Peter J. Williams

Carleton University

Ottawa, Ontario, Canada

75.1 INTRODUCTION

Frost penetration may be defined as "the movement of the freezing front into the ground during freezing" (Harris et al. 1988), and may be measured experimentally or calculated by mathematical models. The almost trivial definition suggests wrongly that the depth of frost penetration is easy to determine.

Soil water usually freezes at lower temperatures than 0°C. The cause of this freezing point depression is that soil water has a lower free energy than free water at atmospheric temperature because it contains dissolved salts and because of adsorption and capillary forces of the soil matrix. The freezing of water in a soil occurs over a range of temperatures, because the more strongly bound soil water (which has a lower free energy) tends to freeze at a lower temperature (Williams and Smith 1989). There is a great variability of frost penetration depth in space and time. A good understanding of the ground thermal regime is required for a reliable extrapolation.

For frost to penetrate the soil surface, the temperature must fall below 0°C. In turn, the resulting thermal properties and heat fluxes will govern further frost penetration. In the field, therefore, the controlling factors will be climate, geographic position, soil type and soil moisture, and surface cover (vegetation, snow). These aspects in the soil should be recorded for a meaningful characterization of frost penetration.

75.2 PHYSICAL PROBING AND PENETROMETER TESTS

Soil samples taken by augering, or drilling down the soil profile, may be examined for ice content and the maximum depth of frost penetration assessed. A transect of boreholes can be obtained in this way, and a frost penetration profile can be drawn. Disadvantages of this method are that sampling is time consuming and destructive.

An alternative procedure of physical probing is to drill a hole through the frozen zone and to lower an L-shaped rod to the bottom of the hole. The rod is raised with the short leg of the L digging into the sidewall of the hole. The freezing front is identified by a sudden increase in resistance (Tobiasson and Atkins 1975).

In measurements using a penetrometer (Rickard and Brown 1972), the strength of the soil is measured. Frozen soil has a higher strength than the same soil in an unfrozen state. If in a vertical array of measurements a sudden, continuing lower soil strength is measured, it is assumed that the soil is unfrozen. The physical probing and penctrometer methods give very rough indications of the position of the actual freezing front.

75.3 VISUAL FROST GAUGES

75.3.1 Materials and Procedure

This method has been developed in Sweden (Gandahl 1957, Gandahl and Bergau 1957) and uses substances, such as fluorescein sand or a bromthymol (= methylene) blue solution in water, that change color upon freezing. Fluorescein sand changes from green in the unfrozen to pale yellow in the frozen state (Rickard and Brown 1972). In the case of methylene blue, a tube is filled with the solution, a dissolved solid such as $NaHCO_3$ (for pH stabilization), and styrofoam balls (Banner and van Everdingen 1979). Methylene blue loses its blue color and becomes colorless upon freezing.

75.3.2 COMMENTS

1   The price of these visual frost gauges is low and construction and operation are easy. Gandahl's device, as well as a visual frost gauge developed in 1965 by the Laboratoire Central des Ponts et Chaussées (1987) in France, measured the position of thawed and frozen material with an accuracy of ±5 cm. Rickard and Brown's gauge had an accuracy of ±2 cm.

2   The access tube for the frost tube has to protrude above the maximum snow level. This causes the solution initially to freeze too soon, because of heat loss via the upper end of the tube. Automated or remote readings are impossible. Access for daily readings leads to packing of snow. Elevated boardwalks affect the radiative heat exchange. Together with the daily extraction of the tubes for readings, all these circumstances affect the thermal regime in the soil.

3   The dissolved solids required for operation of the gauge cause a freezing point depression (for a 0.1% $NaHCO_3$ solution, this is about $-0.79°C$ [Banner and van Everdingen 1979]). The visual frost gauges indicate the depth where the soil temperature equals that of the depressed freezing point of the dissolved salt solution in the frost gauge. This freezing point depression may be greater than that of the soil water solution around the frost gauge, and may lead to a lag in frost indication.

75.4 TEMPERATURE MEASUREMENTS

Several temperature sensors can be appropriately modified to determine the depth of frost penetration in soils. The accuracy of location of the freezing front will depend on the gradient of temperature in the ground. Information on the materials and procedures (principles) of soil temperature measurement is given in Chapter 64.

75.4.1 Resistance Thermometers

According to Hansen (1963), the most precise temperature-sensing devices are electrical resistance thermometers. The instrumentation is robust, stable, and portable. When long cables are required they must be thick to ensure their low resistance and low temperature coefficient of resistance (Judge 1973). The response time is slow. The price of metallic resistance thermometers is high. Protection of the leads from moisture is essential. In temperature measurements with semiconductor resistance thermometers (thermistors; see Chapter 64), a very good precision (to 0.01°C) can be reached. Thermistors are available in a wide range of resistances up to very high values. They are reliable, inexpensive, and the simplest temperature measuring apparatus to use (Judge 1973). Drift can be limited to 0.02°C. Epoxy-coated thermistors showed a drift up to 0.1°C. Thermistors are susceptible to thermal and mechanical shock (Beck 1956 in Judge 1973). Thermal shock can be avoided by using heat sinks during soldering and not vulcanizing sensor pods in cables. Self heating of the thermistor occurs when the heat generated by the passing of an electrical current cannot be carried away by the surrounding medium. The power applied to the sensor should therefore not be too high (e.g., <10 mW, Thermometrics Inc. 1982). For research purposes glass bead thermistors are often used (e.g., at the National Research Council of Canada).

75.4.2 Thermocouples

In frost depth determination by temperature measurements with thermocouples, copper and constantan wires are usually used, because these are resistant to corrosion in moist environments. They are suitable for temperatures below 0°C (Johnston 1973, Tobiasson and Atkins 1975). Thermocouples are inherently simple, flexible, rugged, and easy to handle and install. They respond quickly to rapid transient temperature changes, cover a wide range of temperature, and their output is reasonably linear over portions of that range. Thermocouples are not subject to self-heating problems. Thermocouples are reasonably cheap (Hansen 1963) and ideal as a thermopile (multiple junctions) when greater precision is required, as, for example, for the direct measurement of temperature gradients (Judge 1973). However, compared to other temperature measuring instruments the accuracy of thermocouples is relatively low (up to 0.1 K), particularly when they are connected with long cables. The readings of thermocouples may be prone to instability due to stray currents.

75.4.3 Procedure

When used in the field, temperature sensors such as resistance thermometers, thermistors, and thermocouples are assembled in gauges. The sensors are connected by cables with an electrical device which transduces the electrical signal into a digital signal, which can be read on a readout. The signal can be registered automatically by a datalogger. After measuring the temperature at different levels in the soil, temperature profiles are constructed and the level at which the temperature is 0°C is determined. This level is assumed to be equal to the freezing front. This is a reasonable assumption for coarse-grained soils, in which the freezing point depression is small and may often be neglected. In fine-grained soils the freezing point depression may be considerable, due to the presence of salts and capillary and adsorption forces in the soil. This may result in a slowing down of the rate of frost front penetration in a freezing soil. As a result, there may be a considerable difference between the depth of the 0°C isotherm and the depth of the freezing front.

75.4.4 Comments

If the exact location of the freezing front is needed when using thermocouples, thermistors, or diodes, the freezing point of the soil solution has to be determined as well. If the soil is relatively dry, the soil water has a low free energy and allowance must be made for the perhaps large effect this may have on the temperature of initial freezing of the soil. Which method of temperature measurement is the best in a certain case depends on the kind of problem, the accuracy that is needed, the technical skill, and the financial resources available.

75.5 USE OF MODELS TO CALCULATE THE DEPTH OF FROST PENETRATION

75.5.1 Analytical Models

The simplest model to calculate the depth of frost penetration assumes that all the heat transfer in a freezing soil consists of latent heat (Stefan 1890 in Jumikis 1955). It describes the situation in which the temperature is a function of position and not of time (steady state). Based on these principles the equation was derived:

$$Z = \sqrt{\frac{2K(T_f - T_s)t}{L}} \tag{75.1}$$

where Z = depth of frost penetration (m)
K = thermal conductivity ($Js^{-1}m^{-1}K^{-1}$)
T_f = freezing temperature (K) ($\approx 0°C$)
T_s = temperature at the ground surface (K)
t = time of frost penetration (s)
L = latent heat of fusion (Jm^{-3})

In the calculation of the frost depth by the model of Stefan a large error is generated because the heat that is stored in the soil during the day (or the summer) and which is transferred upward in the soil during the night (or the winter) is not accounted for.

Neumann developed a model which is based on the continuity of heat flow in the soil. It takes into account the latent heat of fusion (when water freezes) and the heat set free by cooling of the unfrozen soil (Jumikis 1955). In general, it is assumed that the depth of frost penetration is proportional to the square root of frost penetration time:

$$Z = j\sqrt{t} \tag{75.2}$$

where Z = depth of frost penetration (m)
t = frost penetration time (s)
j = constant of proportionality ($ms^{-1/2}$)

The calculations of the model result in the transcendental equation in j:

$$\frac{Q_L\rho_s\sqrt{\pi}wj}{2} = b_{sf}(T_f - T_s)\frac{\exp\dfrac{-j^2}{4a_{sf}}}{G\left(\dfrac{j}{2}\sqrt{a_{sf}}\right)} - b_{uf}(T_0 - T_f)\frac{\exp\dfrac{-j^2}{4a_{uf}}}{1 - G\left(\dfrac{j}{2}\sqrt{a_{uf}}\right)} \qquad (75.3)$$

in which

$$b_{sf} = \frac{K_{sf}}{\sqrt{a_{sf}}}$$

$$b_{uf} = \frac{K_{uf}}{\sqrt{a_{uf}}}$$

where Q_L = latent heat of fusion per unit mass (Jkg^{-1})
       ρ_{sf} = density of frozen soil (kg m^{-3})
       w   = moisture content (dimensionless)
       T_f = freezing temperature (K) ($\approx$0°C)
       T_s = temperature at the ground surface (K)
       a_{sf}, a_{uf} = thermal diffusivity of frozen and unfrozen layers, respectively (m^{2}s^{-1})
       $G(\frac{j}{2}\sqrt{a_{uf}})$ = Gauss' probability function
       T_0 = initial temperature in the soil mass (K)
       K_{sf}, K_{uf} = thermal conductivity of resp. frozen and unfrozen layer (Js^{-1}m^{-1}K^{-1})

The transcendental Equation 75.3 can be written as c·j = f(j)

where c   = a constant
       f(j)   = a function of j

The left-hand side of the equation is a straight line through the origin (c·j). The right-hand side is a transcendental curve. Both graphs are plotted in one figure. The abscissa of the intersection of the graph gives the value for j, which is the solution to the equation. With Equation 75.2 (Z = j√t) for each time of frost penetration t, the depth of frost penetration Z can be calculated.

The Neumann model is more realistic than the Stefan model. It assumes an original uniform soil temperature (temperature gradient of 0°C m^{-1}). In reality, this is never the case. A problem arises as to what the initial temperature T_0 is in Equation 75.3. As a solution, T_0 may be substituted by the highest temperature measured in a soil profile, or the temperature at a standardized depth (e.g., 2 m below the soil surface).

This model ignores the upward flow of water and accompanying flow of heat to the freezing front. Soil properties as K and a are considered to be constants while they are clearly functions of the water/ice content of the frozen soil which is itself dependent on negative temperature (Williams and Smith 1989). The method is useful to estimate the depth of frost penetration at a certain time at a certain place. A disadvantage of the analytical models is that much data are required.

75.5.2 Empirical Models

To reduce the required information, empirical models were developed. Aldrich and Paynter (1966) give a model in which the term $(T_f - T_s)t$ of Stefan's Equation 75.1 is replaced by the freezing index F, the absolute value summation of the average daily temperatures of each day that the average temperature was below 0°C. In fine-grained soils a small error is generated by this approximation because T_f may be slightly below 0°C due to the presence of salts and capillary and adsorption forces in the soil. Another error is generated because of the difference between air temperatures (measured usually 1 to 2 m above the soil surface) and the temperature of the ground surface. Often, a correction factor n is introduced to account for this error. The error is small as long as the soil is covered with snow. In empirical models the value of Z in the model of Neumann (75.2) is approximated by multiplying the right-hand side of the equation by a correction factor α, which depends on the mean annual temperature, the surface temperature during the freezing period, the heat capacity C, and the latent heat Q_L. Although these models lack a physical base they gave reasonably good results.

75.5.3 Numerical Models

Since the introduction of computers, numerical models have received gradually more attention. Smith and Tvede (1977) developed a one-dimensional model which calculates the depth of frost penetration under highways. In soils under agricultural land, loss of heat by evaporation has to be taken into account, and the problem should at least be solved two dimensionally. In the model of Benoit and Mostaghimi (1985), the released latent heat and the heat flux from the unfrozen zone by conduction as well as by liquid flow are taken into account. The model also accounts for a layer of snow on the soil surface:

$$\overline{K}_{sf} \frac{dT_{sf}}{dZ_{sf}} = K_{uf} \frac{dT_{uf}}{dZ_{uf}} + LK_w \frac{dh}{dZ_{uf}} + C_{uf}\, dT_{uf} \qquad (75.4)$$

where $\overline{K}_{sf}$ = average thermal conductivity through the combined frozen soil and snow depth thickness $(Js^{-1}m^{-1}K^{-1})$

dT_{sf} = temperature difference between the surface of the soil snow system and the freezing front (K)

dZ_{sf} = depth or thickness of the combined frozen soil snow layers (m)

dZ_{uf} = depth of unfrozen soil to point of stable temperature (m)

K_{uf} = thermal conductivity of unfrozen soil $(Js^{-1}m^{-1}K^{-1})$

L = latent heat of fusion $(Js^{-1}m^{-3})$

k_w = (unsaturated) hydraulic conductivity of unfrozen soil (ms^{-1})

C_{uf} = heat capacity of the (unfrozen) soil $(Js^{-1}m^{-3}K^{-1})$

dT_{uf} = temperature difference between freezing front and that at depth of stable temperature (K)

h = total water potential (m)

where

$$\overline{K}_{sf} = \left(\frac{1}{Z_{sf}} \sum_{i=1}^{n} \frac{dZ_i}{K_i} \right)^{-1} \qquad (75.5)$$

where i $= 1 \ldots n =$ number of soil layer
     Z_i $=$ thickness of soil layer i (m)
     K_i $=$ thermal conductivity of soil layer i ($Js^{-1}m^{-1}K^{-1}$)

The inputs of the model are the maximum and minimum daily air temperatures; the snow depth; the thermal conductivity for snow, frozen soil, and unfrozen soil; hydraulic conductivity; total potential; and soil temperature.

After initialization, daily maximum and minimum air temperatures and snow depths are the only inputs required. In a sensitivity analysis, the model predictions show the greatest sensitivity to errors in soil moisture content, hydraulic conductivity, and thermal conductivity of the frozen soil.

The predicted frost depths agreed well with measured frost depths when freezing occurred. When thawing occurred the predictions differed more from the measurements. The limitation of this model is that much information is required, and that an error-sensitive parameter like the unsaturated hydraulic conductivity is assumed to be constant in the unfrozen soil. This assumption is only true if the spatial steps are very small; otherwise, the hydraulic conductivity varies too much. It should be noted that much research is underway into models describing soil freezing. These are becoming very complex and the models are not yet easily accessible.

REFERENCES

Aldrich, H. P., Jr. and Paynter, H. M. 1966. Depth of frost penetration in non-uniform soil. U.S. Army Cold Regions Research and Engineering Laboratories, Special Report 104 11 pp.

Banner, J. A. and van Everdingen, R. O. 1979. Frost gauges and freezing gauges. National Hydrology Research Institute Paper No. 3, Inland Waters Directorate Technical Bulletin No. 110, Ottawa, Canada. 18 pp.

Benoit, G. R. and Mostaghimi, S. 1985. Modelling soil frost depth under three tillage systems. Trans. ASAE 28: 1499–1505.

Gandahl, R. 1957. Bestämning av tjälgräns i mark med enkel typ av tjälgränsmätare, A frost depth indicator (in Swedish). Statensväginstitut Rapport 30, Stockholm, 1–15.

Gandahl, R. and Bergau, W. 1957. Two methods for measuring the frozen zone in the soil. Pages 32–34 in Proceedings of the Fourth International Conference on Soil Mechanics and Foundation Engineering, London.

Hansen, B. L. 1963. Instruments for temperature measurements in permafrost. Pages 356–358 in

Proceedings of the First International Conference on Permafrost, Lafayette, IN.

Harris, S. A., French, H. M., Heginbottom, J. A., Johnston, G. H., Ladanyi, B., Sego, D. C., and Van Everdingen, R. O. 1988. Glossary of permafrost and related ground-ice terms. NRCC Technical Memorandum No. 142. 156 pp.

Johnston, G. H. 1973. Ground temperature measurements using thermocouples. Pages 1–12 in R. J. E. Brown, Ed. Proceedings, Seminar on the Thermal Regime and Measurements in Permafrost, Saskatoon, 2 and 3 May 1972. NRCC Technical Memorandum 108.

Jumikis, A. R. 1955. The frost penetration problem in highway engineering. Rutgers University Press, New Brunswick, NJ. 162 pp.

Laboratoire Central des Ponts et Chaussées 1987. L'indicateur de profondeur de gel (IPG). Ministère de l'équipement du logement de l'aménagement du territoire et des transports, Laboratoire Central des Ponts et Chaussees, Méthode d'essai LPC No. 29.

Rickard, W. and Brown, J. 1972. The performance of a frost tube for the determination of soil freezing and thawing depths. Soil Sci. 113: 149–154.

Smith, M. W. and Tvede, A. 1977. The computer simulation of frost penetration beneath highways. Can. Geotech. J. 14: 167–179.

Thermometrics Inc. 1982. Thermistors. Thermometrics Inc. Cat No. 181-B, 808 U.S. Highway #1, Edison, NJ.

Tobiasson, W. and Atkins, R. 1975. Frost penetration measurements at the USAF intrusion sensor site, Rome, New York 1973–1974. U.S. Army Cold Regions Research and Engineering Laboratory, Special Report 235. 47 pp.

Williams, P. J. and Smith, M. W. 1989. The frozen earth. Fundamentals of geocryology. Cambridge University Press, Cambridge, 306 pp.

Index

INDEX